Computer Processing of Remotely-Sensed Images

An Introduction

Fourth Edition

Paul M. Mather

University of Nottingham

and

Magaly Koch

Center for Remote Sensing
Boston University

A John Wiley & Sons, Ltd., Publication

Wiley-Blackwell is an imprint of John Wiley & Sons, formed by the merger of Wiley's global Scientific, Technical and Medical business with Blackwell Publishing.

Registered office: John Wiley & Sons Ltd, The Atrium, Southern Gate, Chichester, West Sussex, PO19 8SQ, UK

Editorial offices:
9600 Garsington Road, Oxford, OX4 2DQ, UK
The Atrium, Southern Gate, Chichester, West Sussex, PO19 8SQ, UK
111 River Street, Hoboken, NJ 07030-5774, USA

For details of our global editorial offices, for customer services and for information about how to apply for permission to reuse the copyright material in this book please see our website at www.wiley.com/wiley-blackwell.

Library of Congress Cataloging-in-Publication Data

Mather, Paul M.
Computer processing of remotely-sensed images : an introduction / Paul M. Mather and Magaly Koch. – 4th ed.
p. cm.
Includes index.
ISBN 978-0-470-74239-6 (cloth) – ISBN 978-0-470-74238-9 (pbk.) 1. Remote sensing – Data processing. 2. Remote-sensing images. 3. Image processing – Digital techniques. I. Koch, Magaly. II. Title.
G70.4.M38 2011

621.36′78028566 – dc22

2010018060

A catalogue record for this book is available from the British Library.

This book is published in the following electronic formats: ePDF 9780470666500; Wiley Online Library 9780470666517

Set in 10/12pt Times by Laserwords Private Limited, Chennai, India

Printed and bound in Singapore by Markono Print Media Pte Ltd

First Impression 2011

Contents

Preface to the First Edition

Environmental remote sensing is the measurement, from a distance, of the spectral features of the Earth's surface and atmosphere. These measurements are normally made by instruments carried by satellites or aircraft, and are used to infer the nature and characteristics of the land or sea surface, or of the atmosphere, at the time of observation. The successful application of remote sensing techniques to particular problems, whether they be geographical, geological, oceanographic or cartographic, requires knowledge and skills drawn from several areas of science. An understanding of the way in which remotely sensed data are acquired by a sensor mounted onboard an aircraft or satellite needs a basic knowledge of the physics involved, in particular environmental physics and optics. The use of remotely-sensed data, which are inherently digital, demands a degree of mathematical and statistical skill plus some familiarity with digital computers and their operation. A high level of competence in the field in which the remotely-sensed data are to be used is essential if full use of the information contained in those data is to be made. The term 'remote-sensing specialist' is thus, apparently, a contradiction in terms for a remote-sensing scientist must possess a broad range of expertise across a variety of disciplines. While it is, of course, possible to specialise in some particular aspect of remote sensing, it is difficult to cut oneself off from the essential multidisciplinary nature of the subject.

This book is concerned with one specialised area of remote sensing, that of digital image processing of remotely sensed data but, as we have seen, this topic cannot be treated in isolation and so Chapter 1 covers in an introductory fashion the physical principles of remote sensing. Satellite platforms currently or recently in use, as well as those proposed for the near future, are described in Chapter 2, which also contains a description of the nature and sources of remotely sensed data. The characteristics of digital computers as they relate to the processing of remotely-sensed image data is the subject of Chapter 3. The remaining five chapters cover particular topics within the general field of the processing of remotely-sensed data in the form of digital images, and their application to a range of problems drawn from the Earth and environmental sciences. Chapters 1–3 can be considered to form the introduction to the material treated in later chapters.

The audience for this book is perceived as consisting of undergraduates taking advanced options in remote sensing in universities and colleges as part of a first degree course in geography, geology, botany, environmental science, civil engineering or agricultural science, together with postgraduate students following taught Masters courses in remote sensing. In addition, postgraduate research students and other research workers whose studies involve the use of remotely-sensed images can use this book as an introduction to the digital processing of such data. Readers whose main scientific interests lie elsewhere might find here a general survey of this relatively new and rapidly-developing area of science and technology. The nature of the intended audience requires that the formal presentation is kept to a level that is intelligible to those who do not have the benefit of a degree in mathematics, physics, computer science or engineering. This is not a research monograph, complete in every detail and pushing out to the frontiers of knowledge. Rather, it is a relatively gentle introduction to a subject which can, at first sight, appear to be overwhelming to those lacking mathematical sophistication, statistical cunning or computational genius. As such it relies to some extent on verbal rather than numerical expression of ideas and concepts. The author's intention is to provide the foundations upon which readers may build their knowledge of the more complex and detailed aspects of the use of remote-sensing techniques in their own subject rather than add to the already extensive literature which caters for a mathematically-orientated readership. Because of the multidisciplinary nature of the intended audience, and since the book is primarily concerned with techniques, the examples have been kept simple, and do not assume any specialist knowledge of geology, ecology, oceanography or other branch of Earth science. It is expected that the reader is capable of working out potential applications in his or her own field, or of following up the references given here.

It is assumed that most readers will have access to a digital image processing system, either within their own department or institution, or at a regional or national remote sensing centre. Such processors normally have a built-in software package containing programs to carry out most, if not all, of the operations described in this book. It is hoped that the material presented here will provide such readers with the background necessary to make sensible use of the facilities at their disposal. Enough detail is provided, however, to allow interested readers to develop computer programs to implement their own ideas or to modify the operation of a standard package. Such ventures are to be encouraged, for software skills are an important part of the remote sensing scientist's training. Furthermore, the development and testing of individual ideas and conjectures provides the opportunity for experiment and innovation, which is to be preferred to the routine use of available software. It is my contention that solutions to problems are not always to be found in the user manual of a standard software package.

I owe a great deal to many people who have helped or encouraged me in the writing of this book. Michael Coombes of John Wiley & Sons, Ltd. took the risk of asking me to embark upon the venture, and has proved a reliable and sympathetic source of guidance as well as a model of patience. The Geography Department, University of Nottingham, kindly allowed me to use the facilities of the Remote Sensing Unit. I am grateful also to Jim Cahill for many helpful comments on early drafts, to Michael Steven for reading part of Chapter 1 and for providing advice on some diagrams, and to Sally Ashford for giving a student's view and to George Korybut-Daszkiewicz for his assistance with some of the photography. An anonymous referee made many useful suggestions. My children deserve a mention (my evenings on the word-processor robbed them of the chance to play their favourite computer games) as does my wife for tolerating me. The contribution of the University of Nottingham and the Shell Exploration and Development Company to the replacement of an ageing PDP11 computer by a VAX 11/730-based image processing system allowed the continuation of remote sensing activities in the university and, consequently, the successful completion of this book. Many of the ideas presented here are the result of the development of the image processing software system now in use at the University of Nottingham and the teaching of advanced undergraduate and Masters degree courses. I am also grateful to Mr J. Winn, Chief Technician, Mr. C. Lewis and Miss E. Watts, Cartographic Unit and Mr M.A. Evans, Photographic Unit, Geography Department, University of Nottingham, for their invaluable and always friendly assistance in the production of the photographs and diagrams. None of those mentioned can be held responsible for any errors or misrepresentations that might be present in this book; it is the author's prerogative to accept liability for these.

Paul M. Mather,
March 1987
Remote Sensing Unit,
Department of Geography,
The University of Nottingham,
Nottingham NG7 2RD, UK

Preface to the Second Edition

Many things have changed since the first edition of this book was written, more than 10 years ago. The increasing emphasis on scientific rigour in remote sensing (or Earth observation by remote sensing, as it is now known), the rise of interest in global monitoring and large-scale climate modelling, the increasing number of satellite-borne sensors in orbit, the development of Geographical Information Systems (GISs) technology, and the expansion in the number of taught Masters courses in GIS and remote sensing are all noteworthy developments. Perhaps the most significant single change in the world of remote sensing over the past decade has been the rapid increase in and the significantly reduced cost of computing power and software available to students and researchers alike, which allows them to deal with growing volumes of data and more sophisticated and demanding processing tools. In 1987 the level of computing power available to researchers was minute in comparison with that which is readily available today. I wrote the first edition of this book using a BBC Model B computer, which had 32 kb of memory, 100 kb diskettes and a processor which would barely run a modern refrigerator. Now I am using a 266 MHz Pentium II with 64 Mb of memory and a 2.1 Gb disk. It has a word processor that corrects my spelling mistakes (though its grammar checking can be infuriating). I can connect from my home to the University of Nottingham computers by fibre optic cable and run advanced software packages. The cost of this computer is about 1% of that of the VAX 11/730 that is mentioned in the preface to the first edition of this book.

Although the basic structure of the book remains largely unaltered, I have taken the opportunity to revise all of the chapters to bring them up to date, as well as to add some new material, to delete obsolescent and uninteresting paragraphs, and to revise some infelicitous and unintelligible passages. For example, Chapter 4 now contains new sections covering sensor calibration, plus radiometric and topographic correction. The use of artificial neural networks in image classification has grown considerably in the years since 1987, and a new section on this topic is added to Chapter 8, which also covers other recent developments in pattern recognition and methods of estimating Earth surface properties. Chapter 3, which provides a survey of computer hardware and software, has been almost completely re-written. In Chapter 2 I have tried to give a brief overview of a range of present and past sensor systems but have not attempted to give a full summary of every sensor, because details of new developments are now readily available via the Internet. I doubt whether anyone would read this book simply because of its coverage of details of individual sensors.

Other chapters are less significantly affected by recent research as they are concerned with the basics of image processing (filtering, enhancement and image transforms), details of which have not changed much since 1987, though I have added new references and attempted to improve the presentation. I have, however, resisted the temptation to write a new chapter on GIS, largely because there are several good books on this topic that are widely accessible (for example Bonham-Carter (1994) and McGuire *et al*. (1991)), but also because I feel that this book is primarily about image processing. The addition of a chapter on GIS would neither do justice to that subject nor enhance the reader's understanding of digital processing techniques. However, I have made reference to GIS and spatial databases at a number of appropriate points in the text. My omission of a survey of GIS techniques does not imply that I consider digital image processing to be a 'stand-alone' topic. Clearly, there are significant benefits to be derived from the use of spatial data of all kinds within an integrated environment, and this point is emphasized in a number of places in this book. I have added a significant number of new references to each of the chapters, in the hope that readers might be encouraged to enjoy the comforts of his or her local library.

I have added a number of 'self-assessment' questions at the end of each chapter. These questions are not intended to constitute a sample examination paper, nor do they provide a checklist of 'important' topics (the implication being that the other topics covered in the book are unimportant). They are simply a random set of questions – if you can answer them then you probably understand the contents of the chapter. Readers should use the Mather's Image Processing System (MIPS) software described in Appendices A and B to try out the methods mentioned in these questions. Data sets are also available on the accompanying CD, and are described in Appendix C.

Perhaps the most significant innovation that this book offers is the provision of a CD containing software and images. I am not a mathematician, and so I learn by trying out ideas rather than exclusively by reading or listening. I learn new methods by writing computer programs and applying them to various data sets. I am including a small selection of the many programs that I have produced over the past 30 years, in the hope that others may find them useful. These programs are described in Appendix B. I have been teaching a course on remote sensing for the last 14 years. When this course began there were no software packages available, so I wrote my own (my students will remember NIPS, the Nottingham Image Processing System, with varying degrees of hostility). I have completely re-written and extended NIPS so that it now runs under Microsoft Windows 95. I have renamed it to MIPS, which is rather an unimaginative name, but is nevertheless pithy. It is described in Appendix A. Many of the procedures described in this book are implemented in MIPS, and I encourage readers to try out the methods discussed in each chapter. It is only by experimenting with these methods, using a range of images, that you will learn how they work in practice. MIPS was developed on an old 486-based machine with 12 Mb of RAM and a 200 Mb disk, so it should run on most PCs available in today's impoverished universities and colleges. MIPS is not a commercial system, and should be used only for familiarisation before the reader moves on to the software behemoths that are so readily available for both PCs and UNIX workstations. Comments and suggestions for improving MIPS are welcome (preferably by email) though I warn readers that I cannot offer an advisory service nor assist in research planning!

Appendix C contains a number of Landsat, SPOT, AVHRR and RADARSAT images, mainly extracts of size 512 512 pixels. I am grateful to the copyright owners for permission to use these data sets. The images can be used by the reader to gain practical knowledge and experience of image processing operations. Many university libraries contain map collections, and I have given sufficient details of each image to allow the reader to locate appropriate maps and other back-up material that will help in the interpretation of the features shown on the images.

The audience for this book is seen to be advanced undergraduate and Masters students, as was the case in 1987. It is very easy to forget that today's student of remote sensing and image processing is starting from the same level of background knowledge as his or her predecessors in the 1980s. Consequently, I have tried to restrain myself from including details of every technique that is mentioned in the literature. This is not a research monograph or a literature survey, nor is it primarily an exercise in self-indulgence and so some restriction on the level and scope of the coverage provided is essential if the reader is not to be overwhelmed with detail and thus discouraged from investigating further. Nevertheless, I have tried to provide references on more advanced subjects for the interested reader to follow up. The volume of published material in the field of remote sensing is now very considerable, and a full survey of the literature of the last 20 years or so would be both unrewarding and tedious. In any case, online searches of library catalogues and databases are now available from networked computers. Readers should, however, note that this book provides them only with a background introduction – successful project work will require a few visits to the library to peruse recent publications, as well as practical experience of image processing.

I am most grateful for comments from readers, a number of whom have written to me, mainly to offer useful suggestions. The new edition has, I hope, benefited from these ideas. Over the past years, I have been fortunate enough to act as supervisor to a number of postgraduate research students from various countries around the world. Their enthusiasm and commitment to research have always been a factor in maintaining my own level of interest, and I take this opportunity to express my gratitude to all of them. My friends and colleagues in the Remote Sensing Society, especially Jim Young, Robin Vaughan, Arthur Cracknell, Don Hardy and Karen Korzeniewski, have always been helpful and supportive. Discussions with many people, including Mike Barnsley, Giles Foody and Robert Gurney, have added to my knowledge and awareness of key issues in remote sensing. I also acknowledge with gratitude the help given by Dr Magaly Koch, Remote Sensing Center, Boston University, who has tested several of the procedures reported in this book and included on the CD. Her careful and thoughtful advice, support and encouragement have kept me from straying too far from reality on many occasions. My colleagues in the School of Geography in the University of Nottingham continue to provide a friendly and productive working environment, and have been known occasionally to laugh at some of my jokes. Thanks especially to Chris Lewis and Elaine Watts for helping to sort out the diagrams for the new edition, and to Dee Omar for his patient assistance and support. Michael McCullagh has been very helpful, and has provided a lot of invaluable assistance. The staff of John Wiley & Sons has been extremely supportive, as always. Finally, my wife Rosalind deserves considerable credit for the production of this book, as she has quietly undertaken many of the tasks that, in fairness, I should have carried out during the many evenings and weekends that I have spent in front of the computer. Moreover, she has never complained about the chaotic state of our dining room, nor about the intrusive sound of Wagner's music dramas. There are many people, in many places, who have helped or assisted me; it is impossible to name all of them, but I am nevertheless grateful. Naturally, I take full responsibility for all errors and omissions.

Paul M. Mather
Nottingham, June 1998

Preface to the Third Edition

In the summer of 2001 I was asked by Lyn Roberts of John Wiley & Sons, Ltd. to prepare a new edition of this book. Only minor updates would be needed, I was told, so I agreed. A few weeks later was presented with the results of a survey of the opinions of the 'great and the good' as to what should be included in and what should be excluded from the new edition. You are holding the result in your hands. The 'minor updates' turned into two new chapters (a short one on computer basics, replacing the old Chapter 3, and a lengthier one on the advanced topics of interferometry, imaging spectroscopy and lidar, making a new Chapter 9) plus substantial revisions of the other chapters. In addition, I felt that development of the MIPS software would be valuable to readers who did not have access to commercial remote sensing systems. Again, I responded to requests from postgraduate students to include various modules that they considered essential, and the result is a Windows-based package of 90 000+ lines of code.

Despite these updates and extensions both to the text of the book and the accompanying software, my target audience is still the advanced undergraduate taking a course in environmental remote sensing. I have tried to introduce each topic at a level that is accessible to the reader who is just becoming aware of the delights of image processing while, at the same time, making the reasonable assumption that my readers are, typically, enthusiastic, aware and intelligent, and wish to go beyond the basics. In order to accommodate this desire to read widely, I have included an extensive reference list. I am aware, too, that this book is used widely by students taking Masters level courses. Some of the more advanced material, for example in Chapters 6, 8 and 9, is meant for them; for example the new material on wavelets and developments in principal components analysis may stimulate Masters students to explore these new methods in their dissertation work. The first three chapters should provide a basic introduction to the background of remote sensing and image processing; Chapters 4–8 introduce essential ideas (noting the remark above concerning parts of Chapters 6 and 8), while Chapter 9 is really for the postgraduate or the specialist undergraduate.

I am a firm believer in learning by doing. Reading is not a complete substitute for practical experience of the use of image processing techniques applied to real data that relates to real problems. For most people, interest lies in the meaning of the results of an operation in terms of the information that is conveyed about a problem rather than in probing the more arcane details of particular methods, though for others it is the techniques themselves that fascinate. The level of mathematical explanation has therefore been kept to a minimum and I have attempted to use an approach involving examples, metaphors and verbal explanation. In particular, I have introduced a number of examples, separate from the main text, which should help the reader to interpret image-processing techniques in terms of real-world applications.

Many of these examples make use of the MIPS software that is provided on the CD that accompanies this book. MIPS has grown somewhat since 1999, when the second edition of this book was published. It has a new user interface, and is able to handle images of any size in 8-, 16- or 32-bit representation. A number of new features have been added, and it is now capable of providing access to many of the techniques discussed in this book. I would appreciate reports from readers of any difficulties they experience with MIPS, and I will maintain a web site from which updates and corrections can be downloaded. The URL of this web site, and my email address, can be found in the file `contactme.txt` which is located in the root directory of the CD.

Many of the ideas in this book have come from my postgraduate students. Over the past few years, I have supervised a number of outstanding research students, whose work has kept me up to date with new developments. In particular, I would like to thank Carlos Vieira, Brandt Tso, Taskin Kavzoglu, Premelatha Balan, Mahesh Pal, Juazir Hamid, Halmi Kamarrudin and Helmi Shafri for their tolerance and good nature. Students attending my Masters classes in digital image processing have also provided frank and valuable feedback. I would also like to acknowledge the valuable assistance provided by Rosemary Hoole and Karen Laughton of the School of Geography, University of Nottingham. The help of Dr Koch of Boston University, who made many useful comments on the manuscript and the MIPS software as they have progressed, is also gratefully acknowledged, as is the kindness of Professor J. Gumuzzio and his group at the

Autonomous University of Madrid in allowing me access to DAIS images of their La Mancha study site. The DAIS data were recorded by DLR in the frame of the EC funded programme 'Access to Research Infrastructures', project 'DAIS/ROSIS – Imaging Spectrometers at DLR', Co. Ro. HPRI-CT-1999-0075. Dr Koch has also provided a set of four advanced examples, which can be found in the **`Examples`** folder of the accompanying CD. I am very grateful to her for this contribution, which I am sure significantly enhances the value of the book. My wife continues to tolerate what she quietly considers to be my over-ambitious literary activities, as well as my predilection for the very loudest bits of Mahler, Berlioz, Wagner and others. Colleagues and students of the School of Geography, University of Nottingham, have helped in many ways, not least by humouring me. Finally, I would like to thank Lyn Roberts, Kiely Larkins, and the staff of John Wiley who have helped to make this third edition a reality, and showed infinite patience and tolerance.

A book without errors is either trivial or guided by a divine hand. I can't believe that this book is in the latter category, and it is possible that it isn't in the former. I hope that the errors that you do find, for which I take full responsibility, are not too serious and that you will report them to me.

Paul M. Mather
Nottingham, August 2003

Preface to the Fourth Edition

Almost 25 years have passed since the publication of the first edition of this book. Back in 1987, image processing and geographical information systems (GIS) required dedicated computers that lived in air-conditioned rooms. At that time, I had access to a VAX 11/730, which had about the same computing power as a mobile phone of today. We had four megabytes of RAM and it took 12 hours to carry out a maximum likelihood classification on a 512 × 512 Landsat four-band MSS image. Imagery with the highest multispectral spatial resolution (20 m) was acquired by the SPOT-1 satellite, and there was not a great deal of software to process it. In those days an ability to write your own programmes was essential.

Over those 25 years, new companies, new products and new applications have come into being. Integrated software packages running on fast desktop machines, processing data that may have up to 256 bands, or a spatial resolution of 50 cm, with stereo capability for DEM production, are all seen as a natural if rapid development over the period 1987–2010. Yet the basic principles of image processing remain the same, and so does the structure of this book. However, a new chapter, number ten, has been added. It covers the topic of Environmental Geographical Information Systems and Remote Sensing. I wrote in the preface to the second edition (1999) that I had resisted the temptation to include a chapter on GIS because there were several good GIS texts already available. However, developments in the fields of image processing and GIS mean that nowadays no environmental remote sensing course is complete without some reference to environmental GIS, and no environmental GIS course can be considered adequate without a discussion of the uses of remotely sensed data. In fact, over the past two or three years there has been a 'coming together' of GIS, remote sensing and photogrammetry. Nowadays most software vendors describe their products not as 'Image Processing for Remote Sensing' but as geospatial, integrated, and so on. They can interact with GIS either directly or through a common data format. This book has, therefore, to include material on GIS from a remote sensing point of view. Being aware of my inadequacies in the field of GIS, I asked Dr Magaly Koch of Boston University's Centre for Remote Sensing to help with the new Chapter 10 and with other chapters, as she has the benefit of considerable practical experience in the areas of remote sensing and GIS. I am very grateful to her for accepting this invitation and for making a valuable contribution to the book as a whole.

Perhaps the most important change in the appearance of the book is the introduction of colour illustrations and tables. This development is of tremendous value in a book on image processing. The use of colour supplements is a half-way house, and I always found it more trouble than it was worth. I am grateful to the publishers, and in particular to Fiona Woods, for negotiating a new full colour edition. In retrospect, trying to describe the effects of, for example, a decorrelation stretch while using only greyscale illustrations would have been a challenge to any of the great literary figures of the past, and I am sure that my powers of descriptive writing are but a shadow of theirs.

As far as other changes are concerned, I have rewritten most of Chapter 3. In the third edition Chapter 3 described the workings of the MIPS software package, which was included on a CD that accompanied the book. MIPS is still available, and can now be downloaded from the publisher's web site (www.wiley.com/go/mather4). Chapter 3 now deals with aspects of computing and statistics which should be appreciated and understood by all users of computers. In particular, I feel strongly that users of remote sensing and GIS software packages should have a good understanding of the way their favoured package handles data; many people have been surprised to find that sometimes software does not always do what it says on the box. In comparison, economists and computer scientists have tested spreadsheet software such as Excel almost to destruction (McCullough, 1998, 1999). It is particularly important for the user to be aware of the nature and type of data being processed, as 8-, 10-, 12- and 16-bit data values are now in widespread use. Yet most graphics cards still limit displayed data to 8 bits per primary colour (red, green and blue) and the way that data are converted from their original format (8-, 10-, 11- or 16-bit) to fit the requirements of the graphics hardware is a matter of considerable interest, as the appearance of the image depends to a greater or lesser extent on this conversion process.

Other topics that are new to the fourth edition (apart from Chapters 3 and 10) include: updated information on new satellites and sensors (Chapter 2), new sections on change detection, and image fusion (Chapter 6), a description of the Frost filter (Chapter 7), a new section on independent components analysis (Chapter 8) plus descriptions of recent developments in image classification (support vector machines and decision trees, object-oriented classification and bagging and boosting). Other chapters have been updated, with old references removed and replaced by new ones, though several key papers from the 1970s to 1980s remain in the bibliography where they provide an uniquely intelligible description of a specific topic.

Some may query the size of the bibliography. However, modern bachelors' and Masters' programmes are more focused on assessed project work than was the case in the 1980s, and the key papers in the recent literature are catalogued in the following pages in order to provide guidance to students who are undertaking projects. I have tried to ensure that most journal paper references postdate 1994, because many electronically accessible journals are available only from the mid-1990s on, and I know that both students and researchers like to read their favourite journals online at their desk. The importance of wide reading of original papers should not be underestimated; relying solely on textbooks is not a good idea.

I have already noted that the MIPS software can be downloaded from www.wiley.com/go/mather4. That web site also contains compressed files of PowerPoint figures and tables which can be downloaded by lecturers and teachers. These files, one for each chapter, are compressed using the popular zip method. The four case studies included on the CD that accompanied the third edition of this book are also available on the web site as pdf files (with accompanying data sets).

I would like to acknowledge the help I have received from Rachel Ballard, Fiona Woods, Izzy Canning and Sarah Karim of John Wiley & Sons, Ltd. and Gayatri Shanker of Laserwords, who have looked after me during the production of this book with great patience and forbearance. Special thanks go to Alison Woodhouse, who copy-edited the manuscript in a masterly way. I must also thank my wife, who has been a great support over the past few years while the 'book project' was developing and before. My son James helped enormously in converting diagrams to colour. There are many other people who have had an input into the writing and production of this book, ranging from librarians to graphics designers and copy editors. I thank all of them for their assistance. Finally, a note to the reader: remote sensing is a captivating and intellectually demanding subject, but it is also fun. I have enjoyed teaching and researching in remote sensing for the best part of 30 years and I hope that my enjoyment of and enthusiasm for the subject comes across in the text and maybe inspires you to take a keen interest in whatever aspect of the subject that you are studying.

Needless to say, I take full responsibility for any errors that may be found in the book. I would be grateful if readers could advise me by email of any such errors or misrepresentations. My current email address can be found in the Contacts link on the book's web page.

Paul M. Mather
Nottingham, January, 2010.

In the last decade, technological advances have led to a seemingly overwhelming amount of Earth observation information acquired at increasingly finer spatial, spectral and temporal scales. At the same time data processing and assimilation systems are continuously offering new and innovative ways of capturing, processing and delivering almost in real time this steady stream of geospatial data to end users, who may range from expert users to the untrained person requiring higher-level or value-added image products for decision making and problem solving.

This trend in geospatial technology development has naturally closed or at least narrowed the gap between remote sensing and GIS systems. RS data can provide both image maps for background display and estimates of the characteristics or values of environmental variables. These products can most profitably be used in conjunction with other environmental data such as rainfall or geology in the context of GIS.

I am honoured to contribute a new chapter, Chapter 10, on remote sensing and GIS integration to this new edition and would like to thank Prof. Paul Mather for giving me the opportunity to do so, and for guiding me through the process of 'co-writing a book'. The new chapter draws from my experience in applying remote sensing and GIS techniques over a number of years while working on numerous research projects mainly conducted in Boston University's Center for Remote Sensing. In this respect I am very grateful to Prof. Farouk El-Baz, director of the Center, who has been my adviser and colleague for the past two decades. I would also like to acknowledge the contributions and revisions of my colleagues Dr Francisco Estrada-Belli (Boston University), Dr Michael DiBlasi (Boston University), Dr Luisa Sernicola (University of Naples 'L'Orientale') and my PhD student Mr Ahmed Gaber (Tohoku University) in the case studies included in the new chapter 10. I hope that the combination of theoretical treatment and practical application of geospatial techniques will inspire the reader to create remote sensing and GIS-based solutions to a variety of environmental problems.

Magaly Koch
Boston, January 2010

List of Examples

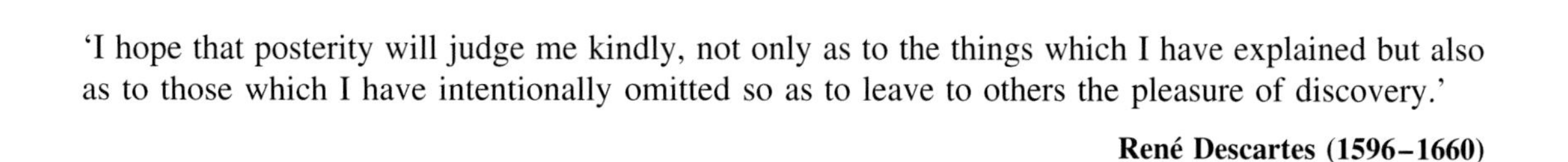

'I hope that posterity will judge me kindly, not only as to the things which I have explained but also as to those which I have intentionally omitted so as to leave to others the pleasure of discovery.'

René Descartes (1596–1660)

'I am none the wiser, but I am much better informed.'

Queen Victoria (1819–1901)

(After being addressed by Lord Rutherford on the state of modern physics)

'Education is an admirable thing, but it is well to remember from time to time that nothing that is worth knowing can be taught.'

Oscar Wilde (1856–1900)

Web Page

A web page is associated with this book. Follow this link: www.wiley.com/go/mather4 and you will find:

1. The MIPS image processing system software, including some test images,
2. PowerPoint Slide Shows containing all the figures and tables from each chapter,
3. A set of four example applications of remote sensing and GIS, including datasets.

1 Remote Sensing: Basic Principles

Electromagnetic radiation is just basically mysterious.

B.K. Ridley, *Time, Space and Things*, 2nd edition. Cambridge University Press, Cambridge, 1984.

1.1 Introduction

The science of remote sensing consists of the analysis and interpretation of measurements of electromagnetic radiation (EMR) that is reflected from or emitted by a target and observed or recorded from a vantage point by an observer or instrument that is not in contact with the target. Earth observation by remote sensing (EO) is the interpretation and understanding of measurements made by airborne or satellite-borne instruments of EMR that is reflected from or emitted by objects on the Earth's land, ocean or ice surfaces or within the atmosphere, together with the establishment of relationships between these measurements and the nature and distribution of phenomena on the Earth's surface or within the atmosphere. Figure 1.1 shows in schematic form the various methods of computer processing (blue boxes) that generate products (green boxes) from remotely-sensed data. This book deals with the methods of computer processing of remotely sensed data (the green and blue boxes) as well as providing an introduction to environmental geographical information systems (E-GISs) (Chapter 10) which make use of remotely-sensed products.

Remotely-sensed images are often used as image maps or backcloths for the display of spatial data in an E-GIS. Methods of improving the appearance of an image (termed *enhancement* procedures) are dealt with in Chapters 4, 5 and 7. Chapter 8 is an introduction to *pattern recognition* techniques that produce labelled images in which each type of land use, for example is represented by a numeric code (for example 1 = broadleaved forest, 2 = water, and so on.). These labelled images can provide free-standing information or can be combined with other spatial data within an E-GIS. Properties of earth surface materials, such as soil moisture content, sea surface temperature (SST) or biomass can be related to remotely sensed measurements using statistical methods. For instance, a sample of measurements of soil moisture content can be collected close to the time of satellite overpass and the corresponding ground reflectance or ground temperature that are recorded by the satellite's instruments can be related via regression analysis to the ground measurements. This sample relationship can then be applied to the entire area of interest. These bio-geophysical variables are used in environmental modelling, often within an E-GIS. Elevation models are another form of remotely-sensed spatial information that is used in E-GIS. Digital elevation models can be derived from optical imagery using two sensors, for example one pointing down and one pointing obliquely backwards (this is the case with the ASTER sensor, discussed in Chapter 2). Another way of producing elevation models is by the use of synthetic aperture radar (SAR) interferometry, which is discussed in Chapter 9. The increasing cooperation between remote sensing specialists and E-GIS users means that more products are available to E-GIS users and the more spatial information is combined with remotely sensed data to produce improved results. This is an example of synergy (literally, working together).

A fundamental principle underlying the use of remotely sensed data is that different objects on the Earth's surface and in the atmosphere reflect, absorb, transmit or emit electromagnetic energy (EME) in different proportions across the range of wavelengths known as the electromagnetic spectrum, and that such differences allow these objects to be identified uniquely. Sensors mounted on aircraft or satellite platforms record the magnitude of the energy flux reflected from or emitted by objects on the Earth's surface. These measurements are made at a large number of points distributed either along a one-dimensional profile on the ground below the platform or over a two-dimensional area below or to one side of the ground track of the platform. Figure 1.2a shows an image being collected by a nadir-looking sensor.

Data in the form of one-dimensional profiles are not considered in this book, which is concerned with

Computer Processing of Remotely-Sensed Images: An Introduction, Fourth Edition Paul M. Mather and Magaly Koch

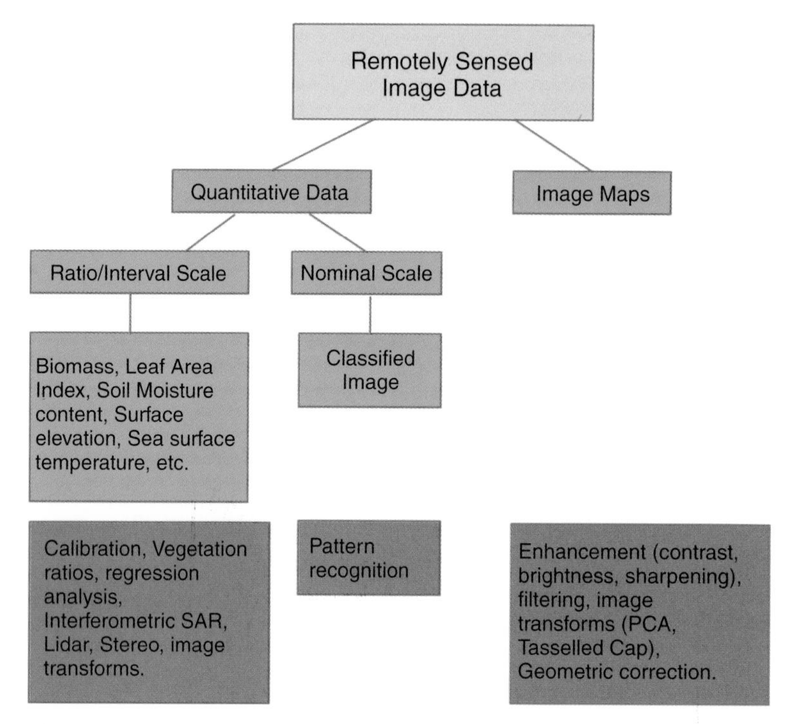

Figure 1.1 *Uses of remotely-sensed data. The green boxes show the products derived from remotely-sensed data, such as image maps and classified images. The blue boxes show the computer processing techniques that are used to derive these products. Image maps are frequently used as backdrops in a GIS, whereas the process of pattern recognition produces labelled (nominal scale) images showing the distribution of individual Earth surface cover types. Quantitative measures such as vegetation indices are derived from calibrated data, and are often linked via regression analysis to Earth surface properties such as sea-surface temperature or soil moisture content. The computer processing techniques to extract and analyse remotely-sensed data are presented in the remainder of this book.*

the processing of two-dimensional (spatial) data collected by imaging sensors. Imaging sensors are either nadir – (vertical) or side-looking. In the former case, the ground area to either side of the point immediately below the satellite or aircraft platform is imaged, while in the latter case an area of the Earth's surface lying to one side or other of the satellite or aircraft track is imaged. The most familiar kinds of images, such as those collected by the nadir-looking thematic mapper (TM) and enhanced thematic mapper plus (ETM+) instruments carried by US Landsat satellites numbered 5 and 7 (6 never reached orbit), and by the HRV instrument (which can be side-looking or nadir-pointing) on board the French/Belgian/Swedish SPOT satellites, are scanned line by line (from side to side) as the platform moves forwards along its track. This forward (or along track) motion of the satellite or aircraft is used to build up an image of the Earth's surface by the collection of successive scan lines (Figure 1.2a).

Two kinds of scanners are used to collect the EMR that is reflected or emitted by the ground surface along each scan line. Electromechanical scanners have a small number of detectors, and they use a mirror that moves back and forth to collect electromagnetic energy across the width of the scan line (AB in Figure 1.2a). The electromagnetic energy reflected by or emitted from the portion of the Earth's surface that is viewed at a given instant in time is directed by the mirror onto these detectors (Figure 1.2b). The second type of scanner, the push-broom scanner, uses an array of solid-state charge-coupled devices (CCDs), each one of which 'sees' a single point on the scan line (Figure 1.2c). Thus, at any given moment, each detector in the CCD array is observing a small area of the Earth's surface along the scan line. This ground area is called a pixel. A remotely-sensed image is made up of a rectangular matrix of measurements of the flux or flow of EMR emanating from individual pixels, so that each pixel value represents the magnitude of upwelling EMR for a small ground area (though it will be seen later that there is 'interference' from neighbouring pixels). This upwelling radiation contains information about (i) the nature of the Earth-surface material present in the pixel area, (ii) the topographic position of the pixel area (i.e. whether it is horizontal, on a sunlit slope or on a shaded slope) and (iii) the state of the atmosphere through which the EMR has to pass. This account of image acquisition is a very simplified one, and more detail is provided in Chapter 2. The nature of Earth-surface

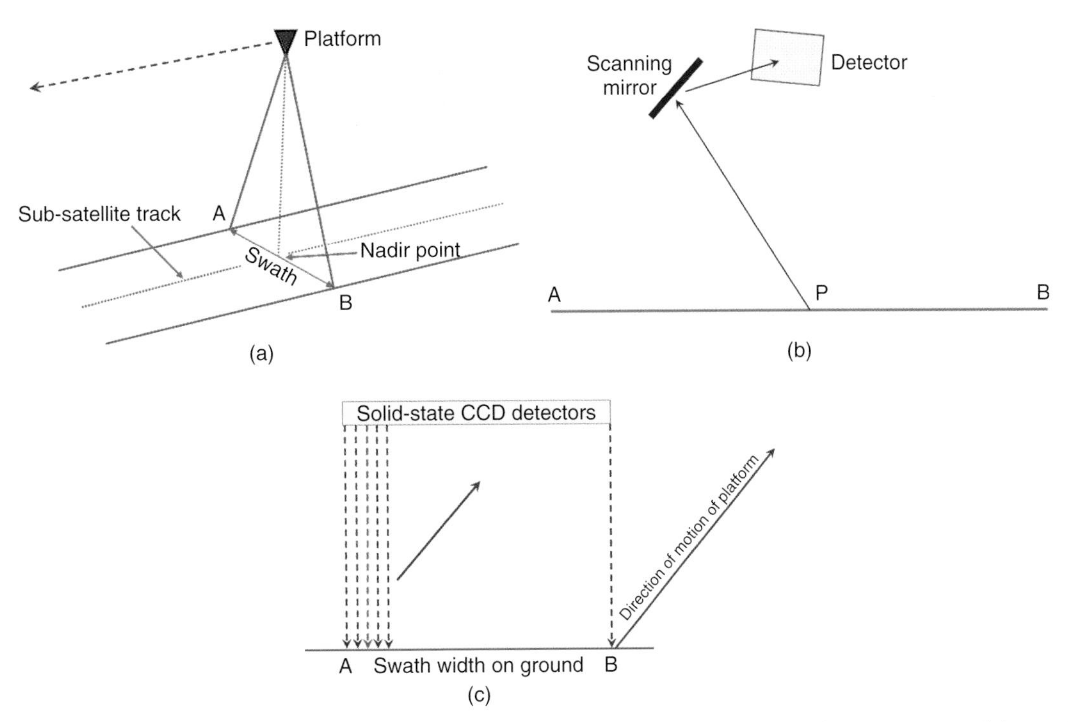

Figure 1.2 *(a) A sensor carried onboard a platform, such as an Earth-orbiting satellite, builds up an image of the Earth's surface by taking repeated measurements across the swath AB. As the satellite moves forward, so successive lines of data are collected and a two-dimensional image is generated. The distance AB is the swath width. The point immediately below the platform is the nadir point, and the imaginary line traced on the Earth's surface by the nadir point is the subsatellite track. (b) Upwelling energy from point P is deflected by a scanning mirror onto the detector. The mirror scans across a swath between points A and B on the Earth's surface. (c) An array of solid state (CCD) detectors images the swath AB. The image is built up by the forward movement of the platform.*

materials and their interaction with EMR is covered in Section 1.3. Topographic and atmospheric interactions are described in Sections 4.7 and 4.4, respectively.

The magnitude of the radiance reflected or emitted by the small ground area represented by a pixel is a physical measurement that is converted to a number, usually an integer (a whole number) lying within a specified range, such as 0–255 (8 bits) or 0–65 535 (16 bits). Remotely-sensed images thus consist of rectangular arrays of numbers, and because they are numerical in nature so computers are used to display, enhance and manipulate them. The main part of this book deals with techniques used in these types of processing. Spatial patterns evident in remotely-sensed images can be interpreted in terms of geographical variations in the nature of the material forming the surface of the Earth. These Earth surface materials may be vegetation, exposed soil and rock or water surfaces. Notice that the characteristics of these materials are not detected directly by remote sensing. Their nature is inferred from the properties of the EMR that is reflected, scattered or emitted by these materials and recorded by the sensor. Another characteristic of digital image data is that they can be calibrated in order to provide estimates of physical measurements of properties of the target such as radiance, reflection or albedo. These values are used, for example in models of climate or crop growth. Examples of the uses of remotely-sensed image data in Earth science and environmental management can be found in Calder (1991). Kaufman *et al.* (1998) demonstrate the wide variety of applications of remote sensing data collected by the instruments on the American Terra satellite. A number of web sites provide access to image libraries. Perhaps the most accessible of these is NASA's Earth Observatory. The Internet contains an ever-changing but large number of Earth-observation images that can be best discovered by using a search engine. Many national space agencies maintain good web sites, for example NASA (USA), CNES (France) and DLR (Germany).

Aerial photography is a familiar form of EO by remote sensing. Past generations of air photographs differ from digital images in that they are analogue in nature. Analogue means: using some alternative physical representation to display some property of interest. For example, a photographic film represents a range of light levels in terms of the differential response of silver halide particles in the film emulsion. Analogue images cannot be processed by computer unless they are converted to digital form, using a scanning device. Computer scanners operate much in the same way as those carried by satellites in that they view a small area of the photograph, record the proportion of incident light that is reflected back by that small area and convert that proportion to a number, usually in the range 0 (no reflection, or black) to 255 (100% reflection, or white). The numbers between 0 and 255 represent increasingly lighter shades of grey.

Nowadays, analogue cameras are rarely used and digital cameras are most often chosen for use in aerial photography. Images acquired by such cameras are similar in nature to those produced by the pushbroom type of sensor mentioned above. Instead of a film, a digital camera has a two-dimensional array of charge-coupled devices (CCD) (rather than a one-dimensional CCD array, as used by the SPOT satellite's HRV instrument, mentioned above). The amount of light from the scene that impinges on an individual CCD is recorded as a number in the range 0 (no light) to 255 (detector saturated). A value of 255 is typically used as the upper bound of the range but a different value may be selected depending on the camera characeteristics. A two-dimensional set of CCD measurements produces a greyscale image. Three sets of CCDs are used to produce a colour image, just as three layers of film emulsion are used to generate an analogue colour photograph. The three sets of CCDs measure the amounts of red, green and blue light that reach the camera. Nowadays, digital imagery is relatively easily available from digital cameras, from scanned analogue photographs, as well as from sensors carried by aircraft, including unmanned drones (see Zhou *et al.*, 2009) and satellites.

The nature and properties of EMR are considered in Section 1.2, and are those which concern its interaction with the atmosphere, through which the EMR passes on its route from the Sun (or from another source such as a microwave radar) to the Earth's surface and back to the sensor mounted onboard an aircraft or satellite. Interactions between EMR and Earth surface materials are summarized in Section 1.3. It is by studying these interactions that the nature and properties of the material forming the Earth's surface are inferred.

The topics covered in this book are dealt with to a greater or lesser extent in a number of textbooks, research monographs and review articles. A good library will provide paper or electronic access to a selection of recent texts that include: Adams and Gillespie (2006), Campbell (2006), Chuvieco (2008), Drury (2004), Elachi and van Zyl (2006), Gao (2009), Gonzales and Woods (2007), Jensen (1996), Landgrebe (2003), Liang (2004), Liang *et al.* (2008), Lillesand, Kiefer and Chipman (2008), Milman (1999), Olsen (2007), Rees (2001), Richards and Jia (2006), Schowengerdt (2006), Smith (2001), Warner, Nellis and Foody (2009) and Weng and Quattrochi (2007). Sanchez and Canton (1998) discuss both remote sensing and space telescope images, while Rees (2001) covers basic principles in a very understandable way. Trauth *et al.* (2007) contains MATLAB code and explanation of some of the topics mentioned in this book, including image processing, DEM manipulation and geostatistics. Some of the material listed above is not written specifically for a remote sensing audience, but nevertheless, contain useful and often additional reading for those readers wishing to follow up a particular topic for a thesis or dissertation. The above list of books and articles may seem to be lengthy, but some readers may prefer to avoid an overtly mathematical approach and thus select one source rather than another.

1.2 Electromagnetic Radiation and Its Properties

1.2.1 Terminology

The terminology used in remote sensing is sometimes understood only imprecisely, and is therefore occasionally used loosely. A brief guide is therefore given in this section. It is neither complete nor comprehensive, and is meant only to introduce some basic ideas. The subject is dealt with more thoroughly by Bird (1991a, b), Chapman (1995), Hecht (2001), Kirkland (2007), Martonchik, Bruegge and Strahler (2000), Nicodemus *et al.* (1977), Rees (2001), Schaepman-Strub *et al.* (2006), Slater (1980), Smith (2001) and the references listed at the end of the preceding section.

EMR transmits energy. As the name implies, EMR has two components. One is the electric field and the other is the magnetic field. These two fields are mutually perpendicular, and are also perpendicular to the direction of travel (Figure 1.3). There is no 'right way up' – EMR can be transmitted with a horizontal electric field and a vertical magnetic field, or vice versa. The disposition of the two fields is described by the polarization state of the EMR, which can be either horizontal or vertical. Polarization state is used in microwave remote sensing (Section 2.4).

Energy is the capacity to do work. It is expressed in joules (J), a unit that is named after James Prescott Joule, an English brewer whose hobby was physics. *Radiant energy* is the energy associated with EMR. The rate of transfer of energy from one place to another (for example from the Sun to the Earth) is termed the *flux* of energy, the word *flux* being derived from the Latin word meaning 'flow'. It is measured in watts (W), after James Watt (1736–1819), the Scottish inventor who was instrumental in designing an efficient steam engine while he was working as a technician at Glasgow University (he is also credited with developing the first rev counter). The interaction between EMR and surfaces such as that of the Earth can be understood more clearly if the concept of

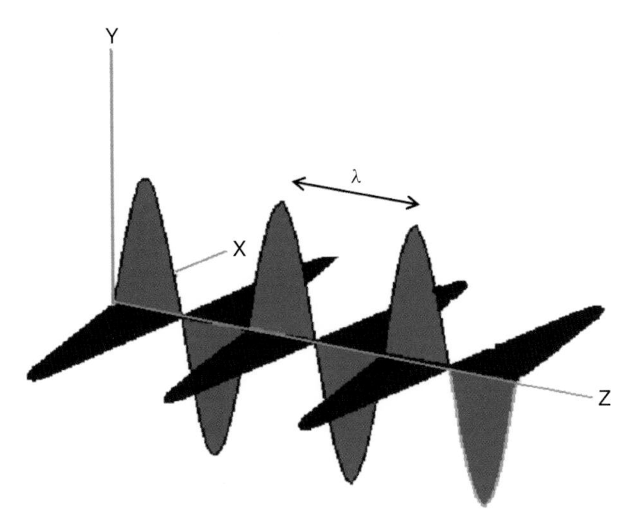

Figure 1.3 *Electromagnetic wave. The wavelength of the electromagnetic energy is represented by the Greek letter lambda (λ). Adapted from a figure by Nick Strobel, from http://www.astronomynotes.com/light/s2.htm. Accessed 24 May 2010.*

radiant flux density is introduced. Radiant flux is the rate of transfer of radiant (electromagnetic) energy. Density implies variability over the two-dimensional surface on which the radiant energy falls, hence radiant flux density is the magnitude of the radiant flux that is incident upon or, conversely, is emitted by a surface of unit area (measured in watts per square metre or $\mathrm{W\,m^{-2}}$). The topic of emission of EMR by the Earth's surface in the form of heat is considered at a later stage. If radiant energy falls (i.e. is incident) upon a surface then the term *irradiance* is used in place of radiant flux density. If the energy flow is away from the surface, as in the case of thermal energy emitted by the Earth or solar energy that is reflected by the Earth, then the term *radiant exitance* or *radiant emittance* (measured in units of $\mathrm{W\,m^{-2}}$) is appropriate.

The term *radiance* is used to mean the radiant flux density transmitted from a unit area on the Earth's surface as viewed through a unit solid (three-dimensional) angle (just as if you were looking through a hole at the narrow end of an ice-cream cone). This solid angle is measured in *steradians*, the three-dimensional equivalent of the familiar radian (defined as the angle subtended at the centre of a circle by a sector which cuts out a section of the circumference that is equal in length to the radius of the circle). If, for the moment, we consider that the irradiance reaching the surface is back-scattered in all upward directions (Figure 1.4), then a proportion of the radiant flux would be measured per unit solid viewing angle. This proportion is the radiance (Figure 1.5). It is measured in watts per square metre per steradian ($\mathrm{W\,m^{-2}\,sr^{-1}}$). The concepts of the radian and steradian are illustrated in Figure 1.6.

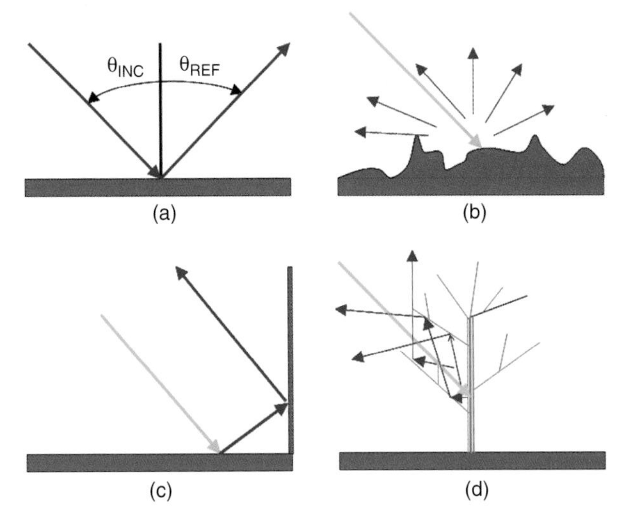

Figure 1.4 *Types of scattering of electromagnetic radiation. (a) Specular, in which incident radiation is reflected in the forward direction, (b) Lambertian, in which incident radiation is equally scattered in all upward directions, (c) corner reflector, which acts like a vertical mirror, especially at microwave wavelengths and (d) volume scattering, in which (in this example) branches and leaves produce single-bounce (primary) and multiple-bounce (secondary) scattering.*

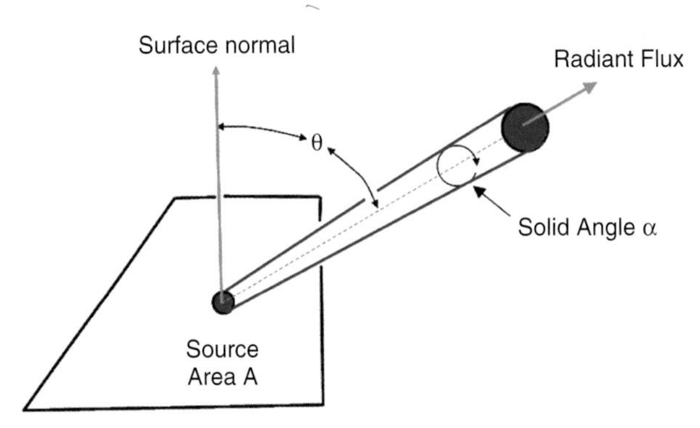

Figure 1.5 *Radiance is the flux of electromagnetic energy leaving a source area A in direction θ per solid angle α. It is measured in watts per square metre per steradian ($\mathrm{W\,m^{-2}\,sr^{-1}}$).*

Reflectance, ρ, is the dimensionless ratio of the radiant emittance of an object and the irradiance. The reflectance of a given object is independent of irradiance, as it is a ratio. When remotely-sensed images collected over a time period are to be compared it is common practice to convert the radiance values recorded by the sensor into reflectance factors in order to eliminate the effects of variable irradiance over the seasons of the year. This topic is considered further in Section 4.6.

The quantities described above can be used to refer to particular wavebands rather than to the whole electromagnetic spectrum (Section 1.2.3). The terms are then preceded by the adjective *spectral;* for example the spectral

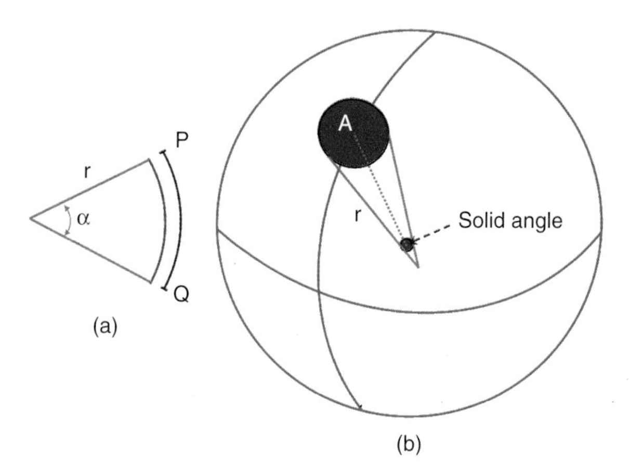

Figure 1.6 *(a) The angle α formed when the length of the arc PQ is equal to the radius of the circle r is equal to 1 radian or approximately 57°. Thus, angle $\alpha = PQ/r$ radians. There are 2π radians in a circle (360°). (b) A steradian is a solid three-dimensional angle formed when the area A delimited on the surface of a sphere is equal to the square of the radius r of the sphere. A need not refer to a uniform shape. The solid angle shown is equal to A/r^2 steradians (sr). There are 4π steradians in a sphere.*

radiance for a given waveband is the radiant flux density in that waveband (i.e. spectral radiant flux density) per unit solid angle. Terms such as *spectral irradiance*, *spectral reflectance* and *spectral exitance* are defined in a similar fashion.

1.2.2 Nature of Electromagnetic Radiation

An important point of controversy in physics over the last 300 years has concerned the nature of EMR. Newton, while not explicitly rejecting the idea that light is a wave-like form of energy (the wave theory) inclined to the view that it is formed of a stream of particles (the corpuscular theory). The wave–corpuscle dichotomy was not to be resolved until the early years of the twentieth century with the work of Planck, Einstein and others. The importance to remote sensing of the nature of EMR is fundamental, for we need to consider radiation both as a waveform and as a stream of particles. The wave-like characteristics of EMR allow the distinction to be made between different manifestations of such radiation (for example microwave and infrared (IR) radiation) while, in order to understand the interactions between EMR and the Earth's atmosphere and surface, the idea that EMR consists of a stream of particles is most easily used. Building on the work of Planck, Einstein proposed in 1905 that light consists of particles called *photons*, which, in most respects, were similar to other sub-atomic particles such as protons and neutrons. It was found that, at the subatomic level, both wave-like and particle-like properties were exhibited, and that phenomena at this level appear to be both waves and particles. Erwin Schrödinger (1867–1961) wrote as follows in *Science, Theory and Man* (New York, Dover, 1957):

> In the new setting of ideas, the distinction [between particles and waves] has vanished, because it was discovered that all particles have also wave properties, and vice-versa. Neither of the concepts must be discarded, they must be amalgamated. Which aspect obtrudes itself depends not on the physical object but on the experimental device set up to examine it.

Thus, from the point of view of quantum mechanics, EMR is both a wave and a stream of particles. Whichever view is taken will depend on the requirements of the particular situation. In Section 1.2.5, the particle theory is best suited to explain the manner in which incident EMR interacts with the atoms, molecules and other particles which form the Earth's atmosphere. Readers who, like myself, were resistant in their formative years to any kind of formal training in basic physics will find Gribben (1984) to be readable as well as instructive, while Feynman (1985) is a clear and well-illustrated account of the surprising ways that light can behave.

1.2.3 The Electromagnetic Spectrum

The Sun's light is the form of EMR that is most familiar to human beings. Sunlight that is reflected by physical objects travels in most situations in a straight line to the observer's eye. On reaching the retina, it generates electrical signals that are transmitted to the brain by the optic nerve. The brain uses these signals to construct an image of the viewer's surroundings. This is the process of vision, which is closely analogous to the process of remote sensing; indeed, vision is a form – perhaps the basic form – of remote sensing (Greenfield, 1997). A discussion of the human visual process can be found in Section 5.2. Note that the process of human vision involves image acquisition (essentially a physiological process) and image understanding (a psychological process), just as EO by remote sensing does. Image interpretation and understanding in remote sensing might therefore be considered to be an attempt to simulate or emulate the brain's image understanding functions.

Visible light is so called because the eye detects it, whereas other forms of EMR are invisible to the unaided eye. Sir Isaac Newton (1643–1727) investigated the nature of white light, and in 1664 concluded that it is made up of differently coloured components, which he saw by passing white light through a prism to form a rainbow-like spectrum. Newton saw the *visible spectrum*, which ranges from red through orange, yellow and green to blue, indigo and violet. Later, the astronomer Friedrich Wilhelm (Sir William) Herschel (1728–1822) demonstrated the existence of EMR with wavelengths beyond those of the visible spectrum; these he called *infrared*, meaning *beyond the red*. It was subsequently found that EMR also exists beyond the violet end of the visible spectrum, and this form of radiation was given the name *ultraviolet*. (Herschel, incidentally, started his career as a band-boy with the Hanoverian Guards and later came to live in England.)

Other forms of EMR, such as X-rays and radio waves, were discovered later, and it was eventually realized that all were manifestations of the same kind of radiation which travels at the speed of light in a wave-like form, and which can propagate through empty space. The speed of light (c_0) is 299 792 458 m s^{-1} (approximately 3×10^8 m s^{-1}) in a vacuum, but is reduced by a factor called the index of refraction if the light travels through media such as the atmosphere or water. EMR reaching the Earth comes mainly from the Sun and is produced by thermonuclear reactions in the Sun's core. The set of all electromagnetic waves is called the *electromagnetic spectrum*, which includes the range from the long radio waves, through the microwave and IR wavelengths to visible light waves and beyond to the ultraviolet and to the short-wave X and γ rays (Figure 1.7).

Symmetric waves can be described in terms of their *frequency* (f), which is the number of waveforms passing a fixed point in unit time. This quantity used to be known as *cycles per second* (cps) but nowadays the preferred term is Hz (Hertz, after Heinrich Hertz (1857–1894), who, between 1885 and 1889 became the first person to broadcast and receive radio waves). Alternatively, the concept of *wavelength* can be used (Figure 1.8). The wavelength is the distance between successive peaks (or successive troughs) of a waveform, and is normally measured in metres or fractions of a metre (Table 1.1). Both frequency and wavelength convey the same information and are often used interchangeably. Another measure of the nature of a waveform is its period (T). This is the time, in seconds, needed for one full wave to pass a fixed point. The relationships between wavelength, frequency and period are given by:

$$f = c/\lambda$$

$$\lambda = c/f$$

$$T = 1/f = \lambda/c$$

In these expressions, c is the speed of light. The velocity of propagation (v) is the product of wave frequency and wavelength, that is

$$v = \lambda f$$

The *amplitude* (A) of a wave is the maximum distance attained by the wave from its mean position (Figure 1.8). The amount of energy, or intensity, of the waveform

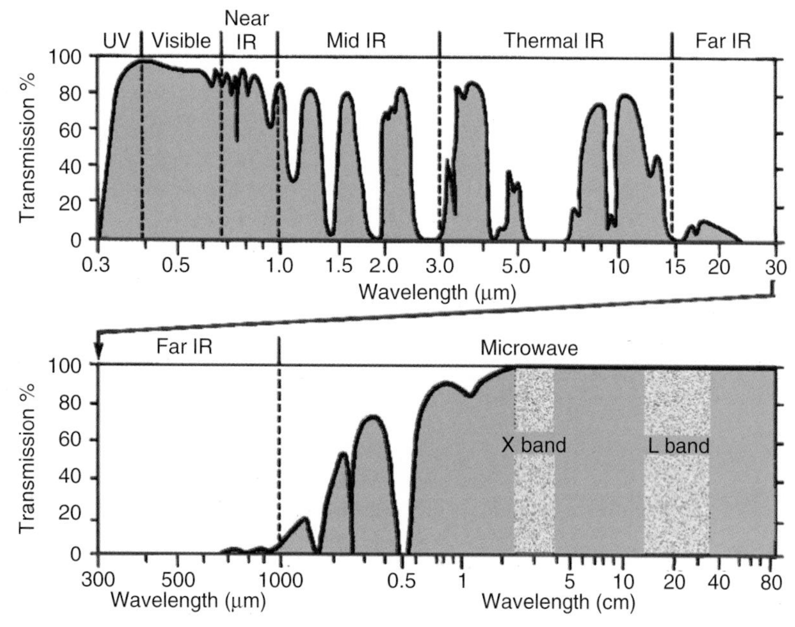

Figure 1.7 *The electromagnetic spectrum showing the range of wavelengths between 0.3 μm and 80 cm. The vertical dashed lines show the boundaries of wavebands such as ultraviolet (UV) and near-infrared (near IR). The shaded areas between 2 and 35 cm wavelength indicate two microwave wavebands (X band and L band) that are used by imaging radars. The curve shows atmospheric transmission. Areas of the electromagnetic spectrum with a high transmittance are known as atmospheric windows. Areas of low transmittance are opaque and cannot be used to remotely sense the Earth's surface. Reprinted from AFH Goetz and LC Rowanm 1981, Geologic Remote Sensing, Science,* ***221****, 781–791, Figure 1.*

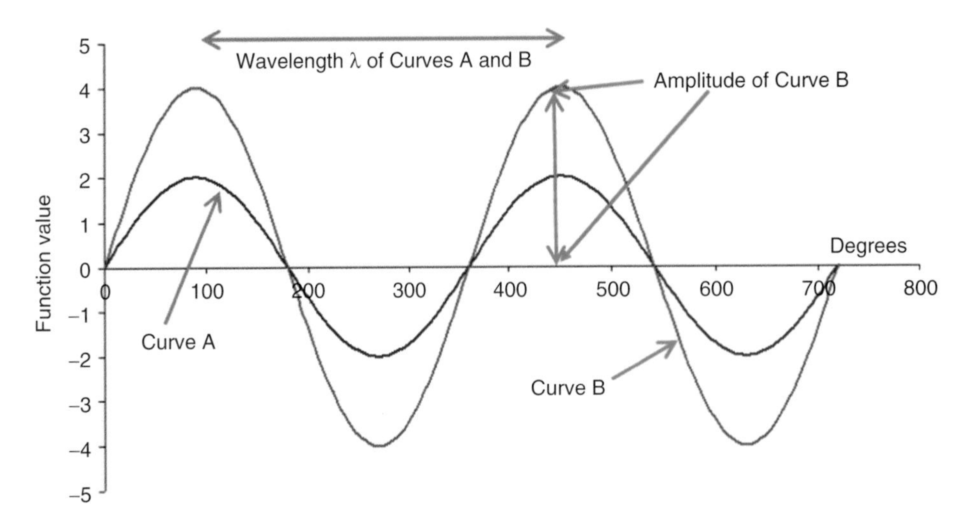

Figure 1.8 *Two curves (waveforms) A and B have the same wavelength (360° or 2π radians, x-axis). However, curve A has an amplitude of two units (y-axis) while curve B has an amplitude of four units. If we imagine that the two curves repeat to infinity and are moving to the right, like traces on an oscilloscope, then the frequency is the number of waveforms (0–2π) that pass a fixed point in unit time (usually measured in cycles per second or Hertz, Hz). The period of the waveform is the time taken for one full waveform to pass a fixed point. These two waveforms have the same wavelength, frequency and period and differ only in terms of their amplitude.*

Table 1.1 *Terms and symbols used in measurement.*

Factor	Prefix	Symbol	Factor	Prefix	Symbol
10^{-18}	atto	a	–	–	–
10^{-15}	femto	f	10^{15}	peta	P
10^{-12}	pico	p	10^{12}	tera	T
10^{-9}	nano	n	10^{9}	giga	G
10^{-6}	micro	μ	10^{6}	mega	M
10^{-3}	milli	m	10^{3}	kilo	k

is proportional to the square of the amplitude. Using the relationships specified earlier we can compute the frequency given the wavelength, and vice versa. If, for example wavelength λ is 0.6 μm or 6×10^{-7} m, then, since velocity v equals the product of wavelength and frequency f, it follows that:

$$v = 6 \times 10^{-7} f$$

so that:

$$f = \frac{c_0}{v} = \frac{3 \times 10^{8}}{6 \times 10^{-7}}\ \text{Hz}$$

that is

$$f = 0.5 \times 10^{15}\text{Hz} = 0.5\ \text{PHz}$$

1 PHz (petahertz) equals 10^{15} Hz (Table 1.1). Thus, EMR with a wavelength of 0.6 μm has a frequency of 0.5×10^{15} Hz. The *period* is the reciprocal of the frequency, so one wave of this frequency will pass a fixed point in 2×10^{-15} s. The amount of energy carried by the waveform, or the squared amplitude of the wave, is defined for a single photon by the relationship

$$E = hf$$

where E is energy, h is a constant known as Planck's constant (6.625×10^{-34} J s) and f is frequency. Energy thus increases with frequency, so that high frequency, short-wavelength EMR such as X-rays carries more energy than does longer-wavelength radiation in the form of visible light or radio waves.

While EMR with particular temporal and spatial properties is used in remote sensing to convey information about a target, it is interesting to note that both time and space are defined in terms of specific characteristics of EMR. A second is the duration of 9 192 631 770 oscillations of the caesium radiation (in other words, that number of wavelengths or cycles are emitted by caesium radiation in 1 s; its frequency is approximately 9 GHz or a wavelength of around 0.03 m). A metre is defined as 1 650 764.73 vacuum wavelengths of the orange–red light emitted by krypton-86.

Visible light is defined as electromagnetic radiation with wavelengths between (approximately) 0.4 and 0.7 μm. We call the shorter wavelength end (0.4 μm) of the visible spectrum 'blue' and the longer wavelength end (0.7 μm) 'red' (Table 1.2). The eye is not uniformly sensitive to light within this range, and has its peak sensitivity at around 0.55 μm, which lies in the green part of the visible spectrum (Figure 1.7 and Figure 1.9). This peak in the response function of the human eye

Table 1.2 *Wavebands corresponding to perceived colours of visible light.*

Colour	Waveband (μm)	Colour	Waveband (μm)
Red	0.780 to 0.622	Green	0.577 to 0.492
Orange	0.622 to 0.597	Blue	0.492 to 0.455
Yellow	0.597 to 0.577	Violet	0.455 to 0.390

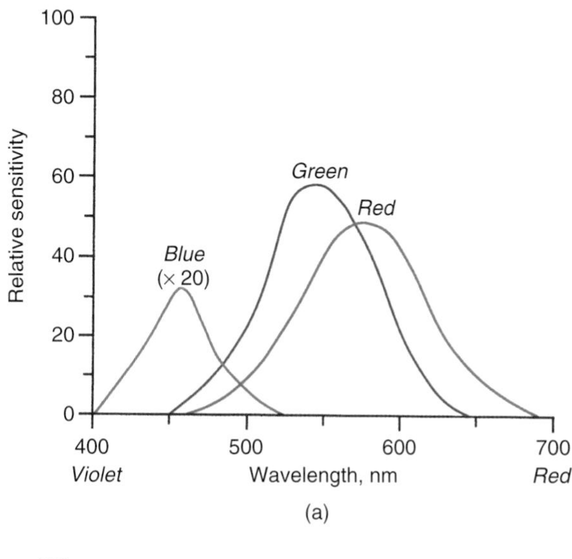

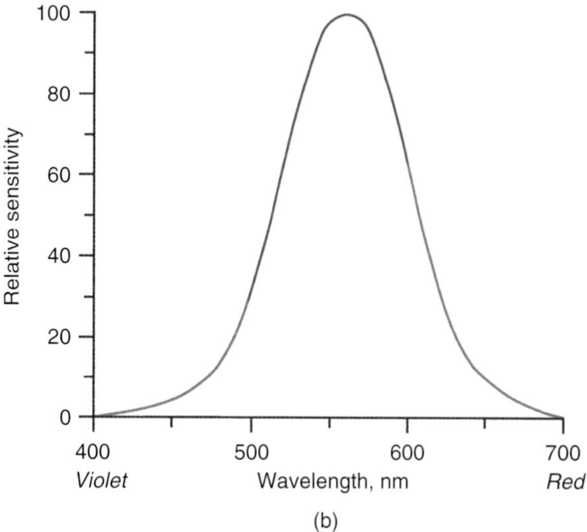

Figure 1.9 *(a) Response function of the red-, green- and blue-sensitive cones on the retina of the human eye. (b) Overall response function of the human eye. Peak sensitivity occurs near 550 nm (0.55 μm).*

corresponds closely to the peak in the Sun's radiation emittance distribution (Section 1.2.4).

The process of atmospheric scattering, discussed in Section 1.2.5 below, deflects light rays from a straight path and thus causes blurring or haziness. It affects the blue end of the visible spectrum more than the red end, and consequently the blue wave and is not used in many remote-sensing systems.

Figure 1.10a–c shows three greyscale images collected in the blue/green, green and red wavebands respectively by a sensor called the Thematic Mapper that is carried by the American Landsat-5 and Landsat-7 satellites (Chapter 2). The different land cover types reflect energy in the visible spectrum in a differential manner, although the clouds and cloud shadows in the upper centre of the image are clearly visible in all three images. Various crops in the fields round the village of Littleport (north of Cambridge in eastern England) can be discriminated, and the River Ouse can also be seen as it flows northwards in the right hand side of the image area. It is dangerous to rely solely on visual interpretation of images such as these. This book is about the processing and manipulation of images, and we will see that it is possible to change the colour balance, brightness and contrast of images to emphasize (or de-emphasize) particular targets. Digital image processing should be an aid to interpretation, but the user should always be aware of enhancements that have been carried out.

EMR with wavelengths shorter than those of visible light (less than 0.4 μm) is divided into three spectral regions, called γ rays, X-rays and ultraviolet radiation. Because of the effects of atmospheric scattering and absorption (Section 4.4; Figure 4.11), none of these wavebands is used in satellite remote sensing, though low-flying aircraft can detect γ-ray emissions from radioactive materials in the Earth's crust. Radiation in these wavebands is dangerous to life, so the fact that it is mostly absorbed or scattered by the atmosphere allows life to exist on Earth. In terms of the discussion of the wave–particle duality in Section 1.2.2, it should be noted that γ radiation has the highest energy levels and is the most 'particle-like' of all EMR, whereas radio frequency radiation is most 'wave-like' and has the lowest energy levels.

Wavelengths that are longer than the visible red are subdivided into the IR, microwave and radio frequency wavebands. The IR waveband, extending from 0.7 μm to 1 mm, is not a uniform region. Short-wavelength infrared (SWIR) or near-infrared (NIR) energy, with wavelengths between 0.7 and 0.9 μm, behaves like visible light and can be detected by special photographic film. IR radiation with a wavelength of up to 3.0 μm is primarily of solar origin and, like visible light, is reflected by the surface of the Earth. Hence, these wavebands are often known as the *optical* bands. Figure 1.11a shows an image of the area shown in Figure 1.10 collected by the Landsat TM sensor in the NIR region of the spectrum (0.75 – 0.90 μm). This image is considerably clearer than the visible spectrum images shown in Figure 1.10. We will see in Section 1.3.2 that the differences in reflection

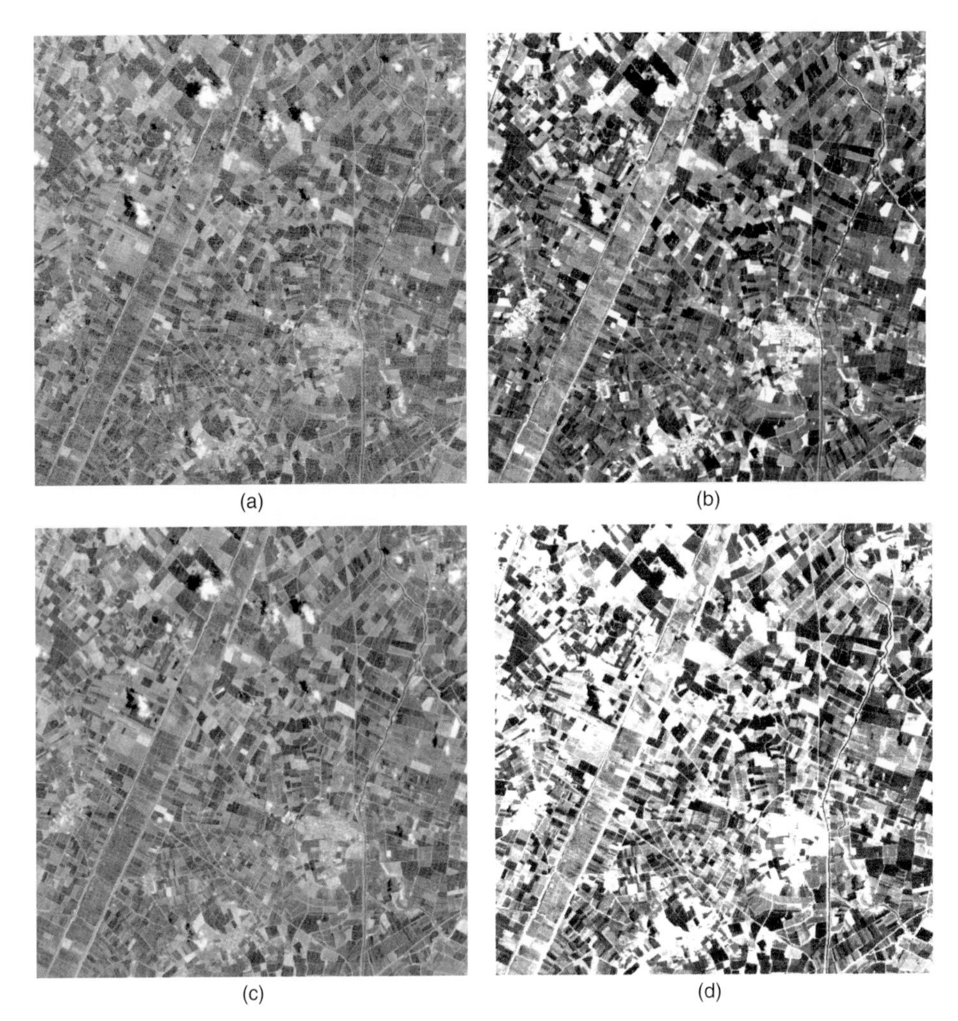

Figure 1.10 *Images collected in (a) band 1 (blue–green), (b) band 2 (green) and (c) band 3 (red) wavebands of the optical spectrum by the Thematic Mapper sensor carried by the Landsat-5 satellite. Image (d) shows the three images (a–c) superimposed with band 1 shown in blue, band 2 in green and band 3 in red. This is called a natural colour composite image. The area shown is near the town of Littleport in Cambridgeshire, eastern England. The diagonal green strip is an area of fertile land close to a river. Original data © ESA 1994; Distributed by Eurimage.*

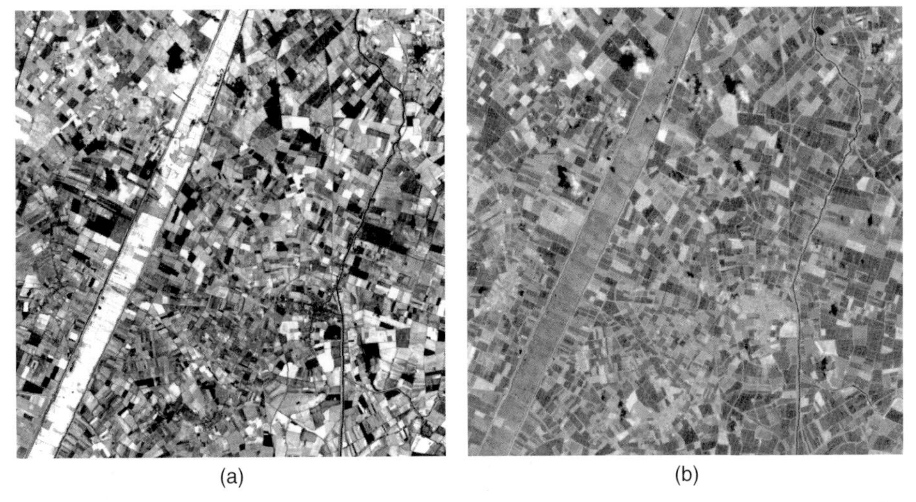

Figure 1.11 *Image of ground reflectance in (a) the 0.75–0.90 μm band (near infrared) and (b) the middle infrared (2.08–2.35 μm) image of the same area as that shown in Figure 1.10. These images were collected by the Landsat-5 Thematic Mapper (bands 4 and 7). Original data © ESA 1994; Distributed by Eurimage.*

between vegetation, water and soil are probably greatest in this NIR band. An image of surface reflection in the Landsat TM mid-IR waveband (2.08 – 2.35 μm) of the same area is shown in Figure 1.11b.

In wavelengths longer than around 3 μm, IR radiation emitted by the Earth's surface can be sensed in the form of heat. The amount and wavelength of this radiation depends on the temperature of the source (Section 1.2.4). Because these longer IR wavebands are sensed as heat, they are called the *thermal infrared* (TIR) wavebands. Much of the TIR radiation emitted by the Earth is absorbed by, and consequently heats, the atmosphere thus making life possible on the Earth (Figure 1.7). There is, however, a 'window' between 8 and 14 μm which allows a satellite sensor above the atmosphere to detect thermal radiation emitted by the Earth, which has its peak wavelength at 9.7 μm. Note, though, that the presence of ozone in the atmosphere creates a narrow absorption band within this window, centred at 9.5 μm. Boyd and Petitcolin (2004) consider remote sensing in the region 3.0 – 5.0 μm. There are a number of regions of high transmittance in this middle IR band, which is really a transition between reflected visible and NIR radiation and emitted TIR radiation.

Absorption of longer-wave radiation by the atmosphere has the effect of warming the atmosphere. This is called the natural greenhouse effect. Water vapour (H_2O) and carbon dioxide (CO_2) are the main absorbing agents, together with ozone (O_3). The increase in the carbon dioxide content of the atmosphere over the last century, due to the burning of fossil fuels, is thought to enhance the greenhouse effect and to raise the temperature of the atmosphere above its natural level. This could have long-term climatic consequences. An image of part of Western Europe acquired by the Advanced Very High Resolution Radiometer (AVHRR) carried by the US NOAA-14 satellite is shown in Figure 1.12. The different colours show different levels of emitted thermal radiation in the 11.5 – 12.5 μm waveband. Before these colours can be interpreted in terms of temperatures, the effects of the atmosphere as well as the nature of the sensor calibration must be considered. Both these topics are covered in Chapter 4. For comparison, a visible band image of Europe and North Africa produced by the Meteosat-6 satellite is shown in Figure 1.13. Both images were collected by the UK NERC-funded satellite receiving station at Dundee University, Scotland. Both images were originally in greyscale but were converted to colour using a procedure called density slicing, which is considered in detail in Section 5.4.1.

That region of the spectrum composed of EMR with wavelengths between 1 mm and 300 cm is called the microwave band. Most satellite-borne sensors that operate in the microwave region use microwave radiation with wavelengths between 3 and 25 cm. Radiation at these wavelengths can penetrate cloud, and the microwave band is thus a valuable region for remote sensing in temperate and tropical areas where cloud cover restricts the collection of optical and TIR images.

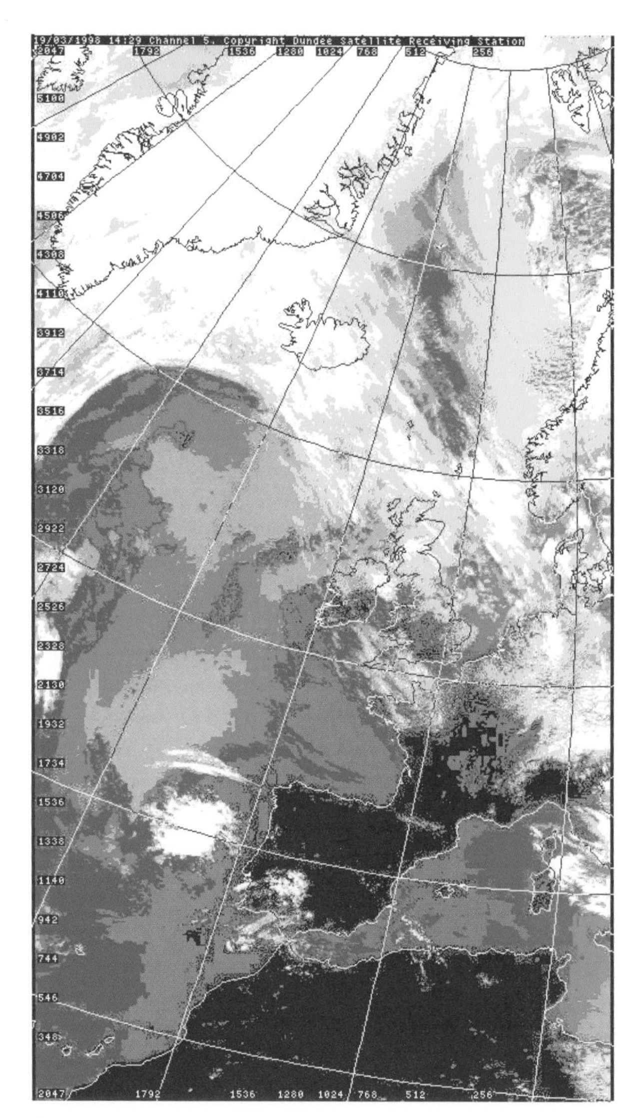

Figure 1.12 *NOAA AVHRR band 5 image (thermal infrared, 11.5–12.5 μm) of western Europe and NW Africa collected at 14.20 on 19 March 1998. The image was downloaded by the NERC Satellite Receiving Station at Dundee University, UK, where the image was geometrically rectified (Chapter 4) and the latitude/longitude grid and digital coastline were added. Darker colours (dark blue, dark green) areas indicate greater thermal emissions. The position of a high-pressure area (anticyclone) can be inferred from cloud patterns. Cloud tops are cold and therefore appear white. The NOAA satellite took just over 15 minutes to travel from the south to the north of the area shown on this image. The colour sequence is (from cold to warm): dark blue–dark green–green–light cyan–pink–yellow–white. © Dundee Satellite Receiving Station, Dundee University.*

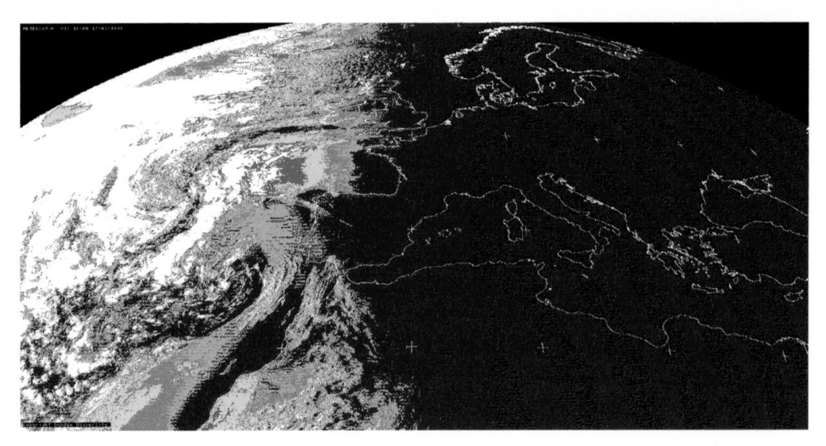

Figure 1.13 *Portion of a Meteosat-6 visible channel image of Europe and North Africa taken at 18.00 on 17 March 1998, when the lights were going on across Europe. Image received by Dundee University, UK. The colour sequence is black–dark blue–cyan–green–yellow–red–white. The banding pattern on the right side of the image (black stripes) is probably electronic noise. © Dundee Satellite Receiving Station, Dundee University.*

Some microwave sensors can detect the small amounts of radiation at these wavelengths that is emitted by the Earth. Such sensors are called passive because they detect EMR that is generated externally, for example by emittance by or reflectance from a target. Passive microwave radiometers such as the SMMR (scanning multichannel microwave radiometer) produce imagery with a low spatial resolution (Section 2.2.1) that is used to provide measurements of sea-surface temperature and wind speed, and also to detect sea ice.

Because the level of microwave energy emitted by the Earth is very low, a high-resolution imaging microwave sensor generates its own EMR at centimetre wavelengths, transmits this energy towards the ground and then detects the strength of the return signal that is scattered by the target in the direction of the sensor. Devices that generate their own electromagnetic energy are called *active* sensors to distinguish them from the *passive* sensors that are used to detect and record radiation of solar or terrestrial origin in the visible, IR and microwave wavebands. Thus, active microwave instruments are not dependent on an external source of radiation such as the Sun or, in the case of thermal emittance, the Earth. It follows that active microwave sensors can operate independently by day or by night. An analogy that is often used is that of a camera. In normal daylight, reflected radiation from the target enters the camera lens and exposes the film. Where illumination conditions are poor, the photographer employs a flashgun that generates radiation in visible wavebands, and the film is exposed by light from the flashgun that is reflected by the target. The microwave instrument produces pulses of energy, usually at centimetre wavelengths, that are transmitted by an antenna or aerial. The same antenna picks up the reflection of these energy pulses as they return from the target.

Microwave imaging sensors are called imaging radars (the word *radar* is an acronym, derived from *ra*dio *d*etection *a*nd *r*anging). The spatial resolution (Section 2.2.1) of imaging radars is a function of their antenna length. If a conventional ('brute force') radar is used, then antenna lengths become considerable as spatial resolution increases. Schreier (1993b, p. 107) notes that if the radar carried by the Seasat satellite (launched in 1981) had used a 'brute force' approach then its 10 m long antenna would have generated images with a spatial resolution of 20 km. A different approach, using several views of the target as the satellite approaches, reaches and passes the target, provides a means of achieving high resolution without the need for excessive antenna sizes. This approach uses the SAR principle, described in Section 2.4, and all satellite-borne radar systems have used the SAR principle. The main advantage of radar is that it is an all-weather, day–night, high spatial resolution instrument, which can operate independently of weather conditions or solar illumination. This makes it an ideal instrument for observing areas of the world such as the temperate and tropical regions, which are often cloud-covered and therefore inaccessible to optical and IR imaging sensors.

A radar signal does not detect either colour information (which is gained from analysis of optical wavelength sensors) or temperature information (derived from data collected by TIR sensors). It can detect both surface roughness and electrical conductivity information (which is related to soil moisture conditions). Because radar is an active rather than a passive instrument, the characteristics of the transmitted signal can be controlled. In particular, the wavelength, depression angle and polarisation of the signal are important properties of the radiation source used in remote sensing. Radar wavelength (Table 1.3)

Table 1.3 Radar wavebands and nomenclature.

Band designation	Frequency (MHz)	Wavelength (cm)
P	300–1000	30–100
L	1000–2000	15–30
S	2000–4000	7.5–15
C	4000–8000	3.75–7.5
X	8000–12 000	2.5–3.75
K_u	12 000–18 000	1.667–2.5
K	18 000–27 000	1.111–1.667
K_a	27 000–40 000	0.75–1.111

determines the observed roughness of the surface, in that a surface that has a roughness with a frequency less than that of the microwave radiation used by the radar is seen as smooth. An X-band (circa 3 cm wavelength) image of the area around the Richat structure in Mauretania is shown in Figure 1.14. Radar sensors are described in more detail in Section 2.4.

Beyond the microwave region is the radio band. Radio wavelengths are used in remote sensing, but not to detect Earth-surface phenomena. Commands sent to a satellite utilize radio wavelengths. Image data is transmitted to ground receiving stations using wavelengths in the microwave region of the spectrum; these data are recorded on the ground by high-speed tape-recorders while the satellite is in direct line of sight of a ground receiving station. Image data for regions of the world that are not within range of ground receiving stations are recorded by onboard tape-recorders or solid-state memory and these recorded data are subsequently transmitted together with currently scanned data when the satellite is within the reception range of a ground receiving station. The first three Landsat satellites (Section 2.3.6) used onboard tape recorders to supplement data that were directly transmitted to the ground. The latest Landsat (number 7) relies on the TDRS (Tracking and Data Relay Satellite) system, which allows direct broadcast of data from Earth resources satellites to one of a set of communications satellites located above the Equator in geostationary orbit (meaning that the satellite's orbital velocity is just sufficient to keep pace with the rotation of the Earth). The signal is relayed by the TDRS system to a ground receiving station at White Sands, NM, USA. European satellites use a similar system called Artemis, which became operational in 2003.

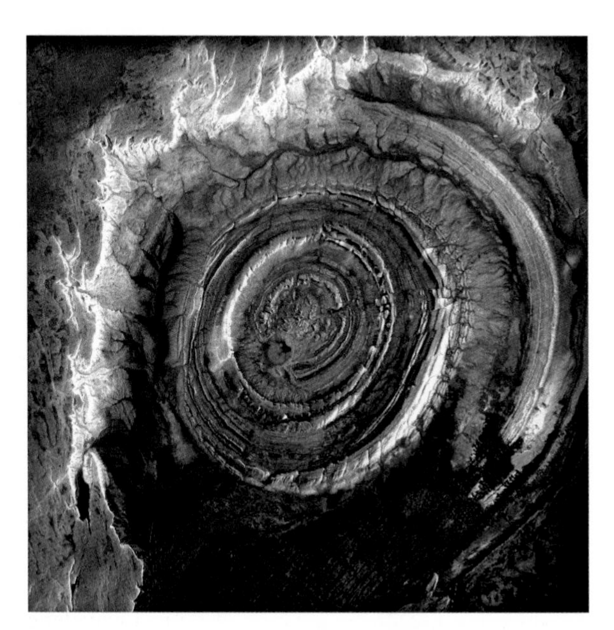

Figure 1.14 *X-band Synthetic aperture radar (SAR) image of the Richat geological structure in Mauretania collected by the Italian satellite COSMO-Skymed 1 on 8 October 2007. The structure is about 60 km in width. COSMO-Skymed (**CO**nstellation of small **S**atellites for **M**editerranean basin **O**bservation) plans to have five satellites in orbit eventually. The third was launched in October, 2008. (http://www.telespazio.it/GalleryMatera.html). COSMO-SkyMed Product © ASI-Agence Spatiale Italiana (YEAR) – All Rights Reserved.*

1.2.4 Sources of Electromagnetic Radiation

All objects whose temperature is greater than absolute zero, which is approximately −273 °C or 0 K (Kelvin), emit radiation. However, the distribution of the amount of radiation at each wavelength across the spectrum is not uniform. Radiation is emitted by the stars and planets; chief of these, as far as the human race is concerned, is the Sun, which provides the heat and light radiation needed to sustain life on Earth. The Sun is an almost-spherical body with a diameter of 1.39×10^6 km and a mean distance from Earth of 150×10^6 km. Its chief constituents are hydrogen and helium. The conversion of hydrogen to helium in the Sun's core provides the energy that is radiated from the outer layers. At the edge of the Earth's atmosphere the power received from the Sun, measured over the surface area of the Earth, is approximately 3.9×10^{22} MW which, if it were distributed evenly over the Earth, would give an incident radiant flux density of 1367 W m^{-2}. This value is known as the *solar constant*, even though it varies throughout the year by about ±3.5%, depending on the distance of the Earth from the Sun (and this variation is taken into account in the radiometric correction of remotely-sensed images; see Section 4.6). Bonhomme (1993) provides a useful summary of a number of aspects relating to solar radiation. On average, 35% of the incident radiant flux is reflected from the Earth (including clouds and atmosphere), the atmosphere absorbs 17%, and 47% is absorbed by the materials forming the Earth's surface. From the Stefan–Boltzmann Law (below) it can be shown that the Sun's temperature is 5777 K if the solar constant is 1367 W m^{-2}. Other estimates of the Sun's temperature range from 5500 to 6200 K. The importance of establishing the surface temperature of the Sun lies in the fact that the distribution of energy emitted in the different regions of the electromagnetic spectrum depends upon the temperature of the source.

If the Sun were a perfect emitter, it would be an example of a theoretical ideal, called a *blackbody*. A blackbody transforms heat energy into radiant energy at the maximum rate that is consistent with the laws of thermodynamics (Suits, 1983). *Planck's Law* describes the spectral exitance (i.e. the distribution of radiant flux density with wavelength, Section 1.2.1) of a blackbody as:

$$M_\lambda = \frac{c_1}{\lambda^5(\exp[c_2/\lambda T] - 1)}$$

where

$c_1 = 3.742 \times 10^{-16}$ W m^{-2}
$c_2 = 1.4388 \times 10^{-2}$ m K
λ = wavelength (m)
T = temperature (Kelvin)
M_λ = spectral exitance per unit wavelength

Curves showing the spectral exitance of blackbodies at temperatures of 1000, 1600 and 2000 K are shown in Figure 1.15. The total radiant energy emitted by a blackbody is dependent on its temperature, and as temperature increases so the wavelength at which the maximum spectral exitance is achieved is reduced. The dotted line in Figure 1.15 joins the peaks of the spectral exitance curves. It is described by Wien's Displacement Law, which gives the wavelength of maximum spectral exitance (λ_m) in terms of temperature:

$$\lambda_m = \frac{c_3}{T}$$

and

$$c_3 = 2.898 \times 10^{-3} \text{ m K}$$

The total spectral exitance of a blackbody at temperature T is given by the Stefan–Boltzmann Law as:

$$M = \sigma T^4$$

In this equation, $\sigma = 5.6697 \times 10^{-8}$ W m^{-2} K^{-4}.

The distribution of the spectral exitance for a blackbody at 5900 K closely approximates the Sun's spectral exitance curve, while the Earth can be considered to act like a blackbody with a temperature of 290 K (Figure 1.16). The solar radiation maximum occurs in the visible spectrum, with maximum irradiance at 0.47 μm. About 46% of the total energy transmitted by the Sun falls into the visible waveband (0.4–0.76 μm).

Wavelength-dependent mechanisms of atmospheric absorption alter the actual amounts of solar irradiance that reach the surface of the Earth. Figure 1.17 shows the spectral irradiance from the Sun at the edge of the atmosphere (solid curve) and at the Earth's surface (broken line). Further discussion of absorption and scattering can be found in Section 1.2.5. The spectral distribution of radiant energy emitted by the Earth (Figure 1.16) peaks in the TIR wavebands at 9.7 μm. The amount of terrestrial emission is low in comparison with solar irradiance. However, the solar radiation absorbed by the atmosphere is balanced by terrestrial emission in the TIR, keeping the temperature of the atmosphere approximately constant. Furthermore, terrestrial TIR emission provides sufficient energy for remote sensing from orbital altitudes to be a practical proposition. The characteristics of the radiation sources used in remote sensing impose some limitations on the range of wavebands available for use. In general, remote sensing instruments that measure the spectral reflectance of solar

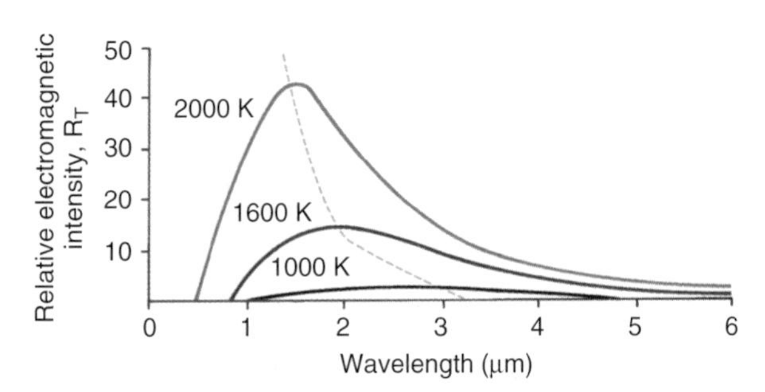

Figure 1.15 Spectral exitance curves for blackbodies at temperatures of 1000, 1600 and 2000 K. The dotted line joins the emittance peaks of the curves and is described by Wien's Displacement Law (see text).

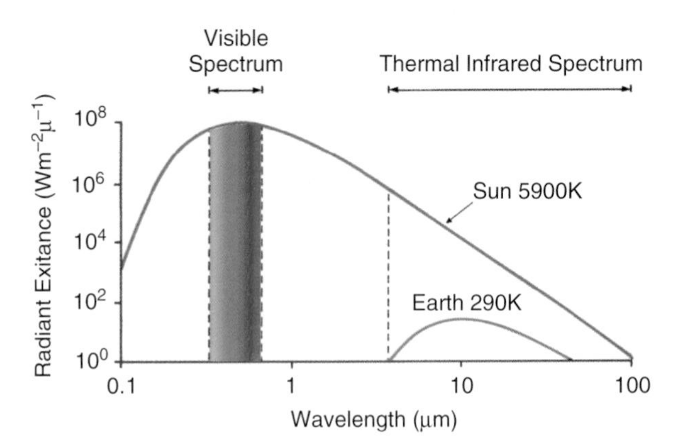

Figure 1.16 Spectral exitance curves for blackbodies at 290 and 5900 K, the approximate temperatures of the Earth and the Sun.

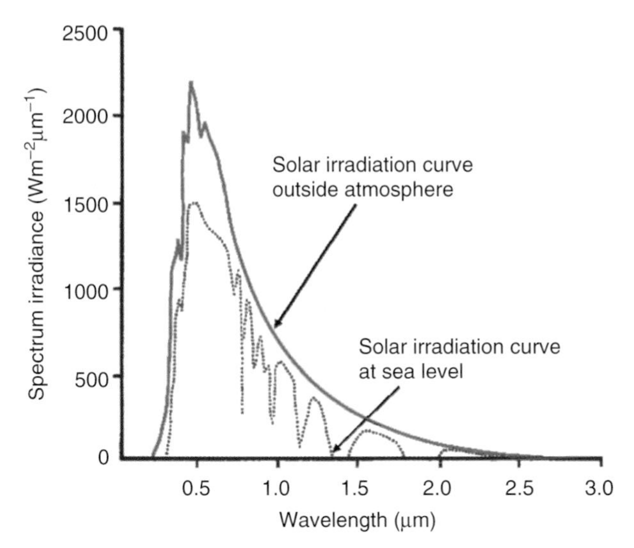

Figure 1.17 *Solar irradiance at the top of the atmosphere (solid line) and at sea-level (dotted line). Differences are due to atmospheric effects as discussed in the text. See also Figure 1.7.* Based on Manual of Remote Sensing, *Second Edition, ed. R.N. Colwell, 1983, Figure 5.5; Reproduced with permission from American Society for Photogrammetry and Remote Sensing, Manual of Remote Sensing.*

radiation from the Earth's surface are restricted to the wavelengths shorter than 2.5 µm. Instruments to detect terrestrial radiant exitance operate in the spectral region between 3 and 14 µm. Because of atmospheric absorption by carbon dioxide, ozone and water vapour, only the 3–5 and 8–14 µm regions of the TIR band are useful in remote sensing. An absorption band is also present in the 9–10 µm region. As noted earlier, the Earth's emittance peak occurs at 9.7 µm, so satellite-borne thermal sensors normally operate in the 10.5–12.5 µm spectral region. The 3–5 µm spectral window can be used to detect local targets that are hotter than their surroundings, for example forest fires. Since the 3–5 µm region also contains some reflected solar radiation it can only be used for temperature sensing at night.

Wien's Displacement Law (Figure 1.15) shows that the radiant power peak moves to shorter wavelengths as temperature increases, so that a forest fire will have a radiant energy peak at a wavelength shorter than 9.7 µm. Since targets such as forest fires are sporadic in nature and require high-resolution imagery the 3–5 µm spectral region is used by aircraft-mounted thermal detectors. This is a difficult region for remote sensing because it contains a mixture of reflected and emitted radiation, the effects of which are not easy to separate.

The selection of wavebands for use in remote sensing is therefore seen to be limited by several factors: primarily (i) the characteristics of the radiation source, as discussed in this section, (ii) the effects of atmospheric absorption and scattering (Section 1.2.5), and (iii) the nature of the target. This last point is considered in Section 1.3.

1.2.5 Interactions with the Earth's Atmosphere

In later chapters, we consider measurements of radiance from the Earth's surface made by instruments carried by satellites such as Landsat and SPOT that operate in the optical wavebands, that is, those parts of the electromagnetic spectrum with properties similar to those of visible light. It was noted at the beginning of this chapter that one aim of remote sensing is to identify the nature, and possibly the properties, of Earth surface materials from the spectral distribution of EMR that is reflected from, or emitted by, the target and recorded by the sensor. The existence of the atmosphere causes problems, because EMR from the Sun that is reflected by the Earth (the amount reflected depending on the reflectivity or albedo of the surface) and detected by the satellite or aircraft-borne sensor must pass through the atmosphere twice, once on its journey from the Sun to the Earth and once after being reflected by the surface of the Earth back to the sensor. During its passage through the atmosphere, EMR interacts with particulate matter suspended in the atmosphere and with the molecules of the constituent gases. This interaction is usually described in terms of two processes. One, called *scattering*, deflects the radiation from its path while the second process, *absorption*, converts the energy present in EMR into the internal energy of the absorbing molecule. Both absorption and scattering vary in their effect from one part of the spectrum to another. Remote sensing of the Earth's surface is impossible in those parts of the spectrum that are seriously affected by scattering and/or absorption, for these mechanisms effectively render the atmosphere opaque to incoming or outgoing radiation. As far as remote sensing of the Earth's surface is concerned, the atmosphere '. . . appears no other thing to me but a foul and pestilential congregation of vapours' (*Hamlet*, Act 2, Scene 2). Atmospheric absorption properties can, however, be useful. Remote sensing of the atmosphere uses these properties and a good example is the discovery and monitoring of the Antarctic ozone hole.

Regions of the spectrum that are relatively (but not completely) free from the effects of scattering and absorption are called *atmospheric windows;* EMR in these regions passes through the atmosphere with less modification than does radiation at other wavelengths (Figure 1.7). This effect can be compared to the way in which the bony tissues of the human body are opaque to X-rays, whereas the soft muscle tissue and blood are transparent. Similarly, glass is opaque to ultraviolet radiation but is transparent at the visible wavelengths. Figure 1.17 shows a plot of wavelength against the magnitude of incoming radiation transmitted through the atmosphere; the window regions are those with a high transmittance. The same information is shown in a different way in Figure 1.7.

The effect of the processes of scattering and absorption is to add a degree of haze to the image, that is, to reduce the contrast of the image, and to reduce the amount of radiation returning to the sensor from the Earth's surface. A certain amount of radiation that is reflected from the neighbourhood of each target may also be recorded by the sensor as originating from the target. This is because scattering deflects the path taken by EMR as it travels through the atmosphere, while absorption involves the interception of photons or particles of radiation. Our eyes operate in the visible part of the spectrum by observing the light reflected by an object. The position of the object is deduced from the assumption that this light has travelled in a straight line between the object and our eyes. If some of the light reflected towards our eyes from the object is diverted from a straight path then the object will appear less bright. If light from other objects has been deflected so that it is apparently coming to our eyes from the direction of the first object then that first object will become blurred. Taken further, this scattering process will make it appear to our eyes that light is travelling from all target objects in a random fashion, and no objects will be distinguishable. Absorption reduces the amount of light that reaches our eyes, making a scene relatively duller. Both scattering and absorption, therefore, limit the usefulness of some portions of the electromagnetic spectrum for remote sensing purposes. They are known collectively as attenuation or extinction.

Scattering is the result of interactions between EMR and particles or gas molecules that are present in the atmosphere. These particles and molecules range in size from microscopic (with radii approximately equal to the wavelength of the EMR) to raindrop size (100 μm and larger). The effect of scattering is to redirect the incident radiation, or to deflect it from its path. The atmospheric gases that primarily cause scattering include oxygen, nitrogen and ozone. Their molecules have radii of less than 1 μm and affect EMR with wavelengths of 1 μm or less. Other types of particles reach the atmosphere both by natural causes (such as salt particles from oceanic evaporation or dust entrained by aeolian processes) or because of human activities (for instance, dust from soil erosion caused by poor land management practices, and smoke particles from industrial and domestic pollution). Some particles are generated by photochemical reactions involving trace gases such as sulfur dioxide or hydrogen sulfide. The former may reach the atmosphere from car exhausts or from the combustion of fossil fuels. Another type of particle is the raindrop, which tends to be larger than the other kinds of particles mentioned previously (10–100 μm compared to 0.1–10 μm radius). The concentration of particulate matter varies both in time and over space. Human activities, particularly agriculture and industry, are not evenly spread throughout the world, nor are natural processes such as wind erosion or volcanic activity. Meteorological factors cause variations in atmospheric turbidity over time, as well as over space. Thus, the effects of scattering are spatially uneven (the degree of variation depending on weather conditions) and vary from time to time. Remotely-sensed images of a particular area will thus be subjected to different degrees of atmospheric scattering on each occasion that they are produced. Differences in atmospheric conditions over time are the cause of considerable difficulty in the quantitative analysis of time sequences of remotely-sensed images.

The mechanisms of scattering are complex, and are beyond the scope of this book. However, it is possible to make a simple distinction between selective and non-selective scattering. Selective scattering affects specific wavelengths of EMR, while non-selective scattering is wavelength independent. Very small particles and molecules, with radii far less than the wavelength of the EMR of interest, are responsible for *Rayleigh scattering*. The effect of this type of scattering is inversely proportional to the fourth power of the wavelength, which implies that shorter wavelengths are much more seriously affected than longer wavelengths. Blue light (wavelength 0.4–0.5 μm) is thus more powerfully scattered than red light (0.6–0.7 μm). This is why the sky seems blue, for incoming blue light is so scattered by the atmosphere that it seems to reach our eyes from all directions, whereas at the red end of the visible spectrum scattering is much less significant so that red light maintains its directional properties. The sky appears to be much darker blue when seen from a high altitude, such as from the top of a mountain or from an aeroplane, because the degree of scattering is reduced due to the reduction in the length of the path traversed through the atmosphere by the incoming solar radiation. Scattered light reaching the Earth's surface is termed diffuse (as opposed to *direct*) irradiance or, more simply, *skylight*. Radiation that has been scattered within the atmosphere and which reaches the sensor without having made contact with the Earth's surface is called the *atmospheric path radiance*.

Mie scattering is caused by particles that have radii between 0.1 and 10 μm, that is approximately the same magnitude as the wavelengths of EMR in the visible, NIR and TIR regions of the spectrum. Particles of smoke, dust and salt have radii of these dimensions. The intensity of Mie scattering is inversely proportional to wavelength, as in the case of Rayleigh scattering. However, the exponent ranges in value from -0.7 to -2 rather than the -4 of Rayleigh scattering. Mie scattering affects shorter wavelengths more than longer wavelengths, but the disparity is not as great as in the case of Rayleigh scattering.

Non-selective scattering is wavelength-independent. It is produced by particles whose radii exceed 10 μm. Such particles include water droplets and ice fragments present in clouds. All visible wavelengths are scattered by such particles. We cannot see through clouds because all visible wavelengths are non-selectively scattered by the water droplets of which the cloud is formed. The effect of scattering is, as mentioned earlier, to increase the haze level or reduce the contrast in an image. If contrast is defined as the ratio between the brightest and darkest areas of an image, and if brightness is measured on a scale running from 0 (darkest) to 100 (brightest), then a given image with a brightest area of 90 and a darkest area of 10 will have a contrast of 9. If scattering has the effect of adding a component of upwelling radiation of 10 units then the contrast becomes 100 : 20 or 5. This reduction in contrast will result in a decrease in the detectability of features present in the image. Figure 1.18 shows relative scatter as a function of wavelength for the 0.3–1.0 μm region of the spectrum for a variety of levels of atmospheric haze.

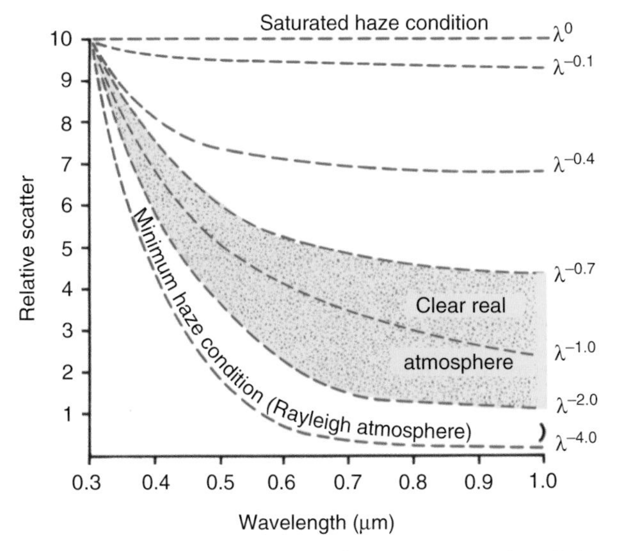

Figure 1.18 *Relative scatter as a function of wavelength for a range of atmospheric haze conditions. Based on R.N. Colwell (ed.), 1983,* Manual of Remote Sensing, *Second Edition, Figure 6.15. Reproduced with permission from American Society for Photogrammetry and Remote Sensing, Manual of Remote Sensing.*

Absorption is the second process by which the Earth's atmosphere interacts with incoming EMR. Gases such as water vapour, carbon dioxide and ozone absorb radiation in particular, regions of the electromagnetic spectrum called absorption bands. The processes involved are very complex and are related to the vibrational and rotational properties of the molecules of water vapour, carbon dioxide or ozone, and are caused by transitions in the energy levels of the atoms. These transitions occur at characteristic wavelengths for each type of atom and at these wavelengths absorption rather than scattering is dominant. Remote sensing in these absorption bands is thus rendered impossible. Fortunately, other regions of the spectrum with low absorption (high transmission) can be used. These regions are called 'windows', and they cover the 0.3–1.3 μm (visible/NIR), 1.5–1.8, 2.0–2.5 and 3.5–4.1 μm (middle IR) and 7.0–15.0 μm (TIR) wavebands. The utility of these regions of the electromagnetic spectrum in remote sensing is considered at a later stage.

1.3 Interaction with Earth-Surface Materials

1.3.1 Introduction

Electromagnetic energy reaching the Earth's surface from the Sun is reflected, transmitted or absorbed. Reflected energy travels upwards through, and interacts with, the atmosphere; that part of it which enters the field of view of the sensor (Section 2.2.1) is detected and converted into a numerical value that is transmitted to a ground receiving station on Earth. The amount and spectral distribution of the reflected energy is used in remote sensing to infer the nature of the reflecting surface. A basic assumption made in remote sensing is that specific targets (soils of different types, water with varying degrees of impurities, rocks of differing lithologies or vegetation of various species) have an individual and characteristic manner of interacting with incident radiation that is described by the spectral response of that target. In some instances, the nature of the interaction between incident radiation and Earth-surface material will vary from time to time during the year, such as might be expected in the case of vegetation as it develops from the leafing stage, through growth to maturity and, finally, to senescence.

The spectral response of a target also depends upon such factors as the orientation of the Sun (solar azimuth, Figure 1.19), the height of the Sun in the sky (solar elevation angle), the direction that the sensor is pointing relative to nadir (the look angle) and the state of health of vegetation, if that is the target. Nevertheless, if the assumption that specific targets are characterized by an individual spectral response were invalid then Earth Observation (EO) by remote sensing would be an impossible task. Fortunately, experimental studies in the field and in the laboratory, as well as experience with multispectral imagery, have shown that the assumption is generally a reasonable one. Indeed, the successful development of remote sensing of the environment over the last few decades bears witness to its validity. Note that the term *spectral signature* is sometimes used to describe the spectral response curve for a target. In view of the dependence of spectral response on the factors mentioned above, this term is inappropriate for

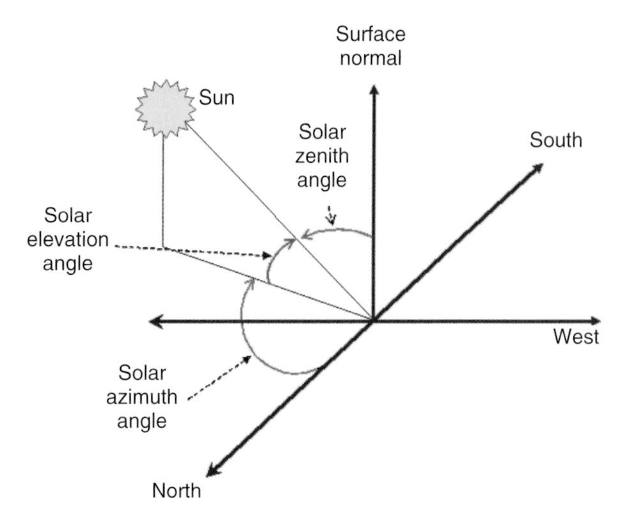

Figure 1.19 *Solar elevation and azimuth angles. The elevation angle of the Sun – target line is measured upwards from the horizontal plane. The zenith angle is measured from the surface normal, and is equal to (90 – elevation angle)°. Azimuth is measured clockwise from north.*

it gives a misleading impression of constancy. Remote sensing scientists are not alone in trying to estimate the nature of surface materials from their spectral reflection properties; an English newspaper, the *Daily Telegraph* (16 March 2009), reported that a thief routinely used Google Earth to identify and steal lead sheathing on the roofs of churches and public buildings.

In this section, spectral reflectance curves of vegetation, soil, rocks and water are examined in order to emphasize their characteristic features. The results summarized in the following paragraphs must not be taken to be characteristic of all varieties of materials or all observational circumstances. One of the problems met in remote sensing is that the spectral reflectance of a given Earth-surface cover type is influenced by a variety of confusing factors. For example, the spectral reflectance curve of a particular agricultural crop such as wheat is not constant over time, nor is it the same for all kinds of wheat. The spectral reflectance curve is affected by factors such as soil nutrient status, the growth stage of the vegetation, the colour of the soil (which may be affected by recent weather conditions), the solar azimuth and elevation angles and the look angle of the sensor. The topographic position of the target in terms of slope orientation with respect to solar azimuth and slope angle also has an effect on the reflectance characteristics of the target, as will the state of the atmosphere. Methods for dealing with some of these difficulties are described in Sections 4.5 and 4.7. Hence, the examples given in this section are idealized models rather than templates.

Before turning to the individual spectral reflectance features of Earth surface materials, a distinction must be drawn between two kinds of reflectance that occur at a surface. *Specular* reflection is that kind of reflection in which energy leaves the reflecting surface without being scattered, with the angle of incidence being equal to the angle of reflectance (Figure 1.4a). Surfaces that reflect *specularly* are smooth relative to the wavelength of the incident energy. *Diffuse* or Lambertian reflectance occurs when the reflecting surface is rough relative to the wavelength of the incident energy, and the incident energy is scattered in all directions (Figure 1.4b). A mirror reflects specularly while a piece of paper reflects diffusely. In the visible part of the spectrum, many terrestrial targets are diffuse reflectors, whereas calm water can act as a specular reflector. At microwave wavelengths, however, some terrestrial targets are specular reflectors, while volume reflectance (scattering) can occur at optical wavelengths in the atmosphere and the oceans, and at microwave wavelengths in vegetation (Figure 1.4d).

A satellite sensor operating in the visible and NIR spectral regions does not detect all the reflected energy from a ground target over an entire hemisphere. It records the reflected energy that is returned at a particular angle (see the definition of radiance in Section 1.2.1). To make use of these measurements, the distribution of radiance at all possible observation and illumination angles (called the *bidirectional reflectance distribution function*; BRDF) must be taken into consideration. Details of the BRDF are given by Slater (1980) who writes:

> ...the reflectance of a surface depends on both the direction of the irradiating flux and the direction along which the reflected flux is detected.

Hyman and Barnsley (1997) demonstrate that multiple images of the same area taken at different viewing angles provide enough information to allow different land cover types to be identified as a result of their differing BRDF. The MISR (Multi-Angle Imaging SpectroRadiometer) instrument, carried by the American Terra satellite, collects multi-directional observations of the same ground area over a timescale of a few minutes, at nadir and at fore and aft angles of view of 21.1°, 45.6°, 60.0° and 70.5° and in four spectral bands in the visible and NIR regions of the electromagnetic spectrum. The instrument therefore provides data for the analysis and characterisation of reflectance variation of Earth surface materials over a range of angles (Diner *et al.*, 1991). Chopping *et al.* (2003) use a BRDF model to extract information on vegetation canopy physiognomy. The European Space Agency's (ESA's) Compact High Resolution Imaging Spectrometer (CHRIS), carried by a small satellite named PROBA (Project for On Board Autonomy) can image an 18.6 km^2 area with its high resolution mode at multiple angles. See http://directory.eoportal.org/get_announce.php?an_id=7299. Furthermore, its agile platform can be tilted during acquisition so multi-angle observations can be acquired. See Guanter, Alonso and Moreno (2005) for more details of the PROBA/CHRIS mission and its first results.

It follows from the foregoing that, even if the target is a diffuse reflector such that incident radiation is scattered in all directions, the assumption that radiance is constant for any observation angle θ measured from the surface normal does not generally hold. A simplifying assumption is known as *Lambert's Cosine Law*, which states that the radiance measured at an observation angle θ is the same as that measured at an observation angle of 0° adjusted for the fact that the projection of the unit surface at a view angle of θ is proportional to $\cos\theta$ (Figure 1.20). Surfaces exhibiting this property are called 'Lambertian', and a considerable body of work in remote sensing either explicitly or implicitly assumes that Lambert's Law applies. However, it is usually the case that the spectral distribution of reflected flux from a surface is more complex than the simple description provided by Lambert's Law, for it depends on the geometrical conditions of measurement and illumination. The topic of correction of images for sun and view angle effects is considered further in Chapter 4.

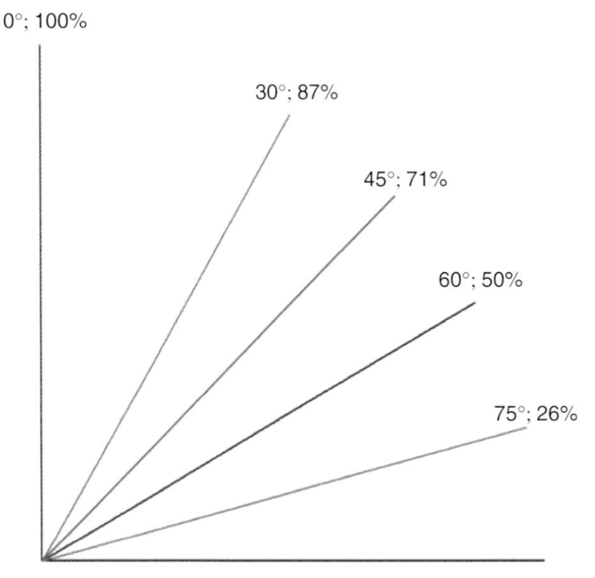

Figure 1.20 *Lambert's cosine law. Assume that the illumination angle is 0°. A range of view angles is shown, together with the percentage of incoming radiance that is scattered in the direction of the view angle.*

1.3.2 Spectral Reflectance of Earth Surface Materials

In this section, typical spectral reflectance curves for characteristic types of Earth-surface materials are discussed. The remarks in Section 1.3.1 should not be overlooked when reading the following paragraphs. The Earth-surface materials that are considered in this section are vegetation, soil, bare rock and water. The short review by Verstraete and Pinty (1992) is recommended. Hobbs and Mooney (1990) provide a useful survey of remote sensing of the biosphere. Aplin (2004, 2005) covers progress in remote sensing in ecology.

1.3.2.1 Vegetation

The reflectance spectra of 3 pixels selected from a DAIS imaging spectrometer data set (Section 9.3) covering a small area of La Mancha in central Spain (Figure 1.21a) show that real-world vegetation spectra conform to the ideal pattern, though there is significant variation, especially in the NIR region. Two of the three curves in Figure 1.21b show relatively low values in the red and the blue regions of the visible spectrum, with a minor peak in the green region. These peaks and troughs are caused by absorption of blue and red light by chlorophyll and other pigments. Typically, 70–90% of blue and red light is absorbed to provide energy for the process of photosynthesis. The slight reflectance peak in the green waveband between 0.5 and 0.6 μm is the reason that actively growing vegetation appears green to the human eye. Non-photosynthetically active vegetation lacks this 'green peak'.

For photosynthetically active vegetation, the spectral reflectance curve rises sharply between about 0.65 and 0.76 μm, and remains high in the NIR region between 0.75 and 1.35 μm because of interactions between the internal leaf structure and EMR at these wavelengths. Internal leaf structure has some effect between 1.35 and 2.5 μm, but reflectance is largely controlled by leaf-tissue water content, which is the cause of the minima recorded near 1.45 and 1.95 μm. The status of the vegetation (in terms of photosynthetic activity) is frequently characterized by the position of a point representative of the steep rise in reflectance at around 0.7 μm. This point is called the red edge point, and its characterization and uses are considered in Section 9.3.2.3.

As the plant senesces, the level of reflectance in the NIR region (0.75–1.35 μm) declines first, with reflectance in the visible part of the spectrum not being affected significantly. This effect is demonstrated by the reflectance spectrum shown in orange on Figure 1.21. The slope of the curve from the red to the NIR region of the spectrum is lower, as is the reflectance in the area of the 'infrared plateau'. However, changes in reflectance in the visible region are not so apparent. As senescence continues, the relative maximum in the green part of the visible spectrum declines as pigments other than chlorophyll begin to dominate, and the leaf begins to lose its greenness and to turn yellow or reddish, depending on species. The

(a)

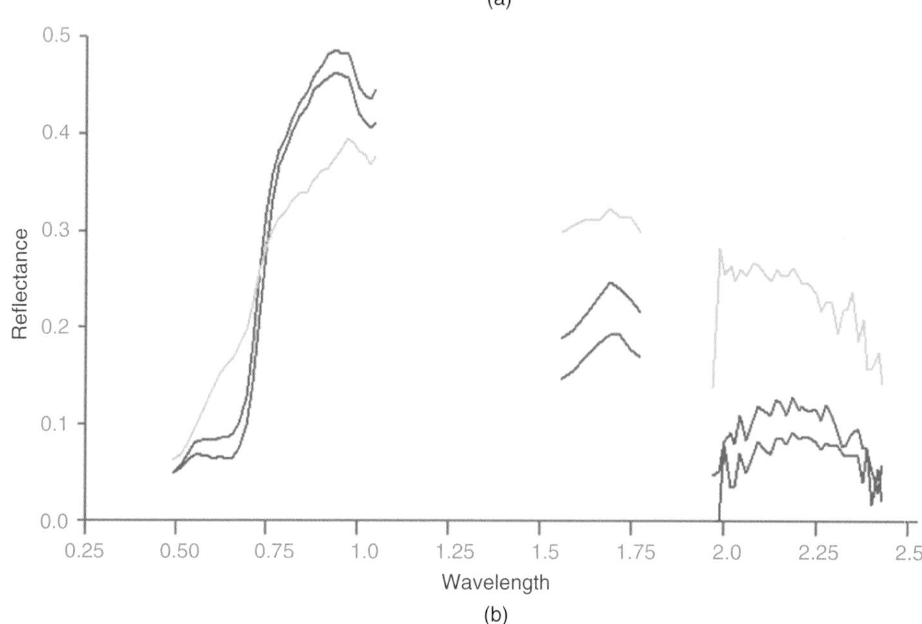

(b)

Figure 1.21 *(a) DAIS image of part of La Mancha, Central Spain. (b) Reflectance spectra in the optical wavebands of three vegetation pixels selected from this image. The two green curves represent the typical spectral reflectance curves of active vegetation. The 2 pixels which these curves represent were selected from the bright red area in the bottom left of image (a) and the similar area by the side of the black lagoon. The third reflectance curve, shown in orange, was selected from the orange area in the top right of image (a). The spectral reflectance plots are discontinuous because parts of the atmosphere absorb and/or scatter incoming and outgoing radiation (see Figure 1.7 and Chapter 4). Reproduced with permission from the German Space Agency, DLR.*

wavelength of the red edge point also changes. Stress caused by environmental factors such as drought or by the presence or absence of particular minerals in the soil can also produce a spectral response that is similar to senescence. Areas of vegetation showing adverse effects due to the presence (or absence) of certain minerals in the soil are called geobotanical anomalies, and their distribution has been used successfully to determine the location of mineral deposits (Goetz, Rock and Rowan, 1983).

The shape of the spectral reflectance curve is used to distinguish vegetated and non-vegetated areas on remotely-sensed imagery. Differences between species can also be considerable, and may be sufficient to permit their discrimination, depending on the number, width and location of the wavebands used by the sensor (Section 2.2). Such discrimination may be possible on the basis of relative differences in the spectral reflectance curves of the vegetation or crop types. Absolute reflectance values (Section 4.6) may be used to estimate physical properties of the vegetation, such as leaf area index (LAI) or biomass production. In agriculture, the estimation of crop yields is often a significant economic requirement. Ratios of reflectance values in two or more spectral bands are widely used to characterize vegetation (Section 6.2.4). It is important to remember, however, the points made in Section 1.3.1; there is no single, ideal spectral reflectance curve for any particular vegetation type, and the recorded radiance from a point on the ground will depend upon the viewing and illumination angles, as well as other variables. The geometry of the crop canopy will strongly influence the BRDF (Section 1.3.1), while factors such as the transmittance of the leaves, the number of leaf layers, the actual arrangement of leaves on the plant and the nature of the background (which may be soil, or leaf litter or undergrowth) are also important. In order to distinguish between some types of vegetation, and to assess growth rates from remotely-sensed imagery, it is necessary to use *multi-temporal imagery*, that is imagery of the same area collected at different periods in the growing season.

1.3.2.2 Geology

Geological use of remotely-sensed imagery relies, to some extent, upon knowledge of the spectral reflectance curves of vegetation, for approximately 70% of the Earth's land surface is vegetated and the underlying rocks cannot be observed directly, and differences in soil and underlying bedrock can be seen in the distribution of vegetation species, numbers of species and vigour. Even in the absence of vegetated surfaces, weathering products generally cover the bedrock. It was noted in the preceding section that geobotanical anomalies might be used to infer the location of mineral deposits. Such anomalies include peculiar or unexpected species distribution, stunted growth or reduced ground cover, altered leaf pigmentation or yellowing (chlorosis) and alteration to the phenological cycle, such as early senescence or late leafing in the spring. It would be unwise to suggest that all such changes are due to soil geochemistry; however, the results of a number of studies indicate that the identification of anomalies in the vegetation cover of an area can be used as a guide to the presence of mineral deposits. If the relationship between soil formation and underlying lithology has been destroyed, for example by the deposition of glacial material over the local rock, then it becomes difficult to make associations between the phenological characteristics of the vegetation and lithology of the underlying rocks.

In semi-arid and arid areas such as the Great Sandy Desert of Western Australia (Figure 1.22), the spectral reflectance curves of rocks and minerals may be used directly in order to infer the lithology of the study area, though care should be taken because weathering crusts with spectra that are significantly different from the parent rock may develop. Laboratory studies of reflectance spectra of minerals have been carried out by Hunt and co-workers in the United States (Hunt, 1977, 1979; Hunt and Ashley, 1979; Hunt and Salisbury, 1970, 1971; Hunt, Salisbury and Lenhoff, 1971). Spectral libraries, accessible over the Internet from the Jet Propulsion Laboratory (the ASTER Spectral Library and the US Geological Survey Digital Spectral Library), contain downloadable data derived from the studies of Hunt, Salisbury and others. A new version (numbered 2.0) of the ASTER Spectral Library is now available (Baldridge *et al.*, 2009) and can be found at http://speclib.jpl.nasa.gov/. These studies demonstrate that rock-forming minerals have unique

Figure 1.22 *Landsat-7 Thematic Mapper image of the Great Sandy Desert of Western Australia. Despite its name, not all of the desert is sandy and this image shows how the differing spectral reflectance properties of the weathered rock surfaces allow rock types to be differentiated. Field work would be necessary to identify the specific rock types, while comparison of the spectral reflectance properties at each pixel with library spectra (such as those contained in the ASTER spectral library, for example Figure 1.23) may allow the identification of specific minerals. The image was collected on 24 February 2001 and was made using shortwave-infrared, near-infrared and red wavelengths as the red, green and blue components of this false-colour composite image. Image courtesy NASA/USGS.*

spectral reflectance curves. The presence of absorption features in these curves is diagnostic of the presence of certain mineral types. Some minerals, for example quartz and feldspars, do not have strong absorption features in the visible and NIR regions, but can be important as dilutants for minerals with strong spectral features such as the clay minerals, sulfates and carbonates. Clay minerals have a decreasing spectral reflectance beyond 1.6 μm, while carbonate and silicate mineralogy can be inferred from the presence of absorption bands in the mid-IR region, particularly 2.0–2.5 μm. Kahle and Rowan (1980) show that multi-spectral TIR imagery in the 8–12 μm region can be used to distinguish silicate and non-silicate rocks. Two examples of library spectra (for limestone and basalt) are shown in Figure 1.23. The data for these figures were derived from the ASTER Spectral Library.

Some of the difficulties involved in the identification of rocks and minerals from the properties of spectral reflectance curves include the effects of atmospheric scattering and absorption, the solar flux levels in the spectral regions of interest (Section 1.2.5) and the effects of weathering. Buckingham and Sommer (1983) indicate

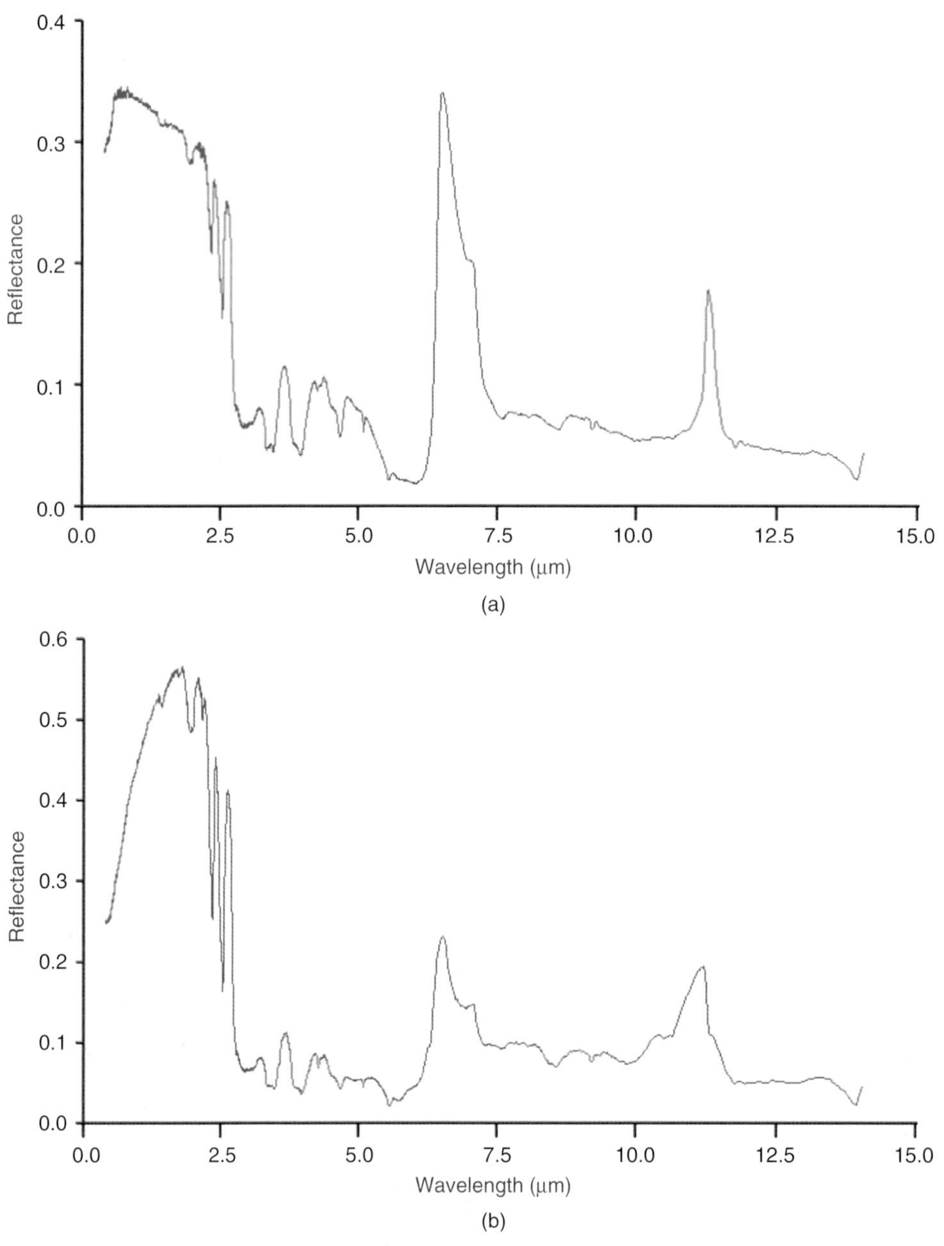

Figure 1.23 *Reflectance and emittance spectra of (a) limestone and (b) basalt samples. Data from the ASTER spectral library through the courtesy the Jet Propulsion Laboratory, California Institute of Technology, Pasadena, California, © California Institute of Technology. All rights reserved.*

that the nature of the spectral reflectance of a rock is determined by the mineralogy of the upper 50 μm, and that weathering, which produces a surface layer that is different in composition from the parent rock, can significantly alter the observed spectral reflectance.

The use of multiband (or hyperspectral) imaging spectrometers mounted on aircraft and satellites can now measure the spectra of ground surface materials at a large number of closely spaced points. The interpretation of these spectra requires a detailed knowledge of the chemistry of the materials concerned. Imaging spectrometers are described in Chapter 9. Clark (1999) gives an accessible survey of the use of imaging spectrometers in identifying surface materials. Introductions to geological remote sensing are Drury (2004), Goetz (1989), Gupta (2003), Prost (2002) and Vincent (1997).

1.3.2.3 Water Bodies

The characteristic spectral reflectance curve for water shows a general reduction in reflectance with increasing wavelength in the visible wavebands, so that in the NIR the reflectance of deep, clear water is virtually zero. This is shown schematically in Figure 1.24a, in which the penetration depth of red, green and blue light is indicated by the arrows depicted in those colours. The black arrow represents NIR radiation, which is absorbed by the first few centimetres of water. Figure 1.25 shows an image of the Tanzanian coast south of Dar-es-Salaam. The red component of the

Figure 1.25 *Image of the coast of Tanzania south of Dar-es-Salaam. Shades of red in the image show spatial variations in near-infrared reflectance, while shades of green in the image show variations in the reflectance of red light. Blue shades in the image show variations in the reflectance of green light. This representation is usually termed 'false colour'. Black shows no reflection in any of the three wavebands, whereas lighter colours show higher reflectance. The water in this area is clear and the reflection from the sea bed is visible, showing the extent of the continental shelf. This image was taken by Landsat's Multispectral Scanner (MSS), which is described in Chapter 2. Image courtesy of NASA/USGS.*

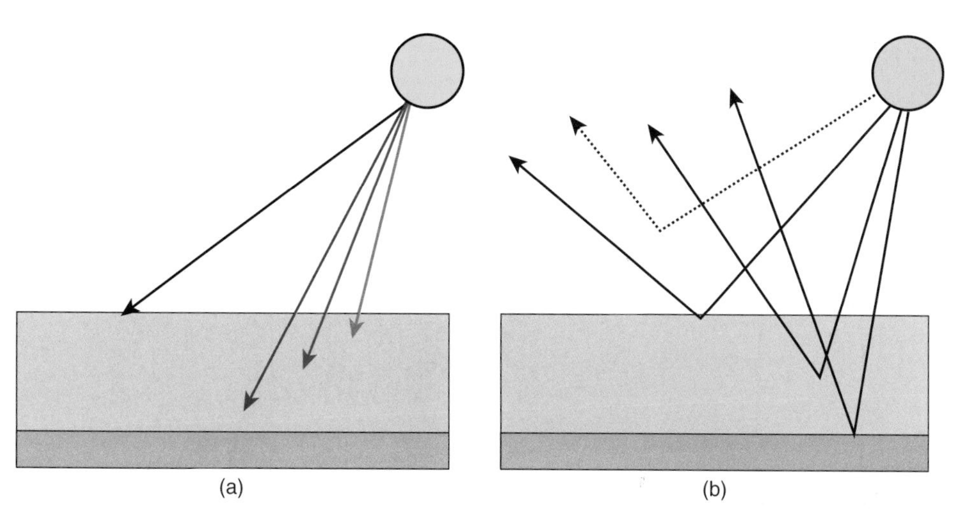

Figure 1.24 *(a) Showing the differential penetration depths of red, green and blue light in clear, calm water. The black line shows longer (infrared) wavelengths that are totally absorbed by the first few centimetres of water. (b) The solid arrows show, from right to left, bottom reflectance (the water depth is assumed to be less than the depth of penetration of blue light), volume reflectance (caused by light being scattered by suspended sediment particles, phytoplankton, dissolved organic matter and surface reflectance. The dotted line shows the path taken by light from the Sun that interacts with the atmosphere in the process of atmospheric scattering (Chapter 4). Electromagnetic radiation scattered by the atmosphere may be considerably greater in magnitude than that which is backscattered by surface reflectance, volume reflectance and bottom reflectance.*

image shows variations in the NIR reflectance of the land and water surface, the green component of the image shows variations in visible red reflectance and the blue component of the image shows variations in the visible green reflectance. The increase in reflection from the water from the deep, offshore region (where the ocean is seen as black) to the inshore region, where the colour changes to blue then blue–green, is clearly apparent. The edge of the continental shelf is shown as an abrupt change of colour, from light blue to black. The red colour of the land surface shows the presence of vegetation, which reflects strongly in the NIR, as noted in Section 1.3.2.1.

The spectral reflectance of water is affected by the presence and concentration of dissolved and suspended organic and inorganic material, and by the depth of the water body. Thus, the intensity and distribution of the radiance upwelling from a water body are indicative of the nature of the dissolved and suspended matter in the water, and of the water depth. Figure 1.24b shows how the information that oceanographers and hydrologists require is only a part of the total signal received at the sensor. Solar irradiance is partially scattered by the atmosphere, and some of this scattered light (the path radiance) reaches the sensor. Next, part of the surviving irradiance is reflected by the surface of the water body. This reflection might be specular under calm conditions, or its distribution might be strongly influenced by surface waves and the position of the sun relative to the sensor, giving rise to *sunglint*. Once within the water body, EMR may be absorbed by the water (the degree of absorption being strongly wavelength-dependent) or selectively absorbed by dissolved substances, or back-scattered by suspended particles. This latter component is termed the volume reflectance. At a depth of 20 m only visible light (mainly in the blue region) is present, as the NIR component has been completely absorbed. Particulate matter, or suspended solids, scatters the downwelling radiation, the degree of scatter being proportional to the concentration of particulates, although other factors such as the particle-size distribution and the colour of the sediment are significant. Over much of the observed low to medium range of concentrations of suspended matter a positive, linear relationship between suspended matter concentration and reflectance in the visible and NIR bands has been observed, though the relationship becomes non-linear at increasing concentrations. Furthermore, the peak of the reflectance curve moves to progressively longer wavelengths as concentration increases, which may lead to inaccuracy in the estimation of concentration levels of suspended materials in surface waters from remotely-sensed data. Another source of error is the inhomogeneous distribution of suspended matter through the water body, which is termed *patchiness*.

The presence of chlorophyll is an indication of the trophic status of lakes and is also of importance in estimating the level of organic matter in coastal and estuarine environments. Whereas suspended matter has a generally broadband reflectance in the visible and NIR, chlorophyll exhibits absorption bands in the region below 0.5 μm and between 0.64 and 0.69 μm. Detection of the presence of chlorophyll therefore requires an instrument with a higher spectral resolution (Section 2.2.2) than

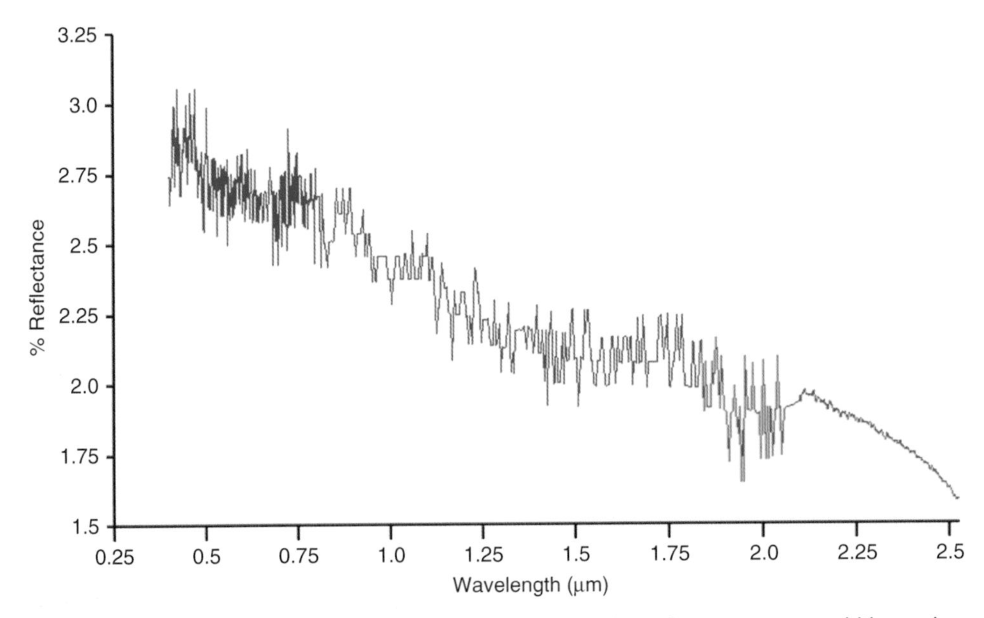

Figure 1.26 *Reflectance spectrum of tap water from 0.4 to 2.55 μm. Data from the ASTER spectral library through the courtesy the Jet Propulsion Laboratory, California Institute of Technology, Pasadena, California, © California Institute of Technology. All rights reserved.*

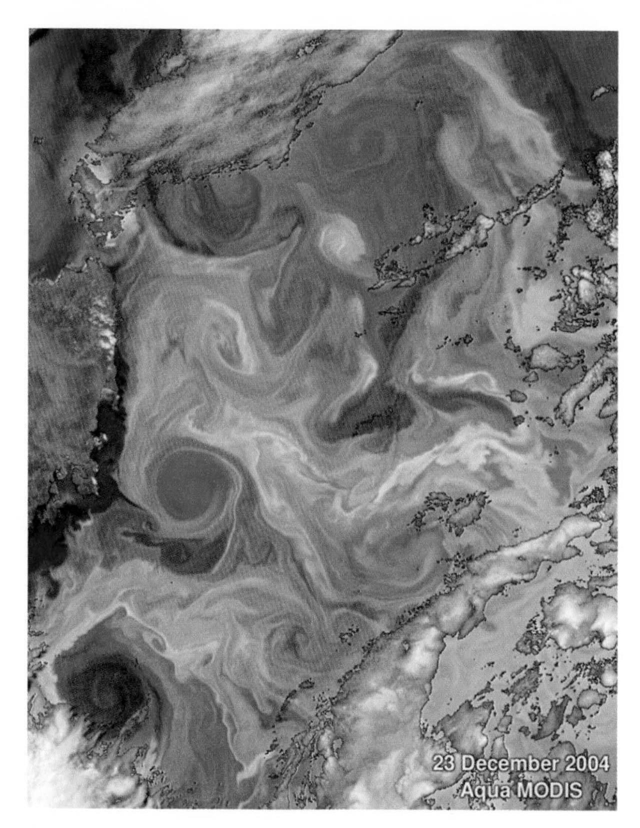

Figure 1.27 *This image of the ocean east of Tasmania in December, 2004 depicts subtle differences in water colour that result from varying distributions of such scattering and absorbing agents in the water column as phytoplankton, dissolved organic matter, suspended sediment, bubbles, and so on. The ocean colours shown above result from independently scaling the satellite-derived normalised water-leaving radiances (nLw) at 551, 488 and 412 nm and using the results as the red, green and blue components of the image, respectively. Differences in the colours may also partially reflect differences in atmospheric components or levels of sun and sky glint or differences in the path that light takes through the MODIS instrument. The MODIS instrument is described in Section 2.3. Source: http://oceancolor.gsfc.nasa.gov/cgi/image_archive.cgi?c =CHLOROPHYLL. Image courtesy of NASA/USGS.*

Figure 1.28 *Surface temperature image of the seas around the Galapagos and Cocos Islands. This heat map, produced through ESA's Medspiration project, shows the sea surface temperatures around Galapagos Islands and Cocos Island in the Pacific Ocean for 18 March 2007 using data from the AATSR sensor carried by the ENVISAT satellite (Section 2.2.1). Reproduced with permission from http://dup.esrin.esa.it/news/inews/inews_130.asp.*

would be required to detect suspended sediment. Furthermore, the level of backscatter from chlorophyll is lower than that produced by suspended sediment; consequently, greater radiometric sensitivity is also required. Ocean observing satellites (Section 2.2.4) carry instruments that are 'tuned' to specific wavebands that match reflectance peaks or absorption bands in the spectra of specific materials such as chlorophyll.

Although the spectral reflectance properties of suspended matter and chlorophyll have been described separately, it is not uncommon to find both are present at one particular geographical locality. The complications in separating-out the contribution of each to the total observed reflectance are considerable. Furthermore, the suspended matter or chlorophyll may be unevenly distributed in the horizontal plane (the patchiness phenomenon noted above) and in the vertical plane. This may cause problems if the analytical technique used to determine concentration levels from recorded radiances is based on the assumption that the material is uniformly mixed at least to the depth of penetration of the radiation. In some cases, a surface layer of suspended matter may ride on top of a lower, colder, layer with a low suspended matter concentration, giving rise to considerable difficulty if standard analytical techniques are used. Reflection from the bed of the water body can have unwanted effects if the primary aim of the experiment is to determine suspended sediment or chlorophyll concentration levels, for it adds a component of reflection to that resulting from backscatter from the suspended or dissolved substances. In other instances, collection of the EMR reflected from the sea bed might

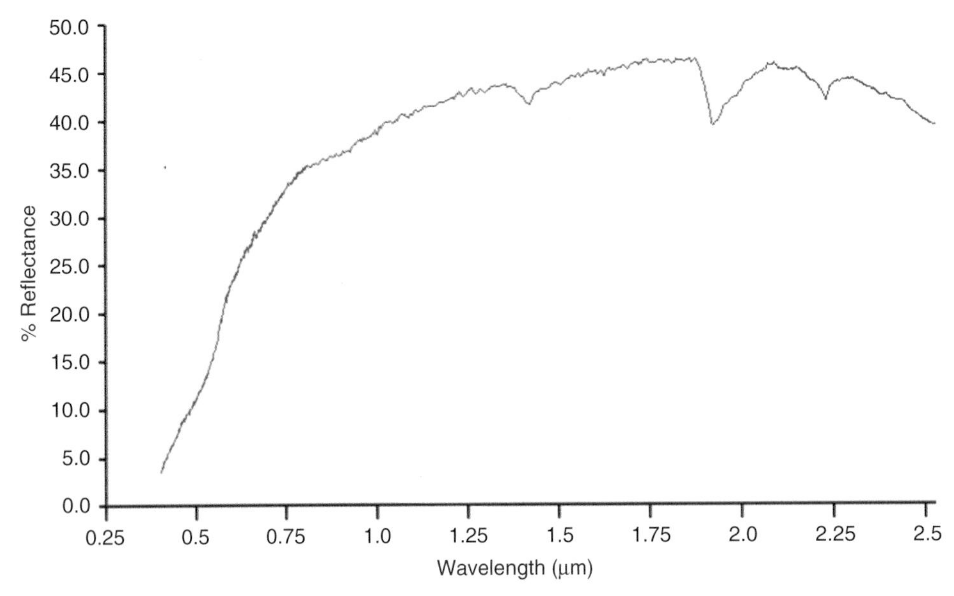

Figure 1.29 *Reflectance spectrum of a brown fine sandy loam soil from 0.4 to 2.5 μm. Note that the y-axis is graduated in percentage reflection. Soil spectra vary with the mineralogical properties of the soil and also its moisture status. The latter varies temporally and spatially. Data from the ASTER spectral library through the courtesy the Jet Propulsion Laboratory, California Institute of Technology, Pasadena, California, © California Institute of Technology. All rights reserved.*

be the primary focus of the exercise, in which case the presence of organic or inorganic material in the water would be a nuisance.

The reflectance spectrum of clear water can be inferred from Figure 1.24a. The magnitude of reflected radiation is low, with absorption increasing with wavelength so that in the NIR region of the spectrum absorption is almost total, and water appears black in the image. Indeed, one of the methods of estimating the atmospheric contribution to the signal received at the sensor is to look at the recorded radiance for a deep clear lake in the NIR. The departure from zero is an estimate of the atmospheric path length (Figure 1.24b). The spectral reflectance curve for tap water is shown in Figure 1.26.

One of the main interests of satellite oceanography is in the derivation of images showing patterns in the oceans and seas of the world. Ocean colour is a term used to describe the variability in colour in images. Figure 1.27 is an example. Ocean colour is related to oxygen consumption and can help in, for example guiding fishing fleets to their prey. A second, important, area of satellite oceanography is the study of SST distribution. Figure 1.28 is an example of a SST image.

1.3.2.4 Soils

The spectral reflectance curves of soils (Figure 1.29) are generally characterized by a rise in reflectivity as wavelength increases – the opposite, in fact, of the shape of the spectral reflectance curve for water. Reflectivity in the visible wavebands is affected by the presence of organic matter in the soil, and by the soil moisture content, while at 0.85–0.93 μm there is a ferric iron absorption band. As ferric iron also absorbs ultraviolet radiation in a broad band, the presence of iron oxide in soils is expressed visually by a reddening of the soil, the redness being due to the absorption of the wavelengths shorter (and longer) than the red. Between 1.3–1.5 and 1.75–1.95 μm water absorption bands occur, as mentioned in Section 1.2.5. Soil reflectance in the optical part of the electromagnetic spectrum is usually greatest in the region between these two water absorption bands, and declines at wavelengths longer than 2 μm with clay minerals, if present, being identifiable by their typical narrow-band absorption features in the 2.0–2.5 μm region. Irons, Weismiller and Petersen (1989) provide a comprehensive survey of factors affecting soil reflectance. Huete (1989) gives a summary of the influence of the soil background on measurements of vegetation spectra, while Huete (2004) is a good review of the remote sensing of soils and soil processes.

1.4 Summary

In the opening paragraph of Section 1.3.2.1, a basic principle of applied remote sensing is set out. This states that individual Earth-surface cover types are distinguishable in terms of their spectral reflection and emission characteristics. Changes in the spectral response of objects can also be used as an indication of changes in the properties of the object, for example the health or growth stage of a plant or the turbidity of a water body. In this chapter, the basic principles of EMR are reviewed briefly, and the relevant terminology is defined. An understanding of these basic principles is essential if the methods described in the remainder of this book are to be applied sensibly. Chapter 2 provides details of the characteristics of sensors that are used to measure the magnitude of EMR that is reflected from or emitted by the Earth surface. It is from these measurements that the spectral reflectance curves of Earth surface elements can be derived. Further details of the derivation of absolute spectral reflectance values from remotely-sensed measurements are provided in Chapter 4, while the use of data from imaging spectrometers, which can record data in tens or hundreds of bands, is described in Chapter 9. However advanced the processing techniques are, their results cannot be properly understood if the user does not have a good working knowledge of the material covered in Chapter 1. Readers are encouraged to consult the main references given in the text, and to familiarize themselves with the reflection spectra of natural targets, for example by using the MIPS software that can be downloaded from the publisher's web site at URL www.wiley.com/go/mather4. MIPS contains a small spectral library which is accessed from the *Plot|Plot Library Spectrum* menu. It includes a selection of spectra from the ASTER spectral library (with the permission of the Jet Propulsion Laboratory, California Institute of Technology, Pasadena, CA, USA).

2 Remote Sensing Platforms and Sensors

2.1 Introduction

This chapter contains a description of the nature and characteristics of digital images of the Earth's surface produced by aircraft and satellite-borne sensors operating in the visible, infrared and microwave regions of the electromagnetic spectrum. The properties of a selection of representative sensors operating in each of these spectral regions are also described, and examples of typical applications are discussed. No attempt is made to provide full details of all planned, current and past platforms and sensors as that subject would require a book in itself that, furthermore, would be out of date before it was written. However, some surveys are valuable historical documents, for example Verger *et al.* (2003). Others, such as the Committee on Earth Observation Satellites (CEOS) *CEOS EO Handbook – Earth Observation Satellite Capabilities and Plans* (CEOS, 2008) are comprehensive and reasonably up to date. See also Wooster (2007) and Petrie (2008). Examples of the most widely used satellite-borne imaging instruments, such as the High Resolution Visible (HRV) (carried by the SPOT satellite), Thematic mapper (TM)/Enhanced Thematic Mapper Plus (ETM+) (Landsat), Advanced Spaceborne Thermal Emission and Reflection Radiometer (ASTER) (Terra) and Synthetic Aperture Radar (SAR) (Radarsats-1 and -2, and ERS-2) are used to illustrate those sections of this chapter that deal with instruments for remote sensing in the optical, near-infrared (NIR), thermal infrared (TIR) and microwave wavebands. The trend towards small satellites carrying a single instrument is also considered, together with the growing number of commercial high-resolution satellite systems. Three developing areas of remote sensing using imaging spectrometers, interferometric SAR and lidar sensors are described in more detail in Chapter 9. These datasets are largely, though not exclusively, collected by sensors carried by aircraft rather than by satellites. A new and interesting means of collecting remotely-sensed data is by mounting instruments on unmanned aerial vehicles (UAVs) which can remain airborne for long periods and provide a stable platform. A Special Issue of *IEEE Transactions on Geoscience and Remote Sensing*, devoted to the use of UAVs in remote sensing is edited by Zhou *et al.* (2009).

Remote sensing of the surface of the Earth has a long history, going back to the use of cameras carried by balloons and pigeons in the eighteenth and nineteenth centuries but in its modern connotation the term *remote sensing* can be traced back to the aircraft-mounted systems that were developed during the early part of the twentieth century, initially for military purposes. Airborne film camera systems are still a source of remotely-sensed data (Lillesand, Kiefer and Chipman, 2008) and spaceborne film camera systems, initially used in low Earth-orbit satellites for military purposes, have also been used for civilian remote sensing from space; for example the National Aeronautics and Space Administration (NASA) Large Format Camera (LFC) flown on the American Space Shuttle in October, 1984. Astronauts onboard the International Space Station (ISS) also take photographs of Earth, using any of a variety of hand-held cameras, both analogue (film) and digital. See http://eol.jsc.nasa.gov/default.htm for details of astronaut photography. Analogue photography is now much less important than it used to be, and most photography using aircraft or spacecraft now uses digital cameras.

Although analogue photographic imagery still has its uses, this book is concerned with the processing of image data collected by scanning systems that ultimately generate digital image products. Analogue cameras and non-imaging (profiling) instruments such as radar altimeters are thereby excluded from direct consideration, although hard copy products from these systems can be converted into digital form by scanning and the techniques described in later chapters can be applied.

The general characteristics of imaging remote sensing instruments operating in the optical wavebands (with wavelengths less than about 3 μm) – namely, their spatial, spectral, temporal and radiometric resolution – are the subject of Section 2.2. In Section 2.3, the properties of images collected in the optical, NIR and TIR regions of the electromagnetic spectrum are described.

Computer Processing of Remotely-Sensed Images: An Introduction, Fourth Edition Paul M. Mather and Magaly Koch

The properties of microwave imaging sensors are outlined in Section 2.4.

Spatial, spectral, temporal and radiometric resolution are properties of remote sensing instruments. A further important property of the remote sensing system is the temporal resolution of the system, that is the time that elapses between successive dates of imagery acquisition for a given point on the ground. This revisit time may be measured in minutes if the satellite is effectively stationary with respect to a fixed point on the Earth's surface (i.e. in geostationary orbit, but note that not all geostationary orbits produce fixed observation points; see Elachi, 1988) or in days or weeks if the orbit is such that the satellite moves relative to the Earth's surface. The Meteosat satellite is an example of a geostationary platform, from which imaging instruments view an entire hemisphere of the Earth from a fixed position above the equator (Figure 1.13). The National Oceanic and Atmospheric Administration (NOAA) (Figure 1.12), Landsat (Figures 1.10 and 1.11) and SPOT satellites are polar orbiters, each having a specific repeat cycle time (or temporal resolution) of the order of hours (NOAA) or days/weeks (Landsat, SPOT). Both the Terra and Aqua satellites are in an orbit similar to that of Landsat. Since Terra's equatorial crossing time is 10.30 and Aqua's is 13.30, it is possible to measure short-term variations in oceanic and terrestrial systems. The temporal resolution of a polar orbiting satellite is determined by the choice of orbit parameters (such as orbital altitude, shape (e.g. circular or elliptic) and inclination), which are related to the objectives of the particular mission. Satellites that are used for Earth observing missions normally have a near-circular polar orbit, though the Space Shuttle flies in an equatorial orbit and some meteorological satellites use a geostationary orbit. Bakker (2000) and Elachi and van Zyl (2006, Appendix B) give details of the mathematics of orbit determination. The relationship between the orbit period T and the orbit radius r is given by Elachi's equation B-5:

$$T = 2\pi r \sqrt{\frac{r}{g_s R^2}}$$

in which g_s is the acceleration due to gravity ($0.00981\,\text{km}\,\text{s}^{-2}$), R is the Earth's radius (approximately 6380 km) and h is the orbital altitude (note that $r = R + h$). If, for example $h = 705\,\text{km}$ then $T \approx 6052\ \text{s} \approx 98.82\,\text{min}$, or more than double the time that Shakespeare's Puck took in *A Midsummer Night's Dream*. These calculations refer to the orbits of Landsat-4, -5 and -7 (Figure 2.1). Thus, by varying the altitude of a satellite in a circular orbit the time taken for a complete orbit is also altered; the greater the altitude the longer the orbital period.

The angle between the orbital plane and the Earth's equatorial plane is termed the *inclination* of the orbit, which is usually denoted by the letter i. Changes in the orbit are due largely to precession, caused mainly by the slightly non-spherical shape of the Earth. If the orbital precession is the same as the Earth's rotation round the Sun then the relationship between the node line and the Sun is always the same, and the satellite will pass over a given point on the Earth's surface at the same Sun time each day. Landsat and SPOT have this kind of orbit, which is said to be *Sun-synchronous*. Figure 2.1 illustrates an example of a circular, near-polar, Sun-synchronous orbit, that of the later Landsat satellites (numbered 4–7). Many satellites carrying Earth-observing instruments use a near-polar, Sun-synchronous orbit because the fact that the Sun's azimuth angle is the same for each date of observation means that shadow effects are reduced (the shadows are always in approximately the same direction). However, the Sun's zenith angle (Figure 1.19) changes throughout the year, so that seasonal differences do occur. Some applications, such as geology, may benefit from a the use of an orbit that is not Sun-synchronous, because different views of the region of interest taken at different Sun azimuth positions may reveal structural features on the ground that are not visible at one particular Sun azimuth angle (Figure 1.19).

As noted earlier, not all Earth observing platforms are in near-polar, Sun-synchronous orbits. The Space Shuttle has an equatorial orbit, which describes an S-shaped curve on the Earth's surface between the approximate latitudes of 50°N and 50°S. The orbit of the International Space Station is similar. A map showing the position and track of the ISS is maintained by NASA at http://spaceflight.nasa.gov/realdata/tracking/.

Thus, the orbit selected for a particular satellite determines not just the time taken to complete one orbit (which is one of the factors influencing temporal resolution) but also the nature of the relationship between the satellite and the solar illumination direction. The temporal resolution is also influenced by the swath width, which is the length on the ground equivalent to one scan line. Landsat TM and ETM+ have a swath width of 185 km whereas the Advanced Very High Resolution Radiometer (AVHRR) sensor carried by the NOAA satellites has a swath width of approximately 3000 km. The AVHRR can therefore provide much more frequent images of a fixed point on the Earth's surface than can the TM, though the penalty is reduced spatial resolution (1.1 km at nadir, compared with 30 m). A pointable sensor such as the SPOT HRV can, in theory, provide much more frequent temporal coverage than the orbital pattern of the SPOT satellite and the swath width of the HRV would indicate.

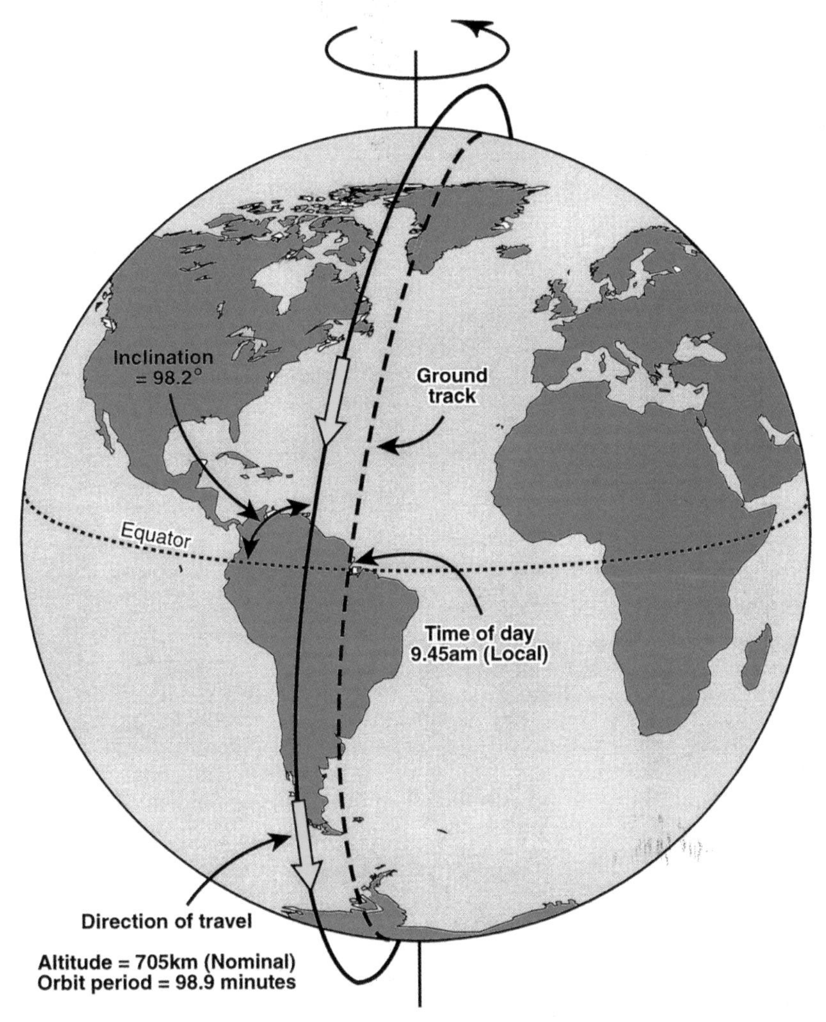

Figure 2.1 *Example of a Sun-synchronous orbit. This is the Landsat-4, -5 and -7 orbit, which has an equatorial crossing time of 09.45 (local Sun time) in the descending node. The satellite travels southwards over the illuminated side of the Earth. The Earth rotates through 24.7° during a full satellite orbit. The satellite completes just over 14.5 orbits in a 24-hour period. The Earth is imaged between 82° N and S latitude over a 16-day period. Landsat-7 orbit is similar. Landsat-7 travels about 15 minutes ahead of Terra, which is in the same orbit, so that sensors onboard Terra view the Earth under almost the same conditions as does the Landsat-7 ETM+.Based on Figure 5.1,* Landsat 7 Science Data Users Handbook, *NASA Goddard Spaceflight Center, Greenbelt, Maryland. http://landsathandbook.gsfc.nasa.gov/handbook/handbook_htmls/chapter5/chapter5.html (accessed 4 January 2009).*

2.2 Characteristics of Imaging Remote Sensing Instruments

The characteristics of imaging remote sensing instruments operating in the visible and infrared spectral region can be summarized in terms of their spatial, spectral and radiometric resolutions. Other important features are the manner of operation of the scanning device that collects the image (electromechanical or electronic) and the geometrical properties of the images produced by the system. The interaction between the spatial resolution of the sensor and the orbital period of the platform determines the number of times that a given point on the Earth will be viewed in any particular time period.

The fact that remote sensing imaging systems have different resolutions (spatial, spectral, radiometric and temporal) is related to the use that is made of the data that are collected by these systems. For observations of dynamic systems, such as the atmosphere and the oceans, a high temporal resolution is required but, because such systems are being observed at a continental or global scale, a coarse spatial resolution is appropriate. Meteosat images the Earth every 15 minutes at a resolution of 5 km at nadir. These images are at a global scale. NOAA satellites have a repeat cycle of 24 hours (almost the whole

Earth is imaged in one day at a spatial resolution at nadir of 1.1 km). The observations could be described as at a continental scale. Medium-resolution systems such as Landsat and SPOT have a repeat cycle of up to 26 days but the pixel size of 20–30 m allows regional studies to be carried out. High resolution systems can have a spatial resolution of less than 1 m. Their most appropriate use is in local studies. Spectral resolution, that is the number and location of the spectral bands viewed by the sensor, should be appropriately selected for the intended target. As noted in Chapter 1, Earth surface materials have different spectral reflectance characteristics and the bands selected for oceanographic observation would differ from a satellite system primarily concerned with land cover mapping. The number of bands is also important; for example the Hyperion instrument carried by Earth Observing-1 (EO-1) has 256 spectral bands. The measurements in these bands would be useful if one's aim was to compare the spectrum of a given pixel with a library spectrum such as those shown in Figure 1.21. Finally, the radiometric resolution of a sensor determines the level of detail that can be seen. It also allows for the monitoring of extremes, for example snow and cloud tops at one end of the range to deep clear water at the other. It follows from this discussion that the different resolutions of an Earth observing system should be selected with the aims of the investigation in mind.

2.2.1 Spatial Resolution

The *spatial resolution* of an imaging system is not an easy concept to define. It can be measured in a number of different ways, depending on the user's purpose. In a comprehensive review of the subject, Townshend (1980) uses four separate criteria on which to base a definition of spatial resolution. These criteria are the geometrical properties of the imaging system, the ability to distinguish between point targets, the ability to measure the periodicity of repetitive targets, and the ability to measure the spectral properties of small targets. These properties are considered briefly here; a fuller discussion can be found in Billingsley (1983), Forshaw *et al.* (1983), Simonett (1983) and Townshend (1980).

The most commonly used measure, based on the geometric properties of the imaging system, is its *instantaneous field of view* (IFOV). The IFOV is defined as the area on the ground that, in theory, is viewed by the instrument from a given altitude at any given instant in time. The IFOV can be measured in one of two ways, as the angle α or as the equivalent distance XY on the ground in Figure 2.2.

Note that Figure 2.2 is a cross-section, and that the line XY is, in fact, the diameter of a circle. The actual, as distinct from the nominal, IFOV depends on a number of

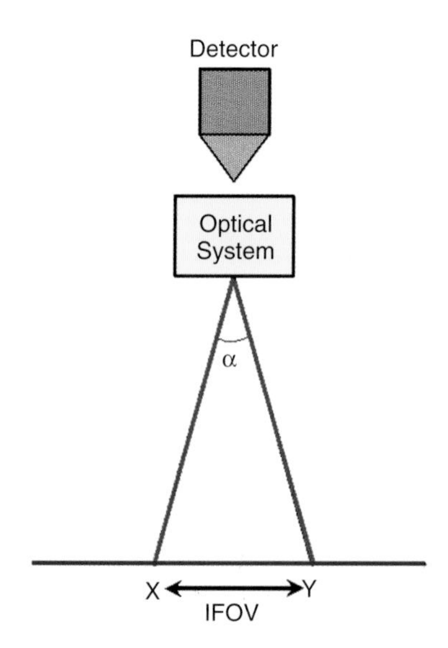

Figure 2.2 *Angular instantaneous field of view (IFOV), α, showing the projection XY on the ground. Note: XY is the diameter of a circle.*

factors. No satellite has a perfectly stable orbit; its height above the Earth will vary, often by tens of kilometres. For instance, Landsats-1 to -3 had a nominal altitude of 913 km, but the actual altitude of these satellites varied between 880 and 940 km. The IFOV becomes smaller at lower altitudes and increases as the altitude increases so, although the spatial resolution of the Landsat-1 to -3 Multi-spectral Scanner (MSS) (Section 2.3.6.1) is generally specified as 79 m, the actual resolution (measured by the IFOV) varied between 76 and 81 m.

The IFOV is the most frequently cited measure of resolution, though it is not necessarily the most useful. In order to explain why this is so, we must consider the way in which radiance from a point source is expressed on an image. A highly reflective point source on the ground does not produce a single bright point on the image but is seen as a diffused circular region, due to the properties of the optics involved in imaging. A cross-section of the recorded or imaged intensity distribution of a single point source is shown in Figure 2.3, from which it can be seen that the intensity of a point source corresponds to a Gaussian-type distribution. The actual shape will depend upon the properties of the optical components of the system and the relative brightness of the point source. The distribution function shown in Figure 2.3 is called the *point spread function* or PSF (Moik, 1980; Slater, 1980; Billingsley, 1983). Richter (1997) considers the role of the PSF in calibrating high spatial resolution imagery. Sensor calibration is considered in Chapter 4. Huang, Davis and Townshend (2002) and Townshend *et al.* (2000) discuss the effects of the PSF in determining the radiance from a given pixel. As Figure 2.3 shows, some of the radiance reaching the sensor from

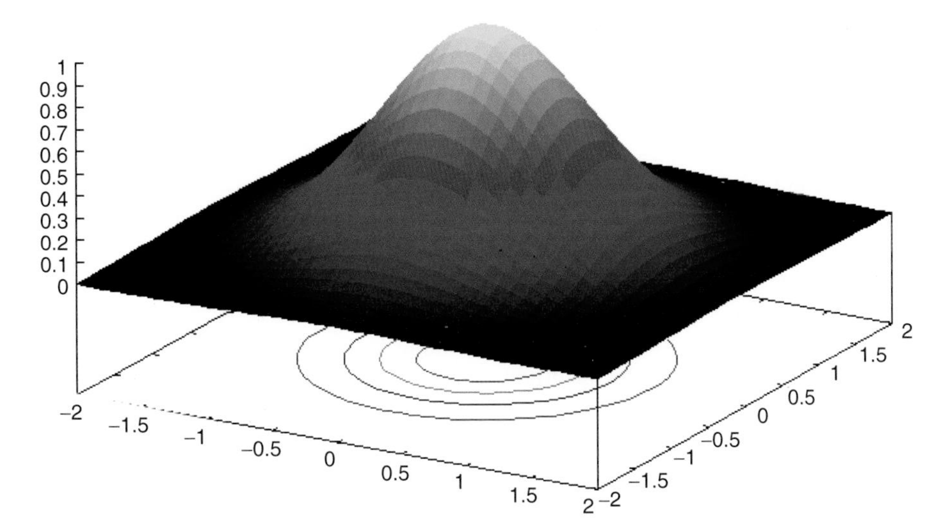

Figure 2.3 *Point spread function. The area of the pixel being imaged runs from* $-0.5 \leq x \leq 0.5$ *and* $-0.5 \leq y \leq 0.5$*, that is centred at (0, 0) but the energy collected by the sensor is non-zero outside this range. The ideal point spread function would be a square box centred at (0, 0) with a side length of 1.0.*

the pixel of side 1 centred at (0, 0) will come from neighbouring pixels; this is called the adjacency effect. Kavzoglu (2004) shows how the two-dimensional PSF of Landsat's ETM+ sensor is derived from the along-track and across-track PSFs.

An alternative measure of IFOV is, in fact, based on the PSF (Figure 2.4) and the 30 m spatial resolution of the Landsat-4 and -5 TM (Section 2.3.6.2) is based upon the PSF definition of the IFOV. The IFOV of the Landsat MSS using this same measure is 90 m rather than 79 m (Townshend, 1980, p. 9). The presence of relatively bright or dark objects within the IFOV of the sensor will increase or decrease the amplitude of the PSF so as to make the observed radiance either higher or lower than that of the surrounding areas. This is why high-contrast features such as narrow rivers and canals are frequently visible on Landsat ETM+ images, even though their width is less than the sensor's spatial resolution of 30 m. Conversely, targets with dimensions larger than the Landsat ETM+ IFOV of 30 m may not be discernible if they do not contrast with their surroundings. The blurring effects of the PSF can be partially compensated for by image processing involving the use of the Laplacian function (Section 7.3.2). Other factors causing loss of contrast on the image include atmospheric scattering and absorption, which are discussed in Chapters 1 and 4.

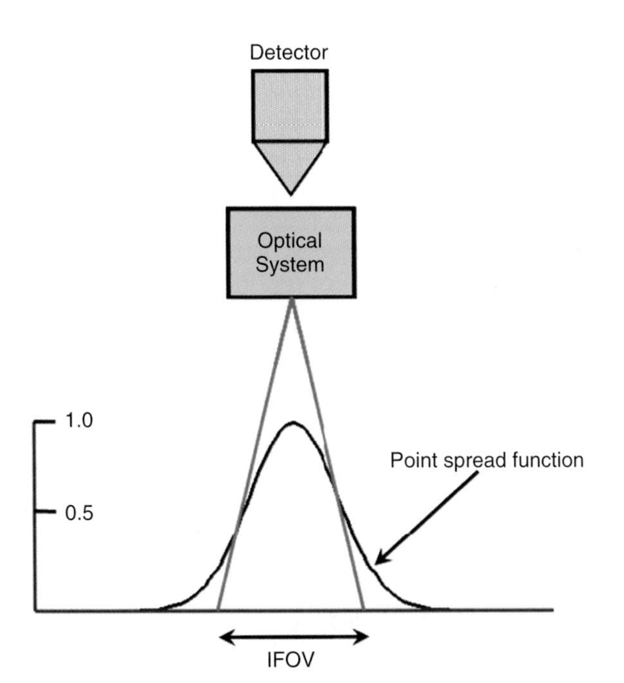

Figure 2.4 *Instantaneous field of view defined by the amplitude of the point spread function.*

The definition of spatial resolving power based on the IFOV is therefore not a completely satisfactory one. As it is a geometrical definition, it does not take into account the spectral properties of the target. If remote sensing is based upon the detection and recording of the radiance of targets, the radiance being measured at a number of discrete points, then a definition of spatial resolution that takes into account the way in which this radiance is generated might be reasonable. This is the basis of the definition of the *effective resolution element* or ERE, which is defined by Colvocoresses (cited by Simonett (1983)) as

> the size of an area for which a single radiance value can be assigned with reasonable assurance that the response is within 5% of the value representing the actual relative radiance.

Colvocoresses estimated the ERE for the Landsat MSS system as 86 and 35 m for the TM. These values might be more relevant than the IFOV for a user interested in classification of multispectral images (Chapter 8).

Other methods of measuring the spatial resolving power of an imaging device are based upon the ability of the device to distinguish between specific targets. There are two such measures in use, and both are perhaps more easily defined for photographic sensors. The first method uses the fact that the PSF of a point source is a bright central disc with bright and dark rings around it. The Rayleigh criterion assumes two equally-bright point sources and specifies that the two sources will be distinguishable on the image if the bright central disc of the one falls on the first dark ring of the PSF of the other. The minimum separation between the point sources to achieve this degree of separation on the image is a measure of the spatial resolving power of the imaging system. The second method assumes that the targets are not points but linear and parallel objects with a known separation that is related to their spatial frequency. If the objects contrast strongly with their background, then one could consider moving them closer together until the point is reached where they are no longer distinguishable. The spatial frequency of the objects such that they are just distinguishable is a measure of spatial resolving power. This spatial frequency is expressed in terms of line pairs per millimetre on the image or as cycles per millimetre. In order to calculate the spatial resolving power by this method it is usual to measure the contrast of the targets and their background. The measure most often used is *modulation* (M), defined as:

$$M = \frac{E_{max} - E_{min}}{E_{max} + E_{min}}$$

E_{max} and E_{min} are the maximum and minimum radiance values recorded over the area of the image. For a nearly homogeneous image, M would have a value close to zero while the maximum value of M is 1.0. Note that a perfectly homogeneous image would have an undefined modulation value since the calculation would involve division by zero. Returning to the idea of the parallel linear targets, we could find the ratio of the modulation measured from the image (M_I) to the modulation of the objects themselves (M_O) This ratio is the modulation transfer factor. A graph of this factor against spatial frequency shows the *modulation transfer function* or MTF. The spatial frequency at which the MTF falls to a half of its maximum value is termed the *effective instantaneous field of view* or EIFOV. The EIFOV of Landsat MSS imagery has been computed to be 66 m while the 30 m resolution of the Landsat TM is computed from the spatial frequency at the point where the MTF has 0.35 of its maximum value (Townshend, 1980, p. 12), whereas the IFOV measure gives resolutions of 79 and 45 m respectively. The EIFOV measure is based on a theoretical target rather than on real targets, and as such gives a result that is likely to exceed the performance of the instrument in actual applications. Townshend and Harrison (1984) describe the calculation and use of the MTF in estimating the spatial resolving power of the Landsat-4 and -5 TM instrument (Section 2.3.6.2).

The IFOV, ERE and EIFOV should not be confused with the pixel size. A digital image is an ordered set of numeric values, each value being related to the radiance from a ground area represented by a single cell or pixel. The pixel dimensions need not be related to the IFOV, the ERE or the EIFOV. For instance, the size of the pixel in a Landsat MSS image is specified as 56 by 79 m. The IFOV at the satellite's nominal altitude of 913 km is variously given as 79, 76.2 and 73.4 m while the ERE is estimated as 87 m (Simonett, 1983, Table 1.2). Furthermore, pixel values can be interpolated over the cells of the digital image to represent any desired ground spacing, using one of the resampling methods described in Chapter 4. The ground area represented by a single pixel of a Landsat MSS image is thus not necessarily identical to the spatial resolution as measured by any of the methods described above.

The discussion in the preceding paragraphs, together with the description given below of the way in which satellite-borne sensors operate, should make it clear that the individual image pixel is not 'sensed' uniquely. This would require a stop/start motion of the satellite and, in the case of the electromechanical scanner, of the scan mirror. The value recorded at the sensor corresponding to a particular pixel position on the ground is therefore not just a simple average of the radiance upwelling from that pixel. There is likely to be a contribution from areas outside the IFOV, which is termed the 'environmental radiance' (Richter *et al.*, 2006) (Section 4.4). Disregarding atmospheric effects (which are also discussed in Section 4.4), the radiance attributable to a specific pixel sensed over an area of terrestrial vegetation is in reality the sum (perhaps the non-linear sum) of the contributions of the different land cover components within the pixel, plus the contribution from radiance emanating from adjacent areas of the ground. Fisher (1997) and Cracknell (1998) examine the concepts underlying the idea of the pixel in some detail, and readers should be aware of the nature of the image pixel in remote sensing before attempting to interpret patterns or features seen on images.

Spatial resolving power is an important attribute of remote sensing systems because differing resolutions are relevant to different problems; indeed, there is a hierarchy of spatial problems that can use remotely-sensed data, and there is a spatial resolution appropriate to each problem.

To illustrate this point, consider the use of an image with a spatial resolution (however defined) of 1 m. Each element of the image, assuming its pixel size was 1 × 1 m, might represent the crown of a tree, part of a grass verge by the side of a suburban road, or the roof of a car. This imagery would be useful in providing the basis for small-scale mapping of urban patterns, analysis of vegetation variations over a small area, or the monitoring of crops in small plots. At this scale it would be difficult to assess the boundaries of, or variation within, a larger spatial unit such as a town; a spatial resolution of 10 m might be more appropriate to this problem. A 10 m resolution would be a distinct embarrassment if the exercise was concerned with the mapping of sea-surface temperature patterns in the Pacific Ocean, for which data from an instrument with a spatial resolution of 500 m or larger could be used. For continental or global-scale problems a spatial resolution of 1 and 5 km respectively would produce data that contained the information required (and no more) and, in addition, was present in manageable quantities. One point to consider concerns the statistical variance of pixel values lying within objects of interest, such as agricultural fields. The spatial resolution must be high enough for the fields to be seen and characterized, but not so high that the statistical variance of pixels within the fields becomes larger than the variation between the mean pixel values of different fields. In the east of England, which has relatively large fields, a spatial resolution of 20–30 m is acceptable, whereas a spatial resolution of 1 m would be embarrassing if the aim was to classify the land cover of the fields into categories such as wheat, barley and pasture. Conversely, if the aim of the exercise is operational precision farming then a resolution of 1 m or less should be sufficient to determine the within-field variability. These figures are illustrative only; the actual values depend on average field size. Nevertheless, the illusion that higher spatial resolution is necessarily better is commonplace; one should always ask 'better for what?' See Atkinson and Curran (1997), Atkinson and Aplin (2004), Hengl (2006), Quattrochi and Goodchild (1997), Thomson (2009) and Wickham and Riitters (1995) for further discussion of this point in a wider context.

The concept of spatial and temporal scale is fundamental to any remote sensing/geographical information science (GIS) project. As noted earlier in this section, the term 'spatial resolution' is open to several interpretations and there is a fundamental need to match the scale of the relevant image to other GIS data layers and to the scope of the problem being considered. Aplin (2006), a special issue of *International Journal of Remote Sensing* (2006), Atkinson and Aplin (2004) and Foody and Curran (1994) consider the problem of spatial and temporal scales in remote sensing. The problem of *scaling-up* or *scaling-down* (i.e. adjusting results or data at one spatial scale to those at another scale) is considered by contributors to van Gardingen, Foody and Curran (1997), for example Barnsley, Barr and Tsang (1997), and Foody and Curran (1994).

2.2.2 Spectral Resolution

The second important property of an optical imaging system is its *spectral resolution*. Microwave SAR images collected by instruments onboard satellites such as ERS-1 and -2, TeraSAR-X, Advanced Land Observation Satellite (ALOS)/PALSAR and Radarsat-1 and -2 (Section 2.4) are generally recorded in a single waveband, as are panchromatic images collected by sensors such as WorldView, IKONOS, SPOT HRV and Landsat-7 ETM+. A panchromatic (literally 'all colours') image is a single band image that measures upwelling radiance in the visible wavebands. In addition to collecting a panchromatic image of the target area, most medium spatial resolution sensors operating in the visible and infrared bands collect multispectral or multiband images, which are sets of individual images that are separately recorded in discrete spectral bands, as shown in Figures 1.10 and 1.11. High spatial resolution satellites such as IKONOS and QuickBird have both panchromatic (high resolution) and multispectral capabilities. The term spectral resolution refers to the width of these spectral bands measured in micrometres (μm) or nanometres (nm). The following example illustrates two important points, namely, that (i) the position in the spectrum, width and number of spectral bands determines the degree to which individual targets (vegetation species, crop or rock types) can be discriminated on the multispectral image and (ii) the use of multispectral imagery can lead to a higher degree of discriminating power than any single band taken on its own.

The reflectance spectra of vegetation, soils, bare rock and water are described in Section 1.3. Differences between the reflectance spectra of various rocks, for example might be very subtle and the rock types might therefore be separable only if the recording device were capable of detecting the spectral reflectance of the target in a narrow waveband. A wide-band instrument would simply average the differences. Figure 2.5a is a plot of the reflection from a leaf from a deciduous tree against wavelength (dotted line). Figure 2.5b shows the reflectance spectra for the same target as recorded by a broad-band sensor such as the Landsat-7 ETM+. Broad-band sensors will, in general, be unable to distinguish subtle differences in reflectance spectra, perhaps resulting from disease or stress. To provide for the more reliable identification of particular targets on a remotely-sensed image the spectral resolution of the sensor must match as closely as possible the spectral reflectance curve of

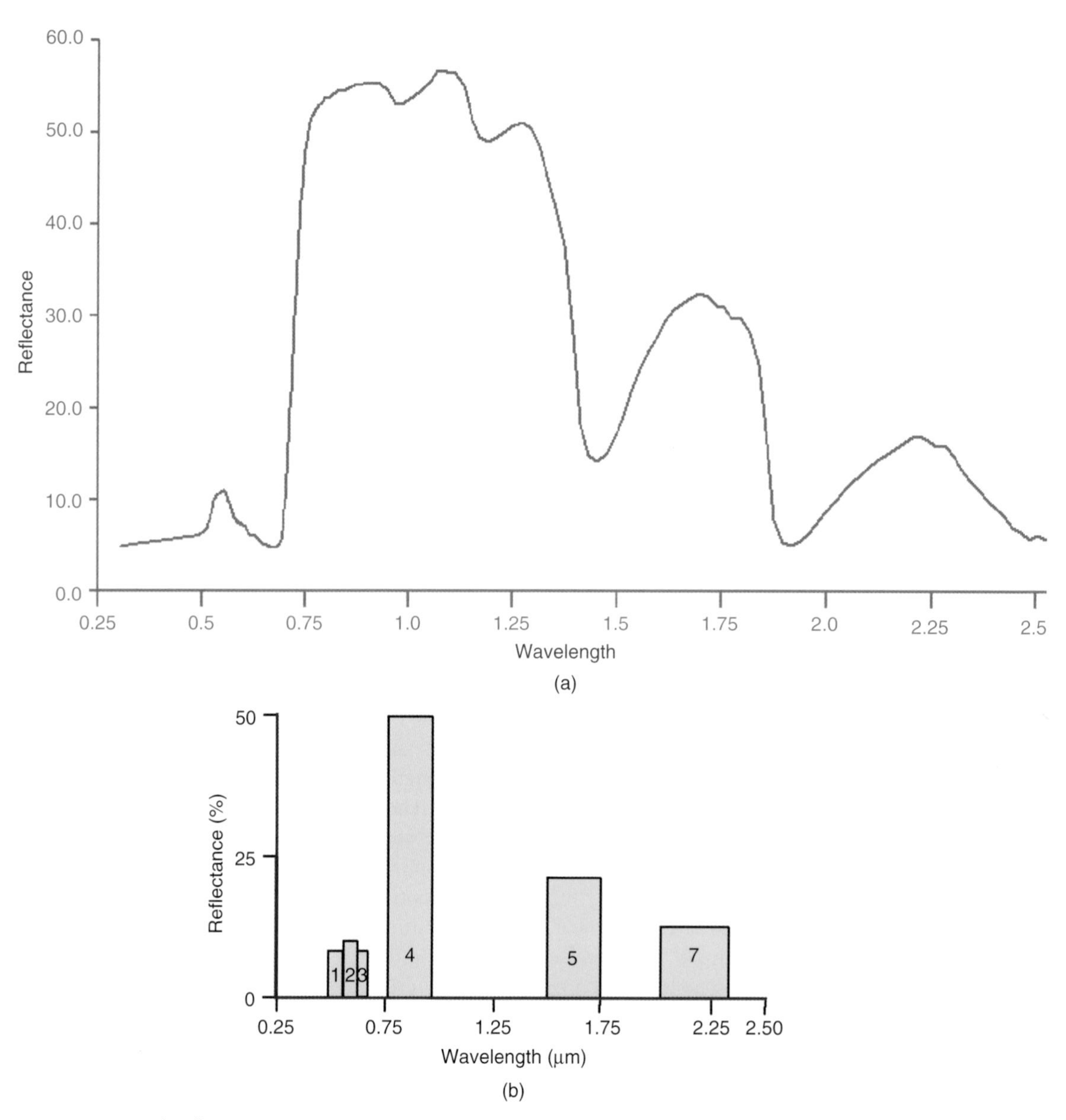

Figure 2.5 *(a) Spectral reflectance curve for a leaf from a deciduous tree. (b) The reflectance spectrum shown in Figure 2.5a as it would be recorded in Landsat ETM+ bands 1–5 and 7. Data from the ASTER spectral library through the courtesy the Jet Propulsion Laboratory, California Institute of Technology,Pasadena, California, © California Institute of Technology. All rights reserved.*

the intended target. This principle is demonstrated by the design of the Coastal Zone Colour Scanner (CZCS) carried by the Nimbus-7 satellite, a design that is common to most ocean-observing satellites, namely, the width and positions of the spectral bands are determined by the spectral reflectance curve of the target. Of course, other considerations must be balanced, such as frequency of coverage or temporal resolution and spatial resolution (Section 2.2.1), as well as practical factors.

Only in an ideal world would it be possible to increase the spectral resolution of a sensor simply to suit the user's needs. There is a price to pay for higher resolution. All signals contain some noise or random error that is caused by electronic noise from the sensor and from effects introduced during transmission and recording. The signal-to-noise ratio (SNR) is a measure of the purity of a signal. Increasing the spectral resolution reduces the SNR of the sensor output because the magnitude of the radiance (the signal strength) in narrower spectral bands is less than that of wider bands while the inherent noise level remains constant. Smith and Curran (1996, 1999) provide details of SNR calculations, and estimate values for the AVHRR, Landsat TM and SPOT HRV as 38 : 1, 341 : 1 and 410 : 1 respectively. See also Atkinson *et al.* (2005, 2007).

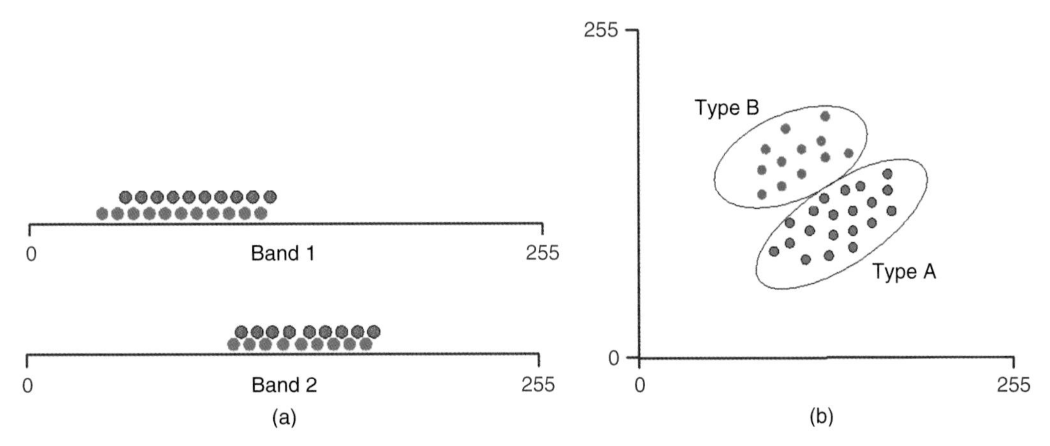

Figure 2.6 *Hypothetical land cover types A and B measured on two spectral bands. (a) Shown as plots on separate bands; the two cover types cannot be distinguished. (b) Shown as a plot on two bands jointly; now the two cover types can be separated in feature space.*

A compromise must be sought between the twin requirements of narrow bandwidth (high spectral resolution) and a low SNR. The *pushbroom* type of sensor is a linear array of individual detectors with one detector per scan line element. The forward movement of the sensor forms the image one scan line at a time (Figure 1.2b). This arrangement provides for a longer 'look' at each scan line element, so more photons reflected from the target are collected, which results in a better SNR than does an electromechanical scanner employing a single detector which observes each scan line element sequentially (Figure 1.2a). The time available to measure the energy emanating from each point along the scan line (termed the dwell time or integration time) is greater for pushbroom scanners, because the individual detector 'sees' more photons coming from a given point than the detector in an electromechanical scanner, which looks at the point only for a short time. Hence, narrower bandwidths and a larger number of quantization levels are theoretically possible without decreasing the SNR to unacceptable levels. Further discussion of different kinds of sensors is contained in Section 2.3.

Justification for the use of multiple rather than single measures of the characteristics of individual Earth surface objects helps to discriminate between groups of different objects. For example, Mather (1976, p. 421) contends that:

> ...The prime justification of adopting a multivariate approach [is] that significant differences (or similarities) may well remain hidden if the variables are considered one at a time and not simultaneously.

In Figure 2.6a the measurements of the spectral reflectance values for individual members of two hypothetical land-cover types are plotted separately for two spectral bands, and there is an almost complete overlap between them. If the measurements on the two spectral bands are plotted together, as in Figure 2.6b, the difference between the types is clear, as there is no overlap. The use of well-chosen and sufficiently numerous spectral bands is a requirement, therefore, if different targets are to be successfully identified on remotely-sensed images.

2.2.3 Radiometric Resolution

Radiometric resolution or radiometric sensitivity refers to the number of digital quantization levels used to express the data collected by the sensor. In general, the greater the number of quantization levels the greater the detail in the information collected by the sensor. At one extreme one could consider a digital image composed of only two levels (Figure 2.7a) in which level 0 is shown as black and level 1 as white. As the number of levels increases to 16 (Figure 2.7b) so the amount of detail visible on the image increases. With 256 levels of grey (Figure 2.7c) there is no discernible difference, though readers should note that this is as much a function of printing technology as of their visual systems. Nevertheless, the eye is not as sensitive to changes in intensity as it is to variations in hue.

The number of grey levels is commonly expressed in terms of the number of binary (base 2) digits (bits) needed to store the value of the maximum grey level.[1] Just two binary digits, 0 and 1, are used in a base 2 number rather than the 10 digits (0–9) that are used in base 10 representation. Thus, for a two-level or black/white representation, the number of bits (binary digits) required per pixel is 1 (giving two states – 0 and 1), while for 4, 16, 64 and 256 levels the number of bits required is 2, 4, 6 and 8 respectively. Thus, '6-bit' data has 64 possible

[1]Base 2 representation of numbers is considered at greater length in Chapter 3.

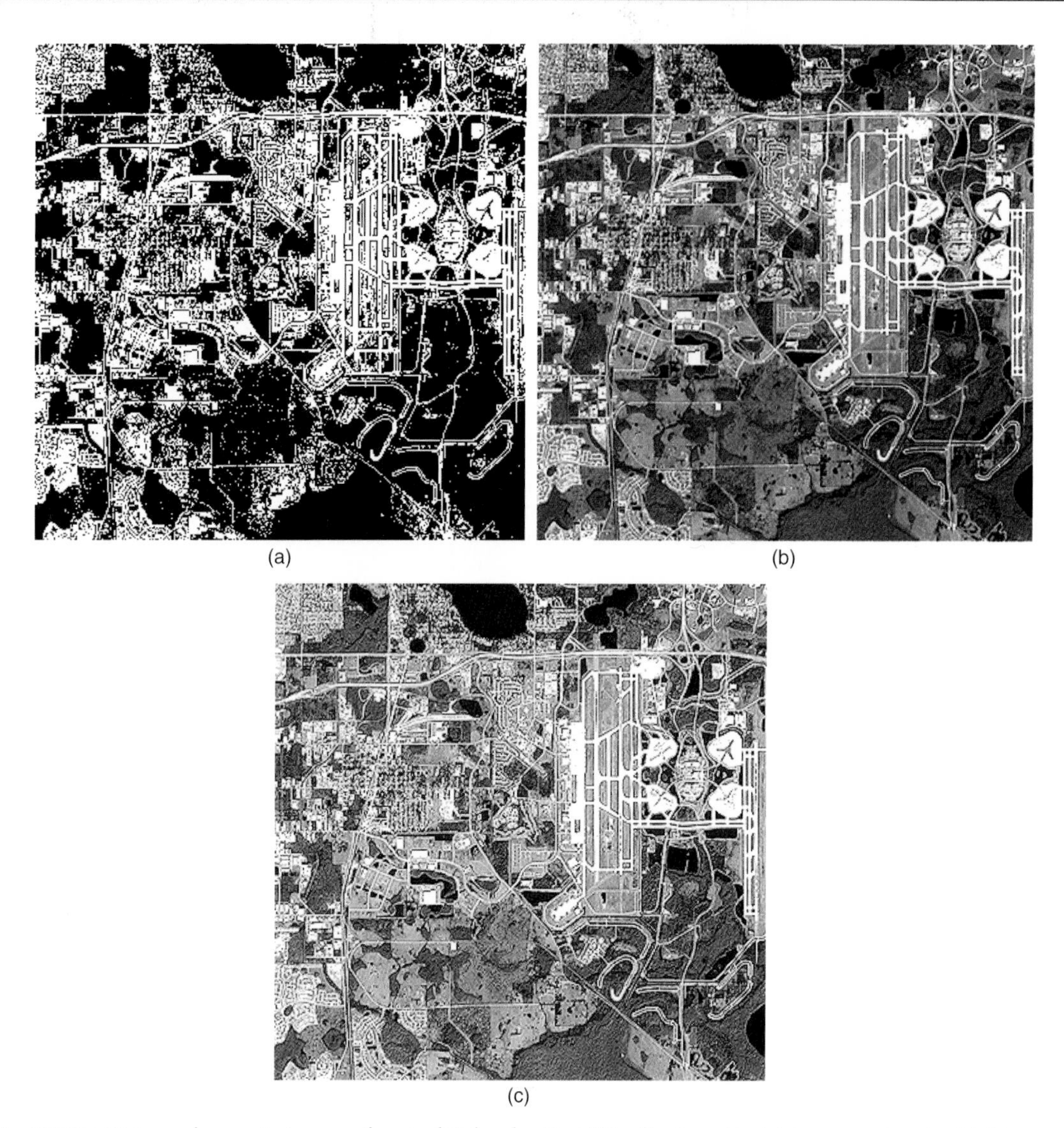

Figure 2.7 *SPOT HRV panchromatic image of part of Orlando, FL, USA, displayed in (a) two grey levels, (b) 16 grey levels and (c) 256 grey levels. Permission to use the data was kindly provided by SPOT image, 5 rue des Satellites, BP 4359, F 331030, Toulouse, France.*

quantization levels, represented by the whole number or integer values 0–63 inclusive (000000–111111 in base 2 notation), with '0' representing black, '63' representing white, and '31' indicating mid-grey. These 'numbers' are enclosed in quotation marks to show that they are not indicating a direct measurement of ground-leaving radiance. They are the steps or quanta into which a range of physical values is divided, hence the term 'quantization', which – in this context – means 'breaking down a continuous range into a discrete number of levels'. If, for instance, a sensor can detect radiance in the range 0.0–10.0 $\mathrm{W\,m^{-2}\,sr^{-1}\,\mu m^{-1}}$ and the SNR of the instrument is such that this range can be divided into 256 levels (for reasons explained in the next paragraph) then level '0' would represent 0.0 $\mathrm{W\,m^{-2}\,sr^{-1}\,\mu m^{-1}}$ and level '255' would represent 10.0 $\mathrm{W\,m^{-2}\,sr^{-1}\,\mu m^{-1}}$. The intermediate levels ('1'–'254') represent equal steps of $(10.0 - 0.0)/255.0 = 0.00392\,\mathrm{W\,m^{-2}sr^{-1}\,\mu m^{-1}}$.

Needless to say, the number of levels used to express the signal received at a sensor cannot be increased simply to suit the user's preferences. The SNR of the sensor, described above, must be taken into consideration. The step size from one level to the next cannot be less than the noise level, or else it would be impossible to say whether a change in level was due to a real change in the radiance of the target or to a change in the magnitude of the noise. A low-quality instrument with a high noise level would necessarily, therefore, have a lower radiometric resolution compared with a high-quality, high SNR instrument. Slater (1980) discusses this point in some detail.

Fourty and Baret (1998) note that estimates of relationships between leaf properties and reflection and transmittance properties are quite sensitive to the radiometric resolution of the instrument used. Bernstein *et al.* (1984) used a measure known as entropy to compare the amount of information (in terms of bits per pixel) for 8-bit and 6-bit data for two images of the Chesapeake Bay area collected by Landsat-4's TM and MSS sensors. The entropy measure, H, is given by:

$$H = -\sum_{i=0}^{k} p(i) \log_2 p(i)$$

where k is the number of grey levels (for example 64 or 256) and $p(i)$ is the probability of level i, which can be calculated from:

$$p(i) = \frac{F(i)}{nm}$$

In this formula, $F(i)$ is the frequency of occurrence of each grey level from 0 to $k - 1$, and nm is the number of pixels in the image. Moik (1980, p. 296) suggests that the use of this estimate of $p(i)$ will not be accurate because adjacent pixel values will be correlated (the phenomenon of spatial autocorrelation, as described by Cliff and Ord 1973) and he recommends the use of the frequencies of the first differences in pixel values; these first differences are found by subtracting the value at a given pixel position from the value at the pixel position immediately to the left, with obvious precautions for the leftmost column of the image. Bernstein *et al.* (1984) do not indicate which method of estimating entropy they use but, since the same measure was applied to both the 6-bit and 8-bit images, the results should be comparable. The 8-bit resolution image added, on average, 1–1.5 bits to the entropy measure compared with a 6-bit image of the same scene. These results are shown in Table 2.1. Moik (1980, Table 9.1) used both measures of entropy given above; for the first measure (based on the levels themselves) he found average entropy to be 4.4 bits for Landsat MSS (6-bit) and, for the second measure (based on first differences) he found average entropy to be 4.2 bits. This value is slightly higher than that reported by Bernstein *et al.* (1984) but still less than the entropy values achieved by the 8-bit TM data. It is interesting to note that, using data compression techniques, the 6-bit data could be re-expressed in 4 or so bits without losing any information, while the 8-bit data conveyed, on average, approximately 5 bits of information. Given the enormous amounts of data making up a multiband image (nearly 300 million pixel values for one Landsat TM image) such measures as entropy are useful both for comparative purposes and for indicating the degree of redundancy present in a given dataset.

Entropy measures can also be used to compare the performance of different sensors. For example, Kaufmann *et al.* (1996) use entropy to compare the German MOMS sensor performance with that of Landsat TM and the SPOT HRV, and Masek *et al.* (2001) use the same measure to compare images of a test area acquired simultaneously by Landsat-5's TM and Landsat-7's ETM+ instruments.

The concepts of spatial, radiometric and temporal resolution are discussed concisely by Joseph (2000).

2.3 Optical, Near-infrared and Thermal Imaging Sensors

The aim of this section is to provide brief details of a number of sensor systems operating in the visible and infrared regions of the electromagnetic spectrum, together with sample images, in order to familiarize the reader with the nature of digital image products and their applications. Further details of the instruments described in this section, as well as others that are planned for

Table 2.1 *Entropy by band for Landsat TM and MSS sensors based on Landsat-4 image of Chesapeake Bay area, 2 November 1982 (scene E-40109-15140). See text for explanation.*

Band number	Thematic mapper		Multispectral scanner	
	Waveband (μm)	Bits per pixel	Waveband (μm)	Bits per pixel
1	0.45–0.52	4.21	0.5–0.6	2.91
2	0.52–0.60	3.78	0.6–0.7	3.57
3	0.63–0.69	4.55	0.7–0.8	4.29
4	0.76–0.90	5.19	0.8–1.1	3.63
5	1.55–1.75	5.92	–	–
6	10.5–12.5	3.53	–	–
7	2.08–2.35	5.11	–	–
Average	–	4.61	–	3.60

Based on Table 1 of Bernstein *et al.* (1984).

Example 2.1: Along-Track Scanning Radiometer (ATSR)

This image from the ATSR is a single-band greyscale image that has been processed by a technique called pseudo-colour (Section 5.4.2). The image covers the 11.0 μm region in the TIR (see Figure 1.7). Cold areas are shown in blue and purple while warmer areas are shown in yellow and red. Land is cooler than sea. The islands in the upper area of the image are Ibiza, Mallorca and Menorca, and the land at the bottom of the image is Algeria. The image shows a thermal eddy and an intricate pattern of temperature variations. The image was acquired on 9 May 1992 and it covers an area of $512 \times 512\,km^2$. Image source: http://www.atsr.rl.ac.uk/images/sample/atsr-1/index.shtml.

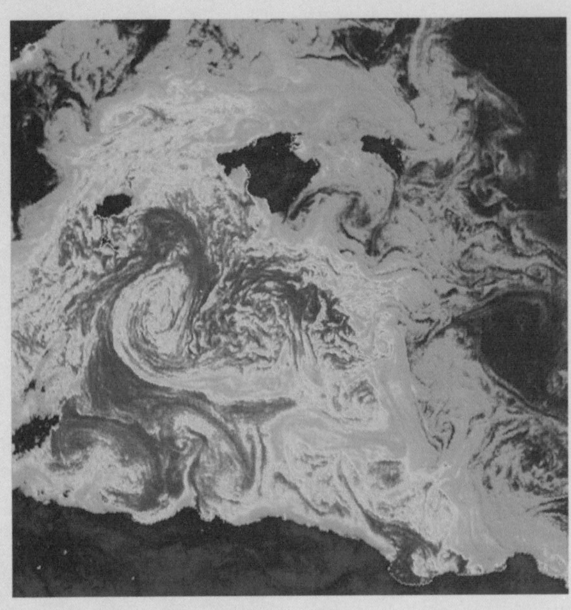

Example 2.1 Figure 1. ATSR image of the Mediterranean Sea between the Balearic Islands and the Algerian coast. This is a thermal infrared image that has been processed by the pseudocolour transform. Image © European Space Agency and Rutherford Appleton Laboratory (RAL).

the next decade, are available from the CEOS via the Internet. A comprehensive survey, now possibly only of historical interest, is provided by Joseph (1996). In this chapter, attention is focused on sensors that in current use, especially the Landsat and SPOT systems, which have generated large and widely used datasets dating from 1972 and 1986, respectively. Sufficient detail is provided for readers to inform themselves of the main characteristics of each system. More information is provided in the references cited. Readers can also access the image files stored on the publisher's website at http://www.wiley.com/go/mather4. The test images are bundled in with the MIPS software in a zip file called **mips.zip**. When you have unzipped this file then you should have a subfolder called **mips/images** containing a set of small images that can be used to view examples of remotely-sensed images acquired by sensors operating in the visible, NIR and thermal regions of the spectrum.

2.3.1 Along-Track Scanning Radiometer (ATSR)

The Along-Track Scanning Radiometer (ATSR) was developed by a consortium of British universities and research institutes led by the UK Rutherford-Appleton Laboratory (RAL). It has three TIR channels centred at 3.7 μm, 10.8 μm and 12 μm, plus a NIR channel centred at 1.6 μm. These thermal channels correspond to bands 3, 4 and 5 of the NOAA AVHRR. The primary purpose of this instrument is to make accurate measurements of global sea surface temperatures for climate research purposes. It was flown on the ERS-1 satellite, which was launched in July 1991. ATSR-2 is a development of ATSR, and has additional channels centred at wavelengths of 0.555, 0.659 and 0.865 μm that are intended for land applications. It is carried by the ERS-2 satellite, which was launched in April 1995. ATSR is an interesting instrument (Harries *et al.*, 1994) as it provides two views of the target from different angles – one from nadir and one in the forward direction. Each view includes an internal calibration system; hence, the data can be corrected for atmospheric effects (which can be estimated from the two views) and calibrated more precisely. The thermal sensor has a nominal accuracy of ±0.05 K. The sensor produces images with a spatial resolution of 1 km for a 500 km swath width. See Example 2.1 for more information on ATSR.

An extended version of ATSR-2, called the Advanced Along Track Scanning Radiometer (AATSR), is carried by the European Space Agency (ESA)'s ENVISAT-1 satellite, which was launched in 2002. The AATSR has seven channels (0.55, 0.67, 0.86, 1.6, 3.7, 11 and 12 μm). The spatial resolution of the visible and near-infrared (VNIR) channels (bands 1–4) is 500 m (with a swath width of 500 km). The infrared channels (bands 5–7) have a spatial resolution of 1 km. An AATSR thermal image is shown in Figure 1.28.

2.3.2 Advanced Very High Resolution Radiometer (AVHRR) and NPOESS VIIRS

The AVHRR, which is carried by the American NOAA series of satellites, was intended to be a meteorological observing system. The imagery acquired by AVHRR has, however, been widely used in land cover monitoring at global and continental scales. Two NOAA satellites are in orbit at any one time, giving morning, afternoon, evening and night-time equatorial crossing times of approximately 07.30, 14.00, 19.30 and 02.00 respectively, though it should be noted that the illumination and view geometry is not the same for all of these overpasses. The NOAA satellite orbit repeats exactly after 9 days. The orbital height is 833–870 km with an inclination of 98.7° and a period of 102 min. The AVHRR/3 instrument, introduced in 1998, has six channels and a spatial resolution, at nadir, of 1.1 km. Data with a 1.1 km nominal resolution are known as *local area coverage* (LAC) data. A lower resolution sample of the LAC data, called *global area coverage* (GAC) data is recorded for the entire 102-min orbit. About 11 min of LAC data can be tape-recorded on each orbit, and so LAC data are generally downloaded to ground receiving stations such as that at the UK Natural Environment Research Council (NERC) ground station at Dundee University in Scotland, which provided the image shown in Figure 1.12. The swath width of the AVHRR is of the order of 3000 km, and so spatial resolution at the edges of the image is considerably greater than the nominal (nadir) figure of 1.1 km. Correction of the geometry of AVHRR is mentioned further in Section 4.3. The effects of off-nadir viewing are not only geometrical for, as noted in Chapter 1, the radiance observed for a particular target depends on the angles of illumination and view. Thus, bidirectional reflectance factors should be taken into consideration when using AVHRR imagery.

The TIR channels are used for sea-surface temperature determination and cloud mapping, whereas the VNIR channels are used to monitor land surface processes, such as snow and ice melt, as well as vegetation status using vegetation indices such as the Normalized Difference Vegetation Index (NDVI) (Section 6.2.4). Band 3B has a shorter wavelength than bands 4 and 5; it can therefore be used to monitor hotter targets such as forest fires (see Figure 1.15) (Chuvieco and Martin, 1994; Chuvieco *et al.*, 2004). An illustration of a sea-surface temperature image can be seen in Figure 1.28 The definitive source of information on the AVHRR sensor is Cracknell (1997).

NOAA-19, launched in February, 2009, is the last in the Television Infrared Observation Satellite (TIROS) series. It will carry the AVHRR/3 and the HIRS/4 instruments. HIRS/4 is used to collect atmospheric profiles in 1 visible, 7 middle IR and 12 long-wave IR bands, from which total atmospheric ozone levels, cloud height and coverage and surface radiance can be derived (Table 2.2). The TIROS programme is to be replaced by the National Polar-orbiting Operational Environmental Satellite System (NPOESS). Imaging capabilities are provided by the Visible/Infrared Imager Radiometer Suite (VIIRS), which has nine VNIR channels, eight mid-infared bands and four TIR bands. The spatial resolution of the instrument at nadir is 400 m., the swath is 3000 km wide and the radiometric resolution is 11 bits. See Çommittee on Earth Studies, Space Studies Board. Commission on Physical Sciences, Mathematics, and Applications, National Research Council (2000), Hutchison and Cracknell (2005), Qu *et al.* (2006), Murphy *et al.* (2006) and Townshend and Justice (2002)

Table 2.2 *The AVHRR/3 Instrument carried by the NOAA satellites.*

Channel number	Wavelengths (μm)	Typical uses
1	0.58–0.68	Daytime cloud and surface mapping
2	0.725–1.00	Land–water boundaries
3A	1.58–1.64	Snow and ice detection
3B	3.55–3.93	Night cloud mapping, sea surface temperature
4	10.30–11.30	Night cloud mapping, sea surface temperature
5	11.50–12.50	Sea surface temperature

Based on http://noaasis.noaa.gov/NOAASIS/ml/avhrr.html.

for more details of NPOESS and VIIRS. Once NPOESS is operational, there will be three satellites in orbit, each with a different overpass time, giving data suitable for the study of diurnal variations. The VIIRS will replace the AVHRR and the MODerate Resolution Imaging Spectrometer (MODIS) instrument (see next section).

Other coarse resolution sensors with thermal bands are the AATSR (Section 2.3.1) and MODIS (Section 2.3.3). VEGETATION (Section 2.3.7.2) carries coarse resolution visible and NIR bands for land surface monitoring.

2.3.3 MODIS

MODIS is a wide field of view instrument that is carried by both the US Terra and Aqua satellites. As noted earlier, the Terra satellite has an equatorial crossing time of 10.30 while the Aqua satellite has an equatorial crossing time of 13.30, so that two MODIS instruments can be used to collect information relating to diurnal variations in upwelling radiance that relate to the characteristics and conditions of land surface, oceanic and atmospheric variables. It is a conventional radiometer, using a scanning mirror, and measures in 36 spectral bands in the range 0.405–14.835 μm (Table 2.3; note that the bands are not listed contiguously in terms of wavelength). Bands 1–19 are collected during the daytime segment of the orbit, and bands 20–36 are collected during the night-time segment. The spatial resolution of bands 1 and 2 is 250 m. Bands 3–7 have a spatial resolution of 500 m, while the remaining bands (8–36) have a spatial resolution at nadir of 1 km.

Estimates of land surface characteristics are derived from the higher-resolution bands (1–7) and ocean/atmosphere measurements are estimated from the coarser resolution (1 km) bands. The swath width is 2330 km, so that each point on the Earth's surface is observed at least once every two days. MODIS data are relayed to White Sands, NM, USA, via the Tracking and Data Relay Satellite (TDRS), and thence to the Goddard Space Flight Centre for further processing.

MODIS data are available in standard form, as calibrated radiances. Forty-one further 'products' derived from MODIS data are being made available. These products include surface reflectance, land surface temperature and emissivity, land cover, leaf area index, sea ice cover, suspended solids in ocean waters and sea surface temperature, thus demonstrating the wide range of applications (relating to terrestrial, atmospheric, oceanographic and cryospheric systems). The processing steps required to generate these products are described by Masuoka *et al.* (1998). Example 2.2 shows a MODIS vegetation product (showing the spatial distribution of two vegetation indices (Section 6.2.4) over North and Central America. Lobser and Cohen (2007) discuss the use of orthogonal transformations (Chapter 6) of MODIS data to measure land surface characteristics. Lunetta *et al.* (2006) use multitemporal MODIS data to measure change in land cover types while Miller and McKee (2004) attempt to map suspended sediment using MODIS 250 m resolution data. Justice *et al.* (2006) discuss fire products obtained from MODIS data (c.f. Section 2.3.2 – AVHRR). Salomonson *et al.* (2002) provide an overview of MODIS's performance up to that date.

2.3.4 Ocean Observing Instruments

A number of dedicated ocean observing satellites have been placed in orbit in the past 25 years. Measurements from instruments carried by these satellites have been used to study the marine food chain and the role of ocean in biogeochemical cycles such as the carbon cycle (Esaias *et al.*, 1998). These measurements have to be very accurate, as the radiance upwelling from the upper layers of the oceans typically does not exceed 10% of the signal recorded at the sensor, the rest being contributed by the atmosphere via scattering (Section 1.2.5). It is thus essential that the procedures for atmospheric correction, that is removal of atmospheric effects (Section 4.4), from the measurements made by ocean observing instruments should be of a high quality. The sensors that are described briefly in this section are the CZCS, the Sea-viewing Wide Field-of-view Sensor (SEAWiFS), Oceansat-2 and MODIS. The last of these is described elsewhere (Section 2.3.3) in more detail.

The CZCS instrument was carried by the Nimbus-7 satellite between 1978 and 1986, and was primarily designed to map the properties of the ocean surface, in particular, chlorophyll concentration, suspended sediment distribution, *gelbstoffe* (yellow stuff) concentrations as a measure of salinity and sea surface temperatures. Chlorophyll, sediment and *gelbstoffe* were measured by five channels in the optical region, four of which were specifically chosen to target the absorption and reflectance peaks for these materials at 0.433–0.453 μm, 0.510–0.530 μm, 0.540–0.560 μm and 0.660–0.680 μm. The 0.7–0.8 μm band was used to detect terrestrial vegetation and sea-surface temperatures were derived from the TIR band (10.5–12.5 μm). The spatial resolution of CZCS was 825 m and the swath width was 1566 km. The sensor could be tilted up to 20° to either side of nadir in order to avoid glint, which is light that is specularly reflected from the ocean surface. The Nimbus satellite had a near-polar, Sun-synchronous orbit with an altitude of 955 km. Its overpass time was 1200 local time, and the satellite orbits repeated with a period of 6 days. The CZCS experiment was finally terminated in December 1986.

Table 2.3 *MODIS wavebands and key uses. Bands 13 and 14 operate in high low gain mode. Bands 21 and 22 have the wavelength range but band 21 saturates at about 500 K, whereas band 22 saturates at about 335 K.*

Band number	Range	Primary use
1	0.620–0.670	Absolute land cover transformation, vegetation chlorophyll
2	0.841–0.876	Cloud amount, vegetation land cover transformation
3	0.459–0.479	Soil/vegetation differences
4	0.545–0.565	Green vegetation
5	1.230–1.250	Leaf/canopy differences
6	1.628–1.652	Snow/cloud differences
7	2.105–2.155	Cloud properties, land properties
8	0.405–0.420	Chlorophyll
9	0.438–0.448	Chlorophyll
10	0.483–0.493	Chlorophyll
11	0.526–0.536	Chlorophyll
12	0.546–0.556	Sediments
13h	0.662–0.672	Atmosphere, sediments
13l	0.662–0.672	Atmosphere, sediments
14h	0.673–0.683	Chlorophyll fluorescence
14l	0.673–0.683	Chlorophyll fluorescence
15	0.743–0.753	Aerosol properties
16	0.862–0.877	Aerosol properties, atmospheric properties
17	0.890–920	Atmospheric properties, cloud properties
18	0.931–0.941	Atmospheric properties, cloud properties
19	0.915–0.965	Atmospheric properties, cloud properties
20	3.660–3.840	Sea surface temperature
21	3.929–3.989	Forest fires and volcanoes
22	3.929–3.989	Cloud temperature, surface temperature
23	4.020–4.080	Cloud temperature, surface temperature
24	4.433–4.498	Cloud fraction, troposphere temperature
25	4.482–4.549	Cloud fraction, troposphere temperature
26	1.360–1.390	Cloud fraction (thin cirrus), troposphere temperature
27	6.535–6.895	Mid troposphere humidity
28	7.175–7.475	Upper troposphere humidity
29	8.400–8.700	Surface temperature
30	9.580–9.880	Total ozone
31	10.780–11.280	Cloud temperature, forest fires and volcanoes, surface temperature
32	11.770–12.270	Cloud height, forest fires and volcanoes, surface temperature
33	13.185–13.485	Cloud fraction, cloud height
34	13.485–13.785	Cloud fraction, cloud height
35	13.785–14.085	Cloud fraction, cloud height
36	14.085–14.385	Cloud fraction, cloud height

Derived from information obtained from http://www.sat.dundee.ac.uk/modis.html.

Example 2.2: MODIS

These images of North and Central America were derived from MODIS data by the process of mosaicing, which involves the fitting together of a set of adjacent images. In this case, the images have been transformed using the NDVI (top), and the Enhanced Vegetation Index or EVI (bottom). The pixel values in Vegetation Index images are correlated with properties of the vegetation cover, such as canopy architecture, density and vigour. Time sequences of such images can be used to monitor changes in biomass and other vegetation-related factors at a regional, continental or global scale. Methods of calculating vegetation indices are described in Section 6.2.4.

The images above represent the state of vegetation in North America during the period 5–20 March 2000. Black areas represent water (for example the Great Lakes and the Atlantic and Pacific Oceans). Areas with low values of NDVI and EVI are shown in white, and progressively browner and then greener colours over land indicate higher values of the index, meaning that vegetation is more vigorous, or denser, or less stressed. Where the temporal frequency of observations is sufficiently high, for example 24 hours, a single 'cloud free' image can be derived weekly from seven individual images 'cloud-free' pixel at each position from the seven available.

Vegetation extent and health are both important inputs to climate models, which are used to predict changes in the global climate. Image source: http://visibleearth.nasa.gov/cgi-bin/viewrecord?2091

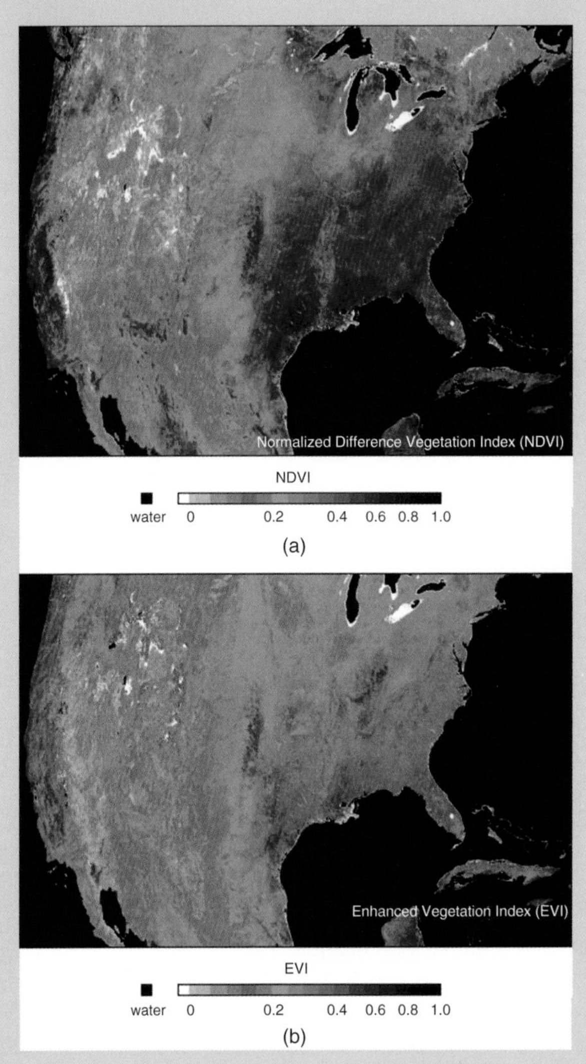

Example 2.2 Figure 1. *MODIS images of North and Central America showing Vegetation Index values. The top image shows the Normalized Difference Vegetation Index (NDVI) while the bottom image shows the Enhanced Vegetation Index (EVI). Images courtesy of NASA's Earth Observatory. http://earthobservatory.nasa.gov/IOTD/view.php?id=696.*

A more recent ocean-observing instrument was the Ocean Colour and Temperature Sensor (OCTS) that was carried aboard the Japanese Advanced Earth Observing Satellite (ADEOS). ADEOS was launched in August 1996 but lasted only until June 1997, when a catastrophic failure occurred. Next, the SEAWiFS was launched in 1997 on a commercial satellite operated by Orbital Sciences Corporation (Orbview). SEAWiFS provides data in eight bands (centred at 412, 443, 490, 510, 555, 670, 705 and 865 nm). The bandwidth of channels 1–6 is 20 nm, and that of bands 7 and 8 is 40 nm. The sensor can be tilted forward or backwards by up to 20° in order to reduce glint. LAC is used to acquire data at a spatial resolution of 1.1 km at nadir, whereas GAC data has a spatial resolution of 4.5 km at nadir.

The Ocean Surface Topography Mission (OSTM) traces its roots back to the TOPEX/Poseidon mission, a joint French/United States altimetry system. Altimetry uses a microwave beam to measure the height of the sea surface to within about 3 cm and thus enable the global study of ocean currents and anomalies such as the El Niño phenomenon (Fu and Cazenava, 2000). The TOPEX/Posiedon system was replaced by Jason-1 in 2001 and Jason-2 in 2008. Jason-1 and -2 follow the same orbit (height 1336 km, inclination 66°) and are operated by a consortium consisting of NASA, NOAA, Centre National d'Etudes Spatiales (CNES) (the French Space Agency) and EUMETSAT, the European Organisation for the exploitation of Meteorological Satellites. Jason-1 and 2 fly in the same orbit, separated by a time interval of approximately 1 min, to allow near-simultaneous observations and thus provide the means of calibrating the datasets produced by the two systems. Once calibration is accomplished the Jason satellites will occupy separate orbits that will provide double the global coverage that Jason-1 was able to provide.

The Indian Space Agency launched Oceansat-2 in September, 2009. It replaces Oceansat-1, which was launched in 1999 with a 5 year design life. Oceansat-2 carries an updated MSS called the Ocean Colour Monitor (OCM) which operates in eight bands, between 412 and 867 nm, collecting data with a spatial resolution of 360 m across a swath of 1420 km. The primary aim of the OCM is to measure ocean colour, particularly phytoplankton and chlorophyll concentrations. This information will be used in a project to manage fisheries. Oceansat-2 also carries a scatterometer, which is a microwave instrument used to measure wind speed, and an Italian instrument that uses global positioning system (GPS) to measure atmospheric conditions. The OCM is tiltable by up to 20° along-track.

As noted in Section 2.3.3, two MODISs are carried by the NASA Terra and Aqua satellites. MODIS has 36 bands in the visible and infrared spectral regions with spatial resolution varying from 250 m for visible bands to 1 km for TIR bands (Table 2.3). Bands 8–16 are similar to the spectral wavebands used by CZCS, OCTS and SEAWiFS. Esaias *et al.* (1998) provide an overview of the capabilities of MODIS for scientific observations of the oceans. They note that the MODIS sensors have improved onboard calibration facilities, and that the MODIS bands that are most similar to the SEAWiFS bands are narrower than the corresponding SEAWiFS bands, giving greater potential for better atmospheric correction (Section 4.4).

A survey of remote sensing applications in biological oceanography by Srokosz (2000) provides further details of these and other sensors that collect data with applications in oceanography. Somewhat more recent surveys are supplied by the texts by Martin (2004) and Robinson (2004), and the *ASPRS Manual* edited by Gowe (2006).

2.3.5 IRS LISS

The Indian Government has an active and successful remote sensing programme. The first Indian remote sensing (IRS) satellite, called Bhaskara-1, was launched in 1979. Later developments led to the IRS programme. IRIS-1A (1988) and IRS-1B (1991) carried the LISS-1 and LISS-2 sensors. LISS is an acronym of Linear Imaging Self-scanning Sensor. LISS-1 was a multispectral instrument with a 76 m resolution in four wavebands. LISS-2 used the same four wavebands, in the VNIR (0.45–0.52 μm, 0.52–0.59 μm, 0.62–0.68 μm and 0.77–0.86 μm), but collected data at a spatial resolution of 36 m. IRS-1C (launched 1995) carried an improved LISS, numbered 3, plus a 5 m resolution panchromatic sensor. LISS-3 includes a mid-infrared band in place of the blue–green band (channel 1 of LISS-1 and LISS-2). The waveband ranges for LISS-3 are 0.52–0.59 μm, 0.62–0.68 μm, 0.77–0.86 μm and 1.55–1.70 μm. The spatial resolution improves from 36 to 25 m in comparison with LISS-2. The panchromatic sensor (0.50–0.75 μm) provides imagery with a spatial resolution of 5 m. IRS-1D carried a similar payload, but did not reach its correct orbit. However, some useful imagery is being obtained from its sensors. Figure 2.8 shows a false-colour (infrared/red/green) composite IRS-1D LISS III image of the Krishna delta on the east coast of India. The panchromatic sensor is, like the SPOT HRV instrument, pointable so that oblique views can be obtained, and off-track viewing provides the opportunity for image acquisition every 5 days, rather than the 22 days for nadir viewing.

IRS-1C also carries the Wide Field Sensor, WIFS, which produces images with a pixel size of 180 m in two bands (0.62–0.68 μm and 0.77–0.86 μm). The swath width is 774 km. Images of a fixed point are produced

Figure 2.8 False-colour (infrared/red/green) composite IRS1D LISS III image of the Krishna delta on the east coast of India. Coastal features such as beach sand, shoals, mudflats, waterlogged areas and salt affected regions inclusive of salt pans, marsh land, cropland, mangroves and suspended sediments are all clearly discernible. LISS-3 (Section 2.3.5) is a four band (0.52 – 0.59, 0.62 – 0.68, 0.77 – 0.86 and 1.55 – 1.70 μm) multispectral sensor that provide 23.5 m resolution coverage. The 23.5 m resolution imagery is resampled to produce a 20 m pixel size. Reproduced with permission from National Remote Sensing Centre of India.

every 24 days, though overlap of adjacent images will produce a view of the same point every 5 days.

Resourcesat-1 (IRS-P6) was launched in 2003 (Seshadri *et al.*, 2005). It carries a LISS-4 sensor, capable of imaging in multispectral or panchromatic mode with a spatial resolution of 5.8 m. The multispectral bands are 0.52–0.59 μm, 0.62–0.68 μm and 0.76–0.86 μm.

Up to date details of the IRS and space programmes is available at http://directory.eoportal.org/get_announce.php?an_id=10251.

2.3.6 Landsat Instruments

2.3.6.1 Landsat Multispectral Scanner (MSS)

Landsat-1, originally called Earth Resources Technology Satellite One (ERTS-1), was the first civilian land observation satellite. It was launched into a 919 km Sun-synchronous orbit on 23 July 1972 by NASA, and operated successfully until January 1978. A second, similar, satellite (Landsat-2) was placed into orbit in January 1975. Landsats-3, -4 and -5 followed in 1978, 1982 and 1984, respectively. A sixth was lost during the launch stage. The Landsat-7 launch took place on 15 April 1999. Landsat-2 and -3 had orbits similar to Landsat-1, but the later satellites use a lower, 705 km, orbit, with a slightly different inclination of 98.2° compared with the 99.09° of Landsats 1–3. The Landsat orbit parameters are such that its instruments are capable of imaging the Earth between 82°N and 82°S latitudes. A special issue of the journal *Photogrammetric Engineering and Remote Sensing* (volume 63, number 7, 1997) is devoted to an overview of the Landsat programme.

Landsat satellites numbered 1–5 inclusive carried the MSS which is described here for historical reasons, as is no longer operational, though a substantial archive of imagery remains available. The MSS was a four-band instrument, with two visible channels in the green and red wavebands, respectively, and two NIR channels (0.5–0.6 μm, 0.6–0.7 μm, 0.7–0.8 μm and 0.8–1.1 μm). These channels were numbered 4–7 in Landsats 1–3 because the latter two satellites carried a second instrument, the Return Beam Vidicon (RBV), which generated channels 1–3. The RBV was a television-based system. It operated by producing an instantaneous image of the scene, and scanning the image, which was stored on a photosensitive tube. Landsat-4 and -5 did not carry the RBV and so the MSS channels were renumbered 1–4. Since Landsat-4 and -5 operate in a lower orbit than Landsat-1 to -3, the optics of the MSS were altered slightly to maintain the swath width of 185 km and a pixel size of 79 m (along track) × 57 m (across track).

The MSS was an electromechanical scanner, using an oscillating mirror to direct reflected radiation onto a set of six detectors (one set for each waveband). The six detectors each record the magnitude of the radiance from the ground area being scanned, which represents six adjacent scan lines (Figure 2.9). The analogue signal from the detectors is sampled at a time interval equivalent to

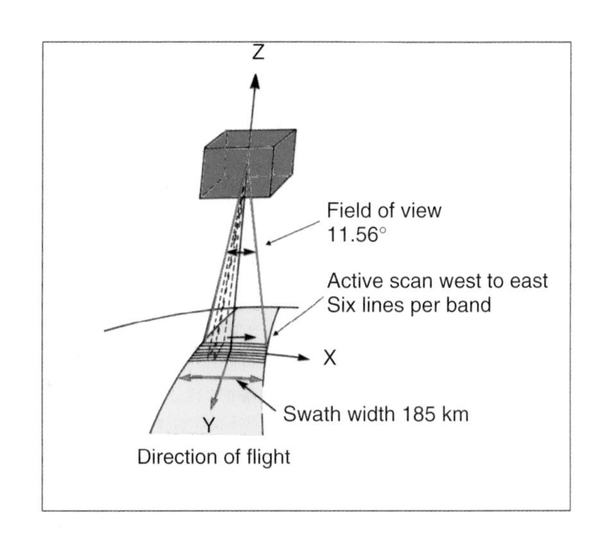

Figure 2.9 Landsat Multispectral Scanner (MSS) operation.

Example 2.3: Landsat MSS

Example 2.3 Figure 1 shows three Landsat MSS images of Mount St Helens dating from 15 September 1973, 22 May 1983 and 31 August 1988. Image processing methods such as image subtraction (Section 6.2.2) are used to obtain a digital representation of areas that have changed over time. The analyst must remember that quantitative measures of change include all aspects of the differences between the images being compared: atmospheric differences (Sections 1.2.5 and 4.4), instrument calibration variations (Section 4.5), as well as other factors such as illumination differences caused by seasonal variations (Section 4.6).When pointable sensors such as the SPOT HRV are the source of the imagery, then viewing geometry effects should also be taken into account.

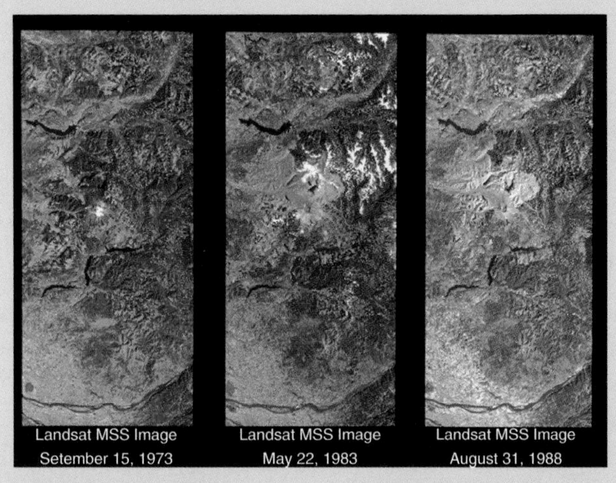

Example 2.3 Figure 1. *Landsat MSS. Landsat MSS images of Mount St. Helens dating from 15 September 1973, 22 May 1983 and 31 August 1988. Source: http://www.gsfc.nasa.gov/gsfc/images/earth/landsat/helen.gif. Courtesy: NASA.*

The analysis of the nature, direction and drivers of change is useful in many areas of environmental management. Changes in land cover, for example can be both a response to and a factor contributing to climate change. Variations in the termini of Alpine glaciers are an indication of local temperatures over a number of years. Change analysis is also an important aspect of urban and regional planning. The long (almost 40 years) archive of Landsat MSS images, together with the moderate spatial resolution and wide-area coverage of the MSS, make it one of the most useful sources of data for decadal change analysis studies.

a distance of 57 m along-scan, and converted to digital form before being relayed to a ground receiving station. A distance of 57 m represents some over-sampling, as the IFOV of the system is equivalent to 79 m on the ground. The detectors are active only as the mirror scans in the forward direction. Satellite velocity is such that it moves forward by an amount equal to 6×79 m during the reverse scan cycle, so that an image is built up in each band in sets of six scan lines. The detectors deteriorate at different rates, so MSS images may show a phenomenon called six-line banding. Methods of removing this kind of 'striping' effect are discussed further in Section 4.2.2.

Landsat MSS images were collected routinely for a period of over 20 years, from the launch of ERTS-1 in July 1972 to November 1997, when the Australian ground receiving station acquired its last MSS image. Although somewhat old-fashioned by today's standards, the MSS performed well and exceeded its design life. The MSS archive provides a unique historical record, and is used in studies of temporal change, for example urban growth, land degradation and forestry, as illustrated in Example 2.3, which looks at change in the Mount St Helen's area of Washington state over a period of 13 years.

As from April, 2008, Landsat MSS imagery has been made available free of charge. Ordering details can be found at http://landsat.usgs.gov/products_data_at_no_charge.php.

2.3.6.2 *Landsat Thematic Mapper (TM)*

The TM instrument is the primary imaging sensor carried by Landsat-4 and -5. Landsat-4 (launched 16 July 1982) was switched off in August 1993 after failure of the data downlinking system. Landsat-5 continues to operate, though only a direct data downlink facility is available. Like the MSS, the TM uses a fixed set of detectors for each band and an oscillating mirror. TM has 16, rather than 4, detectors per band (excluding the TIR channel) and scans in both the forward and the reverse directions. It has seven, rather than four, channels covering the visible, near- and mid-infrared and the TIR, and has a spatial resolution of 30 m. The thermal channel uses four detectors and has a spatial resolution of 120 m. The data are quantized onto a 0–255 range. In terms of spectral and radiometric resolution, therefore, the TM design represents a considerable advance on that of the MSS.

The TM wavebands are as follows: channels 1–3 cover the visible spectrum (0.45–0.52 μm, 0.52–0.60 μm and 0.63–0.70 μm, representing visible blue–green, green and red). Channel 4 has a wavelength range of 0.75–0.90 μm in the NIR. Channels 5 and 7 cover the mid-infrared (1.55–1.75 μm and 2.08–2.35 μm), while channel 6 is the TIR channel (10.4–12.5 μm). The rather disorderly channel numbering is the result of the late addition of the 2.08–2.35 μm band.

Data from the TM instruments carried by the Landsat-4 and -5 satellites and the ETM+ carried by Landsat-7 are transmitted to a network of ground receiving stations. The European stations are located near Fucino, Italy and Kiruna, Sweden. Data are also transmitted via the system of TDRS, which are in geostationary orbits. At least one of the satellites making up the TDRS constellation is in line of sight of Landsats-4, -5 and -7. TDRS transmits the image data to a ground station at White Sands, NM, USA, from where it is relayed to the data processing facility at Norman, OK, USA, using the US domestic communications satellite DOMSAT. Following a problem with the ETM+ instrument in mid-2003, Landsat-5 TM data are again being downlinked to ground receiving stations, almost 20 years after the satellite was launched. Examples of Landsat TM imagery are provided in Figures 1.10 and 1.11.

2.3.6.3 *Enhanced Thematic Mapper Plus (ETM+)*

Landsat-6, which was lost on launch in October 1993, was carrying a new version of the Landsat TM called the ETM. Landsat-7, which was launched on 15 April 1999, and which operated successfully until mid-2003 when an imaging defect occurred, carries an improved version of ETM, called the ETM+. This instrument measures upwelling radiance in the same seven bands as the TM, and has an additional 15 m resolution panchromatic band. The spatial resolution of the TIR channel is 60 m rather than the 120 m of the TM thermal band. In addition, an onboard calibration system allows accurate (±5%) radiometric resolution (Section 2.2.3). Landsat-7 has substantially the same operational characteristics as Landsat-4 and -5, namely, a Sun-synchronous orbit with an altitude of 705 km and an inclination of 98.2°, a swath width of 185 km, and an equatorial crossing time of 10.00. The orbit is the same as that of Terra (equatorial crossing time 10.30), Aqua (equatorial crossing time 13.30) and EO-1. EO-1 carries an Advanced Land Imager (ALI) which is compared to the ETM+ by Neuenschwander, Crawford and Ringrose (2005). See also Ungar *et al.* (2003) and US Geological Survey (2003) for details of the EO-1 mission.

A sketch of the main components of the Landsat-7 satellite is shown in Figure 2.10, and the orbital characteristics are shown in Figure 2.1.

2.3.6.4 *Landsat Data Continuity Mission*

A Landsat follow-on satellite system, the Landsat Data Continuity Mission or LDCM is in preparation, with a planned launch date of December 2012. It will carry the Operational Land Imager or OLI. This instrument has nine bands, shown in Table 2.4. Currently, ETM+ data are being provided by Landsat-5 and Landsat-7 though, as noted above, the Landsat-7 ETM+ instrument has developed a fault which affects data on the right and left hand sides of the image. Powell *et al.* (2007) discuss alternative Landsat-like moderate resolution sensors and their applications. Data from these alternative sensors (such as SPOT High Resolution Geometric (HRG) and Resourcesat LISS-4) can be used to patch up the Landsat ETM+ data.

Given that Landsat-5 is 26 years old at the time of writing, that Landsat-7 data suffer from severe problems, and that the earliest launch date for a Landsat Follow-on Mission is 2012, it seems inevitable that a data gap will occur. Data from other sensors, mentioned at the end of the previous paragraph, may be used to fill in the gap.

2.3.7 SPOT Sensors

2.3.7.1 *SPOT HRV*

The SPOT programme is funded by the governments of France, Belgium and Sweden and is operated by the French Space Agency, CNES, located in Toulouse, France. SPOT-1 was launched on 22 February 1986. It carried an imaging sensor called the HRV instrument that is capable of measuring upwelling radiance in three channels (0.50–0.59 μm, 0.61–0.68 μm and 0.79–0.89 μm) at a spatial resolution of 20 m, or in a single panchromatic channel (0.51–0.73 μm) at a spatial resolution of 10 m. All channels are quantized to a 0–255 scale. HRV does not use a scanning mirror like Landsat MSS and TM. Instead, it uses a linear array of charge-coupled devices, or CCDs, so that all the pixels in an entire scan line are imaged at the same time

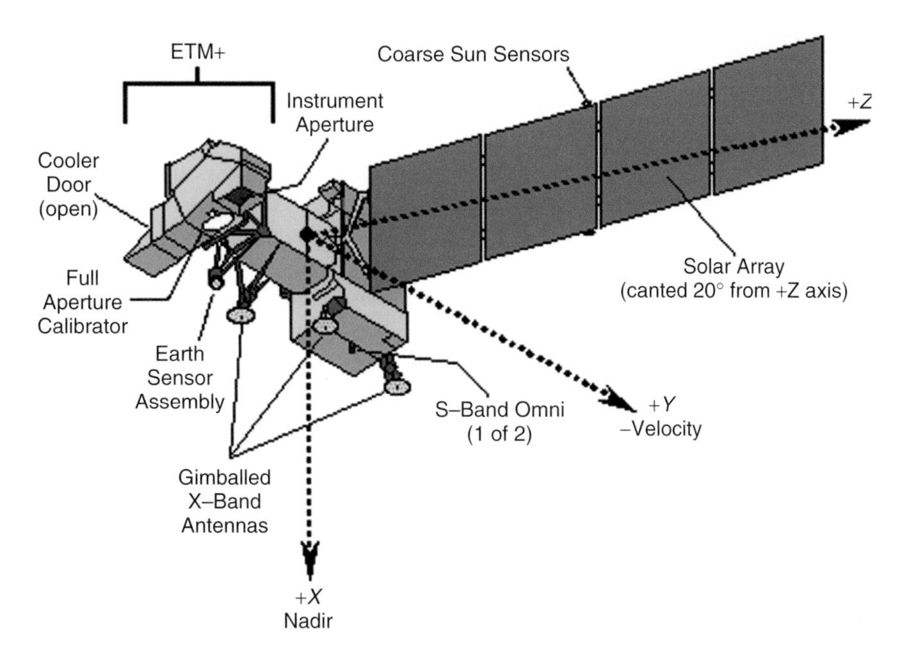

Figure 2.10 *Schematic diagram of the Landsat-7 spacecraft, showing the solar array that supplies power, and the ETM+ instrument. The spacecraft travels in the + Y direction, with the + X direction pointing towards the Earth's centre. Attitude and altitude are controlled from the ground using thruster motors (not shown). Based on Figure 2.2,* Landsat 7 Science Data Users Handbook, *NASA Goddard Spaceflight Center, Greenbelt, MD, USA. http://landsathandbook.gsfc.nasa.gov/handbook.html (accessed 10 April 2009).*

Table 2.4 *Landsat Data Continuity Mission: Operational Land Imager bands.*

Band number	Band name	Min. lower band edge (nm)	Max. upper band edge (nm)	Ground sampling distance (m)
1	Coastal/ aerosol	433	453	28–30
2	Blue	450	515	28–30
3	Green	525	600	28–30
4	Red	630	680	28–30
5	NIR	845	885	28–30
6	SWIR 1	1560	1660	28–30
7	SWIR 2	2100	2300	28–30
8	Panchromatic	500	680	14–15
9	Cirrus	1360	1390	28–30

(Figure 1.2b). As it has no moving parts the CCD push-broom system might be expected to last longer than the electro-mechanical scanners carried by Landsats-1 to -7, though all of the Landsat sensors except the ETM+ on Landsat-7 have exceeded their design lives by substantial amounts. A more important consideration is the fact that, since all of the pixel data along a scan line are collected simultaneously rather than sequentially, the individual CCD detector has a longer 'look' at the pixel area on the ground, and collects more ground-leaving photons per pixel than an instrument based on an electromechanical scanner such as ETM+. This increased dwell time means that the signal is estimated more accurately and the image has a higher SNR (Section 2.2.2).

The SPOT orbit is near polar and Sun-synchronous at an altitude of 832 km, an inclination of 98.7°, with a period of 101.5 minutes. The field of view of the HRV sensor is 4.13° and the resulting swath width is 60 km. Two identical HRV instruments are carried, so when they are working in tandem the total swath width is 117 km (there is a 3 km overlap when both sensors point at nadir). The orbital repeat period is 26 days with an equatorial crossing time of 10:30. However, the system potentially has a higher revisit capability because the HRV sensor is pointable. It can be moved in steps of 0.6° to a maximum of ±27° away from nadir. This feature allows HRV to collect image data within a strip 475 km to either side of nadir. Apart from improving the sensor's revisit capability, the collection of oblique views of a given area (from the left and right sides respectively) provides the opportunity to generate digital elevation models (DEMs) of the area, using photogrammetric methods. However, the use of non-nadir views introduces problems when physical values are to be derived from the quantized counts (Section 4.6).

SPOT-1 was retired at the end of 1990, following the successful launch of SPOT-2 in January 1990. SPOT-3 followed in September 1993. All three carry an identical instrument pack. SPOT-3 was lost following a technical error. SPOT-4 was successfully launched on 24 March 1998. The HRV instrument on SPOT-4 is extended to provide an additional 20 m resolution channel in the mid-infrared region (1.58–1.75 µm) and the new instrument is called the High Resolution Visible Infrared (HRV-IR). This sensor can be used in multi-spectral mode (X), or panchromatic mode (M), or in a combination of X and M modes. SPOT-4 carries an onboard tape recorder, which provides the capacity to store 20 scenes for downloading to a ground station.

The SPOT-5 satellite was launched in May 2002. It carries a four-band multispectral imager called High Resolution Geometric (HRG) with an enhanced panchromatic channel, and a new stereo instrument. Details of these instruments are given in Table 2.5. The spatial resolution of HRG is 10 m (20 m in the short wave infrared (SWIR) band), and panchromatic images can be obtained at 5 or 2.5 m spatial resolution, compared with 10 m for SPOT-1–4. A new instrument, called High Resolution Stereoscopic (HRS) simultaneously collects images from two different angles. The first image looks forward of nadir along the line of flight at an angle of 20° while the second image looks backwards at the same angle. The spatial resolution of HRS is 10 m and its field of view is 8°. The HRV instrument on SPOT-1 to SPOT-4 collected stereo imagery in two stages. At the first stage the HRV sensor was tilted to one side or other of the subsatellite track to capture an off-nadir view of the area of interest. At the second, later, stage a second image of the same area was collected from a later, different, orbit again using the pointing capability of the instrument. The disadvantage of this two-stage approach is that surface conditions may change between the dates

Table 2.5 *Spatial resolution and swath widths for the SPOT-5 instruments HRG (High Resolution Geometric), Vegetation-2 and HRS (High Resolution Stereoscopic) instruments carried by SPOT-5. Note that 2.5 m panchromatic imagery is obtained by processing the 5 m data using a technique called* 'Supermode' *(see text for details).*

Spectral band (µm)		Spatial resolution (m)		
		HRG	Vegetation-2	HRS
P	0.49–0.69	2.5 or 5	–	10
B0	0.43–0.47	–	1000	–
B1	0.50–0.59	10		–
B2	0.61–0.68	10	1000	–
B3	0.79–0.89	10	1000	–
SWIR	1.58–1.75	20	1000	–
Swath width (km)		60	2250	120

of acquisition of the two images, thus causing problems with the coregistering process (Section 4.3.4) and reducing the quality of the DEM that is produced from the stereo pair. HRS uses the *along-track stereo* principle to collect two images from different angles at the same time. The ASTER sensor (Section 2.3.8) uses a slightly different system, collecting simultaneous backward and nadir views in order to generate a stereo image pair. The relative (within the image) accuracy of the elevation values derived from the HRS along-track stereo image pairs is given as 5–10 m, with an absolute (compared with ground observations) accuracy of 10–15 m, making possible the generation of DEM at a map scale of 1:50 000. Cuartero, Felicisimo and Ariza (2005) compare DEMs derived from SPOT and ASTER (Section 2.3.8) data.

This high level of geometric accuracy is due to some extent to the employment of a system for accurate orbit determination called Doppler orbitography and radiopositioning integrated by satellite (DORIS), which was designed and developed by the CNES, the Groupe de Recherches de Géodésie Spatiale (Space Geodesy Research Group) and the French mapping agency (Institut Géographique National or IGN). DORIS consists of a network of transmitting stations, a receiver on board the satellite and a control centre. A receiver on board the satellite measures the Doppler shift of the signal emitted by the transmitting stations at two frequencies (400 MHz and 2 GHz). The use of two frequencies is needed to estimate the delay in propagation of radio waves caused by the ionosphere. The data are stored in the instrument's memory, downloaded to the ground each time the satellite is within range of a receiving station and then processed to determine the precise orbital trajectory. The orbital position of the satellite can be calculated to an accuracy of a few centimetres using this approach.

A development of DORIS, called Détermination Immédiate d'Orbite par DORIS Embarqué (DIODE or real-time orbit determination using onboard DORIS), is carried by SPOT-4 and -5; this device can, as its name suggests, provide immediate orbital information that can be downloaded to a ground station together with the image data. It is used to calculate the position of each pixel on the ground surface, and to express these results in terms of a map projection such as Universal Transverse Mercator (UTM). This is the procedure of *geometric correction* of remotely-sensed images, and it is considered in more detail in Section 4.3. A good guide to map projections is provided by Fenna (2006).

The HRG instrument produces panchromatic images at a resolution of either 2.5 or 5 m. The physical resolution of the instrument is 5 m, but 2.5 m data are generated by a processing technique that the SPOT Image Company calls 'Supermode'. Supermode provides an interesting example of the application of advanced image processing

Example 2.4: Landsat ETM+

The TM and ETM+ instruments have seven spectral bands that cover the visible, NIR and the SWIR regions of the electromagnetic spectrum. Rocks, minerals and surface crusts can be discriminated in these regions (Section 1.3.2.2) and so the use of TM/ETM+ imagery for geological mapping and exploration is an important application area. Techniques such as band ratioing (Section 6.2.4) and colour enhancement (e.g. the decorrelation stretch, Section 6.4.3) are commonly used techniques. Since each of the 30 × 30 m pixels making up the image is unlikely to cover a homogeneous area, techniques such as linear spectral unmixing (Section 8.5.1) can be used to try to identify the proportions of different materials that are present within each pixel area.

Deposits of gold and silver (which are found in Tertiary volcanic complexes) have been exploited in the area shown on Example 2.4 Figure 1 for many years, at places such as Tonopah, Rhyolite and Goldfield in Nevada, United States. Associated with these mineral deposits are hydrothermally altered rocks, which contain iron oxide and/or hydroxyl-bearing minerals. These minerals show absorption bands in the VNIR and SWIR region of the spectrum. Both the TM and ETM+ sensors can detect radiance in these spectral regions (bands 5 and 7). The image has been subjected to a hue-saturation-intensity (HSI) transformation to enhance its colour (Section 6.5).

Example 2.4 Figure 1. *Landsat ETM+ image of the Goldfield/Cuprite area, NV, USA. The colours in this image are related to variations in lithology over this semi-arid area. Image source: http://edcdaac.usgs.gov/samples/goldfield.html. Courtesy United States Geological Survey/NASA. The image has been subjected to a hue-saturation-intensity colour transform (Section 6.5).*

Further reading: Abrams, M., Ashley, R., Rowan, L., Goetz, A., and Kahle, A. (1977) Mapping of hydrothermal alteration in the Cuprite Mining District, Nevada, using aircraft scanner images for the spectral region 0.46 to 2.36 μm. *Geology*, **5**, 713–718.

methods, some of which are described in later chapters of this book. The procedure is described in Example 2.4 and Figure 2.11.

2.3.7.2 Vegetation

SPOT-4 and -5 carry a sensor, Vegetation or VGT, developed jointly by the European Commission, Belgium, Sweden, Italy and France. VGT operates in the same wavebands as HRV-IR except that the swath width is 2250 km, corresponding to a field of view of 101°, with a pixel size at nadir of 1 km. This is called the 'direct' or 'regional' mode. 'Recording' or 'world-wide' mode produces averaged data with a pixel size of around 4 km. In this mode, VGT generates datasets for the region of the Earth lying between latitudes 60°N and 40°S (Arnaud, 1994). The 14 daily orbits will ensure that regions of the Earth at latitudes greater than 35° will be

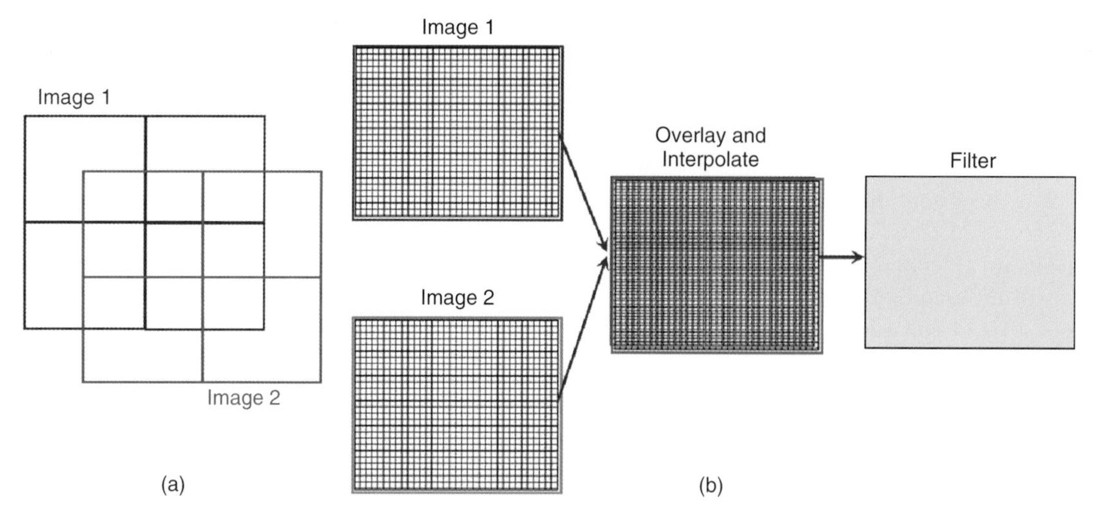

Figure 2.11 *Illustrating SPOT-5 'Supermode'. (a) Two images, each with a spatial resolution of 5 m, are offset by 2.5 m in the x and y directions. (b) These two images (left side) are overlaid and interpolated to 2.5 m pixels (centre), then filtered (right) to produce a Supermode image with a resolution of 2.5 m. See Example 2.5 for more details. Permission to use the data was kindly provided by SPOT image, 5 rue des Satellites, BP 4359, F 331030, Toulouse, France.*

imaged daily, whereas equatorial areas will be imaged every other day. The combination of the Vegetation and HRV-IR sensors on the same platform means that datasets of significantly different spatial resolutions are obtained simultaneously.

VGT data complement those produced by the NOAA AVHRR, which is widely used to generate global datasets (Townshend and Skole, 1994). VGT has the advantage – for land applications – of three spectral bands in the optical region plus one in the infrared part of the spectrum. VGT should also be compared to the MODIS sensor that is carried by both the Terra and Aqua satellites.

2.3.7.3 SPOT Follow-on Programme

SPOT-5 is the last of the line. Its mission was planned to last until 2007 but is still ongoing. Research and development in small satellites and lightweight sensors mean that it is now more cost-effective to use small satellites as described in Section 2.3.9. The attraction of low-cost small satellites has led the French and Italian governments to agree to place a constellation of small satellites into orbit. The French contribution is the Pléiades[2] programme, which will produce two small satellites operating in the optical wavebands. The Italian contribution consists of five small SAR satellites called COSMO/Skymed. SAR sensors are described in Section 2.4. As of late 2009, three of the COSMO/Skymed constellation are already in orbit.

There will be two Pléiades satellites, if plans come to fruition. The first is scheduled for a 2010 launch, and the second in mid-2011. Like COSMO/Skymed, Pléiades will be a joint military/civilian system. It will have a panchromatic sensors, with a nadir spatial resolution of 70 cm, and a four-band multispectral system, operating in blue, green, red and NIR wavebands at a resolution of 2.8 m. The instrument will be very versatile, capable of looking in any direction. A video is available at http://smsc.cnes.fr/PLEIADES/GP_mission.htm, which illustrates the impressive versatility of the optical system. Stereo pairs will be generated from panchromatic imagery and the geometric accuracy of the imagery collected by Pléiades is estimated as sub-metre. The Pléiades home page is at http://smsc.cnes.fr/PLEIADES/index.htm.

The ESA has plans for a Sentinel Programme, running from 2011 to 2016, as part of the European GMES programme. Sentinel-1 consists of a pair of satellites, the first of which is expected to be launched in 2010. Both will carry imaging SAR instruments. Details are provided below (Section 2.4.1). Sentinel-2 (2012) also consists of a pair of satellites, but carrying optical (visible and SWIR) instruments called Multi-Spectral Imager (MSI). Spatial resolution will range from 10 to 60 m in 13 bands. A description is provided by Martimor *et al.* (2007). The MSI could be considered to be a SPOT follow-on instrument, as it provides data in the same spectral region at a similar spatial resolution.

[2]Readers familiar with Greek mythology will be aware that the original Pléiades were the seven daughters of Atlas who, on being pursued for nefarious purposes by Orion, were turned into doves by Zeus and transported to heaven. Orion was later sent up to heaven, but the outcome of their meeting in that place is not known.

2.3.8 Advanced Spaceborne Thermal Emission and Reflection Radiometer (ASTER)

The Japanese-built ASTER is one of the imaging sensors carried by the American Terra satellite. ASTER is a moderate-resolution multispectral sensor, acquiring data in three spectral regions using a separate radiometer for each region. It also has a stereoscopic capability. The spectral regions in which images are acquired are the VNIR, the SWIR and the TIR (Table 2.6). Images collected in the three spectral regions differ in terms of their spatial resolution. For the VNIR sensor, spatial resolution is 15 m. The SWIR bands have a spatial resolution of 30 m, while TIR images have a 90 m spatial resolution. The band numbers, spectral ranges and absolute accuracies of the 14 bands are shown in Table 2.6. Unfortunately, bands 5–9 are no longer functioning, as of January 2009. The SWIR bands started exhibiting anomalous data values in late April 2008 and stopped functioning in January 2009. ASTER SWIR data acquired since April 2008 are basically not useable, and thus cannot replace Landsat-5's TM data should that satellite cease to function before the introduction of the Landsat Follow-on Mission (Section 2.3.6.4).

The ASTER instrument has a number of interesting features. First, the spatial resolution of the VNIR bands is higher than that of either the Landsat ETM+ or the SPOT HRV sensor. Like the HRV sensor, ASTER is 'pointable', that is it can collect images of areas lying to either side of the subsatellite track. The VNIR radiometer can scan up to $\pm 24^{\circ}$ off-nadir, and the other two radiometers can point up to $\pm 8.55^{\circ}$ off-nadir. The inclusion of several bands in both the SWIR and TIR regions makes ASTER a unique spaceborne instrument. The multiple bands in these two spectral regions should prove to be valuable in geological studies.

ASTER has two spectral bands covering the 0.78–0.86 μm region. One is labelled as band 3N and the other is band 3B. Band 3N looks vertically down, while band 3B looks backwards at an angle of 27.6°. Images in these two NIR bands can be used to generate stereo pairs, which can be used for the production of DEMs. The fact that the two images are captured simultaneously in 'along track' mode reduces problems caused by changes in surface reflectance that are associated with 'across-track' mode stereo images such as those produced by SPOT-1 to SPOT-4 HRV. In across-track mode the two images making up the stereo pair are acquired on different orbits, which may be separated in time by weeks or months, depending on cloud cover and instrument availability. Welch *et al.* (1998) and Lang and Welch (1996) discuss the methods used to generate the '*Standard Data Product DEM*' from ASTER band 3B and 3N data (Figure 2.12). They cite results obtained from test sites in the United States and Mexico to show that the root mean square (RMS) error of the elevation measurements is between ±12 and ±30 m, provided that the two images can be co-registered to an accuracy of ±0.5 to ±1.0 pixels using image correlation procedures (Section 4.3.2). Toutin (2002) provides details of an experiment to measure the accuracy of an ASTER DEM, and also describes the methods used by the NASA EOS Land Processes DAAC to generate ASTER DEMs (which are distributed via the United States Geological Survey (USGS) EROS Data Centre in Sioux Falls, SD, USA). The same author (Toutin, 2008) reviews the use of ASTER DEMs in the earth sciences. Global ASTER DEMs (ASTER GDEM) at 30 m resolution are now freely available for the entire globe – see http://www.gdem.aster.ersdac.or.jp.

ASTER data products are available at different levels of processing. Level 0 is full resolution unprocessed and unreconstructed data. Level 1A is equivalent to Level 0 plus ancillary data (radiometric and geometric calibration coefficients). Radiometric and geometric corrections are applied to the Level 1B data, which are supplied in the form of quantized counts. These counts can be converted into apparent radiance using the calibration factors listed in Table 2.7. Levels 2–4 include products such as DEM derived from Band 3N/3B pairs, surface temperatures and radiances and images derived from processing operations such as decorrelation stretch (Section 6.4.3).

An account of ASTER data products is provided by Abrams (2000). The sensor is described by Yamaguchi *et al.* (1998). Cuartero, Felicisimo and Ariza (2005) compare ASTER DEM accuracy with that of DEMs obtained from the SPOT system (see above). Gao and Liu (2008) compare and evaluate ASTER and Landsat ETM+ data in a study of land degradation and find that ETM+ outperforms ASTER in a supervised classification (Chapter 8). This result was attributed to the spectral resolution of the six shortwave infrared bands, which overlap considerably. *IEEE Transactions on Geoscience and Remote Sensing* (2005) is a Special Issue on the ASTER Instrument. Marçal *et al.* (2005) examine the use of ASTER data in land cover classification (Chapter 8). *Remote Sensing of Environment* (2005) is a special issue on scientific results from ASTER.

2.3.9 High-Resolution Commercial and Small Satellite Systems

The release of technology from the military to the commercial sector in the United States during the 1990s allowed a number of American private-sector companies to launch and service their own satellites, for example GeoEye's IKONOS satellite (see below). These satellites each carry a single sensor that produces either a panchromatic image at a spatial resolution of 1 m or better, or a multispectral image, typically

Example 2.5: SPOT-5 Supermode

The SPOT-5 satellite carries the HRG pointable imaging sensor, which produces panchromatic images at a resolution of 5 m using a CCD array with 12 000 elements. In fact, there are two such arrays. Each is programmed to collect a separate image of the area being viewed, with the second image being offset by half a pixel in the horizontal and vertical directions, as depicted in Figure 2.11(a). A new image, with a spatial resolution of 2.5 m, is derived from these overlapped cells. Example 2.5 Figure 1

Example 2.5 Figure 1. *Left-hand SPOT HRG image with spatial resolution of 5 m. Reproduced with permission of SPOT Image, Toulouse.*

Example 2.5 Figure 2. *Right-hand SPOT HRG image with spatial resolution of 5 m. Reproduced with permission of SPOT Image, Toulouse.*

with four spectral bands and a spatial resolution of 4 m. Images from these sensors have brought digital photogrammetry and digital image processing much closer together. The data from the panchromatic band for two adjacent IKONOS orbits can be used to generate DEMs (*ISPRS Journal of Photogrammetry and Remote Sensing*, 2006) as well as for topographic mapping. At the same time, the use of small and relatively cheap satellites to carry a single remote sensing instrument has been pioneered by, among others, Surrey Space Technology Ltd (SSTL), a spin-off company associated with the University of Surrey, United Kingdom, and now owned by EADS/Astrium. The SSTL Disaster Monitoring Constellation or DMC is an example of a group of satellites operating synergistically. A second example is the RapidEye constellation, operated by the German company, RapidEye AG. In this section a brief summary of (i) small independent satellites and (ii) constellations of small satellites is provided. The pace of change is considerable, so to keep up to date readers should visit the Internet and search for 'small satellites'. Recent surveys of small satellites and their applications include Kramer and Cracknell (2008), Sandau, Röser and Valenzuela (2008) and Xue *et al.* (2008).

The first high-resolution commercial optical remote sensing system was IKONOS. The first IKONOS launch failed, but in late 1999 a successful launch placed IKONOS into orbit. (The word '*ikon*' means 'image' or 'likeness' in Greek.) High resolution is the main distinguishing characteristic of the imagery produced by IKONOS for, in panchromatic mode, it can acquire imagery at a spatial resolution of 1 m and multispectral imagery in four channels with a spatial resolution of 4 m. The sensor can be tilted both along and across shows one view of the target area taken by the first HRG sensor. Example 2.5 Figure 2 is the second 5 m resolution panchromatic image from the HRG sensor that is used in the Supermode process. This image is displaced horizontally and vertically by half a pixel with respect to the first image. This displacement produces four pixels measuring 2.5 × 2.5 m nesting within the original 5 × 5 m pixel. Interpolation (resampling) is used to compute values to be placed in the 2.5 m pixels. Resampling is described in section 4.3.3. The third stage of the Supermode process uses a filtering procedure (chapter 7) called deconvolution. SPOT Image uses a method based on the discrete wavelet transform, which is described in section 6.7. Example 2.5 Figure 3 shows the end product.

Example 2.5 Figure 3. *Composite of Example 2.5 figures 1 and 2. The images in Example 2.5 figure 1 and 2 have a spatial resolution of 5 m, but the latter image is offset by 2.5 m. This Supermode image has been resampled and filtered to produce an image with an apparent spatial resolution of 2.5 m. Reproduced with permission of SPOT Image, Toulouse.*

track, and the spatial resolutions cited are valid for off-nadir pointing angles of less than 26°. The panchromatic band covers the spectral region 0.45–0.90 μm, while the four multispectral channels collect data in the blue (0.45–0.53 μm), green (0.52–0.61 μm), red (0.64–0.72 μm) and NIR (0.77–0.88 μm) bands. The satellite flies in a near-circular, Sun-synchronous, polar orbit at a nominal altitude of 681 km and with an inclination angle of 98.1°. The descending nodal time (when the satellite crosses the equator on a north-south transect) is 1030. Data from the IKONOS system are quantized using an 11-bit (0–2047) scale (Section 3.2.1) and are available in a variety of forms ranging from standard system-corrected to geometrically corrected (Section 4.3) and stereo, for the production of DEM. The precision geocorrected imagery is claimed to have a map scale equivalent of 1 : 2400. The use of high-resolution data in updating topographic maps is described by Holland, Boyd and Marshall (2006). A special issue of *ISPRS Journal of Photogrammetry and Remote Sensing* (volume 60, number 3, pages 131–224) is devoted to topographic mapping from high-resolution instruments.

Because of their high spatial resolution, IKONOS images are used for small-area investigations, where they can replace high-altitude air photography to some extent. The fact that the sensor is pointable means that geocorrection involves more complex processing than is required for medium and low-resolution imagery from sensors such as the Landsat ETM+ and NOAA AVHRR. Expert photogrammetric knowledge is needed for these operations. An IKONOS panchromatic image of central London is shown in Figure 2.13. The detail even in this reproduction is clear, and applications in urban planning and change detection, as well as military reconnaissance, are apparent. Dial *et al.* (2003) review applications of IKONOS data.

Table 2.6 *ASTER spectral bands. The ASTER dataset is subdivided into three parts VNIR (Visible and Near Infra-Red), SWIR (Short Wave Infra-Red) and TIR (Thermal Infra-Red). The spatial resolution of each subset is: VNIR 15 m, SWIR 30 m and TIR 90 m. The swath width is 60 km. Data in bands 1–9 are quantized using 256 levels (8 bits). The TIR bands use 12 bit quantization.*

Spectral region	Band number	Spectral range (μm)	Absolute accuracy	Cross-track pointing (°)
VNIR	1	0.52–0.60	$\leq \pm 4\%$	± 24
	2	0.63–0.69		
	3N	0.78–0.86		
	3B	0.78–0.86		
SWIR	4	1.600–1.700	$\leq \pm 4\%$	± 8.55
	5	2.145–2.185		
	6	2.185–2.225		
	7	2.235–2.285		
	8	2.295–2.365		
	9	2.360–2.430		
TIR	11	8.475–8.825	$\leq \pm 3$ K (200–240 K)	
	12	8.925–9.275	$\leq \pm 2$ K (240–270 K)	
	13	10.25–10.95	$\leq \pm 1$ K (270–340 K)	
	14	10.95–11.65	$\leq \pm 2$ K (340–380 K)	

Source: *ASTER Users' Guide, Part 1 (General), Version 3.1*, March 2001, ERSDAC, Japan, and Abrams (2000).

Figure 2.12 *Digital elevation model derived from ASTER data. The area shown covers a part of Southern India for which more conventional digital mapping is unavailable. The ASTER DEM image has been processed using a procedure called density slicing (Section 5.2.1). Low to high elevations are shown by the colours green through brown to blue and white. Original data courtesy NASA/USGS.*

Table 2.7 *Maximum radiance for different gain settings for the ASTER VNIR and SWIR spectral bands.*

Maximum radiance (W m^{-2} sr^{-1} μm^{-1})				
Band no.	High gain	Normal gain	Low gain 1	Low gain 2
1	170.8	427	569	N/A
2	179.0	358	477	
3N	106.8	218	290	
3B	106.8	218	290	
4	27.5	55.0	73.3	73.3
5	8.8	17.6	23.4	103.5
6	7.9	15.8	21.0	98.7
7	7.55	15.1	20.1	83.8
8	5.27	10.55	14.06	62.0
9	4.02	8.04	10.72	67.0

QuickBird was launched in October 2001, and is owned and operated by the American DigitalGlobe company. Like IKONOS, it carries a single instrument capable of producing panchromatic images with a spatial resolution of between 0.61 and 0.73 m, plus multispectral imagery with a spatial resolution of 2.44 and 2.88 m, depending on the angle of tilt of the sensor (which ranges up to 25° off-nadir). The sensor can be tilted along or across track, to produce stereo imagery and to ensure a revisit capability of between one and three and a half

Figure 2.13 IKONOS panchromatic image of central London, showing bridges across the River Thames, the London Eye Ferris wheel (lower centre) and Waterloo railway station (bottom centre). The image has a spatial resolution of 1 m. IKONOS satellite imagery courtesy of GeoEye. Reproduced with permission from DigitalGlobe.

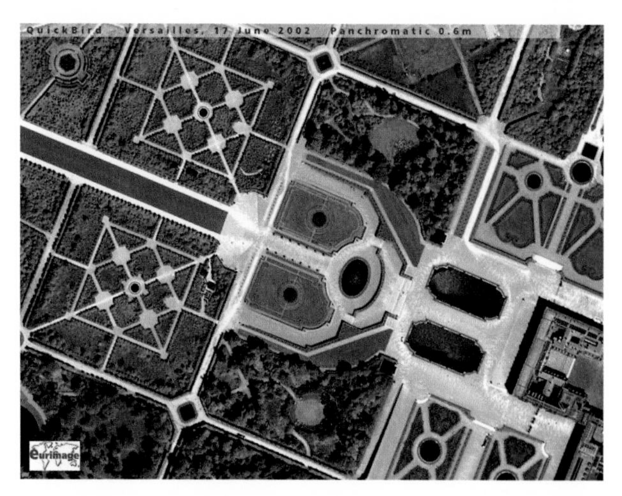

Figure 2.14 QuickBird panchromatic image of the gardens of the Palace of Versailles, near Paris, France. The proclamation of Kaiser Wilhelm I as emperor of Germany was made in the great Hall of Mirrors in the Palace of Versailles following the defeat of France in the Franco-Prussian War of 1871. Image courtesy of DigitalGlobe. © Copyright. All rights reserved.

days. Imagery is available in basic (system corrected), standard (geometrically corrected to a map projection; Section 4.3) and orthorectified forms. Orthorectification is a form of geometric correction that takes into account the relief of the terrain. Figure 2.14 shows a QuickBird panchromatic image of the gardens of the Palace of Versailles, near Paris.

Other commercial high resolution satellite sensors include WorldView (launched 2007) which has the highest resolution of any panchromatic sensor (0.5 m) in orbit. In fact, US Government regulations prevent the sale to non-US citizens of imagery with a resolution of less than 50 cm. Like IKONOS, the radiometric resolution is 11 bits. The spectral response curves for the WorldView-1 and -2 panchromatic instruments are given in Anon (undated). Unlike Worldview-1, WorldView-2, launched in October 2009, carries a multispectral sensor with eight bands (blue, green, red, NIR, red edge, coastal, yellow and NIR-2). These multispectral data have a spatial resolution of 1.8 m at nadir. The panchromatic sensor has a resolution of 0.46 m. All bands have a radiometric resolution of 11 bits and a swath width of 16.4 km. Geolocation of Worldview-2 imagery is assisted by the fact that the platform is very stable; the operators claim that the use of attitude sensors and GPS will make the imagery accurate to 6.5 m (or 2.0 m if ground control points are used). These figures exclude the effects of terrain and of off-nadir viewing.

A British satellite, Topsat, was launched in 2005. It was built by QinetiQ Ltd (the camera system was supplied by the RAL) and funded by the British National Space Centre (BNSC). It carries a multispectral sensor with a spatial resolution of 2.5 m at nadir. Topsat is small – it was built as a technology demonstrator to show that Earth observation can be successful even from a box-sized spacecraft (0.8 m side length). Figure 2.15 shows a TopSat image of the London orbital motorway (M25) where it crosses the Thames Estuary, downstream from London.

Other high-resolution satellite systems that are worth further consideration are FORMOSAT-2, KOMPSAT-2 and Compact High Resolution Imaging Spectrometer (CHRIS). Data from the first two of these satellites is marketed by SPOT Image, and details can be found on their web site. Both produce panchromatic stereo at a resolution of 1 m (KOMPSAT) and 2 m (FORMOSAT) as well as multispectral (4 m KOMPSAT, 8 m FORMOSAT). CHRIS is an ESA experimental project that has produced a good deal of scientific data. It collects data at five different viewing angles – two backwards, two forwards and nadir. The spectral range is 410–1050 nm, and the number of spectral bands is variable. It is one of two spaceborne hyperspectral sensors (the other is NASA's Hyperion). Hyperspectral sensors are considered in more detail in Section 9.3.

So far in this section we have considered single high-resolution dedicated satellites operating in the optical region of the spectrum. One of the problems of optical remote sensing in many areas of the world is that of cloud cover. One solution is to use several satellites spaced out

Figure 2.15 TopSat multispectral image of the Thames near Dartford. The inset shows an enlargement of the area around and including the high-level Queen Elizabeth II bridge taking the M25 (strictly speaking, the A282) over the Thames. Imagery courtesy of TopSat consortium, copyright QinetiQ.

around the same orbit. The UK company SSTL builds relatively cheap but effective satellites using commercial off the shelf (COTS) technology. The company operates the DMC, which currently has six member countries (Algeria, Nigeria, Spain, Turkey, the United Kingdom and China). Each DMC satellite carries a MSS capable of a spatial resolution of 32 m in three spectral bands (NIR, red, green). The latest version of the imaging instrument (DMC-2) has an improved spatial resolution of 22 m. The United Kingdom and Spain currently operate these 22 m resolution systems. The wide ground swath of 600 km enables a revisit of the same area almost anywhere in the world at least every 4 days with just a single satellite. Figure 2.16b shows a 1024 × 1024-pixel extract from a DMC 30 m resolution multispectral image of a part of the south-west United States. The SSTL web site at http://www.dmcii.com/products_sensor.htm contains a review of the sensor, called Surrey Linear Imager-6 (SLIM-6). See also Chander *et al.* (2009b), who assess SLIM-6 data quality and compare the characteristics of the instrument with those of Landsat-7 ETM+. They conclude that

> Indications are that the SSTL UK-DMC SLIM-6 data have few artifacts and calibration challenges, and these can be adjusted or corrected via calibration and processing algorithms.

A second example of a constellation of small satellites is RapidEye, operated by a German company of the same name. The system was built by a consortium of public and private agencies, including the German Space Agency, DLR. Five satellites follow a similar orbital path, so that each point on the Earth's surface between 75°N and S is imaged once per day. The satellites were built by SSTL, and each carries a multispectral sensor which records five bands: blue, green, red, red edge and NIR. (The concept of the red edge is discussed in Section 9.3.2.3.)

2.4 Microwave Imaging Sensors

As noted in Section 1.2.3, the microwave region of the electromagnetic spectrum includes radiation with wavelengths longer than 1 mm. Solar irradiance in this region is negligible, although the Earth itself emits some microwave radiation. Imaging microwave instruments do not, however, rely on the detection of solar or terrestrial emissions (though passive microwave radiometers do detect terrestrial emissions). Instead, they generate their own source of energy, and are thus examples of *active* sensing devices. In this section, the properties of SAR imaging systems carried by the ERS-1/2, ENVISAT, TerraSAR-X, COSMO/Skymed and Radarsat satellites are presented. General details of imaging radar systems are described in Sections 1.2.3 and 9.2.

SAR instruments are more complex than those operating in the optical and TIR regions of the spectrum. In this context, an aperture is an aerial or antenna such as the ones that the reader may have seen above the air traffic

Figure 2.16 Extract from an image acquired by the Algerian AlSAT satellite of the Colorado River in Arizona. This is a 1024 × 1024 pixel (33 × 33 km) extract from the full 600 × 600 km scene. AlSat is a member of SSTL.s Disaster Monitoring Constellation (DMC). Reproduced with premission from AlSAT-1 Image of the Colorado River (DMC Consortium).

A Brief History of Radar

The history of the development of radar makes a fascinating story. The detection of distant targets by radio waves is an idea that dates back to the early twentieth century, when a German scientist, Huelsmayer, patented the concept. Two Americans, Taylor and Young, rediscovered Huelsmayer's idea in 1922, but nothing came of it. By the early 1930s, the British Government was becoming concerned by the prospect of aerial bombardment of cities by the Luftwaffe, and the UK Air Ministry started to investigate ways of locating incoming bombers. Some rather odd ideas emerged, such as that of the Death Ray. The UK Air Ministry offered a prize of £1000 in 1934 for the first person to kill a sheep at 180 m using a Death Ray. The prize was never awarded; unfortunately, the number of sheep that fell in the cause of science (and patriotism) was never recorded. TIR and acoustic devices were considered, but it was Watson-Watt who, in 1935, demonstrated that he could locate an aircraft using the BBC radio transmitter near Daventry in the English Midlands. The Americans had the same idea and did, in fact, detect an aircraft using radar a month before Watson-Watt's success. Watson-Watt became a victim of his own invention when, after the Second World War, he was caught by a radar speed trap.

Early radars such as those used in the 'Battle of Britain' in August/September 1940 operated at long wavelengths (13 m) and were hampered by radio interference. A major breakthrough came in 1939–1940 with the invention of the cavity magnetron by Randall and Boot at Birmingham University. Although the United States was still neutral in 1940, the British Prime Minister, Winston Churchill, authorized the transfer of a cavity magnetron to the United States, where – in response – the Rad Lab was established at MIT, attracting scientists such as Luis Alvarez (who was later to become even more famous for his theory that dinosaurs were annihilated by a meteorite impact at about 65 Ma). One American scientist described the cavity magnetron as the most valuable cargo ever to reach America's shores.

The MIT Rad Lab and the British radar research laboratories used the cavity magnetron to develop shorter-wavelength radars. The British developed the H2S radar, which operated at a 10 cm wavelength and which was fitted to bombers. This targeting radar was the first terrain-scanning radar in the world. Its effectiveness was demonstrated on 1 November 1941, when a 10 cm radar fitted to a Blenheim bomber was able to locate a town at a distance of 50 km.

Radar research and development proceeded rapidly during the Second World War, with naval and airborne radar being widely used by the British and American navies for detection of enemy vessels and surfaced submarines, and for gunnery control. Radars were also carried by aircraft. By 1945, radar was an essential element of the air defence system, targeting and naval operations.

control centre at an airport, or at a military base such as Fylingdales in North Yorkshire. The first airborne radar systems carried radars that were similar in principle to these ground-based radars. The antenna was mounted on the side or on the top of the aircraft, and was pointed to one side of the aircraft at 90° to the direction of forward travel. A pulse of microwave energy was emitted by the antenna and the reflected backscatter from the target was detected by the same antenna, with the result being displayed on a cathode ray tube. This kind of radar is known as a side-looking airborne radar or SLAR. The text box 'Radar History' provides a brief summary of the historical development of radar.

The spatial resolution achieved by a SLAR is proportional to the length of the antenna. For satellite-borne radars, this relationship results in a practical difficulty. In order to achieve a spatial resolution of, say, 30 m from an orbital height of 700 km the antenna length would be several kilometres. If, on the other hand, it were possible to move a 10 m antenna along the orbital path, recording the backscatter from a given target at each position, then it would be possible to simulate an antenna length of several kilometres and achieve an acceptable level of spatial resolution.

One obvious difficulty in understanding how a SAR works is summarized by the question: 'How does the sensor record whether it is approaching, passing or moving away from the target?' The answer is: it doesn't. Instead of recording positions relative to every possible target, which would be difficult, the SAR records the amplitude and the phase of the return signal (Figures 1.8 and 9.1). From these two properties of the return signals, it is possible to calculate the position of the object relative to the antenna. This calculation is based on a concept that is well within our everyday experience, namely the Doppler principle. We can tell whether a police car or an ambulance is approaching us or moving away from us by comparing the tone of the siren at one instant with that at the next instant. If the vehicle is approaching, then the siren's note appears to rise in pitch. If it is moving away from us, then the note seems to drop. The same idea is used in a SAR (the calculations are, however, quite formidable).

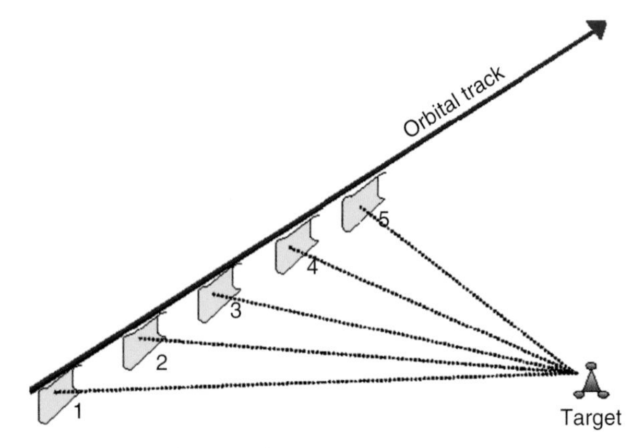

Figure 2.17 *The 'synthetic aperture' is generated by the forward motion of the platform through positions 1–5 respectively. The Doppler principle is used to determine whether the antenna is looking at the target from behind or ahead.*

The Doppler principle can thus be used to increase the spatial resolution of a satellite or airborne imaging radar by allowing the calculation of the motion of the satellite platform relative to a target (Figure 2.17). In addition, the absolute distance between the antenna and the target can be calculated. This distance is called the range. Microwave energy travels at the speed of light and so, if the time taken between the transmission of a pulse and the reception of the return is known, then the distance is $ct/2$ where c is the speed of light, and t is the time taken from transmission of a pulse to the reception of its return. Division by two is undertaken as the pulse travels from the antenna to the target and back again. Radar is called a ranging instrument because it can measure distance from the sensor to the target. Lidar (Section 9.4) is another example of a ranging instrument. However, lidar measures only the distance from the sensor to the target. Radar can also measure some characteristics of the target as well.

We saw in Chapter 1, and in some of the examples presented earlier in this chapter, that optical sensors measure electromagnetic radiation that is reflected by the target. The chemical composition of the target (for example chlorophyll in leaves or minerals in a soil) affects the nature of the reflected radiation in the optical region because each chemical element, or mixture of elements, absorbs some of the incoming radiation at specific wavelengths. The return signal estimates the reflectance spectrum of the target. Techniques for the identification of targets by remote sensing using optical wavelengths assume that similar targets (such as specific vegetation types) have similar reflectance spectra (Chapter 8). The interaction between electromagnetic radiation and a target in the microwave region does not generate information about the types of chemical element or mixtures of elements that are present in the target, but is related to the geometry and surface roughness of that target (relative to the wavelength of the microwave energy pulse) and what are known as the ' dielectric properties' of the target, which generally are closely correlated with its moisture status.

Radar wavelength bands are described by codes such as 'L-band' or 'C-band' that came into use during the Second World War for security purposes. Unfortunately, several versions of the code were used, which confused the Allies as much as the opposition. Table 2.8 shows the commonly accepted delimitation of radar wavelengths. Radar wavelength also has a bearing on the degree of penetration of the surface material that is achieved by the microwave pulses. At L-band wavelengths (approximately 23 cm), microwave radiation can penetrate the foliage of trees and, depending on the height of the tree, may reach the ground. Backscatter occurs from the leaves, branches, trunks and the ground surface. In areas of dry alluvial or sandy soils, L-band radar can penetrate the ground for several metres. The same is true for glacier ice. Shorter-wavelength C-band radiation can penetrate the canopies of trees, and the upper layers of soil and ice. Even shorter wavelength X-band SAR mainly 'sees' the top of the vegetation canopy and the soil and ice surface.

Properties of the microwave radiation used by a SAR, other than wavelength, are important. The *polarization* of the signal has an effect on the nature and magnitude of the backscatter. Figure 1.3 illustrates the concept of polarization of an electromagnetic wave. In a polarized radar, the antenna can transmit and receive signals in either horizontal (H) or vertical (V) mode. If it both transmits and receives in horizontal polarization mode it is designated as 'HH'. If both transmit and receive modes are vertical then the radar system is designated 'VV'. HV and VH modes are also used. HH and VV modes are said to be 'like polarized', whereas VH and HV modes are 'cross-polarized'. The SIR-C SAR carried by the Space Shuttle in 1994 provided polarimetric

Table 2.8 *Radar wavebands and nomenclature.*

Band designation	Frequency (MHz)	Wavelength (cm)
P	300–1000	30–100
L	1000–2000	15–30
S	2000–4000	7.5–15
C	4000–8000	3.75–7.5
X	8000–12 000	2.5–3.75
K_u	12 000–18 000	1.667–2.5
K	18 000–27 000	1.111–1.667
K_a	27 000–40 000	0.75–1.111

radar images in the C and L bands, plus an X-band image. The Advanced Synthetic Aperture Radar (ASAR) instrument on the European ENVISAT can transmit and receive in combinations of H and V polarizations. The Canadian Radarsat-2 also carries a fully polarimetric SAR, as does the Advanced Land Observation Satellite (ALOS) PALSAR and the recently-launched TerraSAR-X (Germany) and COSMO-Skymed (Italy). Freeman *et al.* (1994) discuss the use of multifrequency and polarimetric radar for the identification and classification of agricultural crops (Chapter 8). Mott (2007) is devoted to the analysis of polarimetric radar. Lee and Pottier (2009) is another substantial guide to the use of polarimetric SAR.

Another important property of an imaging radar is the instrument's *depression angle*, which is the angle between the direction of observation and the horizontal (Figure 2.18). The angle between the direction of observation and the surface normal (a line at right angles to the slope of the Earth's surface) is the *incidence angle*, which is also shown on Figure 2.18. The local incidence angle depends on the slope of the ground surface at the point being imaged (Figure 2.18b). Generally, the degree of backscatter increases as the incidence angle decreases. Different depression angles are suited to different tasks; for example ocean and ice monitoring SAR systems use lower depression angles than do SAR systems for land monitoring.

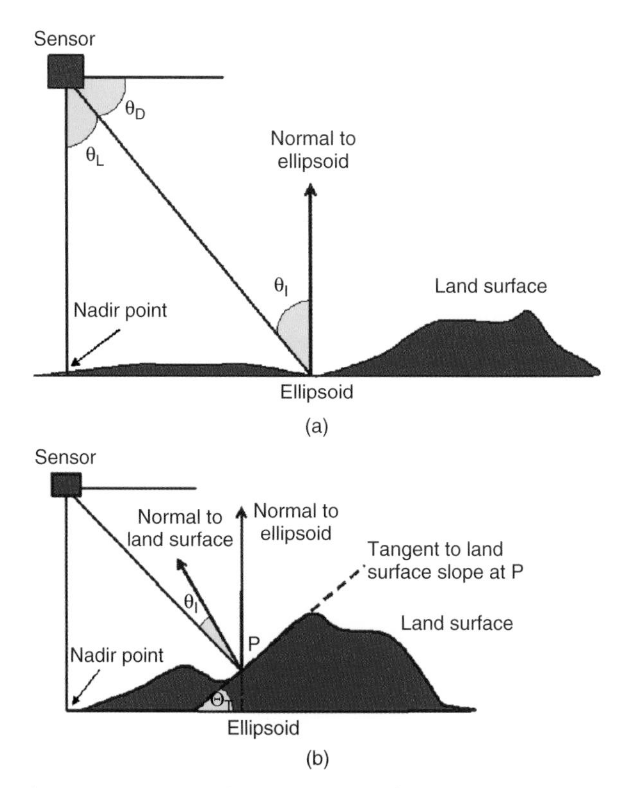

Figure 2.18 *(a) Angles in imaging radar: θ_D is the depression angle (with respect to the horizontal), θ_L is the look angle, which is the complement of the depression angle and θ_I is the incidence angle. θ_L depends on local topography and is equal to the look angle only when the ground surface is horizontal. In (b) the ground slope at P is θ_T and the local incidence angle is θ_I. The normal to the land surface is the line through P that is perpendicular to the tangent to the land surface slope at P.*

The view direction of a radar sensor is also important in detecting geological features with particular orientations, so that radar images of the same terrain with different 'look angles' will show different features (Blom, 1988; Gauthier *et al.*, 1998; Koch and Mather, 1997; Lowman, 1994).

Imaging radars are side-looking rather than nadir-looking instruments, and their geometry is complicated by foreshortening (the top of a mountain appearing closer to the sensor than does the foot of the mountain) and shadow, caused by the 'far side' of a mountain or hill being invisible to the side-looking radar sensor (see Schreier, 1993a, for a thorough description of the workings of a SAR) (Figure 2.19a,b). Furthermore, the interaction between microwave radiation and the ground surface generates a phenomenon called *speckle*, which is the result of interference resulting from the coherent integration of the contributions of all the scatterers in the pixel area (Quegan and Rhodes, 1994). Speckle magnitude is proportional to the magnitude of the backscattered signal, and is rather more difficult to remove from the image than is additive noise. Filtering, which is used to remove unwanted features such as speckle noise, is the topic of Chapter 7. A recent development is the use of wavelets (Section 6.7) to remove speckle filter noise; see, for example Xie, Pierce and Ulaby (2002). See also Section 1.2.3 for an introduction to imaging radar principles, and Section 4.6 for a brief summary of calibration issues. Other basic sources are Leberl (1990), Lewis, Henderson and Holcomb (1998), Rees (2001), Ulaby, Moore and Fung (1981–1986) and Ulaby and Elachi (1990). The textbooks by Kingsley and Quegan (1992), Oliver and Quegan (2004) and Woodhouse (2006) provide a sound introduction to radar systems in remote sensing. Lu, Kwoun and Rhykus (2007) provides a retrospect and prospect. A number of web-based tutorials are recommended – these are listed in Table 2.9.

Software for processing ESA SAR imagery is provided by the Next European Space Agency Synthetic Aperture Radar Toolbox (NEST), which is described as an open source toolbox for reading, post-processing, analysing and visualising the large archive of data (from Level 1) from ESA SAR missions including ERS-1 and -2, ENVISAT and in the future Sentinel-1. In addition, handling of products from third party missions like JERS-1,

```
FILENAME =L71201023_02320000619_B10.FSTFILENAME =L71201023_02320000619_B20.FST
FILENAME =L71201023_02320000619_B30.FSTFILENAME =L71201023_02320000619_B40.FST
FILENAME =L71201023_02320000619_B50.FSTFILENAME =L72201023_02320000619_B70.FST
L71201023_02320000619_B10.FST
L71201023_02320000619_B20.FST
L71201023_02320000619_B30.FST
L71201023_02320000619_B40.FST
L71201023_02320000619_B50.FST
L72201023_02320000619_B70.FST
REV L7A
Number of bands referenced in the header is 6
Band identifiers are 1 2 3 4 5 7
Image filenames for VNIR/SWIR
Band 1 Filename: L71201023_02320000619_B10.FST
Band 2 Filename: L71201023_02320000619_B20.FST
Band 3 Filename: L71201023_02320000619_B30.FST
Band 4 Filename: L71201023_02320000619_B40.FST
Band 5 Filename: L71201023_02320000619_B50.FST
Band 6 Filename: L72201023_02320000619_B70.FST
Images in this file set:
Width: 8311
Depth 7621
Radiometric record
==================
GAINS AND BIASES IN ASCENDING BAND NUMBER ORDER
-6.199999809265137 0.775686297697179
-6.400000095367432 0.795686274883794
-5.000000000000000 0.619215662339154
-5.099999904632568 0.965490219639797
-1.000000000000000 0.125725488101735
-0.349999994039536 0.043725490920684
Geometric record
================
GEOMETRIC DATA MAP PROJECTION =UTM ELLIPSOID =WGS84 DATUM =WGS84
USGS PROJECTION PARAMETERS = 0.000000000000000
0.000000000000000
0.000000000000000 0.000000000000000 0.000000000000000
0.000000000000000 0.000000000000000 0.000000000000000
0.000000000000000 0.000000000000000 0.000000000000000
0.000000000000000 0.000000000000000 0.000000000000000
0.000000000000000
USGS MAP ZONE =31
UL = 0005136.6558W 540441.0389N 247500.000 5999100.000
UR = 0025703.6628E 540824.0187N 496800.000 5999100.000
LR = 0025711.8748E 520506.2184N 496800.000 5770500.000
LL = 0004051.3183W 520139.1763N 247500.000 5770500.000
CENTER = 0010525.9263E 530550.3352N 372150.000 5884800.000 4156 3811
OFFSET =-3391 ORIENTATION ANGLE =0.00
SUN ELEVATION ANGLE =57.2 SUN AZIMUTH ANGLE =147.8
```

lost without any cost being incurred. Some image transform methods, including principal components analysis (PCA) (Section 6.4), the discrete Fourier transform (Section 6.6) and the discrete wavelet transform (Section 6.7) can be used to compress images. These methods exploit the fact that some redundancy exists in a multispectral image set, and they re-express the data in such a way that large reductions in the volume of transformed data represent only small losses of information.

Other methods are *run length encoding* and *Huffman coding*. Run length encoding involves the re-writing of the records of image pixels in terms of expressions of the form (l_i, g_i) where l_i is the number of pixels of value g_i

that occur sequentially. Thus, a sequence of pixel values along a scan line might be 1 1 1 2 2 2 2. This sequence could be encoded as (3, 1) (4, 2) without losing any information. A special character indicates the end of a scan line. Obviously the degree of compression that results from this type of encoding depends on the existence of homogeneous sections of image – in other words, if there are no sequences of equal values then there will be no compression; in fact, if there are no sequences of equal values then there will be expansion rather than compression. Run length encoding is used in fax transmission and may be useful in compressing classified images (Chapter 8) in which individual pixel values are replaced by labels. The *quadtree*, which is described next, may be a better choice, as it is a two-dimensional compression scheme.

A quadtree is a form of two-dimensional data structure or organization that is used in some raster GIS (Figure 3.9). Its major limitation is that the image to be encoded must be square and the side length must be a power of 2. However, the image can be padded with zeros to ensure that this condition is met. The square image is firstly subdivided into four component square subimages of size $2^{n-1} \times 2^{n-1}$. If any of these subimages is homogeneous (meaning that all of the pixels within the sub-image have the same value) then it is not 'quartered' any further. Conversely, those sub-images that are not homogeneous are again divided into four equal parts and the process repeated (in computing terms, the procedure is recursive). When the quadtree operation is completed the individual components, which may be of differing sizes, are given identifying numbers called Morton numbers, and these Morton numbers are stored in ascending order of magnitude to form a *linear quadtree*. For images such as classified images, which generally contain significantly large homogeneous regions, the use of quadtree encoding will result in a substantial saving of storage space. If the image is inhomogeneous then the amount of storage required to store the quadtree may be greater than that required for the raw, uncompressed image. Kess, Steinwand and Reichenbach (1996) use quadtrees to compress the non-land areas of the Global Land 1-km AVHRR data set. They find that the quadtree representation produces a reduction in data volume to 6.72% of the original data size, which is better than that achieved by JPEG, GZIP or LZW compress methods. More details of quadtree-based calculations can be found in Mather (1991), while the definitive reference is Samet (1990).

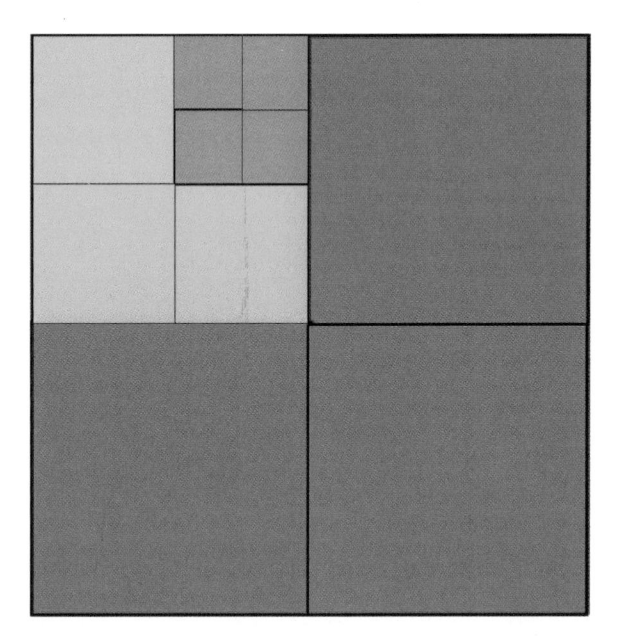

Figure 3.9 *Quadtree decomposition of a raster image. The full image, each dimension of which must be a power of 2, is divided into four equal parts (quads). Each quad is the subdivided into four equal parts in a recursive way. The subdivision process terminates when the pixel values in each sub-(sub-...) quadrant are all equal. The procedure works best for images containing large, homogeneous patches. The illustration shows a three-level decomposition; usually, the number of levels of decomposition is substantially higher than this.*

3.2.3 System Processing

Data collected by remote sensing instruments carried by Earth-orbiting satellites are transmitted to ground receiving stations using high-bandwidth radio. These transmitted data are in raw format. They must be processed before delivery to a user. If we use the Landsat-7 ETM+ instrument as an example, we note that the image data are collected by an optomechanical scanner (Figure 1.2). This scanner uses a mirror to direct radiance upwelling from the ground to a set of 16 detectors, each of which records radiance values for a single scan line (recall from Section 2.3.6.2 that the Landsat TM scanner collects data for 16 scan lines simultaneously). Each of the 16 detectors has seven components, one for each spectral band. The system operates in both forward and reverse mirror directions, so the raw format data has sets of 16 scan lines stored alternately in opposite directions. Without any preprocessing, these data would be difficult to use. Raw format data has other undesirable characteristics. The pixels forming each scan line are not properly aligned geometrically, and artefacts caused by electronic noise in the system may be present. Nor are the data calibrated to radiance units. Most remote sensing image data are delivered to the user in the form of quantized counts, often – but not always – on a scale requiring 8 bits of storage (i.e. levels 0–255). For some applications, it is necessary to convert from quantized counts to radiance units and so this radiometric correction is a vital stage of pre-processing (Section 4.6).

The first level of Landsat ETM+ system processing is called Level 1R. It involves removal of coherent noise, of banding effects caused by the use of 16 detectors (which can generate patterns with a period of 16 scan lines down the image) and calibration of the pixel values to radiance units. The calibration coefficients are provided in the header records of Level 1R data. However, Level 1R data are not corrected for pixel misalignments, so they must undergo a process called geometric correction (Section 4.3) before they are usable. Level 1G Landsat ETM+ data are radiometrically corrected (Section 4.6), like Level 1R, and the pixels are relocated or resampled (Section 4.3.3) so that the image conforms to a map projection. At Level 1G, the image geometry is corrected using system information, such as orbital height of the satellite, the direction of its forward motion, and its attitude (in terms of pitch, roll and yaw). There is no geometric calibration against known ground points.

Table 3.3 *Edited extract from ASTER metadata file, generated by MIPS.*

GROUP VNIRBAND1DATA	
IMAGEDATAINFORMATION1	(4980, 4200, 1)
GROUP IMAGESTATISTICS1	
MINANDMAX1	(60, 255)
MEANANDSTD1 25.771908)	(129.753067,
MODEANDMEDIAN1	(121, 157)
GROUP DATAQUALITY1	
NUMBEROFBADPIXELS1	(0, 0)
GROUP PROCESSINGPARAMETERS1	
CORINTEL1	"N/A"
CORPARA1	"N/A"
RESMETHOD1	"NN"
MPMETHOD1	"UTM"
PROJECTIONPARAMETERS1 6356752.300000, 0.999600, 0.000000, −0.052360, 0.000000, 500000.000000, 0.000000, 0.000000, 0.000000, 0.000000, 0.000000, 0.000000)	(6378137.000000,
UTMZONECODE1	30
GROUP UNITCONVERSIONCOEFF1	
INCL1	0.676000
OFFSET1	−0.676000
CONUNIT1	"W/m2/sr/um"

The definitions of the parameters are given in ASTER Level 1 Data Products Specifications (GDS Version) Version 1.3 produced by the Japanese Earth Remote Sensing Data Centre, dated 25 June 2001. Each of the metadata entities is described as a group, and each group may contain descriptive fields plus values. The portion of the metadata contained in this table refers to the visible and near infrared (VNIR) band 1. The first group, VNIRBAND1DATA, contains one field, which tells us that the image consists of 4980 scan lines each containing 4200 pixels, represented as 1 byte (8 bits) per pixel. Group IMAGESTATISTICS1 provides the minimum, maximum, mean, standard deviation, mode and median of the image pixel values. Group DATAQUALITY tells us that there are no bad pixels. The PROCESSINGPARAMETERS1 group contains details of the method of relocating (resampling) the pixels during system corrections (NN indicates the Nearest Neighbour method, described in Section 4.3.3), the map projection (UTM) and UTM zone (30). The projection parameters for the UTM projection are in the following order: (i) semi-major axis of the ellipsoid, (ii) semi-minor axis of the ellipsoid, (iii) scale factor at the central meridian, (iv) not used – zero, (v) longitude of the central meridian, (vi) latitude of the projection origin, (vii) false easting in the same units as the semimajor axis and (viii) false northing in the same units as the semiminor axis. The remaining fields are set to zero. The final group shown here is UNITCONVERSIONCOEFF1. The two values in this group are the calibration coefficients that are used to convert pixel values to radiance units. The third field of this group indicates that, after conversion, the radiance values are expressed in units of W m^{-2} sr^{-1} μm^{-1}.

However, the pixel positions are said to be accurate to within 250 m for flat areas near sea level. Level 1G data are the standard Landsat-7 ETM+ product. Images derived from higher-level processing (involving the use of ground control points and digital elevation models to refine the image geometry) are offered as 'value-added' products by data providers and by third-party commercial companies. Details of Landsat system data processing can be found in the *Landsat Science Users' Data Handbook* (Irish, 2008).

The processing chain for SPOT HRV data is similar to that described for Landsat-7 ETM+ images. SPOT Image defines three levels of 'system corrections', beginning with Level 1A (which provides radiometric corrections to equalize detector responses), and proceeding to Level 1B, which provides a similar geometric correction to that offered by Landsat ETM+ Level 1G, in addition to radiometric equalization of the detectors. SPOT Image also provides Level 2A pre-processing, which transforms the image data onto a cartographic projection. Like Landsat-7 ETM+ Level 1B data, neither Level 1B nor Level 2A SPOT HRV processing involves any ground control, so the accuracy of the resulting correction depends on the level of error in the determination of the satellite's orbital parameters (see Section 2.3.7.1 for a description of DORIS, used to obtain accurate orbit parameters for the SPOT-5 satellite). None of these corrections includes the effects of the land surface topography, so the geometric accuracies cited by the data providers refer only to flat areas near sea level. Location errors in mountainous areas will be considerably greater.

3.3 Numerical Analysis and Software Accuracy

Mention was made in the introduction to this chapter of the software error that caused the failure of the first Ariane-5 launch in 1997. A Google Scholar search for the term 'software accuracy' will reveal that this failure is not an isolated case, and that many innocent-looking computer routines contain within themselves the seeds of their own destruction. The study of the design of computer algorithms and the evaluation of potential errors in numerical algorithms is called numerical analysis. Typically, these algorithms involve the use of real values rather than integers, although the conversion from real to integer form is not without its problems, as we have already seen. Operations carried out on remotely-sensed images may involve the use of unsigned integers (using 8, 16 or 32-bit representation). An example is the transformation of raw data values to lookup table entries, which is discussed above in the context of pseudocolour images (illustrated in Figure 3.7). There is no rounding error here, nor is it possible for the lookup table entry to lie outside the range 0–255. However, some transformation and classification methods, dealt with in Chapters 6 and 8 respectively, do involve the conversion of integer image pixel values to real or floating point form. Table 3.2 illustrates the properties of integer and real numbers.

There are two principal sources of error in real (floating point) operations on a computer, assuming that the program is correctly coded. One is due to the inherent properties of the computer. The other is in the choice of algorithm used to implement a specific operation. For example, the correlation coefficient can be computed in at least three different ways. One is almost certain to produce severe errors. Before this topic is pursued, however, we will consider numerical error caused by the way that computers store real numbers. These numbers are stored (on Intel-based machines) as a sign bit, a 7-bit exponent and a 24-bit mantissa, giving 32-bits in total. In base 10 notation, we could write 12 345.0 as the mantissa and 2 as the exponent to deconstruct the number $0.1\,234\,500 \times 10^2$. A 24-bit mantissa gives approximately seven decimal digits of accuracy. The range of a 32-bit real number is of the order of $\pm 3.4 \times 10^{38}$, which would seem to be enough for most purposes. However, attempts may be made – particularly with huge GIS datasets – to store numbers bigger than this. The result is called overflow, and may not be detected by the computer operating system. The converse is underflow which is attempting to store a number that is not detectably different from zero. Overflow and underflow are usually, but not always, treated by programmers and program designers so that their occurrence is minimized. Another source of error is the representation, in base 2 form, of a base 10 number. Thus, the base 10 number 0.1 cannot be accurately represented in base 2 (McCullough, 1998). Asking a computer 'is 0.05 + 0.05 equal to 0.1?' should produce the (un)expected answer 'No'.

If overflow/underflow and representational errors do not entrap the unwary, then subtraction of large, nearly-equal real numbers may well do so. A Fortran program using single-precision (32-bit) real numbers to perform the subtraction operation 100 000.1 – 100 000 gave the answer 0.101 563 rather than 0.100 000. This may sound like an insignificant difference, but imagine how error could build up if the data set contains millions of pixel values and each recorded a small discrepancy from its true value.

The subtraction of large nearly-equal real numbers links into the second main cause of numerical error in computing, namely, that due to the poor choice of algorithm. Consider one of the most commonly used statistical measures – the variance, s^2. The classical definition of s^2 (the definitional formula) is given by $s^2 = \frac{1}{n}\sum(x - \overline{x})^2$. This formula is inconvenient as it requires two passes through what might be a very large data set. The mean is computed during the first pass and the variance at the second pass. A one-pass formula, which is logically equivalent to the definitional formula, is often used. In this one-pass formula, s^2 is defined as $s^2 = \frac{1}{n^2}\left[n\sum x^2 - (\sum x)^2\right]$, so that $\sum x$ and $\sum x^2$ can be calculated simultaneously during one pass through the data. Recall that subtracting near-equal and large quantities can lead to significant error, and it will become apparent that this one-pass formula should not be used to calculate the variance of a remotely-sensed image. Some published algorithms, for example Alley (1995), explicitly use this second method.

Other quantities such as the covariance and correlation matrices can be calculated using the one-pass method but this is not recommended. An alternative one-pass algorithm, known as Welford's method, gives a result that is of similar accuracy to the two-pass method. It is an example of an algorithm that is designed with the properties of floating-point arithmetic in mind, and reveals the truth of the old saying that methods useful for hand calculators should not be transferred to the digital computer without some thought. Welford's method is described in Knuth (1998, p. 232). The original paper is Welford (1962).

So far, we have considered underflow, overflow and the subtraction of large but nearly equal values as sources of error. There is one other major source of potential error and that is the analysis of problems that are inherently unstable or ill conditioned. Several of the techniques described in Chapters 6 and 8 make use of an inverse matrix, therefore methods of determining the inverse matrix are considered here. A square matrix generally possesses an inverse. The inverse of matrix $\mathbf{X}$ is written as $\mathbf{X}^{-1}$ and these two matrices are related by the expression $\mathbf{XX}^{-1} = \mathbf{I}$, where $\mathbf{I}$ is the identity matrix which has the same number of rows and columns as $\mathbf{X}$ and $\mathbf{X}^{-1}$ and contains zeros everywhere except along the principal diagonal (which runs from the top left cell (1, 1) to the bottom right cell (n, n)). The principal diagonal of the identity matrix contains 1s. Techniques such as PCA (Chapter 6), spectral unmixing and maximum likelihood classification (Chapter 8) require the computation of the matrix of interband covariances. Computational error will result in inaccuracy in the results, so care should be taken to ensure that a 'good' algorithm is used, such as the Welford approach that is discussed above.

Even a 'good' algorithm will perform poorly if the input matrix does not possess certain properties. Firstly, individual subsets of columns (bands) of the original data from which the covariance matrix is calculated should have low correlations. Orthogonal matrices (which have a zero correlation between their columns) are best from a numerical analysis point of view. The visible bands of the Landsat TM sensor are certainly intercorrelated, as visual inspection of Figure 1.10 shows. The level of intercorrelation may be such that the number of independent sources of information is less than the number of bands. As an example, think of six children as sources of information. If all six speak different languages then we have six different sources of information. If two speak English, two speak German and two speak Russian then the number of independent sources is three. Because the Landsat TM bands are intercorrelated they do not each contribute one unit of information, and this can lead to difficulties in calculating the inverse of the covariance matrix. Matrices based on intercorrelated data sets are said to be ill-conditioned. Small changes in the covariance values lead to disproportionate changes in the inverse covariance matrix. Table 3.4 illustrates this point. A 4 × 4 data matrix $\mathbf{X}$ whose elements are whole numbers is input to a Gauss–Jordan matrix inversion function and the inverse is computed. A check is performed by calculating $\mathbf{XX}^{-1}$, which should produce a 4 × 4 identity matrix, $\mathbf{I}$. Inspection of Table 3.4 shows that the diagonal elements of the matrix $\mathbf{XX}^{-1}$ are close in value to 1.0 and the off-diagonal elements are almost zero, so the inverse matrix $\mathbf{X}^{-1}$ seems to be a reliable estimate of the true inverse matrix. Next, one element of $\mathbf{X}$ is changed, that is the element at the intersection of row 3 and column 3 of $\mathbf{X}$ (written $x(3, 3)$). The value 10 is changed to 9.99. We might reasonably expect that the inverse matrix changes only slightly, but it does not. The first element of the first column has changed from 67.9996 to 71.0418 and most of the other elements have changed by about 5%. When large-scale changes in output values result from slight perturbations in the input matrix $\mathbf{X}$ then the matrix $\mathbf{X}$ is said to be ill-conditioned. This problem cannot be detected by simply looking at the data matrix. There are two methods that can be used to check on the condition of a matrix. One uses the determinant and the second uses the ratio of the largest to the smallest eigenvalue.

All square matrices have a positive number associated with them. This number is called the determinant of the matrix and it ranges upwards from zero. We need not concern ourselves with how the determinant is derived from the matrix; all we need at this stage is to know that the closer the determinant gets to zero the more likely the matrix is to be ill-conditioned. A warning or error message could be delivered to the user if the determinant approaches zero.

The widely used method of PCA is dealt with in Chapter 6. PCA attempts to discover how many independent sources of information or dimensions of variability exist within the data (recall the example used earlier of six children speaking three or six languages). To achieve this aim, PCA makes use of what is called the eigenvalue transform, which generates a set of numbers (one for

Table 3.4 *Example of computational error in matrix inversion. The element (3,3) of the* Initial Data Matrix *is changed from 10.0 to 9.99 and the solution* (Inverse Matrix) *changes considerably (by more than 5%) as a result. In the two cases, the result of multiplying the input matrix* (Initial Data Matrix *or the* Perturbed Data Matrix) *by the computed inverse is shown. The resulting matrix (listed as* Initial Data Matrix × Inverse *or* Perturbed Matrix × Inverse) *should approximate to the* Identity Matrix *(consisting of values of 1.0 along the principal diagonal and 0.0 elsewhere).*

Initial Data Matrix			
5.00000	7.00000	6.00000	5.00000
7.00000	10.0000	8.00000	7.00000
6.00000	8.00000	10.0000	9.00000
5.00000	7.00000	9.00000	10.0000
Inverse Matrix			
67.9 996	−40.9 998	−16.9 999	9.99994
−40.9 998	24.9 999	9.99 994	−5.99997
−16.9 999	9.99994	4.99 998	−2.99999
9.99994	−5.99997	−2.99999	1.99999
Initial Data Matrix × Inverse			
0.999 996	−3.433 228E−05	−1.907 349E−05	−7.629 395E-06
−1.907 349E−06	0.999 977	−3.814 697E−06	0.00000
9.536 743E−07	8.583 069E−06	1.00000	1.907 349E-06
−2.861 023E−06	9.536 743E−07	−2.861 023E−06	0.999 998
Perturbed Matrix			
5.00000	7.00000	6.00000	5.00000
7.00000	10.0000	8.00000	7.00000
6.00000	8.00000	9.99000	9.00000
5.00000	7.00000	9.00000	10.0000
Inverse Matrix			
71.0 418	−42.7 893	−17.8 947	10.5 368
−42.7 893	26.0 525	10.5 263	−6.31 576
−17.8 947	10.5 263	5.26 314	−3.15 788
10.5 368	−6.31 576	−3.15 788	2.09 473
Perturbed Matrix × Inverse			
0.999 990	1.335 144E−05	1.716 614E−05	−1.144 409E−05
−4.291 534E−06	0.999 974	−8.106 232E−06	−4.768 372E−06
−4.529 953E−06	−4.053 116E−06	1.00000	−5.245 209E−06
−2.861 023E−06	1.907 349E−06	−9.536 743E−07	0.999 996

Source of Initial Data Matrix: Kennedy and Gentle (1980, pp. 34–35). The results in this table were calculated using a Gauss–Jordan Fortran routine, operating in single-precision mode, taken from Mather (1976, p. 498).

each row/column so that a 4 × 4 matrix would have four numbers, called eigenvalues. If the eigenvalues of the correlation matrix (which is simply the standardized covariance matrix) are all equal and sum to the matrix size (four in this example) then the matrix is perfectly conditioned and has four independent sources of information. Generally, this happy state rarely prevails, and there are some large eigenvalues and some small eigenvalues. If any of the eigenvalues is equal to zero then the correlation matrix is said to be singular – it does not have an inverse. The ratio of the largest to the smallest eigenvalue can give a value, called the condition number, that could tell us how badly conditioned the covariance/correlation matrix actually is. When all eigenvalues are equal, then the condition number is 1.0 and we have an orthogonal matrix which has mutually independent columns. If the first eigenvalue is very large relative to the other eigenvalues and the smallest eigenvalue has a very small value then the condition number is very large and results derived from the use of the covariance matrix are thus suspect. There is no general agreement on what constitutes a large condition number, unfortunately.

Where the covariance matrix is computed from a sample of data, as is the case with the maximum likelihood classifier (Chapter 8), then the condition number of the covariance matrix may be improved by increasing the size of the sample. Other techniques, such as PCA (Chapter 6), compute covariance/correlation matrixes from the whole data set so increasing sample size is not an option.

In summary, we must take care when carrying out digital image processing operations on remotely sensed images. Software such as ERDAS, ENVI and MATLAB could help by place greater emphasis on ensuring the accuracy of results by building-in checks on potential sources of error, as outlined above. It would be interesting to speculate how results based on ill-conditioned matrices would vary from one software product to another. Comparative analyses of spreadsheets such as EXCEL abound on the Internet, and a similar operation on remotely sensed software packages might prove enlightening. Non-mathematical readers may feel that the danger is exaggerated, whereas computer scientists worry unduly about error. Readers need to be aware of the possibility of errors in numerical computing but few, I expect, will follow the example of Alston S. Householder, a famous computer scientist, who allegedly refused to fly because aircraft are designed using floating point arithmetic.

Further reading on the fascinating subject of computer error includes Nataraj (undated), Higham (2002), Kennedy and Gentle (1980) and McCullough (1998, 1999). For a general introduction to mathematics in remote sensing, see Milman (1999). Most computer science undergraduate texts contain chapters on errors of the kinds discussed above, and a search of the Internet will bring up numerous and fascinating documents.

3.4 Some Remarks on Statistics

Many students dislike statistics, largely because they do not see the point of testing rather badly chosen examples comprising the specification and examination of null hypotheses and the checking of assumptions such as that of normality. However, to use some of the techniques of mixture modelling and classification (Chapter 8), some idea of sampling theory and confidence limits may be found to be useful. For example, the application of the maximum likelihood classifier to determine the class to which a test pixel belongs requires the calculation of a measure that is related to the likelihood of that pixel being a member of a given class, say pasture or forest. If the maximum likelihood value for any class for the test pixel is computed to be 0.25, for example we could very well ask the question 'given a sample of size n and a p- dimensional set of pixel values (such as the $p = 6$ reflective Landsat ETM+ bands), what is the probability that one or more of the test pixels will generate a value as high as 0.25 from random sampling alone?' In other words, how reliable is the likelihood value estimated from the sample? Thus, if we took repeated and different random samples of pixels representing pasture, and if the probability of the test pixel was computed for all random samples, then it is extremely unlikely that the test pixel values would produce a likelihood of 0.25 every time. If the pasture was homogeneous (i.e. if the standard deviation of the values of pixels representing pasture was small) then we would expect some variation around 0.25, with the variation increasing as the standard deviation of the pasture pixels increased. A value of 0.25 might occur (or be exceeded) more than 95% of the time in cases of low standard deviation but may occur or be exceeded only 65% in the case of high standard deviations. The values associated with these percentages are known as confidence limits. Clearly we should have more confidence in the former result rather than in the latter. This is an example of classic inferential statistics which is based on various assumptions. Very few researchers report the confidence levels of their classified images.

It is usually the case that the statistical population(s) of interest is normally distributed (or multivariate normal in the case in which there is more than one variable of interest). The test described above on the single test pixel is based on the assumption that (a) the pixel values for each land cover class (wheat, barley, turnips, pasture, deciduous woodland, coniferous woodland, bare soil, etc.) are each multivariate-normally distributed. The multivariate normal distribution is used to work out the probability that the test pixel belongs to each class in turn. If the classes are not multivariate–normally distributed then the calculation of probabilities will be incorrect and the wrong conclusion may be drawn. We can wriggle out of this one by saying that we aren't too concerned about getting the correct value as long as the resulting rank order of the class membership probabilities is correct. This is probably the case if each class has a unimodal distribution but if the distribution is bimodal or multimodal then we cannot be assured that the probabilities (or, more accurately, the pseudo-probabilities) are in the correct rank order and misclassification will occur. Nor can we use pseudo-probabilities to compute the confidence limits (as described earlier in this section). It is, therefore, dangerous to use traditional statistical methods like maximum likelihood or discriminant analysis on data sets that are not normally distributed.

A second aspect of statistics that should be given some thought is that of sample size. Again, this largely applies to studies involving statistical methods of classification but it has some importance in other areas of image processing. A sample should be representative of the population that it purports to represent, and usually a random or stratified random sampling scheme is adopted. Collecting ground reference data for calibration and validation purposes is time-consuming and costly, especially if there are many land cover classes in the study area, and the temptation is to take sample elements in rectangular clusters. This temptation should be avoided because n sample pixels that are close together do not provide n separate sources of information; the phenomenon of spatial autocorrelation sees to that. So we need a well-distributed set of reference data (Plourde and Congalton, 2003). How many sample elements do we need? This is a question that has occupied more than one mind over several decades. In classical statistics I was taught that statistical theory assumed a large sample size and that the value 30 was generally adopted as the minimum sample size for a univariate test. If we have p dimensions then the sample size should be $30p$ in order to sample each dimension (spectral band) properly. Some researchers have queried this figure (e.g. Van Niel, McVicar and Datt, 2005, who suggest a figure nearer $2p$ or $4p$). Others have suggested that some classification algorithms can work with smaller sample sizes (e.g. Pal and Mather, 2005, 2006). See also Foody, McCullagh and Yates (1995), Foody and Mathur (2004a), Foody *et al.* (2006). The so-called Hughes phenomenon claims that, for a fixed sample size m and for p dimensions, the accuracy of a classifier rises initially and then diminishes as p increases. The point at which this 'curse of dimensionality' comes into play seems to depend on sample size. However, some non-statistical techniques, such as decision trees and support vector machines, are apparently more tolerant to small sample sizes than are methods based on statistics such as the maximum likelihood classifier.

In the preceding paragraph, the number of dimensions (spectral bands) was seen to have an important effect on the outcome of certain classification operations. It is equally true to say that high dimensionality plays a role in those image transforms (Chapter 8) that are based on the covariance matrix. There is little guidance in the literature on the sample size needed for various statistical calculations, but it is clear that the number of parameters to be estimated from the data increases nonlinearly with p. Assume that we are computing the correlation or covariance matrix. If $p = 2$ then we need to estimate three parameters $-r_{11}, r_{22}$ and r_{12}. If we are computing the variance-covariance matrix of hyperspectral data (Chapter 9) then p could be as large as 256, giving 32 896 parameters to be estimated from the sample data. Different computational approaches may have to be adopted for hyperspectral data. Techniques such as PCA (Chapter 6), which uses the variance–covariance matrix, may well need to be reviewed as the many arithmetic computations required for techniques such as PCA may lead to unacceptable rounding errors. For example, some have implemented artificial neural network (ANN, Chapter 8) approaches to the computations of PCA as a result of the very large number of error-prone computations required when p becomes large.

Where a mean vector or covariance matrix is estimated from a sample of reference data (either training data or test data), one must always be aware of the disproportionate effect that outliers in an impure sample have on the estimates. A procedure for down-weighting estimates is described in Section 8.4.1 (in the context of estimating statistical parameters from a sample of data) and in Sections 7.2.1 and 7.2.2, where the moving average filter is compared to the median.

Further reading includes Milman (1999) and Landgrebe (2003). The latter includes useful material on the properties of high-dimensional space associated with hyperspectral data, which is discussed in Chapter 9, while the former is more concerned with the computational aspects of remote sensing data processing. Oliveira and Stewart (2006) and Ralston and Rabinowitz (2000) provide a useful computer science perspective on error in digital computing and on choice of suitable algorithms.

3.5 Summary

Appearances can be misleading, according to folk wisdom. This is nowhere more true than in the case of the manipulation and display of images. In the case of remotely-sensed images it is unlikely that deception would be employed or condoned, though anyone who is unfamiliar with the procedures used to transform image data that is represented by more than 8 bits per pixel onto the 0–255 range may well be misled. The differences between false colour, true colour and pseudocolour images should also be recognized. One pitfall to be avoided by the aware reader is the use of data that has been transformed, perhaps non-linearly, onto an 8-bit scale in subsequent calculations such as band ratios and vegetation indices (Section 6.2.4). A band ratio such as the Normalized Difference Vegetation Index should be computed from the image data themselves, and not from scaled values stored in a display memory.

The topic of data formats is also discussed in this chapter. Extraction of data from a storage medium such as a CD-ROM or DVD is often the first problems faced by a potential user of remotely sensed data.

Again, it is important that the user understands the characteristics of different formats and the properties of the metadata that are provided together with the image data. Differences between levels of processing must always be recognized, as should the properties of 'lossy' and 'lossless' data compression.

The subject of error in digital computing is one that has fascinated me since the mid-1960s when computing as we know it today was in its infancy. Mathematics becomes something else when it is turned into computer algorithms. The work of computer scientist Donald Knuth has always been of interest to me and has led me to a position of scepticism rather than innate trust. Many advanced topics such as PCA and maximum likelihood classification cannot be taught in a satisfactory way without a discussion of the errors involved in the eigenvalue transformation or the inverse matrix problem. Remote sensing software needs to be more rigorously tested – as things stand, there is no reason to lack faith in the various packages that are available but a more open attitude towards computational error needs to be promulgated before results are to be trusted implicitly.

There is an equal amount of interest in statistical topics, such as sampling strategy, sample size, parameter estimation and accuracy assessment. Attention needs to be paid to these topics because an entire analysis may be based on faulty data, for example the presence of outliers in the sample data. One of the aims of this chapter is to ensure that readers are aware of the pitfalls that lie ahead and take due notice of them when selecting computer processing techniques, particularly with high-dimensional data.

Readers interested in computational aspects of remote sensing and image processing might take a look at Plaza and Chang (2008), which contains contributions on topics such as high-performance computing, GRID computing and real-time processing of hyperspectral data (i.e. data in many tens or even hundreds of spectral bands – see Section 9.3). Qu *et al.* (2007) has chapters on data products from various sensors such as MODIS, calibration issues and data formats, specifically HDF. Nachtegael *et al.* (2007) is oriented towards image processing in general but contains a section on remote sensing problems. Ghosh and Pal (2002) is similar.

4 Preprocessing of Remotely-Sensed Data

4.1 Introduction

In their raw form, as received from imaging sensors mounted on satellite platforms, remotely-sensed data generally contain flaws or deficiencies with respect to a particular application. The correction of deficiencies and the removal of flaws present in the data is termed *preprocessing* because, quite logically, such operations are carried out before the data are used for a particular purpose. Despite the fact that some corrections are carried out at the ground receiving station (Section 3.2.3), there is often still a need on the user's part for some further preprocessing. The subject is thus considered here before methods of image display and analysis are examined in later chapters.

It is difficult to decide what should be included under the heading of 'preprocessing', since the definition of what is, or is not, a deficiency in the data depends to a considerable extent on the use to which those data are to be put. If, for instance, a detailed map of the distribution of particular vegetation types or a bathymetric chart is required then the geometrical distortion present in an uncorrected remotely-sensed image will be considered to be a significant deficiency. On the other hand, if the purpose of the study is to establish the presence or absence of a particular class of land use (such as irrigated areas in an arid region) then a visual analysis of a suitably processed false-colour image will suffice and, because the study is concerned with determining the presence or absence of a particular land-use type rather than its precise location, the geometrical distortions in the image will be seen as being of secondary importance. A second example will show the nature of the problem. An attempt to estimate reflectance of a specific target from remotely-sensed data will be hindered, if not completely prevented, by the effects of interactions between the incoming and outgoing electromagnetic radiation and the constituents of the atmosphere. Correction of the imagery for atmospheric effects will, in this instance, be considered to be an essential part of data preprocessing whereas, in some other case (for example discrimination between land-cover types in an area at a particular point in time), the investigator will be interested in relative, rather than absolute, pixel values and thus atmospheric correction would be unnecessary. Measurements of change over time using multitemporal image sets will, in the case of optical imagery, require correction for atmospheric variability, and it will also be necessary to register the images forming the multitemporal sequence to a common geographical coordinate system. In addition, corrections for changes in sensor calibrations will be needed to ensure that like is compared with like.

Because of the difficulty of deciding what should be included under the heading of preprocessing methods, an arbitrary choice has been made. Correction for geometric, radiometric and atmospheric deficiencies, and the removal of data errors or flaws, is covered here despite the fact that not all of these operations will necessarily be applied in all cases. This point should be borne in mind by the reader. It should not be assumed that the list of topics covered in this chapter constitutes a menu to be followed in each and every application. The preprocessing techniques discussed in the following sections should, rather, be seen as being applicable in certain circumstances and in particular cases. The investigator should decide which preprocessing techniques are relevant on the basis of the nature of the information to be extracted from the remotely-sensed data.

The preprocessing techniques described in Section 4.2 are concerned with the removal of data errors and of unwanted or distracting elements of the image. In reality, of course, data errors such as missing scan lines cannot be removed; the data in error are simply replaced with some other data that are felt to be better estimates of the true but unknown values. Similarly, unwanted or distracting elements of the image (such as the banding present on Landsat TM and enhanced thematic mapper plus (ETM+) images, as discussed in Sections 2.3.6 and 4.2.2) can

Computer Processing of Remotely-Sensed Images: An Introduction, Fourth Edition Paul M. Mather and Magaly Koch

only be eliminated or reduced by modifying all the data values in the image. These errors are caused by detector imbalance.

Many actual and potential uses of remotely-sensed data require that these data conform to a particular map projection so that information on image and map can be correlated, for example within a geographical information system (GIS). Two examples will demonstrate the importance of this requirement. In Chapter 8 we see that the classification of a remotely-sensed image is best achieved by establishing the nature of ground cover categories by field work and/or by air-photo and map analysis. In order that the information so derived can be related to the remotely-sensed image, some method of transforming from the scan-line/pixel coordinate reference system of the image to the easting/northing coordinate system of the map is required. Second, if remotely-sensed data are to be used in association with other data within the context of a GIS then the remotely-sensed data and products derived from such data (for example a set of classified images) will need to be expressed with reference to the geographical coordinates to which the rest of the data in the information system conform. In both these cases, there is a need for data preprocessing of a kind known as geometric correction. The same arguments can be put forward if the study involves measurements made on images produced on different dates; if information extracted from the two images is to be correlated then they must be registered, that is, expressed in terms of the same geographical coordinate system. Where an image is geometrically corrected so as to have the coordinate and scale properties of a map, it is said to be *georeferenced*. Geometric correction and registration of images is the topic of Section 4.3.

Atmospheric effects on electromagnetic radiation (due primarily to scattering and absorption) are described in Section 1.2.5. These effects add to or reduce the true ground-leaving radiance, and act differentially across the spectrum. If estimates of radiance or reflectance values are to be successfully recovered from remote measurements then it is necessary to estimate the atmospheric effect and correct for it. Such corrections are particularly important (i) whenever estimates of ground-leaving radiances or reflectance rather than relative values are required, for example in studies of change over time or (ii) where the part of the signal that is of interest is smaller in magnitude than the atmospheric component. For example, the magnitude of the radiance upwelling from oceanic surfaces is generally very low, often being much less than the atmospheric path radiance (the radiance scattered into the field of view of the sensor by the gaseous and particulate components of the atmosphere). If any useful information about variations in radiance upwelling from the ocean surface is to be obtained from a remotely-sensed image then the component of the signal received at the sensor that emanates from the ocean surface must be separated from the larger atmospheric component (Figure 1.23). It is fair to say that no single method of achieving this aim has yet been established, and it is also true that most of the techniques that are in use today and which produce even approximately realistic results tend to be complex in nature. In my experience, simple techniques cannot solve complex problems. The more complex techniques are well beyond the scope of this book and will thus not be considered in any detail. Section 4.4 provides an introductory review of atmospheric correction techniques.

Sections 4.5–4.7 are concerned with the radiometric correction of images. If images taken in the optical and infrared bands at different times (multitemporal images) are to be studied then one of the sources of variation that must be taken into account is differences in the angle of the Sun. A low Sun-angle image gives long shadows, and for this reason might be preferred by geological users because these shadows may bring out subtle variations in elevation. A high Sun angle will generate a different shadow effect. If the reflecting surface is Lambertian (which is, in most cases, a considerable oversimplification) then the magnitude of the radiant flux reaching the sensor will depend on the Sun and view angles. For comparative purposes, therefore, a correction of image pixel values for Sun elevation angle variations is needed. This correction is considered in Section 4.5. The calibration of images to account for degradation of the detectors over time is the topic of Section 4.6. Such corrections are essential if multitemporal images are to be compared, for changes in the sensor calibration factors will obscure real changes on the ground. The effects on recorded radiance levels of terrain slope and orientation are reviewed briefly in Section 4.7.

The material in this chapter is introductory in scope. Research applications will require more elaborate methods of preprocessing. For example, orbital geometry models of geometric correction (Section 4.3.1) may use advanced photogrammetric principles (Konecny, 2003), while the use of the more sophisticated atmospheric correction procedures (Section 4.4) requires a knowledge of higher-level physics. The level of presentation adopted here is intended to provide a basic level of appreciation rather than a full physical understanding. More advanced treatments are provided by Slater (1980), Elachi and van Zyl (2006), Rees (2001) and by various contributors to Asrar (1989).

4.2 Cosmetic Operations

Two topics are discussed in this section. The first is the correction of digital images that contain either partially or entirely missing scan lines. Such defects can be due to errors in the scanning or sampling equipment, in the transmission or recording of image data, or in the reproduction of the media containing the data, such as CD or DVD. Whatever their cause, these missing scan lines are normally seen as horizontal black or white lines on the image, represented by sequences of pixel values such as zero or 255 (in an 8-bit image; Figure 3.4). Their presence intrudes upon the visual examination and interpretation of images and also affects statistical calculations based on image pixel values. Methods to replace missing values with estimates of their true (but unknown) values are reviewed in Section 4.2.1. This is followed by a brief discussion of methods of 'de-striping' imagery produced by electromechanical scanners such as those carried by Landsat (TM and ETM+). As noted in Chapter 2, these scanners collect data for several scan lines simultaneously. The Landsat TM and ETM+ instruments record 16 scan lines for each spectral band on each sweep of the scanning mirror. The radiance values along each of these scan lines are recorded by separate detectors. In a perfect world, each detector would produce the same output if it received the same input. As we know, the world is far from perfect and so, over time, the responses of the detectors making up the set of 16 change at different rates. A systematic pattern is superimposed upon the image, repeating every 16 lines. Techniques to remove this pattern are discussed in Section 4.2.2. Note that they cannot be used with images recorded using solid state (pushbroom) scanners such as the HRV carried by the SPOT satellites because each individual pixel across a scan line is recorded by the corresponding detector in the sensor. Hence, each column of pixels in a SPOT HRV image is recorded by the same detector. With 6000+ columns in an image, the problem of correcting for variations in the detectors is rather more severe than that presented here for electromechanical scanners.

4.2.1 Missing Scan Lines

When missing scan lines occur on an image (Figure 4.1) there is, of course, no means of knowing what values would have been present had the scanner or data recorder been working properly; the missing data have gone for ever. It is, nevertheless, possible to attempt to estimate what those values might be by looking at the image data values in the scan lines above and below the missing values. This approach relies upon a property of spatial data that is called spatial autocorrelation. The word 'auto' means 'self', thus autocorrelation is the relationship between one value in a series and a neighbouring value or values in the same series. Temporal autocorrelation is usually present in a series of hourly readings of barometric pressure, for example. The value at 11.00 tends to be very similar to the value at 10.00 unless the weather conditions are quite abnormal. Spatial autocorrelation is the correlation of values distributed over a two-dimensional or geographical surface. Points that are close in geographical space tend to have similar values on a variable of interest (such as rainfall or height above sea-level). The observation that many natural phenomena exhibit spatial autocorrelation is the basis of the estimation of missing values on a scan line from the adjacent values. This section deals with missing lines or parts of lines; Kuemmerle, Damm and Hostert (2008) give details of a method to detect and correct single-band missing pixels in Landsat TM and ETM+ images.

Figure 4.1 *Illustrating dropped scan lines on a Landsat MSS false colour composite image (bands 7, 5 and 4) of south Wales and north Devon. Original data courtesy of NASA and USGS.*

The simplest method (*Method 1*) for estimating a missing pixel value along a dropped scan line involves its replacement by the value of the corresponding pixel on the immediately preceding scan line. If the missing pixel value is denoted by v_{ij}, meaning the value v of pixel i on scan line j, then the algorithm is simply:

$$v_{ij} = v_{ij-1}$$

Method 1 has the virtue of simplicity. It also ensures that the replacement value is one that exists in the near neighbourhood of pixel (i, j). We will consider an averaging method next, and will see that, where the assumption of positive spatial autocorrelation does not hold, the average of two adjacent pixel values will produce an estimated replacement value that is quite different from either, whereas method 1 produces an estimate that is similar to at least one of its neighbours. Method 1 will need modification whenever the missing line (j) is the first line of an image. In that instance, the value $v_{i,j+1}$ could be used.

Method 2 is slightly more complicated; it requires that the missing value be replaced by the average of the corresponding pixels on the scan lines above and below the defective line, that is:

$$v_{ij} = (v_{ij-1} + v_{ij+1})/2$$

(taking the result to the nearest integer if the data are recorded as integer counts). Where the missing line is the first or last line of the image then Method 1 can be used. As indicated earlier, if nearby pixel values are not highly correlated then the averaging method can produce hybrid pixels that are unlike their neighbours on the scan lines immediately above or below. This is likely to happen only in those cases where the missing line coincides with the position of a boundary such as that between two distinct land-cover types, or between land and water.

Method 3 is the most complex. It relies on the fact that two or more bands of imagery are often available. Thus, Landsat TM produces seven bands, ETM+ produces eight (including the 15 m resolution panchromatic band) and SPOT HRV produces three bands of imagery. If the pixels making up two of these bands are correlated on a pair-by-pair basis then high correlations are generally found for bands in the same region of the spectrum. For instance, the Landsat ETM+ bands 1 and 2 in the blue–green and green wavebands of the visible spectrum are normally highly correlated. The missing pixels in band k might best be estimated by considering contributions from (i) the equivalent pixels in another, highly correlated, band and (ii) neighbouring pixels in the same band, as in the case of the two algorithms already described. If the neighbouring, highly correlated, band is denoted by the subscript r then the algorithm can be represented by

$$v_{i,j,k} = M\left[v_{i,j,r} - (v_{ij+1,r} + v_{ij-1,r})/2\right] + (v_{i,j+1,k} + v_{i,j-1,k})/2$$

The symbol M in this expression is the ratio of the standard deviation of the pixel values in band k and the standard deviation of the pixel values in band r. This algorithm was first described by Bernstein *et al.* (1984) and is examined, together with the two algorithms outlined above, by Fusco and Trevese (1985). The conclusion of the latter authors is that the use of a second correlated band both reduces error and better preserves the geometric structures present in the image. They present some further results and elaborations of the basic algorithm, and readers wishing to go more deeply into the matter are referred to their paper.

The location of missing scan lines might not at first sight seem a topic worthy of serious consideration, for they are usually manifestly obvious when a defective image is examined visually. However, to locate such missing lines interactively using a cursor is a tedious task. The spatial autocorrelation property of images might be used as the basis for formulating a strategy that might allow missing scan lines to be located semi-automatically. If the average of the pixel values along scan line i is computed (with i running from 1 to n, where n is the number of scan lines in the from image) then it might be reasonable to expect that the average of scan line i differs from the average of scan lines $i+1$ and $i-1$ by no more than a value e. The parameter e would be determined by looking at the frequency distribution of the scan line averages over a number of images, or for a representative (and non-defective) part of the image under consideration. Step 1 would then involve locating all those scan lines with average values that deviated by more than e from the average of the preceding scan line. The first scan line of the image could either be omitted or compared with the second scan line. At the end of step 1 we cannot be sure that the unexpectedly deviant behaviour of the scan lines picked out by this comparative method is the result of missing values. Step 2 thus involves a search along each of the scan lines picked out at step 1 for unexpected sequences of values. These unexpected sequences are most likely to be strings of extreme values, either 0 or 255 in 8-bit images. The beginnings and ends of such sequences are marked. At this stage the image can be displayed and a cursor used to mark the start of the suspect sequence. The operator is then able to check that the scan lines or portions of scan lines are indeed defective. Step 3 consists of the application of one of the three methods described earlier, which allow the defective value to be replaced by an estimate of its true but unknown value. Note that isolated aberrant values such as speckle noise on synthetic aperture radar (SAR) images are removed by the use of filters such as the Lee filter or the median filter. These methods are described in Chapter 7.

4.2.2 Destriping Methods

The presence of a systematic horizontal banding pattern is sometimes seen on images produced by electromechanical scanners such as Landsat's TM (Figure 4.2).

This pattern is most apparent when seen against a dark, low-radiance background such as an area of water. The reasons for the presence of this pattern, known as

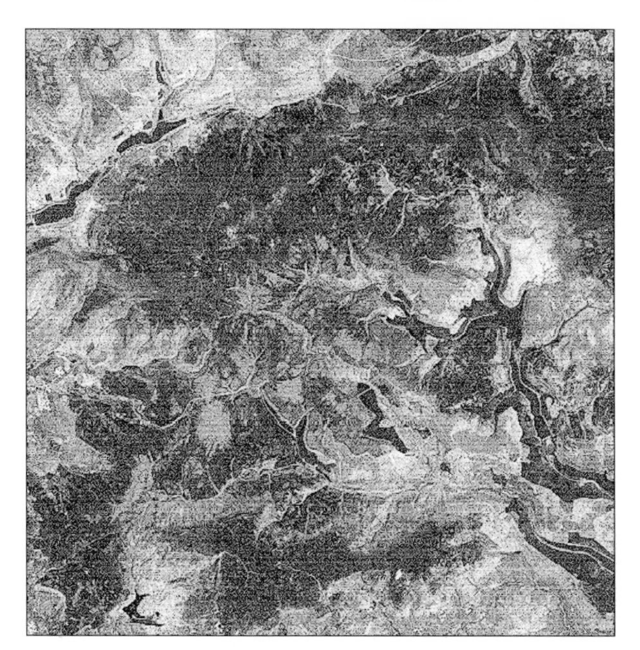

Figure 4.2 *Horizontal banding effects can be seen on this Landsat-4 TM band 1 image of part of the High Peak area of Derbyshire, UK. The banding is due to detector imbalance. As there are 16 detectors per band, the horizontal banding pattern repeats every 16th scan line. The image has been contrast-stretched (Section 5.3) in order to emphasise the banding effect. See Section 4.2.2 for more details. Original data courtesy of NASA and USGS.*

banding, are given in Section 2.3.6.1. It is effectively caused by the imbalance between the detectors that are used by the scanner. This banding can be considered to be a cosmetic defect (like missing scan lines) in that it interferes with the visual appreciation of the patterns and features present on the image. If any statistical analysis of the pixel values is to be undertaken then the problem becomes somewhat more difficult. The pixel values recorded on a CD or DVD are by no means 'raw' data, for they have been subjected to radiometric and geometric correction procedures at the ground receiving station, as described in Section 3.2.3. Hence, there does not seem to be much force in the argument that 'raw data' should not be interfered with. If we take as our starting point the assumption that the image data should be internally consistent (that is areas of equal ground-leaving radiance should be represented by equal pixel values in the image, assuming no other complicating factors) then some kind of correction or compensation procedure would appear to be justified. Two reasons can thus be put forward in favour of applying a 'destriping' correction: (i) the visual appearance and interpretability of the image is thereby improved and (ii) equal pixel values in the image are more likely to represent areas of equal ground-leaving radiance, other things being equal.

Two methods of destriping Landsat imagery are considered in this section. For the sake of simplicity, they are illustrated with reference to Landsat MSS images (which have only six detectors per band) rather than to Landsat TM or ETM+ images, which have 16 detectors per spectral band. Both methods are based upon the shapes of the histograms of pixel values generated by each of the detectors; these histograms are calculated from lines 1, 7, 13, 19, ... (histogram 1), lines 2, 8, 14, 20, ... (histogram 2) and so on until six histograms have been computed (in the case of Landsat MSS) or 16 (in the case of Landsat TM or ETM+).

The first method characterizes the relationship between the scene radiance r_{in} that is received at the detector and the value r_{out} that is output by the sensor system in terms of a linear function. The second method is non-linear in the sense that the relationship between r_{in} and r_{out} is not characterized in terms of a single linear function; a piecewise function made up of small linear segments is used instead. Methods based on low-pass filtering (such as those described by Crippen (1989) and Pan and Chang (1992) are mentioned in Chapter 7.

4.2.2.1 Linear Method

The first method uses a linear expression to model the relationship between the input and output values. The underlying idea is quite simple, though it is based upon the assumption that each of the six detectors 'sees' a similar distribution of all the land-cover categories that are present in the image area. If this assumption is satisfied, and the proportion of pixels representing water, forest, grassland, bare rock, and so on, is approximately the same for each detector, then the histograms generated for a given band from the pixel values produced by the n detectors should be identical (n is the number of detectors used by the scanning instrument, for example 6 for Landsat MSS and 16 for Landsat TM or ETM+). This implies that the means and standard deviations of the data measured by each detector should be the same. Detector imbalance is considered to be the only factor producing differences in means and standard deviations of the subsets of data collected by each detector. To eliminate the striping effects of detector imbalance, the means and standard deviations of the n histograms are equalized, that is, forced to equal a chosen value. Usually the means of the n individual histograms are made to equal the mean of all of the pixels in the image, and the standard deviations of the n individual histograms are similarly forced to be equal to the standard deviation of all of the pixels in the image. Example 4.1 provides a guide to the calculations that are involved.

If the linear method were to be applied on a pixel-by-pixel basis to an image of any great size then it would be

Example 4.1: Destriping – Linear Method

This example demonstrates the way in which the calculations involved in destriping an image using the linear method (Section 4.2.2.1) are carried out. In effect, we partition the image data into k subsets, where k is the number of detectors, and calculate the mean and standard deviation for each subset. Second, the values of the mean and standard deviation of the full data set are required. Given these values, a bias term and an offset term are computed for each subset. The bias term is a multiplier and the offset is a constant to be added. These biases and offsets are applied to the subsets in turn, and the net effect is that all of the subset means are equal to the overall mean and the subset standard deviations are all equal to the overall standard deviation. Given that striping results from detector imbalance (i.e. differences in mean and standard deviation from detector to detector), this equalizing procedure should eliminate striping.

We first need to compute the value of the overall variance V, given by the expression:

$$V = \frac{\sum n_i(\overline{x}_i^2 + v_i)}{\sum n_i} - \overline{X}^2$$

For simplicity, we assume that the number of detectors used in the imaging instrument is 6. The columns in the table below show:

(i) detector number (1–6)
(ii) number of pixels recorded by each detector
(iii) pixel values recorded by each detector
(iv) means of the pixel values for each detector
(v) standard deviations of the pixel values for each detector
(vi) variances of the pixel values for each detector.

(i) Detector number	(ii) Number of pixels	(iii) Pixel values	(iv) Individual means ($\overline{x}_i$)	(v) Standard deviations (s_i)	(vi) Variances ($v_i = s_{2i}$)
1	5	1 3 2 4 6	3.2	1.720	2.96
2	5	3 6 2 3 8	4.4	2.245	5.04
3	5	4 3 4 2 9	4.4	2.417	5.84
4	5	2 4 3 3 7	3.8	1.720	2.96
5	5	0 2 2 2 6	2.4	1.959	3.84
6	5	4 3 3 3 9	4.4	2.332	5.44

(Continues on next page)

inordinately slow. A little thought will show that, for 8-bit images collected by the Landsat TM and ETM+, there are only 256 possible values for each detector. We could build a table consisting of n columns and 256 rows. The input pixel value is the row number and the n corrected values, one per detector, form the data values for that row. The principle is simple. For a given pixel, the row number in the table is the input pixel value while the output (corrected) value is the ith value on that row, where i is the detector number ($1 \leq i \leq n$). The table is known as a lookup table or a direct address table. The use of lookup table methods can be very effective where 8- or 16-bit integer data are being processed, as is the case with most images, because the number of outcomes is limited and is relatively small, so that precalculation of results for all possible cases becomes feasible. More complicated methods using hash tables are needed if the same idea is to be applied to 32-bit integer data, which can represent integers in the range 0 to 2 147 483 647. An example of the use of hash tables in processing remotely-sensed data is given by Mather (1985). Cormen *et al.* (2001) provide a more in-depth study of data structures and algorithms, including hash tables.

4.2.2.2 *Histogram Matching*

The method of de-striping images produced by electromechanical scanners described in Section 4.2.2.1 is based on the assumption that the output from a detector is a linear

We need the value of the overall mean, $\overline{X}$, which appears in the numerator. This is the sum of the individual detector means (column (iv)) divided by the number of detectors (6), which gives $22.6/6 = 3.76\dot{6}$. The numerator of the equation is the sum of $n_i(\overline{x}_i^2 + v_i)$where $\overline{x}_i$ is the mean value of the 5 pixel values for each detector and the n_i (column (ii)) are all equal to 5. The calculation is as follows:

$$[(3.2^2 + 2.96) \times 5] + [(4.4^2 + 5.04) \times 5] \ldots [(4.4^2 + 5.44) \times 5] = 573.$$

The denominator is the sum of the n_i, which equals $5+5+5+5+5+5=30$, so the first term in the equation is equal to $530/40 = 19.1$.

The second term is the square of the overall mean ($\overline{X}^2$), that is, the square of 3.766 or 14.183. Finally, subtract the second term from the first (19.100 – 14.183) to get the value of the overall variance V, which equals 4.192. The overall standard deviation is the square of V, or 2.216. To apply the destriping correction to the image from which the statistics were derived you must calculate the gains (b_i) and offsets (a_i), mentioned above, from:

$$a_i = \frac{S}{s_i}$$

$$b_i = \overline{X} - a_i \overline{x}_i$$

The corrected pixel values r'_{ij} are then found from the relationship $r'_{ij} = a_i r_{ij} + b_i$, where r_{ij} are the uncorrected pixel values. It is important to ensure that the gain and offset computed from the subset of data collected by detector i are applied to the pixels collected by that detector.

The sampling variability of the subset means and standard deviations (which measures their reliability) increases with the size of the subset, so each subset should be reasonably large. An image size of at least 1024 lines and 1024 pixels per line is suggested.

function of the input value according to the expression:

$$r_{\text{out}} = \mathit{offset} + \mathit{gain} \times r_{\text{in}}$$

Horn and Woodham (1979) observe that

> ... it appears that different gains and offsets are appropriate for different scene radiance [r_{in}] ranges. That is, the sensor transfer curves are somewhat non-linear.

In other words, the linear relationship between r_{in} and r_{out} used in Section 4.2.2.1 is an oversimplification. The method described in this section uses the shape of the cumulative frequency histogram of each detector to find an estimate of the non-linear transfer function. The ideal or target transfer function is taken to be defined by the shape of the cumulative frequency histogram of the whole image, which is easily found by carrying out a class-by-class summation of the n individual detector histograms (e.g. 6 for Landsat MSS or 16 for Landsat TM/ETM+). The histogram for detector 1 is computed from the pixel values on scan-lines 1, 7, 13 ... of the image, while the histogram for detector 2 is derived from the pixel values on scan lines 2, 8, 14, ..., and so on. The histograms are expressed in cumulative form (so that class 0 is the number of pixels with a value of 0, class 1 is the number of pixels with values 0 or 1, and so on). Next, each histogram class frequency is divided by the number of pixels counted in that histogram, thus ensuring that the individual histograms and the target histogram are all scaled between 0 and 1.

At this stage, we have n individual cumulative histograms and one target cumulative histogram, where n is the number of detectors. Our aim is to adjust the individual cumulative histograms so that they match the shape of the target cumulative histogram as closely as possible. This is done by adjusting the class numbers of the individual histograms. Thus, class number k in an individual histogram may be equated with class number j in the target histogram. This means that all pixels scanned by the detector corresponding to that individual histogram, and which have the value k, would be replaced in the destriped image by the value j, which is derived from the target histogram. In order to determine the class number in the target histogram to be equated to class number k in the individual histogram, we find the first class in the target histogram for which the cumulative frequency count equals or exceeds the cumulative frequency value of class k in the individual histogram. The class in the target histogram that is found is class y. An example is given in Table 4.1. The frequency value for cell 3 of an individual histogram is 0.57. This value is compared with the target histogram values until the first class with a value greater than or equal to 0.57 is found. This is class 4 of the target histogram. Class 3 of the detector histogram is thus equated to class 4 of the target histogram, and all pixel values of 3 scanned by that detector are replaced with the

Table 4.1 *Example of histogram matching for de-striping Landsat MSS and TM images.*

Input pixel value	Target histogram value	Detector histogram value	Output pixel value
0	0.09	0.08	0
1	0.18	0.11	1
2	0.33	0.18	2
3	0.56	0.57	4
4	0.60	0.66	4
5	0.76	0.78	6
6	0.95	0.95	6
7	1.00	1.00	7

The target histogram is the cumulative histogram of the entire image or subimage. The detector histogram is the cumulative histogram using values of pixels scanned by one of Landsat MSS's six detectors. The output pixel value to replace a given input value is found by comparison of the two histograms. For example, the detector histogram for input pixel value 3 is 0.57. The first value in the target histogram to equal or exceed 0.57 is that in row four. Hence, the pixel values in the uncorrected image that are generated by this detector are replaced by the value 4.

value 4. The procedure is applied separately to all 256 values for each of the 6 (or 16) detectors. The result is generally a reduction in the banding effect, though much depends on the nature of the image. Wegener (1990) gives a critical review of the Horn and Woodham procedure, and presents a modified form of the algorithm.

The lookup table procedure described at the end of Section 4.2.2.1 can be used to make the application of this method more efficient. Thus, for each detector histogram, a table can be constructed so that the output value corresponding to a given input can be easily read. The input value is the pixel value in the image being corrected while the output value is its equivalent in the destriped image.

4.2.2.3 Other Destriping Methods

The procedures discussed in Sections 4.2.2.1 and 4.2.2.2 operate directly on the image data, which has spatial coordinates of (row, column). Hence, these procedures are said to operate in the 'spatial domain'. A number of methods of transforming an image data set from the spatial domain representation to an alternative frequency domain representation are described in Chapter 6. In particular, the Fourier transform has been widely used to determine the existence of periodicities a data series such as may be caused, for example by a recurring 6- or 16-line horizontal pattern. More recently, the wavelet transform has been introduced as an image transform tool; it is also considered in Chapter 6. Gadallah, Csillag, and Smith (2000) describe a new method based on moment matching. Tsai and Chen (2008) provide a recent review.

4.3 Geometric Correction and Registration

Remotely-sensed images are not maps. Frequently, however, information extracted from remotely-sensed images is integrated with map data in a GIS or presented to consumers in a map-like form (for example gridded 'weather pictures' on TV or in a newspaper). If images from different sources are to be integrated (for example multispectral data from Landsat ETM+ and SAR data from ERS-1 and -2, Radarsat or ASAR) or if pairs of interferometric SAR images are to be used to develop digital elevation models (DEMs) (Section 9.2) then the images from these different sources must be expressed in terms of a common coordinate system. The transformation of a remotely-sensed image so that it has the scale and projection properties of a given map projection is called *geometric correction*. A related technique, called *registration*, is the fitting of the coordinate system of one image to that of a second image of the same area. Accurate image registration is needed if a time sequence of images is used to detect changes in, for example the land cover of an area. The terminology of geometric correction and registration is confusing, and terms are used without being properly defined. In this book we use the phrase geometric correction as a generic term covering all techniques, however approximate, of converting the data for a specified image band from row/column to latitude/longitude (lat/long) format. *Rectification* means the equalization of one image coordinate system to another. For example, a multitemporal series of images could be rectified to the first image in the sequence without any consideration of latitude and longitude, north orientation or reference ellipsoid. The term geometric correction can include *geocorrection*, *geocoding*, *georeferencing* and *orthorectification*. Geocorrection is a shorthand form of 'geometric correction' and has no special attributes. It is an unnecessary term. Georeferencing usually implies that the four corners of the image have geographical coordinates but the individual pixels are not given a lat/long pair. No specific account is taken of ellipsoids, or projections. This is the simplest form of geometric correction and is described below (Sections 4.3.1 and 4.3.2). Geocoding means that the image has all the properties of a map (see definition below) whereas orthorectification means that the terrain elevation has been included in the correction process, implying that all pixels are viewed as if from above. This is the most accurate form of geometric correction and, as the (x, y, z) Earth-centric coordinates are now readily obtainable from global positioning system (GPS) and other locational devices both for aircraft and satellites, and as higher-resolution DEM become more readily available (ASTER global DEMs (GDEM) are available at no cost at 30 m resolution; http://asterweb.jpl.nasa.gov/gdem.asp)

so more remotely-sensed imagery (particularly those products and derivations used in a GIS) that has been orthorectified will become the norm.

A map is defined as:

> ...a graphic representation on a plane surface of the Earth's surface or part of it, showing its geographical features. These are positioned according to pre-established geodetic control, grids, projections and scales (Steigler, 1978).

A map projection is a device for the representation of a curved surface (that of the Earth) on a flat sheet of paper (the map sheet). Many different map projections are in common use (see Snyder, 1982; Fenna, 2006; Frei, Graf and Meier, 1993; Grafarend and Krumm, 2006). Each projection represents an effort to preserve some property of the mapped area, such as uniform representation of areas or shapes, or preservation of correct bearings. Only one such property can be correctly represented, though several projections attempt to compromise by minimizing distortion in two or more map characteristics. The UK Ordnance Survey uses a Transverse Mercator projection. A regular grid, graduated in metres and with its origin to the south-west of the British Isles, is superimposed on the map sheet since lines of latitude and longitude plot as complex curves on the Transverse Mercator projection.

Geometric correction of remotely-sensed images is required when the remotely-sensed image, or a product derived from the remotely-sensed image such as a vegetation index image (Chapter 6) or a classified image (Chapter 8), is to be used in one of the following circumstances (Kardoulas, Bird and Lawan 1996):

1. to transform an image to match a map projection
2. to locate points of interest on map and image
3. to bring adjacent images into registration
4. to overlay temporal sequences of images of the same area, perhaps acquired by different sensors and
5. to overlay images and maps within a GIS.

The advent of high-resolution images obtained from instruments carried by satellites such as QuickBird, IKONOS, SPOT-5 and Resourcesat has brought the topic of geometric correction of remotely-sensed images much closer to the field of photogrammetry. For many years photogrammetrists have used accurate camera models to perform analytical corrections on aerial photographs (Konecny, 2003; Wolf and DeWitt, 2000). The use of stellar navigation and GPS on board satellites has meant that orbital parameters required for an analytical solution are now more readily available. Finally, considerable research effort has been directed towards providing a solution to the problem of terrain or relief correction. A brief account of these developments is given in Section 4.7. It should be made clear that the methods of registration and correction described in this chapter will work satisfactorily only in areas of low relative relief. In hilly or mountainous areas the effects of topography can result in the displacement of pixels from their true or relative geographical position.

If a DEM is available at a suitable scale, then the photogrammetric procedure of orthorectification can produce images conforming to map accuracy standards (see Figure 4.10 for an illustration of locational error caused by the topographic effect). An accessible source is Dowman and Dare (1999). Leprince *et al.* (2007) give a more mathematical overview. Often, orthorectification procedures are applied to medium or high resolution images (Aguilar *et al.*, 2008; Robertson, 2003; Wang and Ellis, 2005). Tucker, Grant and Dykstra (2004) and Tatem, Nayar and Hay (2006) provide details of NASA's Landsat global orthorectified data sets, while Masek *et al.* (2006) illustrate the use of these datasets in land surface cover monitoring. Schläpfer and Richter (2002) describe the orthorectification of airborne hyperspectral images, while Liu and Jezek (2004) show how SAR imagery of the Antarctic continent is generated. Wegmüller *et al.* (2003) also use SAR imagery from ENVISAT's ASAR to generate orthorectified imagery.

The sources of geometric error in moderate spatial resolution imagery with a narrow field of view, such as the imagery produced by Landsat ETM+ and SPOT HRV are summarized in Section 2.3. The main categories are: (i) instrument error, (ii) panoramic distortion, (iii) Earth rotation and (iv) platform instability (Bannari *et al.*, 1995a). Instrument errors include distortions in the optical system, non-linearity of the scanning mechanism and non-uniform sampling rates. Panoramic distortion is a function of the angular field of view of the sensor and affects instruments with a wide angular field of view (such as the AVHRR and VIIRS) more than those with a narrow field of view, such as the Landsat TM and ETM+ and the SPOT HRV. Earth rotation velocity varies with latitude. The effect of Earth rotation is to skew the image. Consider the Landsat satellite as it moves southwards above the Earth's surface. At time t, its ETM+ sensor scans image lines 1–16. At time $t + 1$, lines 17–32 are scanned. But the Earth has moved eastwards during the period between time t and time $t + 1$ therefore the start of scan lines 17–32 is slightly further west than the start of scan lines 1–6. Similarly, the start of scan lines 33–48 is slightly further west than the start of scan lines 17–32. The effect is shown in Figure 4.3. Platform instabilities include variations in altitude and attitude. All four sources of error contribute unequally to the overall geometric distortion present in an image. In this section, we deal with the geometric correction of medium-resolution digital images such as

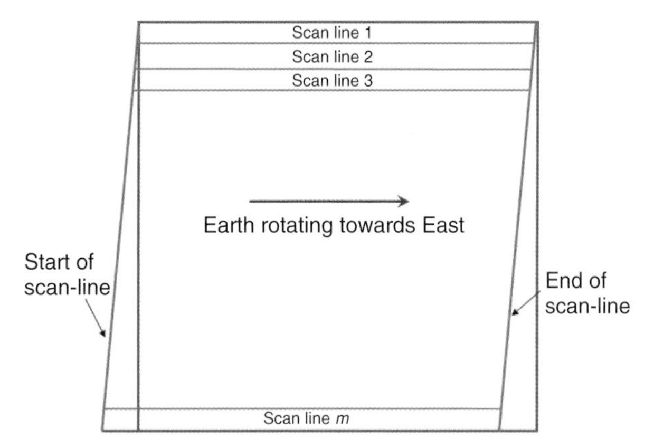

Figure 4.3 *Effects of Earth rotation on the geometry of a line-scanned image. Due to the Earth's eastwards rotation, the start of each swath (of 16 scan lines, in the case of the Landsat-7* ETM+*) is displaced slightly westwards. At the Equator the line joining the first pixel on each scan line (the left margin of the image) is oriented at an angle that equals the inclination angle i of the satellite's orbit. At a latitude of* $(90 - i)^\circ$ *the same line is parallel to the line of latitude* $(90 - i)^\circ$. *Thus, the image orientation angle increases pole-wards. See Section 4.3.1 for further details.*

those acquired by the Landsat TM and SPOT HRV instruments. Correction of wide-angle images derived from the NOAA AVHRR is described by Brush (1985), Crawford, Brooks and Brush (1996), Moreno and Melia (1993) and Tozawa (1983). Geocoding of SAR images is covered by Dowman (1992), Dowman, Laycock and Whalley (1993), Johnsen, Lauknes and Guneriussen (1995) and Schreier (1993a). The use of digital elevation data to correct images for the geometric distortion produced by relief variations is considered by Blaser and Caloz (1991), Itten and Meyer (1993), Kohl and Hill (1988), Palü and Pons (1995), Toutin (1995) and Wong, Orth and Friedmann (1981). Other useful references are Fogel and Tinney (1996), Kropatsch and Strobl (1990), Kwok, Curlander and Pang (1987), Novak (1992), Westin (1990) and Wolberg (1990). Williams (1995) provides an excellent treatment of many aspects covered in this section, including geocoding of SAR and AVHRR imagery. Georeferencing SAR and optical imagery is treated by Hong and Schowengerdt (2005). The text by Wolf and DeWitt (2000) is a good source of information on georeferencing of digital air photographs. Toutin (2004) gives a comprehensive review of the geometric processing of remote sensing images.

The process of geometric correction can be considered to include: (i) the determination of a relationship between the coordinate system of map and image (or image and image in the case of registration); (ii) the establishment of a set of points defining pixel centres in the corrected image that, when considered as a rectangular grid, define an image with the desired cartographic properties; and (iii) the estimation of pixel values to be associated with those points. The relationship between the two coordinate systems (map and image) could be defined if the orbital geometry of the satellite platform were known to a sufficient degree of accuracy. Where orbital parameters are known, methods based upon orbital geometry give high accuracy. The DORIS system, used with the SPOT satellite, is described in Section 2.3.7. Otherwise, orbital models are useful only where the desired accuracy is not high, or where suitable maps of the area covered by the image are not available. A simple method based on nominal orbital parameters is described in Section 4.3.1, while the map-based method is covered in Section 4.3.2. The extraction of the locations of the pixel centre points for the corrected image and the estimation of pixel values to be associated with these output points is considered in Section 4.3.3.

4.3.1 Orbital Geometry Model

Orbital geometry methods are based on knowledge of the characteristics of the orbit of the satellite platform. Bannari *et al.* (1995a) describe two procedures based on the photogrammetric equations of collinearity. These equations describe the properties of the satellite orbit and the viewing geometry, and relate the image coordinate system to the geographical coordinate system. They require knowledge of the geographical coordinates of a number of points on the image. Such points are known as *ground control points* or GCPs.

A simple method of correcting the coordinate system of remotely-sensed images using approximate orbit parameters, described by Landgrebe *et al.* (1974), is used here to illustrate the principles involved. It is not recommended as an operational method, because it is based upon nominal rather than actual orbital parameters, which implies that the accuracy of the geometrically corrected image produced by this technique is not high. Landgrebe *et al.* (1975) suggest that the magnitude of the error is of the order of 1–2%, meaning that if the corrected image is overlaid on a map and both are aligned with reference to a well-defined point, then the error in measured coordinate positions of other points will be 1–2%. Note that the image coordinate system has its origin in the top left corner, at cell (1, 1). The x-axis runs horizontally and increases in value to the right, with the y-axis running vertically, with values increasing downwards. Thus, the x-axis gives the pixel position across the scan line. The y-axis gives the scan line number.

4.3.1.1 Aspect Ratio

Some sensors, such as the Landsat MSS, produce images with pixels that are not square. The Landsat MSS scan lines are nominally 79 m apart, whereas the pixels along each scan line are spaced at a distance of 56 m. Since the instantaneous field of view of the MSS is 79 m there is oversampling in the across-scan direction. As we generally require square rather than rectangular pixels, we can choose 79 m^2 pixels or 56 m^2 pixels to overcome the problem of unequal scale in the x and y directions. Because the across-scan direction is over-sampled, it is more reasonable to choose 79 m^2 pixels. The aspect ratio (the ratio of the $x:y$ dimensions) is 56 : 79 or 1 : 1.41. The first transformation matrix, **M1**, which corrects the image to a 1 : 1 aspect ratio, is therefore

$$\mathbf{M1} = \begin{bmatrix} 0.00 & 1.41 \\ 1.00 & 0.00 \end{bmatrix}$$

This transformation matrix is not required if the image to be geometrically transformed already has square pixels.

4.3.1.2 Skew Correction

Landsat TM and ETM+ images are skewed with respect to the north–south axis of the Earth. Landsat-1 to -3 had an orbital inclination of 99.09° whereas Landsats-4–5 and -7 have an inclination of 98.2°. The satellite heading (the direction of the forward motion of the satellite) at the Equator is therefore 9.09° and 8.2° respectively, increasing with latitude (Figure 2.1). The skew angle θ at latitude L is given (in degrees) by

$$\theta = 90 - \cos^{-1}\frac{\sin(\theta_E)}{\cos(L)}$$

where θ_E is the satellite heading at the Equator and the expression $\cos^{-1}(arg)$ means: the angle whose cosine is arg, that is, the inverse cosine of x. Given the value of θ the coordinate system of the image can be rotated through $\theta°$ anticlockwise so that the scan-lines of the corrected image are oriented in an east–west direction using the transformation matrix **M2**:

$$\mathbf{M2} = \begin{bmatrix} \cos(\theta) & \sin(\theta) \\ -\sin(\theta) & \cos(\theta) \end{bmatrix}$$

The value of L that is normally used in the determination of θ is the centre latitude of the image being corrected. Since the latitude of the satellite is varying continuously this value will be only an approximation.

For latitude $L = 51°$ and using the heading of Landsat-4 and -5 at the equator, θ_E, the value of θ is given by:

$$\theta = 90.00° - \cos^{-1}\frac{\sin(8.20°)}{\cos(51.00°)} = 90.00° - \cos^{-1}\frac{0.14}{0.63}$$

$$= 90.00° - 76.90° = 13.10°$$

and the elements of matrix **M2** are:

$$\mathbf{M2} = \begin{bmatrix} 0.9740 & 0.2286 \\ -0.2286 & 0.9740 \end{bmatrix}$$

4.3.1.3 Earth Rotation Correction

As the satellite moves southwards over the illuminated hemisphere of the Earth, the Earth rotates beneath it in an easterly direction with a surface velocity proportional to the latitude of the nadir or subsatellite point. To compute the displacement of the last line in the image relative to the first line we need to determine (i) the time taken for the satellite sensor to scan the image, and (ii) the eastwards angular velocity of the Earth. The distance travelled by the Earth's surface can then be obtained by multiplying time by velocity. The time taken for the satellite sensor to scan the image can be found if the distance travelled by the satellite and the satellite's velocity are known. Both distance and velocity are expressed in terms of angular measure (i.e. in radians; 1 radian equals approximately 57°; Figure 1.6a). If A is a point on the Earth's surface corresponding to the centre pixel of the first scan line in the image, and if B is the corresponding point for the last scan line in the image, then the curve AB (the line on the Earth's surface joining points A and B) is an arc of a circle centred at the Earth's centre, O. The angle AOB can be calculated because we know that the Earth's equatorial radius (OA or OB) is approximately 6378 km, and an angle (in radian measure) is equal to arc length divided by radius (Figure 1.6a). The arc length AB is the distance between the centre pixels in the first and last scan lines of the image. For Landsat MSS and TM images, AB is 185 km, hence angle AOB (representing the angular distance moved by the Landsat satellite during the capture of one image) equals 185/6378 or 0.029 radians.

The orbital period (the time required for one full revolution) for Landsats 1–3 is 103.267 min (99 min for the later Landsats), so the satellite's angular velocity ω_0 is $2\pi/(103.267 \times 60)$ radians per second, or 0.001014 rad s^{-1}. The problem is now to find the time required for a satellite travelling at this angular velocity to traverse through an angle of 0.029 radians (the angular distance between the first and last scan line, see preceding paragraph). The answer is found by dividing the angular distance to be moved by the angular velocity, and the result of this operation is 0.029/0.001014 = 28.6 s.

Now the question becomes: how far will point B (the centre of the last scan line) move eastwards during the 28.6 s that elapses between the scanning of the first and last scan lines of the MSS or TM image? The answer again depends on latitude. For simplicity, we will take the latitude (L) of the centre of the image. The Earth's surface velocity at latitude L is $V_E(L)$ which is defined as:

$$V_E(L) = R\cos(L)\omega_E$$

R is the Earth's radius, approximately 6378 km, and ω_E (omega-e) is the Earth's angular velocity. Since the Earth rotates once in 23 h, 56 min and 4 s (that is 86 164 s), then its angular velocity is simply $(2\pi/86\,164)\,\text{rad s}^{-1}$, or $0.7292 \times 10^{-4}\,\text{rad s}^{-1}$. If L is 51° then $V_E(L)$ equals $6378 \times 10^{-3} \times 0.6293 \times 0.72921 \times 10^{-4}\,\text{m s}^{-1}$, that is $292.7\,\text{m s}^{-1}$. Now that we know (i) that the time taken to scan an entire Landsat TM/ETM+ or MSS image is 28.6 s and (ii) the Earth's surface velocity, then the calculation of the eastward displacement of the last scan line in the image can be obtained. At 51°N the surface velocity is $292.7\,\text{m s}^{-1}$ so the distance travelled eastwards is $292.7 \times 28.6 = 8371$ m.

These calculations assume that the line AB joining the centres of the first and last scan lines is oriented along a line of longitude whereas, in fact, the Landsat satellites (like SPOT, Terra, Aqua and all Sun-synchronous, polar orbiting platforms) have an orbit that is skewed relative to lines of longitude, as noted in the calculation of matrix **M2** above. The skew angle θ for 51° latitude is 14.54° (see above) so the actual eastwards displacement is $8371 \times \cos(14.54°) = 8103$m. These computations are summarized by the term a_{sk}:

$$a_{sk} = \frac{\omega_E \cos(L)}{\omega_O \cos(\theta)} = 0.0719\frac{\cos(L)}{\cos(\theta)}$$

where ω_E is the Earth's angular velocity, ω_O is the satellite's angular velocity and θ and L are defined above. The transformation matrix **M3** is:

$$\mathbf{M3} = \begin{bmatrix} 1 & 0 \\ a_{sk} & 1 \end{bmatrix}$$

At 51° latitude, **M3** is:

$$\mathbf{M3} = \begin{bmatrix} 0 & 1 \\ 0.04647 & 1 \end{bmatrix}$$

Note that 'fill pixels' are added to the start of each scan-line of a Landsat MSS or TM image by some ground stations to compensate for the Earth rotation effect. If this correction is thought to be sufficient then transformation **M3** can be omitted. Alternatively, since the number of fill pixels is generally given in the header/trailer data associated with each scan-line, these fill pixels can be stripped off and the correction **M3** applied.

The three transformation matrices **M1**, **M2** and **M3** given in Sections 4.3.1.1–4.3.1.3 above are not applied separately. Instead, a composite transformation matrix, **M**, is obtained by multiplying the three separate transformation matrices:

$$\mathbf{M} = \mathbf{M1}\,\mathbf{M2}\,\mathbf{M3}$$

The corrected image coordinate system is related to the raw image coordinate system by

$$\mathbf{x}' = \mathbf{M}\mathbf{x}$$

where $\mathbf{x}'(= x_1', x_2')$ is the vector holding the pixel and line (x and y) coordinates of the corrected pixel and $\mathbf{x}$ $(= x_1, x_2)$ is the original (pixel, line) coordinate. Remember that the origin of the image coordinate system is the top left corner of the image. See Example 4.2 for more details.

4.3.2 Transformation Based on Ground Control Points

The orbital geometry model discussed in Section 4.3.1 is based on nominal orbital parameters. It takes into account only selected factors that cause geometric distortion. Variations in the altitude or attitude of the platform are not considered simply because the information needed to correct for these variations is not generally available. Some satellites such as SPOT-5 now carry instruments that provide precise orbital data (Section 2.3.7), and more complex analytical models than the one described in Section 4.3.1 are used to generate a geometrically corrected image.

An alternative method is to look at the problem from the opposite point of view and, rather than attempt to construct a physical model that define the sources of error and the direction and magnitude of their effects, use an empirical method which compares differences between the positions of common points that can be identified both on the image and on a map of a suitable scale for the same area. For a Landsat ETM+ image, for example a map scale of at least 1 : 25 000 is needed, since the minimum measurable line width on a map is considered to be 1 mm, which translates to 25 m (approximately the size of one Landsat TM/ETM+ pixel) at a map scale of 1 : 25 000. From the differences between the distribution of the common set of points on the image and the distribution of these points on the map, the nature of the distortions present in the image can be estimated, and an empirical transformation to relate the image and map coordinate systems can be computed and applied. This empirical function should be calibrated (using GCPs), applied to the image, and then validated (using a separate test set of GCPs).

The aim of the procedures described in this section is to produce a method of converting map coordinates to image coordinates, and vice versa. Two pieces of information are required. The first is the map coordinates of the image corners. Once the image is outlined on the map, the map coordinates of the pixel centres (at a suitable scale) can be found (Figure 4.4).

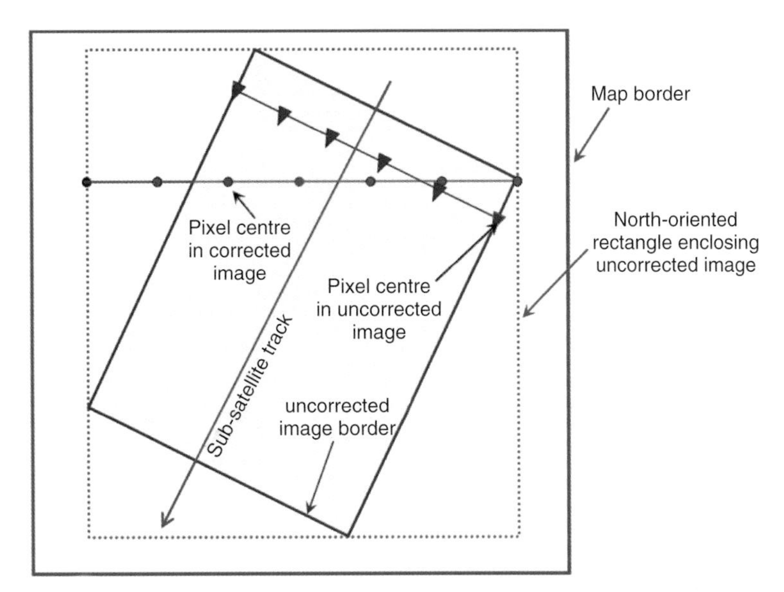

Figure 4.4 *The area of the corrected image is shown by the rectangle that encloses the oblique uncorrected image. The positions of the corners of this enclosing rectangle (in map easting and northing coordinates) can be calculated from the (row, column) image coordinates of the corners of the uncorrected image using a least-squares transformation. Once the corners of the corrected image area are known in terms of map coordinates, the locations of the pixel centres in the corrected image (also in map coordinates) can be determined using simple arithmetic, noting that easting and northing map coordinates are expressed in kilometres. Finally, the pixel centre positions are individually converted to image (row, column) coordinates in order to find the image pixel value to be associated with the pixel position in the corrected image. Not all pixels in the corrected image lie within the area of the uncorrected image. Such pixels receive a zero value.*

The map coordinates of the image corners are found by determining an image-to-map coordinate transformation. The map coordinates of the required pixel centres are converted to image coordinates by a map-to-image coordinate transformation. Both transformations are explained in this section. The final stage, that of associating pixel values with calculated (map) pixel positions, is discussed in the next section under the heading of resampling.

The method relies on the availability of an accurate map of the area at a suitable scale. In some parts of the world maps of sufficient accuracy are not available. It is thus paradoxical that accurate 'image maps' can only be produced for areas for which conventional maps are available using this method unless field surveying techniques, including GPS, are used. The coordinates of selected points, the GCPs, are measured on map and image. GCPs are well defined and easily recognizable features that can be located accurately both on a map and on the corresponding image. They can be located on the ground by the use of GPS rather than by map measurement. Clavet, Lasserre and Pouliot (1993) and Kardoulas, Bird and Lawan (1996) and Toutin (2004) provide details of the use of GPS in locating GCPs for geometric correction. Cook and Pinder (1996) compare the transformation accuracy resulting from the use of control points derived from 1 : 24 000 US Geological Survey maps and from the use of GPS. They summarize the problems involved in the accurate measurement of control points from maps, and conclude that differential GPS provides substantially better results than map digitizing. It is, however, more costly both in terms of equipment and travel time as each control point must be visited and its coordinates measured. Figure 4.5 shows the area around London Heathrow airport. Three 'good' GCP lie within the white circles, which are all motorway junctions. They are good because their position does not change over time (at least, not over the time-scales that we are considering). Bridges over the River Thames would also provide a good choice for locations of GCP. However, it may be less wise to use a point on a reservoir shoreline as a GCP or as a test point, for the water level of the reservoir, and therefore its surface area and size, will vary according to weather conditions. Meanders along rivers also are a poor choice, especially if the map is not up to date.

The symbols (x_i, y_i) refer to the map coordinates of the ith GCP, and the symbols (c_i, r_i) refer to the image column and row image coordinates of the same point. The map coordinates can be expressed in eastings (x) and northings (y) from an arbitrary origin, or in degrees and decimal degrees of longitude and latitude. The image coordinates are expressed in terms of column (c; pixel position along the scan-line) and row (r; scan line number, where scan line 1 is the first scan line of the image (Figure 4.3)). We seek a method that will allow us to convert from (x, y) to (c, r) coordinates and vice versa. To achieve this aim, the method of least squares is used.

Example 4.2: Orbital Geometry Model

This example illustrates the use of the orbital geometry model (Section 4.3.1) in the process of geometric correction of an image. Assume that we have a square image of size 1024 lines and 1024 pixels per line. This image is shown as a square (ABCD) in Example 4.2 Figure 1 (solid line). It is assumed that the image pixels are square so that the transformation matrix $\mathbf{M}_1$ is not required. The Landsat-4 and -5 satellites have an orbital inclination angle of 8.2° so, at a latitude of 51°, angle θ is equal to 13.10°, and the transformation matrices $\mathbf{M}_2$ and $\mathbf{M}_3$ are given by:

$$M_2 = \begin{bmatrix} 0.9740 & 0.2286 \\ -0.2286 & 0.9740 \end{bmatrix}$$

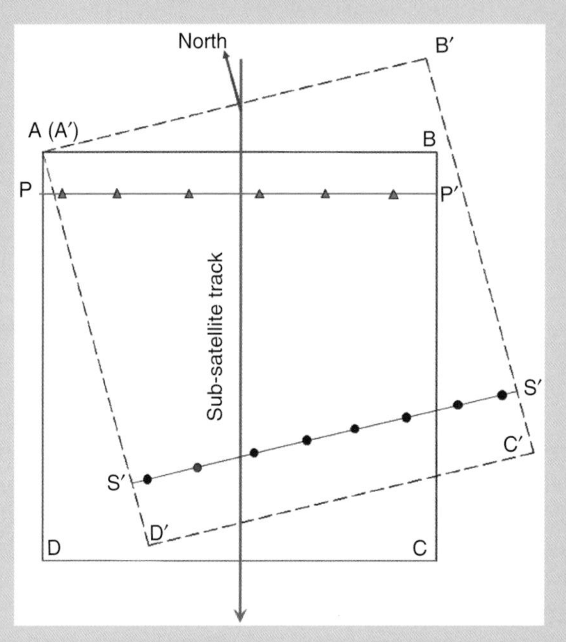

Example 4.2 Figure 1. *An uncorrected (raw) square image is outlined by the solid line joining the points ABCD. After geometric transformation, the map coordinates of the image corners become A′B′C′D′(outlined by dashed line). The subsatellite track is the ground trace of the platform carrying the scanner that collected the raw image. One scan line of the raw image is shown (line PP′) with pixel centres indicated by triangles. The geometrically corrected image (A′B′C′D′) has its columns oriented towards north, with rows running east–west. One scan line in the corrected image is shown (line SS′) with pixel centres indicated by circles. The filled circles show pixel centres that lie within the area of the known, raw image. Note that the angle θ between the subsatellite track and north is equal to 13.0992°.*

(Continues on next page)

To illustrate the method, let us assume that a sample of s_i and t_i values is available, where s and t are any two variables of interest, such as the row number of an image pixel (s) and the easting coordinate of the same point on the corresponding map (t). Furthermore, assume that values of t_i are found relatively easily whereas the values of s_i are difficult to acquire, so it would be advantageous to be able to estimate the value of s given the corresponding value of t. For this reason, we can call s the predicted variable and t the predictor variable. The method of least squares allows the estimation of and value of s (s_i) given the corresponding value of $t(t_i)$ using a function of the form:

$$s_i = a_0 + a_1 t_i + e_i$$

Because only a single predictor variable, t, is used the expression is known as the *univariate* least-squares equation. It allows the difficult-to-measure s_i to be estimated from the value of the corresponding easier-to-measure t_i. The estimated value of s_i is written $\hat{s}$. The terms a_0 and a_1 are computed from a sample of values of s and t using the method of least squares. The criterion used in the least squares procedure is that and

$$M_3 = \begin{bmatrix} 1.00000 & 0.04649 \\ 0.00000 & 1.00000 \end{bmatrix}$$

so that the final transformation matrix **M** is equal to

$$M = \begin{bmatrix} 0.9740 & 0.2719 \\ -0.2266 & 0.9634 \end{bmatrix}$$

The coordinates of the corners of the original image and their transformed equivalents are: [A (0, 0), A′ (0, 0)], [B (1024, 0), B′(997, −232)], [C (1024, 1024), C′(1276, 754)] and [D (0, 1024), D′(278, 986)]. The second stage of the geometric correction operation is to find values for 'new' pixels located at 30 m intervals along the scan lines of the transformed image, that is lines that are 30 m apart and parallel to line A′B′ (or D′E′) in Figure 1. This procedure is termed 'resampling' and is considered in Section 4.3.3. Note that the values for the new pixels can only be obtained in the area where ABCD and A′B′C′D′(Example 4.2 Figure 1) overlap. In practical applications, we would generate a geometrically corrected image of a size sufficient to enclose the whole of the raw image, as shown in Example 4.2 Figure 2.

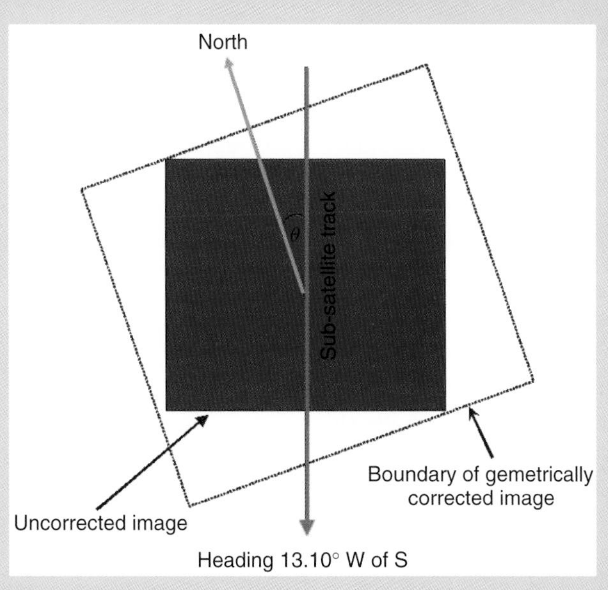

Example 4.2 Figure 2. *In practical applications of geometric correction, the corrected image area is delimited by the rectangle that encloses the uncorrected image. Pixel values within the boundary of the geometrically corrected image but lying outside the area of the uncorrected image are set to zero. The angle θ measures the anticlockwise rotation of 13.10° that is required at latitude 51° to produce a north-oriented image.*

the sum of the squared differences between the true (and usually unavailable) values s_i and the values $\hat{s}_i$ estimated from the univariate least-squares equation (above) is the smallest for all possible value of $\hat{s}_i$. In other words, the coefficients a_0 and a_1 could, in principle, take on any values, such as $a_0 = 8.999$ and $a_1 = -82.192$. However, of all these possible values, those derived using the least squares principle will minimize the criterion $ESS = \sum (s_i - \hat{s}_i)^2$. The term ESS is called the 'explained sum of squares' in least-squares regression analysis. The differences between s and $\hat{s}$ are called 'residuals', and they are represented by the term e_i. Thus, we could omit e_i from the equation and express $\hat{s}$ directly as:

$$\hat{s}_i = a_0 + a_1 t_i$$

The number of predictor variables can be greater than 1. For example, the *bivariate* least squares regression equation would be used if variable s were to be predicted from variables t_1 and t_2. This equation has the form:

$$\hat{s}_i = a_0 + a_1 t_{1i} + a_2 t_{2i}$$

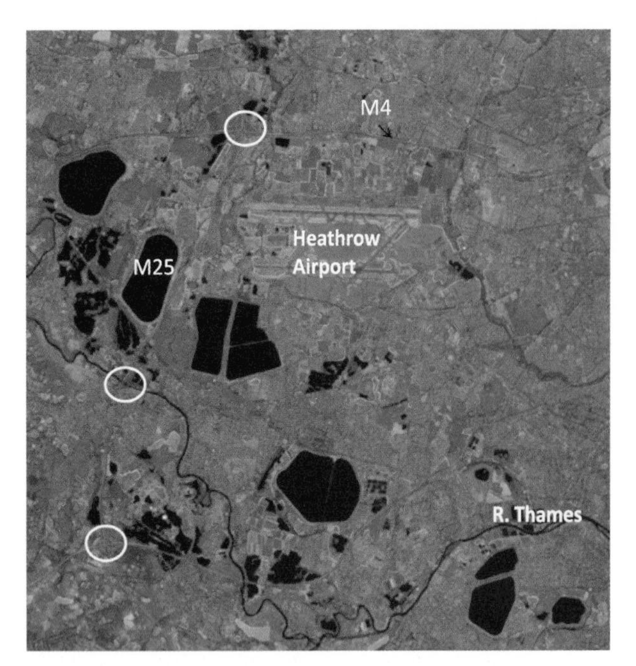

Figure 4.5 Extract of Landsat-5 TM image of the area around Heathrow Airport. The north and south runways are visible, as is the M4 and the M25 (labelled in black). The white circles enclose 'good' ground control points, where the M25 London Orbital Road crosses the M4, the M3 and the River Thames. Poor control points would be located around the edges of the reservoirs (dark blue) because the level, and therefore the spatial extent, of the reservoirs varies according to the weather conditions. Landsat data courtesy NASA/USGS.

where s_i is the ith observation made on variable s, t_{1i} is the ith observation made on variable t_1 and t_{2i} is the ith observation made on variable t_2.

The bivariate linear least squares function is used to find the least squares coefficients for the following four expressions:

1. map x coordinate as a function of image c and r coordinates ($x = f(c, r)$)
2. map y coordinate as a function of image c and r coordinates ($y = f(c, r)$),
3. image c coordinate as a function of map x and y coordinates ($c = f(x, y)$) and
4. image r coordinate as a function of map x and y coordinates ($r = f(x, y)$).

If the coefficients of each of these regression functions are known, it is possible to transform from map (x, y) to image (c, r) coordinates or from image (c, r) to map (x, y) coordinates.

The following bivariate least squares equation relates the map x coordinate to the image r and c coordinates (item (i) in the list above):

$$\hat{x}_i = a_0 + a_1 c_i + a_2 r_i$$

It can be interpreted as follows: the least squares estimate of the map x coordinate of the GCP labelled i can be found from the image column and row coordinates of that GCP (c_i and r_i) if the least-squares coefficients a_0, a_1 and a_2 are known. The values of these coefficients are determined from a sample of values of x, c and r, as described below, and are then applied to all the image pixels in order to estimate map x coordinates. The same operation is performed to find the map y coordinates of all of the pixels in the image. The steps involved are described in the following paragraphs. First, however, some of the technical details involved in the calculations need to be explained.

In mathematical terminology, the bivariate least-squares equation used in the preceding paragraphs is a *first-order polynomial least squares function*. It is first-order because neither of the predictor variables (on the right-hand side of the equation) is raised to a power greater than 1. A first-order function can accomplish scaling, rotation, shearing and reflection but not warping (such as would be necessary to correct for panoramic distortion or for any similar 'bending' effect). A second- or higher-order polynomial can be used to model such distortions, though in practice it is rare for polynomials of order higher than 3 to be used for medium resolution satellite imagery. Where a relatively small image area is being corrected (for instance, a 1024 × 1024 segment) it should be unnecessary to correct for warping at all. Note that the polynomial method does not correct for terrain relief distortions and is therefore applicable only to relatively flat areas. See Kohl and Hill (1988) and other references listed at the end of the introduction to Section 4.5 for details of a modification to the standard polynomial correction that corrects for relief-induced variations using a DEM. Example 4.3 shows polynomial surfaces of orders 1–3.

In general terms, a polynomial function in two variables t and u can be written concisely as:

$$\hat{s} = \sum_{j=0}^{m} \sum_{k=0}^{m-j} a_{jk} t^j u^k$$

where m is the order of the polynomial function. A third-order polynomial, written out in full, becomes:

$$\begin{aligned}\hat{s} = {} & a_{00}t^0u^0 + a_{01}t^0u^1 + \\ & a_{02}t^0u^2 + a_{03}t^0u^3 + \\ & a_{10}t^1u^0 + a_{11}t^1u^1 + \\ & a_{12}t^1u^2 + a_{20}t^2u^0 + \\ & a_{21}t^2u^1 + a_{30}t^3u^0\end{aligned}$$

(do you see the relationship between the subscripts of the coefficients a and the powers to which the corresponding

Example 4.3: Visualizing Polynomial Surfaces

The three surfaces shown in Example 4.3 Figure 1a–c give the z value (vertical axis) at a given point on the base defined by an x and a y coordinate in the range 1–100. The x, y and z coordinates represent one of the four relationships between map eastings and northings and image row and column shown in Section 4.3.2. In this instance, it is assumed that the relationship under study is that between image column (x) and image row (y) and map easting (z) coordinates. Example 4.3 Figure 1a shows a first order or linear surface, which is defined by the relationship $z = a_0 + a_1x + a_2y$. The three coefficients forming the vector $\boldsymbol{a}$ are computed from the GCP data. The result shown in Example 4.3 Figure 1a is a uniformly sloping plane. The fitting procedure is repeated with x and y staying the same but with the z coordinate representing map northing. These two sets of predicted or computed z values can be shifted and rotated with respect to the image easting and northing coordinates to give the sort of relationship shown in Figure 4.8b, where (e', n') are derived from the polynomial surface and (c', r') are measured values at a GCP. Example 4.3 Figure 1b,c show that the relationship between z on the one hand and (x, y) on the other becomes more complex. The second and third order surfaces, also known as the quadratic and cubic, respectively, are warped as well as shifted, scaled and rotated. If GCPs are not measured accurately or if the degree of the fitted surface is too high then the polynomial surface can start to bend and warp in response to errors rather than to a true relationship between x, y and z. A surface order of more than two is rarely required for images collected by sensors with a narrow field of view over a relatively flat area.

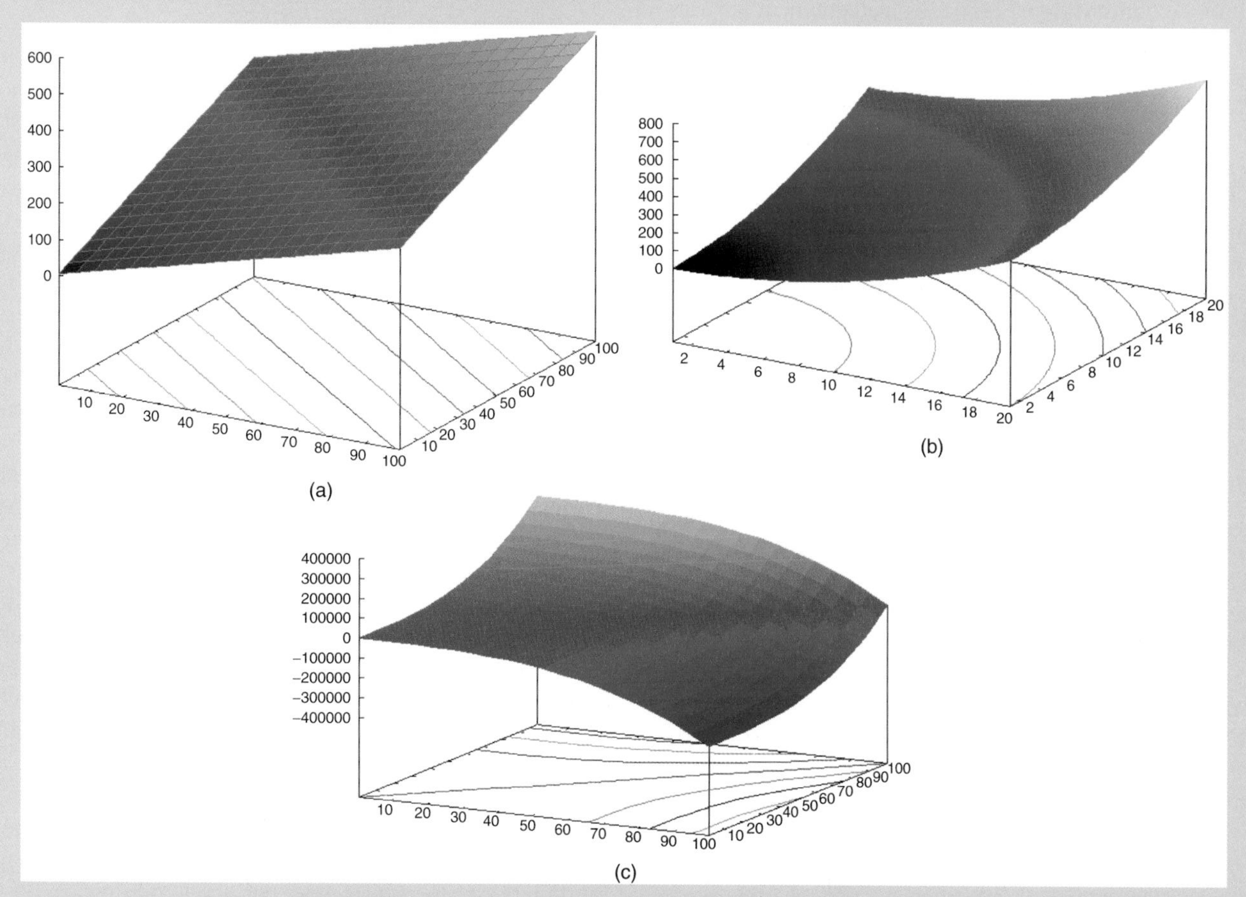

Example 4.3 Figure 1. *(a) first order (b) second order and (c) third order bivariate polynomials. The polynomials are evaluated for a grid of size (100, 100). The numerical labels on the z-axis are arbitrary.*

t and **u** are raised?) The equation can be reduced to a less formidable form as:

$$\hat{s} = a_{00} + a_{10}t + a_{01}u + a_{20}t^2 + a_{11}tu + a_{02}u^2 + a_{30}t^3 + a_{21}t^2u + a_{12}tu^2 + a_{03}u^3$$

The terms s, t and u are replaced by the appropriate terms in expressions (i–iv) above. If, for instance, we wished to estimate y as a function of c and r we would replace s, t and u in the polynomial expansion by y, c and r. There is thus one polynomial function for each of the four coordinate transformations. The first pair gives map coordinates (x, y) in terms of image coordinates (c, r) while the second pair give image coordinates (c, r) in terms of map coordinates (x, y). See Example 4.3 for more details.

Before we consider methods of evaluating polynomial expressions for given sets of (x, y) and (c, r) coordinates we should consider (i) the size of the sample of control points needed to give reliable estimates of the coefficients a_{ii}, (ii) the spatial distribution of the control points and (iii) the accuracy with which they are located (Labovitz and Marvin, 1986). In mathematical terms we need to take a sample of at least n control points where n is the number of coefficients in the polynomial expression. For a first-order polynomial, n is equal to 3. For a second-order polynomial, the value of n is 6 and for a third-order equation n has the value 10. These numbers of control points are necessary purely and simply to ensure that it is *mathematically* possible to evaluate the equations defining the coefficients a_{ij}. It is important to note that the *statistical* requirement, which is concerned not so much with the theoretical feasibility of the calculations but with the reliability of the results, sets a much higher standard. Most conventional statistics texts suggest that a sample size of at least 30 is required to achieve reliable estimates, but experience suggests that 10–15 control points will give acceptable results for a first-order fit and a small image area (up to 1024^2 pixels) medium resolution imagery such as those produced by the Landsat TM/ETM+ and SPOT HRV instruments. More GCPs will be needed in areas of moderate relief, or for images produced by instruments with a wide field of view, where a second order polynomial may be required. Results based on the use of small numbers of GCPs should be treated with caution as they may satisfy the mathematical criteria but may not satisfy the statistical ones. Mather (1995) gives details of a simulation study that emphasizes the importance of adequate GCP sample size.

The second aspect of sampling that should concern users is the spatial distribution of GCPs, a topic that is treated in more detail by Mather (1995). Since the coordinate system being transformed is a two-dimensional one, it seems reasonable to suggest that the control point locations should be represented by a two-dimensional pattern. It may, for instance, be possible to take a large number of control points along a linear transect, such as a main road. The information contained in these control points refers only to one dimension. Results derived from such information could say nothing about variations in the direction perpendicular to the transect line. Another possibility is to locate control points in clusters, for example representing road junctions in small towns that are widely separated. Again, large areas of the image would not be sampled, and results would thus be biased. In extreme cases a condition known mathematically as 'singularity' would be signalled during the least squares computations. This is equivalent in scalar terms to trying to divide by zero. Mather (1976, pp. 124–129) considers this point in detail, and also provides the technical background for two-dimensional least squares problems. Refer also to Section 3.3.

We can conclude from this brief examination not only that control points should be sufficient in number but also that they should be evenly spread, as far as possible, over the image area. This presents problems where substantial parts of the image cover sea or water areas where control points are absent. The same could be said of images covering any relatively featureless region. Unwin and Wrigley (1987) introduce the concept of the *leverage* of a control point, which measures the influence of the point on the overall fit of the polynomial function and allows the user to determine which points require most care and attention.

The accuracy with which control points are measured is also considered by Mather (1995). In his simulation experiment, increasing amounts of random 'noise' were added to the coordinates of the GCPs, and (particularly when the distribution of control points was linear or clustered) the effects of the noise were severe. The residual error is the difference between the value of the map or image coordinate estimated from the equation and the corresponding coordinate measured on the map or image. Residual errors can be calculated for the line/pixel coordinates of the image or the easting/northing coordinates of the map. Using the appropriate polynomial function selected from the four listed above, the map easting coordinate of a control point can be estimated from the image row (line) and column values of the same control point. The difference between the measured and estimated values of the map easting of the control point is the residual value for the map easting. This residual value is expressed in the same units as the map eastings, for example in kilometres. A similar operation can be carried out for the map northings and the image line and pixel coordinates.

Users of some commercial software are encouraged to remove control points that are identified as 'erroneous' by the fact that their associated residual error is considered to be unacceptably high (sometimes 'high' is defined as a value more than one standard deviation distant from the mean). This may seem to be sensible step but, as Morad, Chalmers and O'Regan (1996) show, this is not necessarily the case. Instead of eliminating control points one should seek reasons for the error, which may be the result of erroneous or inaccurate digitization or measurement or the consequence of image distortions such as those induced by high relief. Bolstad, Gessler and Thomas (1990) discuss positional uncertainty in digitizing map data. Since remotely-sensed images are likely to be coregistered with maps in a GIS, it is even more important that the geometric quality of products derived from remotely-sensed image matches the requirements of the project. Overlay operations within a GIS are prone to error if misregistration is present. Users should, therefore, take considerable care in the digitizing and measurement of control point locations because locational errors are magnified by any departures from randomness of the spatial pattern of the control points, and the net result may well be unacceptable.

When using an empirical method such as polynomial least squares to carry out the geometric correction procedure, the user should always be aware of the fact that these residuals are the only measure of goodness of fit that are available. A perfect fit of the polynomial function at the control points tells us nothing about the goodness of fit in the remainder of the area. If a control point (whose location has been checked carefully) is associated with a high residual value is deleted, then the goodness of fit of the least-squares polynomial surface in the neighbourhood of that deleted point is likely to be low. Best practice is to keep some control points in reserve, and use them as check points to measure the distortion in regions of the image that are not close to those control points used to calculate the polynomial function, a procedure that is outlined below in a little more detail. You will find that the residual error values at check points will tend to fall as the order of the polynomial increases from 1 to 2, and possibly 3. Once the order of the polynomial increases beyond three it is probable that the error at check points will increase, simply because the two-dimensional surface that is represented by the polynomial function is becoming increasingly flexible. Since the map and image are only being colocated at the control points, a lot of flexibility is available for distortions of the least-squares surface between the control point locations. Also, the variance of the residuals (used as an overall measure of goodness of fit) will decrease as polynomial order increases, until it becomes zero when the number of coefficients in the polynomial equals the number of control points. This does not imply a 'good fit'; rather, it indicates that the number of GCPs is too small (Example 4.3).

We should also pay attention to the factors to be taken into consideration when control points are selected. If there is a substantial difference in time between the date of imaging and the last full map revision then it would be unwise to select as control points any feature subject to change, for example a point on a river meander, at the edge or corner of a forest plantation or on a coastline. In tidal areas, the land/sea boundary may go back and forth if the region is tidal – sometimes for several hundred metres – and it is thus difficult to correlate a point shown on a map as being on the coastline and the equivalent point on the image. The best control points are intersections of roads or airport runways, edges of dams, or isolated buildings and other permanent features. All are unlikely to change in position, beyond the few centimetres per century produced by continental drift, and all are capable of being located accurately both on map and image. At the 1 : 25 000 map scale or better, control points should be located with and error of 25 m or less, while a cursor can be used to fix the location on the image to within one resolution element. If a zoom function is used to enlarge the image, the exact position of a pair of linear features that intersect at an oblique angle can be estimated to within half a pixel or better.

It would be sensible to test the accuracy of the image–map and map–image coordinate transformations before using them to convert from one system to the other. It seems illogical to generate the least-squares coefficients a_i using a set of (x, y, c, r) control point coordinates and then use the same data to test the adequacy of the calculation. An independent assessment is necessary. As suggested earlier in this discussion, a subset of control points ('test points') that are not used in the least-squares estimation process should be used to assess the goodness of fit of the least-squares transformation by taking each of the test points in turn and converting its c and r image coordinates to map x and y coordinates, then calculating the residual values. The same should be done for the reverse transformation. The standard of the map coordinate residuals and of the image coordinate residuals will give a measure of the goodness of fit. If the residual values are normally distributed, which can be checked by producing a histogram of these values, then 68% of all calculated values should lie within one standard deviation of the mean. The mean residual is zero. If a root mean square error or standard deviation of 0.5 pixel or 10 m, for example is cited then one should be careful not to interpret this statement to mean that *all* of the coordinate transformation results fall within the quoted range.

In some instances, control points may already be available for the required map area and for an earlier image of the region of interest. Rather than go through the tedious procedure of collecting control point information again, the map coordinates of the control points could

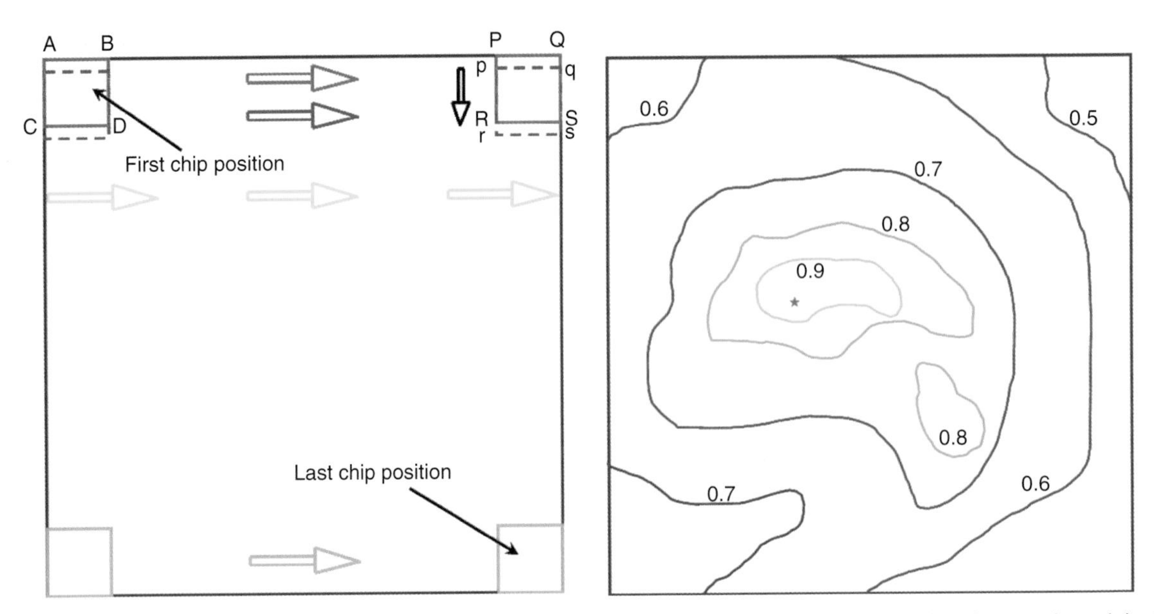

Figure 4.6 (a) Chip ABCD is covering pixels in the top left of the image. The correlation between the chip pixels and the image pixels is calculated, and the chip moves to the right by 1 pixel, and the procedure is repeated. The chip moves right until its right margin is coincident with the right border of the image at PQRS. The chip is moved down one line and back to the left side of the image, and the procedure is repeated. The arrows show the direction of movement of the chip. (b) Isoline map of correlations between image pixels and chip pixels, using the procedure shown in Figure 4.6a. The red star marks the point of maximum correlation.

be reused without difficulty. However, due to fluctuations in the orbit of the satellite, it is unlikely that the control point positions on the second image would be the same as those on the first. Variations in satellite altitude and attitude are quite small, so (i) the pixels in two separate images of the same area collected by the same satellite-borne instrument should have nearly equal sizes and (ii) the linear features whose intersection defined the control point should have approximately the same orientation from image to image of the same area. Benny (1981) made up a set of 'chips' or extracts of the first image of size 19 × 19 pixels, each chip containing a control point. The value of 19 × 19 is not a fixed, nor even a recommended one – much depends on the nature of the image, especially the contrast and the amount of detail. Larger chips are required for images lacking contrast.

The map coordinates of the centre of the chip are known. The problem is to find the image coordinates of the centre of the 19 × 19 chip on the new image. This problem is solved by a correlation procedure. Given that it is possible to estimate the position of the control point on the new image, and maximum deviations from this position in the column and row directions, then a rectangular search area can be defined. The control point is thought to be somewhere inside this search area. The 'chip' is placed over the top left 19 × 19 pixels of this search area and the correlation between the chip pixels and the pixels of the image that underlie the chip pixels is calculated and recorded, together with the row and column coordinates of the image pixel that lies below the centre pixel of the chip (Figure 4.6). The chip is then moved one column to the right and the correlation coefficient is recalculated. Once the chip has reached its furthest right position it is moved down one row and back to its leftmost position. The process continues until the chip has moved to the bottom right-hand corner of the search area.

At this stage, we have (i) the row coordinate of the chip centre, (ii) the corresponding column coordinate and (iii) the correlation value for that position. Values (i) and (ii) could be used as conventional xy coordinates and isolines of correlation could be drawn on this 'map'. The most likely match would be given by the point which had the highest correlation. If the maximum correlation is low (less than about +0.4) then it might be concluded that no match has been found. Benny (1981) describes a 'spiral search' algorithm that differs from the regular search algorithm outlined above. If the regular search algorithm were coded efficiently then it would be as fast as the spiral technique.

An efficient method of conducting the regular search would be to compute the mean of the top left chip for the search chip and the underlying then the deviations from the mean and the standard deviations of the 19 × 19 pixels making up the chip. The statistics for the search chip are computed only once. The correlation coefficient r for the search chip (first time only) and the 19×19 image area lying below the search chip can then be computed from:

$$r = \frac{1}{n}\sum\left(\frac{(x-\overline{x})^2}{S_x}\right)\left(\frac{(y-\overline{y})^2}{S_y}\right)$$

The values x_i are the search chip pixel values and the y_i are the pixels in the search area of the image lying below the chip pixels. The value of r, the correlation between x and y, is placed in the results image at the point lying beneath the centre pixel of the search chip, which is then moved onwards to the right by 1 pixel. The mean of the y_i can be updated quickly using the idea that as the search chip is moved from position 1 to position 2 on line 1 (i.e. moves 1 pixel to the right) then a column of 19 image pixels enters the right side of the region of the image covered by the chip and a column of 19 image pixels on the left side leaves that region. The value of the mean of the y_i is found by subtracting the values of the pixels leaving the region from Σy_i and by adding the values of the pixels entering the region. Since all the calculations in the derivation of r_{xy} except the final division can be performed exactly in integer arithmetic the method is both fast and accurate. This 'moving window' technique is used in Chapter 7 when spatial filtering techniques are considered. See also Section 3.3 for comments on the avoidance of computational error in calculations involving floating-point numbers.

All the available control points could be located by this correlation-based method, which requires the user to provide an estimate of the position of the control point and of the dimensions of the search area for each control point. Benny (1981) goes on to consider a method for the automatic relocation of all control points without user intervention, provided that one control point can be located accurately. He estimates that, for a typical image containing 100 control points, the manual location of the control points would take about 80 hours of effort, plus 1 minute of computer time to carry out the coordinate transformation method described below. For the semiautomatic method, with the user supplying information about the approximate location of each control point, the user man-hours drop from 80 to 8, but computer time goes up to 20 minutes. Finally, the automatic procedure, in which the user supplies information relating to only one control point, the number of man-hours required falls to 0.1 (6 minutes) while the computer time requirement remains at 20 minutes. Automatic identification of GCPs is examined in more detail by Ackermann (1984), Emery, Baldwin and Matthews (2003) and Motrena and Rebordão (1998). Automatic registration of stereo pairs of images, as a prelude to the derivation of DEMs, is described by Al-Rousan *et al.* (1997) and Al-Rousan and Petrie (1998).

Procedures for estimating the coefficients a_{ij} in the least squares functions (i–iv) above relating map and image coordinate systems are now considered. The following description assumes that we wish to estimate the map easting e from the image column and row coordinates r and c for a set of n control points. In practice, all four functions relating map (e, n) and image (r, c) coordinates would be computed (e from r and c, n from r and c, r from e and n and c from e and n). The set of control point map easting coordinates is denoted by the vector **e**, while the powers and cross products of the c and r values are considered to form the matrix **P**. The coefficients a_{ij} are the elements of the coefficients vector **a**. For a second-order fit we would have the system shown in Table 4.2.

The method of least squares is used to find the vector of estimates **e** according to the following model:

$$\mathbf{e} = \mathbf{Pa}$$

P and **e** are explained above, while **a** is a vector of unknown coefficients, which are to be estimated from the

Table 4.2 *Matrix* ***P*** *and vectors* ***e*** *and* ***a*** *required in solution of second-order least-squares estimation procedure.*

$$\mathbf{e} = \begin{bmatrix} e_1 \\ e_2 \\ e_3 \\ e_4 \\ \cdot \\ \cdot \\ \cdot \\ \cdot \\ e_n \end{bmatrix} \quad \mathbf{P} = \begin{bmatrix} 1 & c_1 & r_1 & c_1^2 & c_1r_1 & r_1^2 \\ 1 & c_2 & r_2 & c_2^2 & c_2r_2 & r_2^2 \\ 1 & c_3 & r_3 & c_3^2 & c_3r_3 & r_3^2 \\ 1 & c_4 & r_4 & c_4^2 & c_4r_4 & r_4^2 \\ \cdot & \cdot & \cdot & \cdot & \cdot & \cdot \\ \cdot & \cdot & \cdot & \cdot & \cdot & \cdot \\ \cdot & \cdot & \cdot & \cdot & \cdot & \cdot \\ 1 & c_n & r_n & c_n^2 & c_nr_n & r_n^2 \end{bmatrix} \quad \mathbf{a} = \begin{bmatrix} a_{00} \\ a_{10} \\ a_{01} \\ a_{20} \\ a_{11} \\ a_{02} \end{bmatrix}$$

In this example the map easting vector **e** is to be estimated from the powers and cross-products of the image column (**c**) and row (**r**) vectors which form the matrix **P**. The measurements of **c**, **r** and **e** are measured at n ground control points.

GCP data. The least-squares formula for the evaluation of **a** is:

$$\mathbf{a} = (\mathbf{P}'\mathbf{P})^{-1}\mathbf{P}'\mathbf{e}$$

The elements of vector **a** can be found by a standard subroutine for solving linear simultaneous equations. Such routines work efficiently with well-conditioned equations but can produce significant errors if the equations are ill conditioned. The condition of a matrix such as **P′P** in the expression above can be considered to be a measure of the degree of independence of its columns. Ill-conditioning is indicated by a near-zero determinant, or by a high value of the ratio of the largest and smallest eigenvalues of **P′P**. The degree of independence of a pair of column vectors can be visualized by analogy with the crossing point of two lines. The intersection can be measured accurately if the lines are perpendicular, whereas if the lines cross at a very acute angle the exact point of intersection is difficult to specify (Figure 4.7). In a similar way, the solutions of a set of well-conditioned equations (with the columns of the matrix **P′P** being independent, or nearly so) can be found accurately but the solution vector could be substantially in error if the matrix is badly conditioned. Various studies have indicated that the accuracy of standard procedures (using matrix inversion techniques) for the solution of the least squares equations depends critically on the condition of the matrix. The Gram–Schmidt procedure described by Mather (1976) appears to be a reliable algorithm, and this conclusion is confirmed by later work (Mather, 1995).

It is likely that the matrix **P′P** will be ill-conditioned in our particular case because the columns of **P** are the powers and cross-products of two variables, c and r, for example, $c, r, c^2, cr, r^2, c^3, c^2r, cr^2$ and r^3 in the case of a cubic bivariate polynomial. Hence, standard methods may well produce poor or even misleading results. These will be worsened if the spatial distribution of control points is linear or clustered or if the number of GCPs is small, as noted above. The Gram–Schmidt method is thus to be preferred The application of this method provides estimates of the elements of the coefficients vector **a**, which can then be used to find the vector of map easting coordinate **e**. The coefficients of the other required functions (n estimated from c and r, c estimated from e, and n and r estimated from e and n) can be found in a similar fashion.

Note that the procedures described in this section are applicable only to images obtained from relatively stable platforms such as satellites and for areas with a low relative relief. Scanned images from aircraft contain distortions caused by rapid variations in the aircraft's attitude as measured by pitch, roll and yaw. Such imagery will contain defects such as non-parallel scan lines, which cannot

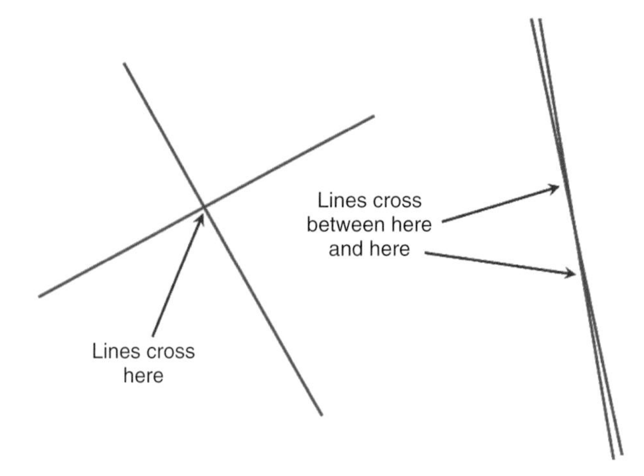

Figure 4.7 *The 'condition' of a matrix in least-squares calculations can be likened to the sharpness of definition of the crossing point of two straight lines. Perpendicular lines have a sharply defined crossing point (left), while the crossing point of two near-coincident lines cannot be well defined (right).*

easily be corrected by polynomial least squares methods. The influence of terrain on the results obtained by the use of the methods described above are summarized earlier in this section. The correction of aircraft imagery, such as the lidar data described in Chapter 9, uses a combination of GPS and inertial navigation systems (INSs) to model the position and attitude of the aircraft.

4.3.3 Resampling Procedures

Once the four transformation equations relating image and map coordinate systems are known, and the results tested, the next step is to find the location on the map of the four corners of the image area to be corrected, and to work out the number of and spacing (in metres) between the pixel centres necessary to achieve the correct map scale. We can now work systematically across the map area, starting at the top left, and locate (in map coordinates e and n) the centre of each pixel in turn. Given the (e, n) location coordinates of a pixel centre on the map we can apply the transformations described in the preceding section to generate (c, r) image coordinates corresponding to the position of the pixel's centre. These (c, r) coordinates are the column and row position in the uncorrected image of the new (geometrically corrected) pixel centres (Figure 4.4 and Figure 4.8a).

It is unlikely that c and r are integers; if they were, it would be possible to take the pixel value at (c, r) (pixel centre position in the uncorrected image, shown by a circle in Figure 4.8a) and transfer it to the corresponding pixel centre in the corrected image, shown by a plus sign in Figure 4.8a. Non-integral values of c and r imply that the corrected pixel centre lies between the columns and rows of the uncorrected image, so that a method of interpolation is needed to estimate the pixel value at (c, r).

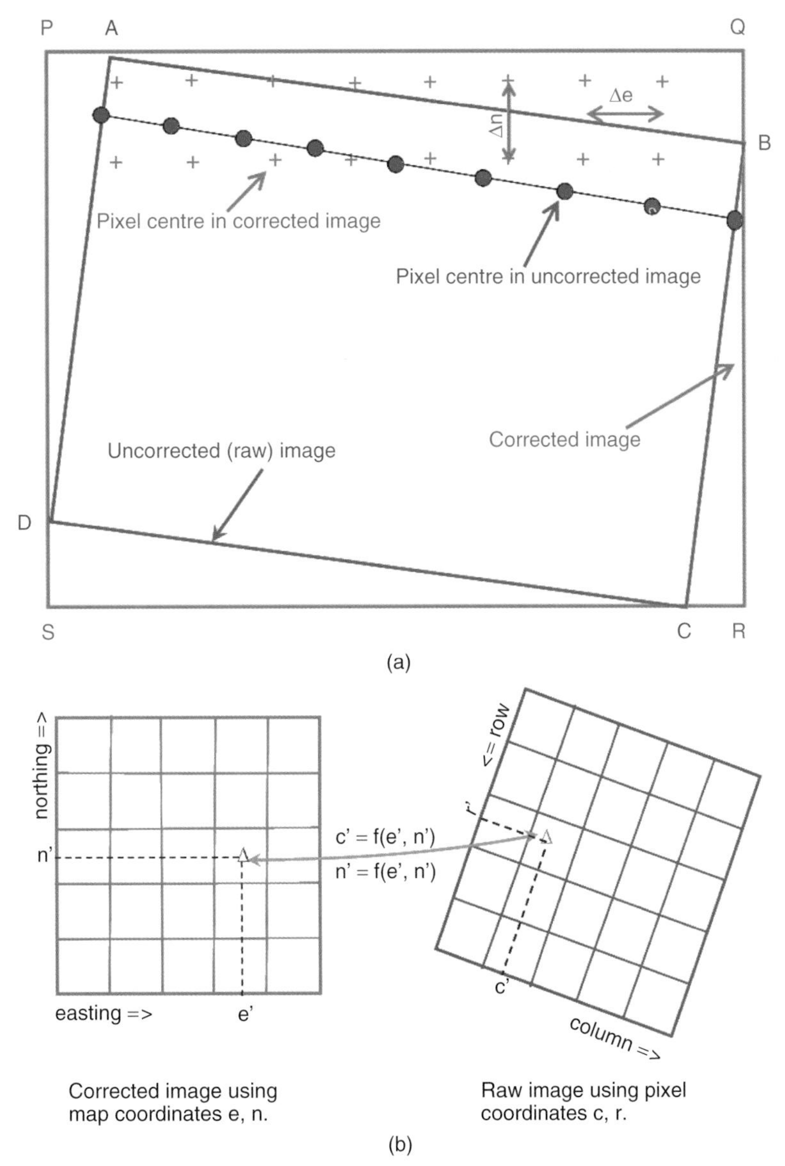

Figure 4.8 *(a) Schematic representation of the resampling process. The extreme east, west, north and south map coordinates of the uncorrected image ABCD define a bounding rectangle PQRS, which is north-orientated with rows running east–west. The elements of the rows are the pixels of the geometrically corrected image expressed in map* (e, n) *coordinates. Their centres are shown by the + symbol and the spacing between successive rows and columns is indicated by* Δe *and* Δn*. The centres of the pixels of the uncorrected image are marked by the symbol o. See text for discussion. (b) The least-squares bivariate polynomial functions take the coordinates at the centre of a pixel with coordinates* (e'n') *in the corrected image and find the coordinates of the corresponding point* (c'r') *in the corrected image. Since the values of the pixels in the uncorrected image are known, one can proceed systematically through the pixels of the corrected image and work out the value to place in each using a procedure called* resampling*. Pixels in the corrected image that have corresponding* (c, r) *locations outside the limits of the uncorrected image (the green rectangle in Figure 4.8a) are given the value zero.*

Figure 4.8b illustrates this point. The coordinates of the pixel at the point (e', n') in the corrected image are transformed using the least-squares polynomial function computed from the GCPs to point (r', c') in the uncorrected image. If r' and c' were integers then the pixel values in the corrected and uncorrected image would be the same. In most cases, r' and c' are non-integral values and lie somewhere between the pixel centres in the uncorrected image. An interpolated value is computed using a procedure called resampling.

Three methods of resampling are in common use. The first is simple – take the value of the pixel in the raw image that is closest to the computed (c, r) coordinates. This is called the *nearest neighbour* method. It has two

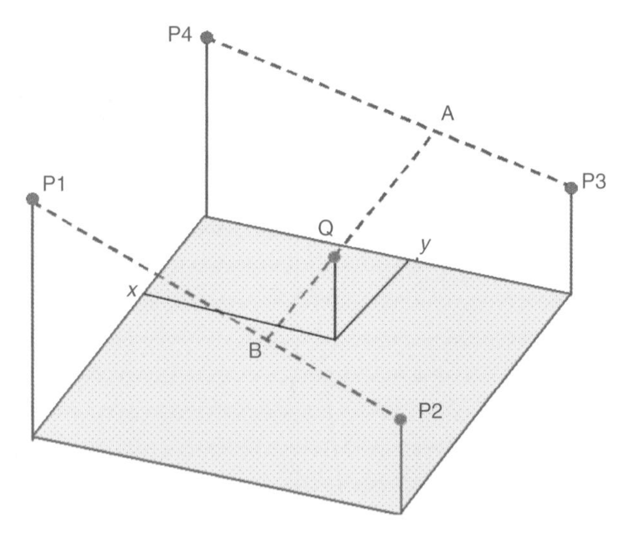

Figure 4.9 Bilinear interpolation. Points P1–P4 represent the centres of pixel in the uncorrected image. The height of the 'pin' at each of these points is proportional to the pixel value. The pixel centre in the geometrically corrected image is computed as (x, y) (point Q). The interpolation is performed in three stages. First, the value at A is interpolated along the line P4–P3, then the value at B is interpolated along the line P1–P2. Finally, the value at Q is interpolated along the line AB.

advantages; it is fast and its use ensures that the pixel values in the output image are 'real' in that they are copied directly from the raw image. They are not 'fabricated' by an interpolation algorithm such as the two that are described next. On the other hand, the nearest neighbour method of interpolation tends to produce a rather blocky effect as some pixel values are repeated.

The second method of resampling is *bilinear interpolation* (Figure 4.9). This method assumed that a surface fitted to the pixel values in the immediate neighbourhood of (c, r) will be planar, like a roof tile. The four pixel centres nearest to (c, r) (i.e points P1–P4 in Figure 4.9) lie at the corners of this tile; call their values v_{ij}. The interpolated value V at (c, r) is obtained from:

$$V = (1-a)(1-b)v_{i,j} + a(1-b)v_{i,j+1} + b(1-a)v_{i+1,j} + abv_{i+1,j+1}$$

where

$a = c - j \qquad j = |c|$

$b = r - i \qquad i = |r|$ and

$|x|$ is the absolute value of the argument x.

Note that the method breaks down if the point (c, r) is coincident with any of the four pixel centres in the uncorrected image and so a test for this should be included in any computer program. Bilinear interpolation results in a smoother output image because it essentially an averaging process. Thus, sharp boundaries in the input image may be blurred in the output image. The computational time requirements are greater than those of the nearest neighbour method.

The third spatial interpolation technique that is in common use for estimating pixel values in the corrected image is called *bicubic* because it is based on the fitting of two third-degree polynomials to the region surrounding the point (c, r). The 16 nearest pixel values in the uncorrected image are used to estimate the value at (c, r) on the output image. This technique is more complicated than either the nearest neighbour or the bilinear methods discussed above, but it tends to give a more natural-looking image without the blockiness of the nearest neighbour or the oversmoothing of the bilinear method, though – as is the case with all interpolation methods – some loss of high-frequency information is involved. The interpolator is essentially a low-pass filter (Chapter 7). The penalty to be paid for these improvements is considerably increased computing requirements.

The cubic convolution method is described by Wolf and DeWitt (2000) as follows: first, take the 4×4 pixel area surrounding the point whose value is to be interpolated. Next, compute two vectors each with four elements. The first contains the absolute differences between the row values (in the full image) and the row value of the point to be interpolated. The second contains the absolute differences for the column values. For example, the 4×4 area may consist of rows 100, 101, 102 and 103 of the full image, with the column values being 409, 410, 411 and 412. The 4×4 set of pixel values forms a matrix **M**. If the point **p** to be interpolated has row/column coordinates of (101.1, 410.5) then the row difference vector is $\mathbf{r} = [1.1, 0.1, 0.9, 1.9]$ and the column difference vector is $\mathbf{c} = [1.5, 0.5, 0.5,$ and $1.5]$.

Next, a row interpolation vector and a column interpolation vector are derived from **r** and **c**. Let v represent any element of **r** or **c**. If the value of v is less than 1.0 then use function f_1 to compute the interpolation value. Otherwise, use function f_2, where f_1 and f_2 are defined as follows:

$$f_1(v) = (a+2)v^3 - (a+3)v^2 + 1$$

$$f_2(v) = av^3 - 5av^2 + 8av - 4a$$

The value of the term a is set by the user, normally in the range -0.5 to -1.0. At this point the elements of **r** and **c** have been converted to row and column interpolation values, respectively. The interpolated value z at point **p** is given by the solution of the matrix equation:

$$z = \mathbf{r}'\mathbf{Mc}$$

The value of a in the row/column interpolation procedure is rarely mentioned, but it has a significant impact on the value returned by the cubic convolution algorithm.

For example, using the same set of data as that presented by Wolf and Dewitt (2000, p. 564), different values of z can be obtained simply by changing the value of a by a small amount. Wolf and DeWitt's matrix **M** is:

$$\mathbf{M} = \begin{bmatrix} 58 & 54 & 65 & 65 \\ 63 & 62 & 68 & 58 \\ 51 & 56 & 59 & 53 \\ 52 & 45 & 50 & 49 \end{bmatrix}$$

for row numbers 618–621 and column numbers 492–621 and column numbers 492–495 of the full image. The row/column coordinates of point **p** are [619.71 493.39]. If the value of a is set to -0.5 then the interpolated value z at point **p** is 60.66. For values of a of -0.6, -0.75 and -1.0 the corresponding z values are 61.13, 61.87 and 63.15. Since the interpolated results are rounded to the nearest integer, small differences do not matter – in this example, the values 60.66 and 61.13 would be rounded to 61. Nevertheless, it would be rather odd of a professional user of the method to concern himself with the arcane algorithmic details of least squares procedures in order to improve computational accuracy while ignoring the choice of a value in the cubic convolution interpolation.

The choice between the three interpolation methods of nearest neighbour, bilinear and cubic convolution depends upon two factors – the use to which the corrected image is to be put, and the computer facilities available. If the image is to be subjected to classification procedures (Chapter 8) then the replacement of raw data values with artificial, interpolated values might well have some effect on the subsequent classification (although, if remote sensing data are to be used alone in the classification procedure, and not in conjunction with map-based data, then it would be more economical to perform the geometric correction after, rather than before, the classification). If the image is to be used solely for visual interpretation, for example in the updating of a topographic map, then the resolution requirements would dictate that the bicubic or cubic method be used. The value of the end product will, ultimately, decide whether the additional computational cost of the bicubic method is justified. Khan, Hayes and Cracknell (1995) consider the effects of resampling on the quality of the resulting image.

4.3.4 Image Registration

Registration of images taken at different dates (multitemporal images) can be accomplished by image correlation methods such as that described in Section 4.3.2 for the location of GCPs. Although least squares methods such as those used to translate from map to image coordinates, and vice versa, could be used to define a relationship between the coordinate systems of two images, correlation-based methods are more frequently employed. A wide range of applications is considered by Goshtasby (2005). A full account of what are termed sequential similarity detection algorithms (SSDAs) is provided by Barnea and Silverman (1972) while Anuta (1970) describes the use of the fast Fourier transform (Chapter 6) in the rapid calculation of the interimage correlations. Eghbali (1979) and Kaneko (1976) illustrate the use of image registration techniques applied to Landsat MSS images. Hong and Zhang (2008b) use wavelet-based methods (see Section 3.7). Townshend *et al.* (1992) point out the problems that may arise if change detection techniques are based on multi-temporal image sets that are not properly registered. This point is discussed further in Section 6.8. As noted earlier, any GIS overlay operation involving remotely-sensed images is prone to error because all such images must be registered to some common geographical reference frame. Misregistration provides the opportunity for error to enter the system.

4.3.5 Other Geometric Correction Methods

The least squares polynomial procedure described in Section 4.3.1.3 is one of the most widely used methods for georeferencing medium-resolution images produced by sensors such as such as the Landsat ETM+, which has a nominal spatial resolution of around 30 m. The accuracy of the resulting geometrically corrected image depends, as we have already noted, on the number and spatial distribution of GCPs. Significant points to note are: (i) the map and image may not coincide in the sense of associating the same coordinates (to a specified number of significant digits) to a given point, even at the control points, because the least squares polynomial produces a global approximation to the unknown correction function and (ii) the method assumes that the area covered by the image to be georeferenced is flat. The effects of terrain relief can produce very considerable distortions in a geometric correction procedure based on empirical polynomial functions (Figure 4.10). The only effective way of dealing with relief effects is to use a mathematical model that takes into account the orbital parameters of the satellite, the properties of the map projection, and the nature of the relief of the ground surface. Where high-resolution images such as those produced by IKONOS and QuickBird are used, the question of accurate, relief-corrected geocoded images becomes critical.

GeoEye, the company which owns and operates the IKONOS system, does not release details of the orbital parameters of the satellite to users (though QuickBird orbital data are available). Instead, they provide the coefficients of a set of *rational polynomials* with their

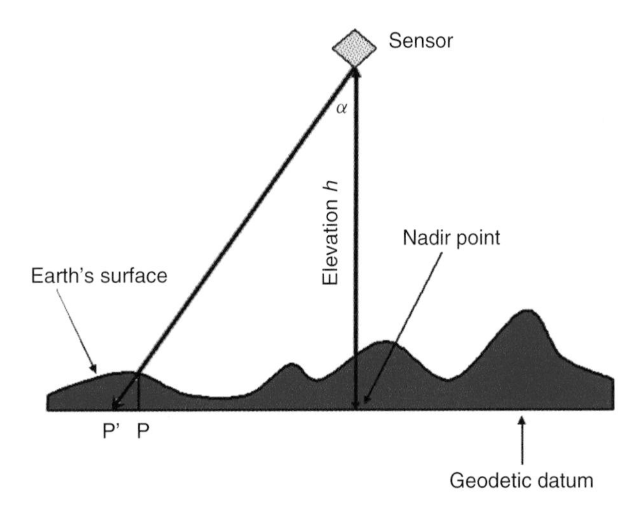

Figure 4.10 *The least-squares polynomial method of geometric correction does not take terrain variations into account. The position of points on a map is given in terms of their location on a selected geodetic datum. At an off-nadir view angle of* α*, point P appears to be displaced to P′. The degree of displacement is proportional to the satellite altitude* h *and the view angle* α*.*

stereo products. The term rational means 'ratios of'. Dowman and Dolloff (2000) provide a useful summary of the characteristics of this georeferencing procedure, while Chen, Teo and Liu (2006) compare the rigorous sensor model and the rational function model for FORMOSAT-2 images, and Arévalo and González (2008) consider geometric rectification of QuickBird imagery. Tao and Hu (2001) describe the rational function model in photogrammetric processing, and Fraser and Yamakawa (2004) evaluate the affine transformation for high-resolution satellite sensor orientation. Di, Ma and Li (2004) also consider rational function models. Cheng and Sustera (2009) examine automatic methods of high-speed, high-accuracy orthorectification and mosaicing of RapidEye imagery (Section 2.3.9). In a well-illustrated paper these authors demonstrate that high accuracy can be achieved via the use of a star tracker, called the Altair HB, which provides attitude information, plus the set of rational polynomial coefficients pertaining to the image in question and, finally, a DEM developed from a spaceborne interferometric SAR elevation mapping mission called SRTM (Section 9.2). With no GCPs available the error in the resulting image is cited as being of the order of 5.7 m in the x direction and 7.5 m in the y direction, or about one pixel. Better results may have been achieved if a map at a larger scale than the 1 : 24 000 map actually used had been available. Although better results were realised when accurate GCPs were used in the correction, the result is still good enough for most purposes.

Geometric correction of SAR images is generally accomplished using orbital models and DEM. The process is complicated by the fact that SAR is a side-looking instrument, producing images in the 'slant range'. Geometric correction involves the conversion from slant range to ground range as well as scaling and north-orientation. The use of a DEM in the geometric correction process allows the consideration of the effects of terrain slope and elevation. A SAR image that is corrected both for viewing geometry and terrain is said to be orthocorrected. Mohr and Madsen (2001) and Sansosti *et al.* (2006) describe the geometric calibration of ERS SAR images, while Hong and Schowengerdt (2005) give details of a technique for the precise registration of radar and optical imagery. The use of SAR images to generate interferograms, which is discussed in Chapter 9, requires accurate geometric correction of the images that are used.

The derivation of DEMs from pairs of radar (SAR) images using interferometric methods is considered in detail in Chapter 9. It should, however, be noted here that DEMs are routinely derived from stereo pairs of optical imagery produced by sensors such as SPOT HRV, ASTER and ALOS PRISM. All of these sensors are described in Chapter 2. Practical aspects of registering the stereo pairs prior to extraction of elevation information are covered by Al-Rousan *et al.* (1997) and Al-Rousan and Petrie (1998), Dowman and Neto (1994), Giles and Franklin (1996), Theodossiou and Dowman (1990), Tokunaga and Hara (1996) and Welch *et al.* (1998). Hirano, Welch and Lang (2003) report on validation and accuracy assessment experiments using ASTER stereo image data.

4.4 Atmospheric Correction

4.4.1 Background

An introductory description of the interactions between radiant energy and the constituents of the Earth's atmosphere is given in Chapter 1. From this discussion one can conclude that a value recorded at a given pixel location on a remotely-sensed image is not a record of the true ground-leaving radiance at that point, for the magnitude of the ground-leaving signal is attenuated due to atmospheric absorption and its directional properties are altered by scattering. Figure 4.11 shows, in a simplified form, the components of the signal received by a sensor above the atmosphere. All of the signal appears to originate from the point P on the ground whereas, in fact, scattering at S_2 redirects some of the incoming electromagnetic energy within the atmosphere into the field of view of the sensor (the atmospheric path radiance) and some of the energy reflected from point Q is scattered

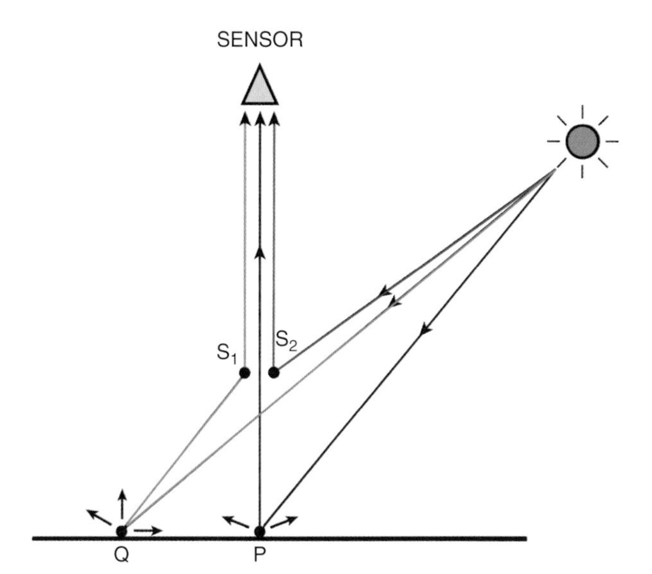

Figure 4.11 *Components of the signal received by an airborne or satellite-mounted sensor. See text for explanation.*

at S_1 so that it is seen by the sensor as coming from P. This scattered energy, called 'environmental radiance', produces what is known as the 'adjacency effect'. To add to these effects, the radiance from P (and Q) is attenuated as it passes through the atmosphere. Other difficulties are caused by variations in the illumination geometry (the geometrical relationship between the Sun's elevation and azimuth angles, the slope of the ground and the disposition of topographic features). These problems are considered in Section 4.5. If the sensor has a wide field of view, such as the NOAA AVHRR, or is capable of off-nadir viewing (for example the SPOT HRV) then further problems result from the fact that the reflectance of a surface will vary with the view angle as well as with the solar illumination angle (this is the bi-directional reflectance property, noted in Chapter 1). Given this catalogue of problems one might be tempted to conclude that quantitative remote sensing is the art of the impossible. However, there is a variety of techniques that can be used to estimate the atmospheric and viewing geometry effects, although a great deal of research work remains to be done in this highly complex area. A good review of terminology and principles is contained in Deschamps, Herman and Tanré (1983), while Woodham (1989) gives a lucid review of problems. The problem of the 'adjacency effect' which is mentioned above is considered by Milovich, Frulla and Gagliardini (1995).

Atmospheric correction might be a necessary preprocessing technique in three cases. In the first, we may want to compute a ratio of the values in two bands of a multispectral image (Section 6.2.4). As noted in Chapter 1, the effects of scattering increase inversely with wavelength, so the shorter-wavelength measurements experience more scattering than the longer-wavelength data. The computed ratio will thus be a biased estimate of the true ratio. In the second situation, a research worker may wish to relate upwelling radiance from a surface to some property of that surface in terms of a physically based model. To do this, the atmospheric component present in the signal recorded by the sensor must be estimated and removed. This problem is experienced in, for example oceanography because the magnitude of the water-leaving radiance (which carries information about the biological and sedimentary materials contained by the upper layers of the ocean) is small compared to the contribution of atmospheric effects. The third case is that in which results or ground measurements made at one time (time 1) are to be compared with results achieved at a later date (time 2). Since the state of the atmosphere will undoubtedly vary from time 1 to time 2, it is necessary to correct the radiance values recorded by the sensor for the effects of the atmosphere. In addition to these three cases, it may well be necessary to correct multispectral data for atmospheric effects even if it is intended for visual analysis rather than any physical interpretation.

The atmosphere is a complex and dynamic system. A full account of the physics of the interactions between the atmosphere and electromagnetic radiation is well beyond the scope of this book, and no attempt is made to provide a comprehensive survey of the progress that has been made in the field of atmospheric physics. Instead, techniques developed by remote sensing researchers for dealing with the problem of estimating atmospheric effects on multispectral images in the 0.4–2.4 μm reflective solar region of the spectrum are reviewed briefly. We begin by summarizing the relationship between radiance received at a sensor above the atmosphere and the radiance leaving the ground surface:

$$L_s = H_{\mathrm{tot}}\rho T + L_{\mathrm{p}}$$

H_{tot} is the total downwelling radiance in a specified spectral band, ρ is the reflectance of the target (the ratio of the downwelling to the upwelling radiance) and T is the atmospheric transmittance. L_{p} is the atmospheric path radiance. The downwelling radiance is attenuated by the atmosphere as it passes from the top of the atmosphere to the target. Further attenuation occurs as the signal returns through the atmosphere from the target to the sensor. Some of the radiance incident upon the target is absorbed by the ground-surface material, with a proportion ρ being reflected by the target. Next, energy reflected by the ground surface from outside the target area is scattered by the atmosphere into the field of view of the sensor. Finally, the radiance reaching the sensor includes a contribution made up of energy scattered within the atmosphere; this is the path radiance term (L_{p}) in the equation. In reality, the situation is more complicated, as Figure 4.11 shows. The path radiance term L_p varies in magnitude inversely with wavelength for scattering increases as wavelength decreases. Hence, L_p will contribute differing amounts to measurements in individual wavebands. In terms of a Landsat TM or ETM+ image the blue-green band (band 1) will generally have a higher L_p component than the green band (band 2).

4.4.2 Image-Based Methods

The first method of atmospheric correction that is considered is the estimation of the path radiance term, L_p and its subtraction from the signal received by the sensor. Two relatively simple techniques are described in the literature. The first is the histogram minimum method, and the second is the regression method. In the first approach, histograms of pixel values in all bands are computed for the full image, which generally contains some areas of low reflectance (clear water, deep shadows or exposures of dark coloured rocks). These pixels will have values very close to zero in near-infrared bands, for example Landsat TM band 4 or SPOT HRV band 3, and should have near-zero values in the other bands in this spectral region. Yet, if the histograms of these other bands are plotted they will generally be seen to be offset progressively towards higher levels. The lowest pixel values (or some combination of the lowest values) in the histograms of visible and near-infrared bands are a first approximation to the atmospheric path radiance in those bands, and these minimum values are subtracted from the respective images. Path radiance is much reduced in mid-infrared bands such as Landsat TM bands 5 and 7, and these bands are not normally corrected.

The regression method is applicable to areas of the image that have dark pixels as described above. In terms of the Landsat ETM+ sensor, pixel values in the near-infrared band (numbered 4) are plotted against the values in the other bands in turn, and a best-fit (least-squares) straight line is computed for each using standard regression methods. The offset a on the x-axis for each regression represents an estimate of the atmospheric path radiance term for the associated spectral band (Figure 4.12).

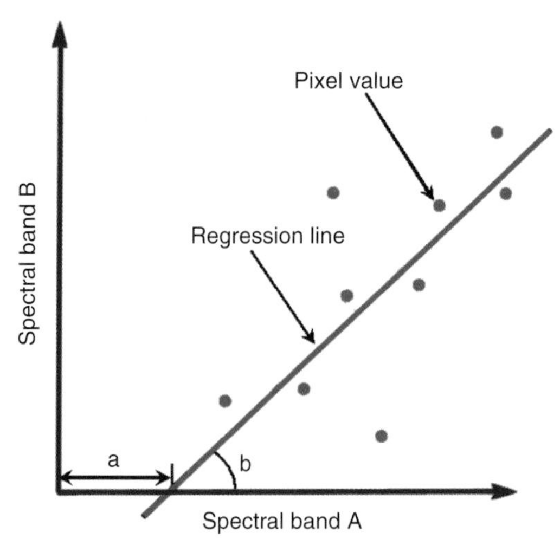

Figure 4.12 *Regression of selected pixel values in spectral band A against the corresponding pixel values in band B. Band B is normally a near-infrared band (such as Landsat ETM+ band 4 or SPOT HRV band 3) whereas band A is a visible/near-infrared band.*

Chavez (1988) describes an elaboration of this 'haze correction' procedure based on the 'histogram minimum' methods described above. Essentially Chavez's method is based on the fact that Rayleigh scattering is inversely proportional to the nth power of the wavelength, the value of n varying with atmospheric turbidity. He defines a number of 'models' ranging from 'very clear' to 'hazy' and each of this is associated with a value of n. A 'starting haze' value is provided for one of the short wavelength bands, and the haze factors in all other bands are calculated analytically using the Rayleigh scattering relationship. The method requires the conversion from pixel values to radiances. The gains and offsets that are used to perform this conversion are the pre-flight calibrations for the Landsat TM. Chavez (1996) gives details of modifications to the method which enhance its performance.

The image-based methods described above (dark object subtraction and regression) simply estimate the contribution to the radiance at a pixel of the atmospheric path radiance. The methods described next are generally used to derive estimates of the radiance reaching the sensor from the target pixel. In other words, they are normally used in conjunction with procedures to correct for illumination effects (Section 4.5) and sensor calibration (Section 4.6). Franklin and Giles (1995) give a systematic account of these procedures, while Aspinall, Marcus and Boardman (2002) consider atmospheric correction of imaging spectrometer data sets (Chapter 9). A full radiometric correction would include conversion of pixel digital value to apparent (at-sensor) radiance, subtraction of the atmospheric contribution, topographic normalization and sensor calibration. The end product is an estimate of the true ground-leaving radiance at a pixel. Such values are required in environmental modelling, and in the measurement of change between two dates of imaging. The use of radiative transfer models and the empirical line method for estimating the atmospheric effect is considered next. Topographic corrections are mentioned briefly in Sections 4.5 and 4.7. Sensor calibration issues are considered in Section 4.6.

4.4.3 Radiative Transfer Models

The methods of atmospheric correction described above rely on information from the image itself in order to estimate the path radiance for each spectral band. However, a considerable body of theoretical knowledge concerning the complexities of atmospheric radiative transfer is in existence, and has found expression in numerical radiative transfer models such as LOWTRAN (Kneizys *et al.*, 1988), MODTRAN (Berk *et al.*, 1999; Guanter, Richter and Kaufmann 2009), ATREM (Gao, Heidebrecht and Goetz, 1993) and 5S/6S (Tanré *et al.*, 1986; Vermote *et al.*, 1997). Operational (as opposed to research) use of these models is limited by the need to supply data relating to the condition of the atmosphere at the time of imaging. The cost of such data collection activities is considerable, hence reliance is often placed upon the use of 'standard atmospheres' such as 'mid-latitude summer'. The use of these estimates results in a loss of accuracy, and the extent of the inaccuracy is not assessable. See Popp (1995) for further elaborations. It is also difficult to apply radiative transfer models to archive data because of lack of knowledge of atmospheric conditions. Richter (1996) shows how spatial variability in atmospheric properties can be taken into account by partitioning the image.

The following example of the use of the *5S* model of Tanré *et al.* (1986) is intended to demonstrate the magnitudes of the quantities involved in atmospheric scattering. Two hypothetical test cases are specified, both using Landsat TM bands 1–5 and 7 for an early summer (1 June) date at a latitude of 51°. A standard mid-latitude summer atmosphere was chosen, and the only parameter to be supplied by the user was an estimate of the horizontal visibility in kilometres (which may be obtainable from a local meteorological station). Values of 5 km (hazy) and 20 km (clear) were input to the model, and the results are shown graphically in Figure 4.13. The apparent reflectance is that which would be detected by a scanner operating above the atmosphere, while the pixel reflectance is an estimate of the true target reflectance which, in this example, is green vegetation. The intrinsic atmospheric reflectance is analogous to the path radiance. At 5 km visibility the intrinsic atmospheric reflectance is greater than pixel reflectance in TM bands 1 and 2. Even at the 20 km visibility level, intrinsic atmospheric reflectance is greater than pixel reflectance in TM band 1. In both cases the intrinsic atmospheric reflectance declines with wavelength, as one might expect. Note that the difference between the apparent reflectance and the sum of the pixel reflectance and the intrinsic atmospheric reflectance is equal to reflectance from pixels neighbouring the target pixels being scattered into the field of view of the sensor (see Figure 4.11). The y-axis in Figure 4.13 shows percent reflectance, that is, the proportion of solar irradiance measured at the top of the atmosphere that is reflected.

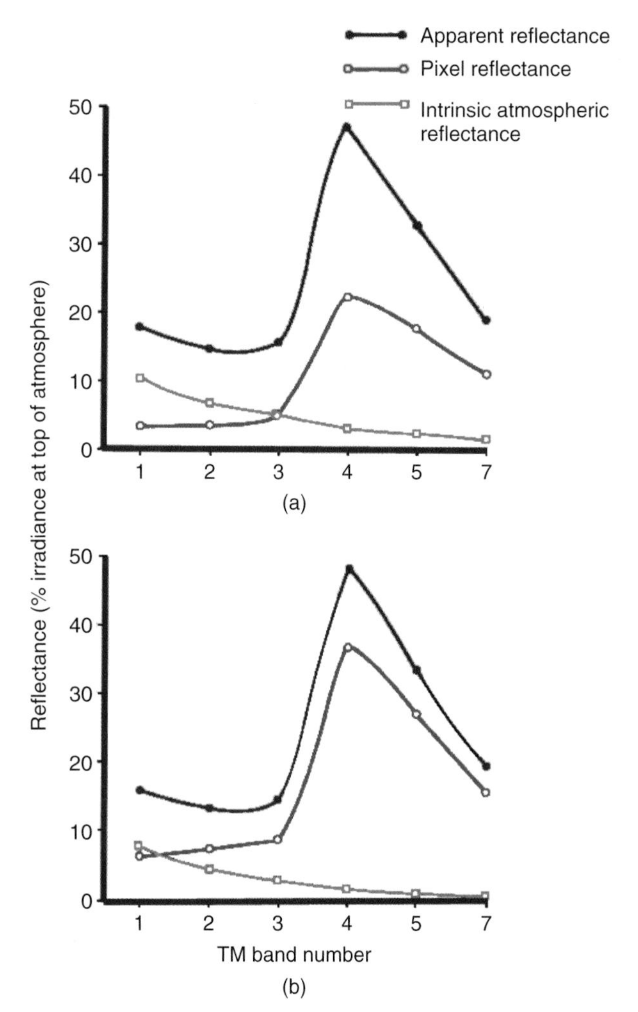

Figure 4.13 Examples of output from atmospheric model (Tanré et al., 1986). Reflectance, expressed as percent irradiance at the top of the atmosphere in the spectral band, is shown for the atmosphere (intrinsic atmospheric reflectance), the target pixel (pixel reflectance) and the received signal (apparent reflectance). The difference between the sum of pixel reflectance and intrinsic atmospheric reflectance and the apparent reflectance is the background reflectance from neighbouring pixels. Examples show results for (a) 5 km visibility and (b) 20 km visibility.

4.4.4 Empirical Line Method

An alternative to the use of radiative transfer models is the empirical line approach (Baugh and Groeneveld, 2008; Smith and Milton, 1999; Karpouzli and Malthus, 2003; Moran *et al.*, 2001; Vaudour *et al.*, 2008). This method is illustrated in Figure 4.14. Two targets – one

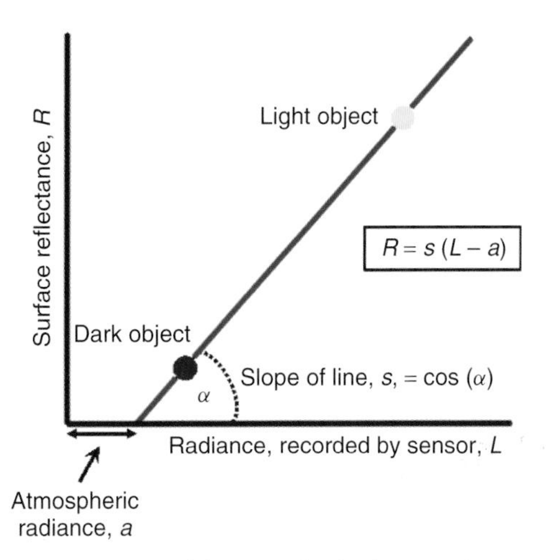

Figure 4.14 *Empirical line method of atmospheric correction. Two targets (light and dark) whose reflectance (*R*) and at-sensor radiance (*L*) are known are joined by a straight line with slope* s *and intercept* a. *The reflectance for any at-sensor radiance can be computed from* $R = s(L - a)$. *Based on Figure 1 of Smith and Milton,* International Journal of Remote Sensing, *1999,* ***20****, 2654. © Taylor and Francis Ltd.*

light and one dark – are selected, and their reflectance is measured on the ground, using a field radiometer, to give values R on the y-axis of Figure 4.14. The radiances recorded by the sensor (shown by the x-axis, L, in Figure 4.14) are computed from the image pixel values using the methods described in Section 4.6. Finally, the slope, s, and intercept, a, of the line joining the two target points are calculated. The conversion equation is $R = s(L - a)$. The term a represents the atmospheric radiance. This equation is computed for all spectral bands of interest. Smith and Milton (1999) report the results of a detailed experiment, and conclude that the method '... is capable of providing acceptable calibration of sensor radiance measurements to estimates of ground reflectance' (Smith and Milton, 1999, p. 2657), though they list a number of theoretical and practical considerations that must be taken into account before reliance is placed on any results. These considerations are discussed in detail by Smith and Milton (1999). Karpouzli and Malthus (2003) also report the results of experiments to estimate surface reflectance using the empirical line method applied to IKONOS data. The absolute differences between calculated band reflectances and corresponding reflectances measured on the ground ranged from 0 to 2.7%. The authors describe these results as '... highly satisfactory' (Karpouzli and Malthus, 2003, p. 1148). Vaudour *et al.* (2008) use the empirical line method to retrieve soil reflectance from SPOT imagery. Yuan and Elvidge (1996) and Hong and Zhang (2008a) compare these and other methods of radiometric normalization. Atmospheric correction of hyperspectral imagery (Section 9.3) is reviewed by Gao, Davis and Goetz (2006).

Some general advice on the need for atmospheric correction in classification and change detection studies is provided by Song *et al.* 2001. These authors suggest that atmospheric correction is not required as long as the training data and the data to be classified are measured on the same relative scale. However, if multitemporal image data are being processed then they must be corrected for atmospheric effects to ensure that they are comparable.

Huang *et al.* (2008) give an example of the use of atmospheric and topographic correction in forest classification. Mitri and Gitas (2004) compare classification performance in classifying burned areas using topographical and non-topographical corrections. Nunes, Marcal and Vaughan (2008) give details of a fast over-land atmospheric correction of visible and near-IR imagery.

4.5 Illumination and View Angle Effects

The magnitude of the signal received at a satellite sensor is dependent on several factors, particularly:

- reflectance of the target
- nature and magnitude of atmospheric interactions
- slope and aspect of the ground target area relative to the solar azimuth
- angle of view of the sensor
- solar elevation angle.

Variations in the spectral reflectance of particular types of Earth-surface cover materials are discussed in Chapter 1, where a general review of atmospheric interactions is also to be found, while Section 4.4 contains an introduction to the correction of image data for atmospheric effects. In this section we consider the effects of: (i) the solar elevation angle, (ii) the view angle of the sensor and (iii) the slope and aspect angles of the target.

In the absence of an atmosphere, and assuming no complicating factors, the magnitude of the radiance reflected from or emitted by a fixed target and recorded by a remote sensor will vary with the illumination angle and the angle of view of the sensor. The reflectance of the target will vary as these two angles alter, and one could therefore consider a function which described the magnitude of the upwelling radiance of the target in terms of these two angles. This function is termed the *bi-directional reflectance distribution function* (or BRDF, described in Chapter 1). When an atmosphere is present an additional complication is introduced, for the irradiance at the target will be reduced as the atmospheric path length increases. The path length (which is the distance that the incoming energy travels through the atmosphere) will increase as the solar elevation angle decreases, and so the degree of atmospheric interference will increase.

The radiance upwelling from the target also has to pass through the atmosphere. The angle of view of the sensor will control the upward path length. A nadir view will be influenced less by atmospheric interactions than would an off-nadir view; for example the extremities of the scan-lines of a NOAA AVHRR image are viewed at angles of up to 56° from nadir, while the SPOT HRV instrument is capable of viewing angles of ±27° from nadir. Amounts of shadowing will also be dependent upon the solar elevation angle. For instance, shadow effects in row crops will be greater at low Sun elevation angles than at high angles. Ranson, Biehl and Bauer (1985) report the results of a study of variations in the spectral response of soybeans with respect to illumination, view and canopy geometry. They conclude that the spectral response depended greatly on solar elevation and azimuth angles and on the angle of view of the sensor, the effect being greater in the visible red region of the spectrum than in the near-infrared. In another study, Pinter *et al.* (1983) show that the spectral reflectance of wheat canopies in MSS and TM wavebands is strongly dependent on the direction of incident radiation and its interaction with vegetation canopy properties, such as leaf inclination and size. Clearly, reflectance from vegetation surfaces is a complex phenomenon which is, as yet, not fully understood and methods for the estimation of the effects of the various factors that influence reflectance are still being evaluated at a research level. It would obviously be desirable to remove such effects by preprocessing before applying methods of pattern recognition (Chapter 8) particularly if results from analyses of images from different dates are to be compared, or if the methods are applied to images produced by off-nadir viewing or wide angle-of-view sensors.

The effects of variation in the solar elevation angle from one image to another of a given area can be accomplished simply if the reflecting surface is Lambertian (Section 1.3.1). This is rarely the case with natural surfaces, but the correction may be approximate to the first order. If the solar zenith angle (measured from the vertical) is θ, the observed radiance is L and the desired view angle is x then the correction is simply:

$$L' = L\frac{\cos(x)}{\cos(\theta)}$$

This formula may be used to standardize a set of multi-temporal images to a standard solar illumination angle. If a suitable DEM is available, and assuming that the image and DEM are registered, then the effects of terrain slope on the angle of incidence of the solar irradiance can be taken into account (Feng, Rivard and Sánchez-Azofeifa, 2003). See Frulla, Milovich and Gagliardini (1995) for a discussion of illumination and view angle effects on the NOAA-AVHRR imagery.

Barnsley and Kay (1990) consider the relationship between sensor geometry, vegetation canopy geometry and image variance with reference to wide field of view imagery obtained from aircraft scanners (though multiangle satellite data are now available from ESA's CHRIS/Proba mission – see Barnsley *et al.*, 2004). The effect of off-nadir viewing is seen as a symmetric increase in reflectance away from the nadir point, so that a plot of the mean pixel value in each column (y-axis) against column number (x-axis) shows a parabolic shape. A first-order correction for this effect uses a least-squares procedure to fit a second-order polynomial of the form:

$$\hat{y}_x = a_0 + a_1 x$$

to the column means. The term $\hat{y}_x$ is the value on the least-squares curve corresponding to image column x. Let $\hat{y}_{\text{nad}}$ be the value on the least squares curve at the nadir point (the centre of the scan line and the minimum point on the parabolic curve). Then the values in column x of the image can be multiplied by the value $\hat{y}_{\text{nad}}/\hat{y}_x$ in order to 'flatten' the curve so that the plot of the column means of the corrected data plot (approximately) as a straight line. Palubinskas *et al.* (2007) give details of a more complicated procedure for radiometric normalization. Helder and Ruggles (2004) describe some radiometric artefacts in the Landsat reflective bands.

4.6 Sensor Calibration

Sensor calibration has one of three aims. First, the user may wish to combine information from images obtained from different sensor systems such as Landsat TM and SPOT HRV. Second, it may be necessary in studies of change to compare pixel values obtained from images that were acquired at different times. Third, remotely-sensed estimates of surface parameters such as reflectance are used in physical and biophysical models. Generally speaking, sensor calibration is normally combined with atmospheric and view angle correction in order to obtain estimates of target reflectance (e.g. Teillet and Fedosejevs, 1995). These authors also provide details of the correction required for variations in solar irradiance over the year. A useful starting point is the review by Duggin (1985). In this section, attention is focused on the Landsat TM/ETM+ optical bands and the SPOT HRV. Calibration of the Landsat TM/ETM+ thermal band (conventionally numbered 6) is discussed by Schott and Volchok (1985), Itten and Meyer (1993), Chander, Markham and Helder (2009a) and Coll *et al.* (2010). Radiometric normalization of sensor scan angle effects is treated by Palubinskas *et al.* (2007) and Paolini *et al.* (2006) report on the use of radiometric correction in change detection studies. Consideration of calibration issues relating to the NOAA-AVHRR is provided by Che and Price (1992) and by Vermote and Kaufman (1995). Gutman and Ignatov (1995) compare pre- and postlaunch calibrations for the NOAA AVHRR, and show how the differences in these calibrations affect the results of Vegetation Index calculations (Chapter 6). A special issue of *Canadian Journal of Remote Sensing* (volume 23, number 4, December 1997) is devoted to calibration and validation issues. Morain and Budge (2004) present a selection of papers delivered at an international workshop on radiometric and geometric calibration.

Atmospheric correction methods are considered in Section 4.4, while corrections for illumination and view angle effects are covered in Section 4.5. In the present section, the topic of sensor calibration is reviewed. Recall from Chapter 1 that the values recorded for a particular band of a multispectral image are counts – they are the representation, usually on a 0–255 scale, of equal-length steps in a linear interpolation between the minimum and maximum levels of radiance recorded by the detector. If more than one detector is used then differences in calibration will cause 'striping' effects on the image (Section 4.2.2). The term 'sensor calibration' refers to procedures that convert from counts to physical values of radiance. The numerical coefficients that are used to calibrate these image data vary over time, and consequently the relationship between the pixel value (count) recorded at a particular location and the reflectance of the material making up the surface of the pixel area will not be constant. Recall that reflectance is the ratio of the radiance reflected or emitted by the target and the incident radiation (irradiance).

In the case of Landsat TM and ETM+, two coefficients are required for conversion of pixel values to radiance. These coefficients are called the gain and the offset. The gain is the slope of the line relating pixel values and radiance. Offset is an additive term, as shown in Figure 4.15.

The determination of these calibration coefficients is not an easy task. Prelaunch calibration factors (coefficients) are obtainable from ground receiving stations and from vendors of image data, and some image processing software packages incorporate these values into their sensor calibration routines. However, studies such as Thome *et al.* (1993) indicate that these calibration factors are time-varying for the Landsat-5 TM sensor, and that substantial differences in the outcome of calibration calculations depend on the values of the calibration factors that are used. The fact that a number of different processing procedures have been applied to Landsat TM data by ground stations further complicates the issue (Moran *et al.*, 1992, 1995). Thome *et al.* (1993, see also Table 2

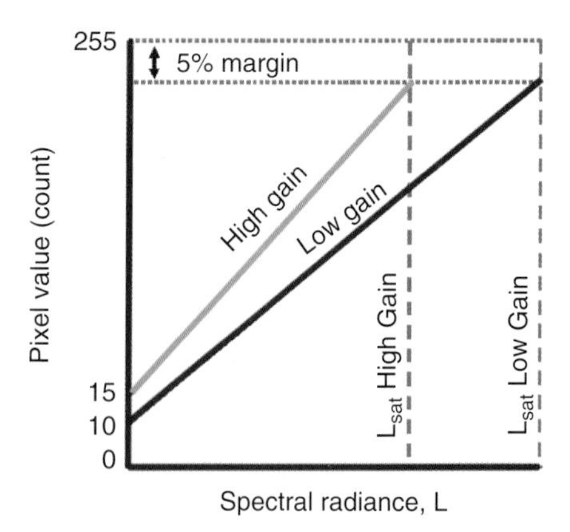

Figure 4.15 *Landsat ETM+ uses one of two gain modes. The spectral radiance reaching the sensor is converted to a digital count or pixel value using high gain mode for target areas which are expected to have a maximum spectral radiance of L_{sat} (High Gain). For other target areas, the maximum radiance is specified as L_{sat} (Low Gain). Each gain setting has an associated offset, measured in counts, which is 10 for low gain and 15 for high gain. Based on Irish (2002), Figure 6.9.*

of Teillet and Fedosejevs, 1995) provide the expressions shown in Table 4.3 for Landsat-5 TM sensor gains in bands 1–5 and 7 (the offset is assumed to be constant over time):

Table 4.3 *Landsat-5 TM calibration coefficients from Thome et al. (1993). G_i is the gain value for band i and D is the number of days since the launch of Landsat-5 (1 March 1984).*

Gain
$G_1 = (-7.84 \times 10^{-5})\,D + 1.409$
$G_2 = (-2.75 \times 10^{-5})\,D + 0.7414$
$G_3 = (-1.96 \times 10^{-5})\,D + 0.9377$
$G_4 = (-1.10 \times 10^{-5})\,D + 1.080$
$G_5 = (7.88 \times 10^{-5})\,D + 7.235$
$G_7 = (7.15 \times 10^{-5})\,D + 15.63$

D is the number of days since 1 March 1984, the date of the launch of the Landsat-5 satellite. The day number in the year is called the Julian Day (JD). You can calculate the JD relative to any year since 1753 using the **Utilities|Julian Dates** module in MIPS. The gain coefficients are used in the equation:

$$L_n^* = (PV - offset)/G_n \qquad (\mathrm{W\,m^{-2}sr^{-1}\mu m^{-1}})$$

In this equation, which differs from the relationship presented in the next paragraph, uses the symbol L_n^* to denote apparent radiance at the sensor, while *PV* is the pixel value, G_n is the sensor gain and the offsets are as follows (values in brackets): TM1 (2.523), TM2

Table 4.4 Landsat-5 TM offset (a_0) and gain (a_1) coefficients.

Band number	a_0	a_1	a_0	a_1
1	−0.1009	0.0636	−0.1331	0.0727
2	−0.1919	0.1262	−0.2346	0.1385
3	−0.1682	0.0970	−0.1897	0.1102
4	−0.1819	0.0914	−0.1942	0.0885
5	−0.0398	0.0126	−0.0398	0.0126
7	−0.0203	0.0067	−0.0203	0.0067

The results of pre-flight calibration are given in the two leftmost columns. The values in the two rightmost columns are derived from observations at White Sands, NM, USA. Note that the gain and offset coefficients for the TM bands 5 and 7 are unchanged. Figures given in units of mW $cm^{-2}sr^{-1}\mu m^{-1}$.

(2.417), TM3 (1.452), TM4 (1.854), TM5 (3.423) and TM7 (2.633). Note that the values of gain and offset given in Table 4.4 refer to the procedure that is described in the following paragraphs.

Hill (1991) and Hill and Aifadopoulou (1990) present a methodology for the calibration of the Landsat TM optical bands and SPOT HRV data. They note that the relationship between radiance and pixel value [PV, sometimes called a digital signal level (DSL) or a digital count (DC)] can be defined for spectral band n of the Landsat TM as follows:

$$L_n^* = a_0 + a_1 PV$$

where a_0 and a_1 are the offset and gain coefficients and L^* is apparent radiance at the sensor measured in units of mW $cm^{-1}sr^{-1}\mu m^{-1}$. The work of Thome *et al.* (1993) on the determination of the values of the gain and offset coefficients is noted above; their gains and offsets use the relationship between L^* and PV described earlier. Other relevant references are Hall *et al.* (1991), Holm *et al.* (1989), Markham, Halthore and Goetz (1992), Moran *et al.* (1990, 1992), Muller (1993), Olsson (1995), Price (1987, 1988), Rao and Chen (1996), Sawter *et al.* (1991), Slater *et al.* (1987), Teillet and Fedosejevs (1995) and Woodham (1989). Table 4.4 shows the Landsat-5 TM calibration coefficients (i) as determined before launch and (ii) as estimated from ground-based measurements at White Sands, New Mexico. Updated radiometric calibration coefficients are provided by Helder *et al.* (2008). Teillet *et al.* (2007) consider the problem of cross-calibration of sensors in the solar reflective wavebands. This aspect of calibration is important whenever data from different sensors are used synergistically. For example, the satellites making up the disaster monitoring constellation (DMC) or the RapidEye system must have instruments that are correctly calibrated so that a target viewed by the sensor onboard UK-DMC2 will produce the same outcome (i.e. radiance in $Wm^{-1}sr^{-1}$) as any other member of the DMC constellation. Also important is the maintenance of internally consistent time series of measures such as NDVI and other vegetation indices derived from AVHRR measurements (Sections 6.2.4 and 6.3.1).

Calibration of Landsat-7 ETM image data is described in Irish (2008). The ETM+ sensor, like the previous TM instruments, operates in one of two modes, termed low and high gain. Where the expected range of upwelling radiances is small, then low-gain mode is used, otherwise high-gain mode is used. The aim of this procedure is to provide the greatest possible image contrast. Figure 4.15 shows graphs of low and high gain modes, and Table 4.4 gives minimum and maximum radiances for the two gain settings for all bands including the panchromatic band, which is labelled as band 8. Irish (2002) gives details of the gain settings used for different targets.

Conversion of Landsat ETM+ counts (Q) to radiance units (L) is accomplished using the following equation:

$$L_\lambda = G_\lambda Q + O_\lambda$$

The gain (G) and offset (O) values for spectral band λ should be read from the metadata provided with the image data. Use of the values contained in the metadata will provide a more accurate estimate of the spectral radiance.

An alternative to the use of absolute calibration procedures for the determination of reflectance factors for Landsat data is to use one of the images in a multitemporal image set as a reference and adjust other images according to some statistical relationship with the reference image. Thus, Elvidge *et al.* (1995) use a regression procedure to perform relative radiometric normalisation of a set of Landsat MSS images. Their procedure assumes that land cover has not changed and that the vegetation is at the same phenological stage in all images. A similar approach is used by Olsson (1993).

The corresponding expression for SPOT HRV uses calibration coefficients derived from the header data provided with the image. Table 4.5 shows a part of the output generated by the MIPS program. The header data refer to a SPOT-1 multispectral (XS) image of the Camargue area of southern France, acquired on 12 January 1987, and gain values a_i for the three multispectral channels are provided. Note that all of the offset values are zero. The apparent radiance of a given pixel is calculated from:

$$L = PV/a_1$$

where a_1 is the gain coefficient, and PV and L are as defined earlier. These coefficients are updated regularly (Begni, 1988; Begni *et al.*, 1988; Moran *et al.*, 1990). These apparent radiance values must be further corrected if imagery acquired from SPOT in off-nadir viewing mode is used (Moran *et al.*, 1990; Muller, 1993).

Table 4.5 *Extract from SPOT header file showing radiometric gains and offsets.*

Scene ID		S1H1870112102714	
Scene centre latitude		N0434026	
Scene centre longitude		E0043615	
Spectral mode (XS or PAN)		XS	
Preprocessing level identification		1B	
Radiometric calibration designator		1	
Deconvolution designator:		1	
Resampling designator:		CC	
Pixel size along line:		20	
Pixel size along column:		20	
Image size in map projection along *y* axis		059792	
Image size in map projection along *x* axis		075055	
Sun calibration operation date		19861115	
This is a multispectral image			
Absolute calibration gains	00.86262	00.79872	00.89310
Absolute calibration offsets	00.00000	00.00000	00.00000

The procedure for conversion of ASTER visible and short wave infrared data (Table 2.4) expressed in counts to radiance units is similar to that described above for SPOT and ETM+ data. The sensor has a number of different gain settings, which can be selected in order to produce a high signal-to-noise ratio for different targets. Table 4.5 shows the maximum radiance values for each band for the high, normal, low 1 and low 2 gain settings. In bands 1–9, pixel values 1–254 represent zero radiance and maximum radiance, respectively. The magnitude of the radiance represented by a change of unity in the quantisation level for band 1, for example increases from 0.6751 (high gain) to 1.6877 (normal gain) to 2.2490 $\text{W m}^{-2}\text{sr}^{-1}\mu\text{m}^{-1}$. Thus, high gain mode can be used for images of regions of low reflectance while low-gain mode is used in regions of high reflectance in order to avoid saturation (which occurs when the radiance received by the sensor exceeds the maximum radiance that can be quantized). The gain setting for any specific image is contained in the metadata. See Table 4.6 for maximum radiance values for ASTER's high, normal and low gain modes.

Given the value of radiance at the sensor (L) it is usual to convert to apparent reflectance, that is, the total reflectance (from target and atmosphere) at the sensor. This value is also known as at-satellite reflectance. Of course, if the image has been corrected for atmospheric effects then the value computed by the equation below is an estimate of actual target reflectance. Conversion to reflectance is accomplished for each individual band using the expression:

$$\rho = \frac{\pi L d^2}{E_s \cos \theta_s}$$

L is the radiance computed as described earlier, E_s is the exoatmospheric solar irradiance (Markham and Barker, 1987; Price, 1988; Table 4.6), d is the relative Earth-Sun distance in astronomical units (the mean distance is 1.0 AU) for the day of image acquisition, and θ_s is the solar zenith angle. The Earth–Sun distance correction factor is required because there is a variation of around 3.5% in solar irradiance over the year. The value of d is provided by the formula:

$$d = 1 - 0.01674 \cos[0.9856(JD - 4)]$$

JD is the 'Julian day' of the year, that is the day number counting 1 January = 1. A utility to calculate the JD is available via the **`Utilities|Julian Dates`** function in MIPS. The module actually calculates the number of days that have elapsed since a reference calendar date since the start of the first millennium, though most readers will wish to compute elapsed days since the start of a given year. The program is useful in computing the time-dependent offset and gain values from the formulae given in Thome *et al.* (1993), described above, which

Table 4.6 Maximum radiance for different gain settings for the ASTER VNIR and SWIR spectral bands.

Maximum radiance (W m^{-2} sr^{-1} μm^{-1})				
Band number	High gain	Normal gain	Low gain 1	Low gain 2
1	170.8	427	569	N/A
2	179.0	358	477	
3N	106.8	218	290	
3B	106.8	218	290	
4	27.5	55.0	73.3	73.3
5	8.8	17.6	23.4	103.5
6	7.9	15.8	21.0	98.7
7	7.55	15.1	20.1	83.8
8	5.27	10.55	14.06	62.0
9	4.02	8.04	10.72	67.0

require the number of days that have elapsed since the launch of Landsat-5 (1 March 1984). Table 4.7 contains details of solar exo-atmospheric spectral irradiance for both the Landsat TM and ETM+ instruments. Comparisons of the radiometric characteristics of these two instruments are provided by Teillet *et al.* (2001), Masek *et al.* (2001) and Vogelmann *et al.* (2001). Thome *et al.* (1997) give details of Landsat TM radiometric calibration, while Thome (2001) discusses the absolute calibration of Landsat ETM+ data. Apparent reflectance is used by Huang *et al.* (2002c) to permit comparison of images collected under different illumination conditions.

Calibration of SAR imagery requires the recovery of the normalised radar cross-section (termed sigma-nought or σ_0 and measured in terms of decibels, dB). The range of σ_0 values is from +5 dB (very bright target) to −40dB (very dark target). Meadows (1995) notes that the purpose of calibration is to determine absolute radar cross-section measurements at each pixel position, and to estimate drift or variation over time in the radiometric performance of the SAR. Calibration can be performed in three ways: by imaging external calibration targets on the ground, by the use of internal calibration data, or by examining raw data quality. Laur *et al.* (2002) give details of calibration of ERS SAR data and the derivation of σ_0. Loew and Mauser (2007) discuss the calibration of SAR data and the removal of terrain effects, which is considered next.

4.7 Terrain Effects

The corrections required to convert ETM+ and SPOT HRV data described in Section 4.6 assume that the area covered by the image is a flat surface that is imaged by a narrow field of view sensor. It was noted in Section 4.6 that apparent reflectance depends also on illumination and view angles, as target reflectance is generally non-Lambertian. That discussion did not refer to the commonly observed fact that the Earth's surface is not generally flat. Variations in reflectance from similar targets will occur if these targets have a different topographic position, even if they are directly illuminated by the Sun. Therefore, the spectral reflectance curves derived from multispectral imagery for what is apparently the same type of land cover (for example wheat or coniferous forest) will contain a component that is attributable to topographic position, and the results of classification analyses (Chapter 8) will be influenced by this variation, which is not necessarily insignificant even in areas of low relief (Combal and Isaka, 2002). Various corrections have been proposed for the removal of the 'terrain illumination effect'. See Li, Daels and Antrop (1996), Proy, Tanré and Deschamps (1989), Teillet, Guindon and Goodenough (1982), Woodham (1989) and Young and Kaufman (1986) for reviews of the problem. Danaher, Xiolaing and Campbell (2001) and Danaher (2002) propose an empirical BRDF correction for Landsat TM and ETM+ images based on the conversion of pixel values to top-of-atmosphere reflectances, as described in the preceding section, and an empirical BRDF model.

Correction for terrain illumination effects requires a DEM that is expressed in the same coordinate system as the image to be corrected. Generally, the image is registered to the DEM, as the DEM is likely to be map-based. The DEM should also be of a scale which is close to that of the image, so that accurate estimates of slope angle and slope direction can be derived for each pixel position in the image. A number of formulae are in common use for the calculation of slope and aspect 'images'

Table 4.7 *Exo-atmospheric solar irradiance for (a) Landsat TM, (b) Landsat ETM+, (c) SPOT HRV (XS) bands and ASTER (Markham and Barker, 1987; Price, 1988; Teillet and Fedosejevs, 1995; Irish, 2008 Thome, personal communication). The centre wavelength is expressed in micrometres (μm) and the exo-atmospheric solar irradiance in mW$cm^{-2}sr^{-1}\mu m^{-1}$. See also Guyot and Gu (1994), Table 2.*

(a) Landsat Thematic Mapper				
Landsat TM/ETM+ band number	Centre wavelength	Centre wavelength (Teillet and Fedosejevs, 1995)	Exo-atmospheric spectral irradiance	Exo-atmospheric spectral irradiance (Teillet and Fedosejevs, 1995)
1	0.486	0.4863	195.70	195.92
2	0.570	0.5706	192.90	182.74
3	0.660	0.6607	155.70	155.50
4	0.840	0.8832	104.70	104.08
5	1.676	1.677	21.93	22.075
7	2.223	2.223	7.45	7.496

(b) Landsat enhanced thematic mapper plus (ETM+)		
Band	Bandwidth (μm)	Exo-atmospheric spectral irradiance
1	0.450–0.515	196.9
2	0.525–0.605	184.0
3	0.630–0.690	155.1
4	0.775–0.900	104.4
5	1.550–1750	22.57
7	2.090–2.350	8.207
8	0.520–0.900	136.8

(c) SPOT high resolution visible (HRV)		
SPOT HRV band no	Centre wavelength	Exo-amospheric spectral irradiance
1	0.544	187.48
2	0.638	164.89
3	0.816	110.14

(d) ASTER		
ASTER band no	Bandwidth (μm)	Exo-atmospheric spectral irradiance
1	0.520–0.600	1846.9
2	0.630–0.690	1546.0
3	0.780–0.860	1117.6
4	1.600–1.700	232.5
5	2.145–2.185	80.32
6	2.185–2.225	74.92
7	2.235–2.285	69.20
8	2.295–2.365	59.82
9	2.360–2.430	57.32

from a DEM and different results may be determined by different formulae (Bolstad and Stowe, 1994; Carara, Bitelli and Carla 1997; Hunter and Goodchild, 1997). A simple method to correct for terrain slope in areas that receive direct solar illumination is simply to use the Lambertian assumption (that the surface reflects radiation in a diffuse fashion, so that it appears equally bright from all feasible observation angles). This cosine correction is mentioned above. It involves the multiplication of the apparent reflection for a given pixel by the ratio of the cosine of the solar zenith angle (measured from the vertical) by the cosine of the incidence angle (measured from the surface normal, which is a line perpendicular to the sloping ground). Teillet, Guindon and Goodenough (1982, p. 88) note that this correction is not particularly useful in areas of steep terrain where incidence angles may approach 90°. Feng, Rivard and Sánchez-Azofeifa (2003) describe a terrain correction for imaging spectrometer data (Chapter 9) based on the assumption of Lambertian behaviour.

Non-Lambertian models include the *Minnaert correction*, which is probably the most popular method of computing a first-order correction for terrain illumination effects (though the method does not include any correction for diffuse radiation incident on a slope). Values of slope angle and slope azimuth angles are needed, so a suitable DEM is required. The Lambertian model can be written as:

$$L = L_N \cos(i)$$

where L is the measured radiance, L_N is the equivalent radiance on a flat surface with incidence angle of zero and i is the exitance angle. The Minnaert constant, k, enters into the non-Lambertian model as follows:

$$L_N = \frac{L\cos(e)}{\cos^k(i)\cos^k(e)}$$

where i, L and L_N are defined as before and e is the angle of exitance, which is equal to the slope angle (β_t in the equation below). The value of $\cos(i)$ is found from the relationship

$$\cos(i) = \cos(\theta_s)\cos(\beta_t) + \sin(\theta_s)\sin(\beta_t)\cos(\phi_s - \phi_t)$$

with θ_s being the solar zenith angle, β_t the slope angle, ϕ_s the solar azimuth angle and ϕ_t the slope azimuth angle. The value of the incidence angle, i, will be in the range $0 - 90°$ if the pixel under consideration receives direct illumination. If the value of i falls outside this range then the pixel lies in shadow.

The value of the Minnaert constant k is the slope of the least-squares line relating $y = \log(L\cos(e))$ and $x = \log(\cos(i)\cos(e))$. Most surfaces have k values between 0 and 1; $k = 1$ implies Lambertian reflectance while $k > 1$ implies a dominance of the specular reflectance component. Once k has been determined then the equivalent radiance from a flat surface can be calculated. However, the value of k depends on the nature of the ground cover, and so would vary over the image even in the absence of any other control. Since a sample of pixels is required in order to estimate the slope of the least-squares line relating $\log(L\cos(e))$ and $\log(\cos(i)\cos(e))$ it might be necessary to segment the image into regions of similar land cover type and calculate a value of k for each type. Often, however, the purpose of performing a terrain illumination correction is to improve the identification of land cover types, hence the problem takes on circular proportions. An iterative approach might be possible, in which classification accuracy assessment (Chapter 8) is used as a criterion. However, a simpler approach would be to calculate an average k value for the whole image. Useful references include Bishop and Colby (2002), Blesius and Weirich (2005), Gitas and Devereux (2006), Gu *et al.* (1999) and Hale and Rock (2003). Riaño *et al.* (2003) provide a useful comparative survey of methods.

Parlow (1996) describes a method for correcting terrain-controlled illumination effects using a simulation model of solar irradiance on an inclined surface. The short wave irradiance model (SWIM) computes both direct and diffuse components of irradiance for given atmospheric conditions and allows the conversion of satellite-observed radiances to equivalent radiances for a flat surface. Note that the Minnaert method, described in the preceding paragraphs, does not consider diffuse illumination. Parlow (1996) shows that correction of the image data for terrain illumination effects produces superior classification performance (Chapter 8). Further references are Conese *et al.* (1993), Costa-Posada and Devereux (1995), Egbert and Ulaby (1972), Hay and Mackay (1985), Hill, Mehl and Radeloff (1995), Huang *et al.* (2008), Jones, Settle and Wyatt (1988), Katawa, Ueno and Kusaka (1986), Mitri and Gitas (2004) and Smith, Lin and Ranson (1980).

4.8 Summary

Methods of preprocessing remotely-sensed imagery are designed to compensate for one or more of (i) cosmetic defects, (ii) geometric distortions, (iii) atmospheric interference and (iv) variations in illumination geometry, to calibrate images for sensor degradation, and to correct image pixel values for the effects of topography. The level of preprocessing required will depend on the problem to which the processed images are to be applied. There is therefore no fixed schedule of preprocessing operations that are carried out automatically prior to the use of remotely-sensed data. The user must be aware of the geometrical properties of the image data and of the effects of external factors (such as the level of, and variations in, atmospheric haze) and be capable of selecting an appropriate technique to correct the defect or estimate the external effect, should that be necessary.

The material covered in this chapter represents the basic transformations that must be applied in order to recover estimates of ground-leaving radiance. The development of models requiring such estimates as input has expanded in recent years. Equally importantly, the use of remote sensing to measure change over time is becoming more significant in the context of global environmental change studies. Multitemporal analysis, the comparison of measurements derived by different sensors at different points in time, and the determination of relationships between target radiance and growth and health characteristics of agricultural crops are examples of applications that require the application of corrections described in this chapter. Procedures to accomplish these corrections are not well formulated at present, and the whole area requires more research and investigation.

5 Image Enhancement Techniques

5.1 Introduction

The ways in which environmental remote sensing satellite and aircraft systems collect digital images of the Earth's surface is described in Chapters 1–3. In these chapters, a remotely-sensed image is characterized as a numerical record of the radiance leaving each of a number of small rectangular areas on the ground (called pixels) in each of a number of spectral bands. The range of radiance values at each pixel position is represented (quantized) in terms of a scale which is normally 8 or more bits in magnitude, depending on the type of scanner that is used and on the nature of any processing carried out at the ground station. Each pixel of a digital multispectral image is thus associated with a set of numbers, with one number per spectral band. For example, Landsat ETM+ provides seven bands of multispectral data, and each pixel can therefore be represented as a group (mathematically speaking, a vector or a set) of seven elements, each expressed on the 0–255 (8 bit) range.

A digital image can therefore be considered as a three-dimensional rectangular array or matrix of numbers, the x- and y-axes representing the two spatial dimensions and the z-axis the quantized spectral radiance (pixel value (PV)). A fourth dimension, time, could be added, since satellite data are collected on a routine and regular basis (every 16 days for Landsat-7, for example). The elements of this matrix are numbers in the range $0–2^{n-1}$ where n is the number of bits used to represent the radiance recorded for any given pixel in the image. As described in Section 3.2, image data that are represented by more than $n = 8$ bits per pixel must be scaled to 8 bit representation before they can be stored in the computer's display memory and viewed on the monitor.

Visual analysis and interpretation are often sufficient for many purposes to extract information from remotely-sensed images in the form of standard photographic prints. If the image is digital in nature, such as the satellite and aircraft-acquired images considered in this book, a computer can be used to manipulate the image data and to produce displays that satisfy the particular needs of the interpreter, who may need, for example to produce an image map for use in a geographical information system (GIS). In this chapter, methods of enhancing digital images are considered. The term *enhancement* is used to mean the alteration of the appearance of an image in such a way that the information contained in that image is more readily interpreted visually in terms of a particular need. Since the choice of enhancement technique is problem-dependent, no single standard method of enhancement can be said to be 'best', for the needs of each user will differ. Also, the characteristics of each image in terms of the distribution of PVs over the 0–255 display range will change from one image to another; thus, enhancement techniques suited to one image (for example covering an area of forest) will differ from the techniques applicable to an image of another kind of area (for example the Antarctic ice-cap).

There are a number of general categories of enhancement technique and these are described in the following sections. As in many other areas of knowledge, the distinction between one type of analysis and another is a matter of personal taste; some kinds of image transformations (Chapter 6) or filtering methods (Chapter 7) can, for instance, reasonably be described as enhancement techniques. In this chapter we concentrate on ways of improving the visual interpretability of an image by one of two methods:

1. Altering image contrast and
2. Converting from greyscale to colour representation.

The first group of techniques consists of those methods which can be used to compensate for inadequacies of what, in photographic terminology, would be called 'exposure'; some images are intuitively felt to be 'too dark', while others are over-bright. In either case, information is not as easily comprehended as it might be if the contrast of the image were greater. In this context, contrast is simply the range and distribution of the PVs over the 0–255 scale used by the computer's display memory. The second category includes those methods that allow the information content of greyscale image to be re-expressed in colour. This is sometimes desirable, for the eye is more sensitive to variations in hue than to changes in brightness.

The chapter begins with a brief description of the human visual system, since the techniques covered in the following sections are fundamentally concerned with the visual comprehension of information displayed in image form.

5.2 Human Visual System

There are a number of theories that seek to explain the manner in which the human visual system operates. The facts on which these theories are based are both physical (to do with the external, objective world) and psychological (to do with our internal, conscious world). Concepts like 'red' and 'blue' are an individual's internal or sensory response to external stimuli. Light reaching the eye passes through the pupil and is focused onto the retina by the lens (Figure 5.1a,b). The retina contains large numbers of light-sensitive photoreceptors, termed rods and cones. These photoreceptors are connected via a network of nerve fibres to the optic nerve, along which travel the signals that are interpreted by the brain as images of our environment.

There are around 100 million rod-shaped cells on the retina, and 5 million cone-shaped cells. Each of these cells is connected to a nerve, the junction being called a synapse. The way in which these cells respond to light is through alteration of a molecule known as a chromophore. Changes in the amount of light reaching a chromophore produce signals that pass through the nerve fibre to the optic nerve. Signals from the right eye are transmitted through the optic nerve to the left side of the brain, and vice versa.

It is generally accepted that the photoreceptor cells, comprising the rods and cones, differ in terms of their inherent characteristics. The rod-shaped cells respond to light at low illumination levels, and provide a means of seeing in such conditions. This type of vision is called *scotopic*. It does not provide any colour information, though different levels of intensity can be distinguished. Cone or photopic vision allows the distinction of colours or hues and the perception of the degree of saturation (purity) of each hue as well as the intensity level. However, photopic vision requires a higher illumination level than does scotopic vision. Colour is thought to be associated with cone vision because there are three kinds of cones, each kind being responsive to one of the three primary colours of light (red, green and blue (RGB)). This is called the tristimulus theory of colour vision. Experiments have shown that the number of blue-sensitive cones is much less than the number of red- or green-sensitive cones, and that the areas of the visible spectrum in which the three kinds of cones respond do, in fact, overlap (Figure 5.2). There are other theories of colour (Fairchild, 2005; Malacara, 2002) but the tristimulus theory is an attractive one not merely because it is simple but because it provides the idea that colours can be formed by adding RGB light in various combinations. Bruce, Green and Georgeson (2003) is a more advanced book covering the physiological and psychological processes of visual perception. Other useful references are Tovée (1996) and the computer vision literature, for example Faugeras (1993) and Shapiro and Stockman (2001). Drury (2004) discusses the properties of the human visual system in relation to the choice of techniques for processing remotely sensed images for geological applications.

A model of 'colour space' can be derived from the idea that colours are formed by adding together differing amounts of RGB. Figure 5.3 shows a geometrical representation of the RGB colour cube. The origin is

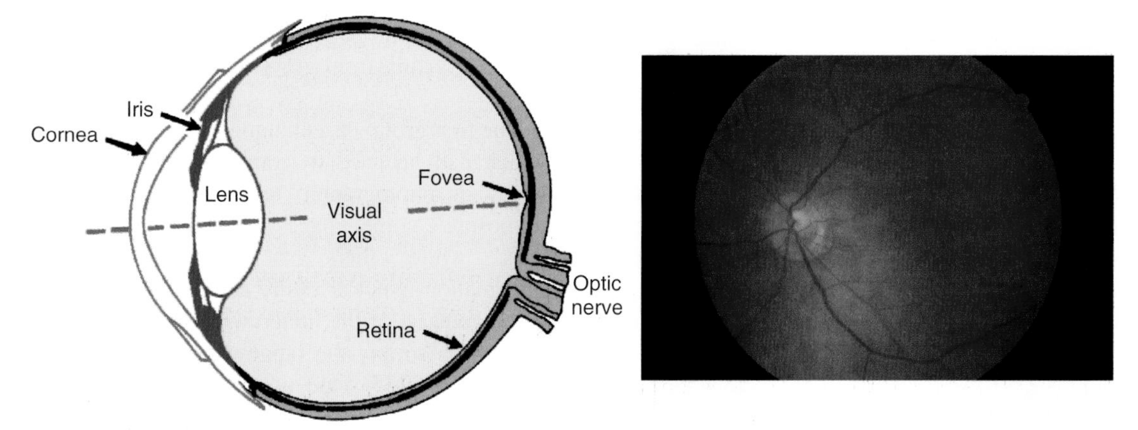

Figure 5.1 *(a) Simplified diagram of the human eye. (b) Senior author's retina. Arteries and veins are clearly visible, and they converge on the optic nerve, which appears in a lighter colour. Rods and cones on the surface of the retina are tiny, and are only visible at a much greater magnification. Courtesy Thomas Bond and Partners, Opticians, West Bridgford, Nottingham.*

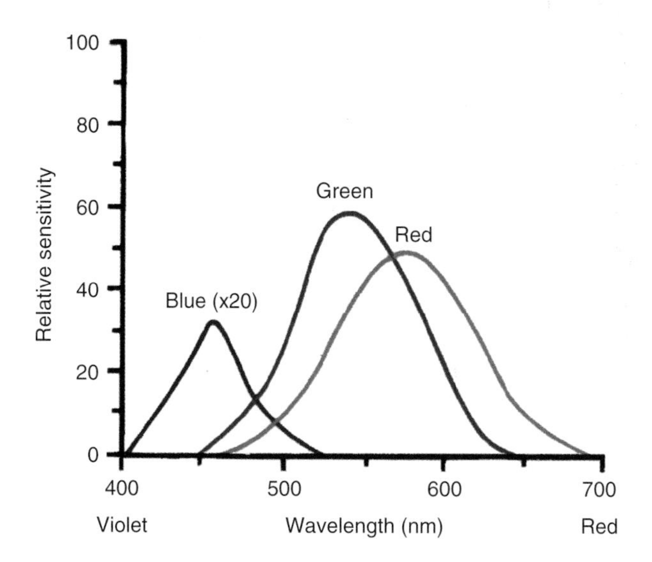

Figure 5.2 *Sensitivity of the eye to red, green and blue light.*

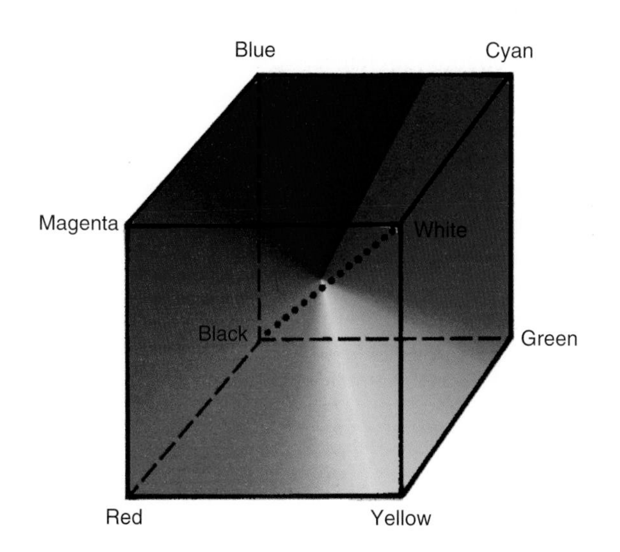

Figure 5.3 *Red–green–blue colour cube.*

at the vertex of the cube marked 'black' and the axes are black–red, black–green and black–blue. A specific colour can be specified by its coordinates along these three axes. Black represents the absence of colour. These coordinates are termed (R, G, B) triples. Notice that white light is formed by the addition of maximum red, maximum green and maximum blue light. The line joining the black and white vertices of the cube represents colours formed by the addition of equal amounts of RGB light; these are shades of grey. Colour television makes use of the RGB model of colour vision. A cathode ray television screen is composed of an array of dots, each of which contains RGB-sensitive phosphors. Colours on the screen are formed by exciting the RGB phosphors in differing proportions. If the proportions of RGB were equal at each point (but varying over the area of the screen) a greyscale image would be seen. A colour picture is obtained when the amounts of RGB at each point are unequal, and so – in terms of the RGB cube – the colour at any pixel is represented by a point that is located away from the black-white diagonal line. Flat screen technology using a liquid crystal display (LCD) and thin film transistor (TFT) allows light from a fluorescent source to pass through RGB liquid crystals with intensity proportional to the voltage at that point. The number of points (representing image pixels) is determined by screen size. Each pixel can be addressed separately and a picture produced in a way that is conceptually similar to the image on a conventional cathode ray tube (CRT) display. LCD screens can be much bigger than CRT; commercial TVs with a diagonal screen size of 52 inches (132 cm) are readily available.

The RGB colour cube model links intuitively with the tristimulus theory of colour vision and also with the way in which a colour television monitor works. Other colour models are available which provide differing views of the nature of our perception of colour. The hue–saturation–intensity (HSI) model uses the concepts of hue, saturation and intensity to explain the idea of colour. Hue is the dominant wavelength of the colour we see; hues are given names such as red, green, orange and magenta. The degree of purity of a colour is given by its saturation. Intensity is a measure of the brightness of a colour. Figure 5.4 shows a geometrical representation of the HSI model. Hue is represented by the top edge of a six-sided cone (hexcone) with red at 0°, green at 120° and blue at 240°, then back to red at 360°. Pure unsaturated and maximum intensity colours lie around the top edge of the hexcone. Addition of white light produces less saturated, paler, colours and so saturation can be represented by the distance from the vertical axis of the hexcone. Intensity (sometimes called value)

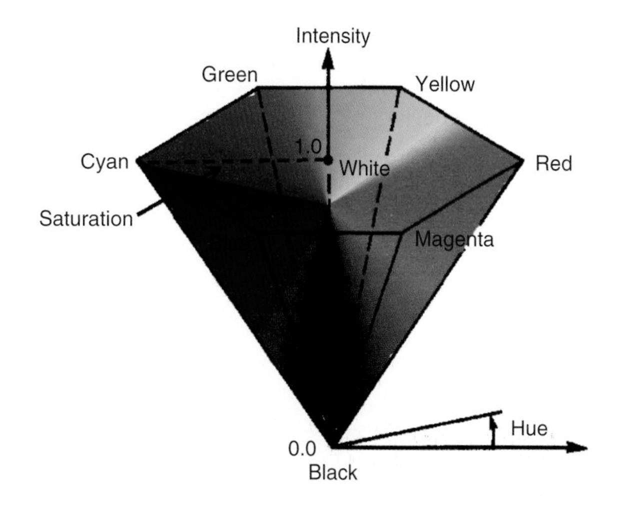

Figure 5.4 *Hue–saturation–intensity (HSI) hexcone.*

is shown as a distance above the apex of the hexcone, increasing upwards as shown by the widening of the hexcone. The point marked black has no hue, nor do any of the shades of grey lying on the vertical axis between the black and white points. All these shades of grey, including white and black, have zero saturation.

The RGB model of colour is that which is normally used in the study and interpretation of remotely-sensed images, and in the rest of this chapter we will deal exclusively with this model. The use of the HSI model is considered in Chapter 6 in common with other image transforms, for the representation of colours in terms of the HSI model can be accomplished by a straightforward transformation of the RGB colour coordinates. The HSI transform can be used to enhance multispectral images in terms of their colour contrast (Section 5.3).

5.3 Contrast Enhancement

The sensors mounted onboard aircraft and satellites have to be capable of detecting upwelling radiance levels ranging from low (for example over oceans) to very high (for example over snow or ice). For any particular area that is being imaged it is unlikely that the full dynamic range from 0 to $(2^n - 1)$ levels of the sensor will be used and the corresponding image is dull and lacking in contrast or over-bright. In terms of the RGB colour cube model of Section 5.2 the PVs are clustered around a narrow section of the black–white axis (Figure 5.3). Not much detail can be seen on such images, which are either underexposed or overexposed in photographic terms. If the range of levels used by the display system could be altered so as to fit the full range of the black-white axis of Figure 5.3 then the contrast between the dark and light areas of the image would be improved while maintaining the relative distribution of the grey levels.

5.3.1 Linear Contrast Stretch

In its basic form the linear contrast-stretching technique involves the translation of the image PVs from the observed range V_{min} to V_{max} to the full range of the display device (generally 0–255, which assumes an 8-bit display memory; see Chapter 3). V is a PV observed in the image under study, with V_{min} being the lowest PV in the image and V_{max} the highest. The PVs are scaled so that V_{min} maps to a value of 0 and V_{max} maps to a value of 255. Intermediate values retain their relative positions, so that the observed PV in the middle of the range from V_{min} to V_{max} maps to 127. Notice that we cannot map the middle of the range of the observed PVs to 127.5 (which is exactly half way between 0 and 255) because the display system can store only the discrete levels 0, 1, 2, ..., 255.

Some dedicated image processing systems include a hardware lookup table (LUT) that can be set so that the colour that you see at a certain pixel position on the screen is a mapping or modification of the colour in the corresponding position in the display memory. The colour code in the display memory remains the same, but the mapping function may transform its value, for example by using the linear interpolation procedure described in the preceding paragraph. The fact that the colour values in the display memory are not altered can be a major advantage if the user has adopted a trial and error approach to contrast enhancement. The mapping is accomplished by the use of a LUT that has 256 entries, labelled 0–255. In its default state, these 256 elements contain the values 0–255. A PV of, say, 56 in the display memory is not sent directly to the screen, but is passed through the LUT. This is done by reading the value held in position 56 in the LUT. In its default (do nothing) state, entry 56 in the LUT contains the value 56, so the screen display shows an image of what is contained in the display memory. To perform a contrast stretch, we first realize that the number of separate values contained in the display memory for a given image is calculated as $(V_{max} - V_{min} + 1)$, which must be 256 or less for an 8-bit display. All LUT output values corresponding to input values of V_{min} or less are set to zero, while LUT output values corresponding to input values of V_{max} or more are set to 255. The range $V_{min} - V_{max}$ is then linearly mapped onto the range 0–255, as shown in Figure 5.5. Using the LUT shown in this figure, any pixel in the image having the value 16 (the minimum PV in the image) is transformed to an output

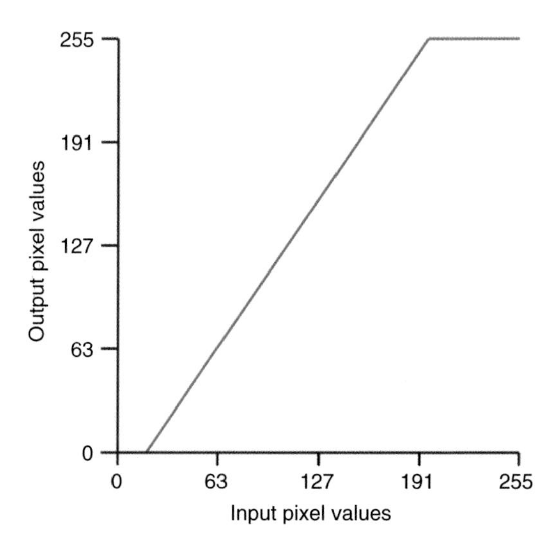

Figure 5.5 *Graphical representation of lookup table to map input pixel values 16–191 on to the full intensity range 0–255. Input values less than 16 are set to 0 on output. Input values of 191 or greater are set to 255 on output. Input values between 16 and 191 inclusive are linearly interpolated to output values 0–255.*

value of 0 before being sent to the digital-to-analogue converter and thence to the display screen. All input values of 191 and more are transformed to output values of 255. The range of input values between 16 and 191 is linearly interpolated onto the full dynamic range of the display device, assumed in this example to be 0–255. If a colour (RGB) image is being stretched then the process is repeated separately for each of the components (R then G then B). Figure 5.6a shows a Landsat-7 ETM+ image of the south-east corner of The Wash in eastern England. The un-stretched image is shown. The histograms of the RGB inputs (corresponding to Landsat ETM+ bands 4, 3 and 2, respectively) are shown in Figure 5.6b. The histograms are calculated from the image PVs, and are simply counts of the number of PVs having the value 0, 1, . . . , 255. Recall that low PVs generate darker shades. The histograms in Figure 5.6b indicate that the image is dark (we can see that from Figure 5.6a but the histogram allows the quantification of the degree of darkness and lightness, and it plays a key role in the methods of image enhancement that are discussed in this chapter).

Figure 5.7a shows the image in Figure 5.6a after a linear contrast stretch in which the highest PV in each of the three bands is stretched to the value 255 and the lowest value is stretched to 0, thus using the full 0–255 dynamic range. Figure 5.7b is the histogram of each channel after the stretch has been applied. The difference between the stretched and raw image is not great.

A slight modification can be used to provide a little more user interaction. The basic technique, as described above, does not take into consideration any characteristic of the image data other than the maximum and minimum PVs. These values may be outliers, located well away from the rest of the image data values. If this were the case, then it could be observed if the image histogram were computed and displayed. It is a relatively straightforward matter to find the 5th and 95th (or any other usually symmetrical pair of) percentile values of the distribution of PVs from inspection of the image histogram. The fifth percentile point is that value exceeded by 95% of the image PVs, while the 95th percentile is the PV exceed by 5% of all PVs in the image. If, instead of V_{max} and V_{min} we use $V_{95\%}$ and $V_{5\%}$ then we can carry out the contrast enhancement procedure so that all PVs equal to or less than $V_{5\%}$ are output as zero while all PVs greater than $V_{95\%}$ are output as 255. Those values lying between $V_{5\%}$ and $V_{95\%}$ are linearly mapped (interpolated), as before, to the full brightness scale of 0–255. Again, this technique is applied separately to each component (RGB) of a false-colour image. Of course, values other than the 5th and 95th percentiles could be used; for example one might elect to choose the 10th and 90th percentiles, or any other pair. The chosen percentage points are usually symmetric around the 50% point, but not necessarily so. Figure 5.8a,b illustrates the application of a 5% linear contrast stretch.

Some image data providers use the value zero as a 'bad pixel' indicator. Others pad a geometrically corrected image (Section 4.3) with 0 pixel. These are called zero-fill pixels. We normally do not want to count these pixels in the image histogram, as their inclusion would bias the histogram. Most software allows 0 pixel to be ignored. Another possibility is that interest centres around a particular part of the 0–255 brightness range, such as 180–250. It is possible to stretch this range so that a PV of 180 maps to zero, and a PV of 250 maps to 255 with values 181–249 being interpolated linearly. Values outside the 180–250 range remain the same. This kind of contrast stretch destroys the relationship between PV and brightness level, but may be effective in visualization of a particular aspect of the information content of the image. Example 5.1 illustrates the use of the linear contrast stretch.

5.3.2 Histogram Equalization

The whole image histogram, rather than its extreme points, is used in the more sophisticated methods of contrast enhancement. Hence, the shape as well as the extent of the histogram is taken into consideration. The first of the two methods described here is called histogram equalization. Its underlying principle is straightforward. It is assumed that in a well-balanced image the histogram should be such that each brightness level contains an approximately equal number of PVs, so that the histogram of these displayed values is almost uniform (though not all 256 available levels are necessarily non-zero). If this operation, called histogram equalization, is performed then the entropy of the image, which is a measure of the information content of the image, will be increased (Section 2.2.3). Because of the nature of remotely-sensed digital images, whose pixels can take on only the discrete values 0, 1, 2, . . . , 255 it may be that there are 'too many' PVs in one class, even after equalization. However, it is not possible to take some of the values from that over-populated class and redistribute them to another class, for there is no way of distinguishing between one PV of 'x' and another of the same value. It is rare, therefore, for a histogram of the PVs of an image to be exactly uniformly distributed after the histogram equalization procedure has been applied.

The method itself involves, firstly, the calculation of the target number of PVs in each class of the equalized histogram. This value (call it n_t) is easily found by dividing N, the total number of pixels in the image, by 256 (the number of histogram classes, which is the number of intensity levels in the image). Next, the histogram of

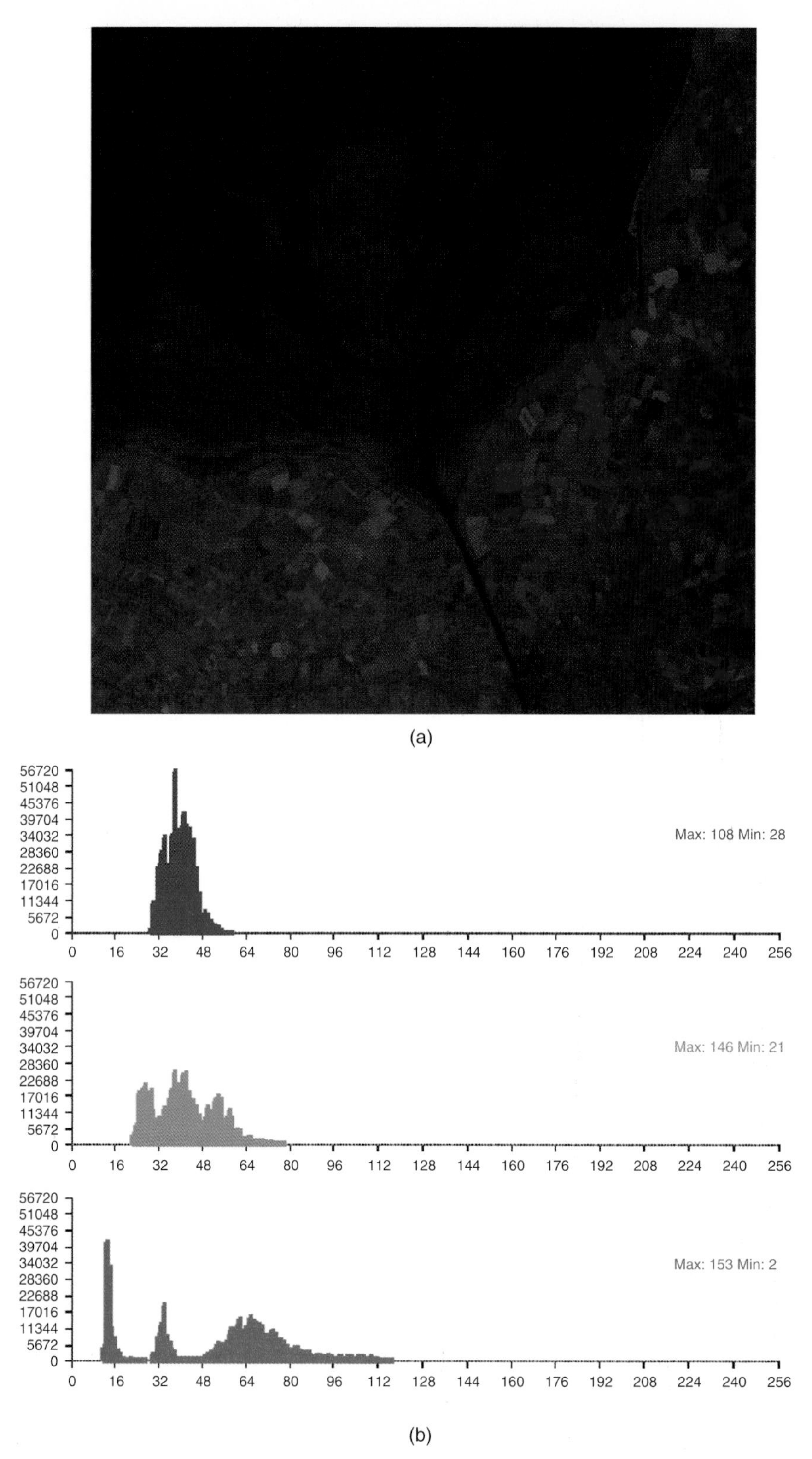

Figure 5.6 *(a) Raw Landsat-7 ETM+ false colour composite image (using bands 4, 3 and 2 as the RGB inputs) of the south-east corner of The Wash, an embayment in eastern England. The River Ouse can be seen entering The Wash. (b) Frequency histograms of the 256 colour levels used in the RGB channels. Landsat data courtesy NASA/USGS.*

(a)

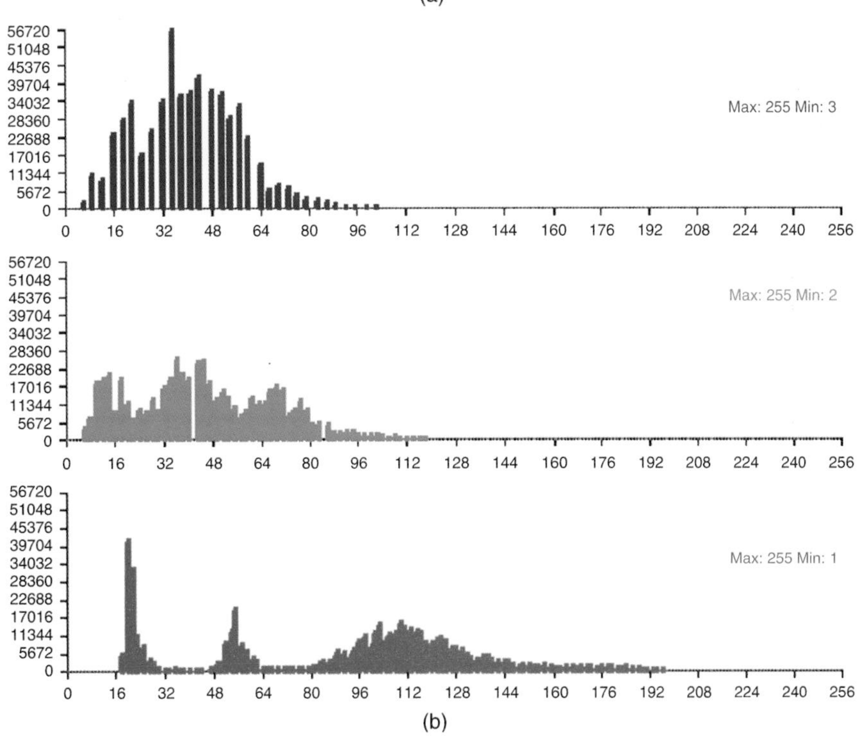

(b)

Figure 5.7 *(a) Image shown in Figure 5.6a after a linear contrast stretch in which the minimum and maximum histogram values in each channel are set to 0 and 255 respectively. (b) The histograms for the stretched image.*

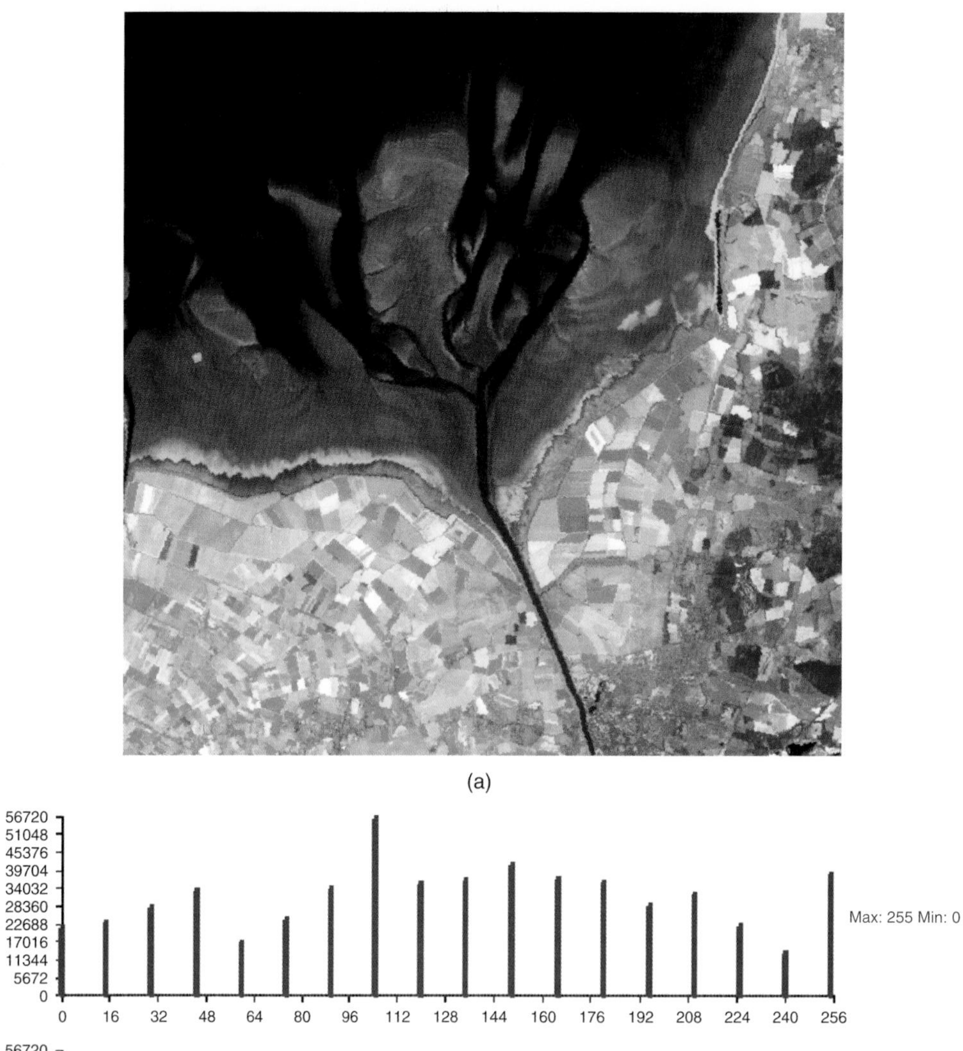

(a)

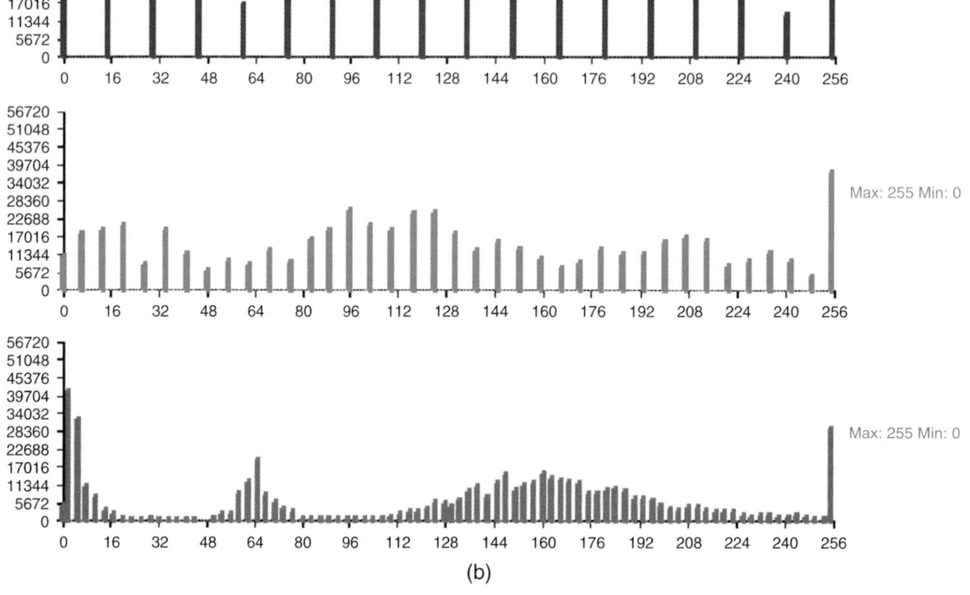

(b)

Figure 5.8 *(a) Linear contrast stretch applied to the image shown in Figure 5.6a. The 5th and 95th percentile values of the cumulative image histograms for the RGB channels are set to 0 and 255 respectively and the range between the 5th and 95th percentiles is linearly interpolated onto the 0–255 scale. (b) Image histograms corresponding to the RGB channels (Landsat TM bands 4, 3 and 2). Landsat data courtesy NASA/USGS.*

Example 5.1: Linear Contrast Stretch

This example uses the image set that is referenced by the image dictionary file *missis.inf*, so you should ensure that this file (and the referenced image files) are available. The image file set consists of seven Landsat TM bands of an area of the Mississippi River south of Memphis, TN, USA. The images have 512 lines of 512 pixels.

Use **View|Display Image** to select **missis.inf** as the current image dictionary file. Display band 4 (**misstm4.img**) in greyscale. The default state of the LUT is to map a pixel intensity value of x (on the range 0–255) to a display value of x, also on the range 0–255. If the PVs in the image are low, then the image will appear dark. Example 5.1 Figure 1 shows an image that is so dark that it may as well have been taken at night. The distribution of image PVs can be seen by displaying the image histogram (**Plot|Histogram**), and it is immediately clear why the image is so dark – its dynamic range is very low (Example 5.1 Figure 2). The extreme values are 4 and 79, though the majority of the PVs are in the range 10–55.

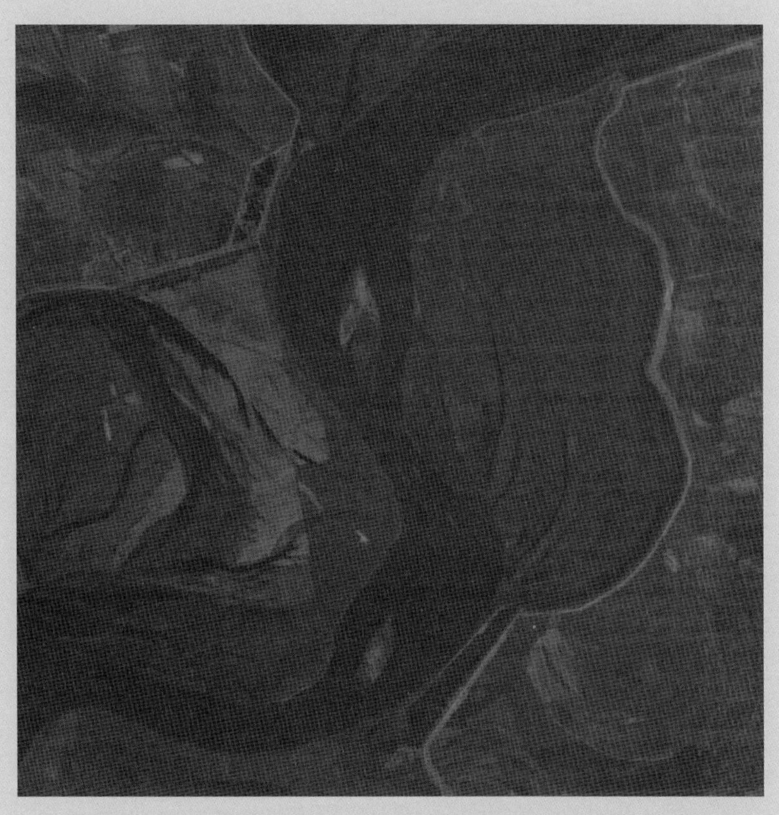

Example 5.1 Figure 1. *Band 4 of Landsat-4 TM image of the Mississippi River near Memphis (details in the file* **missisp.inf***). The dynamic range of the image is very low, and no detail can be seen. The histogram of this image is shown in Example 5.1 Figure 2.*

Now choose **Enhance|Stretch** and, if there is more than one image on the screen, select the appropriate window. Click the button with the caption **Automatic Stretch** and the image shown in Example 5.1 Figure 3 is generated, and displayed in a new window. The Automatic Stretch sets the LUT so that the actual range of the data $-x_{min}$ to $x_{max}-$ is mapped linearly onto the output brightness range of 0–255. Use **Plot|Histogram** to view the histogram of the automatically stretched image (Example 5.1 Figure 4).

Although the image shown in Example 5.1 Figure 3 is much easier to see than the image shown in Example 5.1 Figure 1, it is still rather dark, possibly because x_{min} and x_{max}, the extreme pixel intensity values, are outliers. Try a third experiment. Choose **Enhance|Stretch** again, and this time click **User percentage limits**. An input box will appear, with space for you to enter a lower value and an upper value. If you enter 0 and 100% then the

(Continues on next page)

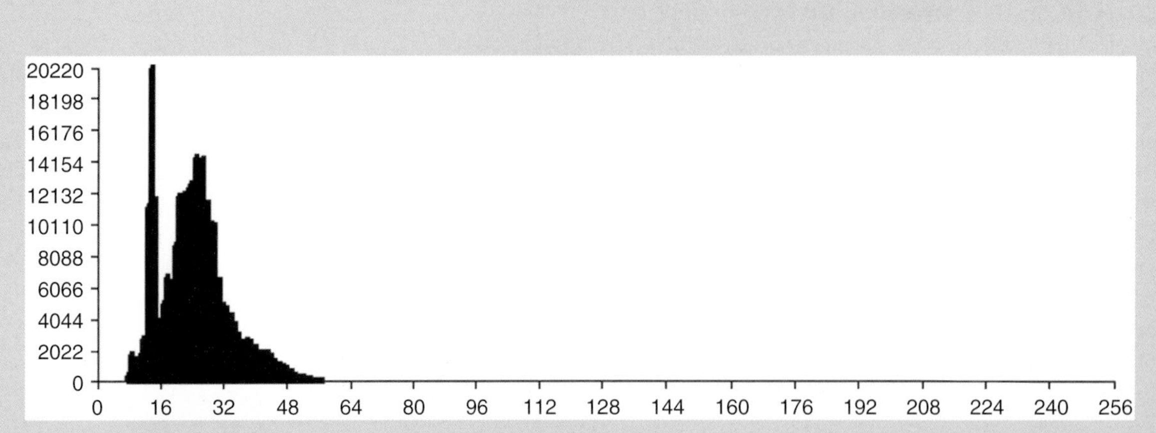

Example 5.1 Figure 2. *Histogram of the image shown in Example 5.1 Figure 1. The narrow peak at a pixel value of 14–15 represents water. The main, wider peak represents land. There are few pixel values greater than 58–60, so the image is dark. The range of pixel values is not great (approximately 8–60) and so contrast is low.*

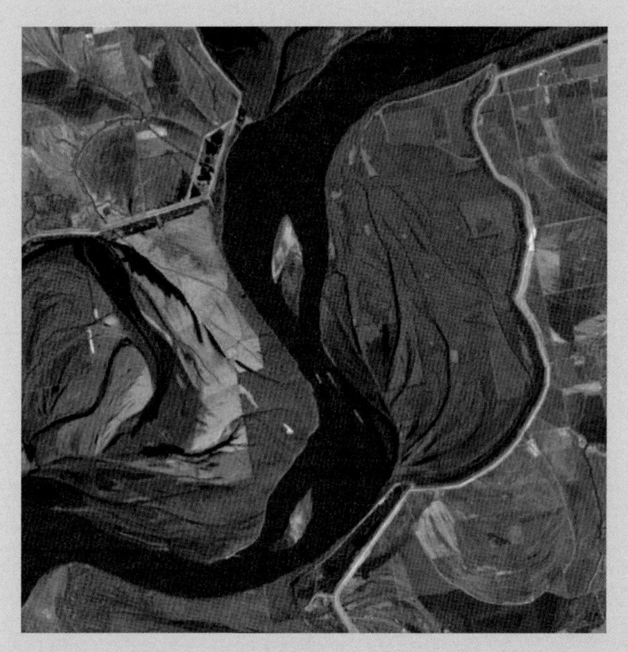

Example 5.1 Figure 3. *The image shown in Example 5.1 Figure 1 after an automatic linear contrast stretch. The automatic stretch maps the dynamic range of the image (8–60 in this case) to the dynamic range of the display (0–255). Compare the histogram of this image (Example 5.1 Figure 4) with the histogram of the raw image (Example 5.1 Figure 2).*

minimum and maximum values in the data, $x_{\min}$ and $x_{\max}$, will be used as the limits of the linear stretch, just as if you had chosen **`Automatic Stretch`**. What is needed is to chop off the extreme low and the extreme high PVs. Look at the histogram of the raw image (Example 5.1 Figure 2) again. The extreme values (from visual inspection) appear to be 8 and 58, yet the lowest PV is 4 and the highest is 79. These relatively few low and high values are distorting the automatic stretch. If we ignore the lowest 5% and the highest 5% of the PVs then we may see an improvement in image brightness. You actually enter 5 and 95 in the two boxes because they represent the percentages of pixels that are *lower* than the specified percentage points – in other words, we specify the 5th and 95th percentiles. The result is shown in Example 5.1 Figure 5, and the corresponding histogram is shown in Example 5.1 Figure 6.

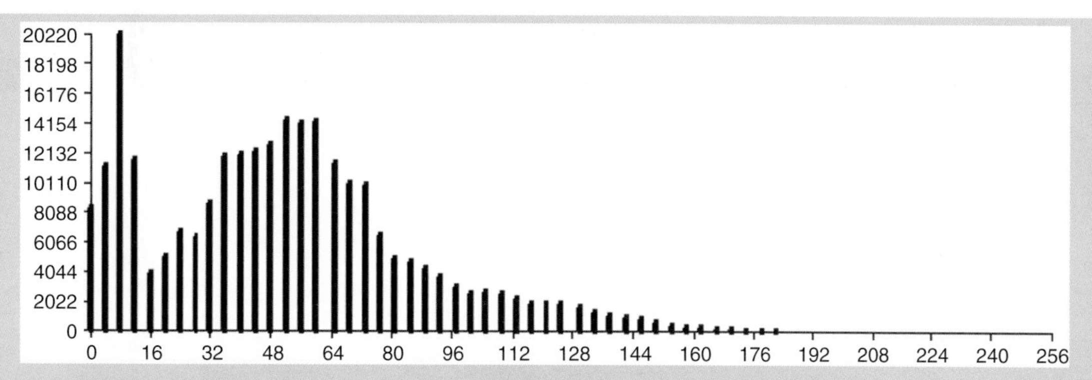

Example 5.1 Figure 4. *Histogram of the contrast-stretched image shown in Example 5.1 Figure 3. Although the lower bound of the dynamic range of the image has been changed from its original value of 8 to 0, the number of pixels with values greater than 182 is relatively low. This is due to the presence of a small number of brighter pixels that are not numerous enough to be significant, but which are mapped to the white end of the dynamic range (255).*

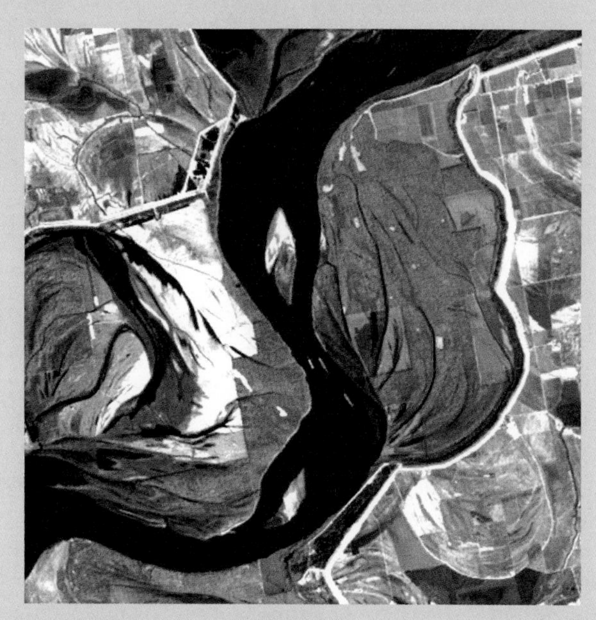

Example 5.1 Figure 5. *The same image as shown in Example 5.1 Figures 1 and 3. This time, a percentage linear contrast stretch has been applied. Rather than map the lowest image pixel value to an output brightness value of zero, and the highest to 255, 2 pixel values are found such that 5% of all image pixel values are less than the first value and 5% are greater than the second. These two values are then mapped to 0 and 255 respectively.*

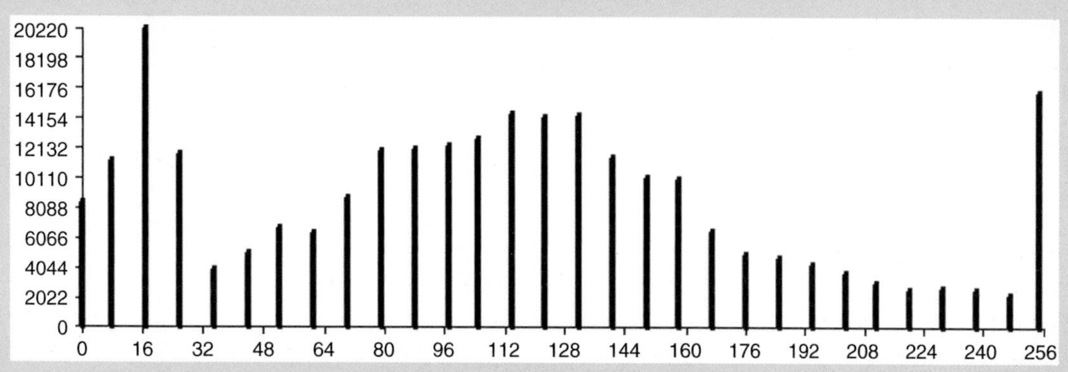

Example 5.1 Figure 6. *Histogram of the image shown in Example 5.1 Figure 5. A percentage linear contrast stretch (using the 5 and 95% cutoff points) has been applied. The displayed image is now brighter and shows greater contrast than the image in Example 5.1 Figure 3.*

(Continues on next page)

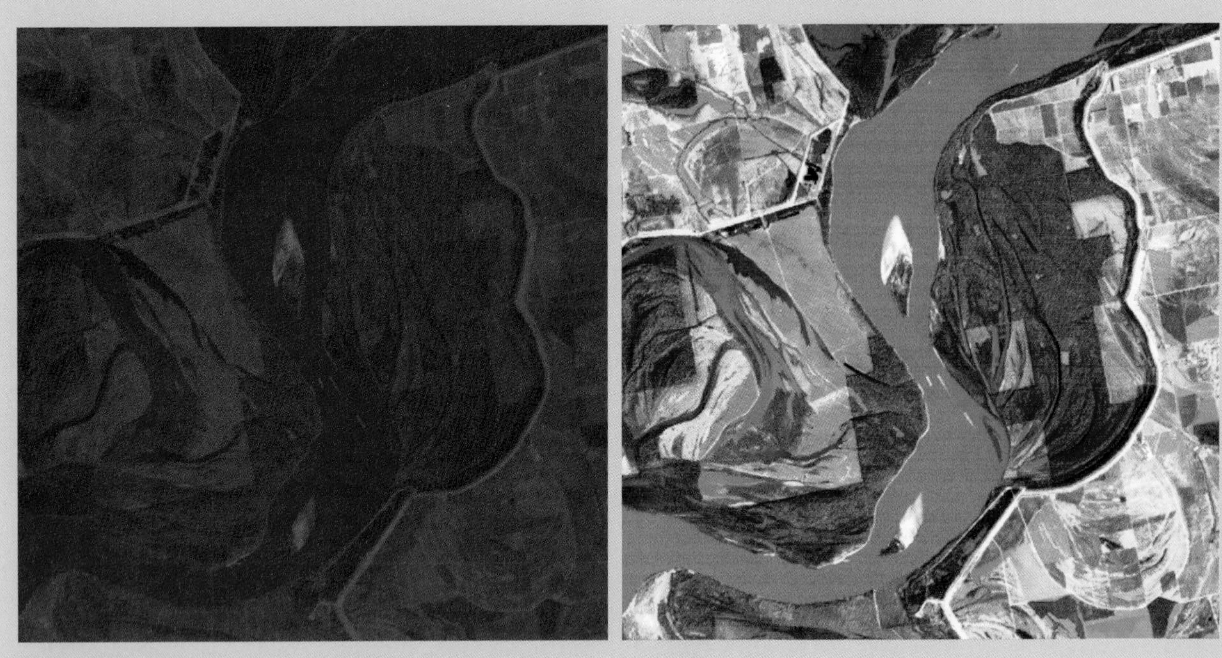

Example 5.1 Figure 7. *Contrast stretching can be applied to all three bands of a natural or false colour composite. (a) raw Mississippi image, and (b) after a histogram equalisation contrast stretch.*

You should relate the visual differences between the images shown in Example 5.1 Figures 1, 3 and 5 to the shape of the corresponding histograms (Example 5.1 Figures 2, 4 and 6) in order to appreciate the relationship between brightness and contrast on the one hand, and the shape and spread of the corresponding histogram.

Now try the following experiments:

- Generate stretched images from the raw image **misstm4.img** using the automatic and manual (**Set DN limits manually**) options. How would you estimate the upper and lower limits of the stretch for the manual option? What happens if you use a range narrower than the 13–42 range used by the 5–95% option? Explain the differences between the output images.
- Try the **Histogram Equalize** and **Gaussian Stretch** options, using a range of standard deviation limits for the latter. Look at the histograms for each output image. What happens if you opt to use a specified part of the histogram rather than its full range? What happens if you increase or decrease the number of standard deviations to either side of the mean in the Gaussian option?
- Which technique produces the 'best' output image? How would you define 'best'?
- Repeat the exercise with a different image, such as **Colo4.img** from the image set referenced by **litcolorado.inf**. Can you say that these techniques always work, or that one method is always better than the rest?

the input image is converted to cumulative form with the number of pixels in classes 0–j represented by C_j. This is achieved by summing the number of pixels falling in classes 0–j of the histogram (the histogram classes are labelled 0–255 so as to correspond to the PVs on an 8-bit scale):

$$C_j = n_0 + n_1 + \ldots + n_j$$

where n_j is the number of pixels taking the greyscale value j. The output level for class j is calculated very simply as C_j/n_t.

The method is not as complicated as it seems, as the example in Table 5.1 demonstrates. The PVs in the raw image (16 levels in the example) are shown in column 1 of Table 5.1, and the number of pixels at each level is given in column 2. The cumulative

Table 5.1 *Illustrating calculations involved in histogram equalization procedure. $N = 262\,144$, $n_t = 16\,384$. See text for explanation.*

Old LUT value	Number in class	Cumulative number	New LUT value
0	1311	1311	0
1	2622	3933	0
2	5243	9176	0
3	9176	18352	1
4	13108	31460	1
5	24904	56364	3
6	30146	86510	5
7	45875	132385	8
8	58982	191367	11
9	48496	239863	14
10	11796	251659	15
11	3932	255591	15
12	3932	259523	15
13	2621	262144	15
14	0	262144	15
15	0	262144	15

number of pixels is listed in column 3. The values in column 4 are obtained by determining the target number of pixels (= total number of pixels divided by the number of classes, that is 262 144/16 = 16 384) and then finding the integer part of C_j divided by n_t, the target number. Thus, input levels 0–2 are all allocated to output level 0, input levels 3 and 4 are allocated to output level 1, and so on. Notice that the classes with relatively low frequency have been amalgamated while the classes with higher frequency have been spaced out more widely than they were originally. The effect is to increase the contrast in the centre of the range while reducing contrast at the margins. Table 5.2 gives the numbers of pixels assigned to each output level.

In this example, which uses only 16 levels for ease of understanding, the output histogram is not uniform. This is not surprising, for the number of pixels at five of the input levels considerably exceeds the target number of 16 384.

The example given in Table 5.1 shows that the effect of the histogram equalization procedure is to spread the range of PVs present in the input image over the full range of the display device; in the case of a colour monitor. This range is normally 256 levels for each of the primary colours (RGB). The relative brightness of the pixels in the original image is not maintained. Also, in order to achieve the uniform histogram the number of levels used is almost always reduced (see for example Table 5.2). This is because those histogram classes with relatively few members are amalgamated to make up the target number, n_t. In the areas of the histogram that have the greatest class frequencies the individual classes are stretched out over a wider range. The effect is to increase the contrast in the densely populated parts of the histogram and to reduce it in other, more sparsely populated areas. If there are relatively few discrete PVs after the equalization process then the result may be unsatisfactory compared to the simple linear contrast stretch.

Sometimes it is desirable to equalize only a specified part of the histogram. For example, if a mask is used to eliminate part of the image (for example water areas may be set to zero) then a considerable number of PVs of zero will be present in the histogram. If there are N zero pixels then the output value corresponding to an input PV of zero after the application of the procedure described above will be N/n_t, which may be large; for instance, if N is equal to 86 134 and n_t is equal to 3192 then all the zero (masked) values in the original image will be set to a value of 27 if the result is rounded to the nearest integer. A black mask will thus be transformed into a dark grey one, which may be undesirable. The calculations described above can be modified so that the input histogram cells between, say, 0 and a lower limit L are not used in the calculations. It is equally simple to eliminate input histogram cells between an upper limit H

Table 5.2 *Number of pixels allocated to each class after the application of the equalisation procedure shown in Figure 5.1a. Note that the smaller classes in the input have been amalgamated, reducing the contrast in those areas, while larger classes are more widely spaced, giving greater contrast. The number of pixels allocated to each non-empty class varies considerably, because discrete input classes cannot logically be split into subclasses.*

Intensity	0	1	2	3	4	5	6	7
Number	9176	22284	0	24904	0	30146	0	0
Intensity	8	9	10	11	12	13	14	15
Number	45875	0	0	58982	0	0	48496	22281

and 255; indeed, any of the input histogram cells can be excluded from the calculations.

The Wash image shown in Figure 5.6a is displayed in Figure 5.9a after a histogram equalization contrast stretch. The histogram of the image in Figure 5.9a is given in Figure 5.9b.

5.3.3 Gaussian Stretch

A second method of contrast-enhancement based upon the histogram of the image PVs is called a Gaussian stretch because it involves the fitting of the observed histogram to a Normal or Gaussian histogram. A Normal distribution gives the probability of observing a value x given the mean $\overline{x}$ is defined by

$$p(x) = \frac{1}{\sigma\sqrt{2\pi}} e^{\frac{-(x-\overline{x})^2}{2\sigma^2}}$$

The standard deviation, σ, is defined as the range of the variable for which the function $p(x)$ drops by a factor of $e^{-0.5}$ or 0.607 of its maximum value. Thus, 60.7% of the values of a normally distributed variable lie within one standard deviation of the mean. For many purposes a Standard Normal distribution is useful. This is a Normal distribution with a mean of zero and a unit standard deviation. Values of the Standard Normal distribution are tabulated in standard statistics texts, and formulae for the derivation of these values are given by Abramowitz and Stegun (1972).

An example of the calculations involved in applying the Gaussian stretch is shown in Table 5.3. The input histogram is the same as that used in the histogram equalization example (Table 5.1). Again, 16 levels are used for the sake of simplicity. Since the Normal distribution ranges in value from $-\infty$ to $+\infty$, some delimiting points are needed to define the end points of the area of the distribution that are to be used for fitting purposes. The range ± 3 standard deviations from the mean is used in the example. Level 1 is, in fact, the probability of observing a value of a Normally distributed variable that is three standard deviations or more *below* the mean; level 2 is the probability of observing a value of a Normally-distributed variable that is between 2.6 and 3 standard deviations below the mean, and so on. These values can be derived from an algorithm based on the approximation specified by Abramowitz and Stegun (1972). Column (i) of the table shows the PVs in the original, un-enhanced image. Column (ii) gives the points on the Standard Normal distribution to which these PVs will be mapped, while column (iii) contains the probabilities, as defined above, which are associated with the class intervals. Assume that the number of pixels in the image is $512 \times 512 = 262\,144$ and the number of quantization levels is 16. The target number of pixels (that is the number of pixels that would be observed if their distribution were Normal) is found by multiplying the probability for each level by the value 262 144. These results are contained in column (iv) and, in cumulative form, in column (v). The observed counts for the input image are shown by class and in cumulative form in columns (vi) and (vii). The final column gives the level to be used in the Gaussian-stretched image. These levels are determined by comparing columns (v) and (vii) in the following manner. The value in column (vii) at level 0 is 1311. The first value in column (v) to exceed 1311 is that associated with level 1, namely, 1398; hence, the input level 0 becomes the output level 1. Taking the input (cumulative) value associated with input level 1, that is 3933, we find that the first element of column (v) to exceed 3933 is that value associated with level 3 (9595) so input level 1 becomes output level 3. This process is repeated for each input level. Once the elements of column (viii) have been determined, they can be written to the LUT and the input levels of column (i) will automatically map to the output levels of column (viii).

The range ± 3 standard deviations as used in the example is not the only one which could have been used. A larger or smaller proportion of the total range of the Standard Normal distribution can be specified, depending on the requirements of the user. Usually, the limits chosen are symmetric about the mean, and the user can provide these limits from a terminal. An example of the Gaussian contrast stretch is given in Figure 5.10.

If Tables 5.1 and 5.2 are compared it will be seen that the Gaussian stretch emphasizes contrast in the tails of the distribution while the histogram equalization method reduces contrast in this region. However, at the centre of the distribution the reverse may be the case, for the target number for a central class may well be larger for the Gaussian stretch than the histogram equalization. In the worked example the target number for each class in the histogram equalization was 16 384; note that the target numbers for classes 5–10 inclusive in the Gaussian stretch exceed 16 384. Input classes may well have to be amalgamated in order to achieve these target numbers. Table 5.2 and Table 5.4 give the number of pixels allocated to each output class after the application of the histogram equalization and Gaussian contrast stretches, respectively. In both cases, the range of levels allocated to the output image exceeds the range of PVs in the input image; this will result in an overall brightening of the displayed image.

The application of contrast-enhancement techniques is discussed above in terms of a single greyscale image. The techniques can be used to enhance a false colour image by applying the appropriate process to the RGB channels separately. Methods of simultaneously 'stretching' the colour components of a false-colour image are dealt

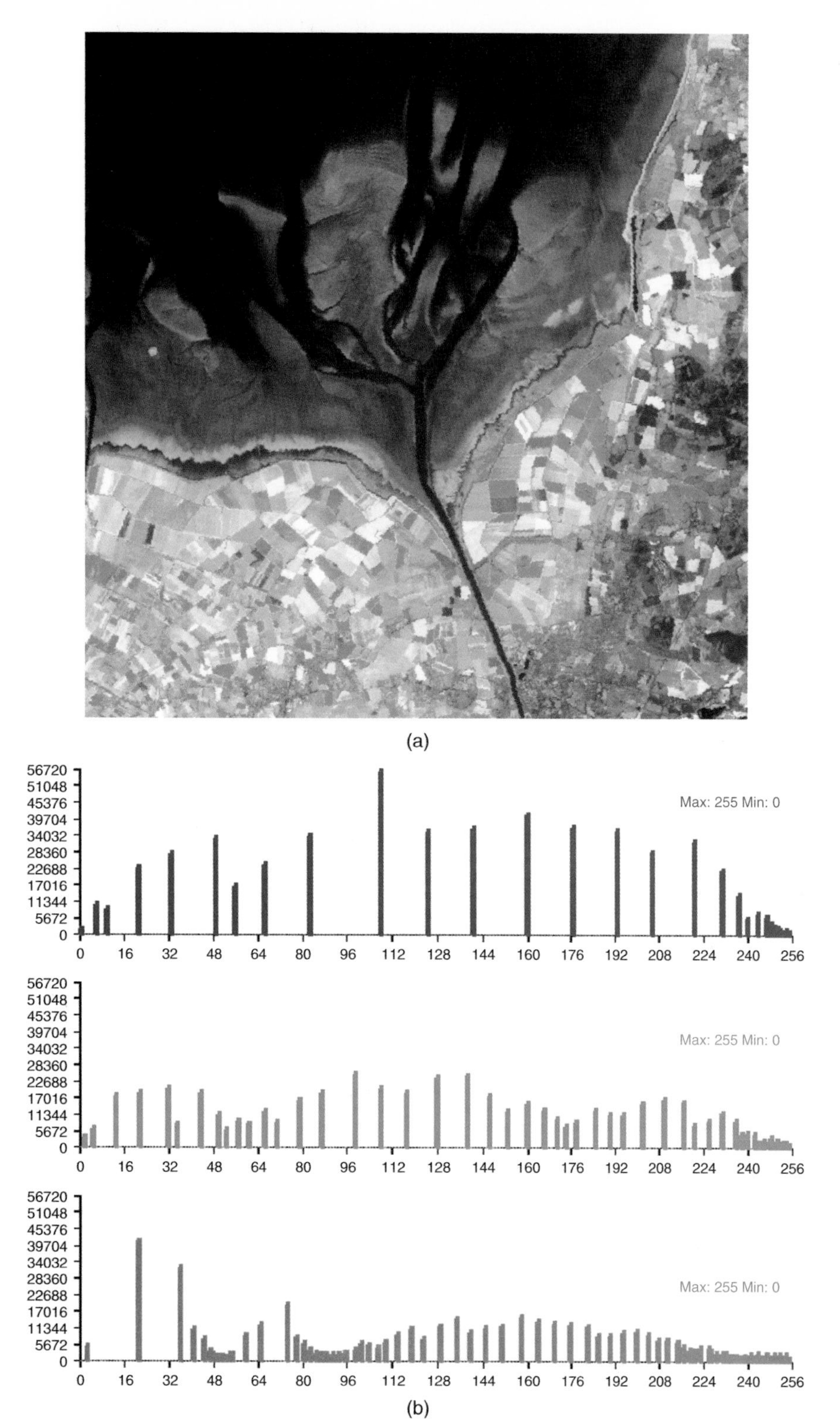

Figure 5.9 *(a) Histogram equalization contrast stretch applied to the image shown in Figure 5.6a. (b) Histogram of the image shown in Figure 5.9a. It is difficult to achieve a completely flat or uniform histogram, and in this case the frequency distribution of the image pixel values is slightly bell shaped. Landsat data courtesy NASA/USGS.*

Table 5.3 *Fitting observed histogram of pixel values to a Gaussian histogram. See text for discussion.*

(i)	(ii)	(iii)	(iv)	(v)	(vi)	(vii)	(viii)
0	<−3.0	0.0020	530	530	1311	1311	1
1	−2.6	0.0033	868	1398	2622	3933	3
2	−2.2	0.0092	2423	3821	5243	9176	3
3	−1.8	0.0220	5774	9595	9176	18352	4
4	−1.4	0.0448	1175	21346	13108	31460	5
5	−1.0	0.0779	20421	41767	24904	56364	6
6	−0.6	0.1156	30303	72070	30146	86510	7
7	−0.2	0.1465	38401	110471	45875	132385	8
8	0.2	0.1585	41555	152026	58982	191367	10
9	0.6	0.1465	38401	190427	48496	239863	11
10	1.0	0.1156	30303	220730	11796	251659	12
11	1.4	0.0779	20421	241151	3932	255591	13
12	1.8	0.0448	11751	252902	3932	259523	14
13	2.2	0.0220	5774	258676	2621	262144	15
14	2.6	0.0092	2423	261099	0	262144	15
15	>3.0	0.0040	1045	262144	0	262144	15

with elsewhere (the HSI transform in Section 6.5 and the decorrelation stretch in Section 6.4.3).

If an image covers two or more spectrally distinctive regions, such as land and sea, then the application of the methods so far described may well be disappointing. In such cases, any of the contrast-stretching methods described above can be applied to individual parts of the range of PVs in the image; for instance, the histogram equalization procedure could be used to transform the input range 0–60 to the output range 0–255, and the same could be done for the input range 61–255. The same procedure could be used whenever distinct regions occur if these regions can be identified by splitting the histogram at one or more threshold points. While the aesthetic appeal of images enhanced in this fashion may be increased, it should be noted that pixels with considerably different radiance values will be assigned to the same displayed or output colour value. The colour balance will also be quite different from that resulting from a standard colour-composite procedure.

5.4 Pseudocolour Enhancement

In terms of the RGB colour model presented in Section 5.2, a greyscale image occupies only the diagonal of the RGB colour cube running from the 'black' to the 'white' vertex (Figure 5.3). In terms of the HSI model, grey values are ranged along the vertical or intensity axis (Figure 5.4). No hue or saturation information is present, yet the human visual system is particularly efficient in detecting variations in hue and saturation, but not so efficient in detecting intensity variations. Three methods are available for converting a grey scale image to colour. The colour rendition as shown in the output image is not true or natural, for the original (input) image does not contain any colour information, and enhancement techniques cannot generate information that is not present in the input image. Nor is the colour rendition correctly described as false colour, for a false colour image is one composed of three bands of information which are represented in visible RGB. The name given to a colour rendition of a single band of imagery is a pseudocolour image. Three techniques are available for converting from greyscale to pseudocolour form. These are the techniques of density slicing, pseudocolour transform and user-specified colour transform. Each provides a method for mapping from a one-dimensional greyscale to a three-dimensional (RGB) colour.

5.4.1 Density Slicing

Density slicing is the representation of a set of contiguous grey levels of a greyscale image by specific colours. The range of contiguous grey levels (such as 0–10 inclusive) is called a 'slice'. The greyscale range 0–255 is normally converted to several colour slices. It is acknowledged that conversion of a greyscale image to pseudocolour is

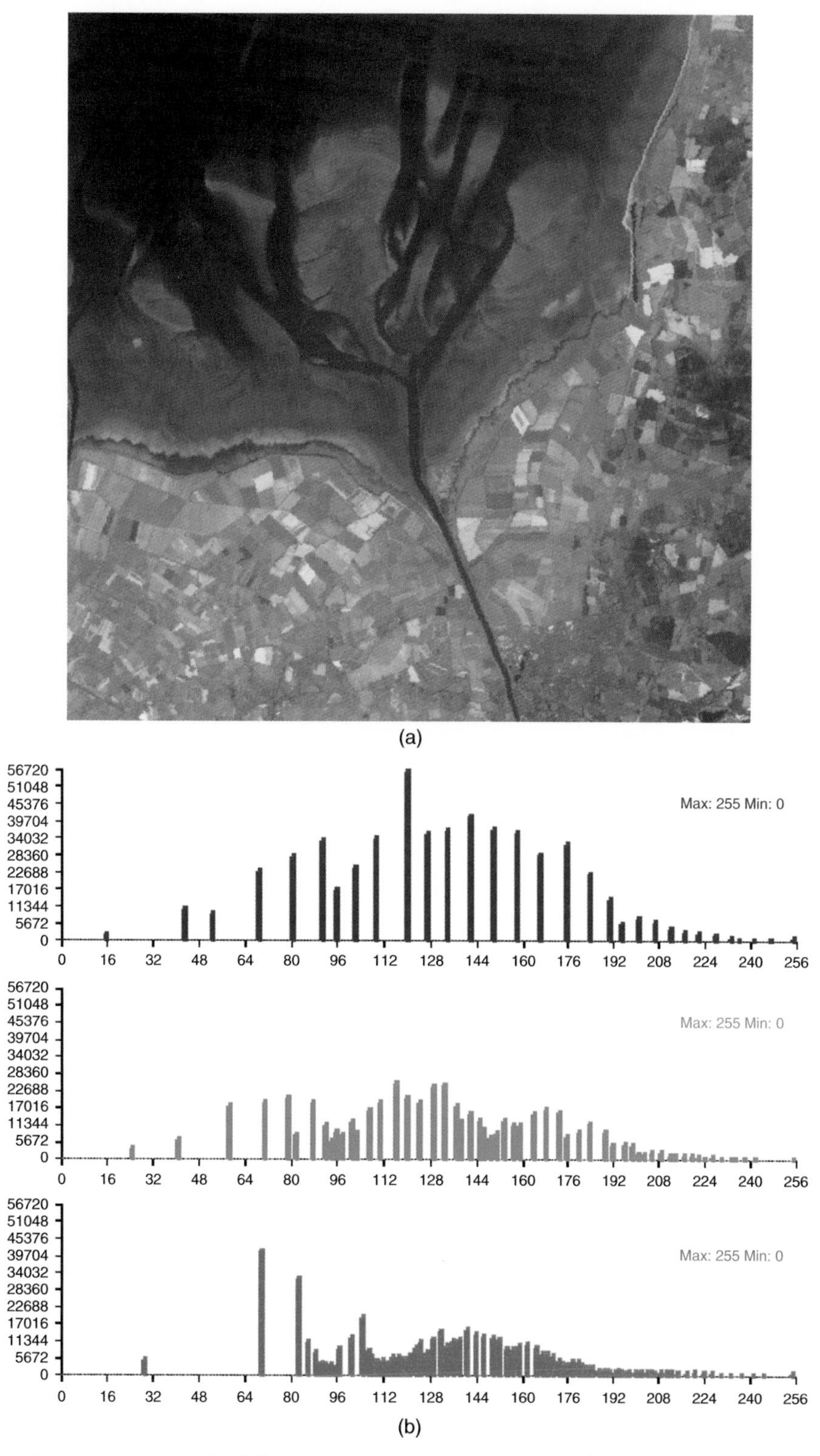

Figure 5.10 *(a) Gaussian contrast stretch of the image shown in Figure 5.6a. (b) Histogram of Gaussian contrast stretched image. Compare with Figure 5.9b – the number of classes at the two ends of the distribution is larger with the Gaussian stretch but the number of classes at the centre of the distribution is reduced. Landsat data courtesy NASA/USGS.*

Table 5.4 Number of pixels at each level following transformation to Gaussian model.

Intensity	(0)	(1)	(2)	(3)	(4)	(5)	(6)	(7)
Number	0	1311	0	7865	9176	13108	24904	30146
Intensity	(8)	(9)	(10)	(11)	(12)	(13)	(14)	(15)
Number	45875	0	58782	48496	11796	3932	3932	2621

(a)

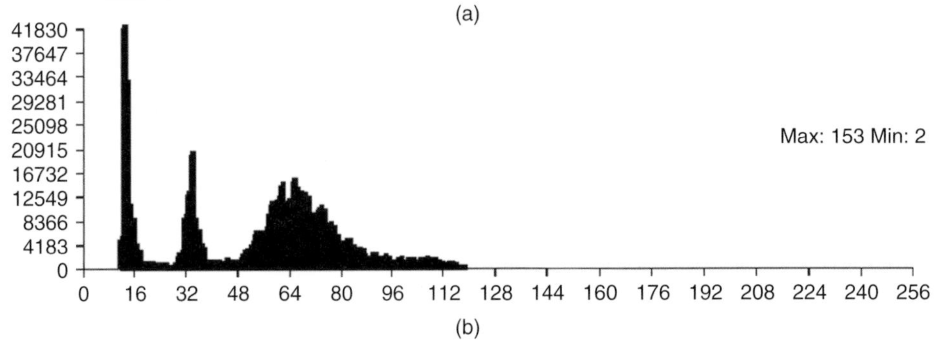

(b)

Figure 5.11 *(a) Landsat ETM+ Band 4 (NIR) image of the south-east corner of The Wash, eastern England. Water absorbs NIR 5 radiation almost completely, whereas growing crops reflect strongly, and appear in light shades of grey. This image is shown in pseudocolour in Figures 5.12 and 5.13. (b) Histogram of Figure 5.11a. Landsat data courtesy NASA/USGS.*

an effective way of highlighting different but internally homogeneous areas within an image, but at the expense of loss of detail. The loss of detail is due to the conversion from a 256-level greyscale image to an image represented in terms of many fewer colour slices. The effect is (i) to reduce the number of discrete levels in the image, for several grey levels are usually mapped onto a single colour and (ii) to improve the visual interpretability of the image if the slice boundaries and the colours are carefully selected.

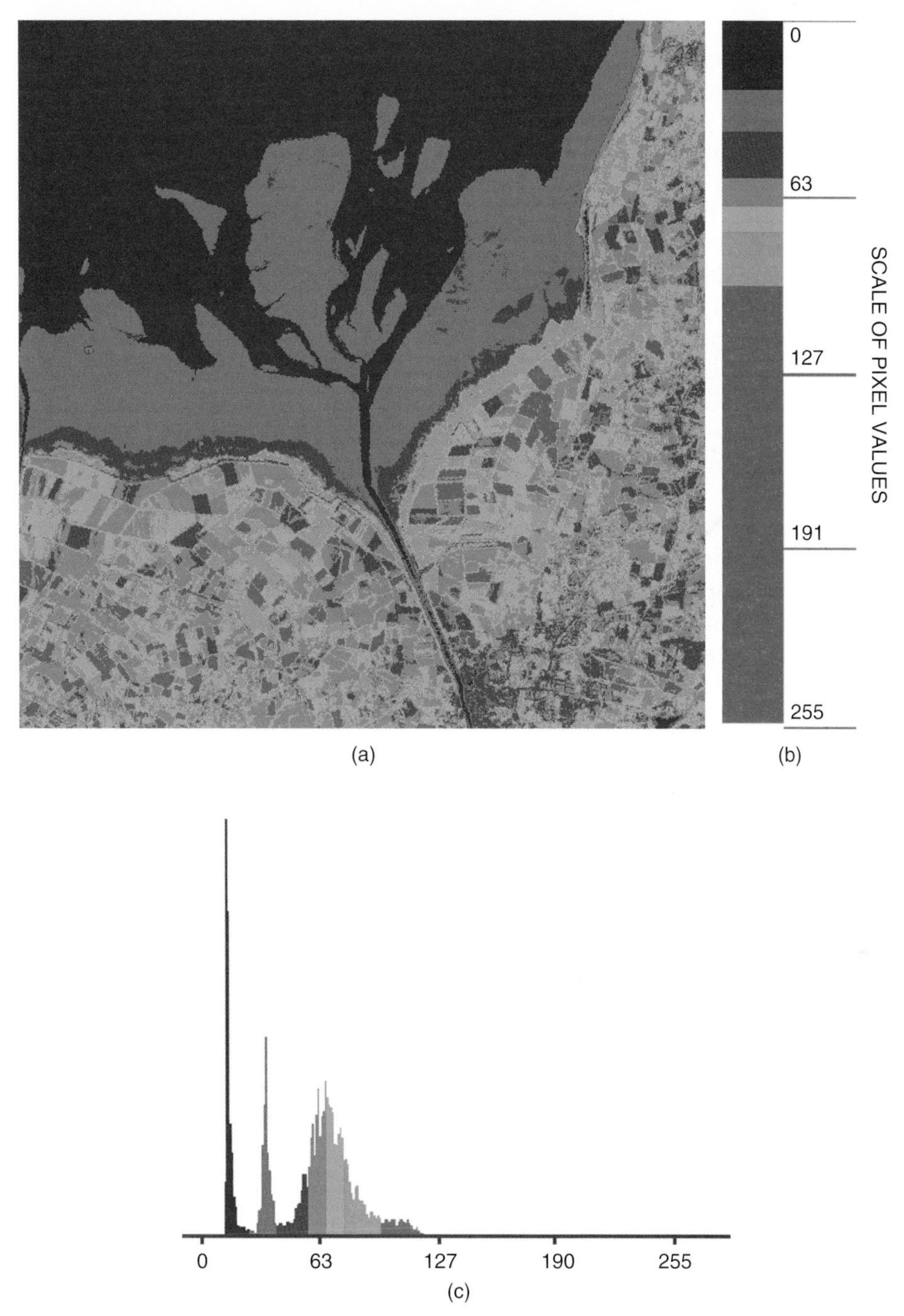

Figure 5.12 (a) Greyscale image of Figure 5.11a converted to colour by slicing the greyscale range 0–255 and allocating RGB values to each slice (b) the density slice colour bar and (c) the image histogram using the slice colours. This rendition is performed manually and at each step a 'slice' of the colour bar (density) is allocated a colour of the user's choice. Here, water is blue and the bright shades of grey are shown in red. Landsat data courtesy NASA/USGS.

In most image-processing systems the user is allowed to specify any colour for the current slice, and to alter slice boundaries in an upwards or downwards direction by means of a joystick or mouse. The slice boundaries are thus obtained by an interactive process, which allows the user to adjust the levels until a satisfactory result has been achieved. The choice of colour for each slice is important if information is to be conveyed to the viewer in any meaningful way, for visual perception is a psychological as well as a physiological process. Random colour selections may say more about the psychology of the perpetrator than about the information in the image. Consider, for example a thermal infrared image of the heat emitted by the Earth. A colour scale ranging from light blue to dark blue, through the yellows and oranges to red would be a suitable choice for most people have an intuitive feel for the 'meaning' of colours in terms of temperature. A scale taking in white, mauve, yellow, black, green and pink might confuse rather than enlighten.

Figure 5.11a shows a greyscale image which is to be converted to pseudocolour using the process of density slicing. This image is band 4 of the Landsat ETM+ false colour image shown in Figure 5.6. Figure 5.11b is the histogram of this image. Figure 5.12(a–c) are the density sliced image, the colour wedge giving the relationship between greyscale and colour, and the histogram of the density sliced image with colours superimposed.

5.4.2 Pseudocolour Transform

A greyscale image has equal RGB values at each pixel position. A *pseudocolour transform* is carried out by changing the colours in the RGB display to the format shown in the lower half of Figure 5.13. The settings shown in the lower part of Figure 5.13 send different colour (RGB) information to the digital to analogue converter (and hence the screen) for the same greyscale PV. The result is an image that pretends to be in colour. It is called a pseudocolour image. Like the density slicing method, the pseudocolour transform method associates each of a set of grey levels to a discrete colour. Usually, the pseudocolour transform uses a lot more colours than the density slice method. If the histogram of the values in the greyscale image is not approximately uniform then the resulting pseudocolour image will be dominated by one colour, and its usefulness thereby reduced. Analysis of the image histogram along the lines of the histogram equalization procedure (Section 5.3) prior to the design of the pseudocolour LUTs would alleviate this problem. Figure 5.13 shows the way in which a greyscale PV is represented as an RGB triple – the PVs (N1, N1, N1) are converted by this 'do-nothing' LUT to the same triple (N1, N1, N1). The greyscale image on the screen is directly equivalent to the greyscale values in the image. The lower part of Figure 5.13 illustrates the same input PVs (N1, N1, N1) but in this case the RGB LUTs are set to transform this triple to the values (N2, N3, N4). The values N2, n3 and N4 define a colour which is dominated by red and green. Figure 5.14a is a pseudocolour image of the Wash area converted to pseudocolour using the colour translation wedge shown in Figure 5.14b (which is, in fact, made up of 49 steps around the top of the HSI hexcone shown in Figure 5.4. There is nothing mysterious about the value 49 – it just happened that way).

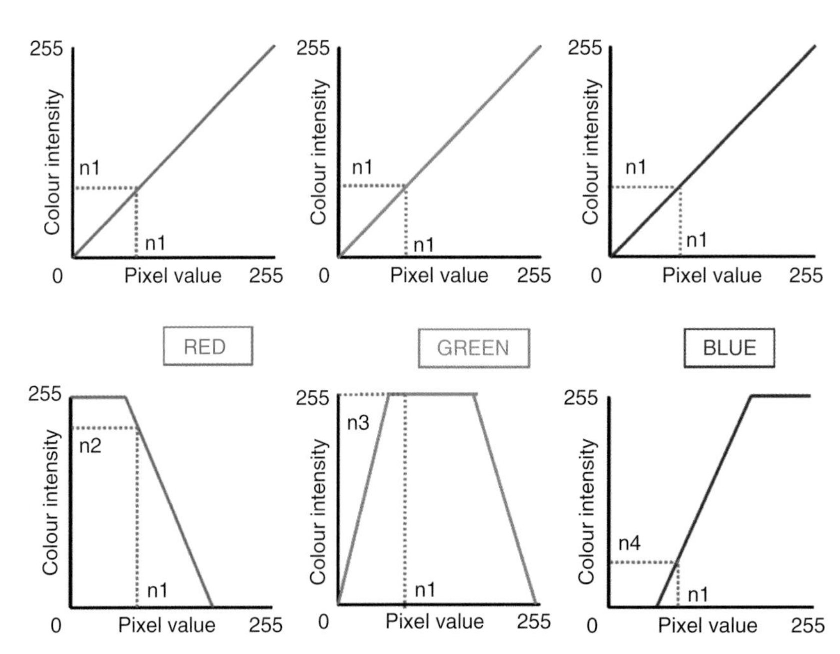

Figure 5.13 Illustrating the pseudocolour transform. A greyscale image is stored in all three (RGB) display memories, and the lookup tables (LUTs) for all three display memory are equivalent, sending equal RGB values to the screen at each pixel position. Thus, the greyscale pixel value N1 is sent to the display memory as the values (N1, N1, N1). The pseudocolour transform treats each display memory separately, so that the same pixel value in each of the RGB display memories sends a different proportion of red, green and blue to the screen. For example, the pixel value N1 in a greyscale image would be seen on screen as a dark grey pixel. If the pseudocolour transform were to be applied, the pixel value N1 would transmit the colour levels (N2, N3 N4) to the display memory, as shown by the purple dotted lines in the lower part of the diagram. The values N2, N3 and N4 would generate a colour that was close to maximum yellow, with a slight bluish tinge.

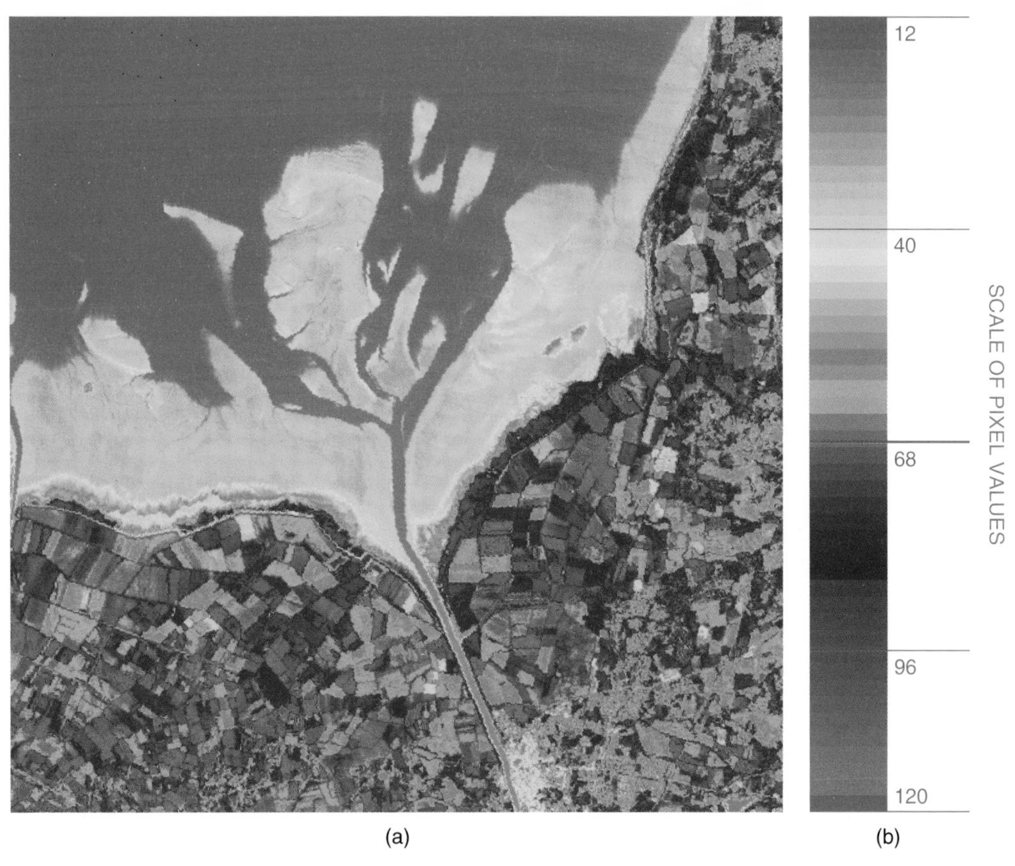

Figure 5.14 *(a) Pseudocolour transformation of the grey scale image shown in Figure 5.11. The range of greylevels and their colour equivalents are shown in (b). Notice that the whole 0–255 range is not used and the transformation is carried out only on pixel values between 12 and 120. These values were found by visual inspection of the histogram shown in Figure 5.12b. The overwhelming majority of pixel values lie in this range. The colour wedge shown in (b) has 49 colours, from red through yellow, green, cyan, blue and magenta. The pixel range of 12–120 is mapped onto these 49 colours to give the image shown in (a), which is more informative than the greyscale equivalent in Figure 5.11a.*

5.5 Summary

Image enhancement techniques include, but are not limited to, those of contrast improvement and greyscale to colour transformations. Other image-processing methods can justifiably be called enhancements. These include: (i) methods for detecting and emphasizing detail in an image (Section 7.4), (ii) noise reduction techniques, ranging from removal of banding (Section 4.2) to filtering (Chapter 7), (iii) colour transforms based on principal components analysis, called the decorrelation stretch (Section 6.3) and (iv) the HSI transform, Section 6.5. All these methods alter the visual appearance of the image in such a way as to bring out or clarify some aspect or property of the image that is of interest to a user. The range of uses to which remotely-sensed images can be put is considerable and so, although there are standard methods of enhancement such as those described here, they should not be applied thoughtlessly but with due regard to the user's requirements and purpose.

6 Image Transforms

6.1 Introduction

An image transform is an operation that re-expresses in a different, and possibly more meaningful, form all or part of the information content of a multispectral or greyscale image. In principle, an image transform is like a good cook, who can take a set of ingredients and turn them into cakes of different types. There is nothing in each cake except the same original ingredients yet they look (and probably taste) different. There are many different ways of looking remotely-sensed images, just as there are many different ways of looking at social, economic or political questions.

A number of different transforms are considered in this chapter. The term 'transform' is used somewhat loosely, for the arithmetic operations of addition, subtraction, multiplication and division are included, although they are not strictly transforms. These operations, which are described in Section 6.2, allow the generation of a derived image from two or more bands of a multispectral or multitemporal image. The derived image may well have properties that make it more suited to a particular purpose than the original. For example, the numerical difference between two images collected by the same sensor on different days may provide information about changes that have occurred between the two dates, while the ratio of the near-infrared (NIR) and red bands of a single-date image set is widely used as a vegetation index that correlates with difficult to measure variables such as vegetation vigour, biomass and leaf area index (LAI).

Vegetation indices are also discussed in Section 6.3. They are based on a model of the distribution of data values on two or more spectral bands considered simultaneously. Two examples of these transformations are the Perpendicular Vegetation Index (PVI), which uses a two-dimensional model of the relationship between vegetation and soil pixels, and the Tasselled Cap transformation, which is based on the optical (visible plus NIR) bands of a multispectral data set.

Section 6.4 provides an introduction to the widely used technique of principal components analysis (PCA), which is a method of re-expressing the information content of a multispectral set of m images in terms of a set of m principal components, which have two particular properties: zero correlation between the m principal components, and maximum variance. The maximum variance property of principal components means that the components are extracted in order of decreasing variance. The first component is that linear combination of spectral bands that has the maximum variance of all possible linear combinations of the same spectral bands. The second principal component is that linear combination of spectral bands that has the maximum variance with respect to the remaining part of the data once the effects of the first principal component have been removed, and so on. The zero correlation property means that principal components are statistically unrelated, or orthogonal. It is usually found that much of the information in the original m correlated bands is expressible in terms of the first p of the full set of m principal components, where p is less than m. This property of PCA is useful in generating a false-colour composite image. If the image set consists of more than three bands then the problem arises of selecting three bands for display in red, green and blue (RGB). Since the principal components of the image set are arranged in order of variance (which is generally assumed to correlate with information, but may also include a noise component) then the first three principal components can be used as the RGB components of a false-colour composite image. No linear combination (i.e. weighted sum) of the original bands can contain more information than is present in the first three principal components. Another use of PCA is in reducing the amount of calculation involved in automatic classification (Chapter 8) by basing the classification on p principal components rather than on m spectral bands. In addition, the p principal component images require less storage space than the m-band multispectral image from which they were derived. Hence, PCA can also be considered to be a data compression transformation.

Rather than maximize the variance of the principal components, we could choose another criterion such as

Computer Processing of Remotely-Sensed Images: An Introduction, Fourth Edition Paul M. Mather and Magaly Koch

maximizing the signal to noise ratio (SNR). The standard principal components procedure, despite popular belief, does not remove noise. If we can estimate the level of noise in the data then we could extract components that are arranged in order of decreasing SNR. This modification of PCA, which we will call noise-adjusted PCA, is described in Section 6.4.2.

Section 6.5 deals with a transformation that is concerned with the representation of the colour information in a set of three coregistered images representing the RGB components of a colour image. Theories of colour vision are summarized in Section 5.2 where it is noted that the conventional RGB colour cube model is generally used to represent the colour information content of three images. The hue–saturation-intensity (HSI) hexcone model is considered in Section 6.5 and its applications to image enhancement and to the problem of combining images from different sources (such as radar and optical images) are described.

The transforms and operations described above act on two or more image bands covering a given area. Section 6.6 introduces a method of examining the information content of a single-band greyscale image in terms of its frequency components. The discrete Fourier transform (DFT) provides for the representation of image data in terms of a coordinate framework that is based upon spatial frequencies rather than upon Euclidean distance from an origin (i.e. the conventional Cartesian or *xy* coordinate system). Image data that have been expressed in frequency terms are said to have been transformed from the image or spatial domain to the frequency domain. The frequency-domain representation of an image is useful in designing filters for special purposes (described in Chapter 7) and in colour coding the scale components of the image.

A related transformation, called the discrete wavelet transform (DWT), represents an attempt to bridge the gap between the spatial and frequency domains, for it decomposes the input signal (which may be one-dimensional, like a spectrum collected by a field spectrometer, or two-dimensional, like an image) in terms of wavelength (1D) or space (2D) *and* scale simultaneously. One major use of wavelets is to remove noise from (i.e. 'denoise') one- and two-dimensional signals. Data from the instruments carried by the Pléiades platform (Section 2.3.7.3) will be downlinked in compressed form, using a wavelet compression algorithm with an average compression factor of 4 (http://directory.eoportal.org/get_announce.php?an_id=8932). Wavelet-based denoising is described in Chapter 9, while the basics of the wavelet transform are considered in Section 6.7.

With the increasing availability of imaging spectrometer data sets (Section 9.3) that are composed of measurements in many tens or hundreds of spectral bands, methods of analysing and extracting information from the shape of spectral reflectance curves have been developed. An example is the derivative operation. The first derivative measures the slope of the spectral reflectance curve, while the second derivative measures the change in slope steepness. Both measures are useful in locating wavelengths of interest on the spectral reflectance curve, for example the position of the 'red edge' in a vegetation spectrum. These and other advanced methods of analysing imaging spectrometer data are described in Chapter 9.

6.2 Arithmetic Operations

The operations of addition, subtraction, multiplication and division are performed on two or more coregistered images of the same geographical area (Section 4.3). These images may be separate spectral bands from a single multispectral data set or they may be individual bands from image data sets that have been collected at different dates. Addition of images is really a form of averaging for, if the dynamic range of the output image is to be kept equal to that of the input images, rescaling (usually division by the number of images added together) is needed. Averaging can be carried out on multiple images of the same area in order to reduce the noise component. Subtraction of pairs of images is used to reveal differences between those images and is often used in the detection of change if the images involved were taken at different dates.

Multiplication of images is rather different from the other arithmetic operations for it normally involves the use of a single 'real' image and a binary image made up of ones and zeros. The binary image is used as a mask, for those image pixels in the real image that are multiplied by zero also become zero, while those that are multiplied by one remain the same.

Division or ratioing of images is probably the arithmetic operation that is most widely applied to images in geological, ecological and agricultural applications of remote sensing, for the division operation is used to detect the magnitude of the differences between spectral bands. These differences may be symptomatic of particular land cover types. Thus, a NIR : red ratio might be expected to be close to 1.0 for an object which reflects equally in both of these spectral bands (for example a cloud top) while the value of this same ratio will be well above one if the NIR reflectance is higher than the reflectance in the visible red band, for example in the case of vigorous vegetation.

6.2.1 Image Addition

If multiple, coregistered, images of a given region are available for the same time and date of imaging then addition (averaging) of the multiple images can be used as a means of reducing the overall noise contribution. A single image might be expressed in terms of the following model:

$$G(x, y) = F(x, y) + N(x, y)$$

where $G(x, y)$ is the recorded image, $F(x,y)$ the true image and $N(x,y)$ the random noise component. $N(x, y)$ is often hypothesized to be a random Normal distribution with a mean of zero, since it is the sum of a number of small, independent errors or factors. The true signal, $F(x,y)$, is constant from image to image. Therefore, addition of two separate images of the same area taken at the same time might be expected to lead to the cancellation of the $N(x,y)$ term for, at any particular pixel position (x,y), the value $N(x,y)$ is as likely to be positive as to be negative. Image addition, as noted already, is really an averaging process. If two images $G_1(i,j)$ and $G_2(i,j)$ are added and if each has a dynamic range of 0–255 then the resulting image $G_{sum}(i, j)$ will have a dynamic range of 0–510. This is not a practicable proposition if the image display system has a fixed, 8-bit, resolution. Hence it is common practice to divide the sum of the two images by two to reduce the dynamic range to 0–255. The process of addition is carried out on a pixel-by-pixel basis as follows:

$$G_{sum}(i, j) = (G_1(i, j) + G_2(i, j))/2$$

The result of the division is normally rounded to the nearest integer. Note that if the operation is carried out directly on images stored on disk then the dynamic range may be greater than 0–255, and so the scaling factor of 2 will have to be adjusted accordingly. Images stored in the computer display memory have a dynamic range of 0–255. However, take care – recall from the discussion in Section 5.3 that the application of a contrast stretch will cause the pixel values in the display memory to be changed, so adding together two contrast-enhanced images does not make sense. Always perform arithmetic operations on images that have not been contrast enhanced. A further difficulty is that images with a dynamic range of more than 0–255 are scaled before being written to display memory. The scaling used is automatic, and depends on the image histogram shape (Section 3.2.1). Any arithmetic operation carried out on images with an extended dynamic range (i.e. greater than 0–255) that have been scaled to 0–255 for display purposes is likely to be meaningless. The operation should be carried out on the full range images stored on disk, not on their scaled counterparts.

If the algorithm described in the previous paragraph is applied then the dynamic range of the summed image will be approximately the same as that of the two input images. This may be desirable in some cases. However, it would be possible to increase the range by performing a linear contrast stretch (Section 5.3.1) by subtracting a suitable offset o and using a variable divisor d:

$$G'(i, j) = (G_1(i, j) + G_2(i, j) - o)/d$$

The values of o and d might be determined on the basis of the user's experience, or by evaluating $G_{\mathrm{sum}}(i,j)$ at a number of points systematically chosen from images G_1 and G_2. Image G' will have a stretched dynamic range in comparison with the result of the straight 'division by 2'.

Other methods of removing noise from images include the use of the DWT to estimate a 'noise threshold' which is then applied to the data series (such as a one-dimensional reflectance spectrum or a two-dimensional image) in order to remove additive noise of the kind described above. The wavelet transform is considered further in Section 6.7 and Chapter 9.

6.2.2 Image Subtraction

The subtraction operation is often carried out on a pair of coregistered images of the same area taken at different times. The purpose is to assess the degree of change that has taken place between the dates of imaging (see, for example Dale, Chandica and Evans, 1996). Image subtraction or differencing is also used to separate image components. In Chapter 9, for example the wavelet transform is used to separate the signal (information) in an image from the noise. The difference between the original and de-noised image can be found by subtracting the denoised image from the original image.

Image differencing is performed on a pixel-by-pixel basis. The maximum negative difference (assuming both images have a dynamic range of 0–255) is $(0-255 =) - 255$ and the maximum positive difference is $(255 - 0 =) + 255$. The problem of scaling the result of the image subtraction operation onto a 0–255 range must be considered. If the value 255 is added to the difference then the dynamic range is shifted to 0–510. Next, divide this range by 2 to give a range of 0–255. Variable offsets and multipliers can be used as in the case of addition (Section 6.2.1) to perform a linear contrast-stretch operation. Formally, the image subtraction process can be written as:

$$G_{\mathrm{diff}}(i, j) = (255 + G_1(i, j) - G_2(i, j)/2$$

using the same notation as previously. If interest is centred on the magnitude rather than the direction of change then the following method could be used:

$$G_{\mathrm{absdiff}}(i, j) = |G_1(i, j) - G_2(i, j)|$$

The vertical bars |.| denote the absolute value (regardless of sign). No difference is represented by the value 0 and the degree of difference increases towards 255. Note the remarks above (Section 6.2.1) on the logic of applying arithmetic operations to images with a dynamic range other than 0–255.

A difference image $G_{\mathrm{diff}}(i,j)$ tends to have a histogram that is Normal or Gaussian in shape with the peak at a count of 127 (if the standard scaling is used, in which zero difference translates to a pixel value of 127), tailing off rapidly in both directions. The peak at a count of 127 represents pixels that have not changed while the pixels in the histogram tails have changed substantially. The image $G_{\mathrm{absdiff}}(i,j)$ has a histogram with a peak at or near zero and a long tail extending towards the higher values. A density-sliced difference image is shown in Figure 6.2c (the histogram of this difference image is shown in Figure 6.1). Differences in TM/ETM+ band 2 are shown. Figures 6.2a,b show respectively the TM band 4, 3, 2 false-colour composite for 1984 and the ETM+ band 4, 3, 2 false-colour composite for 1993. The bright areas of the false-colour composite images (which have been enhanced by a 5–95% linear stretch) are desert while the red areas are vegetation. Areas of red in Figure 6.2c are those where the 1984 band 2 pixel values are lower than the corresponding pixel values in the 1993 image. These negative change areas represent vegetation present in 1993 but not in 1984 at the bottom of the difference image, and also a change from brackish lagoon to deeper clearer water in the top left corner of the same image. Areas of brown, dark green, light green and blue show least change (the areas have remained desert over the period 1984–1993), but yellow areas are those where the pixel values in band 2 for 1993 are greater than those for 1984. These scattered yellow areas are located in the area covered by vegetation in 1984 and may represent ploughed fields or land that has been allowed to revert to desert.

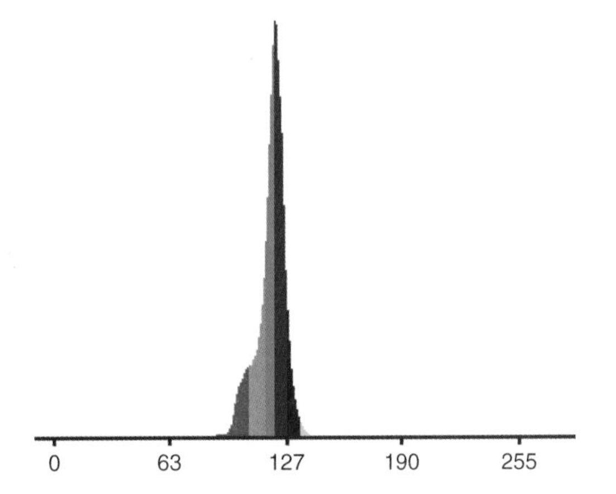

Figure 6.1 Histogram of difference between 1993 and 1984 images of Alexandria, Egypt (see Figure 6.2), after scaling to the 0–255 range. An indicated difference value of 127 equates to a real difference of zero. Histogram x-axis values lower than 127 indicate negative differences and values above 127 indicate positive differences. In practice the modal class is close to, but not exactly, 127 as a result of differences in illumination geometry and atmospheric conditions. The corresponding colour wedge is shown in Figure 6.2d.

Jupp and Mayo (1982) provide an interesting example of the use of image subtraction. They use a four-band Landsat MSS image to generate a classified (labelled) image (Chapter 8) in which a single-band image is generated, with each pixel in the image being given a numerical label to indicate the land cover class to which it belongs. These labels are determined on the basis of the spectral characteristics of each pixel. For example, labels of 1, 2 and 3 could be used to indicate forest, grassland and water. The mean values of the pixels in each class in every band are computed, to produce a table of k rows and p columns, with k being the number of classes and p the number of spectral bands used in the labelling process. A new image set is then generated, with one image per spectral band. Those pixels with the label i in the classified image are given the mean value of class i. This operation results in a set of p 'class mean value' images. A third image set is then produced, consisting of p residual images which are obtained by subtracting the actual pixel value recorded in a given spectral band from the corresponding value in the 'mean' image for that pixel. Residual images can be combined for colour composite generation. The procedure is claimed to assist in the interpretation and understanding of the classified image, as it highlights pixels that differ from the mean value of all pixels allocated the same class. This could be useful in answering questions such as: 'have I omitted any significant land cover classes?' or 'why is class x so heterogeneous?'

6.2.3 Image Multiplication

Pixel-by-pixel multiplication of two remotely-sensed images is rarely performed in practice. The multiplication operation is, however, a useful one if an image of interest is composed of two or more distinctive regions and if the analyst is interested only in one of these regions. Figure 6.3a shows a Landsat-2 MSS band 4 (green) image of part of the Tanzanian coast south of Dar-es-Salaam. Variations in reflectance over the land area distract the eye from the more subtle variations in the radiance upwelling from the upper layers of the ocean. The masking operation can eliminate variations over the distracting land region. The first step is the preparation of the mask that best separates land and water, using the NIR band (Figure 6.3b) since reflection from water bodies in the NIR spectral band is very low, while reflection from vegetated land areas is high, as noted in Section 1.3. A suitable threshold is chosen by visual inspection of the image histogram of the NIR pixel values. A binary mask

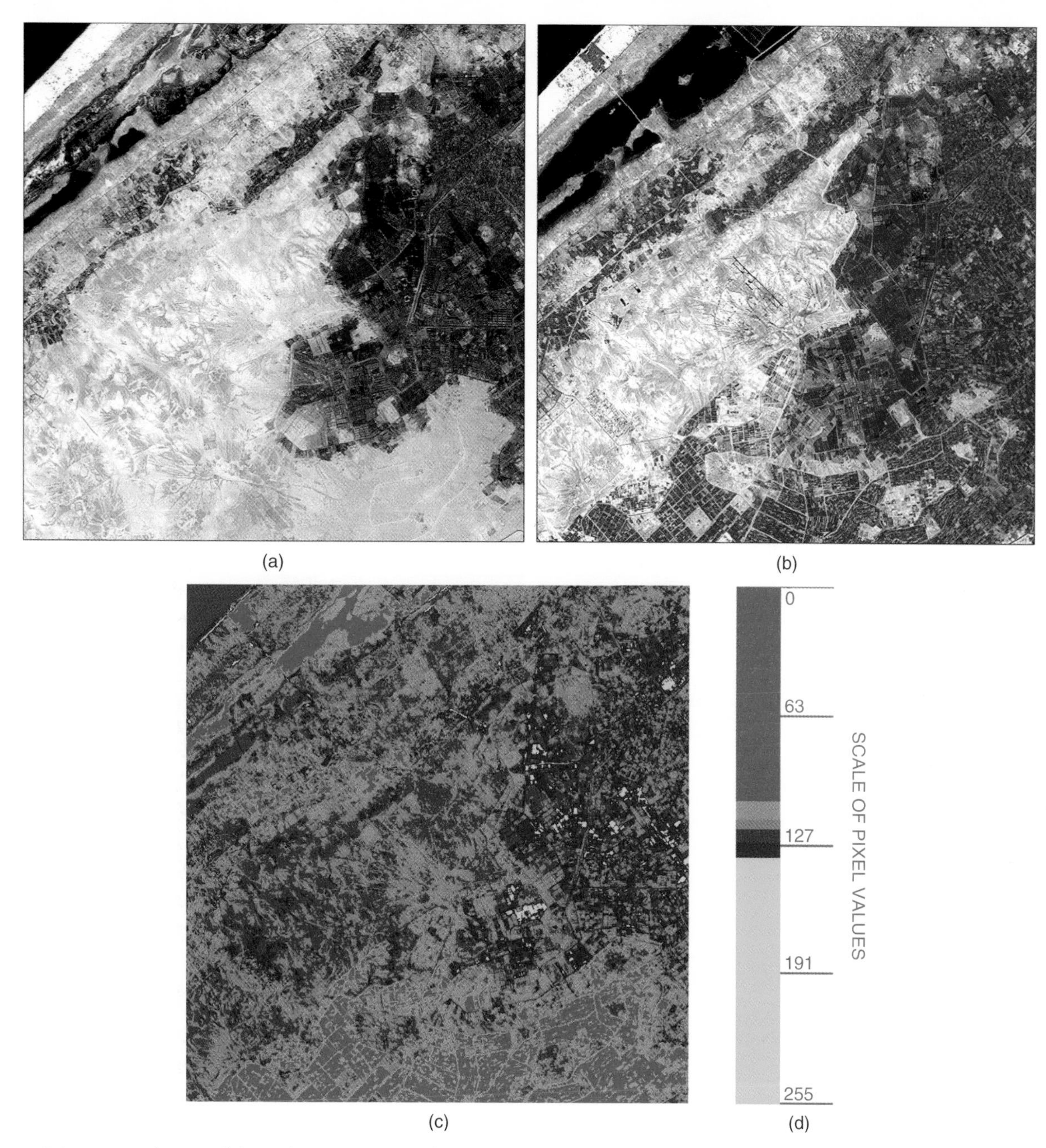

Figure 6.2 *(a) Landsat TM false colour composite (bands 4, 3 and 2) of a (1984) sub-image of Alexandria, Egypt, after a linear contrast stretch. (b) Corresponding ETM+ image for 1993. (c) Density sliced difference image based on band 2 images. (d) Colour wedge for difference image. These colours are also used in the histogram shown in Figure 6.1. Landsat data courtesy NASA/USGS.*

image is then generated from the NIR image by labelling with '255' those pixels that have values below the threshold (Figure 6.3c). Pixels whose values are above the threshold are labelled '1', so the mask image displays as a black-and-white image with the masked area appearing white. The second stage is the multiplication of the image shown in Figure 6.3a and the mask image (Figure 6.3c). Multiplication by 1 is equivalent to doing nothing, whereas multiplication by 0 sets the corresponding pixel in the masked image to 0. Using the above procedure, the pixels in the Tanzanian coast band 4 image that represent land are replaced by zero values, while 'ocean' pixels are unaltered. Application of the density slice procedure produces the image shown in Figure 6.3d. In practice, the pixel values in the two images (mask and band 4 in this example) are not multiplied but are processed by a simple logical function: if the mask pixel is zero then set the corresponding image pixel to zero, otherwise do

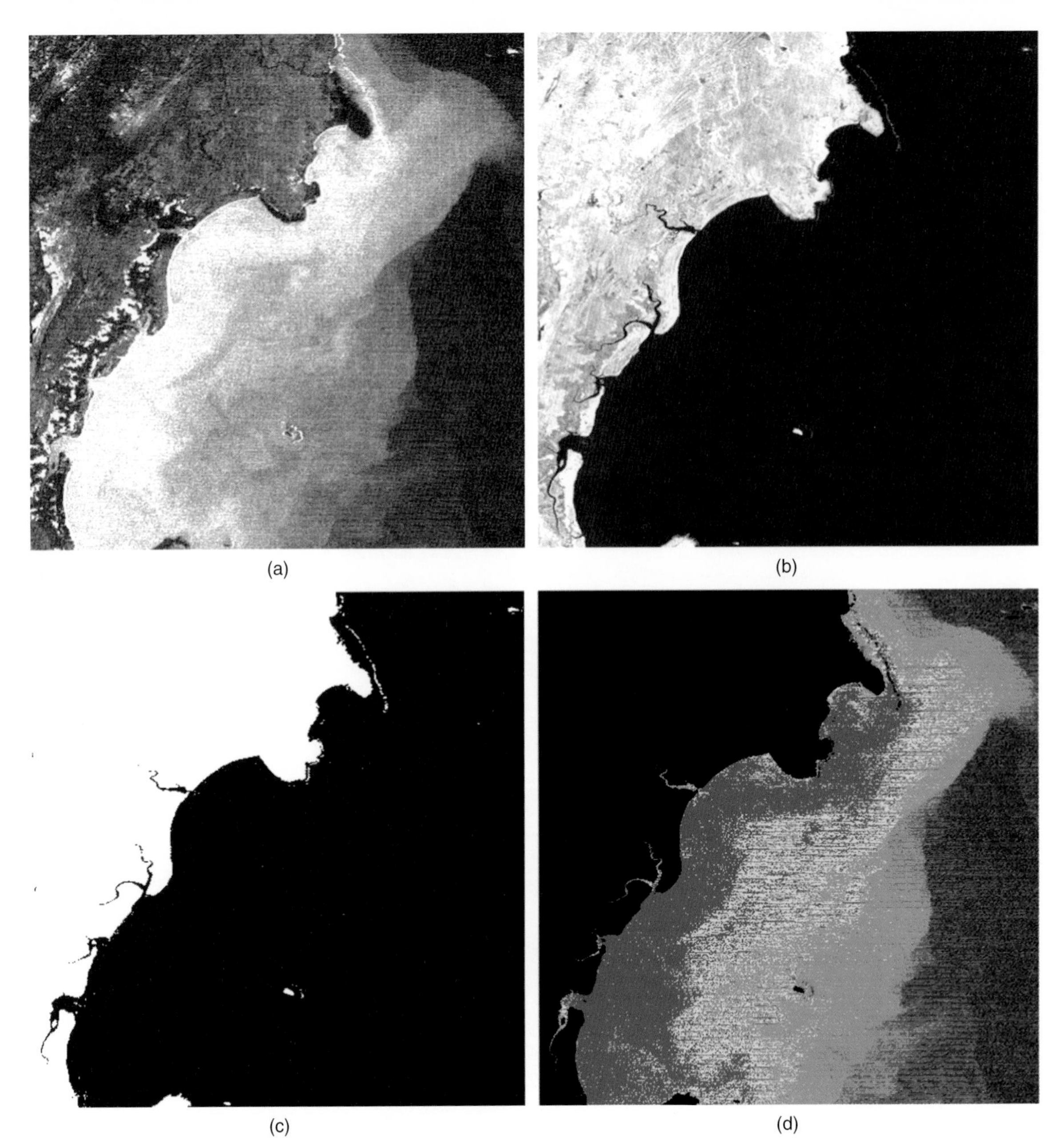

Figure 6.3 *Illustrating the use of image multiplication in creating a land/sea mask that enables the full range of the display to be used to represent variations in green light penetration depth in a region of the Tanzanian coast, south of Dar es Salaam. (a) Landsat MSS Band 4 (green), (b) Landsat MSS Band 7 (near infrared), (c) Land/sea mask created from near infrared band (b). Land is shown in white. (d) Masked and density sliced green band (a). The sixth line banding phenomenon is prominent. Landsat data courtesy NASA/USGS.*

nothing. Some software packages use the two extremes (0 and 255) of the range of pixel values to indicate 'less than' and 'greater than' the threshold, respectively.

6.2.4 Image Division and Vegetation Indices

The process of dividing the pixel values in one image by the corresponding pixel values in a second image is known as ratioing. It is one of the most commonly used transformations applied to remotely-sensed images. There are two reasons why this is so. One is that certain aspects of the shape of spectral reflectance curves of different Earth-surface cover types can be brought out by ratioing. The second is that undesirable effects on the recorded radiances, such as that resulting from variable illumination (and consequently changes in apparent

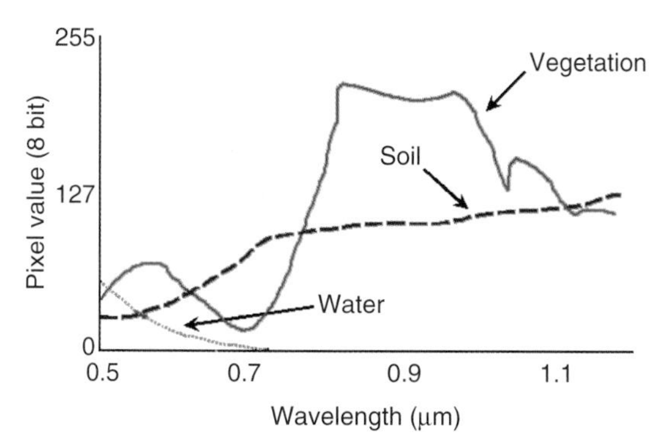

Figure 6.4 The ratio of a pixel value at near-infrared wavelengths (around 1.0 μm) to the corresponding pixel value in the red region of the spectrum (0.6–0.7 μm) will be large if the area represented by the pixel is covered by vigorous vegetation (solid curve). It will be around 1.0 for a soil pixel, but less than 1.0 for a water pixel. In effect, the IR/R ratio is measuring the slope of the spectral reflectance curve between the infrared and red wavelengths. Inspection of the curves shown in this figure shows that the curve for vegetation has a very significant slope in this region.

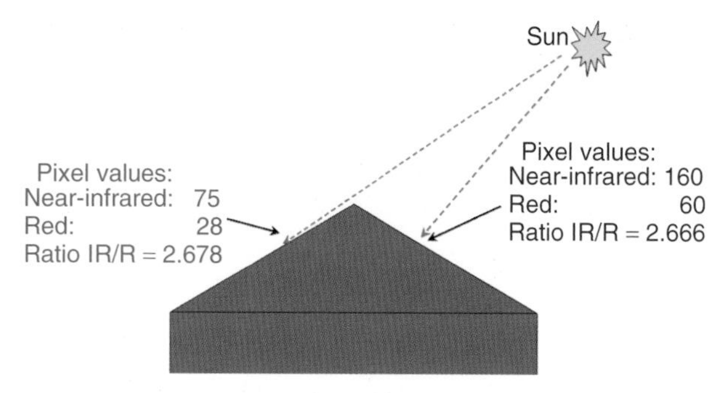

Figure 6.5 Ratio of pixel values in the near-infrared region to the corresponding pixel value in the visible red region of the spectrum. The ratios for the illuminated and shaded slopes are very similar, although the pixel values differ by a factor of more than 2. Hence an image made up of IR : R ratio values at each pixel position will exhibit a much-reduced shadow or topographic effect.

upwelling radiance) caused by variations in topography can be reduced. Figure 6.4 shows the spectral reflectance curves for three cover types. The differences between the curves can be emphasized by looking at the gradient or slope between the red and the NIR bands, for example bands 3 (red) and 4 (NIR) in a Landsat ETM+ image, or bands 3 (NIR) and 2 (red) of the SPOT HRV image set. The shape of the spectral reflectance curve for water shows a decline between these two points, while that for vegetation shows a substantial increase. The spectral reflectance curve for soil increases gradually between the two bands. If a pixel value in the NIR band is divided by the equivalent value in the red band then the result will be a positive real number that exceeds 1.0 in magnitude. The same operation carried out on the curve for water gives a result that is less than 1.0, while the soil curve gives a value somewhat higher than 1.0. The greater the difference between the pixel values in the two chosen bands the greater the value of the ratio.

The two images may as well be subtracted if this were the only result to be derived from the use of ratios. Figure 6.5 shows a hypothetical situation in which the irradiance at point *B* on the ground surface is only 50% of that at *A* due to the fact that one side of the slope is directly illuminated by the Sun. Subtraction of the values in the two bands at point *A* gives a result that is double that which would be achieved at point *B* even if both points are located on the same ground-cover type. However, the ratios of the two bands at *A* and *B* are the same because the topographic effect has been largely cancelled out in this instance. This is not always the case, as shown by the discussion below.

One of the most common spectral ratios used in studies of vegetation status is the ratio of the NIR to the equivalent red band value for each pixel location. This ratio exploits the fact that vigorous vegetation reflects strongly in the NIR and absorbs radiation in the red waveband (Section 1.3.2.1). The result is a greyscale image that can be smoothed by a low-pass filter (Section 7.2) and density-sliced (Section 5.4.1) to produce an image showing variation in biomass (the amount of vegetative matter) and in LAI as well as the state of health (physiological functioning) of plants (Dong *et al.*, 2003; Hansen and Schjoerring, 2003; Lu, 2006; Serrano, Filella and Peñuelas, 2000). Smith, Steven and Colls (2004) discuss the use of vegetation ratios to detect plant stress resulting from gas leaks. A critical analysis of vegetation indices is given by Myneni *et al.* (1995).

More complex ratios involve sums of and differences between spectral bands. For example, the Normalized Difference Vegetation Index (NDVI), defined in terms of the NIR and red (R) bands as:

$$NDVI = \frac{NIR - R}{NIR + R}$$

is preferred to the simple R : NIR ratio by many workers because the ratio value is not affected by the absolute pixel values in the NIR and R bands.

Figure 6.6b, d show the NDVI and IR/R images for the (false colour) image of the Nottingham area (Figure 6.6a). On the basis of visual evidence, the difference between the simple ratio and the NDVI is not great. However, the fact that sums and differences of bands are used in the NDVI rather than absolute values may make the NDVI more appropriate for use in studies where comparisons over time for a single area are involved, since the NDVI might be expected to be influenced to a lesser extent by variations in atmospheric conditions (but see below).

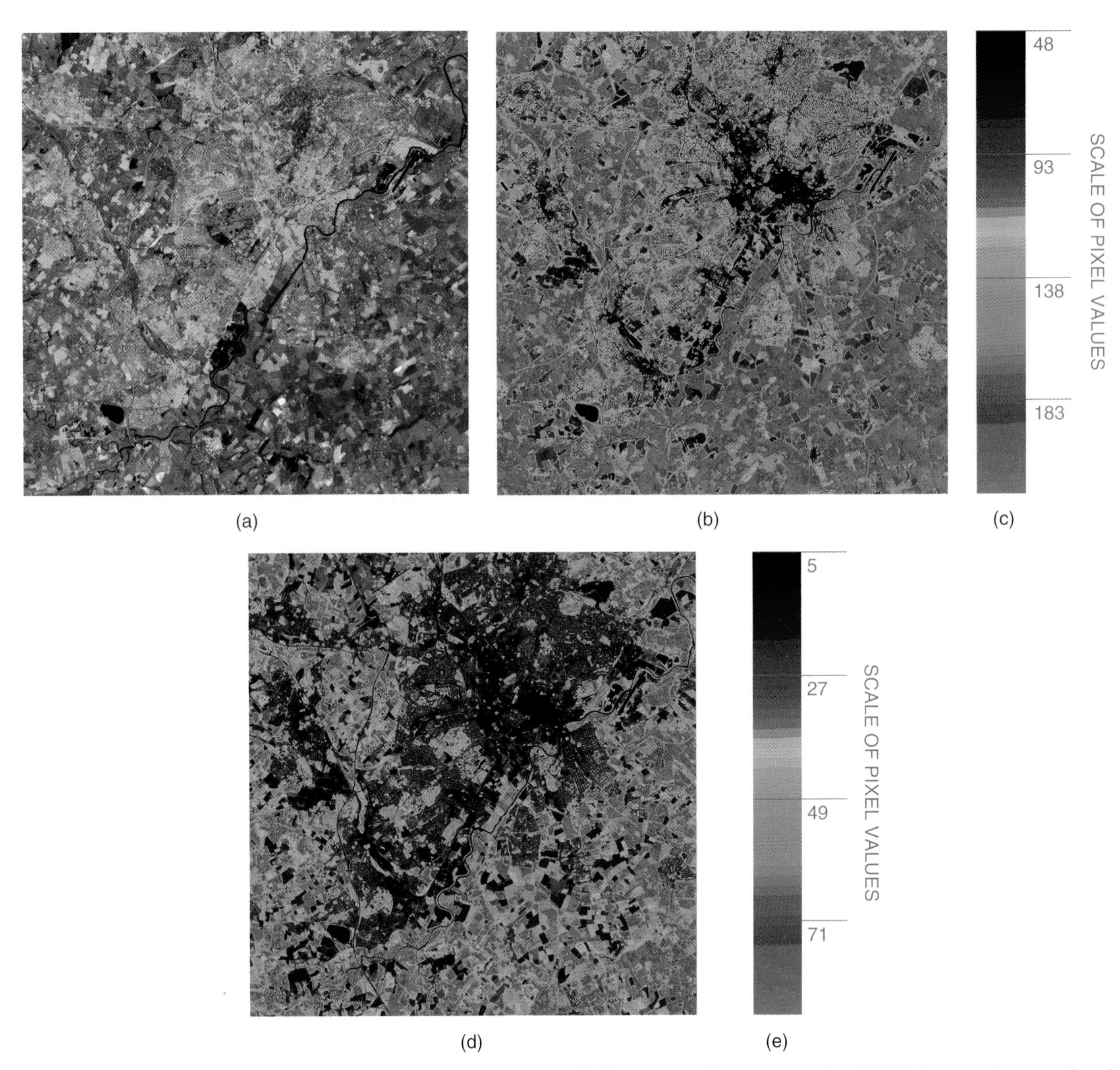

Figure 6.6 *(a) SPOT HRV false colour composite image of the Nottingham area. (b) NDVI image in pseudocolour (c) Pseudocolour wedge for (b), (d) Near-infrared : red ratio image in pseudocolour. (e) colour wedge for (d). Permission to use the data was kindly provided by SPOT image, 5 rue des Satellites, BP 4359, F 331030, Toulouse, France.*

Sensor calibration issues, discussed in Section 4.6, may have a significant influence on global NDVI calculations based on the NOAA AVHRR. Gutman and Ignatov (1995) show how the difference between pre- and post-launch calibrations lead to unnatural phenomena such as the 'greening of deserts'.

The derivation of 'better' vegetation indices is an active research area. Perry and Lautenschlager (1984) argue that various ratios are functionally equivalent. The simple ratio and NDVI, plus other band ratios, are all affected by external factors such as the state of the atmosphere, illumination and viewing angles and soil background reflectance. This is one reason why NDVI-composite images derived from multiple NOAA AVHRR require careful analysis. These images are produced by selecting cloud-free pixels from a number of images collected over a short period (a week or a month) to produce a single image from these selected pixels. Because of the orbital characteristics of the NOAA satellite it is probable that the NDVI values for adjacent pixels have been collected at different illumination and viewing angles. See Section 1.3.1 for a discussion of the bidirectional reflectance properties of Earth surface materials. Further discussion of the problem of estimating vegetation characteristics from remotely-sensed data is contained in Section 9.3, where the use of high spectral resolution (imaging spectrometer) data is discussed.

A class of indices called soil adjusted vegetation indices (SAVIs) has been developed, and there are quite a number to choose from (the original SAVI, the Transformed Soil-Adjusted Vegetation Index, TSAVI, the Global Environment Monitoring Index, GEMI and a number of others). Bannari *et al.* (1995b), Rondeaux (1995) and Rondeaux, Steven and Baret (1996) provide an extensive review. Steven (1998) discusses the Optimized Soil-Adjusted Vegetation Index (OSAVI) and shows that the form:

$$OSAVI = \frac{NIR - R}{NIR + R + 0.16}$$

minimizes soil effects. Readers interested in pursuing this subject should refer to Baret and Guyot (1991), Huete (1989), Pinty, Leprieur and Verstraete (1993) and Sellers (1989). Leprieur, Kerr and Pichon (1996) assess the comparative value of the different vegetation indices using NOAA AVHRR data. Assali and Menenti (2000) and Jakubauskas, Legates and Kastens (2001) illustrate the analysis of temporal sequences of NDVI data using a Fourier-based approach (Section 6.6). Canisius, Turral and Molden (2007) also use a Fourier-based approach using time series of NDVI values derived from AVHRR data, as do Roerink *et al.* (2003). Jönsson and Eklundh (2004) provide a computer program for analysis of time-series of satellite sensor data. McMorrow *et al.* (2004) assess the use of hyperspectral data (Section 9.3) for characterizing upland peat composition. Haboudane *et al.* (2002, 2004) also use hyperspectral data. Carlson and Ripley (1997) relate NDVI to fractional vegetation cover and LAI. Gillies, Carlson and Cui (1997) and Nishida *et al.* (2003) describe a relationship between NDVI, surface temperature and soil moisture content. As NOAA-AVHRR is being phased out and as new sensors such as MODIS become established so the intercalibration of vegetation indices from different sensors becomes an important issue (Sakamoto *et al.*, 2005; Steven *et al.*, 2003; van Leeuwen *et al.*, 2006). Other references include Bannari *et al.* (1995b), Baret and Guyot (1991), Gao *et al.* (2000), Gilabert *et al.*, (2002), Gitelson *et al.* (2002), Haboudane *et al.* (2004), Hobbs and Mooney, (1990), Liang (2004, 2007), Pinty, Leprieur and Verstraete (1993), Tucker (1979) and van der Meer, van Dijk and Westerhof (1995).

If ratio values are to be correlated with field observations of, for example LAI or estimates of biomass, or if ratio values for images of the same area at different times of the year are to be compared, then some thought should be given to scaling of the ratio values. Ratios are computed as 32-bit real (floating-point) values. Scaling is necessary only to convert the computed ratio values to integers on the range 0–255 for display purposes, as explained in Section 3.2.1. If the user wishes to correlate field observations and ratio values from remotely-sensed images then unscaled ratio values should be used. Most image processing software packages calculate the NDVI value at each pixel position and provide one of two options. First, as the NDVI is known to range between + and −1, the calculated NDVI values are arranged so that the lowest possible NDVI value (−1) is recorded in the display memory of the computer as 0 (zero) and the highest possible NDVI value (+1) is recorded as 255. Thus, an NDVI value of zero is transformed to the 8-bit value 127. Two sets of images taken at different times can be compared directly using this scaling method, since the full range of the NDVI is transformed to the full 8-bit range of 0–255. A second option allows the user to scale the range $NDVI_{MIN}$–$NDVI_{MAX}$ (the minimum and maximum NDVI values appearing in the image) onto the range 0–255. Since the values of $NDVI_{MIN}$ and $NDVI_{MAX}$ will generally differ from image to image, the results of this scaling operation are not directly comparable. The issue of scaling 16- and 32-bit integers and 32-bit real values onto a 0–255 range in order to display the image on a standard PC is considered in Section 3.2.

It was noted above that one of the reasons put forward to justify the use of ratios was the elimination of variable illumination effects in areas of topographic slope. This, of course, is not the case except in a perfect world. Assume that the variation in illumination at point (i,j) can be summarized by a variable $c(i,j)$ so that, for the pixel at location (i, j), the ratio between channels q and p is expressed as:

$$R(i, j) = \frac{c(i, j)v(q, i, j)}{c(i, j)(v(p, i, j)} = \frac{v(q, i, j)}{v(p, i, j)}$$

where $v(q, i, j)$ is the radiance from pixel located at point (i, j) in channel q. The term $c(i, j)$ is constant for both bands q and p and therefore cancels out. If there were an additive as well as a multiplicative term then the following logic would apply:

$$R(i, j) = \frac{c(i, j)v(q, i, j) + r(q)}{c(i, j)v(p.i.j) + r(p)}$$

and it would be impossible to extract the true ratio $v(q, i, j)/v(p, i, j)$ unless the terms $r(p)$ and $r(q)$ were known or could be estimated. The terms $r(p)$ and $r(q)$ are the atmospheric path radiances for bands p and q (Section 1.2.5), which are generally unknown. They are also unequal because the amount of scattering in the atmosphere increases inversely with wavelength. Switzer, Kowalik and Lyon (1981) consider this problem and show that the atmospheric path radiances must be estimated and subtracted from the recorded radiances before ratioing. A frequently-used method, described in Section 4.4.2, involves the subtraction of constants $k(p)$ and $k(q)$ which are the minimum values in the histograms of channels p and q respectively; these values might be expected to provide a first approximation to the path radiances in the two bands. Switzer, Kowalik and Lyon (1981) suggest that this 'histogram minimum' method overcorrects the data. Other factors, such as the magnitude of diffuse irradiance (skylight) and reflection from cross-valley slopes, also confuse the issue. It is certainly not safe to assume that the 'topographic effect' is completely removed by a ratio operation. Atmospheric effects and their removal, including the 'histogram minimum' method, are considered in Chapter 4.

The effects of atmospheric haze on the results of ratio analyses are studied in an experimental context by Jackson, Slater and Pinter (1983). These authors find that, for turbid atmospheres, the NIR/red ratio was considerably less sensitive to variations in vegetation status and they conclude that:

> The effect (of atmospheric conditions) on the ratio is so great that it is questionable whether interpretable results can be obtained from satellite data unless the atmospheric effect is accurately accounted for on a pixel-by-pixel basis' (p. 195).

The same conclusion was reached for the NDVI. Holben and Kimes (1986) also report the results of a study involving the use of ratios of NOAA AVHRR bands 1 and 2 under differing atmospheric conditions. They find that the NDVI is more constant than individual bands.

Other problems relate to the use of ratios where there is incomplete vegetation coverage. Variations in soil reflectivity will influence ratio values, as discussed above. The angle of view of the sensor and its relationship with solar illumination geometry must also be taken into consideration if data from off-nadir pointing sensors such as the SPOT HRV or from sensors with a wide angular field of view such as the NOAA AVHRR are used (Barnsley, 1983; Wardley, 1984; Holben and Fraser, 1984). In order to make his data comparable over time and space, Frank (1985) converted the Landsat MSS digital counts to reflectances (as described in Section 4.6) and used a first-order correction for solar elevation angle based on the Lambertian assumption (Section 1.3.1; Figure 1.4; Section 4.5). Useful contributions to the study of vegetation indices are Huete *et al.* (2002), who compare vegetation indices derived from MODIS data at two spatial resolutions. Gitelson *et al.* (2002) review the use of vegetation indices to estimate vegetation fractions. Liang (2005, 2007) gives a thorough review of methods of estimating land surface characteristics, while Haboudane *et al.* (2004) and Gong *et al.* (2003) focus on indices derived from hyperspectral data (Section 9.3). The paper by Kowalik, Lyon and Switzer (1983) is still worth reading for an account of the impact of additive radiance terms to the calculation of vegetation indices. Baret and Buis (2008) provide a recent review.

6.3 Empirically Based Image Transforms

Experience gained during the 1970s with the use of Landsat MSS data for identifying agricultural crops, together with the difficulties encountered in the use of ratio transforms (Section 6.2) and principal component transforms (Section 6.4), led to the development of image transforms based on the observations that (i) scatter plots of Landsat MSS data for images of agricultural areas show that agricultural crops occupy a definable region of the four-dimensional space based on the Landsat MSS bands and (ii) within this four-dimensional space the region occupied by pixels that could be labelled as 'soil' is a narrow, elongated ellipsoid. Pair-wise plots of Landsat MSS bands fail to reveal these structures fully because they give an oblique rather than a 'head-on' view of the sub-space occupied by pixels representing vegetation. Kauth and Thomas (1976) propose a transformation that, by rotating and scaling the axes of the four-dimensional space, would give a more clear view of the structure of the data. They called their transform the *Tasselled Cap* since the shape of the region of the transformed feature space that was occupied by vegetation in different stages of growth appeared like a Scottish 'bobble hat'. Other workers have proposed other transforms; perhaps the best known is the PVI which was based on a similar idea to that of the Tasselled Cap, namely, that there is a definite axis in four-dimensional Landsat MSS space that is occupied by pixels representing soils, ranging from soils of low reflectance to those of high reflectance (see also Baret, Jacquemond and Hanocq, 1993). These two transformations are described briefly in the next two subsections.

6.3.1 Perpendicular Vegetation Index

A plot of radiance measured in the visible red band against radiance in the NIR for a partly vegetated area will result in a plot that looks something like Figure 6.7. Bare soil pixels lie along the line S_1-S_2, with the degree of wetness of the soil being higher at the S_1 end of the 'soil line' than at the S_2 end. Vegetation pixels will lie below and to the right of the soil line, and the perpendicular distance to the soil line was suggested by Richardson and Wiegand (1977) as a measure which was correlated with the green LAI and with biomass. The formula used by Richardson and Wiegand (1977) to define the PVI is based on either Landsat-1–3 MSS

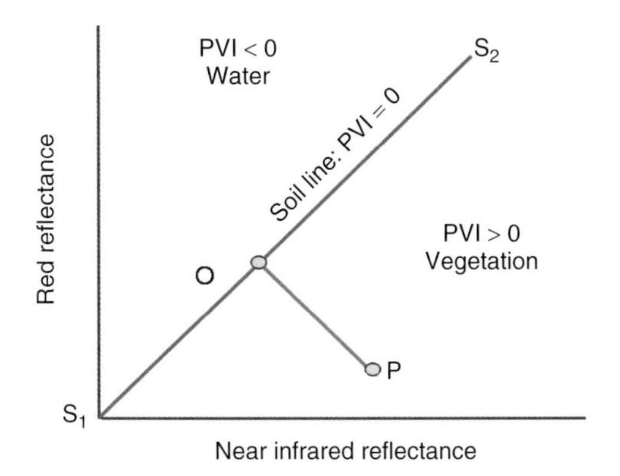

Figure 6.7 *The 'soil line' S_1–S_2 joins the position of the expected red and near-infrared reflectance for wet soils (S_1) with that for dry soils (S_2). Vigorous vegetation shows high reflectance in the near-infrared (horizontal axis) and low reflectance at visible red wavelengths. Point P therefore represents a pixel that has high near-infrared and low visible red reflectance. The PVI measures the orthogonal distance from P to the soil line (shown by line PO).*

band 7 or 6, denoted by PVI7 and PVI6, respectively. Note that this rendition of the PVI is now of historical interest. It is examined here for illustrative purposes only. Bands 6 and 7 of the Landsat MSS covered NIR regions and band 5 covered the visible green waveband.

$$PVI7 = \sqrt{\begin{array}{l}(0.355MSS7 - 0.149MSS5)^2\\ +(0.355MSS5 - 0.852MSS7)^2\end{array}}$$

$$PVI6 = \sqrt{\begin{array}{l}(0.498MSS6 - 0.487MSS5 - 2.507)^2\\ +(2.734 + 0.498MSS5 - 0.543MSS6)^2\end{array}}$$

Neither of these formulae should be used on Landsat MSS images without some forethought. First, the PVI is defined as the perpendicular distance from the soil line (Figure 6.7). Richardson and Wiegand (1977) equation for the soil line is based on 16 points representing soils, cloud and cloud shadows in Hidalgo and Willacy Counties, Texas, USA, for four dates in 1975. It is unlikely that such a small and geographically limited sample could adequately define the soil line on a universal basis. A locally valid expression relating 'soil' pixel values in Landsat MSS bands 5 and 7 (or 5 and 6) of the form $X5 = c + X7$ is needed. Second, the Richardson and Wiegand equation is based on the assumption that the maximum digital count in Landsat MSS bands 4 (green), 5 (red) and 6 (NIR) is 127 with a maximum of 63 in Landsat MSS band 7 (NIR). Landsat MSS images supplied by ESA are normally system-corrected and the pixel values in all four Landsat MSS bands are expressed on a 0–255 scale. Landsat TM and ETM+ images are recorded on a 0–255 scale, so this problem should not arise. The PVI equations listed above do not, however, apply to ETM+ or TM images.

The PVI has been used as an index that takes into account the background variation in soil conditions which affect soil reflectance properties. Jackson, Slater and Pinter (1983) demonstrate that the PVI is affected by rainfall when the vegetation cover is incomplete. However, they considered it to be 'moderately sensitive' to vegetation but was not a good detector of plant stress. The effects of atmospheric path radiance on the PVI were reported as reducing the value of the index by 10–12% from a clear to a turbid atmospheric condition. This is considerably less than the 50% reduction noted for a NIR : red ratio.

The PVI is not now widely used. It is described here so as to introduce the concept of the 'soil line'. Nowadays, the Tasselled Cap or KauthThomas transformation is generally preferred, as it can be modified to deal with data from different sensors. However, its formulation depends on the definition of the soil line using empirical data.

6.3.2 Tasselled Cap (Kauth–Thomas) Transformation

The PVI (Section 6.3.1) uses spectral variations in two of the four Landsat MSS bands, and relates distance from a soil line in the two-dimensional space defined by these two bands as a measure of biomass or green LAI. Kauth and Thomas (1976) use a similar idea except their model uses all four Landsat MSS bands. Their procedure has subsequently been extended to higher-dimensional data such as that collected by the Landsat TM and ETM+ instruments. The simpler four-band combination is considered first. In the four-dimensional feature space defined by the Landsat MSS bands, Kauth and Thomas (1976) suggest that pixels representing soils fall along an axis that is oblique with respect to each pair of the four MSS axes. A triangular region of the four-dimensional Landsat MSS feature space is occupied by pixels representing vegetation in various stages of growth. The Tasselled Cap transform is intended to define a new (rotated) coordinate system in terms of which the soil line and the region of vegetation are more clearly represented. The axes of this new coordinate system are termed 'brightness', 'greenness', 'yellowness' and 'nonesuch'. The brightness axis is associated with variations in the soil background reflectance. The greenness axis is correlated with variations in the vigour of green vegetation while the yellowness axis is related to variations in the yellowing of senescent vegetation. The 'nonesuch' axis has been interpreted by some authors as being related to atmospheric conditions. Due to the manner in which these axes are computed they are statistically uncorrelated, so that they can be represented in the four-dimensional space defined by the four Landsat MSS bands by four orthogonal lines. However, the yellowness and nonesuch functions have not been widely used and the Tasselled Cap transformation has often been used to reduce the four-band MSS data to two functions, brightness and greenness. For further discussion of the model, see Crist and Kauth (1986).

The justification for this dimensionality-reduction operation is that the Tasselled Cap axes provide a consistent, physically based coordinate system for the interpretation of images of an agricultural area obtained at different stages of the growth cycle of the crop. Since the coordinate transformation is defined *a priori* (i.e. not calculated from the image itself) it will not be affected by variations in crop cover and stage of growth from image to image over a time-series of images covering the growing season. The principal components transform (Section 6.4) performs an apparently similar operation; however, the parameters of the principal components transform are computed from the statistical relationships between the individual spectral bands of the specific image being analysed. Consequently, the parameters of the principal components transform vary from one multispectral image set to another as the correlations among the bands depend upon the range and statistical distribution of pixel values in each band, which will differ from an early growing season image to one collected at the end of the growing season.

If the measurement for the jth pixel on the ith Tasselled Cap axis is given by u_j, the coefficients of the ith transformation by $\mathbf{r}_i$ and the vector of measurements on the four Landsat MSS bands for the same pixel by $\mathbf{x}_j$ then the Tasselled Cap transform is accomplished by:

$$u_j = \mathbf{r}_i' \mathbf{x}_j + c$$

In other words, the pixel values in the four MSS bands (the elements of $\mathbf{x}_j$) are multiplied by the corresponding elements of $\mathbf{r}_i$ to give the position of the jth pixel with respect to the ith Tasselled Cap axis, $\mathbf{u}$. The constant c is an offset which is added to ensure that the elements of the vector $\mathbf{u}$ are always positive. Kauth and Thomas (1976) use a value of 32.

The vectors of coefficients $\mathbf{r}_i$ are defined by Kauth and Thomas (1976) as follows:

$$\mathbf{r}_1 = \{0.433, 0.632, 0.586, 0.264\}$$

$$\mathbf{r}_2 = \{-0.290, -0.562, 0.600, 0.491\}$$

$$\mathbf{r}_3 = \{-0.829, 0.522, -0.039, 0.194\}$$

$$\mathbf{r}_4 = \{0.223, 0.012, -0.543, 0.810\}$$

These coefficients assume that Landsat MSS bands 4–6 are measured on a 0–127 scale and band 6 is measured on a 0–63 scale. They are also calibrated for Landsat-1 data and slightly different figures may apply for other Landsat MSS data. The position of the $\mathbf{r}_1$ axis was based on measurements on a small sample of soils from Fayette County, Illinois, USA. The representativeness of these soils as far as applications in other parts of the world is concerned is open to question.

Crist (1983) and Crist and Cicone (1984a, 1984b) extend the Tasselled Cap transformation to data from the six reflective bands of Landsat TM datasets. Data from the Landsat TM thermal infrared channel (conventionally labelled band 6; see Section 2.3.6) are excluded. They found that the brightness function $\mathbf{r}_1$ for the Landsat MSS Tasselled Cap did not correlate highly with the Landsat TM Tasselled Cap equivalent, though Landsat MSS greenness function $\mathbf{r}_2$ did correlate with the Landsat TM greenness function. The TM data was found to contain significant information in a third dimension, identified as wetness. The coefficients for these three functions are given in Table 6.1.

The brightness function is simply a weighted average of the six TM bands, while greenness is a visible/NIR contrast, with very little contribution from bands 5 and 7. Wetness appears to be defined by a contrast between the mid-infrared bands (5 and 7) and the red/NIR bands (3 and 4). The three Tasselled Cap functions can be considered to define a three-dimensional space in which the positions of individual pixels are computed using the coefficients listed in Table 6.1.

The plane defined by the greenness and brightness functions is termed by Crist and Cicone (1984a) the 'plane of vegetation' while the functions brightness and wetness define the 'plane of soils' (Figure 6.8). A Tasselled Cap transform of the 1993 ETM+ image of Alexandria, Egypt, is shown in Figure 6.9.

Several problems must be considered. The first is the now familiar problem of dynamic range compression that, in the case of the Tasselled Cap transform, assumes an added importance. One of the main reasons for supporting the use of the Tasselled Cap method against, for example the principal components technique (Section 6.4) is that the coefficients of the transformation are defined *a priori*, as noted above. However, if these coefficients are applied blindly, the resulting Tasselled Cap coordinates will not lie in the range 0–255 and will thus not be displayable on standard image processing equipment. The range and frequency distribution of Tasselled Cap transform values varies from image to image, however. The problem is to define a method of dynamic range compression that will adjust the Tasselled Cap transform values on to a 0–255 range without destroying interimage comparability. Because the values of the transform are scene dependent it is unlikely that a single mapping function will prove satisfactory for all images but if an image-dependent

Table 6.1 Coefficients for the Tasselled Cap functions 'brightness', 'greenness' and 'wetness' for Landsat Thematic Mapper bands 1–5 and 7.

TM band	1	2	3	4	5	7
Brightness	0.3 037	0.2 793	0.4 343	0.5 585	0.5 082	0.1 863
Greenness	−0.2 848	−0.2 435	−0.5 436	0.7 243	0.0 840	−0.1 800
Wetness	0.1 509	0.1 793	0.3 299	0.3 406	−0.7 112	−0.4 572

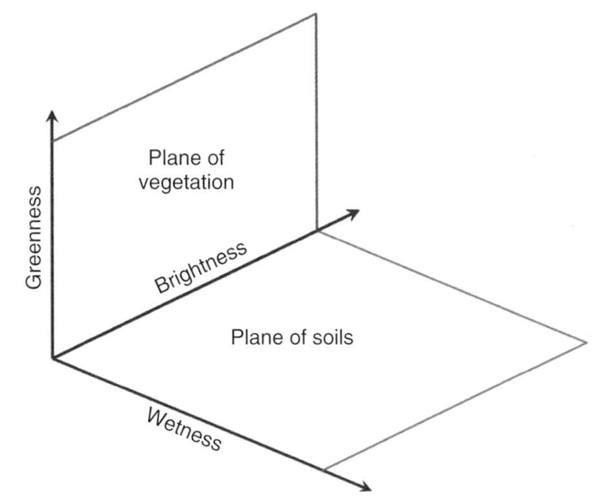

Figure 6.8 The Tasselled Cap transformation defines three fixed axes. Image pixel data are transformed to plot on these three axes (greenness, brightness and wetness) which jointly define the Plane of Vegetation and the Plane of Soils. See text for discussion. Based on Crist, E.P. and Cicone, R.C., 1986, Figure 3. Reproduced with permission from American Society for Photogrammetry and Remote Sensing, Manual of Remote Sensing.

mapping function is selected then interimage comparison will be made more difficult. Crist (personal communication) suggests that the range of Tasselled Cap function values met with in agricultural scenes will vary between 0 and 350 (brightness), −100 and 125 (greenness) and −150 to 75 (wetness). If interimage comparability is important, then the calculated values should be scaled using these limits. Otherwise, the functions could be evaluated for a sample of pixels in the image (for example 25% of the pixels could be selected) and the sample minimum and maximum values ($x_{\min}$ and $x_{\max}$) calculated. The following formula can then be applied to the pixel values x to give the Tasselled Cap values y.

$$y = \frac{x - x_{\min}}{x_{\max} - x_{\min}} \times 255$$

where y is the scaled value (0–255) and x is the raw value. In order to prevent over- and undershoots, a check should be made for negative y values (which are set to zero) or values of y that are greater than 255 (these are set to 255). Problems of scaling are discussed in more detail in Section 3.2.1. As the cost of disk storage has fallen substantially over the past few years, it is now more likely that the output from the Tasselled Cap procedure is represented in the form of 32-bit real values, which can be manipulated in accordance with the requirements of any particular problem.

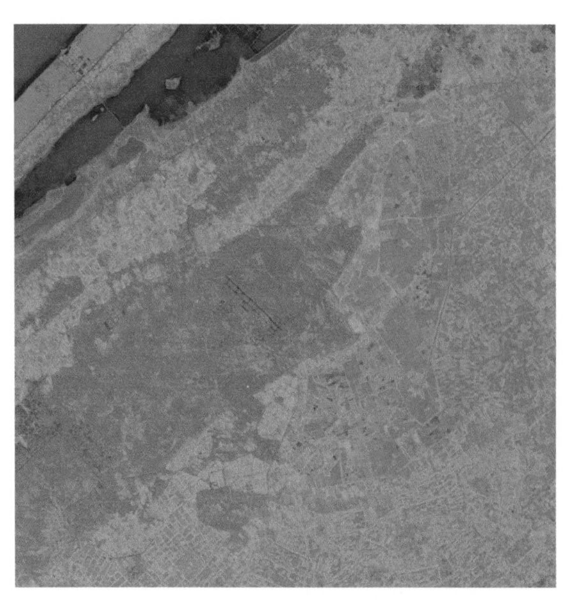

Figure 6.9 Tasselled Cap image derived from 1993 ETM+ image of Alexandria, Egypt, shown in Figure 6.2b. Brightness is shown in red – with the sandy desert area being clearly delineated. Greenness is allocated to the green band, and wetness to the blue band. The water areas are clearly identified in the top left corner of the image, while the agricultural areas (shown in red on Figure 6.2b) are shown in shades of cyan, a mixture of green (greenness) and blue (wetness). More detail of field boundaries and roads and tracks can be seen on this image, compared to Figure 6.2b. Landsat data courtesy NASA/USGS.

A second problem which interferes with the comparison of multidate Tasselled Cap images is the problem of changes in illumination geometry and variations in the composition of the atmosphere. Both of these factors influence the magnitude of the ground-leaving radiance from a particular point, so that satellite-recorded radiances from a constant target will change even though the characteristics of the target do not change. These problems are addressed in a general way in Chapter 4

(Sections 4.4 and 4.5); they are particularly important in the context of techniques that endeavour to provide the means to carry out comparisons between multitemporal images. One solution is proposed by Huang *et al.* (2002c), whose use of apparent reflection is noted in Section 4.6. These authors use procedures similar to those described below to compute Tasselled Cap coefficients from Landsat TM apparent reflectance data.

Like the PVI, the Tasselled Cap transform relies upon empirical data for the determination of the coefficients of the brightness axis (the soil line in the terminology of the PVI). It was noted above that the Kauth and Thomas (1976) formulation of the MSS Tasselled Cap was based on a small sample of soils from Fayette County, Illinois, USA. The TM brightness function is also based on a sample of North American soils, hence applications of the transformation to agricultural scenes in other parts of the world may not be successful if the position of the brightness axis as defined by the coefficients given above does not correspond to the reflectance characteristics of the soils in the local area. Jackson (1983) describes a method of deriving Tasselled Cap-like coefficients from soil reflectance data, using Gram-Schmidt (GS) orthogonal polynomials. He uses reflectance data in the four Landsat MSS bands for dry soil, wet soil, green and senesced vegetation to derive coefficients for three functions representing brightness, greenness and yellowness. Where possible, the coefficients of the orthogonal, Tasselled Cap-like functions should be calculated for the area of study as they may differ from those provided by Crist and Cicone (1984a, 1984b) and listed above. Lobser and Cohen (2007) and Zhang *et al.* (2002) extend the concept of the Tasselled Cap to MODIS data, while Horne (2003) supplies Tasselled Cap coefficients for the four bands of IKONOS multispectral scanner data. Dymond, Mladenoff and Radeloff (2002) illustrate the use of the TM Tasselled Cap in forestry applications. Arbia, Griffith and Haining (2003) examine spatial error propagation when linear combinations of spectral bands are computed, using the case of vegetation indices – including Tasselled Cap – as an example.

6.4 Principal Components Analysis

6.4.1 Standard Principal Components Analysis

Adjacent bands in a multi- or hyperspectral remotely-sensed image are generally correlated. Multiband visible/NIR images of vegetated areas exhibit negative correlations between the NIR and visible red bands and positive correlations among the visible bands because the spectral reflectance characteristics of vegetation (Section 1.3.2.1) are such that as the vigour or greenness of the vegetation increases the red reflectance diminishes and the NIR reflectance increases. The presence of correlations among the bands of a multispectral image implies that there is redundancy in the data. Some information is being repeated. It is the repetition of information between the bands that is reflected in their intercorrelations. If two variables, x and y, are perfectly correlated then measurements on x and y will plot as a straight line sloping upwards to the right (Figure 6.10a). Since the positions of the points shown along line AB occupy only one dimension, the relationships between these points could equally well be given in terms of coordinates on line AB. Even if x and y are not perfectly correlated there may be a dominant direction of scatter or variability, as in Figure 6.10b. If this dominant direction of variability (AB) is chosen as the major axis then a second, minor, axis (CD) could be drawn at right-angles to it. A plot

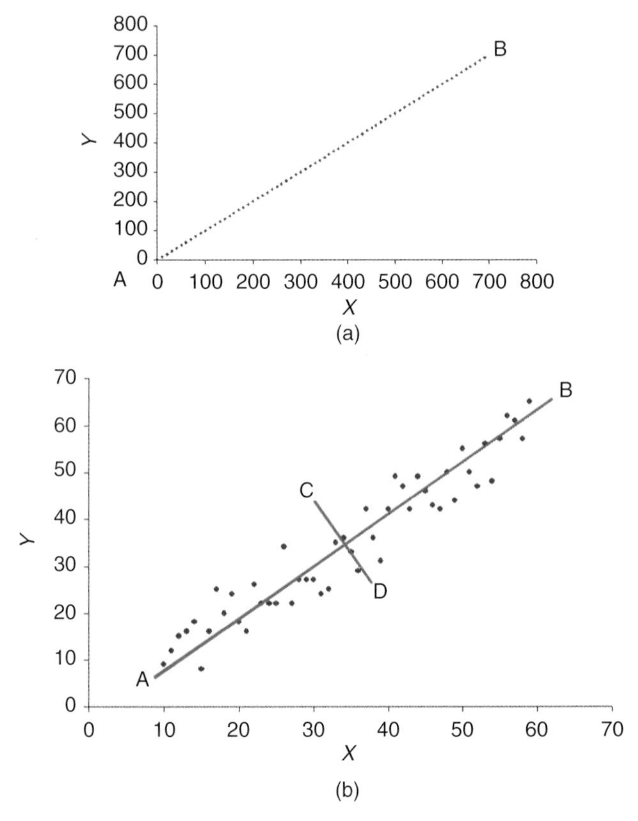

Figure 6.10 (a) Plot of two variables, x and y, which are perfectly correlated ($r = 1.0$). The (x, y) points lie on a straight line between A and B. Although this is a two-dimensional plot, all the points lie on a one-dimensional line. One dimension is therefore redundant. (b) In contrast to the plot shown in (a), this distribution of (x, y) points does not lie along a single straight line between A and B. There is some scatter in a second, orthogonal, direction CD. The distance relationships between the points would be the same if we used AB as the x-axis and CD as the y-axis, though the numerical coordinates of the points would change.

using the axes AB and CD rather than the conventional x and y axes might, in some cases, prove more revealing of the structures that are present within the data. Furthermore, if the variation in the direction CD in Figure 6.10b contains only a small proportion of the total variability in the data then it may be ignored without too much loss of information, resulting in data compression.

This example shows that we must draw a basic distinction between the number of variables (e.g. spectral bands) in an image data set and the intrinsic dimensionality of that data set. In Figure 6.10a the number of variables is two (x and y) but the dimensionality of the data as shown by the scatter of points is one. In Figure 6.10b the dimensionality is again effectively one, although the number of observed variables is, in fact, two. In both examples the use of the single axis AB rather than the x and y axes accomplishes two aims: (i) a reduction in the size of the data set since a single coordinate on axis AB replaces the two coordinates on the x and y axes and (ii) the information conveyed by the set of coordinates on AB is greater than the information conveyed by the measurements on either the x or the y axes individually. In this context information means variance or scatter about the mean; it can also be related to the range of states or levels in the data, as shown in the discussion of entropy in Section 2.2.3.

Multispectral image data sets generally have a dimensionality that is less than the number of spectral bands. For example, it is shown in Section 6.3.2 that the four-band Landsat MSS Tasselled Cap transform produces two significant dimensions (brightness and greenness) while the six-band Landsat TM Tasselled Cap transform defines three meaningful functions (dimensions). The purpose of principal (n.b., not 'principle'!) components analysis is to define the number of dimensions that are present in a data set and to fix the values of the coefficients which specify the positions of that set of axes which point in the directions of greatest variability in the data (such as axes AB and CD in Figure 6.10b). These axes or dimensions of variability are always uncorrelated. A principal components transform of a multispectral image (or of a set of registered multitemporal images; see Sections 6.8 and 6.9) might therefore be expected to perform the following operations:

- estimate the dimensionality of the data set and
- identify the principal axes of variability within the data.

These properties of PCA (sometimes also known as the Karhunen-Loève transform) might prove to be useful if the data set is to be compressed, for example for transmission over a slow connection. Also, relationships between different groups of pixels representing different land cover types may become clearer if they are viewed in the principal axis reference system rather than in terms of the original spectral bands, especially as the variance of the data set is concentrated in relatively fewer principal components. Variance is often associated with information. The data compression property is useful if more than three spectral bands are available. A conventional RGB colour display system relates a spectral band to one of the three colour inputs (RGB). The Landsat TM provides seven bands of data, hence a decision must be made regarding which three of these seven bands are to be displayed as a colour composite image. If the basic dimensionality of the TM data is only three then most of the information in the seven bands will be expressible in terms of three principal components. The principal component images could therefore be used to generate a RGB false-colour composite with principal component number 1 shown in red, number 2 in green and number 3 in blue. Such an image contains more information than any combination of three spectral bands.

The positions of the mutually perpendicular axes of maximum variability in the two-band data set shown in Figure 6.10b can be found easily by visual inspection to be the lines AB and CD. If the number of variables (spectral bands) is greater than three then a geometric solution is impracticable and an algebraic procedure must be sought. The direction of axis AB in Figure 6.10b is defined by the sign of the correlation between variables x and y; high positive correlation results in the scatter of points being restricted to an elliptical region of the two-dimensional space defined by the axes x and y. The line AB is, in fact, the major or principal axis of this ellipse and CD is the minor axis. In a multivariate context, the shape of the ellipsoid enclosing the scatter of data points in a p-dimensional space is defined by the variance-covariance matrix computed from p variables or spectral bands. The variance in each spectral band is proportional to the degree of scatter of the points in the direction parallel to the axis representing that variable, so that it can be deduced from Figure 6.11a that for the circular distribution the variances of variables X and Y (represented by GH and EF) are approximately equal. The covariance defines the shape of the ellipse enclosing the scatter of points. Figure 6.11a shows two distributions. One (green outline) has a high positive covariance while the other (blue outline) has a covariance of zero. The mean of each variable gives the location of the centre of the ellipse (or ellipsoid in a space of dimensionality higher than two). Thus, the mean vector and the variance-covariance matrix define the location and shape of the scatter of points in a p-dimensional space. The information contained in the variances and covariances of a set of variables is used again in the definition of the maximum likelihood classification procedure that is described in Section 8.4.2.3.

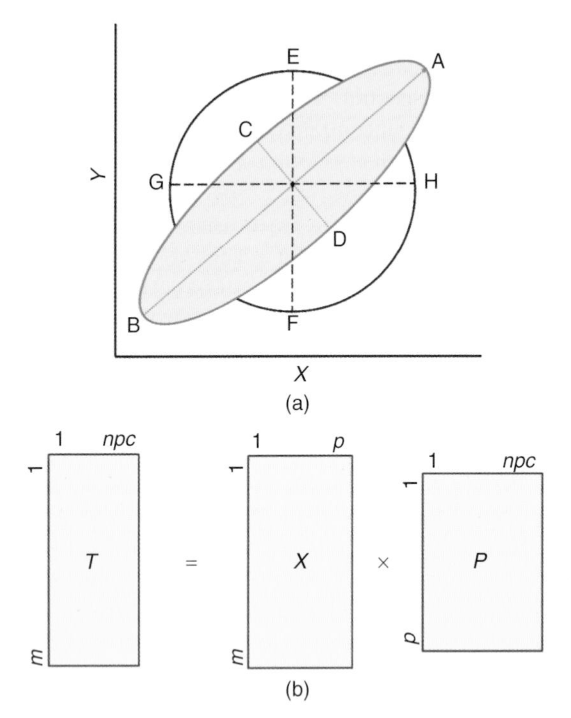

Figure 6.11 *(a) The ellipse is characterized in the two dimensional space defined by variables X and Y by long axis AB and short axis CD, which define the two orthogonal directions of maximum scatter. The circle shows equal scatter in all directions from the centre, so the positions of its axes EF and GH are purely arbitrary – there is no direction of maximum variance. In the case of the ellipse, the direction of slope of line AB indicates that there is a strong positive correlation between the two variables while its shape shows that one variable has a larger variance than the other. The lack of a preferred orientation of scatter in the case of the circle indicates a zero correlation with the two variables X and Y having equal variances. (b) Matrix multiplications in PCA. See text for discussion.*

The relationship between the correlation matrix and the variance–covariance matrix sometimes leads to confusion. If the p variables making up the data set are measured on different and incompatible scales (for example three variables may be measured, respectively, in metres above sea level (elevation), in millibars (barometric pressure) and kilograms (weight)) then unit changes in the variances of these variables are not directly comparable – how many metres are equal to 1000 millibars? The importance of the variance in defining the scatter of points in a particular direction has already been stressed, so it is clear that if the variance is to be used in defining the shape of the ellipsoid enclosing the scatter of points in the p-dimensional space then the scales used to measure each variable must be comparable.

To illustrate this point, consider what would be the outcome (in terms of the shape of the enclosing ellipsoid) if the three variables mentioned earlier were measured in feet above sea level (rather than metres), inches of mercury (rather than millibars) and weight in ounces (rather than kilograms). Not only would the variance of each of the variables be altered but the shape of the enclosing ellipsoid would also change. The degree of change would not be constant in each dimension, and the shape of the second ellipsoid based on imperial units would not be related in any simple fashion to the shape of the first ellipsoid, based on metric units. Consequently, the lengths and orientations of the principal axes would change. It is in these circumstances that the correlation coefficient rather than the covariance is used to measure the degree of statistical association between the spectral bands. The correlation is simply the covariance measured for standardized variables. To standardize a variable the mean value is subtracted from all the measurements and the result is divided by the standard deviation. This operation converts the raw measurements to standard scores or z-scores, which have a mean of zero and a variance of unity. The off-diagonal elements of the correlation matrix are the covariances of the standard scores and the diagonal elements are the variances of the standard scores, which are by definition always unity. Since the shape of the ellipsoid enclosing the scatter of data points is altered in a complex fashion by the standardization procedure it follows that the orientation and lengths of the principal axes, such as AB and CD in Figure 6.10b, will also change. The effects of the choice of the sample variance-covariance matrix $\mathbf{S}$ or the correlation matrix $\mathbf{R}$ on the results of the principal components operation are considered below, and an illustrative example is provided.

The principal components of a multispectral image set are found by algebraic methods that are beyond the scope of this book (see Jolliffe (2002), Mather (1976) or Richards and Jia (2005) for a more detailed discussion, or Jackson (1991) for an easier introduction). The procedure can be illustrated easily, however. What we are seeking is a transformation matrix that will rotate and shift the xy axes of the Figure 6.11a to the positions AB and CD (the long axes of the yellow ellipse). This transformation matrix has p rows and npc columns (Figure 6.11b) where p is the number of images in the multispectral or hyperspectral data set and npc is the number of principal component images that are to be calculated. The image data matrix, $\mathbf{X}$, is postmultiplied by $\mathbf{P}$ to give the matrix of principal component images, $\mathbf{T}$. Matrix $\mathbf{X}$ has m rows and p columns, where m is the number of pixels in each image. There are npc columns in matrix $\mathbf{T}$, each column holding a principal component image with m pixels. The value of npc lies between 1 and p. To find the (i, j)th element of $\mathbf{T}$, multiply the elements of row j of matrix $\mathbf{X}$ by the corresponding elements of column j of matrix $\mathbf{P}$.

The elements of the transformation matrix $\mathbf{P}$ are determined by the principle of maximum variance. That is, the transformation from **X** into **T** must ensure that the variance of Principal Component 1 is the largest of any linear combination of coefficients p_{ij}. But this criterion could be met by setting all the elements of **P** to infinity. So we add the extra stipulation that $\sum_{i=1}^{p} p_{ij}^2 = 1$ or, in other words, the sums of the squares of the elements of any column j of **P** add to 1.0. Fortunately, a matrix procedure called eigenvalue/eigenvector extraction provides exactly what we want. The matrix **P** is the matrix of unit eigenvectors of the correlation matrix **R** or of the variance-covariance matrix **S** of the raw data, **X**. So the PCA operation is simple: calculate **R** or **S**, obtain its eigenvectors **P** and carry out the matrix multiplication **XP** as shown in Figure 6.11b. The eigenvalues **Λ** associated with the eigenvectors give the variance of the corresponding principal component image, The eigenvalues are normally arranged in descending order of magnitude, and it is commonly found that the variances of the first two or three principal component images account for more than 90% of the total variation in **X**, the original image set.

It follows from the previous paragraph that standardized units of variance must be used if the features (bands) are measured on incompatible scales. The calculation of the correlation matrix **R** includes data standardization. If, as may be the case with radiance data rather than DNs, the variables are all measured in terms of the same units (such as $\mathrm{mW\,cm^2\,sr^{-1}}$) then standardization is unnecessary and, indeed, undesirable for it removes the effects of changes in the degree of variability between the bands. In such cases, the matrix **S** of variances and covariances is used.

If the principal component images are to be displayed on a screen, then they must be scaled to fit the range 0–255. This scaling problem has already been discussed earlier in this chapter in the context of band ratios and the Tasselled Cap transform. Given that the principal component images will not generally be integers in the range 0–255, the most effective method of scaling is to store the principal component images as 32-bit real numbers, and use one of the methods of scaling described in Chapter 3 to convert these real numbers to a 0–255 range when required (see Figure 3.6). When comparing principal component images produced by different computer programs, it is always wise to determine the nature of the scaling procedure used in the generation of each image. Reference to Figure 3.6b,c shows how the effective visualizaation of the results of PCA can be achieved.

If all available bands are input to the principal components procedure then, depending on whether the analysis is based on interband correlations or covariances, the information contained in a subset of bands may be under-represented, as a result of the spectral resolution of the sensor. For example, the Landsat ETM+ has one thermal infrared band (out of seven) whereas ASTER has 4 out of 13 (Table 2.4). Relative to the ASTER data set, the information content of the thermal infrared band will be under-represented in the Landsat ETM+ dataset. Siljestrőm *et al.* (1997) use a procedure that they call *selective principal components analysis*, which involves the division of the bands into groups on the basis of their intercorrelations. For TM data, these authors find that bands 1, 2 and 3 are strongly intercorrelated and are distinct from bands 5 and 7. Band 4 stands alone. They carry out PCA separately on bands (1, 2, 3) and bands (5, 7) and use the first principal component from each group, plus band 4, to create a false-colour image, which is then used to help in the recognition of geomorphological features.

Table 6.2 gives the correlation matrix for the six reflective (visible plus NIR and mid-infrared) bands (numbered 1–5 and 7) of a Landsat TM image of the Littleport area shown in Figure 6.12a. Correlations rather than covariances are used as the basis for the principal components procedure because the digital counts in each band do not relate to the same physical units (that is a change in level from 30 to 31, for example, in band 1 does not represent the same change in radiance as a similar change in any other band. See Section 4.6 for details of the differences in calibration between TM bands. High positive correlations among all reflective bands except band 4 (NIR) can be observed. The lowest correlation between any pair of bands (excluding band 4) is +0.660. The correlations between band 4 and the other bands are negative. This generally high level of correlation implies that the radiances in each spectral band except band 4 are varying spatially in much the same way. The negative correlation

Table 6.2 *Correlations among Thematic Mapper reflective bands (1–5 and 7) for the Littleport TM image. The means and standard deviations of the six bands are shown in the rightmost two columns.*

TM band	1	2	3	4	5	7	Mean	Standard deviation
1	1.000	0.916	0.898	−0.117	0.660	0.669	65.812	8.870
2	0.916	1.000	0.917	−0.048	0.716	0.685	29.033	5.652
3	0.898	0.917	1.000	−0.296	0.757	0.819	26.251	8.505
4	−0.117	−0.048	−0.296	1.000	−0.161	−0.474	93.676	24.103
5	0.660	0.716	0.757	−0.161	1.000	0.883	64.258	18.148
7	0.669	0.685	0.819	−0.474	0.883	1.000	23.895	11.327

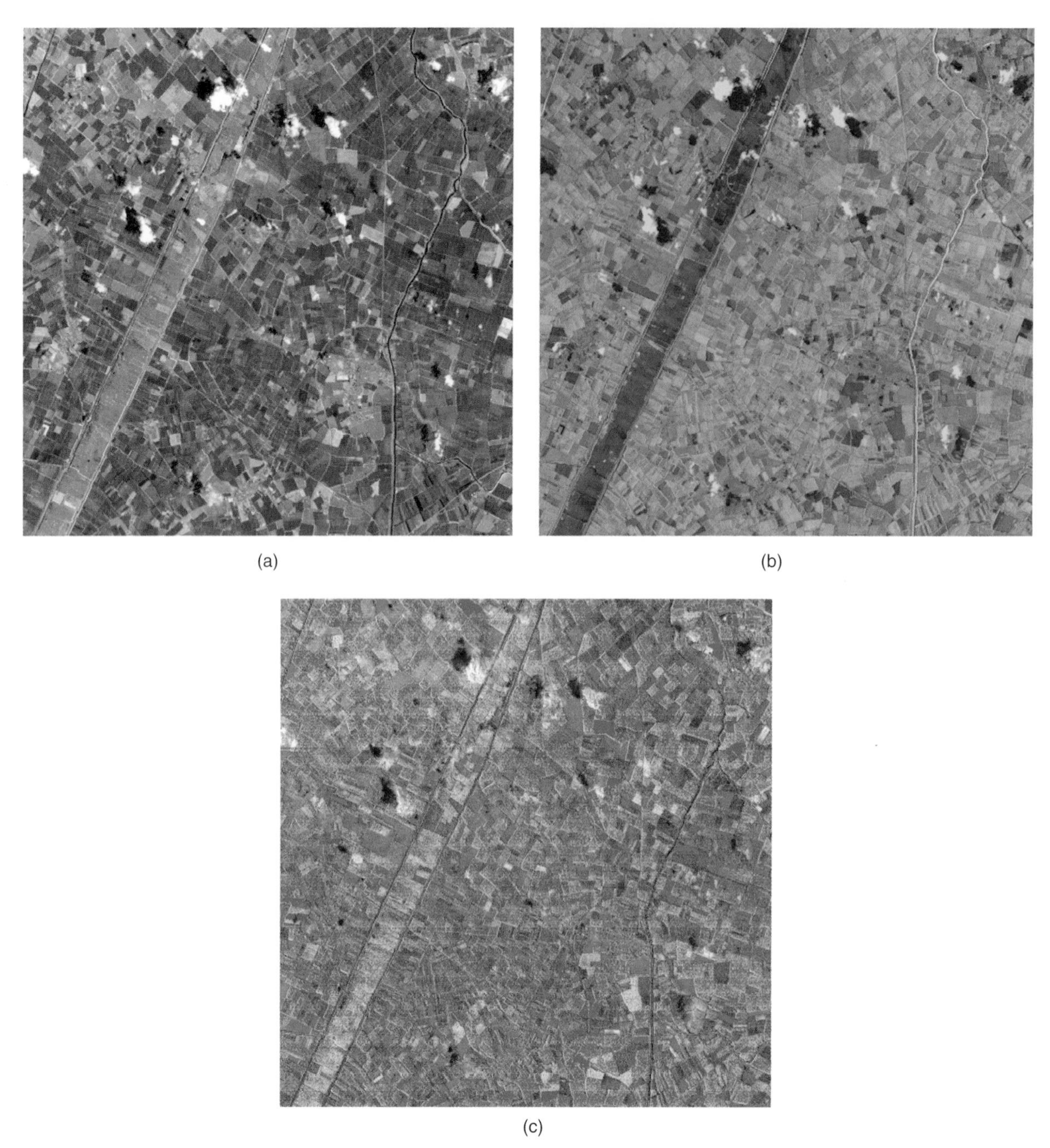

Figure 6.12 *(a) False colour composite image (ETM+ bands 4, 3, 2) of the Littleport area of eastern England. The two parallel lines running up from the lower right corner to the top centre are drainage channels. The R. Ouse is the deep blue line running vertically from the bottom right. Clouds and shadows are apparent mainly in the upper centre of the image. Fields of growing crops and pasture are coloured red. Ploughed fields are cyan, as is the village of Littleport in the bottom centre. Landsat data courtesy NASA/USGS. (b) Principal components 1–3 of the correlation matrix between the six reflective bands of the Littleport TM subimage. The red channel shows principal component 1, which appears to show the negative of brightness as cloud shadow and water appear yellow. The area of land between the two drainage channels is shown in deep purple, though little detail is present. It is difficult to give a name to the three components. (c) Principal components 4–6 of the correlation matrix of the Littleport image. The River Ouse and the two parallel drainage channels are very clear as is some detail of the area between the drainage channels. The image is rather noisy, though not enough to render it unusable.*

between band 4 and the optical bands can be explained by the fact that the area shown in the image is an agricultural one, and the main land cover type is vegetation. The idealized spectral reflectance curve for vegetation (Figure 1.21) shows the contrast between the NIR and red channels. Vegetation ratios exploit this fact, and as the ground cover of vegetation, and the vigour (greenness) of the vegetation increase so the contrast between NIR and red reflectance increases as the NIR reflectance rises and the red reflectance falls. Hence, this negative correlation between band 4 and other bands can be partially explained. Table 6.2 also shows the means and standard deviations of the six bands. Note the significantly higher standard deviations for bands 4, 5 and 7 which show that variability of pixel values is greater in these bands than in bands 1, 2 and 3. The mean pixel values also differ. This is a result of the nature of the ground cover and the calibration of the TM sensor.

The eigenvalues and scaled eigenvectors, or principal component loadings, derived from the correlation matrix measure the concentration of variance in the data in six orthogonal directions (Table 6.3, Figure 6.11a). Over 70% of the variability in the data lies in the direction defined by the first principal component. Column 1 of Table 6.3 gives the relationship between the first principal component and the six TM bands; all bands except the NIR have entries that are greater than 0.87, while the NIR band has an entry of −0.309. This indicates (as was inferred from the correlation matrix) that there is considerable overlap in the information carried by the different channels, and that there is a contrast between the NIR and the other bands. The image produced by the first principal component (Figure 6.12b, red channel) summarizes information that is common to all channels. It can be seen to be a weighted average of five of the six TM bands contrasted with the NIR band.

The second principal component of the Littleport image set (Figure 6.12a, green channel) is dominated by the contribution of the NIR band. There is a small contribution from the three visible bands. Between them, principal components 1 and 2 account for over 88% of the variability in the original six-band data set. A further 8% is contributed by principal component 3 (Figure 6.12b, blue channel), which appears to be highlighting a contrast between the infrared and the visible bands. Visual analysis of the principal component images shown in Figure 6.12b appears to indicate that principal components 1 and 2 may be picking out differences between different vegetation types in the area, while principal component 3 is related to water content of the soil.

Principal components 4–6 (Figure 6.12c) together contain only 3.1% of the variation in the data. If the noise present in the image data set is evenly distributed among the principal components then the lower-order (higher numbered) principal components might be expected to have a lower SNR than the higher-order principal components. On these grounds it might be argued that principal components 4–6 are not worthy of consideration. This is not necessarily the case. While the contrast of these lower-order principal component images is less than that of the higher-order components, there may be patterns of spatial variability present that should not be disregarded, as Figure 6.12c shows. The sixth principal component shown in Figure 6.12c is clearly spatially non-random. Townshend (1984) gives a good example of the utility of low-order principal components. His seventh principal component accounted for only 0.08% of the variability in the data yet a distinction between apple and plum orchards and between peaty and silty-clay soils was brought out. This distinction was not apparent on any other component or individual TM band, nor was it apparent from a study of the principal component loadings. It

Table 6.3 *Principal component loadings for the six principal components of the Littleport TM image. Note that the sum of squares of the loadings for a given principal components (column) is equal to the eigenvalue. The percent variance value is obtained by dividing the eigenvalue by the total variance (six in this case because standardized components are used – see text) and multiplying by 100.*

	PC 1	PC 2	PC 3	PC 4	PC 5	PC 6
TM band 1	0.899	0.242	−0.288	0.223	0.002	−0.002
TM band 2	0.914	0.303	−0.182	−0.143	−0.095	−0.103
TM band 3	0.966	0.033	−0.165	−0.117	0.086	0.134
TM band 4	−0.309	0.924	0.214	−0.006	0.076	−0.001
TM band 5	0.871	0.019	0.470	0.038	−0.124	0.059
TM band 7	0.904	−0.285	0.267	0.009	0.148	−0.094
Eigenvalue	4.246	1.086	0.481	0.085	0.059	0.041
% variance	70.77	18.10	8.02	1.42	0.99	0.68
Cumulative % variance	70.77	88.88	96.90	98.32	99.31	100.00

is important to check principal component images by eye, using one's knowledge of the study area, rather than rely solely upon the magnitudes of the eigenvalues as an indicator of information content, or on inferences drawn from the principal component loadings.

Do not be misled, however, by the coherent appearance of the lower-order principal component images. Figure 6.12c (blue channel) shows the sixth and last principal component derived from the Littleport TM image data set. This principal component accounts for only 0.68% of the total (standardized) variance of the image data set, yet it is interpretable in terms of spatial variation. It should be borne in mind that the information expressed by this principal component has been (i) transformed from a 32-bit real number into a count on the 0–255 range, as described in Section 3.2.1 and (ii) subjected to a histogram equalization contrast stretch. Yet, if the first principal component were expressed on a 0–255 scale and the ranges of the other principal components adjusted according to their associated eigenvalue, then principal component 6 would have a dynamic range of just 0–10.

The use of lower-order (higher numbered) components depends on the aims of the project. If the aim is to capture as much as possible of the information content of the image set in as few principal components as possible then the lower-order principal components should be omitted. If the PCA is based on the correlation matrix is used then the 'eigenvalue greater than 1' (or 'eigenvalue – 1') criterion could be used to determine how many principal components to retain. If an individual standardized band of image data has a variance of one then it might be possible to argue that all retained principal components should have a variance of at least one. If this argument were used then only the first two principal components of the Littleport image set would be retained, and 11% of the information in the image data set would be traded for a reduction of 66.66% in the image data volume. On the other hand, one may wish merely to orthogonalize the data set (that is express the data in terms of uncorrelated principal components rather than in terms of the original spectral bands) in order to facilitate subsequent processing. For example, the performance of the feed-forward artificial neural net classifier, discussed in Chapter 8, may perform better using uncorrelated inputs. In such cases, all principal components should be retained. It is important to realize that the aims of a project should determine the procedures followed, rather than the reverse.

The example used earlier in this section is based on the eigenvalues and eigenvectors of the correlation matrix, **R**. We saw earlier that the principal components of the variance-covariance (**S**) and correlation (**R**) matrices are not related in a simple fashion. Consideration should therefore be given to the question of which of the two matrices to use. Superficially it appears that the image data in each band are comparable, all being recorded on a 0–255 scale. However, reference to Section 8.4.3 shows that this is not so. The counts in each band can be referred to the calibration radiances for that band, and it will be found that:

1. The same pixel value in two different bands equates to different radiance values.
2. If multidate imagery is used then the same pixel value in the same band for two separate dates may well equate to different radiance values because of differences in sensor calibration over time.

The choice lies between the use of the correlation matrix to standardize the measurement scale of each band or conversion of the image data to radiances followed by PCA of the variance-covariance matrix (Singh and Harrison, 1985). However, if multidate imagery is used the question of comparability is of great importance.

Figure 6.13a,b and Tables 6.4 and 6.5 summarize the results of a principal components transform of the Littleport TM image data set based on the variance-covariance matrix (**S**) rather than the correlation matrix (**R**). Compare the results for the analysis based on the correlation matrix, described above and summarized in Table 6.2 and Figure 6.12. The effects of the differences in the variances of the individual bands are very noticeable. Band 4 has a variance of 580.95, which is 47.56% of the total variance, and as a consequence band 4 dominates principal components 1 and 2. As in the correlation example, the first principal component is a contrast between band 4 and the other five bands, but this time the loading on band 4 is the highest in absolute terms rather than the lowest and the percentage of variance explained by the first principal component is 58.22 rather than 70.77. Principal component two, accounting for 35.35% of the total variance rather than 18.10% in the correlation case, is now more like a weighted average. The false-colour composite made up of principal components 1–3 (Figure 6.13a) appears, from visual inspection, to contain more detail than the corresponding image using correlations (Figure 6.12b). The higher-order principal components are much less important even than they are when the analysis is based on correlations, and the corresponding principal component images are noisier. Which result is 'better' depends, of course, on the user's objectives. The aim of PCA should be to generate a set of images that are more useful in some way than are the untransformed images, rather than to satisfy the pedantic requirements of statistical theory. Nevertheless, the point is clear: the principal components of **S**, the variance-covariance matrix, are quite different from the principal components of **R**, the correlation matrix. Methods of dealing with the noise that affects the lower-order (higher numbered) principal component images are considered below.

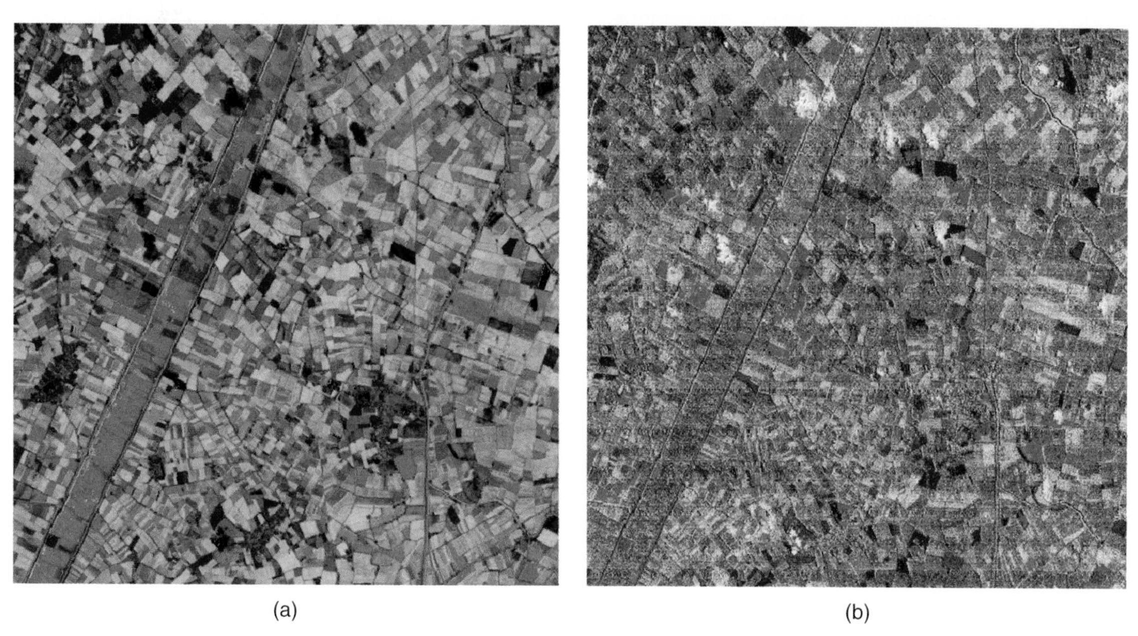

Figure 6.13 *(a) The first three principal components derived from the variance-covariance matrix of the Littleport subimage shown in Figure 6.12a. (b) The corresponding principal components 4–6. See text for further discussion.*

Table 6.4 *Variance-covariance matrix for the Littleport TM image set. The last row shows the variance of the corresponding band expressed as a percentage of the total variance of the image set. The expected variance for each band is 16.66%, but the variance ranges from 2.61% for band 2 to 47.56% for band 4.*

TM band	1	2	3	4	5	7	Mean	Standard deviation
TM 1	78.67	45.90	67.71	−25.04	106.25	67.22	65.812	8.870
TM 2	45.90	31.94	44.09	−6.51	73.45	43.87	29.033	5.652
TM 3	67.71	44.09	72.34	−60.75	116.78	78.86	26.251	8.505
TM 4	−25.04	−6.51	−60.75	580.95	−70.32	−129.52	93.676	24.103
TM 5	106.25	73.45	116.78	−70.32	329.37	181.56	64.258	18.148
TM 7	67.22	43.87	78.86	−129.52	181.56	128.30	23.895	11.327
Percentage variance	6.44	2.61	5.92	47.56	26.96	10.50	16.666	15.90

Table 6.5 *Principal component loadings for the six principal components of the Littleport TM image, based on the covariance matrix shown in Table 6.3a.*

	PC 1	PC 2	PC 3	PC 4	PC 5	PC 6
TM band 1	0.531	0.593	0.575	−0.491	−0.168	−0.018
TM band 2	0.502	0.674	0.477	−0.034	0.139	0.215
TM band 3	0.707	0.530	0.421	0.057	0.178	−0.085
TM band 4	−0.834	0.552	0.000	0.019	−0.001	−0.002
TM band 5	0.664	0.714	−0.211	−0.070	0.006	−0.003
TM band 7	0.684	0.438	−0.064	0.233	−0.055	0.013
Eigenvalue	711.19	431.87	61.20	9.69	5.52	2.05
% Variance	58.22	35.35	5.01	0.79	0.45	0.17
Cumulative % variance	58.22	93.58	98.59	99.38	99.83	100.00

In Section 6.3.2 the relationship between the principal components and the Tasselled Cap transformations is discussed. It was noted that the principal components transform is defined by the characteristics of the inter-band correlation or covariance structure of the image, whereas the Tasselled Cap transformation is based on external criteria, namely, the pre-defined position of the soil brightness, greenness and wetness axes. In other words, the positions of the Tasselled Cap axes are defined mathematically by pre-determined coefficients such as those listed in Table 6.1, whereas the coefficients of the principal components transform are derived from the correlation (**R**) or variance-covariance (**S**) matrix of the data set, which varies from image to image. Because the results of a PCA are image-specific it follows that principal component transformed data values cannot be directly compared between images as the principal components do not necessarily have the same pattern of principal component loadings. While PCA is useful in finding the dimensionality of an image data set and in compressing data into a fewer number of channels (for display purposes, for example) it does not always produce the same components when applied to different image data sets of the same area.

The use of PCA in measuring change in multitemporal image sets and in merging images with different properties such as high-resolution panchromatic and lower resolution multispectral images is considered in Sections 6.8 and 6.9.

Further reading and case studies of the use of PCA applied to remotely-sensed data are provided by Avena, Ricotta and Volpe (1999), who consider the spatial structure of a multispectral data set, Soares Galveno, Pizarro and Neves Epiphanio (2001), who study tropical soils, Huang and Antonelli (2001), who use PCA to compress high-resolution data, Blackburn and Milton (1997), who consider PCA in ecological studies, while Crósta *et al.* (2003) apply PCA to an ASTER image set to identify key alteration minerals. Kaewpijit, Le Moigne and El-Ghazawi (2003) compares PCA and wavelets (Section 6.7) in a study of data compression. Jolliffe (2002) and Jackson (1991) are standard texts – the latter is less demanding of the reader. Hsieh (2008) introduces the concept of non-linear PCA while Chitroub (2005) shows how PCA can be performed by an artificial neural network (Section 8.4.3). This latter method is especially useful with high-dimensional data because the variance-covariance matrix becomes large, and the number of samples per matrix element reduces, thus increasing the error associated with each element of the variance-covariance matrix. In addition, the calculation of the eigenvalues and eigenvectors of the variance-covariance matrix (or the correlation matrix) is subject to rounding error, as described in Chapter 3.

Other references to the use of PCA in remote sensing are given in Sections 6.8 and 6.9 in which the problems of change detection and data fusion, respectively, are presented.

6.4.2 Noise-Adjusted PCA

The presence of noise that tends to dominate over the signal in lower-order (higher numbered) principal component images is mentioned a number of times in the preceding section. Some researchers, for example Green *et al.* (1988) and Townshend (1984), note that some airborne thematic mapper data do not behave in this way. Roger (1996) suggests that the principal components method could be modified so as to eliminate the noise variance. This 'noise-adjusted' PCA would then be capable of generating principal component images that are unaffected by noise. The method presented here and incorporated into the MIPS software is similar in principle to the 'noise-adjusted PCA' of Green *et al.* (1988) and Roger (1996). It is described in more detail in Tso and Mather (2009), while Nielsen (1994) gives a definitive account.

Standard PCA extracts successive principal components in terms of the 'maximum variance' criterion. Thus, if we assume that every variable (feature or spectral band) $\mathbf{x}_j$ entered in the PCA has a mean of zero, then we could define principal component number one as that linear combination of the variables that maximizes the expression $\sum_{i=1}^{p}\sum_{j=1}^{n} x_{ji}^2$, where x_{ji} is an element of the data matrix that consists of n observations (pixels) measured on p spectral bands. The second and subsequent principal components are defined in terms of the same criterion, but after the variance attributable to higher-order (lower numbered) principal components has been removed.

Using this idea, principal components are seen as linear combinations of the original variables $\mathbf{x}_j$, with the coefficients of the linear combination being defined so that the maximum variance criterion is satisfied. A linear combination of variables $\mathbf{x}_j$ is simply a weighted sum of the x_{ij} and the coefficients of the linear combination are the weights. There are an infinite number of possible weights – PCA determines the set that satisfies the desired criterion, that of maximum variance, as explained in Section 6.4.1.

What if we use a criterion other than that of maximum variance? Could there be another set of weights that satisfies the new criterion? This new criterion could be expressed in words as 'maximize the ratio of the signal variance to the noise variance', which we will express mathematically as $\frac{\sigma_S^2}{\sigma_N^2}$ which is the ratio of signal variance (σ_S^2) to noise variance (σ_N^2). Two questions arise at this point: first, is there a procedure that could maximize the new criterion? Second, how do we calculate σ_S^2 and σ_N^2? The answer to the first question is 'yes', and it is described later in this section. The second question is, in fact, more difficult. We can easily compute the sums of squares given by the expression $\sum_{i=1}^{p}\sum_{j=1}^{n} x_{ji}^2$ because the values of $\mathbf{X}(= x_{ji})$ are known. However, we need a method that will separate the individual measurements x_{ij} into two parts, the first part representing the 'signal' part of each x (i.e. x_S) and the second representing the noise contribution (i.e. x_N). This idea could be written as $x = x_S + x_N$. If any two of these terms are known then the other can be calculated. Unfortunately, though, only x – the pixel value x- is known.

One way out of this paradox is to consider the nature of 'noise'. If we can think a single-valued spatial data set (such as a digital elevation model or DEM, Figure 2.12) as consisting of the sum of spatial variability over different scales, then – for a DEM covering a hilly area at a map scale of 1 : 100 000 or smaller – we see patterns that relate to the *regional* disposition of hills and valleys and may be able to draw some conclusions about the structure of the area as a whole; for example we may deduce from the fact that the valleys radiate from a central point, which is also the highest point, that the area has a dome-like structure. Alternatively, if the rivers form a trellis-like pattern, as in parts of the Appalachian Mountains in the eastern United States, then we may conclude that the drainage pattern is controlled by the nature of the underlying rocks, which consist of alternating hard and soft bands.

In both of these instances it is the regional-scale pattern of hills and valleys on which our attention is focussed. No consideration is given to the presence of small hummocks on valley floors, or the numerous tiny gullies that are located near the heads of the main valleys. Those phenomena are too small to be of relevance at a map scale of 1 : 100 000, so they are ignored. They represent 'noise' whereas the regional pattern of hills and valleys represents the 'signal' or information at this spatial scale. One way of measuring the noise might be to take the difference in height between a given cell in the DEM and the cells to the north and east, respectively. If the cell of interest lies on a flat, plateau-like, area with no minor relief features, then these horizontal and vertical differences in elevation would be close to zero. In the hummocky valley bottoms and the gullied valley heads these local differences in height would be larger. Since the hummocks and gullies represent noise at our chosen scale of observation, we might think of using vertical and horizontal differences to quantify or characterizenoise.

The MIPS procedure **`Noise Reduction Transform`** (accessed from the **`Transform`** menu) uses this simple idea to separate values x into the signal (x_S) and noise (x_N) components. First, the covariance matrix $\mathbf{C}$ of the full dataset (consisting of n pixels per band, with p bands) is computed. Next, the horizontal and vertical pixel differences in each of the p bands are found, and their covariance matrices calculated and combined to produce the noise covariance matrix $\mathbf{C}_N$. The covariance matrix of the signal, $\mathbf{C}_S$, is found by subtracting $\mathbf{C}_N$ from $\mathbf{C}$. Now we can define the criterion to be used in the calculation of the coefficients of our desired linear combination; it is simply 'maximize the ratio $\mathbf{C}_S/\mathbf{C}_N$', meaning: our linear combinations of the p bands will be ordered in terms of decreasing SNR rather than in terms of decreasing variance (as in PCA). Mathematical details of how this is done are beyond the scope of this book, but the algorithm is essentially the same as that used in the multiple discriminant analysis technique, described in Mather (1976).

The outcome is a set of coefficients for p linear combinations of the p spectral bands, ranked from 1 (the highest SNR) to p (the lowest SNR). These coefficients are applied to the data in exactly the same way as described above for standard principal components. Now, however, the resulting images should be arranged in order of the ratio of signal to noise variance, rather than in terms of total variance, as shown in Figure 6.14 in which the first and sixth noise-reduced principal components are displayed. To help you compare the results, the image used in this example is the same as that used at the end of Section 6.4.1 (Figures 6.12 and 6.13).

6.4.3 Decorrelation Stretch

Methods of colour enhancement are the subject of Chapter 5.2. The techniques discussed in that chapter include linear contrast enhancement, histogram equalization and the Gaussian stretch. All of these act upon a single band of the false colour image at a time and thus must be applied separately to the RGB components of the image. As noted in Section 6.4.1, PCA removes the correlation between the bands of an image set by rotating the axes of the data space so that they become oriented with the directions of maximum variance in the data, subject to the constraint that these axes are orthogonal. If the data are transformed by PCA to a three-dimensional space defined by the principal axes, and are 'stretched' within this space, then the three contrast stretches will be at right angles to each other. In RGB space the three colour components are likely to be correlated, so the effects of stretching are not independent for each colour (Gillespie *et al.*, 1986).

Decorrelation stretching requires the three bands making up the RGB colour composite images to be subjected to a PCA, a stretch applied in the principal components

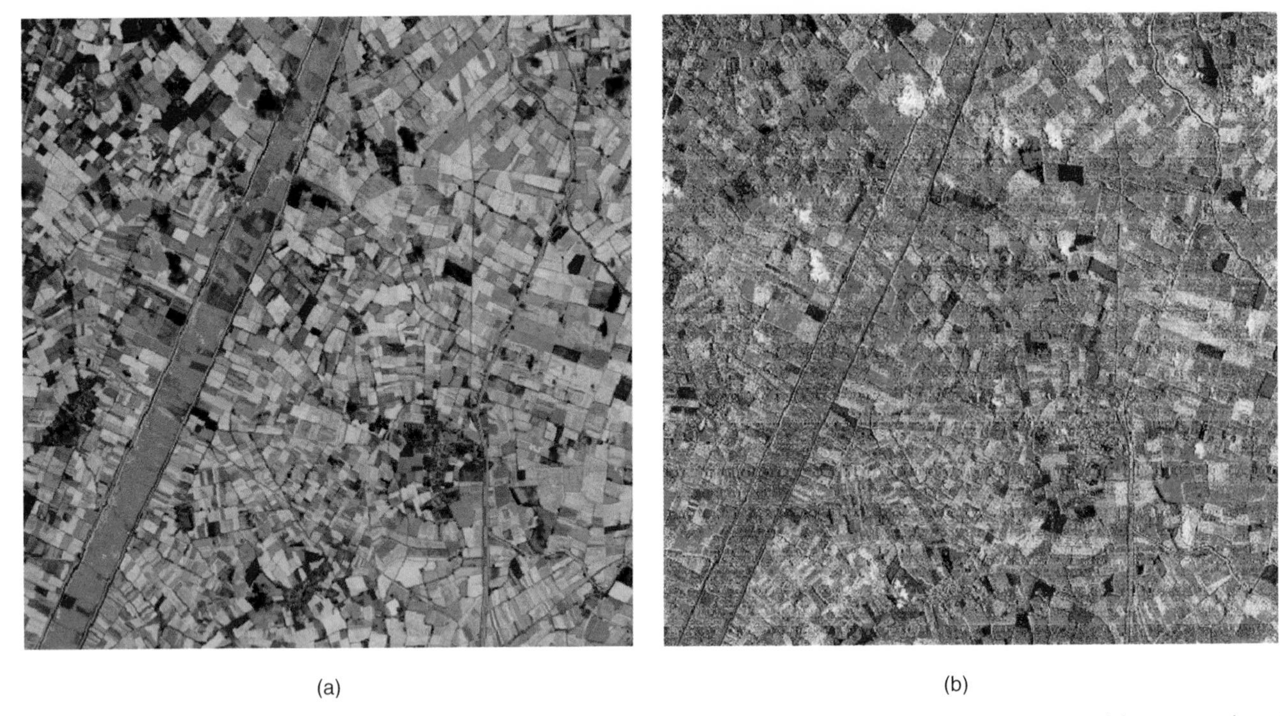

Figure 6.14 *Noise reduction transform for the Littleport image. (a) noise-reduced components 1–3 in RGB and (b) noise-reduced components 4–6 in RGB. Compare with Figures 6.12 and 6.13.*

space, and the result transformed back to the original RGB space. The result is generally an improvement in the range of intensities and saturations for each colour, with the hue remaining unaltered (Section 6.5 provides a more extended discussion of these properties of the colour image). Poor results can be expected when the RGB images do not have approximately Gaussian histograms, or where the image covers large, homogeneous areas.

The decorrelation stretch, like PCA, can be based on the covariance matrix **S** or the correlation matrix **R** (Section 6.4.1). Use of **R** implies that all three bands are given equal weight. If the stretch is based on **S**, each band is weighted according to its variance. The following description uses the notation **R** to indicate *either* the correlation *or* the covariance matrix. Essentially, the principal component images **y** are calculated from the eigenvector matrix **E** and the raw images **X** using the relationship $\mathbf{T} = \mathbf{XP}$, and the inverse transform is $\mathbf{X} = \mathbf{TP}^{-1}$ (Section 6.4.1; Figure 6.11b; Alley, 1995; Campbell, 1996; Jackson, 1991; Rothery and Hunt, 1990). Since **E** is an orthogonal matrix, it follows that $\mathbf{E}' = \mathbf{E}^{-1}$.

Alley (1995) uses a stretching parameter based upon the eigenvalues plus a shift to convert the raw decorrelation stretched values onto a 0–255 scale. Each decorrelation stretched band is given a mean of 127 and a standard deviation of 50. Clipping may be required if transformed values are negative or exceed 255 in value. This configuration does not necessarily produce optimum results. As in the case of standard PCA, and other image transforms which operate in the domain of real numbers, the scaling of the result back to the range 0–255 is logically simple but practically difficult. The method suggested above for the Tasselled Cap transform involves calculating the maximum and minimum of the raw transformed values, then scaling this range onto the 0–255 scale using offsets and stretching parameters. Where the stretch is being applied to an image displayed on screen this method is reasonably fast and usually gives a good result (see Figure 6.15). Further consideration of the procedure is provided by Guo and Moore (1996) and Campbell (1996). The latter author presents a detailed analysis of the decorrelation stretch process. Ferrari (1992), Rowan and Mars (2003) and White (1993) illustrate the use of the method in geological image interpretation. Krause *et al.* (2004) use PCA and decorrelation stretch in a geographical information system (GIS)-based study of coastal north Brazil. The widespread acceptance of the decorrelation stretch procedure is shown by its inclusion as a standard product for data collected by the ASTER sensor carried by the Terra satellite. Figure 6.15c shows a decorrelation-stretched TM image of part of the coastline of eastern England. Detail over the water areas is increased in comparison to the false-colour composite and the HSI transform (see next section), but the HSI stretch gives more detail over land.

Figure 6.15 (a) Landsat TM bands 4, 3, 2 false colour composite of the coastline of The Wash, eastern England, after a 5–95% linear contrast stretch. (b) The Wash image after a HSI transform. The saturation and intensity are stretched linearly and the hue is left unchanged. (c) The Wash image after a decorrelation stretch based on the covariance matrix. Landsat data courtesy NASA/USGS.

6.5 Hue-Saturation-Intensity (HSI) Transform

Details of alternative methods of representing colours are discussed in Section 5.2 where two models are described. The first is based on the RGB colour cube. The different hues generated by mixing RGB light are characterized by coordinates on the RGB axes of the colour cube (Figure 5.3). The second representation uses the HSI hexcone model (Figure 5.4) in which hue, the dominant wavelength of the perceived colour, is represented by angular position around the top of a hexcone, saturation or purity is given by distance from the central, vertical axis of the hexcone and intensity or value is represented by distance above the apex of the hexcone. Hue is what we perceive as colour (such as mauve or purple). Saturation is the degree of purity of the colour, and may be considered to be the amount of white mixed in with the colour. As the amount of white light increases so the colour becomes more pastel-like. Intensity is the brightness or dullness of the colour. It is sometimes useful to convert from RGB colour cube coordinates to HSI hexcone coordinates, and vice versa. The RGB coordinates will be considered to run from 0 to 1 (rather than 0–255) on each axis, while the coordinates for the hexcone model will consist of (i) hue expressed as an angle between 0 and 360° and (ii) saturation and intensity on a 0–1 scale. Note that the acronym IHS (intensity–hue–saturation) is sometimes used in place of HSI.

The application of the transform for colour enhancement is straightforward. The three-band image to be processed is converted to HSI representation, and a linear contrast stretch is applied to the saturation and/or the intensity components. The HSI data are then converted back to RGB representation for display purposes. Figure 6.15b shows an image of part of The Wash coastline of eastern England after a HSI transformation. The red colour of the growing crops in the fields is enhanced relative to the linear contrast stretched version (Figure 6.15a) but some of the detail of the sediment in the water is lost.

The HSI transformation has been found to be particularly useful in geological applications, for example Jutz and Chorowicz (1993) and Nalbant and Alptekin (1995). Further details of the HSI transform are given in Blom and Daily (1982), Foley *et al.* (1997), , Hearn and Baker (1997), Pohl and van Genderen (1998) and Mulder (1980). Terhalle and Bodechtel (1986) illustrate the use of the transform in the mapping of arid geomorphic features, while Gillespie *et al.* (1986) discuss the role of the HSI transform in the enhancement of highly correlated images. Massonet (1993) gives details of an interesting use of the HSI transform in which the amplitude, coherence and phase components of an interferometric image (Chapter 9) are allocated to HSI, respectively, and the inverse HSI transform applied to generate a false colour image that highlights details of coherent and incoherent patterns. Schetselaar (1998) discusses alternative representations of the HSI transform, and Andreadis *et al.* (1995) give an in-depth study of the transform. Phillip and Rath (2002) consider different colour spaces, including HSI, and Pitas (1993) lists *C* routines for the RGB to HSI colour transform. Yet another useful source is Plataniotis and Venetsanopoulos (2000) who also give formulae for the transformation from RGB colour space to HSI colour space and back. Not all the formulae are identical, which is a potential source of confusion.

The HSI transform can be used to combine the spectral detail of an RGB colour composite image with the spatial detail of a geometrically registered panchromatic image in a process called image fusion. This topic is considered in Section 6.9.

6.6 The Discrete Fourier Transform

6.6.1 Introduction

The coefficients of the Tasselled Cap functions, and the eigenvectors associated with the principal components, define coordinate axes in the multidimensional data space containing the multispectral image data. These data are re-expressed in terms of a new set of coordinate axes and the resulting images have certain properties, which may be more suited to particular applications. The Fourier transform operates on a single-band (greyscale) image, not on a multispectral data set. Its purpose is to break down the spatial variation in grey levels into its spatial scale components, which are defined to be sinusoidal waves with varying amplitudes, frequencies and directions. The coordinates k_1 and k_2 of the two-dimensional space defined by the axes U, V in which these scale components are represented are given in terms of frequency (cycles per basic interval). This representation is called the frequency domain (Figure 6.16b) whereas the normal row/column coordinate system in which images are normally expressed is termed the spatial domain (Figure 6.16a). The Fourier transform is used to convert a single-band image from its spatial domain representation to the equivalent frequency domain representation, and vice versa.

The idea underlying the Fourier transform is that the greyscale values forming a single-band image can be viewed as a three-dimensional intensity surface, with the rows and columns defining two axes (x and y in Figure 6.16a) and the grey level intensity value at each pixel giving the third (z) dimension. A series of waveforms of increasing frequency and with different orientations is fitted to this intensity surface and the information associated with each such waveform is calculated. The Fourier transform therefore provides details of (i) the frequency of each of the scale components (waveforms) fitted to the image and (ii) the proportion of information associated with each frequency component. Frequency is defined in terms of cycles per basic interval where the basic interval in the across-row direction is given by the number of pixels on each scan line, while the basic interval in the down-column direction is the number of scan lines. Frequency could be expressed in terms of metres by dividing the magnitude of the basic interval (in metres) by cycles per basic interval. Thus, if the basic interval is 512 pixels each 20 m wide then the wavelength of the fifth harmonic component is $(512 \times 20)/5$ or 2048 m. The first scale component, conventionally labelled zero, is simply the mean grey level value of the pixels making up the image. The remaining scale components have increasing frequencies (decreasing wavelengths) starting with 1 cycle per basic interval, then $2, 3, \ldots, n/2$ cycles per basic interval where n is the number of pixels or scan lines in the basic interval.

This idea can be more easily comprehended by means of an example using a synthetic one-dimensional data series. This series consists of the sum of four sine waves

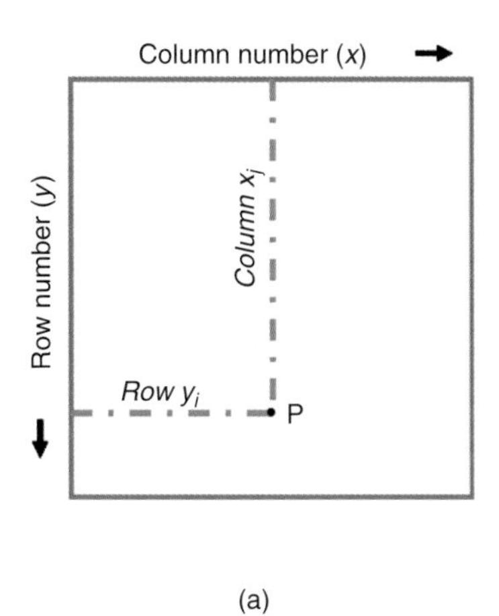

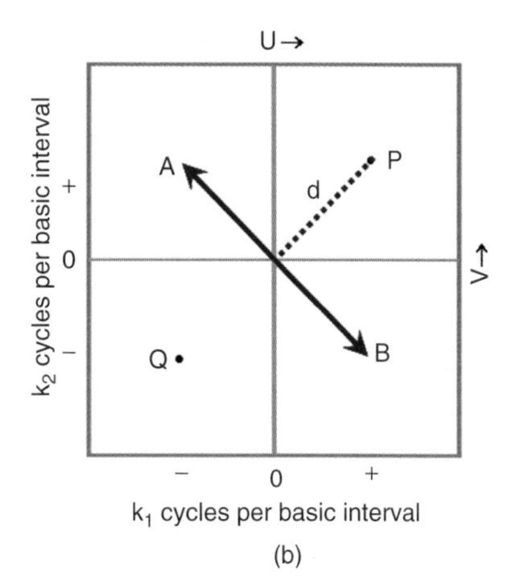

Figure 6.16 *(a) Spatial domain representation of a digital image. Pixel P has coordinates (i, j) with respect to the image row and column coordinate axes, or ($i\Delta x$, $i\Delta y$) metres, where Δx and Δy are the dimensions of the image pixels in the row and column direction, respectively. (b) Frequency-domain representation, showing the amplitude spectrum of the image. The value at P is the amplitude of a sinusoidal wave with frequency $U = k_1$ and $V = k_2$ cycles per basic interval in the u and v directions, respectively. The wavelength of this sinusoidal wave is proportional to the distance d. The orientation of the waveform is along direction AB. Point Q (-U, -V) is the mirror image of point P.*

that differ in frequency and amplitude. Sine wave 1 (Figure 6.17a) has an amplitude of 1.0 (see annotation on the y-axis) and completes one cycle over the basic interval of 0–6.28 radians (360°), so its frequency is 1 Hz. The second, third and fourth sine waves (Figure 6.17b, c and d) have amplitudes of 2.0, 3.0 and 3.0 (note the y-axis annotation) respectively, and frequencies of 2, 3 and 32 Hz. These sine waves were calculated at 256 points along the x-axis, and Figure 6.17e shows the result of adding up the values of sine waves 1–4 at each of these 256 points. It is difficult to discern the components of this composite sine wave by visual analysis. However, the discrete forward Fourier transform (Figure 6.17f) clearly shows the frequencies (horizontal axis) and amplitudes (vertical axis) of the four component sine waves. Note that Figure 6.17f shows only the first 40 harmonics out of a possible $n/2 = 256/2 = 128$ harmonics. These frequency/amplitude diagrams show the *amplitude spectrum* (sometimes called the power spectrum in the engineering literature). Other small frequency components are visible in the amplitude spectrum; these result from the fact that the series is a discrete one, measured at 256 points, which gives only 8 points to define each of the sinusoids in Figure 6.17d. In the following paragraphs we will use the term *scale component* to indicate a significant harmonic (i.e. one with a relatively large amplitude). The sum of the amplitudes is equal to the variance of the data.

6.6.2 Two-Dimensional Fourier Transform

If this simple example were extended to a function defined over a two-dimensional grid then the differences would be that (i) the scale components would be two-dimensional waveforms and (ii) each scale component would be characterized by orientation as well as by amplitude. The squared amplitudes of the waves are plotted against frequency in the horizontal and vertical directions to give a two-dimensional amplitude spectrum, which is interpreted much in the same way as the one-dimensional amplitude spectrum shown in Figure 6.17f, the major differences being:

1. The frequency associated with the point $[k_1, k_2]$ in the two-dimensional amplitude spectrum is given by:

$$k_{12} = \sqrt{k_1^2 + k_2^2}$$

   where the basic intervals given by each axis of the spatial domain image are equal, or by:

$$k_{12} = \sqrt{k_1/n_1\Delta t_1 + k_2/n_2\Delta t_2}$$

   where the basic intervals in the two spatial dimensions of the image are unequal. In the latter case, $n_1\Delta t_1$ and $n_2\Delta t_2$ are the lengths of the two axes, n_1 and n_2 are the number of sampling points along each axis, and Δt_1 and Δt_2 are the sampling intervals. This implies that frequency is proportional to distance from the centre of the amplitude spectrum which is located at the point [0, 0] in the centre of the amplitude spectrum diagram (Rayner, 1971).
2. The angle of travel of the waveform whose amplitude is located at point (k_1, k_2) in the amplitude spectrum is perpendicular to the line joining the point (k_1, k_2) to the centre (DC) point of the spectrum (0, 0). This point is illustrated in Figure 6.18b, which shows the two-dimensional amplitude

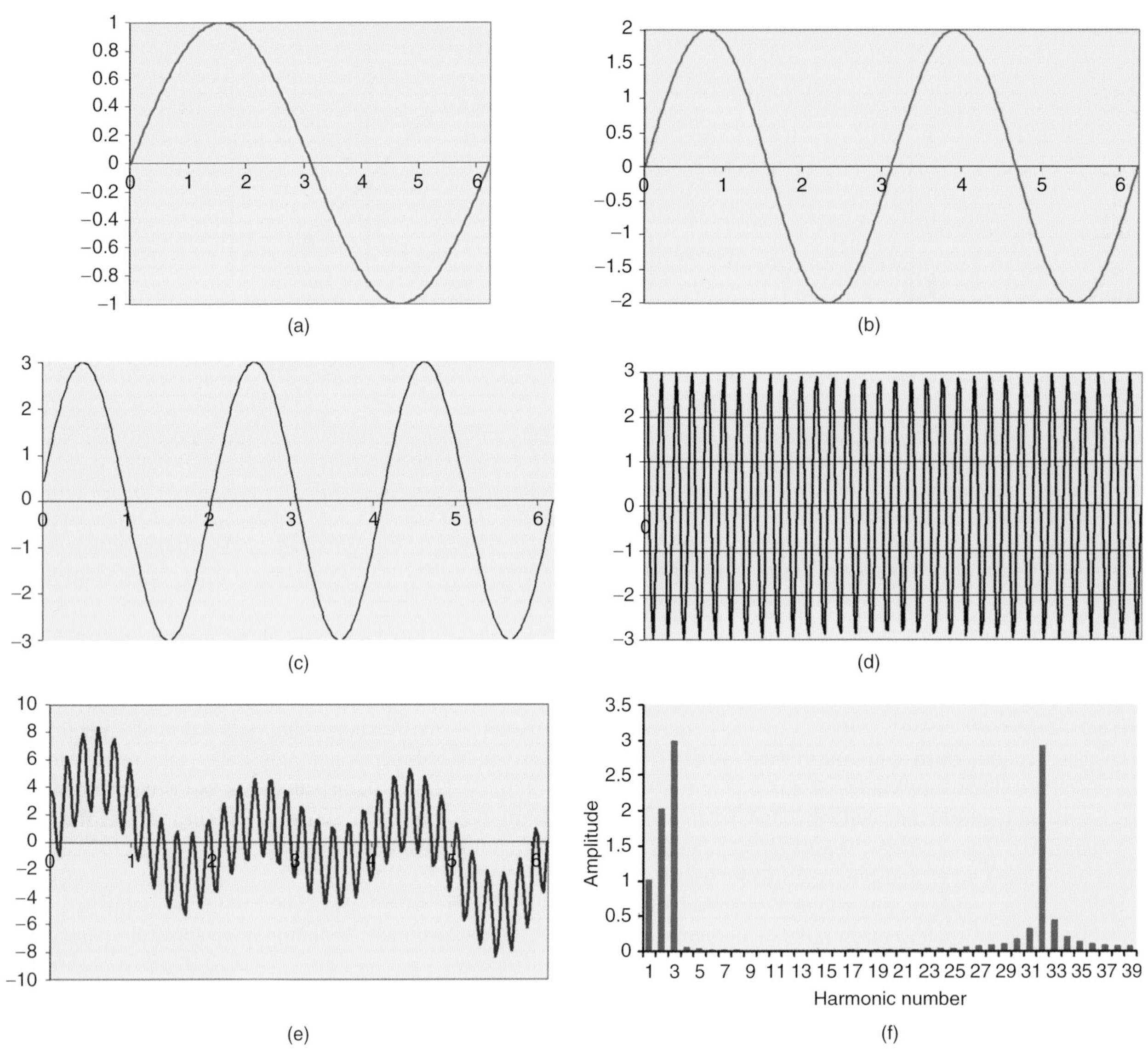

Figure 6.17 (a) Sinusoidal curve with an amplitude (variance) of 1.0 (y-axis) performing one cycle in the basic interval (i.e. -2π) sampled at 256 points (x-axis). (b) As (a) but with two cycles per basic interval and an amplitude of 2.0. (c) As A with three cycles per basic interval and an amplitude of 3.0. (d) As (a) but with 32 cycles per basic interval and an amplitude of 3.0. (e) The sum of the sinusoids (a–d) inclusive. (f) The one-dimensional Fourier transform of (e) showing harmonic number 1 having an amplitude of 1.0, harmonic 2 with an amplitude of 2, harmonic 3 with an amplitude of 3 and harmonic 32 also with an amplitude of 3. There is evidence of leakage around harmonic 32, as adjacent harmonics should be nonzero. This leakage is due to the relatively poor definition of the sinusoid with 32 cycles per basic interval. At a sampling rate of 256 there are only eight points describing each cycle. Only the first 40 harmonics are shown.

spectrum of an 512 × 512 pixel image made up of a set of horizontal lines spaced 16 rows apart. These lines are represented digitally by rows of 1s against a background of 0s (Figure 6.18a). The amplitude spectrum shows a set of symmetric points running horizontally through the origin (centre point or DC). The points are so close that they give the appearance of a line. They represent the amplitudes of the waveforms reconstructed from the parallel, horizontal lines which could be considered to lie on the crests of a series of sinusoidal waveforms progressing down the image from top to bottom. Since the direction of travel could be topbottom or bottomtop the amplitude spectrum is symmetric and so the two points closest to the DC represent the amplitudes of the set of waves whose wavelength is equal to the spacing between the horizontal lines. The two points further out from the DC represent a

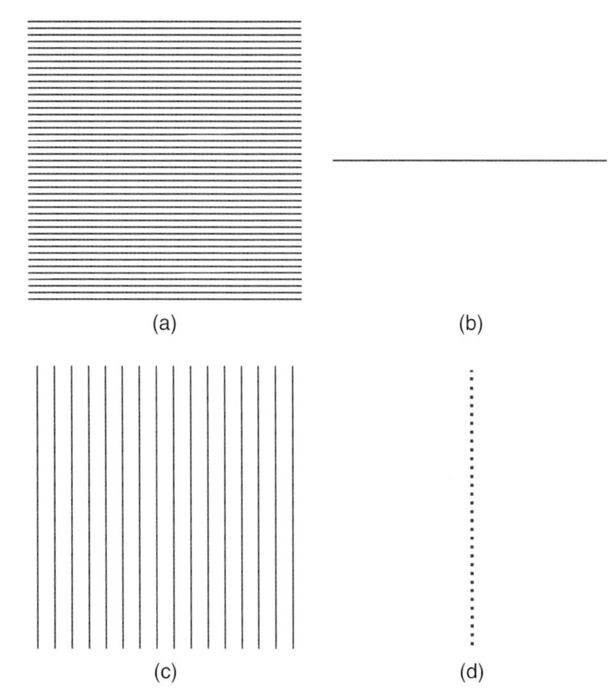

Figure 6.18 *The left-hand images (a and c) show the spatial domain representation of two patterns (horizontal lines spaced 16 rows apart and vertical lines spaced 32 columns apart). The right-hand images show the frequency domain representation of the two patterns. The frequency domain representation used here is the logarithm of the two-dimensional amplitude spectrum.*

spurious waveform, which has a wavelength equal to double the distance between the horizontal lines. Such spurious waveforms represent a phenomenon termed *aliasing* (Rosenfeld and Kak, 1982; Gonzales and Woods, 2007; Figure 6.19) which can be defined as the case in which several different frequency curves fit all or some of the available data points. It is a result of sampling a continuous signal.

Figure 6.20 may help to illustrate some of these ideas. Figure 6.20a shows a two-dimensional sine wave orientated in a direction parallel to the x-axis of the spatial domain coordinate system. Figure 6.20b shows another sine wave, this time oriented parallel to the y-axis, and Figure 6.20c shows the result of adding these two sine waves together to generate an undulating surface.

The two-dimensional amplitude spectrum of a composite sine wave is shown in Figure 6.21b. Figure 6.21a shows the pattern of high and low values (peaks and troughs – the same pattern as in Figure 6.20c but in the form of an image. In an ideal world, with an infinite sample (rather than the 512×512 grid used here) we might expect that the frequency-domain representation would show four high values located above and below and to the right and left of the DC at a distance from the DC that is proportional to the frequency of the sine waves (i.e. the number of times that the complete sine wave is repeated in the x direction (or the y direction, since the two axes are equal). The caption to Figure 6.21 explains why this is not the case, for the amplitude in the diagonal directions of Figure 6.21a is greater then either the vertical or horizontal amplitude.

Calculation of the amplitude spectrum of a two-dimensional digital image involves techniques and concepts that are too advanced for this book. A simplified account will be given here. Fourier analysis is so called because it is based on the work of Jean Baptiste Joseph, Baron de Fourier, who was Governor of Lower Egypt and later Prefect of the Departement of Grenoble during the Napoleonic era. In his book *Theorie Analytique de la Chaleur*, published in 1822, he set out the principles of the Fourier series which has since found wide application in a range of subjects other than Fourier's own, the analysis of heat flow. The principle of the Fourier series is that a single-valued curve (i.e. one which has only a single y value for each separate x value) can be represented by a series of sinusoidal components of increasing frequency. The form of these sinusoidal components is given by

$$f(t) = a_0 + \sum_n a_n \cos n\omega t + \sum_n b_n \sin n\omega t$$

in which $f(t)$ is the value of the function being approximated at point t, ω is equal to $2\pi/T$ and T is the length of the series. The term a_0 represents the mean level of the functions and the summation terms represent the contributions of a set of sine and cosine waves of increasing frequency. The fundamental waveform is that with a period equal to T seconds, or a frequency of ω Hz. The second harmonic has a frequency of 2ω, while 3ω is the frequency of the third harmonic, and so on.

The a_i and b_i terms are the cosine and sine coefficients of the Fourier series. It can be seen in the formula above that the coefficients a_i are the multipliers of the cosine terms and the b_i are the multipliers of the sine terms. Sometimes the sine and cosine terms are jointly expressed as a single complex number, that is, a number which has 'real' and 'imaginary' parts, with the form $(a_i + jb_i)$. The 'a' part is the real or sine component, and b is the imaginary or cosine component. The term j is equal to $\sqrt{-1}$. Hence, the a_i and b_i are often called the real and imaginary coefficients. We will not use complex number notation here. The coefficients a and b can be calculated by a least-squares procedure. Due to its considerable computing time requirements this method is no longer used in practice. In its place, an algorithm called the fast Fourier transform (FFT) is used (Bergland, 1969; Gonzales and Woods, 2007; Lynn, 1982; Pavlidis, 1982; Pitas, 1993; Ramirez, 1985). The advantage of the FFT over the older method can be summarized by the fact

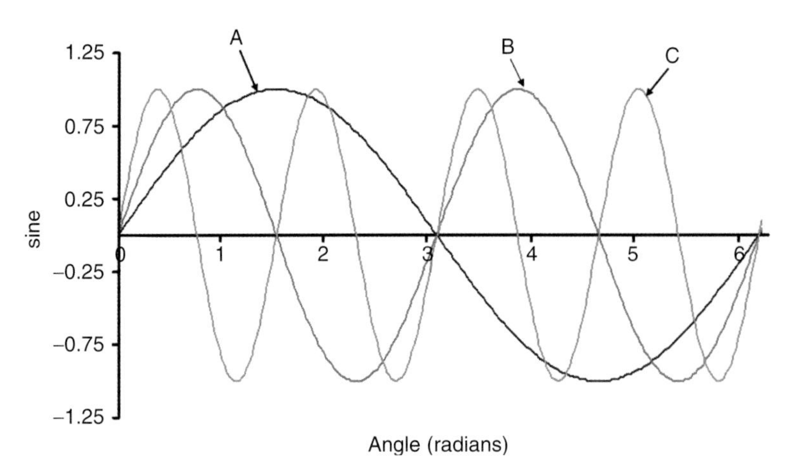

Figure 6.19 Illustrating the concept of aliasing. If we have only three values of x, at 0.0000, 3.1415 and 6.2830 (0, 2π and π radians) then any of a number of sine waves will pass through these points. Here, three are shown. Curve A (blue) completes one cycle over the range 0–6.2830 radians, curve B (magenta) completes two cycles, and curve C (orange) completes four cycles. On the basis of the data alone, it is not possible to say whether all of these curves exist in the data or whether some are artefacts generated by aliasing (i.e. false identity).

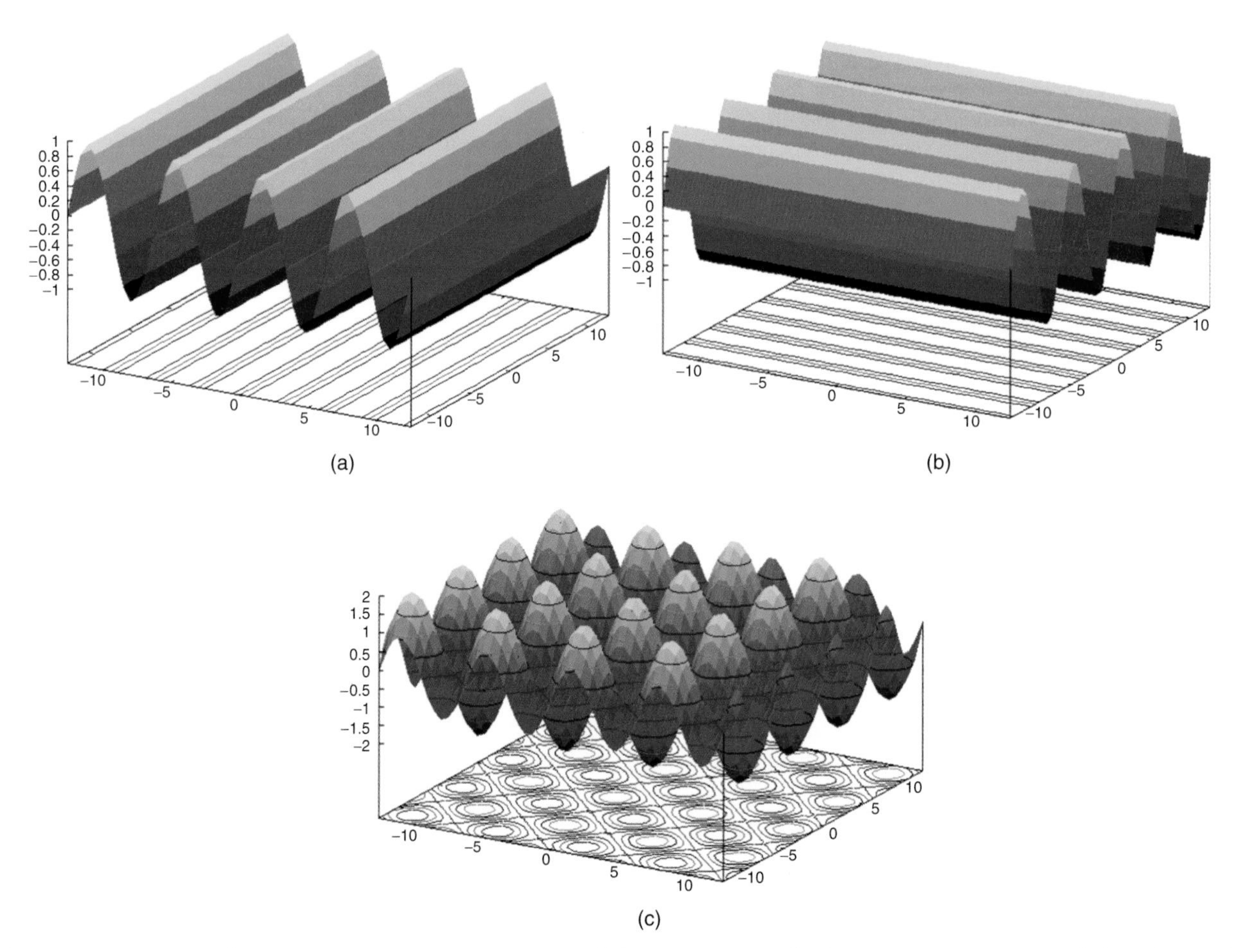

Figure 6.20 (a) Two-dimensional sine wave running parallel to the x-axis. (b) Sine wave running parallel to the y-axis. (c) Composite formed by summing the values of the sine waves shown in (a) and (b).

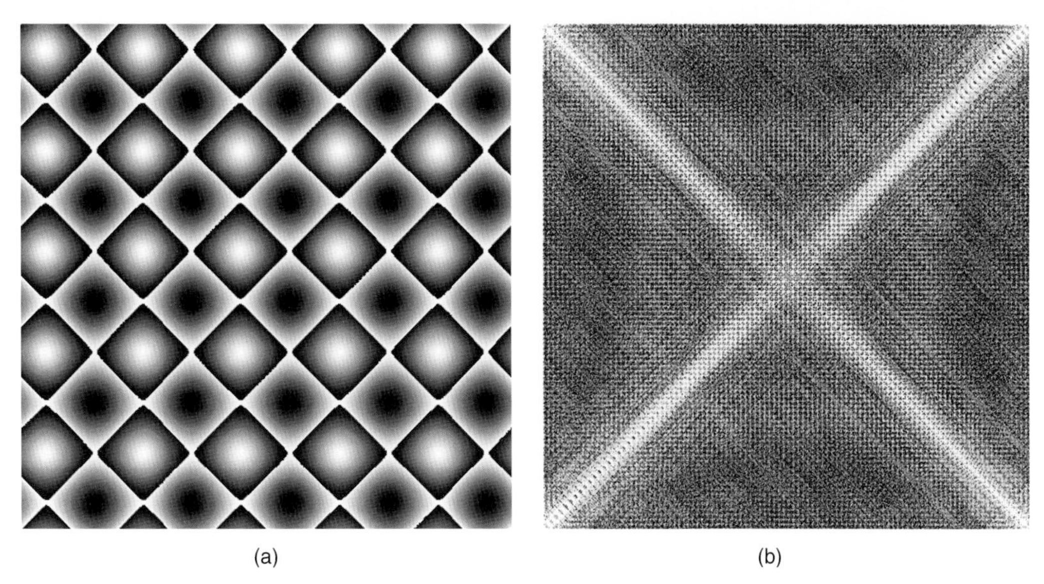

Figure 6.21 (a) Image of composite sine waves shown in Figure 6.20c. (b) The corresponding amplitude spectrum. This example is curious in that one might think that the dominant orientations were vertical and horizontal, but not diagonal. Closer inspection of Figure 6.21a shows that the horizontal/vertical pattern is either dark-intermediate-dark or light-intermediate-light, whereas the diagonal pattern is dark-light-dark or light-dark-light. In other words, the amplitude or range of the diagonal pattern is greater than that of the horizontal/vertical pattern, and this is reflected in the frequency domain representation in (b). If you go back to Figure 6.20c you will appreciate that the diagonal pattern has a greater amplitude than the horizontal or vertical pattern, even though the image is constructed by the addition of waveforms running parallel to the x- and y-axes respectively.

that the number of operations required to evaluate the coefficients of the Fourier series using the older method is proportional to N^2 where N is the number of sample points (length of the series) whereas the number of operations involved in the FFT is proportional to $N \log_2 N$. The difference is brought out by a comparison of columns (ii) and (iii) of Table 6.6. However, in its normal implementation the FFT requires than N (the length of the series) should be a power of 2. Singleton (1979a) gives an algorithm that evaluates the FFT for a series of length N that need not be a power of 2 while Bergland and Dolan (1979) provide a Fortran program listing of an algorithm to compute the FFT in a very efficient manner.

Once the coefficients a_i and b_i are known the amplitude of the ith harmonic at any point (u, v) is computed from

$$A_i = \sqrt{a_i^2 + b_i^2}$$

which is numerically equivalent to the complex number abs($a + bj$), and the phase angle (the displacement of the first crest of the sinusoid from the origin, measured in radians or degrees) is defined as

$$\theta = \tan^{-1}(b_i/a_i)$$

Generally, only the amplitude information is used.

Table 6.6 Number of operations required to compute the Fourier transform coefficients a and b for a series of length N (column (i)) using least-squares methods (column (ii)) and the Fast Fourier Transform (FFT) (column (iii)). The ratio of column (ii) to column (iii) shows the magnitude of the improvement shown by the FFT. If each operation took 0.01 second then, for the series of length N = 8096, then the least-squares method would take 7 days, 18 h and 26 min. The FFT would accomplish the same result in 17 min 45 s.

(i) N	(ii) N^2	(iii) $N \log_2 N$	(iv) (ii)/(iii)
2	4	2	2.00
4	16	8	2.00
16	256	64	4.00
64	4 096	384	10.67
128	16 384	896	18.29
512	262 144	4 608	56.89
8 096	67 108 864	106 496	630.15

The procedure to calculate the forward Fourier transform (it is, in fact, the forward DFT) for a two-dimensional series, such as a greyscale image, involves the following steps:

1. Compute the Fourier coefficients for each row of the image, storing the coefficients a_i and b_i in separate

two-dimensional arrays. The coefficients form two-dimensional arrays of real numbers, or they can be considered to form a single two-dimensional complex array. The former representation is used here.

2. Compute the Fourier transform of the columns of the two matrices composed, respectively, of the a_i and b_i coefficients to give the Fourier coefficients of the two-dimensional image. There are two sets of coefficients, corresponding to the a_i and b_i terms.

Step 2 requires that the two coefficient matrices are transposed; this is a very time-consuming operation if the image is large. Singleton (1979b) describes an efficient algorithm for the application of the FFT to two-dimensional arrays. Example 6.1 illustrates the procedure for calculating the Fourier transform of an image.

6.6.3 Applications of the Fourier Transform

As noted already, the main use of the Fourier transform in remote sensing is in frequency-domain filtering (Section 7.5). For example, Lei *et al.* (1996) use the Fourier transform to identify and characterize noise in MOMS-02 panchromatic images in order to design filters to remove the noise. Hird and McDermid (2009) compare methods of noise reduction, including Fourier transforms, in one-dimensional time series of NDVI data. Other applications include the characterization of particular terrain types by their Fourier transforms (Leachtenauer, 1977), and the use of measures of heterogeneity of the grey levels over small neighbourhoods based on the characteristics of the amplitude spectra of these neighbourhoods. If an image is subdivided into 32 × 32 pixel subimages and if each subimage is subjected to a Fourier transform then the sum of the amplitudes in the area of the amplitude spectrum closest to the origin gives the low-frequency or smoothly varying component while the sum of the amplitudes in the area of the spectrum furthest away from the origin gives the high-frequency, rapidly changing component. These characteristics of the amplitude spectrum have also been used as measures of image texture, which is considered further in Section 8.7.1. Fourier-based methods have also been used to characterise topographic surfaces (Brown and Scholz, 1985), in the calculation of image-to-image correlations during the image registration process (Section 4.3.4), in analysis of the performance of resampling techniques (Shlien, 1979; Section 4.3.3) and in the derivation of pseudocolour images from single-band (mono) images (Section 5.4). De Souza Filho *et al.* (1996) use Fourier-based methods to remove defects from optical imagery acquired by the Japanese JERS-1 satellite. Temporal sequences of vegetation indices (Section 6.2.4) are analysed using one-dimensional Fourier transforms by Olsson and Ekhlund (1994), and a similar approach is used by Menenti *et al.* (1993), Assali and Menenti (2000), Canisius, Turral and Molden (2007) and Roerink *et al.* (2003).

In the derivation of a pseudocolour image the low frequency component (Figure 7.13) is extracted from the amplitude spectrum and an inverse Fourier transform applied to give an image that is directed to the red monitor input. The intermediate or midrange frequencies are dealt with similarly and directed to the green input. The blue input is derived from the high-frequency components, giving a colour rendition of a black-and-white image in which the three primary colours (RGB) represent the low, intermediate and high frequency or scale components of the original monochrome image. Blom and Daily (1982) and Daily (1983) describe and illustrate the geological applications of a variant of this method using synthetic-aperture radar images. Their method is based on the observation that grey level (tonal) variations in synthetic aperture radar images can be attributed to two distinct physical mechanisms. Large-scale features (with low spatial frequency) are produced by variations in surface backscatter resulting from changes in ground surface cover type. High spatial frequencies correlate with local slope variations, which occur on a much more rapid spatial scale of variability. The amplitude spectrum of the SAR image is split into these two components (high and low frequencies) using frequency-domain filtering methods that are described in Section 7.5. The result is two filtered amplitude spectra, each of which is subjected to an inverse Fourier transform to convert from the frequency back to the spatial domain. The low-pass filtered image is treated as the hue component in HSI colour space (Section 6.5) and the high-pass filtered image is treated as the intensity component. Saturation is set to a value that is constant over the image; this value is chosen interactively until a pleasing result is obtained. The authors suggest that the pseudocolour image produced by this operation is easier to interpret than the original greyscale SAR image. Further details of image merging are given in Section 6.9.

6.7 The Discrete Wavelet Transform

6.7.1 Introduction

The idea of representing the information content of an image in the spatial domain and the frequency domain is introduced in Section 6.6. These two representations give different 'views' of the information contained in the image. The DFT is presented in that same section as a technique for achieving the transformation of a grey scale (single-band) image from the spatial domain of (row, column) coordinates to the frequency domain of (vertical frequency, horizontal frequency) coordinates. While the frequency domain representation contains information about the presence of different waveforms making up the grey scale image, it does not tell us *where* in the image a specific waveform with a particular frequency occurs; its is assumed that the 'frequency mix' is the same in all parts of the image. Another disadvantage of the frequency domain representation is the need to assume statistical stationarity, which requires that the mean and variance of the pixel values are constant over all regions of the image. In addition, the DFT assumes that the image repeats itself in all directions to infinity.

The DFT has a number of useful applications in the analysis of remotely-sensed images. These are elaborated in Section 6.6 and 7.5. In this section, the DWT is introduced. It augments, rather than replaces, the DFT because it represents a compromise between the spatial and frequency domain representations. It is impossible to measure exactly both the frequencies present in a grey scale image *and* the spatial location of those frequencies (this is an extension of Heisenberg's Uncertainty Principle). It is, however, possible to transform an image into a representation that combines frequency bands (ranges of frequencies) and specific spatial areas. For example, the Windowed Fourier Transform generates a separate amplitude spectrum for each of a series of sub-regions of the image and thus provides some idea of the way in which the frequency content of the data series changes with time or space. The Windowed Fourier Transform does not have the flexibility of the DWT, which is generally preferred by statisticians.

An outline of the DWT is provided in Section 6.7.1. Details of the use of the DWT in removing noise from the reflectance spectra of pixels (i.e. one-dimensional signals) and images (two-dimensional signals) are given in Section 9.3.2. The derivation of the two-dimensional DWT is described in Section 6.7.2. For a more advanced treatment, see Addison (2002), Mallat (1998) and Starck, Murtagh and Bijaou (1998). Strang (1994) provides a gentler introduction.

6.7.2 The One-Dimensional Discrete Wavelet Transform

There are a number of ways of presenting the concept of the wavelet transform. The approach adopted in this section is more intuitive than mathematical, and it uses the idea of cascading low-pass and high-pass filters (such filters are discussed in Section 7.1). Briefly stated, a low-pass filter removes or attenuates high frequencies or details, producing a blurred or generalized output. A high-pass filter removes the slowly changing background components of the input data, producing a result that contains the details without the background. Examples of low-pass and high-pass filters are shown in Figure 6.22. The wavelet transform can be considered as a sequence of (high-pass, low-pass) filter pairs, known as a filter bank, applied to a data series **x** that could, for example be a reflectance spectrum sampled or digitized over a given wavelength range. Note that the samples are assumed to be equally spaced along the x-axis, and the number of samples (n) is assumed to be a power of 2, that is $n = 2^j$, where j is a positive integer.

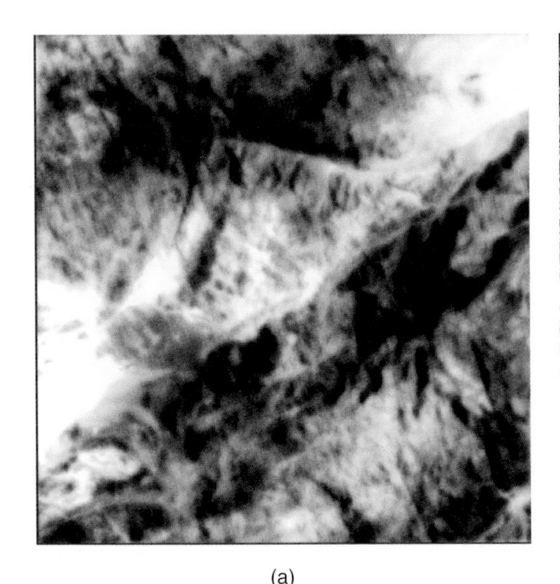

(a)

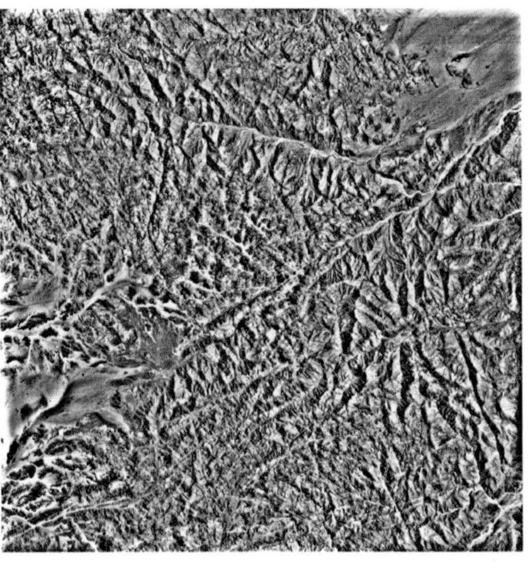

(b)

Figure 6.22 Landsat band 4 greyscale image of part of the Red Sea Hills after (a) a low-pass filter and (b) a high-pass filter have been applied. Filtering is discussed in Chapter 7. A low pass filter removes detail whereas a high-pass filter enhances detail. Landsat data courtesy NASA/USGS.

Example 6.1: The Fourier Transform

This example has the aim of demonstrating how the Fourier amplitude spectrum is computed, and how it is interpreted. The example is continued in Section 7.5, where frequency-domain filtering is discussed.

Example 6.1 Figure 1 shows a false colour image of an area of the Red Sea Hills in Sudan east of Port Sudan. The area experiences an arid climate, so surface materials are mainly bare rock and weathering products such as sand (bright area near the lower left corner). The region was tectonically active in the past and a number of fault lines are visible. The aim of this example is to show how to apply the Fourier transform to convert the representation of the image information from the spatial to the frequency domain.

Example 6.1 Figure 1. *Landsat-5 TM image of part of the Red Sea Hills, Sudan. This false colour image has band 7 in the red channel, band 5 in the green and band 3 in the blue channel. It has been enhanced by a linear 5–95% stretch.*

If you are using MIPS then the steps are as follows. Other systems such as ERDAS, ENVI and MATLAB will have similar commands. First, ensure that the file `sudanhi.inf` is available. Copy it from the CD to your hard drive if necessary, and follow the steps outlined below:

1. Use **`View Display Image`**, select **`sudanhi.inf`**, and display band 5 of this image set as a greyscale image. The dimensions of the image are 1024 × 1024 pixels.
2. Choose **`Filter Fourier Transform Forward Transform`**. The log of the amplitude spectrum of the image is displayed in a new window.

The log of the amplitude spectrum is shown above as Example 6.1 Figure 2. The origin of the frequency domain is the centre of the image, and the units of measurement are *cycles per basic interval* in the u and v (horizontal and vertical directions). The lengths of the u and v axes are both equal to 1024 pixels, which is the basic interval, so the brightness levels of the four nearest pixels to the origin, which is at the centre of the image, that is the pixels that are above, below, left and right of the central pixel (DC) represent the proportion of total image information at the lowest possible spatial frequency of one cycle per basic interval (equal to a wavelength 1024 × 30 m or 30.720 km) in the horizontal and vertical directions. The second closest pixels in the same directions show the information present at spatial frequencies of two cycles per basic interval or a spatial wavelength of 512 × 30 m or 15.360 km. Recall from Figure 6.17 that a complex one-dimensional curve can be reconstructed from its Fourier transform. The same applies to surfaces – the band 5 Sudan image can be reconstructed from the amplitude spectrum shown in Example 6.1 Figure 2.

Example 6.1 Figure 2. *Logarithm of the Fourier amplitude spectrum of the image shown in Example 6.1 Figure 1.*

The amplitude spectrum is interpreted in terms of spatial frequencies or wavelengths only in circumstances in which a particular frequency is to be identified. For example, the Landsat TM and ETM+ sensors gather image data in 16-line sections, and so there may be evidence of a peak in the amplitude spectrum at a spatial wavelength of $16 \times 30 = 480$ m, which corresponds to a frequency of $1024/16 = 64$ (1024 is the vertical or v axis length). Bright points positioned on the v axis at 64 pixels above and below the centre point of the amplitude spectrum would indicate the presence of a strong contribution to image variance at frequencies corresponding to a spacing of 16 lines on the image.

More generally, the shape of the amplitude spectrum can give some indication of the structures present in the image. Figure 6.18 illustrates the shape of the Fourier amplitude spectrum for two extreme cases – of vertical and horizontal lines. Hence, we would expect to see some evidence of directionality in the spectrum if the features in the image area had any preferred orientation. Another use of the amplitude spectrum is to quantify the texture of the image (Section 8.7.1). An image with a fine texture would have a greater proportion of high frequency information compared with an image with a coarse texture. Therefore, the ratio of high frequency components (those that are further from the origin that a given distance d_1, for example) to low frequency components (those closer to the origin than a specified distance d_2) can be used to measure the texture of an image or a sub-image. Usually, a small moving window of size 32 × 32 pixels is passed across the image, the ratio of high to low frequencies is computed, and this value is assigned to the centre pixel in the window.

The amplitude spectrum is also used as a basis for filtering the image (Chapter 7). Filtering involves the selective removal or enhancement of specific frequency bands in the amplitude spectrum, followed by an inverse Fourier transform to convert the filtered information back to the spatial domain. This procedure is illustrated in Chapter 7 using the Sudan image (Example 6.1 Figure 1) as an example.

The steps involved in the one-dimensional DWT are as follows:

1. Apply a low-pass filter to the full data series $\mathbf{x}$ of length n (where n is a power of 2). If n is not a power of 2, pad the series with zeros to extend its length to the next higher power of 2. Take every second element of $\mathbf{x}$ to produce a series $\mathbf{x}_1$ of length $n/2$.
2. Apply a high-pass filter to the full data series $\mathbf{x}$, padding with zeros as in step 1 if necessary. Take every second element of the result, to produce a set of $n/2$ high-pass filter outputs comprising the first level of the detail coefficients, $\mathbf{d}_1$, of length $n/2$.
3. Take the result from step 1, which is a data series $\mathbf{x}_1$ of length $n/2$ that has been smoothed once. Apply a low-pass filter. Take every second element of the result to produce a series $\mathbf{x}_2$ of length $n/4$.
4. Apply a high-pass filter to the output from step 1. Take every second element of the result, to give a vector $\mathbf{d}_2$ of length $n/4$. Vector $\mathbf{d}_2$ forms the second level of detail coefficients.
5. Repeat the low-pass/high-pass filter operation on the output of step 3. Continue filtering and decimating until the length of the resulting data series is one.

These operations are sometimes termed *subband coding* and the sequence of high pass and low pass filters is known as a *quadrature mirror* filter. The operations are shown schematically in Figure 6.23.

The wavelet or detail coefficients formed at steps 2, 4, and so on (i.e. vectors $\mathbf{d}_1, \mathbf{d}_2, \mathbf{d}_3, \ldots, \mathbf{d}_{n/2}$ using the notation introduced above) can be interpreted as follows. The vector $\mathbf{d}_1$ represents the detail (plus noise) in the original image. Vector $\mathbf{d}_2$ characterizes the detail (plus noise) in the once-smoothed image. Vectors $\mathbf{d}_3, \mathbf{d}_4, \ldots, \mathbf{d}_n$ contain the detail (plus noise) derived by filtering the twice-smoothed, thrice-smoothed and n-times smoothed image. There are $n/2$ elements in $\mathbf{d}_1$, $n/4$ elements in $\mathbf{d}_2$, $n/8$ elements in $\mathbf{d}_3$, and so on, as the series length is halved at each step. Thus, for an original series length n of 32, the number of detail coefficients is $16 + 8 + 4 + 2 + 2$. The last two coefficients are somewhat different from the others, but that need not concern us here. Each subset of the $n = 32$-point raw data sequence, with a length of 16, 8, 4 and 2, is derived from an increasingly smoothed series, the effect of down-sampling being to make the series sampled further and further apart. If the location of the samples is plotted, a time-scale diagram such as that shown in Figure 6.24 is produced. If the input data series is formed of the elements of a reflectance spectrum, then the time dimension is replaced by wavelength.

In a data series of length 512, the 256 coefficients at level 1 are derived from the original (raw) series. The 128 coefficients at level 2 come from the series after one smoothing, the 64 coefficients at level 3 are extracted from the twice-smoothed series, and so on. As progressive smoothing implies a continual reduction in high-frequency content, it follows that the representation of the data with the largest high frequency content is level 1, and that with the lowest high-frequency content is level 9. The higher frequencies are sampled at a higher rate than the low frequencies, which is logical as high frequencies vary more quickly in time and space than do low frequencies. An alternative way of representing

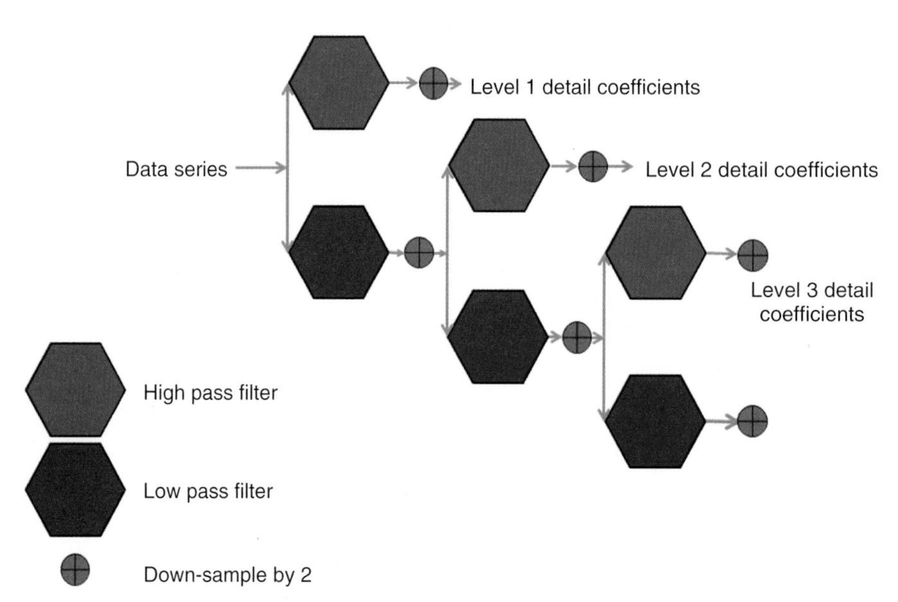

Figure 6.23 *Illustrating the calculation of wavelet coefficients for a one-dimensional series of measurements. Levels 1, 2 and 3 are shown but the procedure continues until down-sampling by two results in a series containing only a two elements.*

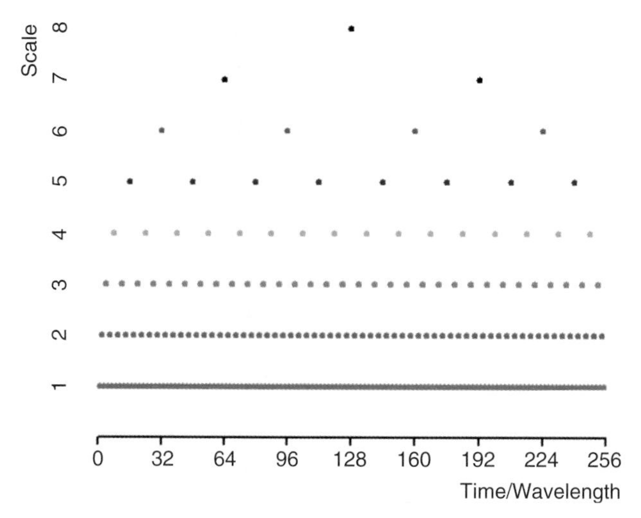

Figure 6.24 *Time-scale graph for a one-dimensional series of length 256 points. The raw data can be considered to be level 0. Dyadic sampling is applied so that at level 1 the number of detail coefficients is 128, reducing to 64 at level 2, and so on. See Figure 6.25 for an alternative representation of this process. Note that when remote sensing data are used then the time axis will normally be replaced by wavelength.*

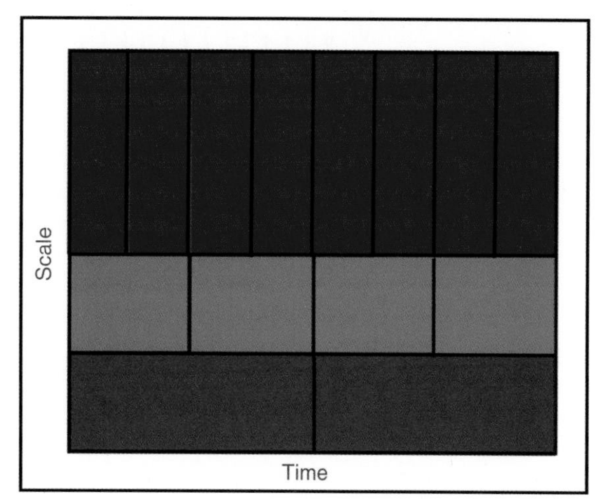

Figure 6.26 *Wavelet detail coefficients are arranged in the time-scale diagram to show their extent (in time, horizontally) and in scale (vertically). Each box corresponds to one coefficient, with the number of coefficients being reduced by a factor of 2 at each level. The highest-level coefficients have the greatest sampling frequency in time, but cover the greatest extent in scale.*

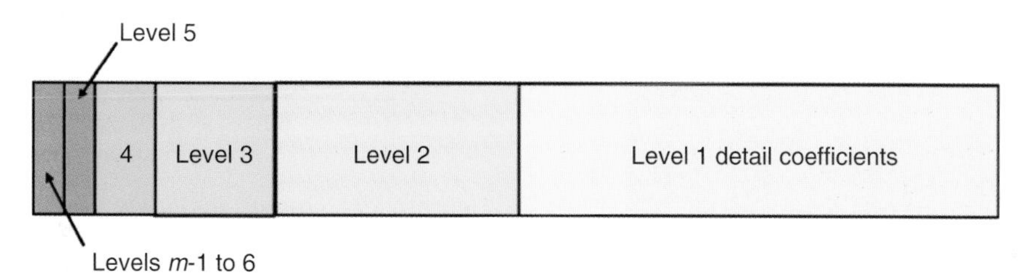

Figure 6.25 *Assume that the data series length is $n = 2^m$. There are m - 1 sets of detail coefficients. The n/2 highest-level (1) detail coefficients are shown on the right. The number of detail coefficients decreases by a factor of 2 at each level, so that the number of level 2 coefficients is n/4, the number of level 3 coefficients is (n/8), and so on.*

the wavelet detail coefficients is to place them in the order shown in Figure 6.25.

The sampling pattern in Figure 6.24 is termed dyadic (from the Greek word for 'two') because the sample size is reduced by a factor of 2 at each level. Some authors refer to this process as decimation.[1] The magnitudes of the dyadic samples can be represented in a time-scale diagram by placing the wavelet detail coefficients in 'layers' from top to bottom, as shown in Figure 6.26.

Probably the best way to discover what wavelets are, and how they work, is to look at some examples. We begin with a very simple sine wave, shown in Figure 6.27. This sine wave repeats 40 times over the basic interval of 512 samples so its frequency is 40 Hz. The coefficients resulting from a DWT applied to the data of Figure 6.27 are shown in Figure 6.28. The number of levels for $n = 512$ is nine. All levels of detail coefficients show a symmetry that mirrors the symmetry of the sine wave.

Figure 6.29 shows the DFT of the same sine wave data shown in Figure 6.27. Since there is only a single frequency (40 Hz) present in the data, the Fourier amplitude spectrum – not unexpectedly – shows a single spike centred at a frequency of 40 Hz. The narrow spread of values around 40 Hz in the Fourier amplitude spectrum is the result of the fact that a continuous function (a sine wave) has been sampled at a specific spacing.

The second example is an extension of the first. Figure 6.30 shows a composite sine wave, formed by adding together two sine waves with frequencies of 40 and 80 Hz, respectively. The DFT of the data shown in

[1]Decimation implies a reduction by a ratio of 1 : 10 (as in the ancient Roman military punishment, revived in 71 AD by the Roman general Crassus after his men had fled before Spartacus's army of slaves, and described by Plutarch in *The Life of Crassus*: '*... five hundred that were the beginners of the flight, he divided into fifty tens, and one of each was to die by lot, thus reviving the ancient Roman punishment of decimation ...*' – had he killed one in every two he would have had no army left).

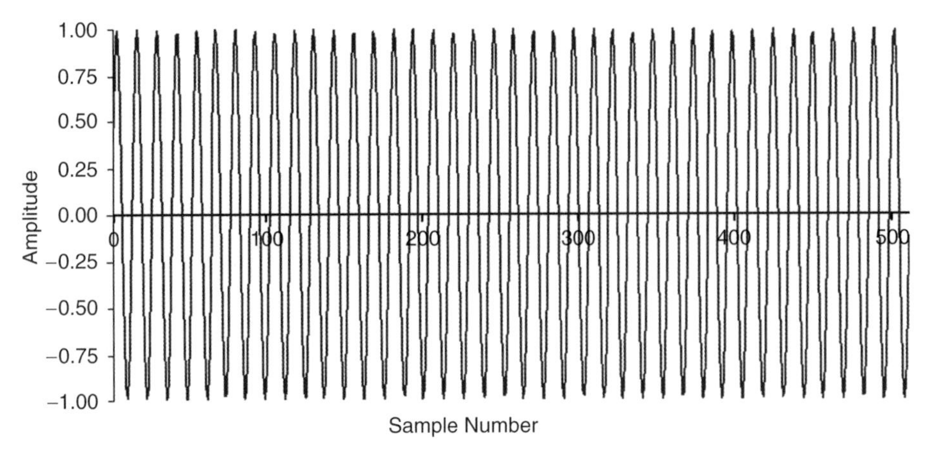

Figure 6.27 *Sine wave with frequency of 40 Hz calculated for 512 points. The amplitude of the sine wave is given by the vertical axis, and sample number is shown on the horizontal axis.*

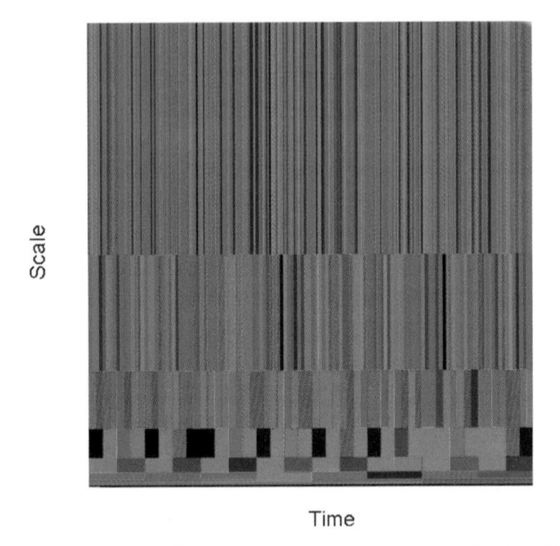

Figure 6.28 *Time (horizontal axis) – scale (vertical axis) display of wavelet detail coefficients for a sine wave of frequency 40 Hz, shown in Figure 6.27. The sine wave was sampled at 512 points, so there are nine levels of detail coefficients.*

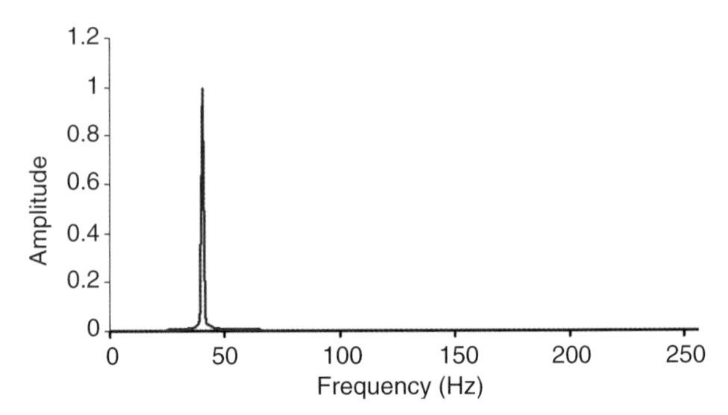

Figure 6.29 *Fourier amplitude spectrum of the sine wave shown in Figure 6.27. The presence of a periodic component with a frequency of 40 Hz is apparent. The horizontal axis is time, and the vertical axis measures variance.*

Figure 6.30 is given in Figure 6.31. Again, the Fourier representation is adequate, as it correctly shows two spikes centred at 40 and 80 Hz. The result of the DWT is not shown, because it also represents the data adequately. A more complicated signal is generated by the Chirp function, which is a sine wave with frequency increasing with time (Figure 6.32), unlike the sine wave data, which are uniform, symmetric and repeat to infinity. The Fourier amplitude spectrum simply indicates a number of frequency components in the approximate range 1–50 Hz. It does not pick out the change in frequency of the sine waves as time (represented by the x-axis) increases (Figure 6.33). However, the higher levels of detail coefficients generated by the DWT (Figure 6.34) show a pattern of increasing frequency from left to right, which gives a more realistic visual representation of the variation in the data set than does the DFT.

If we take the two sine waves shown in Figure 6.30 but, rather than adding them together at each of the 512 sample points, we sample the 40 Hz sine wave at points 1–256 and sample the 80 Hz sine wave at points 257–512, then the result is a separated pair of sine waves (Figure 6.35). Interestingly, the amplitude spectrum of the series shown in Figure 6.35 that is produced by the Fourier transform (see Figure 6.36) is identical to that shown in Figure 6.31, which was generated from the data shown in Figure 6.30. This result shows that the Fourier transform cannot define *where* in the data series a particular frequency occurs. It uses a universal basis function (i.e. based on the full data series) composed of sine and cosine waves repeating to infinity, whereas the DWT uses a localized basis function (as the filters deal with the data 2, 4, 8, 16, ..., points at a time). The DWT is thus said to have compact as opposed to universal support. The DWT result for the data of Figure 6.34 is shown in Figure 6.37, and it is apparent that the higher-level DWT coefficients show a clear distinction between the left and right halves of the data series.

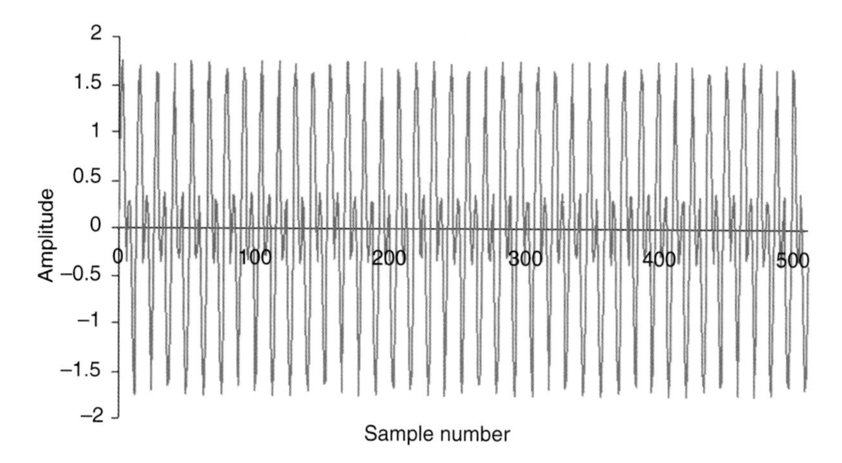

Figure 6.30 *Composite wave formed by adding two sine waves with frequencies of 40 and 80 Hz respectively.*

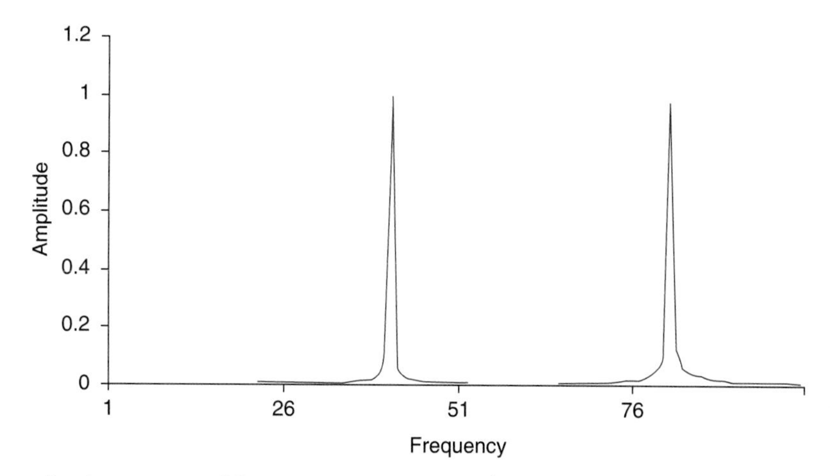

Figure 6.31 *Fourier amplitude spectrum of the composite sine wave shown in Figure 6.30. Both components (with frequencies of 40 and 80 Hz, respectively) are identified.*

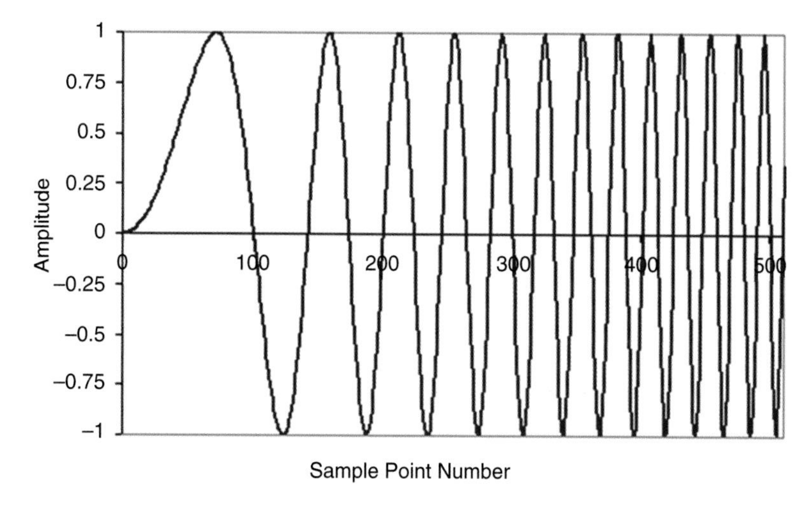

Figure 6.32 *The Chirp function.*

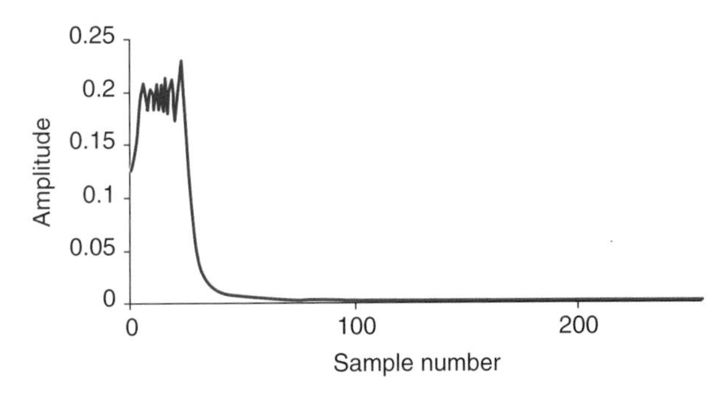

Figure 6.33 Fourier amplitude spectrum of the Chirp function (Figure 6.32).

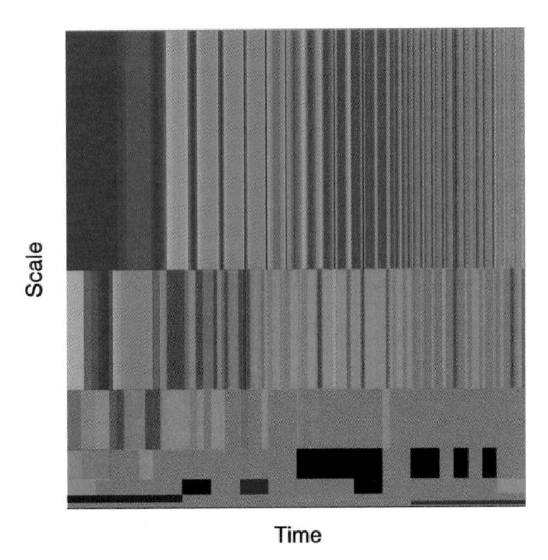

Figure 6.34 Time-scale plot of wavelet transform detail coefficients for Chirp function (Figure 6.32). Time is represented by the x-axis and scale by the y-axis, with the finest scale at the top and the coarsest scale at the bottom.

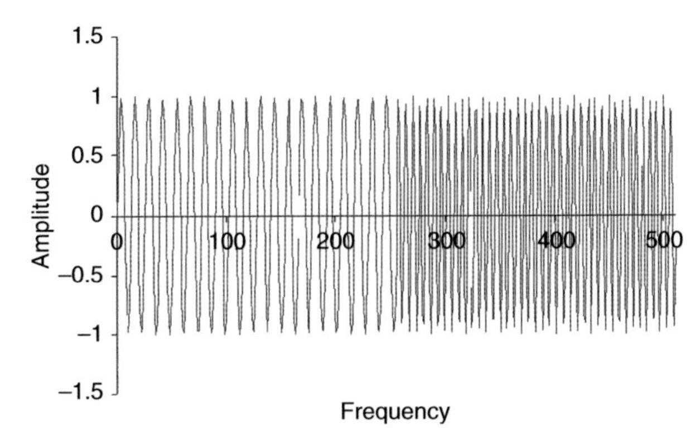

Figure 6.35 The same two sine waves as in Figure 6.30. However, in this example, the first sine wave forms the left half of the series (samples 1–256) and the second sine wave is on the right (samples 257–512).

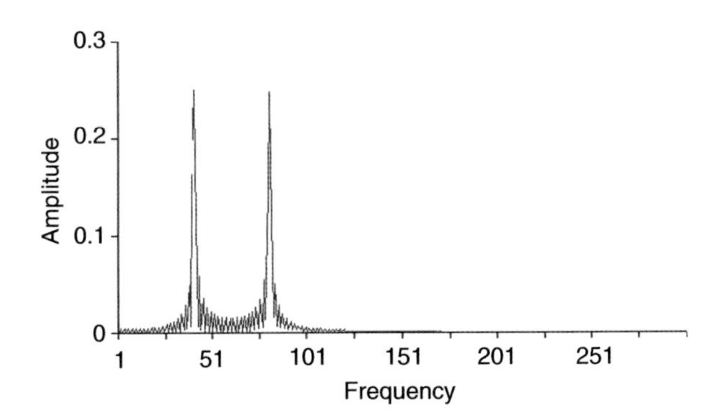

Figure 6.36 Fourier amplitude spectrum of the data shown in Figure 6.35. This spectrum is identical to the one shown in Figure 6.31, though the input data are different.

6.7.3 The Two-Dimensional Discrete Wavelet Transform

The two-dimensional DWT is calculated in the same way as the two-dimensional DFT, that is by performing a one-dimensional DWT along the rows of the image to give an intermediate matrix, and then applying the one-dimensional DWT to the columns of this intermediate matrix. The size of the image is m rows and n columns, where both m and n are powers of 2. The lowest-level detail coefficients – which contain the most detail – form the right-hand end of the one-dimensional DWT output vector (Figure 6.35) while the lowest level detail coefficients are located at the left-hand end of the one-dimensional DWT output. If you think about m such output vectors lying horizontally and being intersected by n vectors lying vertically then the highest level detail in the vertical direction forms the lower part of the output image while the highest level detail in the horizontal direction forms the right-hand part of the output image. The result is that the highest level of 'horizontal' detail coefficients dominates the upper right quadrant of the output image, the highest level 'vertical' detail coefficients dominates the lower left quadrant of the output image, and highest level horizontal and vertical coefficients interact in the lower right quadrant of the output image. The 'once-smoothed' image (reduced in resolution by a factor of 2) is located in the upper left quadrant of the output, where the lowest-level horizontal and vertical detail coefficients interact. The third level has horizontal, vertical and diagonal components with the upper left quadrant containing the thrice-smoothed image (Figure 6.37).

The second-level decomposition acts on the coefficients contained in the upper right quadrant of the first level decomposition. This upper left quadrant is converted by row-column operations to a set of subquadrants. This process can be repeated until the sub-quadrants are only 1 pixel in size, or it can be stopped at any point.

The importance of the two-dimensional DWT is that the direction (horizontal, vertical or diagonal) of the detail

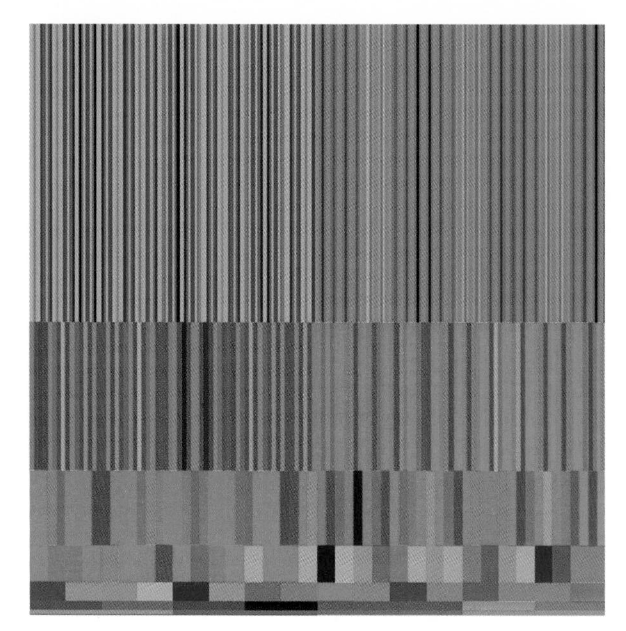

Figure 6.37 *Wavelet time-scale diagram of the two sine waves shown in Figure 6.35.*

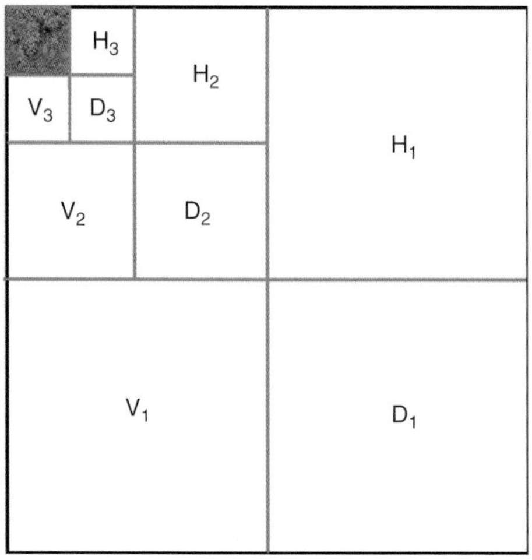

Figure 6.38 *Disposition of 'horizontal' (H), 'vertical' (V) and 'Diagonal' (D) wavelet detail coefficients in the two-dimensional DWT. Three levels of decomposition are shown, indicated by subscripts 1, 2 and 3. The thrice-smoothed image occupies the top left cell of the third level.*

coefficients at each hierarchical level are clearly defined. The highest-level detail coefficients (indicated by H_i, V_i and D_i in Figure 6.38) can be extracted and processed for particular purposes. For example, noise present in the image can be estimated and removed using procedure described in Section 9.3.2.2.2. The information present in the higher-level detail coefficients can be used to characterise the texture present in an image. Texture can range from smooth to rough, and is a measure of the local variability of grey levels (Section 8.7.1).

Applications of wavelets in image processing are given by Ranchin and Wald (1993) and Prasad and Iyengar (1997). Du, Guindon and Cihlar (2002) use wavelets to remove haze in high-resolution images, while wavelets are used to smooth interferograms (Section 9.2) by Braunich, Wu and Kong (2000). Blackburn (2007a) uses the wavelet transform in a study of reflectance spectra. The same author (Blackburn, 2007b) uses wavelets in a study of plant pigments. The use of wavelets as a cosmetic procedure (removal of striping, Section 4.2.2) is described by Torres and Infante (2001). An interesting application uses wavelets to combine panchromatic and multispectral images (a technique known as 'pan sharpening' or 'image fusion'; see also Section 6.9, where the use of the HSI transform in image fusion is described). Lu *et al.* (2007) use one-dimensional wavelets to denoise temporal sequences of MODIS products. The use of two-the dimensional DWT in image 'denoising' is discussed in Chapter 9.

6.8 Change Detection

6.8.1 Introduction

One of the main advantages of remotely-sensed data possesses is that collection of data takes place repeatedly over time. For example, the Landsat-7 satellite has a repetition cycle of 16 days, whereas NOAA AVHRR views the Earth's surface every 24 hours (Chapter 2). Of course, cloud cover is a problem in some areas of the world and several years may pass before the combination of a satellite overpass and a clear day arrives. The use of radar (ASAR, PALSAR, Radarsat-1 and -2, TerraSAR-X and COSMO/Skymed, for instance) can overcome cloud problems but radar imagery is not as well understood as optical images.

The importance of change detection applications in remote sensing is demonstrated by the volume of literature that surveys the available methods and sets out the relative merits of each. The books edited by Lunetta and Elvidge (1998) and Khorram *et al.* (1999) include contributions dealing with methods of detecting and measuring change. Recent reviews are by Coppin *et al.* (2004), Lu *et al.* (2004), Rogan and Chen (2004) and Sui *et al.* (2008). Hansen *et al.* (2008) consider change in tropical forests between 2000 and 2005 using multitemporal remotely-sensed data.

The first problem in change detection is to acquire a pair of images separated by a suitable time period. This time period may range from minutes (in the case of geostationary satellite images, where cloud movement is monitored to get an estimate of wind speed) to days or weeks (when crops are monitored over the growing season) or even years (for the application of interferometric methods of determining elevation, as described in Chapter 9). Where there is a strong seasonal effect, as in the temperate regions of the world such as Western Europe, New Zealand and countries with a Mediterranean climate it is best to ensure that the two image sets to be used in change detection are captured at the same time of year. Even so, differing weather condition in the weeks before data acquisition may generate apparent changes that are due primarily to the antecedent weather.

Differences between two image sets for the same area may thus reflect seasonal changes, changes due to antecedent weather conditions and apparent changes that are introduced by differences in sensor calibration, atmospheric effects and viewing/illumination geometry. Some methods that are based on correlations do not need correction for atmospheric path radiance as the means of all the image bands are set to zero; however, absorption and other effects are not accounted for. To be able to compare like with like, the two image data sets to be compared should be atmospherically and radiometrically corrected, and differences of illumination should also be accounted for. The methods to achieve these corrections are described in Chapter 4. Paolini *et al.* (2006) study the effects of radiometric corrections on change detection, while Song *et al.* (2001) provide useful advice on the need for atmospheric correction in change detection studies.

A further preprocessing step is accurate geometrical correction; resolving differences due to inadequate geometrical correction, for example registering one image to the other, is a vital preprocessing step. Brown, Foody and Atkinson (2007) consider misregistration error in change detection using airborne imagery, a topic that is also reviewed by Gong, Ledrew and Miller (1992) who discuss the effects of misregistration of images (Chapter 4) on the subtraction process, and present methods to reduce these effects, which will appear erroneously on the difference image as areas of change.

In this section, three methods of change detection are described. Descriptions of other methods, which may be more or less suited to a particular application, are provided by Canty (2007) (with ENVI/IDL code). Khorram *et al.* (1999) and contributors to Lunetta and Lyon (2004) and Stehman and Wickham (2006) discuss the important issue of accuracy estimation in change detection. Kontoes (2008) gives an example of the use of change vector analysis (Lambin and Strahler, 1994a, 1994b; Lambin, 1996). Lunetta *et al.* (2006) give an example of change detection using MODIS data, while Serra, Pons and Sauri (2003) use postclassification change detection. Sesnie *et al.* (2008) discuss classification and change detection in complex neotropical environments. Stehman, Sohl and Loveland (2005) examine sampling strategies for change detection. Lu *et al.* (2004) list 31 different methods of change detection and provide more than 13 pages of references. Other sources of information are Chen *et al.* (2003), Collins and Woodcock (1996), Coppin *et al.* (2004), Grey, Luckman and Holland (2003), Johnson and Kasischke (1998), Lambin (1996), Lambin and Strahler (1994a, 1994b), Mas (1999), Siljestrom and Moreno (1995) and Varjo (1996). In addition, the reviews and books mentioned in the opening paragraphs of this section give broad overviews.

Apart from the introductory paragraphs above, this section contains details of three examples of change detection, using image differencing, PCA and canonical correlation change detection methods. All of the examples use the Alexandria 1984 and 1993 image data sets shown in Figure 6.2a, b. The two datasets have been georeferenced but no radiometric or atmospheric corrections have been carried out.

6.8.2 NDVI Difference Image

The first example uses the NDVI images derived from the 1984 and 1993 TM/ETM+ images of Alexandria, Egypt. Bands 4 and 3 are used, and before the NDVI images were calculated, their histograms were displayed and offsets determined so that the histogram minimum method of correcting for atmospheric path radiance could be applied. The offsets for bands 4 and 3 of the 1984 data set are 13 and 24 respectively, while the corresponding offsets for the 1993 data set are 16 and 25. The resulting pseudocolour difference image is included as Figure 6.39. The two NDVI images from which Figure 6.39 was derived can be seen as Figure 6.40a, b.

The most noticeable thing about the two NDVI images and the change image is their relative lack of detail compared with the false colour composites shown in Figure 6.2a, b. Even careful density slicing did not permit the use of more than five colour bands (Figure 6.39b, c). The overall impression is that the use of just two bands (TM/ETM+ bands 4 and 3) does not give a full picture of the changes that have occurred between the two dates, particularly as the measure used (the NDVI) is designed for measuring vegetation status. For example, the airport in the centre of the 1993 image is absent from the 1984 image. This change is not reflected in the NDVI difference image. The method may be more suited to regions with a greater vegetation cover than that found in the Alexandria area.

6.8.3 PCA

The technique of PCA is described in Section 6.4. In that section it was noted that the first PC image derived from a single-date image set is a weighted average of the input bands, while the second PC is a contrast between

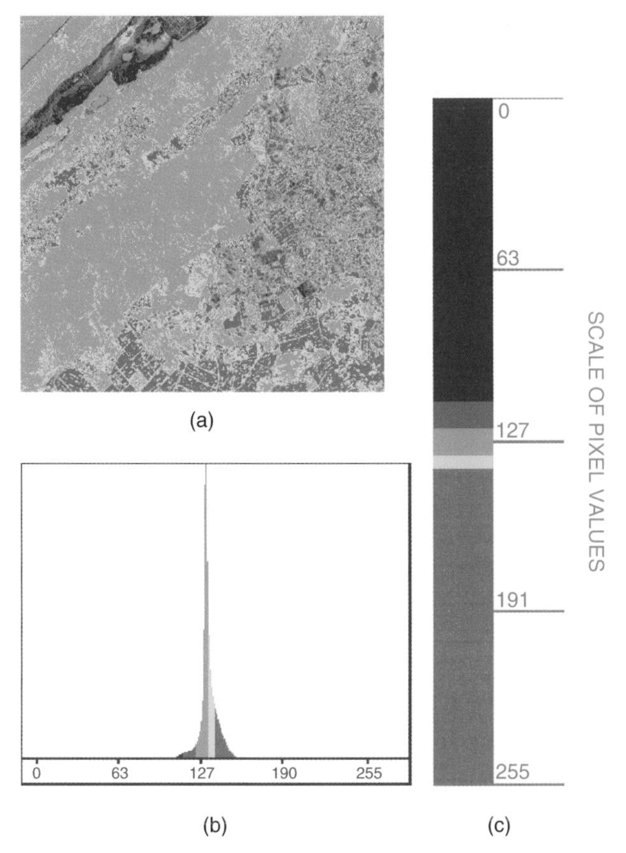

Figure 6.39 *(a) Difference between NDVI93 (Figure 6.40b) and NDVI84 (Figure 6.40a after subtraction of offsets. (b) Colour wedge used in pseudocolour operation on (a). (c) Histogram of (a) using colour palette depicted in (b).*

the NIR and the red bands, pointing to a component that was identifying areas of more extensive or more luxuriant vegetation in contrast to bare soil or water. In change detection studies the two data sets being analysed are combined forming, in this case, a 12-band image set with the Alexandria 1984 data forming columns 1–6 and with the 1993 data set appended as columns 7–12 (considering each image band as a column formed by taking rows 1, 2, 3, . . . in turn and concatenating them). This combined data set was analysed using PCA based on the 12 × 12 correlation matrix. The use of correlation or covariance measures involves subtraction of the mean value in each band from all the pixel values in that band, effectively standardizing the data for its means. This has the effect of removing any additive effect caused, perhaps, by atmospheric path radiance.

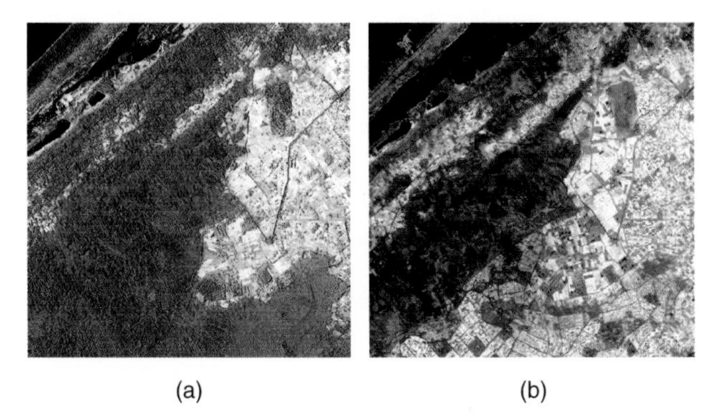

Figure 6.40 *(a) Alexandria NDVI image for 1984. (b) Alexandria NDVI image for 1993. Both are contrast stretched.*

The statistics from the PCA of the combined data sets are listed in Table 6.7 and the first six component images form Figure 6.41. It is clear from the correlation matrix shown in Table 4.5 that the highest correlations are those between adjacent spectral bands (mainly in excess of 0.9). Spectral band 1 for both years (labelled 1 and 7) shows lower correlations with all other bands than the other bands, but the lowest correlations are between spectral bands from different years, the lowest being 0.482 between 1984 band 1 and 1993 band 4 (numbered 10 in Table 4.5). The eigenvalues are typical of those computed from remotely-sensed image data in that the first principal component summarizes almost 80% of the total variation in the 12 images. The second principal component captures another 11% and the first six principal components together account for 99.4% of the total variability. Six component images were calculated using the weights in Table 6.7, which gave the images shown in Figure 6.41.

Study of the weights (eigenvectors) from which the principal component images were calculated and the corresponding images reveals that component 1 (Figure 6.41a) is a weighted average of all 12 bands. It shows the infrastructure present as well as the desert surface. Because all 12 weights have negative values, the areas that are bright in the original images (Figure 6.2) are dark in principal component 1 while dark areas in the original images become bright in the principal component image. Though the weights indicate that this principal component is a weighted average component, the different regions within the area of the image are brought out clearly. As noted earlier, the desert surface is dark while the infrastructure present in both years (1984 and 1993) is bright. The infrastructure developed between 1984 and 1993 appears as mid-grey. There are some interesting patterns over the desert surface and roads and field boundaries are clear in both the bright and mid-grey regions. Overall, principal component 1 appears to provide a good single-band summary of the regions of change in the two images.

The weights associated with principal component 2 show a clear contrast between bands 1–6 (1984) and bands 7–12 (1993), with the 1984 image being bright and the 1993 image being dark. Visual analysis of the corresponding principal component image (Figure 6.41b)

Table 6.7 *Correlation matrix, eigenvalues and eigenvectors of the combined 1984 and 1993 Alexandria images. The first six bands are TM bands for 1984. Bands 7–12 are the six TM bands for 1993. See Figure 6.41 for the first six principal component images.*

Combined 1984 and 1993 correlation matrix												
	1	**2**	**3**	**4**	**5**	**6**	**7**	**8**	**9**	**10**	**11**	**12**
1	1.000											
2	0.964	1.000										
3	0.926	0.988	1.000									
4	0.794	0.894	0.914	1.000								
5	0.665	0.783	0.830	0.902	1.000							
6	0.679	0.797	0.845	0.890	0.989	1.000						
7	0.725	0.711	0.678	0.661	0.614	0.609	1.000					
8	0.708	0.727	0.710	0.711	0.674	0.673	0.981	1.000				
9	0.674	0.714	0.712	0.733	0.727	0.726	0.952	0.988	1.000			
10	0.482	0.577	0.602	0.742	0.815	0.780	0.745	0.801	0.845	1.000		
11	0.572	0.646	0.664	0.737	0.794	0.784	0.856	0.910	0.949	0.915	1.000	
12	0.588	0.652	0.667	0.713	0.757	0.760	0.875	0.929	0.961	0.867	0.987	1.000
Eigenvalues and associated variance (individual and cumulative)												
PC	Eigenvalue	%variance	Cumulative % variance	PC	Eigenvalue	% variance	Cumulative %var					
1	9.555	0.796	0.796	7	0.331	0.276	0.997					
2	1.308	0.109	0.905	8	0.134	0.111	0.998					
3	0.780	0.650	0.970	9	0.782	0.652	0.999					
4	0.134	0.111	0.981	10	0.377	0.314	0.999					
5	0.845	0.704	0.988	11	0.320	0.267	0.999					
6	0.728	0.607	0.994	12	0.209	0.174	1.000					
Eigenvectors (weights) for the first six principal components												
	1	**2**	3	**4**	**5**	**6**						
1	−0.264	0.369	0.405	0.048	−0.257	−0.374						
2	−0.284	0.375	0.205	0.028	0.087	−0.167						
3	−0.287	0.377	0.089	−0.081	0.185	−0.135						
4	−0.292	0.262	−0.173	0.426	0.534	0.475						
5	−0.288	0.152	−0.445	−0.154	−0.337	0.162						
6	−0.288	0.173	−0.416	−0.381	−0.305	0.148						
7	−0.284	−0.243	0.391	0.091	−0.377	0.344						
8	−0.297	−0.253	0.281	0.011	−0.047	0.271						
9	−0.302	−0.269	0.156	−0.086	0.062	−0.397						
10	−0.277	−0.270	−0.333	0.675	−0.226	−0.316						
11	−0.297	−0.304	−0.116	−0.157	0.261	−0.238						
12	−0.295	−0.309	−0.026	−0.377	0.361	−0.184						

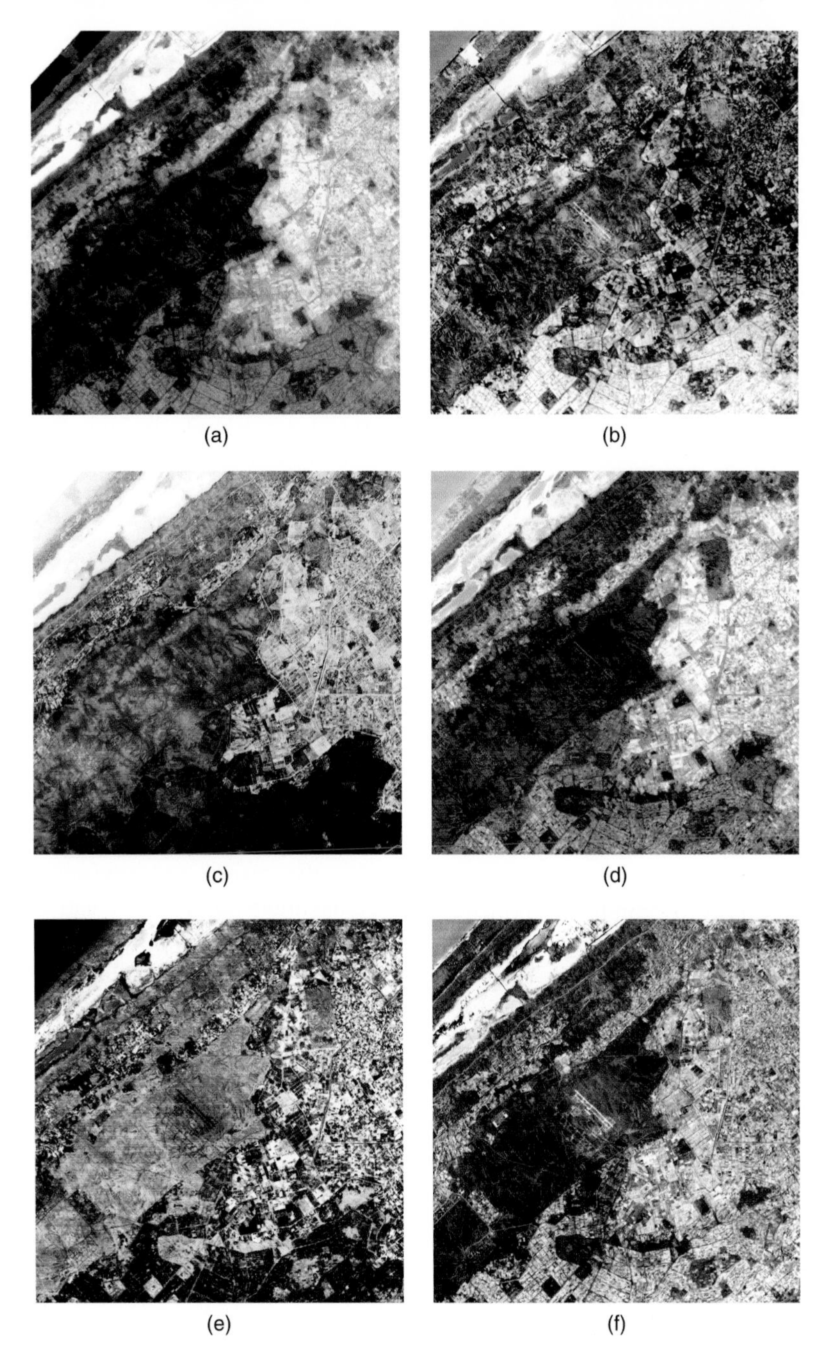

Figure 6.41 (a–f) The first six principal components of the combined 1984 and 1993 Alexandria dataset. See Table 6.7 for the associated statistics.

shows that this distinction is not as clear-cut as one might expect. New developments between 1984 and 1993 show up as lighter areas (including the airport that appears near the image centre). Both the desert surface and the old infrastructure appear dark, with considerable structural detail appearing. What appears to be a building in the extreme top left of the image appears white. This building cannot be seen on principal component 3, which has a pattern of weights that contrast the visible spectrum (with positive weights, thus appearing in lighter shades of grey on the image) with the NIR spectral bands, with negative weights. Older areas of development appear bright, while the darkest areas correspond to new infrastructure.

One could analyse the remaining principal components in a similar way before coming to the conclusion that the dimensions of change (i.e. new infrastructure such as the airport and new buildings, changes in the patterns or structures visible on the desert surface (light coloured in

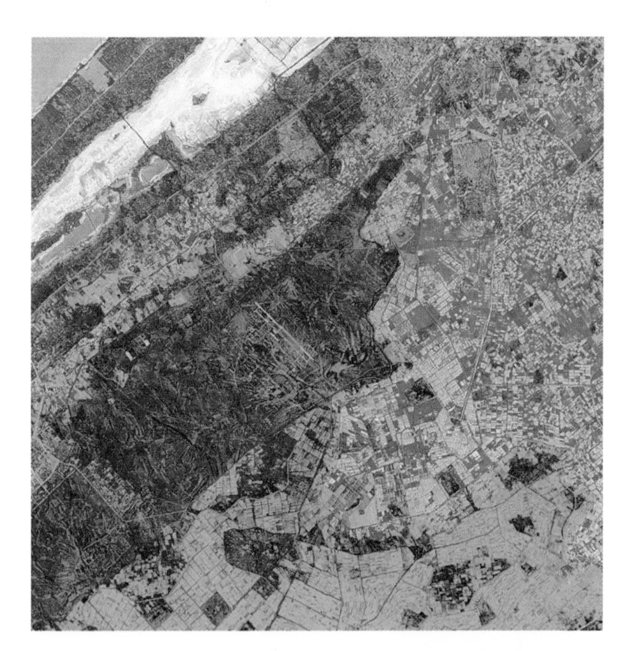

Figure 6.42 Principal components of change for the Alexandria image; principal components 1–3 are shown in red, green and blue respectively. The image has been subjected to a histogram equalization contrast stretch and to a process called sharpening, which is described in more detail in Chapter 7.

Figure 6.2)) are not adequately summarized by a single principal component image. Figure 6.42 shows a much more revealing picture produced by combining principal components 1, 2 and 3 into a false-colour image that has been enhanced by the use of a histogram equalization contrast stretch and by the application of a process called sharpening (described in Chapter 7) that brings out fine details in an image. The composite image shows the extent of development in a pink colour, with new developments being seen in light green. This light green surrounds areas of new development shown in pink and delineates the contiguous area of new development in the lower part of the image. Patterns on the desert surface appear in blue/green and brown tones, and the airport built between 1984 and 1993 standing out in green. The building in the extreme top left corner, mentioned above, is shown in cyan. The colour combination of principal components 1–3 is more informative in terms of the location and pattern of change than any of the principal components alone, and would provide a valuable input to a GIS-based study of change.

Examples of the use of PCA in change detection are provided by Henebry (1997), Koch (2000), Li and Yeh (1998), and Siljestrom and Moreno (1995).

6.8.4 Canonical Correlation Change Analysis

Canonical correlation change analysis (CCCA for short) is a novel image processing technique that is described in Nielsen's (1994) PhD thesis, though the canonical correlation procedure has been used in conventional multivariate analysis for many years (e.g. Cooley and Lohnes, 1962; Timm, 2002; Rencher, 2002). A recent use of the technique in remote sensing is Zhang *et al.* (2007). The mathematics can appear daunting, but the underlying concepts can be understood easily. Consider one of the principal component images shown in Figure 6.41; it has been generated by finding a set of axes that individually maximize the variance remaining in the image set after earlier principal components have been extracted. The first principal component image is derived by taking, for every pixel in the dataset, the sum of (weight (1) times pixel value in band 1) + (weight (2) times pixel value in band 2), and so on for all bands. The weights are shown in Table 6.7, and they are computed so that they have the property of maximum variance. In other words, no other combination of weights could produce an image that had a higher variance than principal component 1, subject to the constraint that the sum of squares of the weights is unity. A principal component is therefore defined as a weighted sum of the raw image data, with the maximum variance property.

CCCA uses two image sets, not one; in this section we have concentrated on the Alexandria images of 1984 and 1993, which we will call **X** and **Y** respectively. If we define the desired property to be maximum correlation not maximum variance, we can set out the problem: find a weighted linear combination of the bands of **X** that has the highest correlation with the corresponding weighted linear combination of the bands of **Y**. In a sense we do two PCAs on **X** and **Y** but use the maximum correlation criterion rather than maximum variance. The first principal component images for **X** and **Y** would then have the highest correlation of any other possible linear combination of **X** and **Y**.

The output from the application of CCCA to the 12-band Alexandria image sets (**X** and **Y**) consists mainly of (i) the values of the canonical correlations between the each of the six canonical images of **X** and the corresponding six canonical images of **Y** and (ii) 12 greyscale images in order band 1–6 of **X** and band 1–6 of **Y** so that in the composite data set the first canonical image pair is numbered 1 and 7, the second 2 and 8, and so on. One might be tempted to think that the canonical image pair with the lowest correlation would show the areas of greatest difference (i.e. change) but the following example may give food for thought.

Table 6.8 shows the results of a CCCA of the two Alexandria data sets. The first three images derived from the corresponding weight vectors (also shown in Table 6.8) all have canonical correlations above 0.5 in magnitude. Little can be said about Table 6.8 other than that; the weight vectors are included to reassure the

Table 6.8 *Canonical correlations and column eigenvectors (weights) for Alexandria 1984 TM and 1993 ETM+ images. See text for discussion. Figures 6.43 and 6.44 show the images corresponding to these weights.*

		Eigenvalues			Canonical correlations		
1		0.768			0.876		
2		0.601			0.775		
3		0.362			0.602		
4		0.229			0.479		
5		0.025			0.158		
6		0.000			0.017		
Matrix of right-hand vectors (TM 1984)							
TM band		**1**	**2**	**3**	**4**	**5**	**6**
	1	−0.063	1.729	−2.445	−5.858	0.652	−1.881
	2	1.817	1.270	1.875	10.787	−5.176	8.564
	3	−1.372	−1.626	1.206	−3.390	6.477	−9.786
	4	−0.854	−0.255	−0.634	−0.407	−2.533	−1.499
	5	−1.059	1.214	−3.155	4.276	6.445	6.030
	6	0.753	−1.687	3.420	−5.640	−6.145	−1.415
Matrix of left-hand vectors (TM 1993)							
ETM band		**1**	**2**	**3**	**4**	**5**	**6**
	1	0.315	1.244	−0.408	−1.962	0.210	5.046
	2	−0.272	2.250	−0.817	−0.110	−3.313	−13.060
	3	0.702	−2.719	1.670	2.875	5.492	7.029
	4	−0.242	−0.028	0.310	1.172	−2.753	2.085
	5	−2.330	1.356	−5.892	1.453	4.601	−1.111
	6	1.091	−1.401	5.537	−3.536	−4.139	0.283

reader that the author is not suffering from delusions. Rather more informative are the canonical correlation images derived from these weight vectors, shown in Figures 6.43 and 6.44.

All the image pairs (e.g. Figure 6.43a, b) are arranged in decreasing order of correlation. Noise is a problem in some of the lower-correlated images. Canonical image 1 of the **X** (1984) dataset highlights infrastructure that is mainly present in 1984 and not necessarily in 1993, whereas image 1 of the **Y** (1993) dataset shows the reverse, that is infrastructure that is mainly present in 1993. For instance the runway of the airport in the middle of the desert was not paved in 1984 but a landing strip was already there (though it is not very apparent on the image). In 1993 the airport was finished and the runway becomes quite visible. That runway only appears in the images corresponding to the **Y** dataset. Band 2 of the **X** and **Y** datasets (Figure 6.43c, d seem to highlight features mainly in the desert surface. Again, band 2 of the **X** dataset shows the desert features in 1984 and band 2 of the **Y** dataset those found in 1993. The lower-order canonical correlation image can be interpreted similarly, but with increasing difficulty as the SNR becomes smaller. Figure 6.44 shows an unusual false colour composite image in which the red band holds the canonical correlation image for the **X** dataset and the corresponding image for the **Y** dataset is shown in green. A null image (composed of zeros) is placed in the blue band so that only combinations of red and green (i.e. shades of red, yellow and green) can be seen. This image, like the other canonical correlation images, shows more horizontal banding noise than do the PCA change images (Figures 6.41 and 6.42). A comparison between Figures 6.42 and 6.44 is left as an exercise for the reader.

This section closes with a succinct review of the mathematics underlying CCCA. We begin by noting that **X** represents the first data set, with $npix \times nl$ rows and

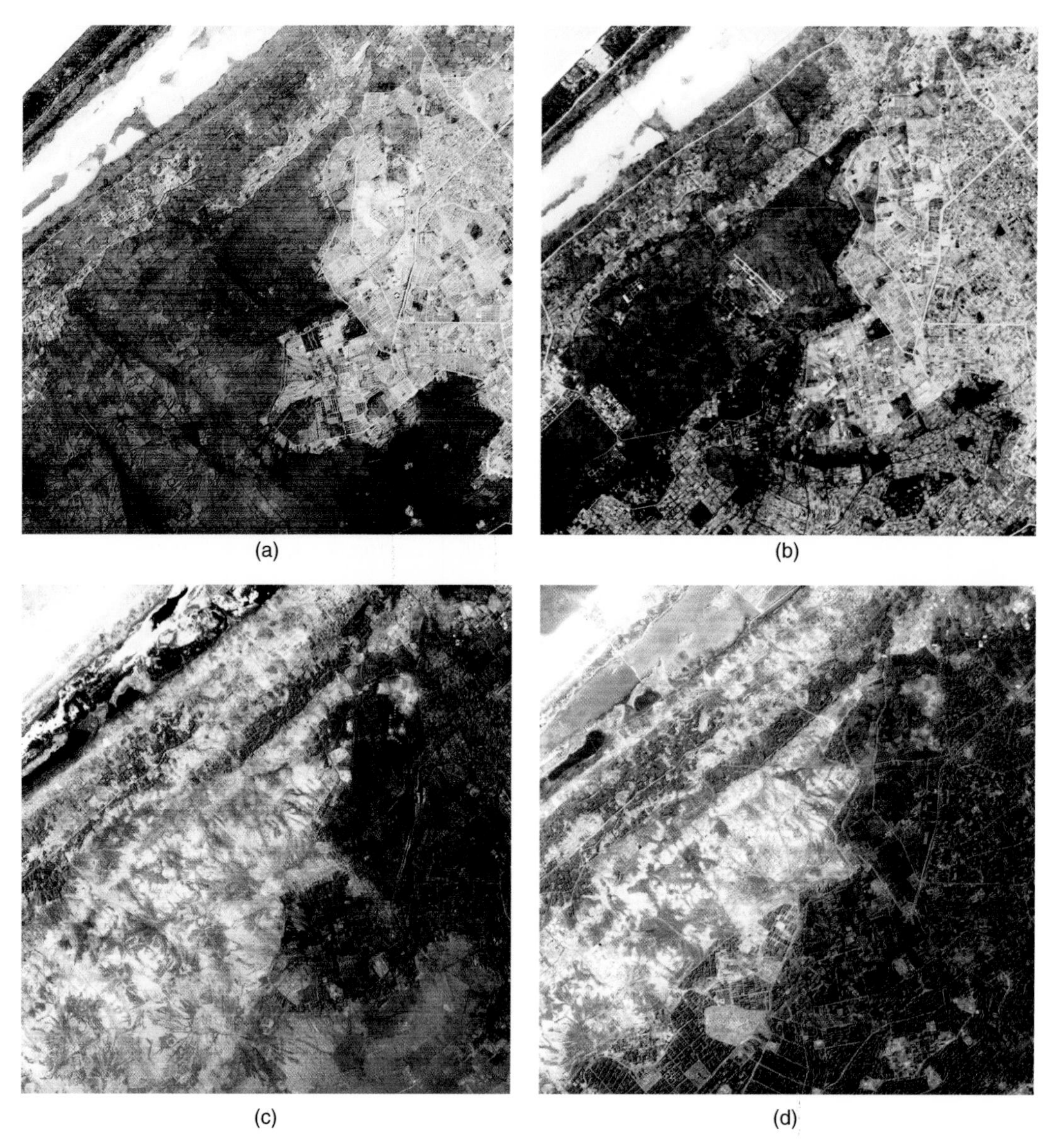

Figure 6.43 *Canonical correlation change images. The left-hand column of images (a), (c), (e), (g), (i) and (k) are the canonical correlation change images computed from the **X** or left-hand data set (the 1984 Alexandria dataset) while the right-hand column (b), (d), (f), (h), (j) and (l) are the corresponding images for the **Y** or right-hand dataset (the 1993 Alexandria dataset). See Table 6.8 for the related statistics. See Figure 6.2 for the raw images.*

nbands columns, where *npix*, *nl* and *nbands* are the number of pixels per row, or the image width, *nl* is the number of scan lines, or the image height, and *nbands* is the number of spectral bands in the image set. Matrix **Y** represents the second data set in a similar fashion. If $\mathbf{R}_{xx}$ is the matrix of correlations between the columns of **X**, $\mathbf{R}_{yy}$ the correlations among the columns of **Y**, and $\mathbf{R}_{xy}$ and $\mathbf{R}_{yx}$ the intercorrelations

$$\mathbf{R} = \mathbf{R}_{yy}^{-1}\mathbf{R}_{yx}\mathbf{R}_{xx}^{-1}\mathbf{R}_{xy}$$

between the columns of **X** and the columns of **Y** then the eigenvalues and eigenvectors of the product matrix **R**, defined as: give the canonical correlations (which are equal to the square roots of the eigenvalues of **R**) and the two sets of weight vectors that are derived from the eigenvectors of **R**. The matrix product can be written as

$$\mathbf{R} = \mathbf{R}_{yy}^{-1}\mathbf{B}$$

where

$$\mathbf{B} = \mathbf{R}_{yx}\mathbf{R}_{xx}^{-1}\mathbf{R}$$

A routine to solve the generalized eigenvalue problem can then be used to compute eigenvalues and eigenvectors. Such routines can be found in libraries such as Linpack and Eispac on the Internet. MATLAB users need to use routine `eig`. The CCCA procedure is incorporated into

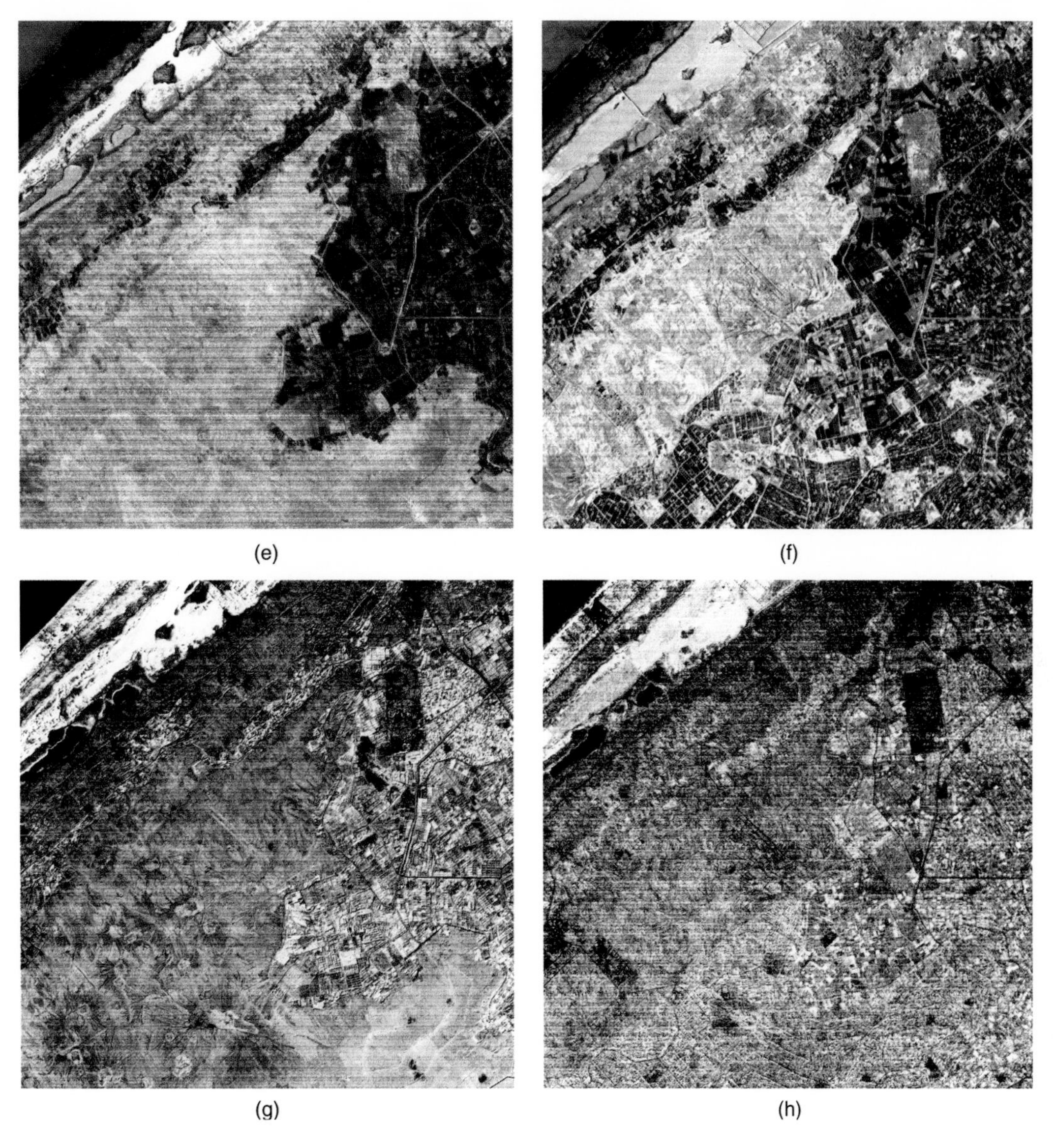

Figure 6.43 (continued)

MIPS, which was used to process the examples given in this section.

6.8.5 Summary

It was pointed out at the beginning of this section that there is a large number of methods of change detection, and one cannot take the results of one comparative analysis to provide a definitive answer to the question 'Which methods is best?'. The characteristics of the image may make one or other method appear 'best', because there is no universal agreement on the meaning of 'best'. In this study, the CCCA and PCA methods performed better than the difference of NDVI images in the sense that infrastructure details at the two dates (1984 and 1993) were clearly visible. For most readers, the PCA method may be easier to comprehend. Note, however, that no radiometric or atmospheric corrections were carried out on the images prior to the analyses, whereas a real-world study would be expected to incorporate both of these corrections, as well as topographic correction in hilly areas where shadow length may change due to differences in solar elevation angles, even if the imagery was collected at the same time of day.

The USGS (United States Geological Survey) has introduced a new web site called Terralook (http://terralook.cr.usgs.gov/) which provides georeferenced time sequences of two or more Landsat MSS, TM and ETM+ images as well as ASTER images, both data sets being global in coverage. Unfortunately, the

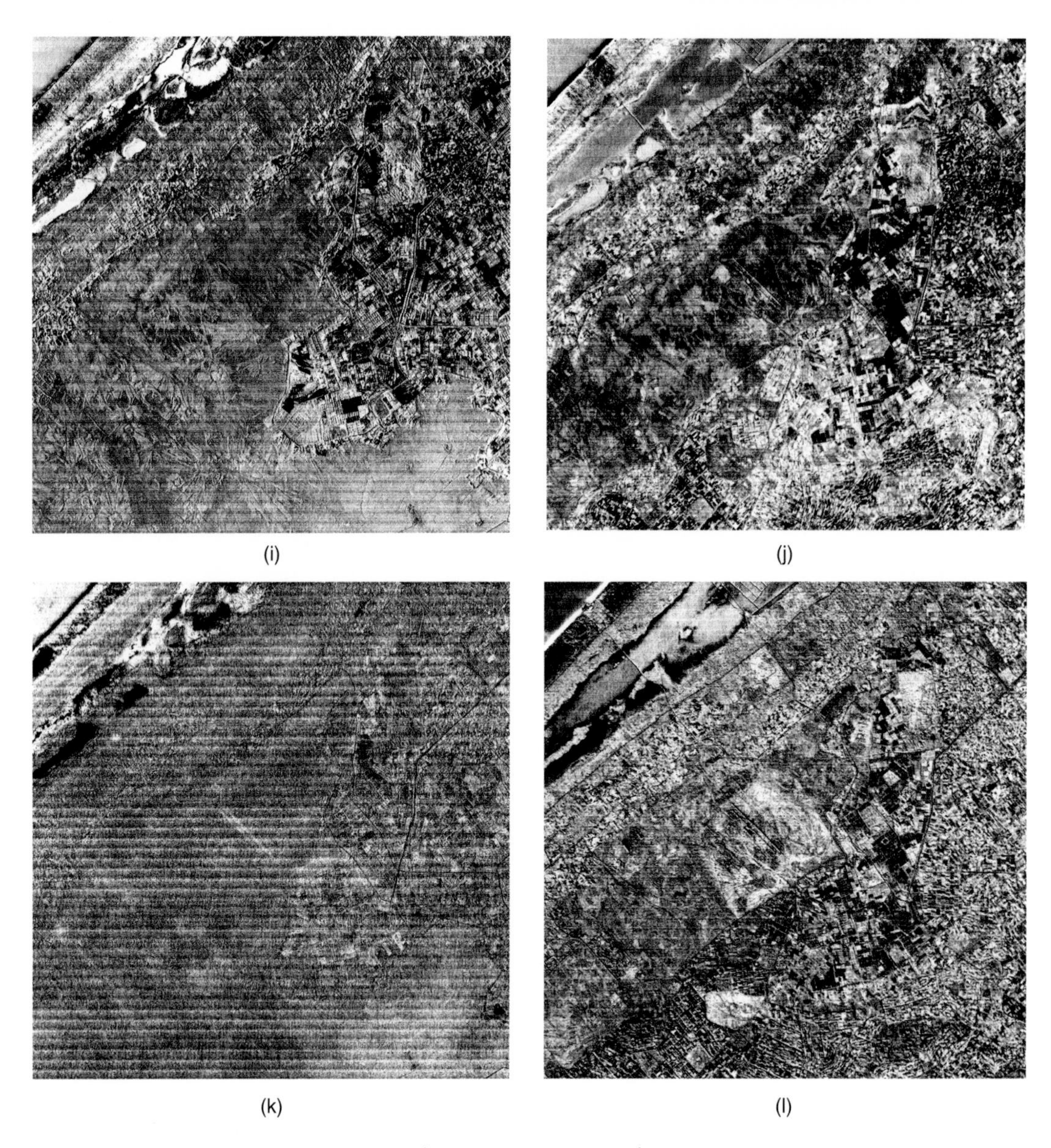

Figure 6.43 (continued)

images are only available in JPEG format which is a lossy compression method (see Chapter 3). Lecturers and instructors may nevertheless find material here for teaching and demonstration purposes.

6.9 Image Fusion

6.9.1 Introduction

Image fusion refers to the combination of image data from different sources with the aim of increasing the information content of the resulting merged image in accordance with the principle that the whole is greater than the sum of the parts. Often the motivation for employing image fusion techniques is to provide a suitable image map for import into a GIS. Visual appearance is therefore of considerable importance. Image fusion includes the operation known as pan-sharpening, in which a high-resolution panchromatic image is fused with a lower-resolution false-colour composite image to produce a fused image which retains the colour (spectral) information of the false colour composite image and combines it with the spatial sharpness of the panchromatic image. A second example is the fusion of a single-band SAR image with a colour composite image. The resulting image, if the fusion is successful, retains the colour information of the false-colour composite and adds the information in the SAR image, which is sensitive to surface roughness and soil moisture variations.

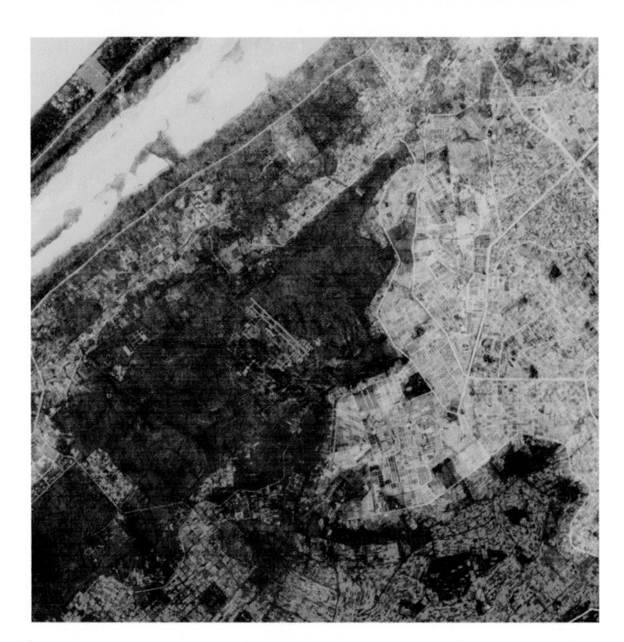

Figure 6.44 Canonical correlation image 1 for 1984 is shown in red and canonical correlation image 1 for 1993 is shown in green. The blue channel is set to a null image. Red areas have changed between 1984 and 1993 (for example the desert surface). Yellow areas are unchanged between the two dates and green areas are those of new developments. The airport stands out clearly, as does the new building in the top left corner. More scanline noise seems to be apparent on this image than on the PCA false colour image (Figure 6.43).

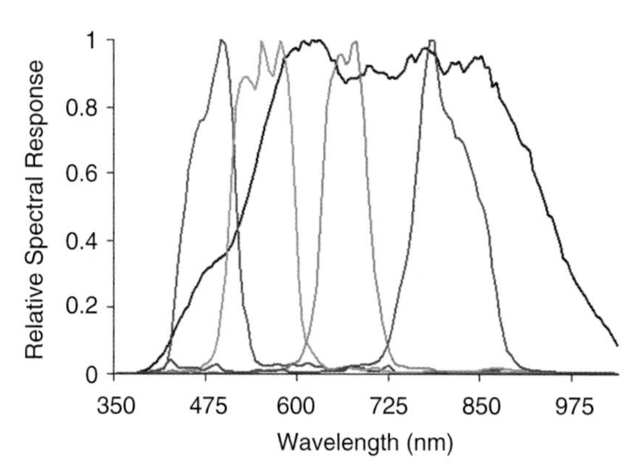

Figure 6.45 Relative spectral response of IKONOS bands. The red, green and blue bands are shown in their respective colours. The NIR band is in purple and the panchromatic band is shown by the black curve. Note how the panchromatic band covers a significant part of the NIR spectrum. In contrast, the SPOT HRV panchromatic band covers only slightly more than the visible spectrum (0.48–0.71 μm). Courtesy of GeoEye, Inc. © 2010. All rights reserved.

Figure 6.46 IKONOS image of part of the UAE in false colour. The image covers an area of 2 × 2 km with a resolution of 4 m. Courtesy of GeoEye, Inc. © 2010. All rights reserved.

A number of books and review articles deal with image fusion, including Du *et al.* (2007), Hyder, Shabazian and Waltz (2002), IEEE Transactions on Geoscience and Remote Sensing (2008), Karathanassi, Kolokousis and Ioannidou (2007), Li, Kwok and Wang (2002), Ling *et al.* (2007), Naik and Murthy (2003), Shan and Stilla (2008), Pohl and van Genderen (1998), Stathaki (2008) and Wald (2002). Of the range of methods outlined in these surveys, four are chosen to illustrate the range and diversity of image fusion techniques. These are the HSI transform (Section 6.5), PCA (Section 6.4), GS orthogonalization (mentioned in Section 6.2.3 in connection with the Tasselled Cap transform), and wavelet-based algorithms (Section 6.7). The first four of these techniques used the ENVI software package, while the wavelet-based method is implemented in MIPS. These methods are illustrated with reference to an IKONOS data set that covers part of the United Arab Emirates. This data set is called the UAE image in the remainder of this section. IKONOS generates a multispectral data set at 4 m resolution and a panchromatic image at a resolution of 1 m. The spectral response of the IKONOS bands is shown in Figure 6.45. The image covers an area of 2 × 2 km; thus the multispectral images have a resolution of 500 × 500 pixels and the panchromatic image is 2000 × 2000 pixels in size (Figures 6.46 and 6.47).

The first stage in the process of pan-sharpening is the geometric rectification of the low-resolution multispectral image and the high-resolution panchromatic image. This stage is accomplished by resampling the low-resolution image to give the same pixel size as the high-resolution image. Resampling is covered in Section 4.4.3 and 4.3.4. The simplest method, nearest neighbour

Figure 6.47 Panchromatic band (resolution 1 m) of the IKONOS UAE image. Courtesy of GeoEye, Inc.

interpolation, tends to produce a blocky saw-tooth effect and so a more complex interpolation scheme is generally employed, even though procedures such as bilinear interpolation tend to smooth the image rather than sharpen it. The end-product is a registered set of false-colour and panchromatic imagery.

The second stage in the process of pan-sharpening is the choice of algorithm. In this section, the four techniques listed above are compared on the basis of both subjective and quasi-objective criteria. A number of studies provide details of these methods, and of other competing techniques; see Aiazzi *et al.* (2007), Blom and Daily (1982), Hong and Zhang (2008b), Kalpoma and Kudoh (2007), Liu (2000), Malpica (2007), Ranchin (2002a, 2002b), Ranchin and Wald (2000), Tu *et al.* (2004) and Zhou, Civco and Silander (1998). The four selected methods are described in the following Sections 6.9.2–6.9.5). The results are then compared using a number of quasiobjective methods in Section 6.9.6. The reader should bear in mind the opening sentences of this section, which specify that the aim of the pan-sharpening method is to provide an image map for importing into a GIS. Other applications are listed by Karathanassi, Kolokousis and Ioannidou (2007).

6.9.2 HSI Algorithm

The HSI transform is useful in two ways: first, as a method of image enhancement and, second, as a means of combining co-registered images from different sources. The first of these applications is described in Section 6.5. In terms of image fusion applications, the forward transform of the geometrically rectified and resampled multispectral image is computed and the HSI components are extracted. The intensity component is replaced by the registered panchromatic image and the result back-transformed to RGB colour space. The method is easy to understand and is widely used. It is less successful than other methods when the panchromatic band does not overlap the visible and NIR bands. In the implementation used here, the pan-sharpened image is reconstituted in 8-bit representation (0–255). The remaining four methods convert the output image to the 11 bit representation of the input IKONOS images.

6.9.3 PCA

Applications of PCA to remotely-sensed data normally result in a first principal component that is a weighted average of al bands in the data set. This common variation can be considered to represent variations in brightness or intensity. Pan sharpening using PCA replaces PC1 with the panchromatic image and then performs an inverse PCA on the result. Inverse PCA is used in the decorrelation stretch procedure (Section 6.4.3). The success of the method depends on the validity of the assumption that PC1 represents brightness or intensity.

6.9.4 Gram-Schmidt Orthogonalization

The GS orthogonalization procedure is simply a method of transforming a matrix so that its columns are orthogonal (i.e. they are uncorrelated). This method is patented by the Eastman Kodak Company and is available in the ENVI software package. It is similar in principle to the PCA method in that the low-resolution multispectral image is orthogonalized (but via the GS procedure rather than using an eigenvalue–eigenvector transform) and the panchromatic band substituted for the first GS vector. An inverse transform is then carried out to generate the final result.

6.9.5 Wavelet-Based Methods

Figure 6.38 shows a three-level wavelet decomposition of an image. The first level decomposition would simply have a degraded image in the top left quadrant. The quadrants labelled H_1, D_1 and V_1 are the first-level horizontal, diagonal and vertical detail images. In its simplest form, wavelet-based pan-sharpening involves the computation of a level 1 transform for each of the three components of the resampled multispectral image and for the panchromatic image. One or more of the H_1, D_1 or V_1 components is then replaced by the corresponding component of the panchromatic image and an inverse wavelet transform applied. The result should incorporate the

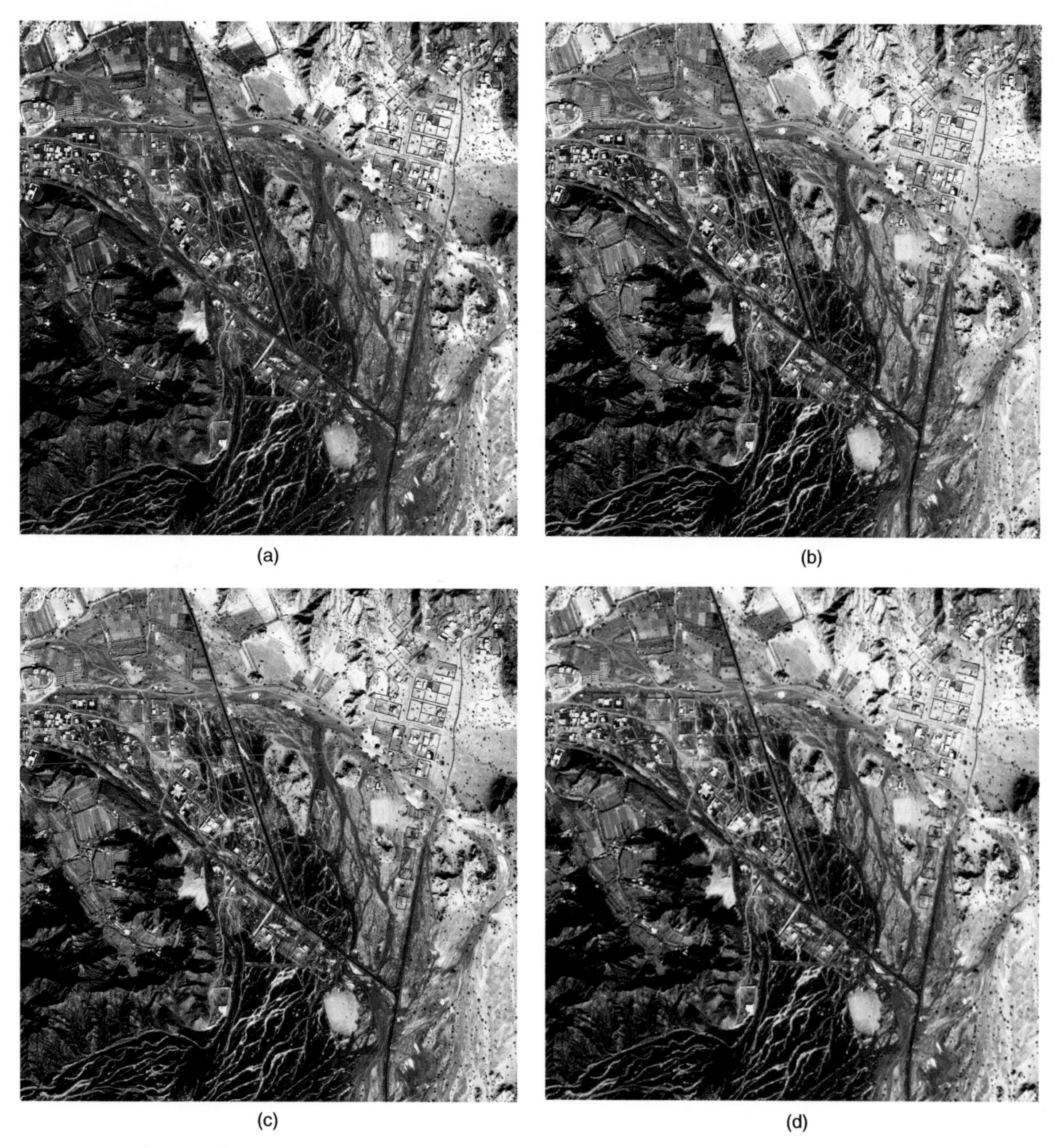

Figure 6.48 Pan-sharpened images using the IKONOS multispectral and panchromatic images shown in Figures 6.46 and 6.47. Note that these images are 2000 × 2000 pixels in size and reproduction on a printed page reduces the apparent resolution. See Figure 6.49 for examples of full-resolution subimages. (a) HSI transform, (b) Principal components transform, (c) Gram-Schmidt transform and (d) wavelet transform.

once-smoothed multispectral image and the panchromatic detail. As implemented in MIPS, a choice of mother wavelets is required. For the purposes of this exercise, a Daubechies-4 wavelet was used and all three quadrants of detail coefficients in the false-colour image were replaced by their panchromatic equivalents (Figure 6.37, upper and lower right and lower left quadrants). See Amolins, Zhang and Dare (2007) and Pajares and de la Cruz (2004) for reviews of the use of wavelets in image fusion. Garguet-Duport *et al.* (1996), Yocky (1996) and Zhou, Civco and Silander (1998) also provide more details.

6.9.6 Evaluation – Subjective Methods

Bearing in mind the fact that it is specified in the opening sentences of this section that the pan-sharpened image is computed for visual analysis, the first criterion to be used in selecting a method is the degree of success achieved by the pan-sharpening process as seen by the user. Printed images are rarely a completely satisfactory method of reproducing what is seen on screen, so the following descriptions might appear to be too imaginative. The explanation for this is the fact that the comparison is based on the images seen on screen. The four pan-sharpened images are reproduced as Figure 6.48a–d. The original 4 m resolution image is shown in Figure 6.46.

The first thing one notices is the relative lack of colour in the HSI transformed image in Figure 6.48a. As explained above, this may be because this image was output in 8-bit format while all of the others are expressed in terms of 11 bits. The remaining (pan-sharpened) are shown in Figure 6.48b–d. At this scale it is not possible to distinguish between the images on grounds of quality. It should be noted that the 8-bit image was subjected to a 5% linear contrast stretch, while the remaining images were converted from 11-bit representation to the 8-bit representation that is required for image display via the use of the equal class frequency transformation. This is an example of what was referred to in Chapter 3, where it was noted that image enhancement procedures could be performed automatically by the software and a misleading impression could thereby be obtained.

Figure 6.49 shows a selection of full-scale subimages, beginning with the 4-m multispectral image and the 1-m panchromatic image. Now it is possible to see that substantial differences between the results of the different techniques are present. In terms of visual appearance, the PCA and GS methods outperform the rest, while the wavelet method produces the worst performance, with the blocky appearance of the nearest-neighbour resampled image being very apparent. The HSI image is rather subdued in comparison to the results of the PCA and GS methods. In the next section, quantitative measures of quality are applied and it is interesting to note how objective methods can mislead whereas the eye rarely does.

6.9.7 Evaluation – Objective Methods

Karathanassi, Kolokousis and Ioannidou (2007) give details of a number of quantitative measures that they use to evaluate a number of pan-sharpening algorithms. Of these, the mean, standard deviation, correlation and entropy are used here. The concept of entropy was introduced in Section 2.2.3; it gives an estimate of the number of bits of information needed to represent the image. Table 6.9 lists the mean values, standard deviations and entropy measures for the 1 m resampled false-colour composite image (it is labelled 'RGB resampled' as the descriptors of the display channels are used rather than the wavebands of the image displayed in that channel. The images (represented by the columns of Table 6.9) are all in 11-bit integer format except the HSI result, which is represented in terms of 8-bit pixels. The means and standard deviations of all of the images approximate the mean and standard deviation of the RGB resampled image. This is as a result of histogram matching being applied to the image product. The minimum and maximum values of the RGB resampled image are taken and the histogram of the pan-sharpened image resulting from the application of the PCA or GS algorithm (discount the HSI as it is only 8 bits) is stretched so that its histogram minimum and maximum values equate to that of the input image. This is to ensure that the overall brightness of the original and pan-sharpened image are approximately equal. This procedure was not applied to the output from the wavelet algorithm, so its mean and standard deviation over all three input bands are higher than the RGB resampled image. However, if we take entropy as an indication of quality then the wavelet results are clearly the best, with an entropy of about 9.25 bits compared with 8.75–8.98 for the other methods, excluding HSI, which has a range of entropy values of about 7.75. Thus, on the basis of a quantitative information measure, the wavelet method is best.

Correlation of images has also been used as a measure of quality. One might think that the higher the correlation gets, the better the fit. In fact the RGB resampled image suffers, as we have seen, from saw-tooth patterns as a result of the use of the nearest neighbour interpolation method. It comes as no surprise, then, to find that the correlation between the RGB values of the entropy image and those of the RGB resampled image are highest, and approach 1.0 whereas the other methods are correlated at a level of 0.9 or so (Table 6.10). However, the reason for the high correlation of the wavelet results is the fact that they too demonstrate a saw-tooth pattern. It would seem that a higher correlation with the RGB resampled image is not a good guide to quality. The worst performer of the techniques used in this experiment has the highest scores on the information measure (entropy) as well as the highest correlation with the false colour resampled multispectral image.

Given that the primary motivation for performing the pan-sharpening operation is to produce an image map that can act as a background layer in a GIS, the subjective method of evaluation – that is look and see – is preferable to any quantitative measure of quality. Of the methods used here, the GS and PCA procedures are both acceptable in that the effects of resampling are not visually apparent and the colours match quite closely those of the original multispectral image.

References to various aspects of data fusion include Amarsaikhan and Douglas (2004) on classification, Cetin

Figure 6.49 *Pan-sharpening illustrated using a 350 × 350 chip from the IKONOS MSS and panchromatic images shown in figures 6.46 and 6.47.* ***(a)*** *IKONOS MSS image (4 m resolution),* ***(b)*** *IKONOS panchromatic image (1 m resolution),* ***(c)*** *Hue, saturation and intensity (HSI) transform,* ***(d)*** *Principal Components transform,* ***(e)*** *Gram-Schmidt transform,* ***(f)*** *Wavelet transform.*

Table 6.9 Summary statistics for the data fusion example. The mean, standard deviation and entropy of the resampled multispectral image (RGB resampled) and for the four fusion methods (Gram-Schmidt, principal components, hue-saturation-intensity and wavelet) are shown. See text for elaboration.

	RGB resampled			Gram-Schmidt			PCA			HSI			Wavelet		
	Mean	SD	Entropy	Mean	SD	Entropy	Mean	SD	Entropy	Mean	SD	Entropy	Mean	SD	Entropy
R	444	125	8.95	444	130	8.98	444	130	8.97	112	54	7.63	512	156	9.26
G	463	124	8.94	463	122	8.90	463	122	8.90	117	59	7.73	545	148	9.20
B	512	113	8.79	512	111	8.75	512	111	8.75	115	60	7.74	490	136	9.06

Table 6.10 Columns show the correlation between the four fusion methods and the red, green and blue bands of the resampled multispectral false colour image.

	Red band (resampled)	Green band (resampled)	Blue band (resampled)
RGB resampled	1.00	1.00	1.00
Gram-Schmidt	0.92	0.90	0.90
PCA	0.92	0.91	0.90
HSI	0.87	0.91	0.91
Wavelet	0.99	0.99	0.98

and Musaoglu (2009) on hyperspectral data fusion, Hyder, Shabazian and Waltz (2002) on multisensor fusion, the IEEE special issue on data fusion IEEE Transactions on Geoscience and Remote Sensing (2008), Kalpoma and Kudoh (2007) on IKONOS image fusion, the PE&RS special issue on remote sensing data fusion Shan and Stilla (2008), Ranchin (2002a, 2002b), Stathaki (2008) on fusion algorithms, Wald (2002), and Warrender and Augusteijn (1999) on fusion of image classifications.

6.10 Summary

A range of image transform techniques is considered in this chapter. Arithmetic operations (addition, subtraction, multiplication and division) have a utilitarian use – for example image subtraction is used routinely in the separation of the high-frequency component of an image during the filtering process (Section 7.5) while addition is used in the method of lineament detection described in Section 9.2. Image division, or ratioing, is one of the most common transformations applied to remotely-sensed images in both geological and agricultural studies for simple band ratios reflect differences in the slopes of the spectral reflectance curves of Earth surface materials. Problems experienced with the use of ratios include the difficulty of separating the effect of atmospheric path radiances, and the choice of dynamic range compression technique. Nevertheless ratio images are a widely used and valuable tool. The empirical transformations considered in Section 6.3 were developed for use with images of agricultural areas. One of the special problems here is that the database on which these transformations (the PVI and the Tasselled Cap transformation) are based is limited and site-specific. Their unrestricted use with images from parts of the world, other than those regions of the United States where they were developed, is questionable.

The PCA (KarhunenLoève) transformation has a long history of use in multivariate statistics. Even so, it is not as well understood by the remote sensing community as it might be. Like most parametric statistical methods it is based on a set of assumptions that must be appreciated, if not completely satisfied, if the methods are to be used successfully. The final transformation techniques covered in this chapter are the Fourier transform and the wavelet transform. The level of mathematics required to understand a formal presentation of these methods is generally well above that achieved by undergraduate students in geography, geology and other Earth sciences. The intuitive explanations given in Sections 6.6 and 6.7 might serve to introduce such readers to the basic principles of the method and allow a fuller understanding of the frequency-domain filtering techniques described in Section 7.5 and the de-noising procedures applied to reflectance spectra and images in Section 9.2.

The reader should appreciate that the presentation in the latter parts of this chapter is largely informal and non-mathematical. Many pitfalls and difficulties are not covered. These will, no doubt, be discovered serendipitously by the reader in the course of project work.

7 Filtering Techniques

7.1 Introduction

The image enhancement methods discussed in Chapter 5 change the way in which the information content of an image is presented to the viewer, either by altering image contrast or by coding a grey-scale image in pseudocolour so as to emphasize or amplify some property of the image that is of interest to the user. This chapter deals with methods for selectively or suppressing information at different spatial scales present in an image. For example, we may wish to suppress the high-frequency noise pattern caused by detector imbalance that is sometimes seen in Landsat MSS and TM images and which results from the fact that the image is electro-mechanically scanned in groups of six lines (MSS) or 16 lines (TM/ETM+) (Sections 2.3.6 and 4.2.2). On the other hand, we may wish to emphasize some spatial feature or features of interest, such as curvilinear boundaries between areas that are relatively homogeneous in terms of their tone or colour, in order to sharpen the image and reduce blurring. The techniques operate selectively on the image data, which are considered to contain information at various spatial scales. The idea that a spatial (two-dimensional) pattern, such as the variation of grey levels in a grey scale image, can be considered as a composite of patterns at different scales superimposed upon each other is introduced in Section 6.6 in the context of the Fourier transform. Large-scale background or regional patterns, such as land and sea, are the basic components of the image. These large-scale patterns can be thought of as 'background' with 'detail' being added by small-scale patterns. Noise, either random or systematic, is normally also present.

In symbolic terms, the information contained in an image can be represented by the following model:

$$
\begin{aligned}
\textit{image data} &= \textit{regional pattern} + \textit{local pattern} + \textit{noise} \\
&= \textit{background} + \textit{foreground (detail)} + \textit{noise} \\
&= \textit{low frequencies} + \textit{high frequencies} + \textit{noise}
\end{aligned}
$$

There is no reason to suppose that noise affects only the foreground or detail, though noise is often described as a high-frequency phenomenon. Noise can be either random or periodic. An example of random noise is the speckle pattern on synthetic-aperture radar (SAR) images. Periodic noise can be the result of a number of factors, such as the use of an electromechanical scanner, or the vibration from an aircraft engine.

The representation of the spatial variability of a feature in terms of a regional pattern with local information and noise superimposed has been widely used in disciplines that deal with spatially distributed phenomena. Patterns of variation are often summarized in terms of generalizations. For example, a geographer might note that, in Great Britain, 'mean annual rainfall declines from west to east' in the knowledge that such a statement describes only the background pattern, upon which is superimposed the variations attributable to local factors. In both geography and geology, the technique of trend surface analysis has been found useful in separating the regional and local components of such spatial patterns (Davis, 1973; Mather, 1976). Lloyd (2006) also considers spatial variation at different scales.

By analogy with the procedure used in chemistry laboratories to separate the components of a suspension, the techniques described in this chapter are known as *filtering*. A digital filter can be used to extract a particular spatial scale component from a digital image. The slowly varying background pattern in the image can be envisaged as a two-dimensional waveform with a long wavelength or low frequency; hence a filter that separates this slowly varying component from the remainder of the information present in the image is called a *low-pass filter*. Conversely, the more rapidly varying detail is like a two-dimensional waveform with a short wavelength or high frequency. A filter to separate out this component is called a *high-pass filter*. These two types of filter are considered separately. Low-frequency information allows the identification of the background pattern, and produces an output image in which the detail has been

Computer Processing of Remotely-Sensed Images: An Introduction, Fourth Edition Paul M. Mather and Magaly Koch

smoothed or removed from the original (input) image (hence low-pass filtering can be thought of as a form of blurring the image). High-frequency information allows us either to isolate or to amplify the local detail. If the high-frequency detail is amplified by adding back to the image some multiple of the high-frequency component extracted by the filter then the result is a sharper, de-blurred image. Anyone who listens to the organ music of JS Bach will be able to identify the low frequency components as the slowly changing low bass notes played by the foot pedals, while the high frequency detail consists of the shorter and much more numerous notes played on the manuals. Bach's music and remotely-sensed images have something in common, for both combine information at different scales (temporal scales in music, spatial scales in images).

Three approaches are used to separate the scale components of the spatial patterns exhibited in a remotely-sensed image. The first is based upon the transformation of the frequency domain representation of the image into its scale or spatial frequency components using the Discrete Fourier Transform (Section 6.6), while the second method is applied directly to the image data in the spatial domain. A third, more recent, development is that of the discrete wavelet transform, which uses both frequency (scale) and spatial representations of the data. The principles of the wavelet transform are discussed in Section 6.7 and applications are summarized in Chapter 9. Fourier-based filtering methods are considered in Section 7.5. In the following two sections the most common spatial-domain filtering methods are described. There is generally a one-to-one correspondence between spatial and frequency-domain filters. However, specific filters may be easier to design in the frequency domain but may be applied more efficiently in the spatial domain. The concept of spatial and frequency-domain representations is shown in Figure 6.21. Whereas spatial-domain filters are generally classed as either high-pass (sharpening) or as low-pass (smoothing), filters in the frequency domain can be designed to suppress, attenuate, amplify or pass any group of spatial frequencies. The choice of filter type can be based either on spatial frequency or on direction, for both these properties are contained in the Fourier amplitude spectrum (Section 6.6).

7.2 Spatial Domain Low-Pass (Smoothing) Filters

Before the topic of smoothing a two-dimensional image is considered, we will look at a simpler expression of the same problem, which is the smoothing of a one-dimensional pattern. Figure 7.1 shows a plot of grey levels along a cross-section from the top left corner (0, 0) to the bottom right corner (511, 511) of the TM band 7 image shown in Figure 1.11b. Figure 7.1a shows the cross-section for the unfiltered image, while Figure 7.1b shows the same cross-section after the application of a low pass (smoothing) filter. Clearly, the level of detail has been reduced and the cross-section curve is more generalized, though the main peaks are still apparent. Figure 7.2 displays another plot showing grey level value (vertical axis) against position across a scan-line of a digital image. The underlying pattern is partially obscured by the presence of local patterns and random noise. If the local variability, and the random noise, were to be removed then the overall pattern would become more clearly apparent and a general description of trends in the data could then be more easily made. The solid line in Figure 7.2 is a plot of the observed pixel values against position along the scan line, while the dotted line and the broken line represent the output from median and moving-average filters respectively. These filters are described below. Both produce smoother plots than the raw data curve, and the trend in the data is more easily seen. Local sharp fluctuations in value are removed. These fluctuations represent the high-frequency component of the data and may be the result of local characteristics or of noise. Thus, low-pass filtering is used by Crippen (1989), Eliason and McEwen (1990) and Pan and Chang (1992) to remove banding effects on remotely-sensed images (Section 4.2.2), while Dale, Chandica and Evans (1996) use a low-pass filter in an attempt to smooth away the effects of image-to-image misregistration.

7.2.1 Moving Average Filter

The moving average filter simply replaces a data value by the average of the given data point and a specified number of its neighbours to the left and to the right. If the coordinate on the horizontal axis of Figure 7.2 is denoted by the index j then the moving-average filtered value at any point j is x'_j. The procedure for calculating x'_j depends on the number of local values around the data point to be filtered that are used in the calculation of the moving average. This number is always an odd, positive integer so that there is a definite central point (thus, the central value in the sequence 1, 2, 3 is 2 whereas there is no specific central value in the sequence 1, 2, 3, 4). The broken line in Figure 7.2 is based on a five-point moving average, defined by

$$x'_j = (x_{j-2} + x_{j-1} + x_j + x_{j+1} + x_{j+2})/5$$

Five raw data values represented by the vector $\boldsymbol{x}$ and centred on point x_j are summed and averaged to produce one output value (x'_j). If a three-point moving average

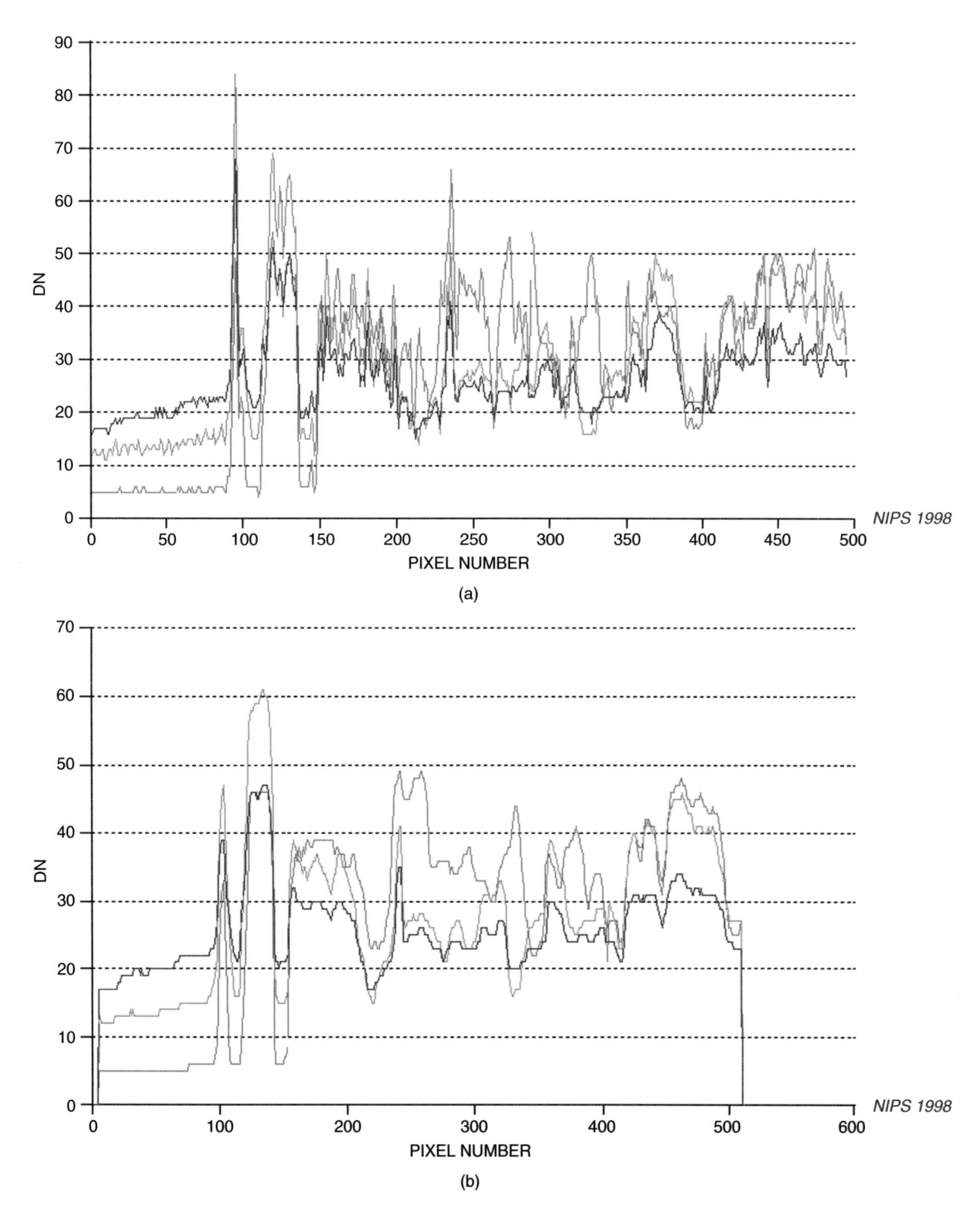

Figure 7.1 *(a) Cross-section of a Landsat TM image, with band 4 shown in red, band 3 in green and band 2 in blue. (b) Cross-section between the same points as used in (a) after the application of a smoothing filter (a 7 × 7 median filter was used to generate this cross-section, as described in Section 7.2.2). The reduction in detail is clearly apparent. Landsat data courtesy NASA/USGS.*

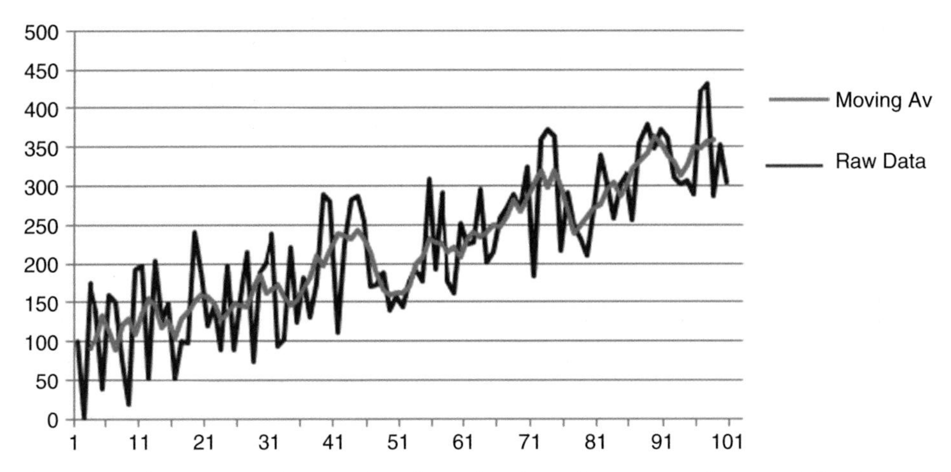

Figure 7.2 *One-dimensional data series showing the effect of and moving average (low-pass) filtering. The original data are shown in dark blue and the data after the application of a 5 × 5 moving average filter are shown in red. Note that the first two and last two data points in the raw data cannot be filtered.*

had been used then three raw data values centred on point j (i.e. points $j-1$, j and $j+1$) would be summed and averaged to give one output value at x'_j.

If the number of data elements included in the averaging process is n, then $[n/2]$ values at the beginning of the input series and $[n/2]$ values at the end of the input series do not have output values associated with them, because some of the input terms x_{j-1}, x_{j-2} and so on will not exist for $j<[n/2]$, just as some of the terms $x_{j+1}, x_{j+2}, \ldots$ will not exist for $j>N-[n/2]$ (the symbol [.] indicates the integer part of the given expression and N is the total number of raw data values in the (input) series that is being filtered). The filtered (output) series is thus shorter than the input series by $n-1$ elements, where n is the length of the filter (three point, five point, etc.). Thus, a moving average curve, such as that shown in Figure 7.2, will have no values at points x_1 and x_2 or at x_{n-1} and x_n. A 5 × 5 filter applied to an image will leave an unfiltered margin of 2 pixels around the four sides of the image. These marginal pixels are usually set to zero.

In calculating a five-point moving average for a one-dimensional series the following algorithm might be used: add up the input (x) values 1–5 and divide their sum by 5 to give x'_3, the first filtered (output) value. Note that filtered values x'_1 and x'_2 cannot be calculated; the reason is given in the preceding paragraph. Next, add raw data values x_2 to x_6 and divide their sum by 5 to give x'_4. This procedure is repeated until output value x'_{n-2} has been computed, where n is the number of input values (again, x'_{n-1} and x'_n are left undefined). This algorithm is rather inefficient, for it overlooks the fact that the sum of $x_2 - x_6$ is easily obtained from the sum of x_1 to x_5 simply by subtracting x_1 from the sum of $x_1 - x_5$ and adding x_6. The terms x_2, x_3 and x_4 are present in both summations, and need not be included in the second calculation. If the series is a long one then this modification to the original algorithm will result in a more efficient program.

A two-dimensional moving average filter is defined in terms of its horizontal (along-scan) and vertical (across-scan) dimensions. Like the one-dimensional moving average filter, these dimensions must be odd, positive and integral. However, the dimensions of the filter need not be equal. A two-dimensional moving average is described in terms of its size, such as 3 × 3. Care is needed when the filter dimensions are unequal to ensure that the order of the dimensions is clear; the Cartesian system uses (x, y) where x is the horizontal and y the vertical coordinate, with an origin in the lower left of the positive quadrant. In matrix (image) notation, the position of an element is given by its row (vertical, y) and column (horizontal, x) coordinates, and the origin is the upper left corner of the matrix or image. The central element of the filter, corresponding to the element x'_j in the one-dimensional case described earlier, is located at the intersection of the central row and column of the $n \times m$ filter window. Thus, for a 3 × 3 window, the central element lies at the intersection of the second row and second column. To begin with, the window is placed in the top left corner of the image to be filtered (Figure 7.3) and the average value of the elements in the area of the input image that is covered by the filter window is computed. This value is placed in the output image at the point in the output image corresponding to the location of the central element of the filter window. In effect, the moving average filter window can be thought of as a matrix with all its elements equal to 1; the output from the convolution of the window and the image is the sum of the products of the corresponding window and image elements divided by the number of elements in the window. If $\mathbf{F}$ $(= f_{ij})$ is the filter matrix, $\mathbf{G}$ $(= g_{ij})$

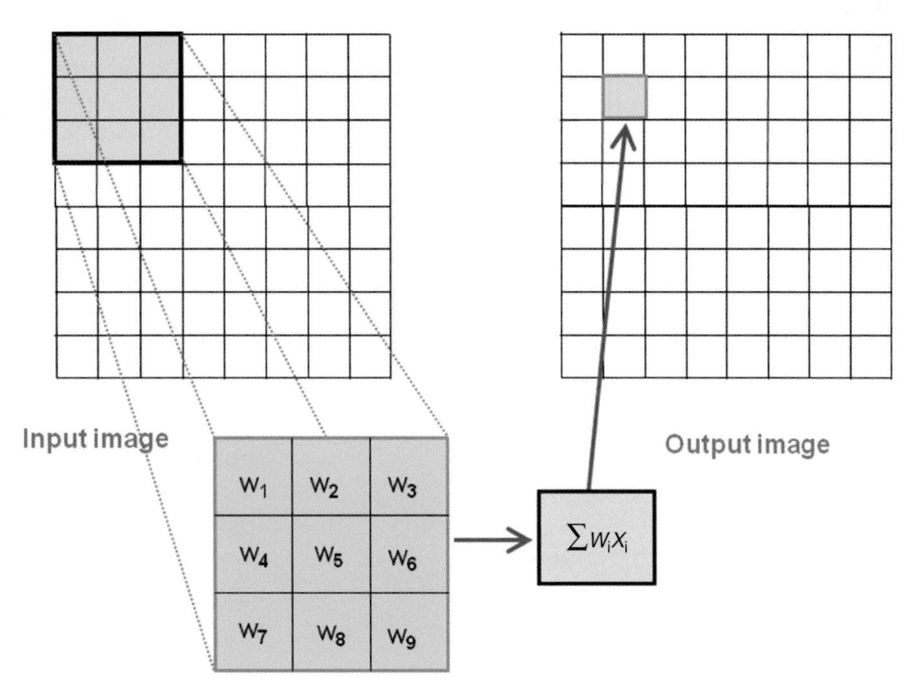

Figure 7.3 *Illustrating the operation of a spatial domain filter. In this example, the filter size is three rows by three columns. The filter size is always an odd, positive integer so that there is a 'central' pixel (on row 2, column 2 in this case). The 3 × 3 matrix of filter weights $w_1 \ldots w_9$) is placed over a 3 × 3 patch of image pixels and the weight is multiplied by the associated pixel value. These products are summed and the result is placed in the position in the output image that corresponds to the position of the central cell of the weight matrix. In the case of a 3 × 3 moving average filter, with nine elements, the weights w_i are all equal to 1/9.*

is the input image and $\mathbf{O}$ ($= o_{ij}$) the output (filtered) image then

$$o_{ij} = \left\{ \sum_{p=-b}^{b} \sum_{q=-c}^{c} g_{p+i,q+j} f_{r+p,s+q} \right\} / mn$$

where

- b integer part of $n/2$
- c integer part of $m/2$
- n number of rows in filter matrix (odd number)
- m number of columns in filter matrix (odd number)
- r central row in filter matrix ($= [n/2]$)
- s central column in filter matrix ($= [m/2]$)
- i, j image pixel underlying element (r, s) of filter matrix (coordinates in row/column order)
- $[e]$ integer part of expression e.

For example, given a 5 × 3 filter matrix the value of the pixel in the filtered image at row 8, column 15 is given by:

$$o_{8,15} = \left\{ \sum_{p=-2}^{2} \sum_{q=-1}^{1} g_{8+p,15+q} f_{3+p,2+q} \right\} / 15$$

with $b = 2$, $c = 1$, $r = 3$ and $s = 2$. Notice that the indices i and j must be in the range $b < i < (N - b + 1)$ and $c < j < (M - c + 1)$ if the image has N rows and M columns numbered from 1 to N and 1 to M respectively. This means that there are b empty rows at the top and bottom of the filtered image and c empty columns at either side of the filtered image. This unfiltered margin can be filled with zeros or the unaltered pixels from the corresponding cells of the input image can be placed there.

The initial position of the filter window with respect to the image is shown in Figure 7.3. Once the output value from the filter has been calculated, the window is moved one column (pixel) to the right and the operation is repeated. The window is moved rightwards and successive output values are computed until the right-hand edge of the filter window hits the right margin of the image. At this point, the filter window is moved down one row (scan-line) and back to the left-hand margin of the image. This procedure is repeated (Figure 7.4). The window is moved rightwards and successive output values are computed until the filter window reaches the right-hand edge of the image. At this point, the filter window is moved down one row (scan line) and back to the left-hand margin of the image. This procedure is repeated until the filter window reaches the bottom right-hand corner of the input image. The output image values form a matrix that has fewer rows and columns than the input image because it has an unfiltered margin corresponding to the

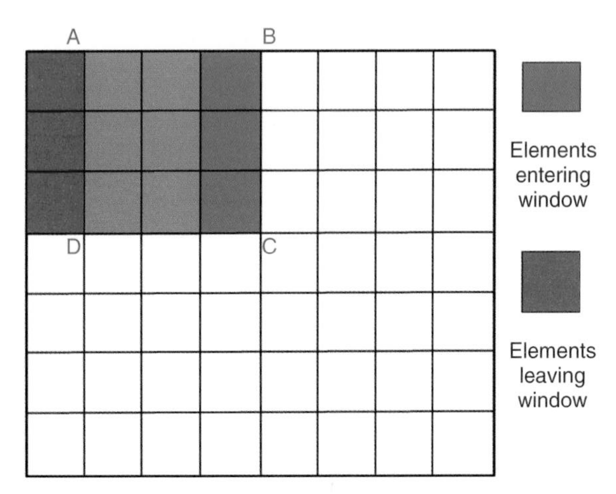

Figure 7.4 *The filter window ABCD has moved one column to the right from its initial position (position 1) in the top left corner of the image and now is in position 2. The elements of the column to be subtracted from the sum calculated at position 1 are indicated by green shading. Those elements to be added to the sum calculated at position 1 are indicated by red shading.*

top and bottom rows and the left and right columns of the input matrix that the filter window cannot reach. Generally, these missing rows and columns are filled with zeroes in order to keep the input and output images the same size.

The effect of the moving average filter is to reduce the overall variability of the image and lower its contrast. At the same time those pixels that have larger or smaller values than their neighbourhood average (think of them as grey level peaks and troughs) are respectively reduced or increased in value so that local detail is lost. Noise components, such as the banding patterns evident in line-scanned images, are also reduced in magnitude by the averaging process, which can be considered as a smearing or blurring operation. In cases where the overall pattern of grey level values is of interest, rather than the details of local variation, neighbourhood grey level averaging is a useful technique.

Examples of 3×3 and 5×5 moving average filter weights are given in Figure 7.5. The input to the filter program takes the form of a two-dimensional array of integers with a divisor that is equal to the product of the two dimensions of the box. Thus, in Figure 7.5a the divisor is 9, while in Figure 7.5b it is 25. If you apply these two filters, you will see that increasing the window size of a moving average filter results in a greater degree of smoothing, since more pixels are included in the averaging process. Example 7.1 illustrates the effects of applying a moving average of 3×3 and 7×7 on an ETM+ image of part of The Wash in eastern England.

1/9	1/9	1/9
1/9	1/9	1/9
1/9	1/9	1/9

(a)

1/25	1/25	1/25	1/25	1/25
1/25	1/25	1/25	1/25	1/25
1/25	1/25	1/25	1/25	1/25
1/25	1/25	1/25	1/25	1/25
1/25	1/25	1/25	1/25	1/25

(b)

Figure 7.5 *Moving average filter weights for (a) a 3×3 filter and (b) a 5×5 filter. The same effect can be achieved by using filter weights of 1, so that the weighted summation in Figure 7.3 becomes a simple addition. The sum is then divided by the product of the two dimensions of the filter, that is 9 in the case of (a) and 25 for (b).*

7.2.2 Median Filter

An alternative smoothing filter uses the median of the pixel values in the filter window (sometimes called the neighbourhood) rather than the mean. The median filter is generally thought to be superior to the moving average filter, for two reasons. First, the median of a set of n numbers (where n is an odd integer) is always one of the data values present in the set. Second, the median is less sensitive to errors or to extreme data values. This can be demonstrated by a simple, one-dimensional, example. If the 9 pixel values in the neighbourhood of, and including the point (x, y) are {3, 1, 2, 8, 5, 3, 9, 4, 27} then the median is the central value (the fifth in this case) when the data are ranked in ascending or descending order of magnitude. In this example the ranked values are {1, 2, 3, 3, 4, 5, 8, 9, 27} giving a median value of 4. The mean is 6.88, which would be rounded up to a value of 7 for display purposes, as most display systems use an integer scale of grey levels, such as 0–255. The value 7 is not present in the original data, unlike the median value of 4. Also, the mean value is larger than 6 of the 9 observed values, and may be thought to be unduly influenced by one extreme data value (27), which is three times larger than the next highest value in the set. Thus, the median filter removes isolated extreme pixel values or spikes, such as the value 27 in the example, which might represent isolated noise pixels. It also follows that the median filter preserves edges better than a moving-average filter, which blurs or smoothes the grey levels in the neighbourhood of the central point of the filter window. Figure 7.2 shows (i) a one-dimensional sequence of values, (ii) the result of applying a moving average of length 5 to the given data and (iii) the result of applying a median filter also of length 5.

It is clear that, while both the median and the moving average filters remove high-frequency oscillations, the median filter more successfully removes isolated spikes and better preserves edges, defined as pixels at which the gradient or slope of grey level value changes markedly. SAR images often display a noise pattern called *speckle* (Section 2.4). This is seen as a random pattern of bright points over the image. The median filter is frequently used to eliminate this speckle without unduly blurring the sharp features of the image (Blom and Daily, 1982; Held *et al.*, 2003).

The mean is relatively easily computed; it involves a process of summation and division, as explained earlier, and considerable savings of computer time can be achieved by methods described in Section 7.2.1. In contrast, the median requires the ranking of the data values lying within the $n \times m$ filter window centred on the image point that is being filtered. The operation of ranking or sorting is far slower than that of summation, for $[n/2]+1$ passes through the data are required to determine the median ($[n/2]$ indicates 'the integer part of the result of dividing n by 2'). At each pass, the smallest value remaining in the data must be found by a process of comparison. Using the values given in the example in the preceding paragraph, the value 1 would be picked out after the first pass, leaving eight values. A search of these eight values gives 2 as the smallest remaining value, and so on for ($[n/2]+1 = [9/2]+1 = 4+1 = 5$) passes. The differences between the summation and ranking methods are amplified by the number of times the operation is carried out; for a 3×3 filter window and a 512×512 image the filter is evaluated $510 \times 510 = 260\,100$ times. Thus, although the median filter might be preferred to the moving average filter for the reasons given earlier, it might be rejected if the computational cost was too high.

Fortunately, a less obvious but faster method of computing the median value for set of overlapping filter windows is available when the data are composed of integers. This fast algorithm begins as usual with the filter window located in the top left-hand corner of the image, as shown in Figure 7.3. A histogram of the $n \times m$ data points lying within the window is computed and the corresponding class frequencies are stored in a one-dimensional array. The median value is found by finding that class (grey level) value such that the cumulative frequency for that class equals or exceeds $[n/2]+1$. Using the data given earlier, the cumulative class frequencies for the first few grey levels are (0) 0; (1) 1; (2) 2; (3) 4 and (4) 5. The value in brackets is the grey level and the number following the bracketed number is the corresponding cumulative frequency, so that no pixels have values of zero, whereas 4 pixels have values of 3 or less. Since n is equal to 9, the value of ($[n/2]+1$) is 5 and no further calculation is necessary; the median is 4. This method is considerably faster than the obvious (brute force) sorting method because histogram calculation does not involve any logical comparisons, and the total histogram need not be checked in order to find the median. Further savings are achieved if the histogram is updated rather than recalculated when the filter window is moved to its next position, using a method similar to that illustrated in Figure 7.4. First, the cumulative frequencies are reduced as necessary to take account of the left-hand column of the window, which is moving out of the filter, and then the cumulative frequencies are incremented according to the values of the new right-hand column of pixel values, which is moving into the window. This part of the procedure is similar to the updating of the sum of pixel values as described in Section 7.2.1 in connection with the moving average filter. If the fast algorithm is used then the additional computational expense involved in computing the median is not significant.

The concept of the median filter was introduced by Tukey (1977) and its extension to two-dimensional images is discussed by Pratt (1978). The fast algorithm described above was reported by Huang, Yang and Tang (1979), who also provide a number of additional references. See also Brownrigg (1984) and Danielsson (1981). Blom and Daily (1982), Chen, Ma and Li-Hui Chen (1999), Chan, Ho and Nikolova (2005) and Rees and Satchell (1997) illustrate the use of the median filter applied to SAR images.

7.2.3 Adaptive Filters

Both the median and the moving average filter apply a fixed set of weights to all areas of the image, irrespective of the variability of the grey levels underlying the filter window. Several authors have considered smoothing methods in which the filter weights are calculated anew for each window position, the calculations being based on the mean and variance of the grey levels in the area of the image underlying the window. Such filters are termed adaptive filters. Their use is particularly important in the attenuation of the multiplicative noise effect known as speckle, which affects SAR images. As noted in Section 7.2.2, the median filter has been used with some success to remove speckle noise from SAR images. However, more advanced filters will, in general,

Example 7.1: Moving Average Filters

The purpose of this example is to evaluate the effects of changing the window size of the two-dimensional moving average filter. The instructions to the image processing software assume you are using MIPS but you could be using any image processing software, as all the usual brands have filtering modules.

Start MIPS and display a greyscale or false colour image of your choice. In this example, I am using a Landsat ETM+ false-colour image of part of The Wash on the east coast of England. The original (contrast stretched) image is shown in Example 7.1 Figure 1. Next, select `Filter User Defined Filter`. Select the vertical and horizontal sizes of the filter to use as 5 × 5. The weights are shown as a 5 × 5 array of 1's with a divisor of 25. These weights and divisor define the 5 × 5 moving average filter. The filtered image (Example 7.1 Figure 2) is shown in a new window.

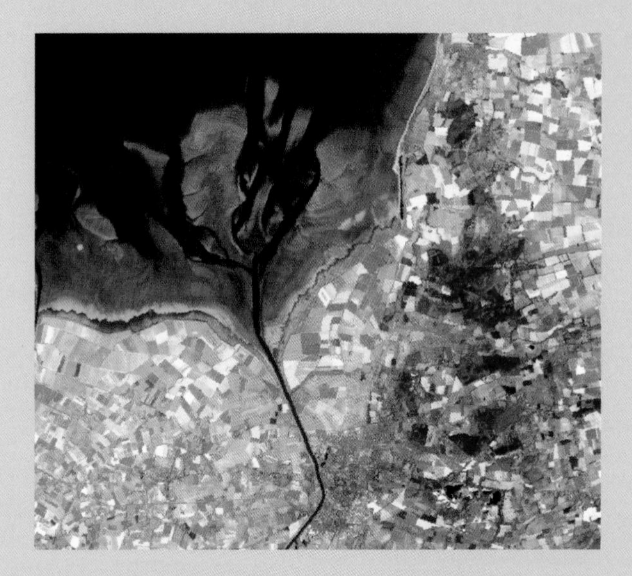

Example 7.1 Figure 1. *Contrast-stretched Landsat ETM+ false colour image of part of the The Wash in eastern England.*

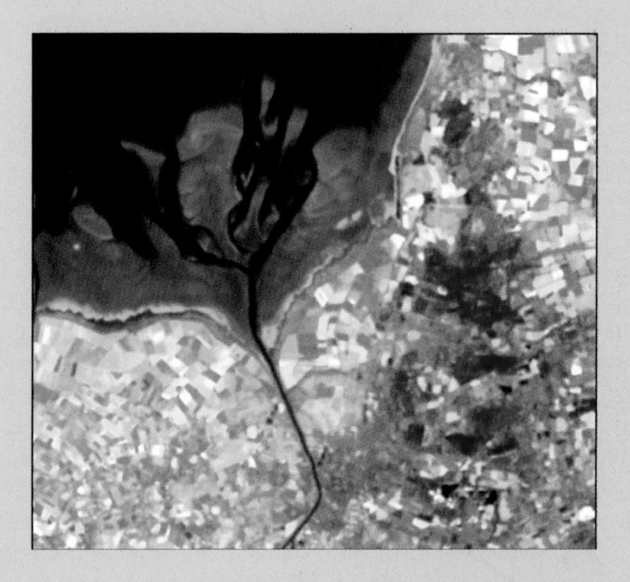

Example 7.1 Figure 2. *The image shown in Example 7.1 Figure 1 after the application of a 5 × 5 moving average filter.*

Finally, choose **Filter User Defined Filter** again, and change the filter size to 7 × 7. Accept the default weights and divisor, and a 7 × 7 moving average filtered image appears on the screen (Example 7.1 Figure 3). You can try other moving average window sizes (they do not have to be square; 3 × 1 or 7 × 3 are acceptable, for example). How would you describe (i) the general effect of the moving average filter and (ii) the specific effects of changing window size?

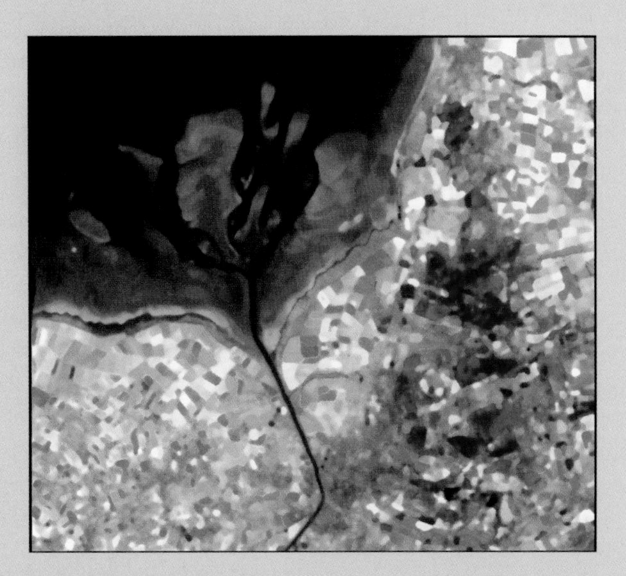

Example 7.1 Figure 3. *The image shown in Example 7.1 Figure 1 after the application of a 7 × 7 moving average filter.*

produce superior results in the sense that they are theoretically capable of removing speckle without significantly degrading the high-frequency component of the SAR image. See the texts by Oliver and Quegan (2004) and Woodhouse (2006) for in-depth discussions of the origins and statistical properties of speckle in SAR imagery.

One of the best-known and simplest speckle suppression filters is the *sigma filter*, proposed by Lee (1983a, 1983b). This filtering method is based on the concept of the Normal distribution. Approximately 95% of the values of observations belonging to a Normal distribution with mean μ and standard deviation σ fall within $\pm 2\sigma$ of the mean value. Lee's method assumes that the grey level values in a single-band SAR image are Normally distributed and, for each overlapping, rectangular window, computes estimates of the local mean $\bar{x}$ and local standard deviation s from the pixels falling within the window. A threshold value is computed from those pixels whose values lie within $\pm 2s$ of the window mean $\bar{x}$. Pixels outside this range are not included in the calculation. The method breaks down when only a few of the pixels in the window have values that are within $\pm 2s$ of the window mean. A parameter k is used to control the procedure. If fewer than k pixels are selected by the threshold (i.e. fewer than k pixels lie in the range $\bar{x} \pm 2s$) then the procedure is aborted, and the filtered pixel value to the left of the current position is used. Alternatively, the average of the four neighbouring pixels replaces the window centre pixel. This modification can cause problems in the first case if the pixel concerned is located on the left margin of the image. In the second case, the filtered values for the pixels to the right of and below the current pixel will need to be calculated before the average can be obtained. Lee *et al.* (2009) present an improved sigma filter for speckle removal. Lopes *et al.* (1990) consider adaptive speckle filters in which the size of the moving window is related to scene homogeneity.

A second widely used speckle filter is the Frost adaptive filter (Frost *et al.*, 1982). This filter, like the sigma filter, uses a moving window the dimensions of which are odd, positive integers. Unlike the moving average filter, the Frost filter coefficients adapt themselves according to the image pixel values that lie below the moving window by using local statistics (i.e. mean and variance of the window pixels). The output of the filter, which replaces the central window pixel value in the image being filtered, is defined as:

$$V = \frac{\sum_1^m G_i W_i}{\sum_1^m W_i}$$

where V is the output from the filter. G represents the m pixel greyscale values underlying the window (indexed from 1 to m by reading across the rows of the moving window so that for a 3×3 image the value of m is 9 and the grey level values are indexed as 1–3 across row 1, 4–6 across row 2 and 7–9 across row 3). The weights are denoted by W and they are arranged in the same order as the elements of G. These weights are computed from $W = \exp(-AT)$ in which A is a constant given by:

$$A = expdamp\left(\frac{var}{mean^2}\right)$$

where *expdamp* is a user-defined exponential damping factor, which takes a default value of 1.0. Take care if you code this formula in a program or a spreadsheet; remember to check whether mean is equal to zero before doing the division. Larger values of *expdamp* preserve edges but diminish the degree of smoothing. Smaller values of *expdamp* result in more smoothing and less well-preserved edges. The values *var* and *mean* are the variance and the mean values of the m pixel values covered by the window. The elements of T are the absolute values of the euclidean distances from the m pixels in the window to the central pixel. The Frost filter is implemented on most image processing systems and readers should experiment using different values of *expdamp* and different window sizes (as is the case with most window-based filters, the degree of smoothing or blurring increases with window size).

Further details of speckle filters are provided by Oliver and Quegan (2004), Lee and Pottier (2009) (this book deals with polarimetric SAR but the underlying principles are the same) Loizou and Pattichis (2008) (the topic of this book is ultrasound, but again the principles are the same; the book includes some MATLAB code). See also Dong, Milne and Forster (2001), Touzi (2002), Lopes, Touzi and Nezry (1990), Lopes *et al.* (1993), Franceschetti and Lanari (1999) and, of course, the original paper by Frost *et al.* (1982). Other references on speckle filtering include Park, Song and Pearlman (1999) who describe an adaptive windowing scheme, while Dong *et al.* (1998) use recursive wavelet transforms and compare the performance of wavelet and spatial filters using quantitative evaluation measures. Rio and Lozano-Garcia (2000) use spatial filtering of Radarsat SAR data as a pre-processing method prior to classification (Chapter 8). Amarsaikhan and Douglas (2004) compare four speckle filters in terms of their ability to preserve textural information (Section 8.7.1). Xiao, Li and Moody (2003) give a good review of speckle filtering, and Chen, Ma and Li-Hui Chen (1999) discuss a modified median filter. Chan, Ho and Nikolova (2005) give details of median filtering for speckle removal. An advanced treatment of algorithms is contained in Arce (2004).

Figure 7.6 shows (a) a multitemporal ERS-1 C-band SAR image of agricultural fields in East Anglia, United Kingdom (the image is distributed with MIPS as the file `\mips\images\east anglia.inf`). The red, green and blue bands are coregistered images collected at different times during the growing season; (b) the result of applying the Frost filter to Figure 7.6a using a filter window size of 5×5 and a damping factor of 1.0 and (c) the difference image after a decorrelation stretch. The filtered image (b) is obviously smoothed and the difference image (c) shows that some edge information has been removed along with the speckle noise.

Other developments in the use of the sigma filter are summarized by Smith (1996), who describes two simple modifications to the standard sigma filter to improve its computational efficiency and preserve fine features. Serkan *et al.* (2008) give details of an adaptive mean filter that preserves edges. Reviews of speckle filtering of SAR images are provided by Desnos and Matteini (1993) and Lee *et al.* (1994). Lee *et al.* (2009) provide details of an updated sigma filter. Wakabayeshi and Arai (1996) discuss an approach to speckle filtering that uses a chi-square test. Martin and Turner (1993) consider a weighted method of SAR speckle suppression, while Alparone *et al.* (1996) present an adaptive filter using local order statistics to achieve the same objective. Order statistics are based on the local grey level histogram, for example the median. More advanced methods of speckle filtering using simulated annealing are described by White (1993). Other references are Beauchemin, Thomson and Edwards (1996), who use a measure of texture (the contrast feature derived from the Grey Level Co-occurrence Matrix, described in Section 8.7.1) as the basis of the filter, and Lopes *et al.* (1993). More recent developments in the suppression of speckle noise are based on the discrete wavelet transform (Sections 6.7 and 9.3.2.2). Xiao, Li and Moody (2003) provide a good review of speckle filtering, while Xie *et al.* (2003), Solbø and Eltoft (2004), Vidal-Pantaleoni and Martí (2004) and Pizurica *et al.* (2001) consider wavelet-based speckle filtering. Other useful reading is: Park, Song and Pearlman (1999), Dong *et al.* (1998), Rio and Lozano-Garcia (2000), Amarsaikhan and Douglas (2004) as well as the specialist texts by Woodhouse (2006) and Oliver and Quegan (2006).

The idea of edge-preserving smoothing, as used in the sigma filter, is also the basis of a filtering method proposed by Nagao and Matsuyama (1979). This method attempts to avoid averaging pixel values that belong to

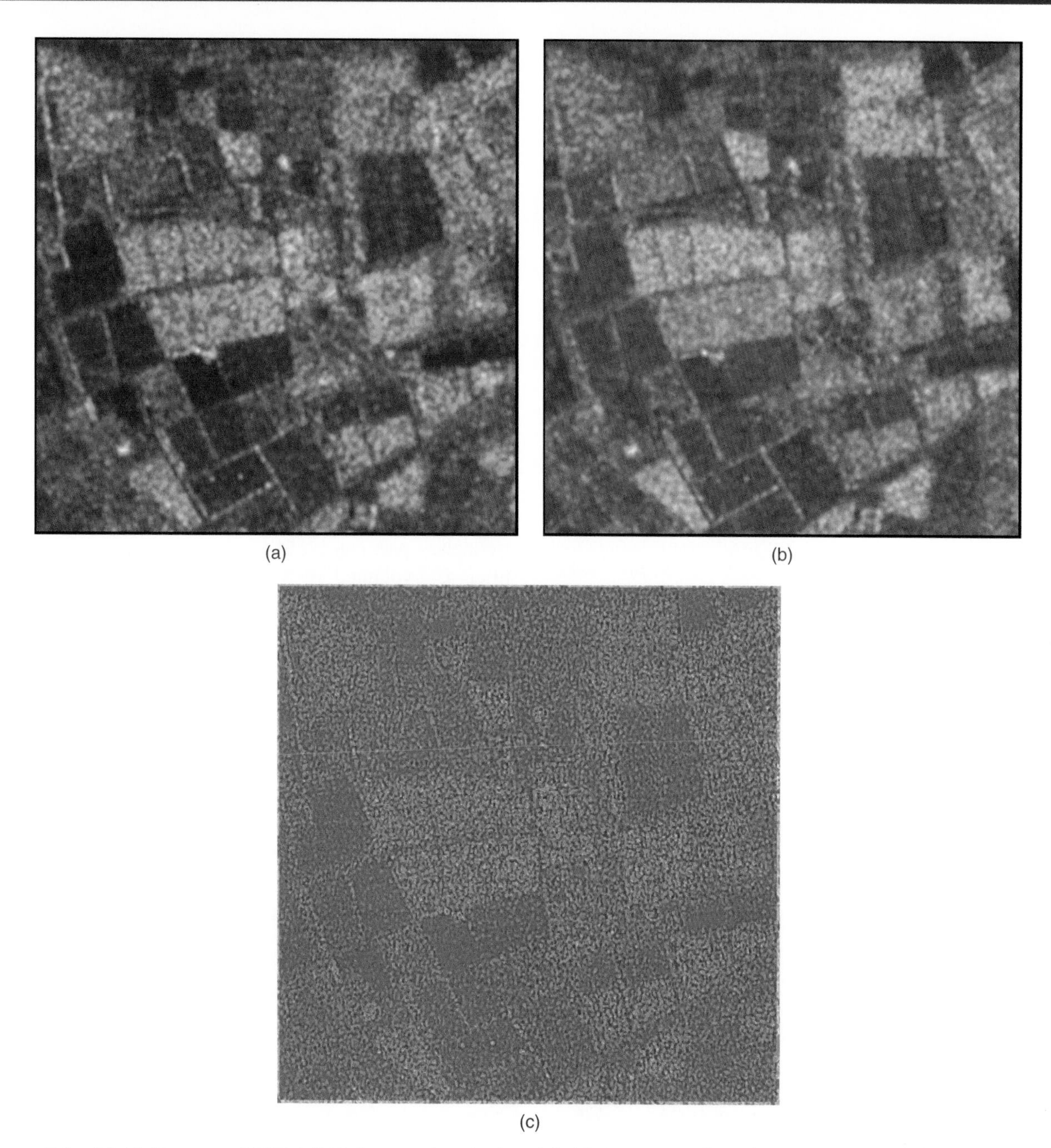

Figure 7.6 *(a) Multitemporal ERS-1 SAR image of agricultural fields in East Anglia. The images were collected at different times during the growing season and have been coregistered (Chapter 4). (b) Image in (a) after the application of a Frost filter using a window size of 5 × 5 and a value of* expdamp *of 1.0. (c) Difference between (a) and (b) after a decorrelation stretch enhancement. It is clear that (c) contains some systematic (edge) information as well as speckle noise. Image (b) is also blurred in comparison with image (a). ERS data courtesy of the European Space Agency.*

different 'regions' that might be present in the image. The boundary between two regions contained within a window area might be expected to be represented by an 'edge' or sharp discontinuity in the grey level values. Hence, Nagao and Matsuyama suggest that a bar be rotated around the centre of the window and the bar at the position with the smallest standard deviation of the pixels' grey scale values be selected as the 'winner', since a small standard deviation indicates the absence of any edges. The centre pixel value is replaced by the average of the pixel values in the winning bar (Figure 7.7). The Nagao–Matsuyama filter is implemented in MIPS (under the **`Filter`** menu item). Figure 7.8a,b show a Landsat TM image of the Gregory Rift Valley in East Africa before and after the application of the Nagao-Matsuyama filter.

0	0	0	0	0
0	1	1	1	0
0	1	1	1	0
0	1	1	1	0
0	0	0	0	0

0	1	1	1	0
0	1	1	1	0
0	0	1	0	0
0	0	0	0	0
0	0	0	0	0

0	0	0	0	0
0	0	0	1	1
0	0	1	1	1
0	0	0	1	1
0	0	0	0	0

0	0	0	0	0
0	0	0	0	0
0	0	1	0	0
0	1	1	1	0
0	1	1	0	0

0	0	0	0	0
1	1	0	0	0
1	1	1	0	0
1	1	0	0	0
0	0	0	0	0

1	1	0	0	0
1	1	1	0	0
0	1	1	0	0
0	0	0	0	0
0	0	0	0	0

0	0	0	1	1
0	0	1	1	1
0	0	1	1	0
0	0	0	0	0
0	0	0	0	0

0	0	0	0	0
0	0	0	0	0
0	0	1	1	0
0	0	1	1	1
0	0	0	1	1

0	0	0	0	0
0	0	0	0	0
0	1	1	0	0
1	1	1	0	0
1	1	0	0	0

Figure 7.7 *Nagao and Matsuyama filters. The 1s and 0s can be interpreted as a logical mask, with 1 meaning 'true' and 0 meaning 'false'. For a given window position, the variance of the pixels in the 'true' positions is calculated. The pixel value transferred to the output image is the mean of the 'true' pixels in the window with the lowest variance.*

7.3 Spatial Domain High-Pass (Sharpening) Filters

The process of imaging or scanning involves blurring, as noted in the discussion of the point spread function (PSF) in Chapter 2. High frequencies are more heavily suppressed than are the low-frequency components of the image. It might therefore seem likely that the visual quality of an image might be improved by selectively increasing the contribution of its high-frequency components. Since the low-pass filters discussed in Section 7.2 involve some form of averaging (or spatial integration) then the use of the 'mathematical opposite' of averaging or integrating, namely the derivative function, might seem to be suited to the process of sharpening or de-blurring an image. However, a simpler way of performing an operation that is equivalent to high-pass filtering is considered before derivative-based methods are discussed.

7.3.1 Image Subtraction Method

According to the model described in Sections 6.6 and 7.1 an image can be considered to be the sum of its low and high frequency components, plus noise. The low-frequency part can be isolated by the use of a low-pass filter as explained in Section 7.2. This low-frequency image can be subtracted from the original, unfiltered, image leaving behind the high frequency component. The resulting image can be added back to the original, thus effectively doubling the high-frequency component.

The addition and subtraction operations must be done with care (Section 6.2). The sum of any 2 pixel values drawn from images each having a dynamic range of 0–255 can range from 0 to 510, so division by 2 is needed to keep the sum within the 0–255 range. The difference between 2 pixel values can range from −255 to

(a)

(b)

Figure 7.8 *Output from the Nagao–Matsuyama filter for a Landsat TM image of part of the Gregory Rift Valley of Kenya. (a) Original image (TM bands 7, 5 and 3 in RGB) and (b) filtered image. Both images were enhanced using a 5–95% contrast stretch. Differences between the two images are small when viewed on a printed page but on-screen viewing shows the removal of some minor detail in (b). Landsat data courtesy NASA/USGS.*

+255; if the result is to be expressed on the range 0–255 then (i) 255 is added to the result and (ii) this value is divided by 2. In a difference image, therefore, 'zero' has a grey scale value of 127. The foregoing assumes that the image pixels contain 8-bit integers; where the radiometric resolution is 11 or 16-bit (or whenever the data are expressed as 32-bit real numbers) procedures to convert to 8-bit representation have to be undertaken (Chapter 3).

7.3.2 Derivative-Based Methods

Other methods of high-pass filtering are based on the mathematical concept of the derivative, as noted earlier. The derivative of a continuous function at a specified point is the rate of change of that function value at that point. For example, the first derivative of position with respect to time (the rate of change of position over time) is velocity, assuming direction is constant. The greater the velocity of an object the more rapidly it changes its position with respect to time. The velocity can be measured at any time after motion commences. The velocity at time t is the first derivative of position with respect to time at time t. If the position of an object were to be graphed against time then the velocity (and hence the first derivative) at time t would be equal to the slope of the curve at the point time $= t$. Hence, the derivative gives a measure of the rate at which the function is increasing or decreasing at a particular point in time or, in terms of the graph, it measures the gradient of the curve.

In the same way that the rate of change of position with time can be represented by velocity, so the rate of change of velocity with time can be found by calculating the first derivative of the function relating velocity and time. The result of such a calculation would be acceleration. Since acceleration is the first derivative of velocity with respect to time and, in turn, velocity is the first derivative of position with respect to time then it follows that acceleration is the second derivative of position with respect to time. It measures the rate at which velocity is changing. When the object is at rest its acceleration is zero. Acceleration is also zero when the object reaches a constant velocity. A graph of acceleration against time would be useful in determining those times when velocity was constant or, conversely, the times when velocity was changing.

In terms of a continuous grey scale image, the analogue of velocity is the rate of change of grey scale value over space. This derivative is measured in two directions – one with respect to x, the other with respect to y. The overall first derivative (with respect to x and y) is the square root of the sum of squares of the two individual first derivatives. The values of these three derivatives (in the x direction, y direction and overall) tell us (i) how rapidly the greyscale value is changing in the x direction, (ii) how rapidly it is changing in the y direction and (iii) the maximum rate of change in any direction, plus the direction of this maximum change. All these values are calculable at any point in the interior of a continuous image. In those areas of the image that are homogeneous, the values taken by all three derivatives (x, y and overall) will be small. Where there is a rapid change in the grey scale values, for example at a coastline in a near-infrared image, the gradient (first derivative) of the image at that point will be high. These lines or edges of sharp change in grey level can be thought of as being represented by the high-frequency component of the image for, as mentioned earlier, the local variation from the overall background pattern is due to high-frequency components (the background pattern is the low-frequency component). The first derivative or gradient of the image therefore identifies the high-frequency portions of the image.

What does the second derivative tell us? Like the first derivative it can be calculated in both the x and y directions and also with respect to x and y together. It identifies areas where the gradient (first derivative) is constant, for the second derivative is zero when the gradient is constant. It could be used, for example to find the top and the bottom of a 'slope' in grey level values.

Images are not continuous functions. They are defined at discrete points in space, and these points are usually taken to be the centres of the pixels. It is therefore not possible to calculate first and second derivatives using the methods of calculus. Instead, derivatives are estimated in terms of differences between the values of adjacent pixels in the x and y directions, though diagonal or corner differences are also used. Figure 7.9 shows the relationship between a discrete, one-dimensional function (such as the values along a scan line of a digital image, as shown in Figure 7.1) and its first and second derivatives estimated by the method of differences. The first differences (equivalent to the first derivatives) are

$$x\, p(i, j) = p(i, j) - p(i - 1, j)$$

$$y\, p(i, j) = p(i, j) - p(i, j - 1)$$

in the x (along-scan) and y (across-scan) directions respectively, while the second difference in the x direction is:

$$\begin{aligned} x_2 p(i, j) &= \Delta x p(i + 1, j) - \Delta x p(i, j) \\ &= [p(i + 1, j) - p(i, j)] \\ &\quad - [p(i, j) - p(i + 1, j)] \\ &= p(i + 1, j) + p(i - 1, j) - 2p(i, j) \end{aligned}$$

Similarly, the second difference in the y direction is:

$$\Delta y_2 p(i, j) = p(i, j + 1) + p(i, j - 1) - 2p(i, j)$$

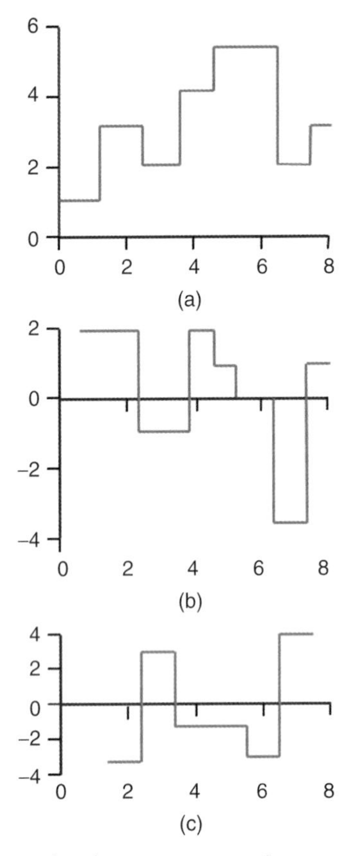

Figure 7.9 Graphical representation of (a) a one-dimensional data series with corresponding (b) first and (c) second-order differences.

The calculation and meaning of the first and second differences in one dimension are illustrated in Table 7.1. A discrete sequence of values, which can be taken as pixel values along a scan line, is shown in the top row of Table 7.1 and the first and second differences are shown in rows 2 and 3. The first difference is zero where the rate of change of greyscale value is zero, positive when 'going up' a slope and negative when going down. The magnitude of the first difference is proportional to the steepness of the 'slope' of the greyscale values, so steep 'slopes' (where grey scale values are increasing or decreasing rapidly) are characterized by first differences that are large in absolute value. The second difference is zero where the first difference values are constant, negative at the foot of a 'slope' and positive at the top of the 'slope'. The extremities of a 'slope' are thus picked out by the second difference.

The computation of the magnitude of the maximum first difference or gradient of a digital image can be carried out by finding Δx and Δy as above and then determining the composite gradient, given by:

$$\Delta xyp(i, j) = \sqrt{\{[\Delta xp(i, j)]^2 + [\Delta yp(i, j)]^2\}}$$

and the direction of this composite gradient is:

$$\theta = tan^{-1}[\Delta yp(i, j) \div \Delta xp(i, j)]$$

Other gradient measures exist. One of the most common is the Roberts Gradient, ΔR. It is computed in the two diagonal directions rather than in the horizontal and vertical directions from:

$$\Delta R = \sqrt{[p(i, j) - p(i+1, j+1)]^2} \sqrt{+[p(i, j+1) - p(i+1, j)]^2}$$

or

$$\Delta R = |p(i, j+1) - p(i+1, j+1)| + |p(i, j+1) - p(i+1, j)|$$

The second form is sometimes preferred for reasons of efficiency as the absolute value (|.|) is more quickly computable then the square root, and raising the inter-pixel difference values to the power of 2 is avoided. The Roberts Gradient function is implemented in the MIPS program, via the **Filter** main menu item. Figure 7.10a shows the histogram of the Landsat ETM+ sub-image of the south-east corner of The Wash in eastern England, after the application of the Roberts Gradient operator. Most of the pixels in the Roberts Gradient image are seen to have values less than 50, so a manual linear contrast stretch was applied, setting the lower and upper limits of the stretch to 0 and 50, respectively. The result is shown in Figure 7.10b, in which the grey level values are proportional to ΔR.

In order to emphasize the high-frequency components of an image a multiple of the gradient values at each pixel location (except those on the first and last rows and columns) can be added back to the original image.

Table 7.1 Relationship between discrete values (f) along a scan line and the first and second differences ($\Delta(f)$, $\Delta_2(f)$)). The first difference (row 2|) indicates the rate of change of the values off shown in row 1. The second difference (row 3) gives the points at which the rate of change itself alters. The first difference is computed from $\Delta(f) = f_i - f_{i-1}$, and the second derivative is found from $\Delta_2(f) = \Delta(\Delta(f)) = f_{i+1} + f_{i-1} - 2f_i$.

f	0	0	0	1	2	3	4	5	5	5	5	4	3	2	1	0
$\Delta(f)$		0	0	1	1	1	1	1	0	0	0	−1	−1	−1	−1	−1
$\Delta_2(f)$		0	1	0	0	0	0	−1	0	0	−1	0	0	0	0	

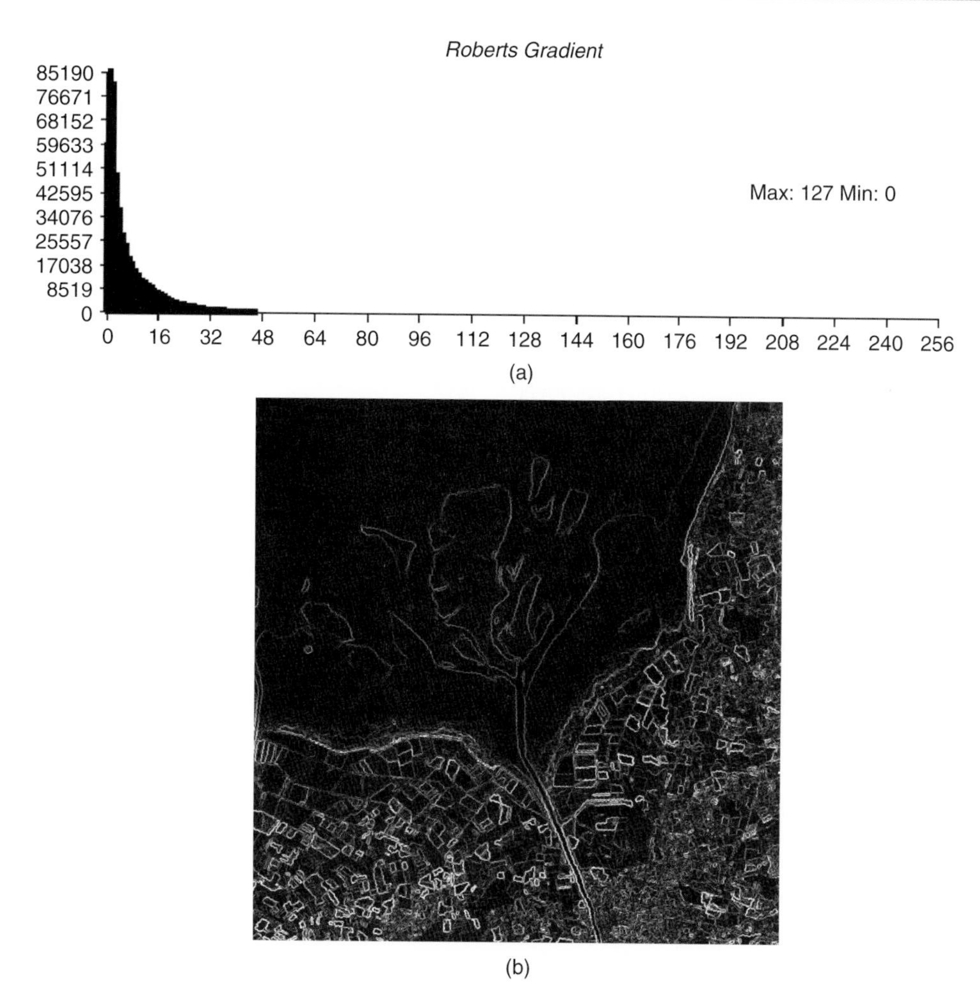

Figure 7.10 *(a) Histogram of the image shown in Figure 7.9b. (b) Landsat ETM+ band 4 image of the south-east corner of The Wash in eastern England, after the application of the Roberts Gradient filter and a linear contrast stretch using specified lower and upper limits of 0 and 50 (see the image histogram, (a)). The field boundaries are brought out very clearly, as is the mouth of the River Ouse. One of the challenges of information extraction from images is to derive vector representations of the objects seen in edge-detecting filters such as the Roberts Gradient.Landsat data courtesy NASA/USGS.*

Normally the absolute values of the gradient are used in this operation. The effect is to emphasize those areas where the greyscale values are changing rapidly. Another possibility is to define a threshold value by inspection of the histogram of gradient values. Where the gradient value at a pixel location exceeds this threshold value the pixel value is set to 255, otherwise the gradient is added back as before. This will over-emphasize the areas of greatest change in grey level.

The second difference function of a digital image is given by

$$\begin{aligned}\Delta xy_2\, p(i,j) &= \Delta x_2 p(i,j) + \Delta y_2 p(i,j)\\ &= [p(i+1,j) + p(i-1,j) + p(i,j+1)\\ &\quad + p(i,j-1)] - 4p(i,j)\end{aligned}$$

In image processing, this function is called the Laplacian operator. Like its one-dimensional analogue shown in Table 7.1 this operator takes on a negative value at the foot of a greyscale 'slope' and a positive value at the crest of a 'slope' (Figure 7.11). The magnitude of the value is proportional to the gradient of the 'slope'. If absolute values are taken, then Laplacian operator will pick out the top and the bottom of 'slopes' in greyscale values. Alternatively, the signed values (negative at the foot, positive at the crest) can be displayed by adding 127 to all values, thus making 127 the 'zero' point on the grey scale. Negative values of the Laplacian will be shown by darker shades of grey, positive values by lighter grey tones. Like the gradient image, the Laplacian image can be added back to the original image though, as noted below, it is more sensible to subtract the Laplacian. The effect is sometimes quite dramatic, though much depends on the 'noisiness' of the image. Any adding-back of high-frequency information to an already noisy image will inevitably result in disappointment. Figure 7.12b shows

Figure 7.11 Section across a greyscale image. The Laplacian operator outputs a positive value at points where the grey level curve reaches a minimum, and negative values at those points where the grey level curve shows a maximum. Subtraction of the output from the Laplacian operator from the grey level curve would over-emphasise both the dark minima at the base of a slope and the bright maximum at the top of the slope.

the result of subtracting the Laplacian image from the original, un-enhanced image of the Painted Desert area of Arizona, USA (Figure 7.12a). The result is much less 'hazy' than the original, and the effects of the linear stretch are considerable.

Rosenfeld and Kak (1982, pp. 241–244) give reasons why this reduction in haziness is be observed. If the discussion of the PSF (Chapter 2) is recalled, it will be realized that the effect of imaging through the atmosphere and the use of lenses in the optical system is to diffuse the radiance emanating from a point source so that the image of a sharp point source appears as a circular blob. Rosenfeld and Kak (1982) show that the Laplacian operator approximates in mathematical terms to the equation known as Fick's Law, which describes the two-dimensional diffusion process. Thus, subtracting the Laplacian from the original image is equivalent to removing the diffused element of the signal from a given pixel. Another possible explanation is that the value recorded at any point contains a contribution from the neighbouring pixels. This is a reasonable hypothesis for the contribution could consist of the effects of diffuse radiance, that is radiance from other pixels that has been scattered into the field of view of the sensor. The Laplacian operator effectively subtracts this contribution.

The weight matrix to be passed across the image to compute the Laplacian is shown in Table 7.2a, while the 'image-minus-Laplacian' operation can be performed directly using the weight matrix shown in Table 7.2b. Other forms of the weight matrix are conceivable; for example diagonal differences rather than vertical and horizontal differences could be used, or the diagonal differences plus the vertical/horizontal differences. A wider neighbourhood could be used, with fractions of the difference being applied. There seems to be little or no reason why such methods should be preferred to the basic model unless the user has some motive based upon the physics of the imaging process.

(a)

(b)

Figure 7.12 Landsat TM image of the Little Colorado River, Painted Desert, Arizona (bands 7, 5 and 3 in RGB). The un-enhanced image is shown in (a). The image shown in (b) has been subjected to the 'image minus Laplacian' operation followed by a 5–95% linear contrast stretch. Landsat data courtesy NASA/USGS.

Table 7.2 (a) Weight matrix for the Laplacian operator. (b) These weights subtract the output from the Laplacian operator from the value of the central pixel in the window.

(a)		
0	1	0
1	−4	1
0	1	0

(b)		
0	−1	0
−1	5	−1
0	−1	0

The basic model of a high-pass image-domain filter involves the subtraction of the pixel values within a window from a multiple of the central pixel. The size of the window is not limited to 2×2 or 3×3 which are used in the derivative-based filters described above. Generally, if the number of pixels in a window is k then the weight given to the central pixel is $(k - 1)$ while all other pixels have a weight of -1. The product of the window weights and the underlying image pixel values is subsequently divided by k. The size of the window is proportional to the wavelengths allowed through the filter. A low-pass filter will remove more of the high-frequency components as the window size increases (i.e. the degree of smoothing is proportional to the window size). A high-pass filter will allow through a broader range of wavebands as the window size increases. Unless precautions are taken, the use of very large window sizes will cause problems at the edge of the image; for instance, if the window size is 101×101 then the furthest left that the central pixel can be located is at row 51, giving a margin of 50 rows that cannot be filtered. For 3×3 filters this margin would be one pixel wide, and it could be filled with zeros. A zero margin 50 pixels wide at the top, bottom, left and right of an image might well be unacceptable. One way around this problem is to ignore those window weights that overlap the image boundary, and compute the filtered value using the weights that fall inside the image area. The value of the central weight will need to be modified according to the number of weights that lie inside the image area. This implies that the bandwidth of the filter will vary from the edge of the image until the point at which all the window weights lie inside the image area.

High-pass filters are used routinely in image processing, especially when high-frequency information is the focus of interest. For instance, Ichoku *et al.* (1996) use the 'image minus Laplacian' filter as part of a methodology to extract drainage-pattern information from satellite imagery. Krishnamurthy, Manalavan and Saivasan (1992) and Nalbant and Alptekin (1995) demonstrate the value of high-frequency enhancement and directional filtering in geological studies. Al-Hinai, Khan and Canaas (1991) use a high-pass filter to enhance images of sand dunes in the Saudi Arabian desert.

7.4 Spatial Domain Edge Detectors

A high-pass filtered image that is added back to the original image is a high-boost filter and the result is a sharpened or de-blurred image. The high-pass filtered image can be used alone, particularly in the study of the location and geographical distribution of 'edges'. An edge is a discontinuity or sharp change in the greyscale value at a particular pixel point and it may have some interpretation in terms of cultural features, such as roads or field boundaries, or in terms of geological structure or relief. We have already noted that the first difference can be computed for the horizontal, vertical and diagonal directions, and the magnitude and direction of the maximum spatial gradient can also be used. Other methods include the subtraction of a low-pass filtered image from the original (Section 7.3.1) or the use of the Roberts Gradient. A method not so far described is the Sobel nonlinear edge operator (Gonzales and Woods, 2007), which is applied to a 3×3 window area. The value of this operator for the 3×3 window defined by:

A	*B*	*C*
D	*E*	*F*
G	*H*	*I*

is given for the pixel underlying the central window weight (E) by the function:

$$S = \sqrt{X^2 + Y^2}$$

where

$$X = (C + 2F + I) - (A + 2D + G)$$

$$Y = (A + 2B + C) - (G + 2H + I)$$

This operation can also be considered in terms of two sets of filter weight matrices. X is given by the following weight matrix, which determines horizontal differences in the neighbourhood of the centre pixel:

−1	0	1
−2	0	2
−1	0	1

while Y is given by a weight matrix which involves vertical differences:

−1	−2	−1
0	0	0
1	2	1

An example of the output from the Sobel filter for a Landsat MSS false colour image of part of the Tanzanian coast is shown in Figure 7.13.

Shaw, Sowers and Sanchez (1982) and Pal and Pal (1993) provide assessment of these techniques of edge-detection, including its role in image segmentation. They conclude that first-differencing methods reveal local rather than regional boundaries, and that increasing the size of a high-pass filter window increases the amount of regional-scale information. The Roberts and Sobel techniques produced a too-intense enhancement of local edges but did not remove the regional patterns. Cheng *et al.* (2001) consider segmentation of colour images.

One of the many uses of edge-detection techniques is in the enhancement of images for the visual identification and analysis of geological lineaments, which are defined as

> mappable, simple or composite linear features whose parts are aligned in a rectilinear or slightly curvilinear relationship and which differ distinctly from the pattern of adjacent features and which presumably reflect a subsurface phenomenon (O'Leary, Friedmann and Pohn, 1976, p. 1467).

The subsurface phenomena to which the definition refers are presumed to be fault and joint patterns in the underlying rock. However, linear features produced from remotely-sensed images using the techniques described in this section should be interpreted with care. For example, the position of what may appear to be lineaments from SAR imagery depends on the SAR's look direction and on the instrument's depression angle. An example of the use of an edge-detection procedure to highlight linear features for geological interpretation is to be found in Moore and Waltz (1983). Sander (2007) reviews the use of remotely-sensed lineaments in groundwater exploration. Tripathi, Gokhale and Siddiqui (2000) use directional morphological transforms to identify lineaments.

Other applications of edge-detection techniques include the determination of the boundaries of homogeneous regions (segmentation) in an image (Quegan and Wright, 1984; Jacquez, Maruca and Fortin, 2002).

(a)

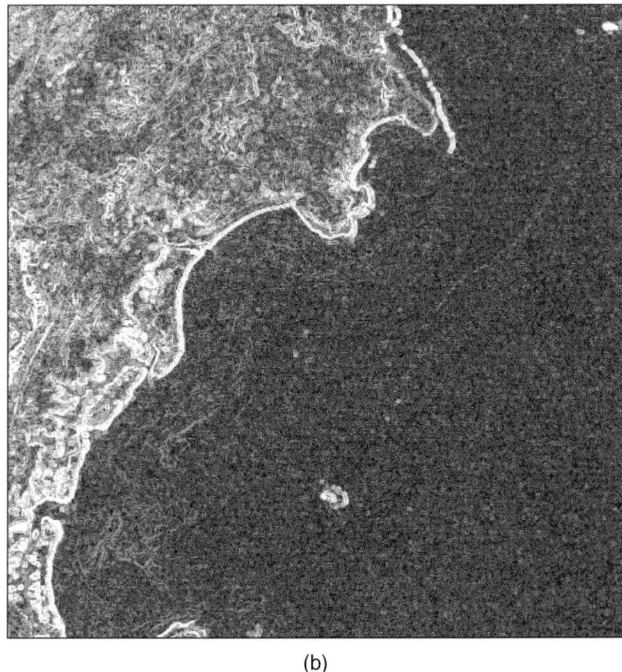

(b)

Figure 7.13 (a) Landsat MSS near-infrared image of the Tanzanian coast south of Dar es Salaam. (b) After the application of the Sobel filter. Both images have been enhanced by a linear contrast stretch. Note how the strength or magnitude of the output from the Sobel filter is related to the degree of contrast between the pixels on either side of the edge in question. Landsat data courtesy NASA/USGS.

A comprehensive review by Brady (1982) considers the topic of image segmentation within the context of image understanding. Pavlidis (1982) is still a useful source. Algorithms for edge detection and region segmentation are discussed by Farag (1992) and Pitas (1993). Reviews of edge detection and linear feature extraction methodologies are provided by Budkewitsch, Newton and Hynes (1994) and Wang (1993). Riazanoff, Cervelle and Chorowicz (1990) describe ways of thinning (*skeletonizing*) lines which have been identified using edge-detection techniques. Such lines are generally defined by firstly applying a high-pass filter, then thresholding the resulting image using edge magnitude or strength to give a binary image. One focus of interest in edge detection is the topic of road detection. Gruen and Li (1995) use a wavelet transform and dynamic programming techniques. Shi and Zhu (2002) consider the problem in terms of high-resolution imagery, while Mena (2003) and Péteri and Ranchin (2007) review the state of the art. The definitive reference is still Marr and Hildreth (1980).

7.5 Frequency Domain Filters

The Fourier transform of a two-dimensional digital image is discussed in Section 6.6. The Fourier transform of an image, as expressed by the amplitude spectrum, is a breakdown of the image into its frequency or scale components. Since the process of digital filtering can be viewed as a technique for separating these components, it might seem logical to consider the use of frequency-domain filters in remote sensing image processing. Such filters operate on the amplitude spectrum of an image and remove, attenuate or amplify the amplitudes in specified wavebands. A simple filter might set the amplitudes of all frequencies less than a selected threshold to zero. If the amplitude spectrum information is converted back to the spatial domain by an inverse Fourier transform, the result is a low-pass filtered image. Any wavelength or waveband can be operated upon in the frequency domain, but three general categories of filter are considered here – low-pass, high-pass and band-pass. The terms low-pass and high-pass are defined in Section 7.1. A band-pass filter removes both the high and low frequency components, but allows an intermediate range of frequencies to pass through the filter, as shown in Figure 7.14. Directional filters can also be developed, because the amplitude spectrum of an image contains information about the frequencies and orientations as well as the amplitudes of the scale components that are present in an image.

The different types of high-, low- and band-pass filters are distinguished on the basis of what are known as their 'transfer functions'. The transfer function is a graph of frequency against filter weight, though the term filter weight should, in this context, be interpreted as 'proportion of input amplitude that is passed by the filter'.

Figure 7.15a shows a cross-section of a transfer function that passes all frequencies up to the value f_1 without alteration. Frequencies higher in value than f_1 are subjected to increasing attenuation until the point f_2. All

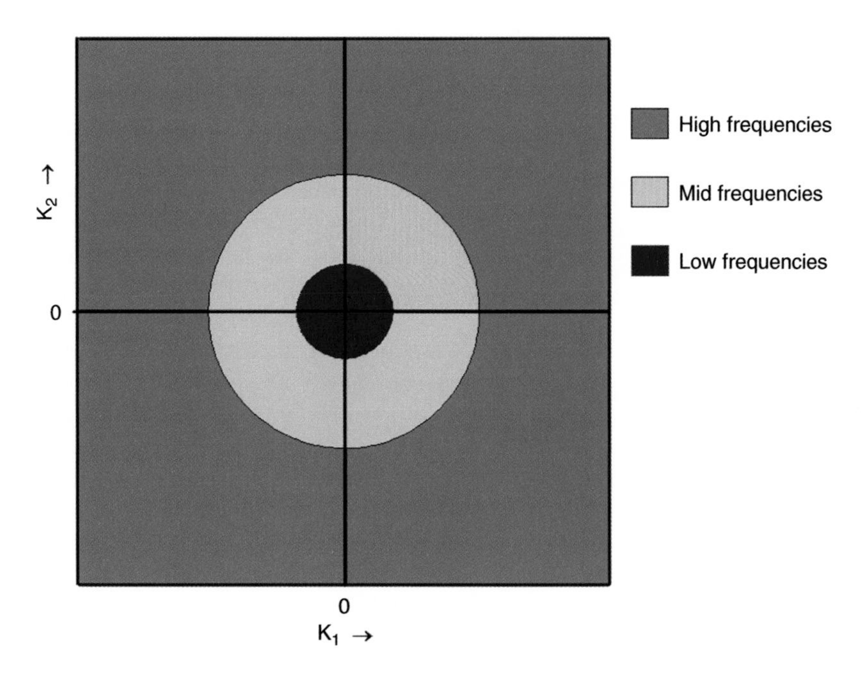

Figure 7.14 *Location of low, mid and high frequency components of the two-dimensional amplitude spectrum.*

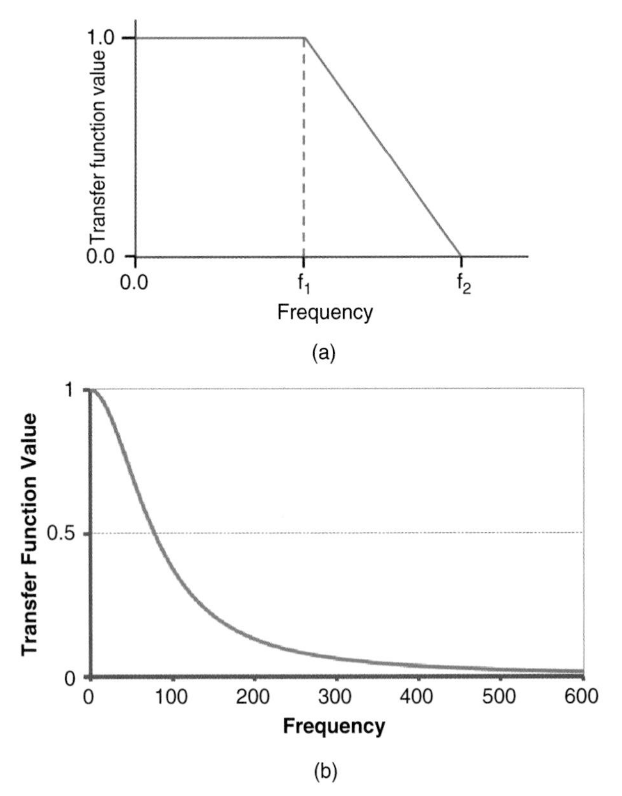

Figure 7.15 *(a) Filter transfer function (one-dimensional slice) that passes unchanged all spatial frequencies lower than f_1, attenuates all frequencies in the range $f_1 - f_2$ and suppresses all frequencies higher than f_2. The degree of attenuation increases linearly in the range $f_1 - f_2$. This filter would leave low frequencies unchanged, and would suppress high frequencies. (b) Transfer function for a low-pass Butterworth filter with cut-off frequency D_0 equal to 50. The shape of the transfer function is smooth, which is an advantage as sharp edges cause 'ringing'.*

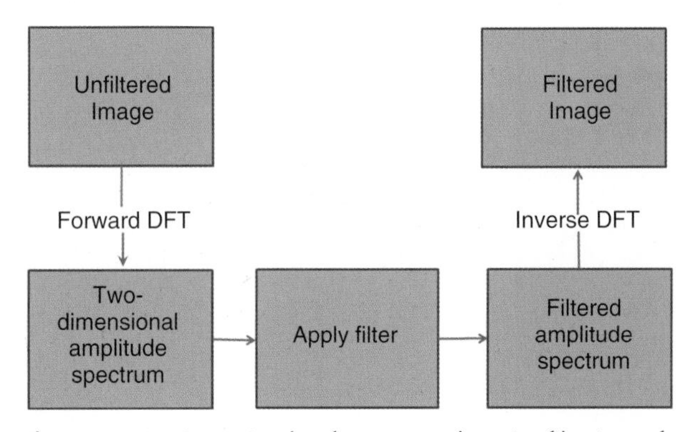

Figure 7.16 *Steps in the frequency domain filtering of a digital image.*

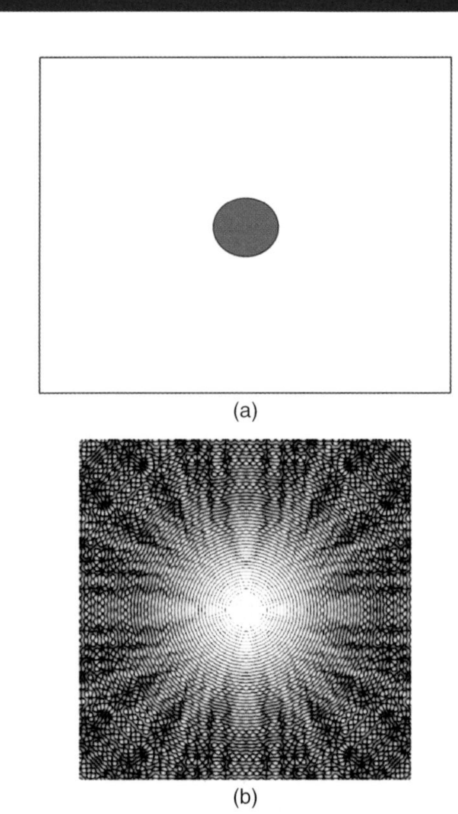

Figure 7.17 *(a) Representation of a frequency-domain low-pass ideal filter. The transfer function of the ideal filter has a sharp edge, in this case at a frequency of 50 Hz. (b) The logarithm of the two-dimensional Fourier amplitude spectrum of (a). Note the concentric circles centred on the origin of the amplitude spectrum. When the inverse transform is applied, these circles are, in effect, superimposed on the forward transform of the image. The result is a pattern of ripples on the transformed image. This phenomenon is called 'ringing'.*

frequencies with values higher than f_2 are removed completely. Figure 7.15b shows the transfer function of a more complex filter, a Butterworth low-pass filter, which is described in more detail below.

Care should be taken in the design of filter transfer functions. As noted earlier, the spatial domain filtered image is derived from the two-dimensional amplitude spectrum image by multiplying the two-dimensional amplitude spectrum by the two-dimensional filter transfer function and then performing an inverse Fourier transform on the result of this calculation (Figure 7.16). Any sharp edges in the filtered amplitude spectrum will convert to a series of concentric circles in the spatial domain, producing a pattern of light and dark rings on the filtered image. This phenomenon is termed *ringing*, for reasons that are evident from an inspection of Figure 7.17a,b. Gonzales and Woods (2007) discuss this aspect of filter design in detail.

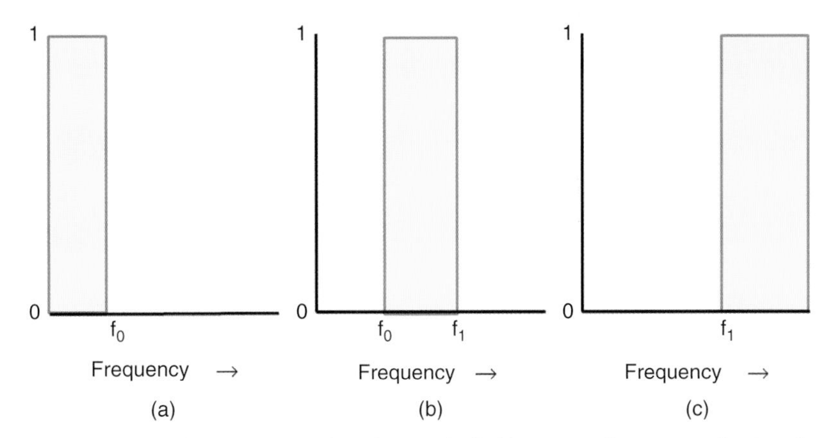

Figure 7.18 *Cross sections of transfer functions for three ideal filters: (a) low-pass, (b) band-pass and (c) high-pass. The amplitude coefficients lying within the shaded area of each filter are unchanged as the transfer function value is 1.0. Those amplitude coefficients lying outside the coloured area are set to zero.*

A cross-section through the transfer function of a *low-pass ideal filter* is shown in Figure 7.18a. The degree of smoothing achieved by the low-pass ideal filter depends on the position of the cut-off frequency, f_0. The lower the value of f_0, the greater the degree of smoothing, as more intermediate and high frequency amplitude coefficients are removed by the filter. The transfer functions for band-pass and high-pass ideal filters are also shown in Figure 7.18. Their implementation in software is not difficult, as the cut-off frequencies form circles of radii f_0 and f_1 around the centre point of the transform (also known as the DC point in the literature of image processing). Figure 7.19a–c illustrate the results of the application of increasingly severe low-pass ideal filters to the TM band 7 image shown in Figure 2.11 using D_0 values of 100, 50 and 5. The degree of smoothing increases as the cut-off frequency decreases. Figure 7.19c shows very little real detail but is, nevertheless, one of the frequency components of the TM image.

Because of their sharp cut-off features, ideal filters tend to produce a filtered image that can be badly affected by the ringing phenomenon, as discussed earlier. Other filter transfer functions have been designed to reduce the impact of ringing by replacing the sharp edge of the ideal

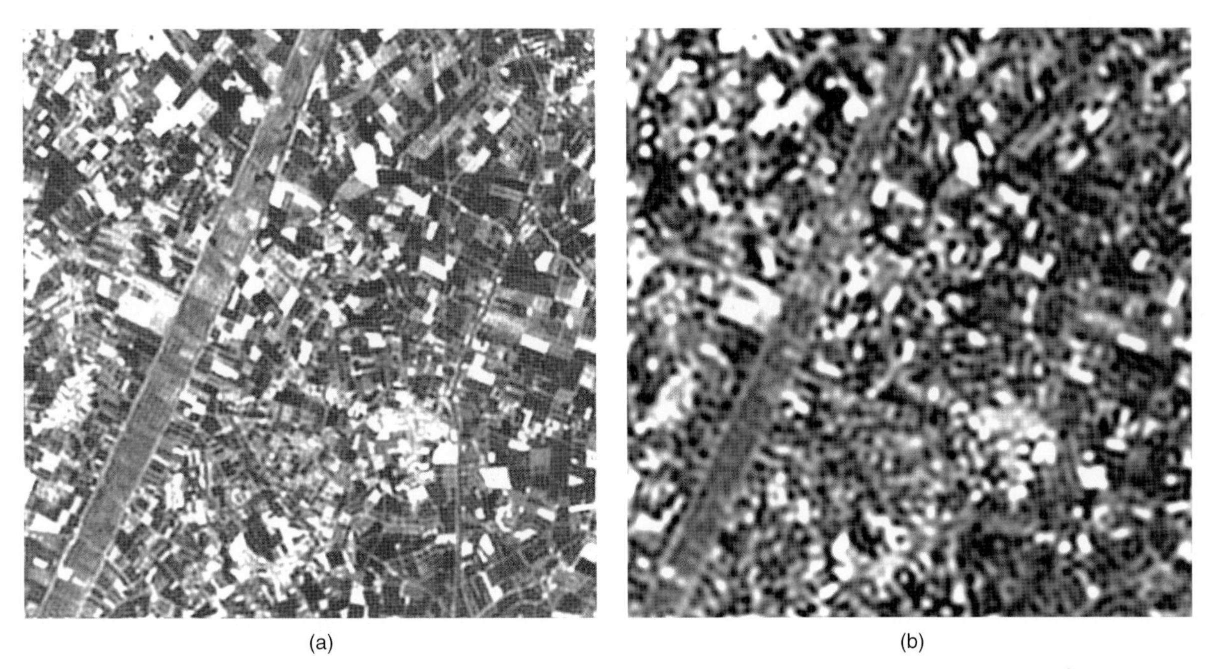

Figure 7.19 *Illustrating the results of the application of increasingly severe low-pass Ideal filters to the Littleport TM band 7 image shown in Figure 1.11b. The filter radii D_0 used in the Ideal filter are (a) 50, (b) 100 and (c) 5. Landsat data courtesy NASA/USGS. Figure 7.19 (c) is located on the next page.*

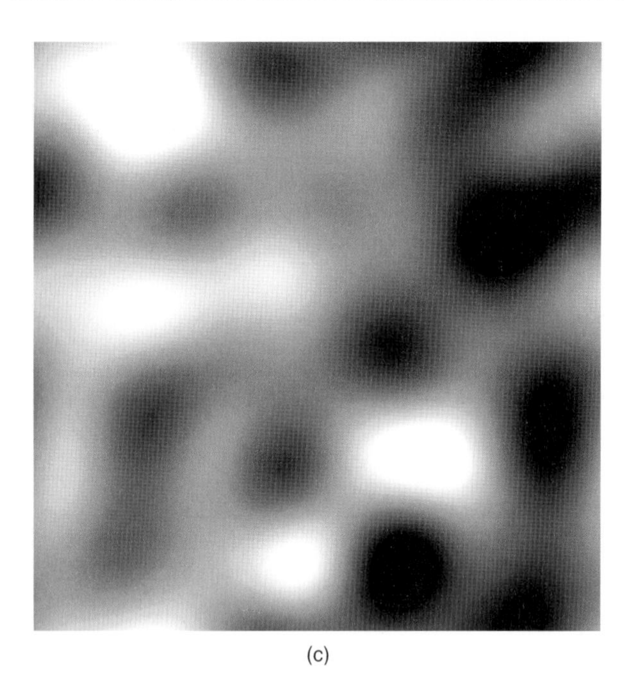

(c)

Figure 7.19 *(continued)*

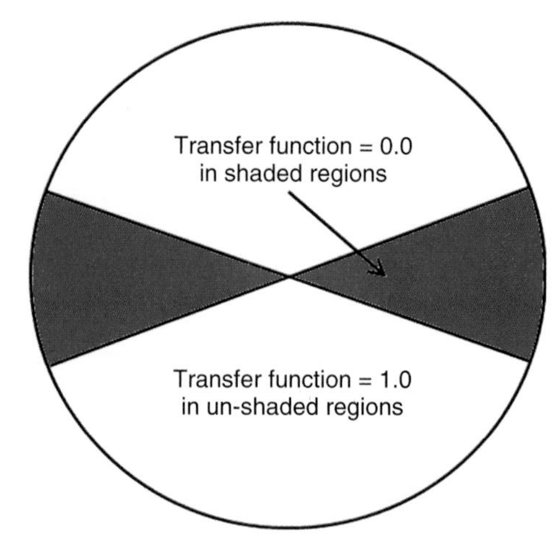

Figure 7.20 *Frequency domain directional filter designed to eliminate all horizontal frequency components. Note that the origin of the (u, v) frequency domain coordinates is the centre of the circle, which has a radius equal to n/2, where n is the dimension of the image (assumed to be square; if the image is rectangular then the circle becomes an ellipse). Thus, for a 512 × 512 image the maximum frequency is 256 Hz.*

filter with a sloping edge or with a function that decays exponentially from the cut-off frequency. An example of this latter type is the Butterworth filter (Figure 7.15b), which is defined by:

$$H(U, V) = 1.0\frac{1.0 + 0.414}{[D(u, v)/D_0]^2}$$

$H(u, v)$ is the value of the filter transfer function for frequencies u and v (remember that the origin of the coordinates u, v is the centre point of the frequency domain representation), $D(u, v)$ is the distance from the origin to the point on the amplitude spectrum with coordinates (u, v) and D_0 is the cut-off frequency, as shown in Figure 7.15b, which is a plot of the value of the transfer function $H(u, v)$ against frequency. This form of the Butterworth filter ensures that $H(u, v) = 0.5$ when $D(u, v)$ equals D_0. Gonzales and Woods (2007) describe other forms of filter transfer function.

Directional filters can be implemented by making use of the fact that the amplitude spectrum contains scale and orientation information (Section 6.6). A filter such as the one illustrated in Figure 7.20 removes all those spatial frequencies corresponding to sinusoidal waves oriented in a east–west direction. Such filters have been used in the filtering of binary images of geological fault lines (McCullagh and Davis, 1972).

High-frequency enhancement is accomplished by firstly defining the region of the amplitude spectrum containing 'high' frequencies and then adding a constant, usually 1.0, to the corresponding amplitudes before carrying out an inverse Fourier transform to convert from the frequency to the spatial domain representation. The transfer function for this operation is shown in Figure 7.21. It is clear that it is simply a variant of the ideal filter approach with the transfer function taking on values of 1 and 2 rather than 0 and 1.

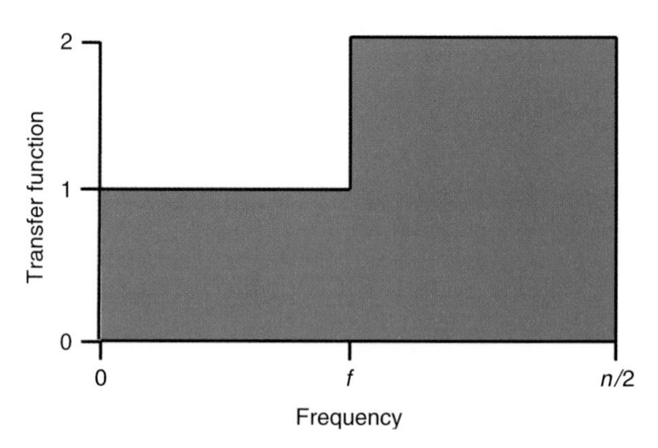

Figure 7.21 *High-frequency boost filter. Spatial frequencies less than f Hz are left unchanged, as the corresponding transfer function value (y-axis) is 1.0. Frequencies higher than f Hz are doubled in magnitude. When the inverse transform is completed (Figure 7.15), the spatial domain representation of the filtered image will show enhanced.*

Example 7.2: Frequency Domain Filtering

The example of the fourier transform in Chapter 6 demonstrated the use of the forward fourier transform to generate an amplitude spectrum. The log of the amplitude spectrum of a Landsat TM image of part of the Red Sea Hills in eastern Sudan is displayed in that example. In this example, a filter is applied to the log of the same amplitude spectrum. Two types of filtering are demonstrated – high-pass and low-pass. Example 7.2 Figure 1 shows the log of the filtered amplitude spectrum of the Sudan image. The black hole in the centre is result of applying a high-pass Butterworth filter with a cut-off frequency of 50 (pixels from centre). The magnitudes of the amplitudes within that circle have been modified using the transfer function described in Section 7.5. Example 7.2 Figure 2 shows the spatial domain image that results from the application of inverse discrete Fourier transform to the filtered amplitude spectrum (Example 7.2 Figure 1). It is clear that much of the tonal information in the original image is low frequency in nature, because the removal of the central disc of the amplitude spectrum (Example 7.2 Figure 1) has eliminated most of the tonal variation, leaving an image that comprises the medium and high frequency components. The high frequency components correspond to sharp edges that may be related to the positions of linear features such as fault lines.

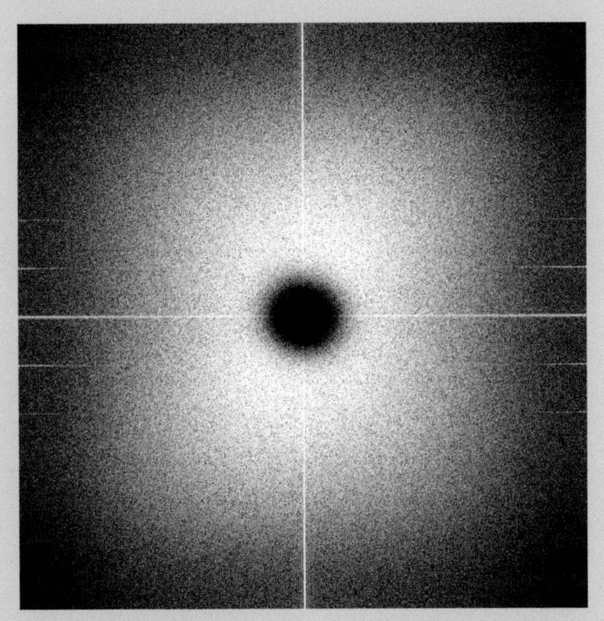

Example 7.2 Figure 1. *Logarithm of the filtered amplitude spectrum of the image shown in Example 6.1 Figure 1. The full amplitude spectrum is shown in Example 6.2 Figure 2. The dark circle in the centre of the filtered amplitude spectrum shows that, in comparison with Example 6.1 Figure 2, the frequency components close to the coordinate centre have been suppressed or attenuated. This figure therefore illustrates a high-pass filter.*

Example 7.2 Figure 3 shows the amplitude spectrum after the application of a Butterworth low-pass filter that suppresses frequencies beyond a cut-off point of 100 (pixels from the centre of the transform). Example 7.2 Figure 4 was reconstructed by applying an inverse Fourier transform to the filtered amplitude spectrum (Example 7.2 Figure 3). Most of the tonal variation is retained, as this varies only slowly across the image from light to dark. However, some of the detail has been removed.

These operations were performed by using the Fourier entry on the MIPS `Filter` menu, and selecting (in sequence): `Forward Transform`, `Filter Butterworth High (or Low) Pass` and `Inverse Transform`. See Section 6.6 for further details of the Fourier transform and Section 7.5 for a discussion of frequency domain filtering.

(Continues on next page)

Example 7.2 Figure 2. *The result of applying the Inverse Discrete Fourier Transform to the filtered amplitude spectrum shown in Example 7.2 Figure 1. The original image is shown in Example 6.1 Figure 1. It is clear that the high-pass filter has removed the background information, leaving behind the high-frequency information (sharp changes in greylevel and edges).*

Example 7.2 Figure 3. *The logarithm of the amplitude spectrum of the image shown in Example 6.2 Figure 1, after the application of a low-pass filter (in this example, a Butterworth low-pass filter with a cut-off frequency of 100 was used).*

Example 7.2 Figure 4. *Image recovered from the amplitude spectrum shown in Example 7.2 Figure 3. This image is effectively the complement of the high-pass filtered image shown in Example 7.2 Figure 2. The detail has been lost (compare Example 6.1 Figure 1) but the overall background pattern of light and dark (together with some major transitions in grey level, which may represent geological faults or fractures).*

Filtering in the frequency domain can be seen to consist of a number of steps, as follows (see also Example 7.2):

1. Perform a forward Fourier transform of the image and compute the amplitude spectrum (Section 6.6).
2. Select an appropriate filter transfer function and multiply the elements of the amplitude spectrum by the appropriate transfer function.
3. Apply an inverse Fourier transform to convert back to spatial domain representation.

Although frequency domain methods are far more flexible that the spatial domain filtering techniques the cost of computing the forward and inverse Fourier transforms has limited its use in the past. As noted in Section 6.6, the two-dimensional Fourier transform requires the transposition of two large matrices holding the intermediate sine and cosine coefficients. This used to be a time-consuming operation when computer memory was limited, but this is no longer the case.

Examples of the use of frequency-domain filtering include de Souza Filho *et al.* (1996), who describe a method to remove noise in JERS-1 imagery. Lei *et al.* (1996) also use frequency-domain filtering methods to clean up MOMS-02 images. Aeromagnetic data is analysed using frequency-domain techniques by Hussein, Rabie and Abdel Nabie (1996). Gonzales and Woods (2007) and Pitas (1993) provide more detailed accounts of frequency domain filtering than the one presented here, though the level of mathematics required to understand their presentations is rather higher than that found here.

7.6 Summary

Filtering of digital images is used to remove, reduce or amplify specific frequency components of an image. The most commonly used filters operate in the spatial domain and can be divided into low-pass or smoothing filters and high-pass or sharpening filters. Uses of smoothing filters include the suppression of noise and other unwanted effects, such as the banding phenomenon, which affects some Landsat ETM+ images. Sharpening filters are used to improve the visual interpretability of the image by, for example de-blurring the signal. Edge and line detection is seen as an extension of the technique of image sharpening. Filtering in the frequency domain is achieved via the application of the principles of the Fourier transform, discussed in Section 6.6. While these methods are inherently more flexible than are spatial domain filters the computational cost of applying them is considerable, and they are often understood less intuitively. Recent developments in computer hardware, especially random access memory and processor speed, mean that frequency-domain methods may become more popular.

8 Classification

What is or is not a cow is for the public to decide.

L. Wittgenstein

8.1 Introduction

This chapter is written with two audiences in mind. The first of these consists of undergraduates following second and third year courses in remote sensing and geographical information system (GIS) who want a gentle, non-mathematical introduction to the ideas behind pattern recognition and classification. They will find that the first few sections can accommodate their needs. The remainder of the chapter is intended for a more advanced audience, including those following Masters courses or researching a topic for a dissertation or a presentation. The fact that the chapter presents a progressive view of the subject should encourage the more advanced reader to 'brush up' on his or her knowledge of the basic geometrical ideas underlying the topic, while at the same time encouraging the less advanced reader to absorb some of the more intricate material that is normally presented at Masters level. Readers requiring a more sophisticated approach should consult contributions to Chen (2007), Landgrebe (2003), Theodoridis and Koutroumbas (2006), Lu and Weng (2007) and Tso and Mather (2009) as well as the *Special Issue on Pattern Recognition* (IEEE Transactions on Geoscience and Remote Sensing, 2007). Wilkinson (2005) provides a useful survey of satellite image classification experiments over the period 1990–2005. Rogan and Chen (2004) present a good general survey of classification methods.

The process of classification consists of two stages. The first is the recognition of categories of real-world objects. In the context of remote sensing of the land surface these categories could include, for example woodlands, water bodies, grassland and other land cover types, depending on the geographical scale and nature of the study. The second stage in the classification process is the labelling of the entities (normally pixels) to be classified. In digital image classification these labels are numerical, so that a pixel that is recognized as belonging to the class 'water' may be given the label '1', 'woodland' may be labelled '2', and so on. The process of image classification requires the user to perform the following steps:

1. Determine *a priori* the number and nature of the categories in terms of which the land cover is to be described.
2. Assign labels to the pixels on the basis of their properties using a decision-making procedure, usually termed a classification rule or a decision rule.

Sometimes these steps are called *classification* and *identification* (or *labelling*), respectively. The classification stage is normally based on a predetermined number of classes that, one hopes, can be observed on the ground at the chosen spatial scale. These are the target or information classes. Clustering, which is described next, produces classes that are more or less spectrally distinct, and these are called spectral classes. They may correspond to information classes, providing the spatial scales match each other. Wemmert *et al.* (2009) describe clustering using data sources (SPOT HRV and QuickBird multispectral) of differing spatial resolutions.

In contrast to the classification procedure, the process of clustering does not require the definition of a set of categories in terms of which the land surface is to be described. Clustering is a kind of exploratory data analysis or data mining procedure, the aim of which is to determine the number (but not initially the identity) of land cover categories that can be separated in the area covered by the image, and to allocate pixels to these categories. Identification of the clusters or categories in terms of the nature of the land cover types is a separate stage that follows the clustering procedure. Several clusters may correspond to a single land-cover type. Methods of relating the results of clustering to real-world categories are described by Lark (1995). Tran, Wehrens and Buydens (2005) present a tutorial on clustering multispectral images within a chemometric context.

These two approaches to pixel labelling are known in the remote sensing literature as *supervised* and *unsupervised* classification procedures, respectively. They can be used to segment an image into regions with similar attributes. Although land cover classification is used above as an example, similar procedures can be applied to clouds, water bodies and other objects present in the image. In all cases, however, the properties of the pixel to be classified are used to label that pixel. In the simplest case, a pixel is characterized by a vector whose elements are its grey levels in each spectral band. This vector represents the spectral properties of that pixel.

A set of grey scale values for a single pixel measured in a number of spectral bands is known as a *pattern*. The spectral bands (such as the seven Landsat ETM+ bands) or other, derived, properties of the pixel (such as context and texture, which are described in later sections of this chapter) that define the pattern are called *features*. The classification process may also include features such as land surface elevation or soil type that are not derived from the image. A pattern is thus a set of measurements on the chosen features for the individual (pixel or object) that is to be classified. The classification process may therefore be considered as a form of pattern recognition, that is, the identification of the pattern associated with each pixel position in an image in terms of the characteristics of the objects or materials that are present at the corresponding point on the Earth's surface.

Pattern recognition methods have found widespread use in fields other than Earth observation by remote sensing; for example military applications include the identification of approaching aircraft and the detection of targets for cruise missiles or speed cameras that read your number plate. Robot or computer vision involves the use of mathematical descriptions of objects 'seen' by a television camera representing the robot eye and the comparison of these mathematical descriptions with patterns describing objects in the real world. In every case, the crucial steps are: (i) selection of a set of features which best describe the pattern at the spatial scale of interest and (ii) choice of a suitable method for the comparison of the pattern describing the object to be classified and the target patterns. In remote sensing applications it is usual to include a third stage, that of assessing the degree of accuracy of the allocation process.

A geometrical model of the classification or pattern recognition process is often helpful in understanding the procedures involved; this topic is dealt with in Section 8.2. The more common methods of unsupervised and supervised classification are covered in Sections 8.3 and 8.4. Supervised methods include those based on statistical concepts and those based on artificial neural networks (ANNs). The methods described in these sections have generally been used on spectral data alone (that is, on the individual vectors of pixel values). This approach is called 'per-point' or 'per-pixel' classification based on spectral data. The addition of features that are derived from the image data has been shown to improve the classification in many cases. Other more recent developments in classification are summarized in Section 8.6. These developments include the use of decision trees (DTs), support vector machines (SVMs) and Independent components analysis (ICA). The use of multiple (hybrid) classifiers is also considered.

Texture is a measure of the homogeneity of the neighbourhood of a pixel, and is widely used in the interpretation of aerial photographs. Objects on such photographs are recognized visually not solely by their greyscale value (tonne) alone but also by the variability of the tonal patterns in the region or *neighbourhood* that surrounds them. Texture is described in Section 8.7.1.

Visual analysis of a photographic image often involves assessment of the context of an object as well as its tone and texture. *Context* is the relationship of an object to other, nearby, objects. Some objects are not expected to occur in a certain context; for example jungles are not observed in Polar regions in today's climatic conditions. Conversely glacier ice is unlikely to be widespread in southern Algeria within the next few years. In the same vein, a pixel labelled 'wheat' may be judged to be incorrectly identified if it is surrounded by pixels labelled 'snow'. The decision regarding the acceptability of the label might be made in terms of the pixel's context rather than on the basis of its spectral reflectance values alone. Contextual methods are not yet in widespread use, though they are the subject of on-going research. They are described in Section 8.8.

The number of spectral bands used by satellite and airborne sensors ranges from the single band of the SPOT HRV in panchromatic to several hundred bands provided by imaging spectrometers (Section 9.3). The methods considered in this chapter are, however, most effective when applied to multispectral image data in which the number of spectral bands is less than 12 or so. The addition of other 'bands' or features such as texture descriptors or external data such as land surface elevation or slope derived from a digital elevation model (DEM) can increase the number of features available for classification. The effect of increasing the number of features on which a classification procedure is based is to increase the computing time requirements but not necessarily the accuracy of the classification. Some form of feature selection process to allow a trade-off between classification accuracy and the number of features is therefore desirable (Section 8.9). The assessment of the accuracy of a thematic map produced from remotely sensed data is considered in Section 8.10.

8.2 Geometrical Basis of Classification

One of the easiest ways to perceive the distribution of values measured on two features is to plot one feature against the other. Figure 8.1 is a plot of catchment area against stream discharge for a hypothetical set of river basins. Visual inspection is sufficient to show that there are two basic types of river basin. The first type has a small catchment area and a low discharge whereas the second type has a large area and a high discharge. This example might appear trivial but it demonstrates two fundamental ideas. The first is the representation of the selected features of the objects of interest (in this case the catchment area and discharge) by the axes of a Euclidean space (termed 'feature space'), and the second is the use of measurements of distance (or, conversely, closeness) in this Euclidean space to measure the resemblance of pairs of points (representing river basins) as the basis of decisions to classify particular river basins as large area/high discharge or small basin/low discharge. The axes of the graph in Figure 8.1 are the x, y axes of a Cartesian coordinate system. They are orthogonal (at right-angles) and define a two-dimensional Euclidean space. Variations in basin area are shown by changes in position along the x-axis and variations in river discharge are shown by position along the y-axis of this space. Thus, the position of a point in this two-dimensional space is directly related to the magnitude of the values of the two features (area and discharge) measured on the particular drainage basin represented by that point.

The eye and brain combine to provide what is sometimes disparagingly called a 'visual' interpretation of a pattern of points such as that depicted in Figure 8.1. If we analyse what the eye/brain combination does when faced with a distribution such as that shown in Figure 8.1 we realize that a 'visual' interpretation is not necessarily a simple one, though it might be intuitive. The presence of two clusters of points is recognized by the existence of two regions of feature space that have a relatively dense distribution of points, with more or less empty regions between them. A point is seen as being in cluster 1 if it is closer to the centre of cluster 1 than it is to the centre of cluster 2. Distance in feature space is being used as a measure of similarity (more correctly 'dissimilarity' as the greater the interpoint distance the less the similarity). Points such as those labelled P and Q in Figure 8.1 are not allocated to either cluster, as their distance from the centres of the two clusters is too great. We can also visually recognize the compactness of a cluster using the degree of scatter of points (representing members of the cluster) around the cluster centre. We can also estimate the degree of separation of the two clusters by looking at the distance between their centres and the scatter of points around those centres. It seems as though a visual estimate of distance (closeness and separation) in a two-dimensional Euclidean space is used to make sense of the distribution of points shown in the diagram. However, we must be careful to note that the scale on which the numerical values are expressed is very important. If the values of the y-coordinates of the points in Figure 8.1 were to be multiplied or divided by a scaling factor, then our visual interpretation of the interpoint relationships would be affected. If we wished to generalize, we could draw a line in the space between the two clusters to represent the boundary between the two kinds of river basin. This line is called a *decision boundary*.

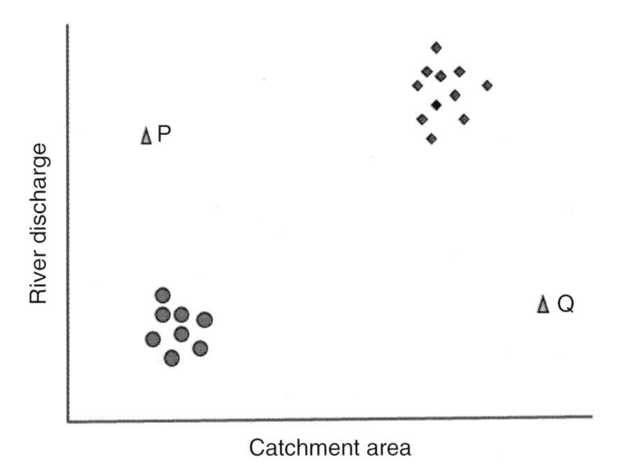

Figure 8.1 *Plot of catchment (watershed) area against river discharge at mouth of catchment. Two distinct groups of river catchments can be seen – small catchments with low river discharge and large catchments with high river discharge. It is difficult to say to which of these groups the catchments represented by points P and Q belong.*

The same concepts – the association of a feature or characteristic of an object with one axis of a Euclidean space and the use of interpoint distance as the basis of a decision rule – can easily be extended to three dimensions. Figure 8.2 shows the same hypothetical set of river basins, but this time they are represented in terms of elevation above sea level as well as area and discharge. Two groupings are evident as before, though it is now clear that the small basins with low discharge are located at higher altitudes than the large basins with high discharge. Again, the distance of each point from the centres of the two clouds can be used as the basis of an allocation or decision rule but, in this three-dimensional case, the decision boundary is a plane rather than a line.

Many people seem to find difficulty in extending the concept of inter-point distance to situations in which the objects of interest have more than three characteristics. There is no need to try to visualize what a four, five or even seven-dimensional version of Figure 8.2 would look like; just consider how straight-line distance is measured in one, two and three dimensional Euclidean spaces in

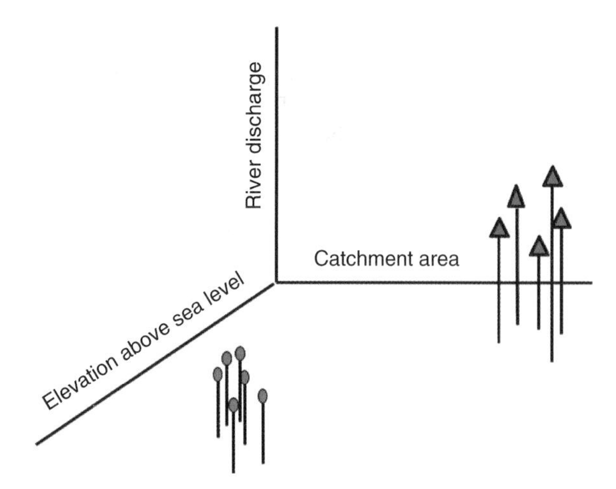

Figure 8.2 *Plot of catchment (watershed) area, river discharge and elevation above sea level for a sample of drainage basins. Two groups of drainage basins are identifiable in this three-dimensional feature space.*

which x, y and z represent the axes:

$$d_{12} = \sqrt{(x_1 - x_2)^2}$$

$$d_{12} = \sqrt{(x_1 - x_2)^2 + (y_1 - y_2)^2}$$

$$d_{12} = \sqrt{(x_1 - x_2)^2 + (y_1 - y_2)^2 + (z_1 - z_2)^2}$$

The squared differences on each axis are added and the square root of the sum is the Euclidean distance from point 1 to point 2. This is a simple application of the theorem of Pythagoras (Figure 8.3). If we replace the terms x_j, y_j and z_j (where j is an index denoting the individual point) by a single term x_{ij}, where i is the axis number and j the identification of the particular point, then the three expressions above can be seen to be particular instances of the general case, in which the distance from point a to point b is:

$$d_{ab} = \sqrt{\sum_{i=1}^{p} (x_{ia} - x_{ib})^2}$$

where d_{ab} is the Euclidean distance between point a and point b measured on p axes or features. There is no reason why p should not be any positive integer value – the algebraic formula will work equally well for $p = 4$ as for $p = 2$ despite the fact that most people cannot visualize the $p > 3$ case. The geometrical model that has been introduced in this section is thus useful for the appreciation of two of the fundamental ideas underlying the procedure of automated classification, but the algebraic equivalent is preferable in real applications because (i) it can be extended to beyond three dimensions and (ii) the algebraic formulae form the basis of computer programs.

It may help to make things clearer if an example relating to a remote sensing application is given at this point.

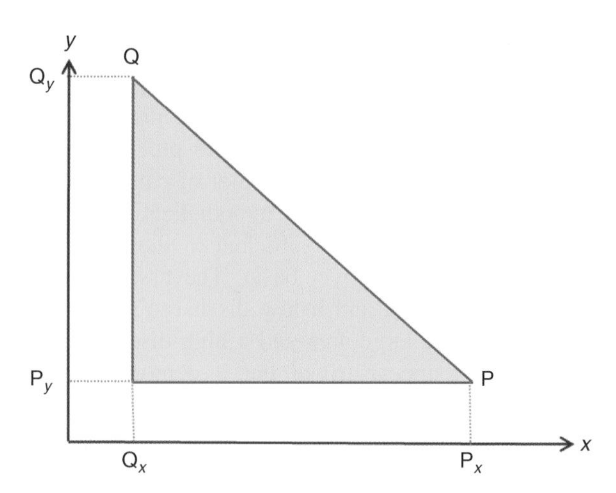

Figure 8.3 *P and Q represent two objects to be compared, such as two trees. We have measurements of height and diameter of the two trees, and these features are represented by axes x and y, respectively. Pythagoras's Theorem is used to calculate the distance PQ in a two-dimensional feature space using the formula* $PQ = \sqrt{(P_x - Q_x)^2 + (P_y - Q_y)^2}$.

The discussion of the spectral response of Earth-surface materials in Section 1.3.2 shows that deep, clear water bodies have a very low reflectance in the near infrared waveband, and their reflectance in the visible red waveband is not much higher. Vigorous vegetation, on the other hand, reflects strongly in the near-infrared waveband whereas its reflectance in the visible red band is relatively low. The red and near infrared wavebands might therefore be selected as the features on which the classification is to be based. Estimates can be made of the pixel grey scale values in each spectral band for sample areas on the image that can be identified *a priori* as 'water', 'cloud top' and 'vigorous vegetation' on the basis of observations made in the field, or from maps or aerial photographs, and these estimates used to fix the mean position of the points representing these three categories in Figure 8.4. The two axes of the figure represent near-infrared and red reflectance, respectively, and the mean position of each type is found by finding the average red reflectance (y coordinate) and near-infrared reflectance (x coordinate) of the sample values for each of the two categories. Fixing the number and position of the large circles in Figure 8.4 represents the first stage in the strategy outlined at the start of this section, namely, the building of a classification.

Step two is the labelling of unknown objects (we could use verbal labels, as we have done up to now, or we could use numerical labels such as '1', '2' and '3'. Remember that these numbers are merely labels. The points labelled a–f in Figure 8.4 represent unclassified pixels. We might choose a decision rule such as 'points will be labelled as members of the class whose centre is closest in feature

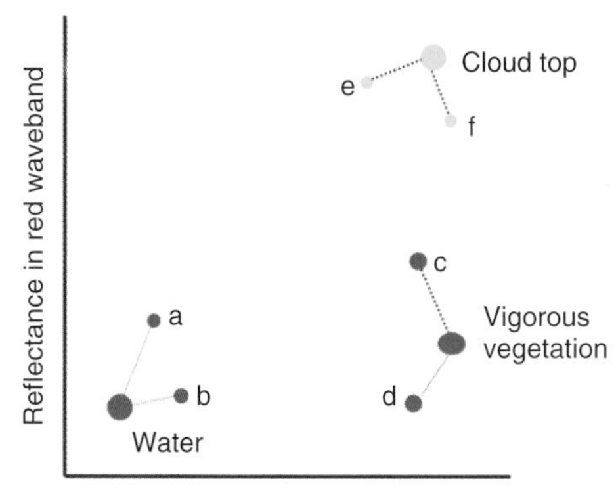

Figure 8.4 *The positions of points representing the average reflectance of vigorous vegetation, water and cloud tops can be estimated from knowledge of their spectral reflectance characteristics (Chapter 1). Points a–f can be allocated to one of these three categories using the criterion of 'minimum distance', that is maximum similarity.*

space to the point concerned'. The distance formula given above could then be used on the points taken one at a time to give the Euclidean straight-line distance from that point (representing a pattern associated with a particular pixel) to each of the centres. Those points that are closer to the mean value for vigorous vegetation are labelled '1' while those closer to the central water point are labelled '2'. Finally, points nearest the cloud top point are labelled '3'. If this procedure is applied to a two-band image, as shown in Figure 8.4, the end product is a matrix of the same dimensions as the image being classified. The elements of this new matrix are numerical pixel labels, which in this example are either 1s, 2s or 3s. If the colour green is associated with the value '1', the colour blue with the value '2' and the colour white with the label '3' then a colour-coded thematic map of the image area would result, in which water would be blue, vigorous vegetation would be green and cloud tops would appear as white, assuming, of course, that the classification procedure was a reliable one. The position of the decision boundaries is given by the set of points that are equidistant from all three class centres.

It will be shown later that the decision rule used in this example ('allocate an unknown pixel to the closest class centroid') is not the only one that can be applied. However, the process of image classification can be understood more clearly if the geometrical basis of the example is clearly understood.

8.3 Unsupervised Classification

It is sometimes the case that insufficient observational or documentary evidence of the nature of the land-cover types is available for the geographical area covered by a remotely-sensed image. In these circumstances, it is not possible to estimate the mean centres of the classes, as described above. Even the number of such classes might be unknown. In this situation we can only 'fish' in the pond of data and hope to come up with a suitable catch. In effect, the automatic classification procedure is left largely, but not entirely, to its own devices – hence the term 'unsupervised clustering'. The relationship between the labels allocated by the classifier to the pixels making up the multispectral image and the land-cover types existing in the area covered by the image is determined after the unsupervised classification has been carried out. Identification of the spectral classes picked out by the unsupervised classifier in terms of information classes existing on the ground is achieved using whatever information is available to the analyst. The term 'exploratory' might be used in preference to 'unsupervised' because a second situation in which this type of analysis might be used can be envisaged. The analyst may well have considerable ground data at his or her disposal but may not be certain (i) whether the spectral classes he or she proposes to use can, in fact, be discriminated given the data available and/or (ii) whether the proposed spectral classes are 'pure' or 'mixed'. As we see in Section 8.4, some methods of supervised classification require that the frequency distribution of points belonging to a single spectral class in the p-dimensional feature space has a single mode or peak. In either case, exploratory or unsupervised methods could be used to provide answers to these questions.

8.3.1 The k-Means Algorithm

An exploratory classification algorithm should require little, if any, user interaction. The workings of such a technique, called the k-means clustering algorithm, are now described by means of an example. Figure 8.5 shows two well-separated groups of points in a two-dimensional feature space. The members of each group are drawn from separate bivariate-normal distributions. It is assumed that we know that there are two groups of points but that we do not know the positions of the centres of the groups in the feature space. Points '1_0' and '2_0' represent a first guess at these positions. The 'shortest distance to centre' decision rule, as described earlier, is used to label each unknown point (represented by a dot in the figure) with a '1' or a '2' depending on the relative Euclidean distance of the point from the initial cluster centres, labelled '1_0' and '2_0'. Thus, the (squared) Euclidean distances to cluster centres 1 and 2 (d^2_{q1} and d^2_{q2}) are computed for each point q, and q is allocated the label '1' if d^2_{q1} is less than d^2_{q2} or the label '2' if d^2_{q2} is less than d^2_{q1}. If the two squared distances are equal, then the point is arbitrarily allocated the label '1'.

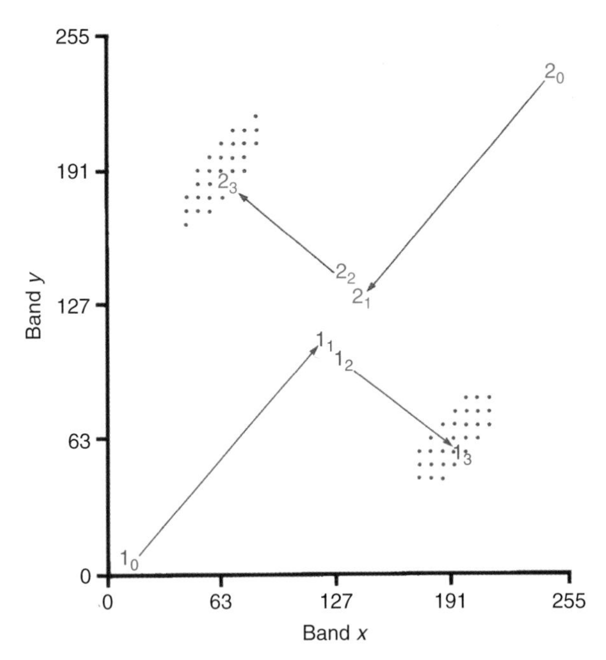

Figure 8.5 *Illustrating the iterative calculation of centroid positions for two well-separated groups of points in a two-dimensional feature space defined by axes labelled Band* x *and Band* y. *Points* 1_0 *and* 2_0 *migrate in three moves from their initial random starting positions to the centres of the two clouds of points that represent the two classers of pixels.*

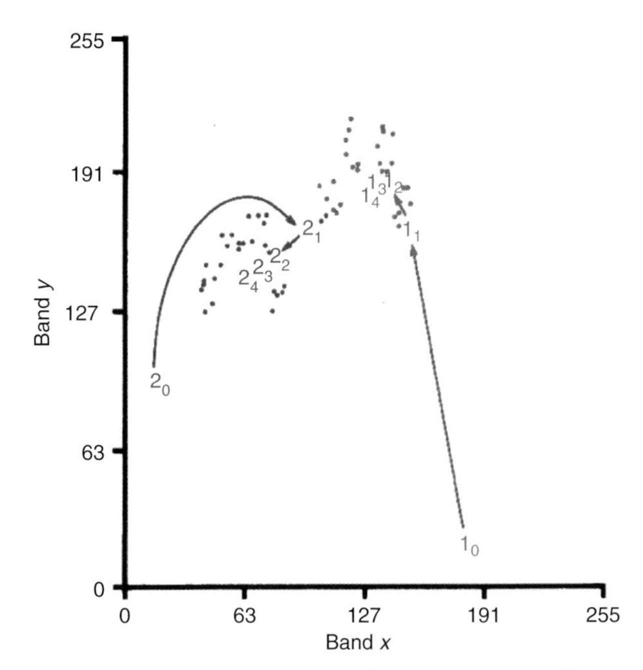

Figure 8.6 *Iterative calculation of centroid positions for two diffuse clusters. See text for discussion.*

At the end of this labelling sequence the mean of the values of all points labelled '1' is computed for each of the axes of the feature space, and the same is done for all points labelled '2' to give the coordinates in the feature space of the centroids of the two groups of points. These new centroids are shown in the diagram as '1_1' and '2_1'. The points are re-labelled again using the shortest-distance-to-mean decision rule, based this time on the new positions of the centroids. Again, the position of the centroid of the points labelled '1' at this second iteration is computed and is shown as '1_2'. The centroid of the set of points labelled '2' is found in a similar fashion and is shown as '2_2'. Distances from all points to these new centres are calculated and another pair of new centroids are found ('1_3' and '2_3'). These centroids are now at the centres of the two groups of points that were artificially generated for this example, and re-labelling of the points does not cause any change in the position of the centroids, hence this position is taken as the final one.

To show that the technique still works even when the two groups of points are not so well separated, as in Figure 8.5, a second pair of groups of points can be generated. This time the coordinates of the points in a two-dimensional feature space are computed by adding random amounts to a preselected pair of centre points to give the distribution shown in Figure 8.6. The start positions of the migrating centroids are selected randomly and are shown on the figure as '1_0' and '2_0' respectively. The same relabelling and recalculation process as that used in the previous example is carried out and the centroids again migrate towards the true centres of the point sets, as shown in Figure 8.6. However, this time the decision boundary is not so clear-cut and there may be some doubt about the class membership (label) of points that are close to the decision boundary.

Since the relabelling procedure involves only the rank orders of the distances between point and centroids, the squared Euclidean distances can be used, for the squares of a set of distance measures have the same rank order as the original distances. Also, it follows from the fact that the squared Euclidean distances are computed algebraically that the feature space can be multidimensional. The procedures in the multidimensional case involve only the additional summations of the squared differences on the feature axes as shown in Section 8.2; no other change is needed. Also note that the user can supply the starting centroid values in the form of mean values for each cluster for each feature. If this starting procedure is used then the method can no longer be described as 'unsupervised'.

8.3.2 ISODATA

In the examples used so far it has been assumed that the number of clusters of points is known in advance. More elaborate schemes are needed if this is not the case. The basic assumption on which these schemes are based is that the clusters present in the data are 'compact' (that is the points associated with each cluster are tightly grouped

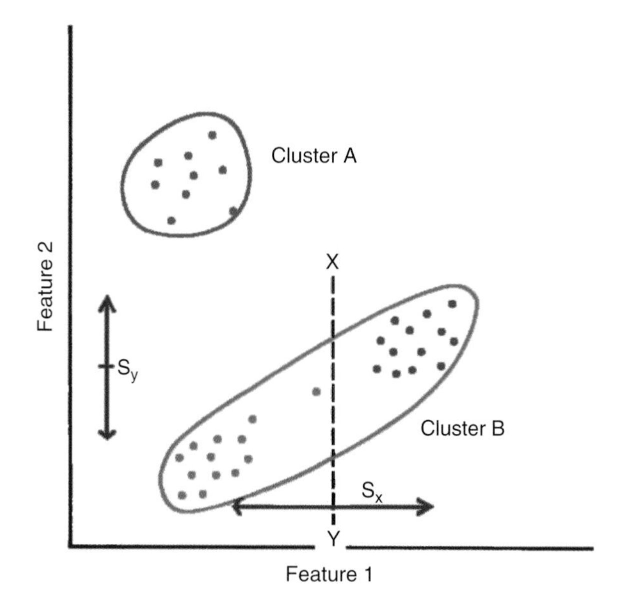

Figure 8.7 *Compact (A) and elongated cluster (B) in a two-dimensional feature space. Cluster B has standard deviations s_x and s_y on features 1 and 2, respectively. Since s_x is larger than a user-specified threshold value, cluster B is split along the line XY.*

around the cluster centre, and thus occupy a spheroidal region of feature space). A measure of the compactness of a cluster can be taken as the set of standard deviations for the cluster measured separately for each feature (Figure 8.7). If any of these feature standard deviations for a particular cluster is larger than a user-specified value then that cluster is considered to be elongated in the direction of the axis representing that feature.

A second assumption is that the clusters are well separated in that their intercentre distances are greater than a preselected threshold. If the feature-space coordinates of a trial number of cluster centres are generated randomly (call the number of centres k_0) then the closest-distance-to-centre decision rule can be used to label the pixels which, as before, are represented by points in feature space. Once the pixels have been labelled then (i) the standard deviation for each feature axis is computed for each of the non-null k_0 clusters and (ii) the Euclidean distances between the k_0 cluster centres are found. Any cluster that has one or more 'large' standard deviations is split in half along a line perpendicular to the feature axis concerned (Figure 8.7) while any clusters that are closer together than a second user-supplied threshold (in terms of their inter-centre distance) are amalgamated. Application of this split-and-merge routine results in a number k_1 of new cluster centre positions, and the pixels are re-labelled with respect to these k_1 centroids. The split and merge function is again applied, and a new number k_2 of centres is found. This process iterates until no clusters are split or merged and no pixels change cluster. This gives the number k_p of clusters and the positions of their centroids in the feature space. At each cycle, any cluster centres that are associated with less than a pre-specified number of pixels are eliminated. The corresponding pixels are either ignored as unclassifiable in subsequent iterations or else are put back into the pool for re-labelling at the next iteration.

This split and merge procedure forms the basis of the ISODATA algorithm (ISODATA is an acronym derived from *I*terative *S*elf-*O*rganizing *D*ata *A*nalysis *T*echniques, with a terminal '*A*' added for aesthetic reasons). Note that some commercial software packages use a basic k-means unsupervised clustering (Section 8.3.1) but call it 'ISODATA'. The ISODATA algorithm can be surprisingly voracious in terms of computer time if the data are not cleanly structured (i.e. do not possess clearly separated and spheroidal clusters). Unless precautions are taken it can easily enter an endless loop when clusters that are split at iteration i are merged again at iteration $i + 1$, then are split at iteration $i + 2$. Little general guidance can be given on the choice of the number of initial cluster centres k_0 or the of the elongation and closeness threshold values to be used. It is often sensible to experiment with small subsets of the image to be classified to get a 'feel' for the data. This algorithm has been in use for many years. A full description is given by Tou and Gonzales (1974) while the standard reference is Duda, Hart and Stork (2000). The account by Bow (2002) includes a flow-chart of the procedure as well as a lengthy example.

The ISODATA procedure as described above is implemented in MIPS. The coordinates of the initial centres in feature space are generated by randomly selecting (x, y) coordinate pairs and choosing to use the vector of pixel values corresponding to the pixel in row y, column x as a starting centre. This strategy helps to reduce the chances of the generation of cluster centres that are located away from any pixel points in feature space. Input includes the 'desired' number of clusters, which is usually fewer than the initial number of clusters, and the maximum number of clusters, together with the splitting and merging threshold distances. The stopping criteria are (i) when the average inter-centre Euclidean distance falls below a user-specified threshold or (ii) when the change in average intercentre Euclidean distance between iteration i and iteration $i - 1$ is less than a user-specified amount. Alternatively, a standard set of defaults values can be selected, though it should be said that these defaults are really no more than guesses. The MIPS implementation of ISODATA outputs the classified image to the screen at each iteration, and the cluster statistics are listed on the lower toolbar, which should be kept visible.

Example 8.1 provides some insights into the workings of the ISODATA procedure as implemented in MIPS. Memarsadeghi *et al.* (2007) present a fast algorithm for

Example 8.1: ISODATA Unsupervised Classification

The aim of this example is to demonstrate the operation of the ISODATA unsupervised classifier. An outline of the workings of this algorithm is provided in the main text. The commands shown here, such as **Classify|Isodata** are specific to the MIPS software. Other packages will have similar commands. However, you should be aware that some versions of ISODATA do not incorporate the split and merge algorithm described in the text. These algorithms are actually implementing the *k-means* method.

We begin by selecting **Classify|Isodata** from the main menu, and then choosing the **INF** file **missis.inf** (provided on the web site download and, if MIPS was properly installed, copied to your hard disk in the **mips/images** folder). Next, specify that you wish to base the classification on bands 1–5 and 7 of this Landsat TM subimage. The subimage size is quite small (512 × 512 pixels) so select the whole of the subimage for analysis. Now we have to decide whether or not to accept the default options for the user-specified parameters (Example 8.1 Table 1). As we do not know yet whether these parameter values are suitable, and given that the subimage is not too big, select **Use Defaults**, just to see what happens. Do not save the starting centres (this option is available for users who wish to restart the procedure using the same starting cluster centres but with different parameter values, as noted below). Instead, select **Generate Randomly** so that the initial cluster centres are selected from the image using random x and y pixel coordinates. The next dialog box asks if you want to save the randomly generated cluster centres. Click **Don't Save**. Note that if you repeat the ISODATAclassification on the same image set using the **Generate Randomly** option, then the final results may well differ, as the final solution depends to a considerable extent on the starting configuration. This is one of the less welcome features of iterative optimization algorithms.

Example 8.1 Table 1 *ISODATA parameters and their effects.*

Parameter number	Parameter description	Default value	Action
1	Starting number of clusters.	20	Affects final number of clusters.
2	Desired number of clusters.	10	Should be half the value of parameter 1.
3	Maximum number of clusters.	50	Stops excessive splitting.
4	Minimum number of pixels in a cluster.	50	Kills off small clusters by declaring them to be 'dead'.
5	Exclusion distance.	200	Any pixels further than the exclusion distance from their nearest centre are declared to be unclassified (label 0). This parameter can be used to encourage more spherical clusters.
6	Closeness criterion.	30	Cluster centres closer than this can be merged. Decrease this value if merging is too voracious.
7	Elongation criterion.	16	Clusters that extend further than this criterion along one axis are split perpendicular to that axis. Increase this value if splitting is excessive.
8	Maximum number of iterations.	35	This is normally sufficient.
9	Maximum number of clusters that can be merged at one time.	2	Use this to increase or decrease the merging tendency.
10	Relative decrease in intercluster–centre distance.	1	Stops ISODATA if the decrease I the value of the intercluster–centre distance becomes less than this.
11	Absolute value of intercluster–centre distance.	5	Stops ISODATA if the value of the intercluster–centre distance becomes less than this.

The Mississippi image now appears on the screen in colour-coded form, with a colour table. Details of the number of clusters and the overall pixel-to-centre distance measure are listed on the lower toolbar. You can continue for a further iteration or quit at this point. If you continue, you will see the colours on the classified image change as clusters are merged or split. Eventually, the change in the intercluster squared Euclidean distance will fall below the threshold, and the iterative process will terminate. Alternatively, the default number of iterations (35) will be executed and the process will again terminate. The result should be similar to the image shown in Example 8.1 Figure 1.

Example 8.1 Figure 1. *Output from the ISODATA procedure applied to the Landsat TM image set referenced by the file* **missis.inf**. *The subimage size is 512 × 512 pixels. See text for elaboration.*

You are then offered the option of performing a hierarchical cluster analysis on the ISODATA results (Example 8.1 Figure 2). A dendrogram is a two-dimensional hierarchical representation of the distance (dissimilarity) relationships among a set of objects, which in this case are the cluster centres. You can cut the dendrogram vertically at any point along the x-axis. A vertical line at the cutting point at, for example a distance of nine units in Example 8.1 Figure 2 represents a four-cluster solution, with ISODATA cluster centres numbered 1, 7, 3, 11, 5, 8 and 12 forming the first composite cluster. The second cluster groups together centres 9 and 14, and the third amalgamates cluster centres 2, 15 and 10. The final grouping consists of ISODATA cluster centres 4, 13 and 6. These relationships are useful in understanding the nature of the ISODATA results, for they show the structure present at different levels of dissimilarity (x-axis). The groupings are also used in the reclassification process, which is described next.

Reclassification is simply the allocation of the same colour code to two or more ISODATA classes. When you select this option you can type the number of the class you have chosen then use the mouse to left-click on a colour in the palette. You can choose the same colour for several classes – for instance, you may decide to colour classes 4, 13 and 6 in blue. These three ISODATA classes have been grouped in the dendrogram, as explained in the preceding paragraph. If you are not satisfied with the result, you can repeat the reclassification exercise.

The final decision to be made in the ISODATA process is how to save the classified image. You can use the **Utilities|Copy to Clipboard|Copy Image** option to place the image on the Windows clipboard, or the **File|Export Image Set** to save the result as a TIFF or bitmap image, or you can use the final ISODATA option, which is to save the image (together with an associated INF file) as a set of labels. These labels are simply the ISODATA class identifiers of the pixels. The same class identifiers are used in the reclassification process. You may wish to save the label image so that you can use a common colour scheme on the results of ISODATA classifications of several images, or the output from ISODATA for a single image set but using different parameters. Note that the bitmap and TIFF representations save the RGB colours associated with each pixel, not the class labels.

Since the ISODATA procedure is started by picking random pixels to act as the initial centres, it is impossible to say what exactly will happen when you run the ISODATA module. The final result should look something like Example 8.1 Figure 1 and the dendogram, showing the relationships between the ISODATA classes, looks like Example 8.1 Figure 2. Remember that the ISODATA classes are identified solely on the basis of spectral

(Continues on next page)

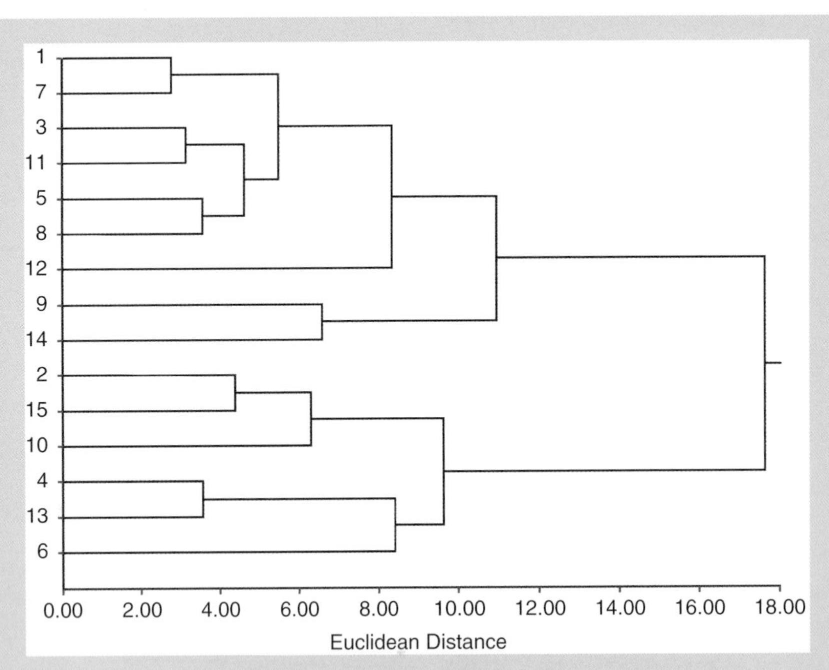

Example 8.1 Figure 2. *Dendrogram showing the hierarchical dissimilarity relationships among the ISODATA cluster centres shown as different colours in Example 8.1 Figure 1. The class labels associated with the cluster centres are shown on the y-axis. The Euclidean distance (x-axis) is a measure of dissimilarity (in that a value of 0.0 indicates perfect similarity). See text for discussion.*

characteristics (modified, perhaps, by topographic effects) and, as yet, they have no 'real-world' interpretation. Using a combination of subject-matter reasoning (i.e. inferring the nature of the spectral classes from geographical location and spectral reflectance properties), and sources of information such as existing maps, it may be possible to assign labels to some of the classes in the classified image. For example, the dark, wide linear feature running from the top right to the bottom left of the image shown in Example 8.1 Figure 1 is probably the Mississippi River.

The dendrogram (Example 8.1 Figure 2), which shows the similarities between the spectral classes at different levels of generalization, can also be used to aid interpretation. Starting at the right-hand side of the dendrogram, we see that at a Euclidean distance of about 18.5 all of the pixels belong to a single class (the Earth's surface). At a dissimilarity level of 18.5, the ISODATA classes split into two groups, one containing six classes and the other consisting of the remaining nine classes. By locating these six and nine classes in the image, we can again use inferential reasoning to label the spectral classes. Moving left or right along the horizontal axis of the dendrogram is equivalent to changing scale.

MIPS outputs a lot of information to the log file, so that you can work out how the clusters are split and merged. The final results for this example are shown in Example 8.1 Table 2. Use the information in column two (number of pixels allocated to this class) to 'weed out' any small, inconsequential, clusters. Next, look at the shape of the spectrum of the mean or centroid of each of the remaining classes. Knowledge of the spectral reflectance characteristics of the major land surface cover types is useful. Finally, the last column gives a measure of the compactness of each cluster. A low value indicates a very homogeneous class, and a high value indicates the reverse. Look at class number four, which is the largest of the classes, with 44 676 of the 262 144 pixels comprising the image. It is very compact (the mean squared distance of 9.54 is calculated by measuring the mean Euclidean distance from each of the 44 676 pixels in the class to the centre point of the class, which is described by the values in columns 3–8. The centroid (mean) value on TM bands 1–5 and 7 shows very low values in the near-infrared region, with moderate values in the visible bands (1–3). Figure 6.4 shows a typical spectral reflectance curve for water, which corresponds quite well to the profile of class number 4. Class 13 is similar to class 5, according to the dendrogram (Example 8.1 Figure 2), and its centroid profile is also typical of water. It may be concluded that classes 4 and 13 represent two different but related water classes, though class 13 contains only about 6 000 pixels.

Other classes are less compact than class 4. Class nine in particular is very diffuse, though it contains more than 23 000 pixels. It is not easy to interpret the centroid values in columns 3–8 in terms of a specific cover

Example 8.1 Table 2 *Summary of the output from the ISODATA unsupervised classification.*

Cluster number	Number of pixels	TM band 1	TM band 2	TM band 3	TM band 4	TM band 5	TM band 7	Mean squared distance
1	15 077	56.2	18.6	20.1	21.3	35.0	15.8	26.66
2	10 883	56.8	19.3	20.5	16.5	18.0	8.1	32.19
3	30 967	55.4	18.0	19.6	24.0	47.0	20.9	13.85
4	44 676	61.8	23.9	27.1	13.9	3.1	1.5	9.54
5	7 261	58.5	20.8	22.6	33.8	45.0	18.4	35.84
6	3 579	70.6	29.6	36.7	22.8	5.1	2.3	56.94
7	13 486	56.6	18.7	20.5	22.4	40.9	18.7	21.27
8	9 547	61.8	22.1	26.0	26.9	43.9	20.4	48.45
9	23 545	66.1	25.4	30.5	36.2	70.5	35.1	172.19
10	3 199	64.2	24.8	29.3	22.4	20.6	9.2	78.39
11	26 836	57.7	19.5	22.1	26.4	52.4	23.8	23.39
12	15 853	59.6	21.9	23.0	43.1	52.4	21.1	58.89
13	6 255	58.1	21.2	21.7	11.2	7.0	3.2	35.18
14	36 024	62.3	22.6	26.6	29.7	58.6	28.8	44.14
15	14 956	55.7	18.3	19.3	19.0	27.3	12.5	26.84

type, but use of the `Reclassify` procedure in MIPS may give some indication of the spatial pattern of class 9, and the dendrogram may also provide some clues.

You can try running the program again, using default parameters. Each time, different pixels are selected to act as starting centres so the result is never the same twice in succession. Sometimes the algorithm spirals out of control and produces a single class. When you feel that you understand how the procedure works, you can experiment with the parameter values. If you do this, then store a set of starting centre coordinates so that you can reuse them each time. By doing this you can eliminate the effects of starting the clustering process from different points and you will therefore isolate the effects of the changes you have made to the default parameters. The main parameters are shown in Example 8.1 Table 1.

ISODATA, and at the same time summarize much of the material in this section.

8.3.3 A Modified k-Means Algorithm

A second, somewhat less complicated, method of estimating the number of separable clusters in a dataset involves a modification of the k-means approach that is outlined in Section 8.3.1 to allow merging of clusters. An overestimate of the expected number of cluster centres ($k_{\max}$) is provided by the user, and pixels are labelled as described in Section 8.3.1, using the closest-distance-to-centre decision rule. Once the pixels have been labelled the centroids of the clusters are calculated and the relabelling procedure is employed to find stable positions for the centroids. Once such a position has been located, a measure of the compactness of each cluster i is found by summing the squared Euclidean distances from the pixels belonging to cluster i to the centroid of cluster i. The square root of this sum divided by the number of points in the cluster gives the root mean squared deviation for that cluster. It is now necessary to find the pair of clusters that can be combined so as (i) to reduce the number of cluster centres by one and (ii) at the same time cause the least increase in the overall root mean square deviation. This is done by computing the quantity P from:

$$P = \frac{n_i n_j}{n_i + n_j} \sum_{k=1}^{p} (y_{ik} - y_{jk})^2 \quad (i = 2, k;\ j = l, i - 1)$$

for every pair of cluster centres (y_i and y_j) where p is the number of dimensions in the feature space and n_i is the number of pixels assigned to cluster i. If clusters $i = r$ and $j = s$ give the lowest value of P then the centroids of clusters r and s are combined by a weighted average procedure, the weights being proportional to the numbers of pixels in clusters r and s. If the number of clusters is still greater than or equal to a user-supplied minimum value

k_{min}, then the re-labelling procedure is then employed to reallocate the pixels to the reduced number of centres and the overall root mean square deviation is computed. The procedure is repeated for every integral value of k (the number of clusters) between k_{max} and k_{min} or until the analyst terminates the procedure after visually inspecting the classified image. As with the ISODATA algorithm, empty clusters can be thrown away at every iteration. Mather (1976) provides a Fortran program to implement this procedure.

The result of an unsupervised classification is a set of labelled pixels, the labels being the numerical identifiers of the classes. The label values run from 1 to the number of classes (k) picked out by the procedure. The class numbered zero (or $k + 1$) can be used as the label for uncategorized pixels. The image made up of the labels of the pixels is displayed by assigning a colour or a grey tone to each label. From a study of the geographical location of the pixels in each class, an attempt is normally made to relate the spectral classes (groups of similar pixels) to corresponding information class (categories of ground cover). Alternatively, a method of hierarchical classification can be used as in Example 8.1 to produce a linkage tree or dendrogram from the centroids of the unsupervised classes, and this linkage tree can be used to determine which spectral classes might best be combined. The relationship between spectral classes and information classes is likely to be tenuous unless external information can be used for, as noted earlier, unsupervised techniques of classification are used when little or no detailed information exists concerning the distribution of ground cover types. An initial unsupervised classification can, however, be used as a preliminary step in refining knowledge of the spectral classes present in the image so that a subsequent supervised classification can be carried out more efficiently. The classes identified by the unsupervised analysis could, for example form the basis for the selection of training samples for use in a supervised technique of classification (Section 8.4). General references covering the material presented above are Bow (2002), Everitt (1993) and Kaufman and Rousseeuw (2005). Example 8.1 gives some practical advice on cluster labelling.

8.4 Supervised Classification

Supervised classification methods are based on external knowledge of the area shown in the image. Unlike some of the unsupervised methods discussed in Section 8.3 supervised methods require input from the user before the chosen algorithm is applied. This input may be derived from fieldwork, air photo analysis, reports or from the study of appropriate maps of the area of interest. Supervised methods are implemented using either statistical or non-statistical algorithms. Statistical algorithms use parameters derived from sample data in the form of training classes, such as the minimum and maximum values on the features, or the mean values of the individual clusters, or the mean and variance-covariance matrices for each of the classes. Non-statistical methods such as ANNs do not rely on statistical information derived from the sample data but are trained on the sample data directly and do not rely on assumptions about the frequency distributions of the image bands. In contrast, statistical methods such as the maximum likelihood (ML) procedure are based on the assumption that the frequency distribution for each class is multivariate normal in form. Thus, statistical methods are said to be *parametric* (because they use estimates of statistical parameters derived from training data) whereas neural methods are *non-parametric*. The importance of this statement lies in the fact that additional non-remotely-sensed data such as slope angle or soil type can more easily be incorporated into a classification using a non-parametric method, because such data are unlikely to follow a multivariate normal frequency distribution. The ML method is described in Section 8.4, as are neural classifiers. Section 8.5 covers mixing models (including ICA) and fuzzy classifiers. Other approaches to classification (Section 8.6) include SVMs, DTs, hybrid classifiers and object-oriented methods. Texture, context and the incorporation of other sources of spatial data are summarized in Sections 8.7 and 8.8. The final two sections focus on feature selection and accuracy assessment. Since all methods of supervised classification use training data samples it is logical to consider the characterization of training data in the next section.

8.4.1 Training Samples

Supervised classification methods require prior knowledge of the number and, in the case of statistical classifiers, certain aspects of the statistical nature of the information classes with which the pixels making up an image are to be identified. The statistical characteristics of the classes that are to be estimated from the training sample pixels depend on which method of supervised classification is used. The simple parallelepiped method requires estimates of the extreme values on each feature for each class, while the k-means or centroid method needs estimates of the multivariate means of the classes. The most elaborate statistical method discussed in this book, the ML algorithm, requires estimates of the mean vector and variance-covariance matrix of each class. Neural classifiers operate directly on the training data, but are strongly influenced by misidentification of training samples as well as by the size of the training datasets. Misidentification of an individual training sample pixel may not have much influence on a statistical classifier, but the impact on a neural classifier could be considerable. The material contained in this section must be interpreted in the light of whichever of these methods is used. It should also be noted that a second, separate set of data – the test dataset – is required in order to assess the accuracy of the classification (Section 8.10). The test data should be gathered using the same rules and guidance as is set out below.

It is of crucial importance to ensure that the *a priori* knowledge of the number and statistical characteristics of the classes is reliable. The accuracy of a supervised classification analysis will depend upon two factors: (i) the representativeness of the estimates of both the number and the statistical nature of the information classes present in the image data and (ii) the degree of departure from the assumptions upon which the classification technique is based. These assumptions vary from one technique to another. These assumptions will be mentioned in the following subsections. In this section we concentrate on the estimation of statistical properties, in particular the mean and variance of each spectral band and the covariances of all pairs of spectral bands. These methods can be used to locate aberrant pixels that can then be eliminated or down-weighted.

The validity of statistical estimates depends upon two factors – the size and the representativeness of the sample. Sample size is not simply a matter of 'the bigger the better' for cost is, or should be, an important consideration. Sample size is related to the number of variables (spectral bands in this case) whose statistical properties are to be estimated, the number of those statistical properties, and the degree of variability present in the class. In the case of a single variable and the estimation of a single property (such as the mean or the variance) a sample size of 30 is usually held to be sufficient. For the multivariate case the size should be at least $30p$ pixels per class where p is the number of features (e.g. spectral bands), and preferably more, though there is evidence that some classifiers – specifically the SVM classifier, described in Section 8.6.1 – work well with small amounts of training data, provided that the support vectors are represented in the training data. However, Su (2009) considers that SVM need a long computer time with large training samples, and proposes a clustering method to reduce training sample size. See Section 8.10 for a more in-depth discussion of the effects of sample size on classification accuracy. Small training samples must be representative, however. If you inspect in Example 8.1 Table 1 you will see that some of these (spectral) classes are very variable and will require a greater number of training samples in order to be properly characterized. Other classes are much more compact so that a smaller number of training samples will be adequate to represent their characteristics.

Training samples are normally located by fieldwork or from air photograph or map interpretation, and their positions on the image found either by visual inspection or by carrying out a geometric correction on the image to be classified. It is not necessary to carry out the procedure of geometric transformation on the full image set to be classified, unless the resulting classified image is to be input to a GIS. All that is required is the set of transform equations that will convert a map coordinate pair to the corresponding image column and row coordinates (Section 4.3). Using these equations, the location on the image of a training sample whose map coordinates are known is a relatively simple matter, provide that the geometric transform is accurate. If geometric correction is required, it is best carried out on a single-band classified image (in which the pixel 'values' are the labels of the classes to which the pixels have been allocated) rather than on the p images to be classified. Not only is this less demanding of computer resources, but it ensures that the radiometric (pixel) values are not distorted by any resampling procedure (Khan, Hayes and Cracknell, 1995). If external data are used in the classification (Section 8.7.2) then geometric correction of image and non-image data to a common reference system is a necessary prerequisite.

The minimum sample size specified in the preceding paragraphs is valid only if the individual members of the training sample are independent, as would be the case if balls were drawn randomly from a bag by an impartial referee. Generally, however, the characteristics of adjacent pixels are not independent – if you were told that pixel a was identified as 'forest' you might be reasonably confident that its neighbour, pixel b, would also be a member of the class 'forest'. If a and b were statistically independent there would be an equal chance that b was a member of any other of the candidate classes, irrespective of the class to which a was allocated. The correlation between nearby points in an image is called *spatial autocorrelation*.

It follows that the number n of pixels in a sample is an over-estimate of the number of fully independent pieces of information in the sample if the pixels making up the training sample are autocorrelated, which may be the case if blocks of pixels are selected rather than scattered, individual pixels. The consequence of autocorrelation is that the use of the standard statistical formulae to estimate the means and variances of the features, and the correlations among the features, will give biased results. Correlations between spectral bands derived from spatially autocorrelated training data will, in fact, be underestimated and the accuracy of the classification will be reduced as a result. Campbell (1981) found that variance–covariance matrices (the unstandardized analogue of the correlation matrix) were considerably greater when computed from randomly selected pixels

Table 8.1 *Variance–covariance matrices for four Landsat MSS bands obtained from random sample (upper figure) and contiguous sample (in parentheses) drawn from same data.*

	MSS 1	MSS 2	MSS 3	MSS 4
MSS 1	1.09 (0.40)	–	–	–
MSS 2	1.21 (0.21)	3.50 (1.01)	–	–
MSS 3	−1.00 (−0.78)	−1.65 (−0.19)	23.15 (14.00)	–
MSS 4	−0.51 (−0.43)	−1.85 (−1.10)	12.73 (9.80)	11.58 (8.92)

Source: Campbell (1981), Table 7.

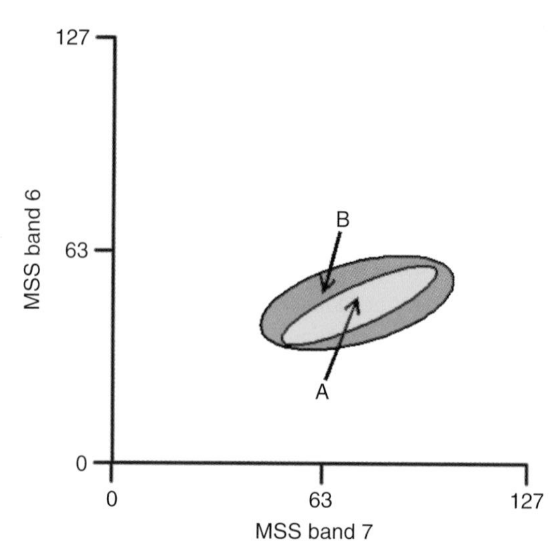

Figure 8.8 *Ellipsoids derived from the variance-covariance matrices for training sets based on contiguous (A) and random (B) samples. Derived from Campbell (1981).*

within a class rather than from contiguous blocks of pixels from the same class (Table 8.1). Figure 8.8 shows the ellipsoids defined by the mean vectors and variance–covariance matrices for Landsat MSS bands 6 and 7 of Campbell's (1981) contiguous and random data (marked A and B respectively). The locations of the centres of the centres of the two ellipses are not too far apart but their orientation, size and shape differ somewhat. Campbell (1981) suggests taking random pixels from within a training area rather than using contiguous blocks, while Labovitz and Matsuoko (1984) prefer a systematic sampling scheme with the spacing between the samples being determined by the degree of positive spatial autocorrelation in the data. Dobbertin and Biging (1996) report that classification accuracy is reduced when images show a high level of spatial autocorrelation. Derived features such as texture might be expected to display a higher degree of spatial autocorrelation than the individual pixel values in the raw images, because such measures are often calculated from overlapping windows. Better results were obtained from randomly selected training pixels than from contiguous blocks of training pixels, a conclusion also reached by Gong, Pu and Chen (1996) and Wilson (1992). The variances of the training samples were also higher when individual random training pixels were used rather than contiguous pixel blocks. The method of automatically collecting training samples, described by Bolstad and Lillesand (1992) may well generate training data that are highly autocorrelated. Stehman, Sohl and Loveland (2005) evaluate various sampling strategies for estimating land cover over large areas. Gallego (2004) is another useful source, while Longley *et al.* (2005) includes a section on spatial sampling. Finally, Plourde and Congalton (2003) consider the impact of sample placement and sampling strategy.

The degree of autocorrelation will depend upon (i) the natural association between adjacent pixels, (ii) the pixel dimensions and (iii) the effects of any data preprocessing. The degree of autocorrelation can be calculated by taking sequences of pixels that are spaced 1, 2, 3, ..., units apart and plotting the correlations between a set of pixels and its first, second, third and subsequent nearest neighbours in the form of a correlogram. A diagram of the kind shown in Figure 8.9 might result, and the autocorrelation distance (in terms of number of pixels) can be read directly from it. As pixel size increases so the autocorrelation distance will diminish. The problem of spatially-autocorrelated samples is considered in more detail in the papers cited above, and in Basu and Odell

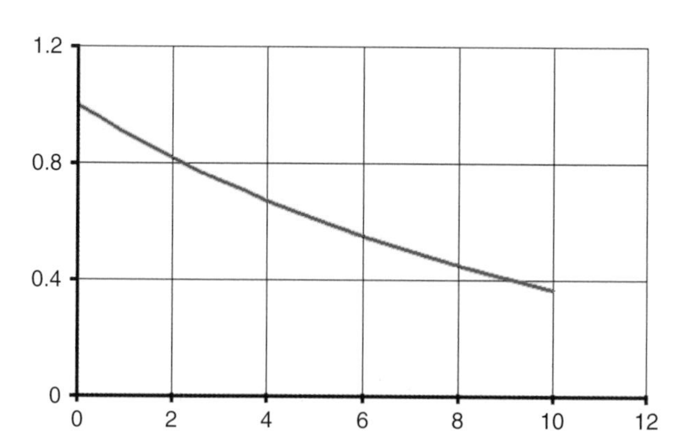

Figure 8.9 *Illustrating the concept of autocorrelation. The diagram shows the correlation between the pixels in an image and their* nth *nearest neighbours in the x direction* ($n = 1, \ldots, 10$). *The correlation at distance (lag) one is computed from two sets of data – the values at pixels numbered* 1, 2, 3, ..., n – *1 along each scan line and the values at the pixels numbered 2, 3, 4, ...,* n *along the same scan lines. The higher the correlation the greater the resemblance between pixels spaced one unit apart on the x-axis. The same procedure is used to derive the 2nd, 3rd, ...,* mth *lag autocorrelation.*

(1974), Craig (1979) and Labovitz, Toll and Kennard (1982). The definitive reference on spatial autocorrelation is still Cliff and Ord (1973). Geostatistical methods are based upon the spatial autocorrelation property, and are discussed briefly in Section 8.5.

Another source of error encountered in the extraction of training samples for use with statistical classifiers such as ML (Section 8.4.2) is the presence in the sample of atypical values. For instance, one or more vectors of pixel measurements in a given training sample may be contaminated in some way; hence, the sample mean and variance–covariance matrix for that class will be in error. Campbell (1980) considers ways in which these atypical values can be detected and proposes estimators of the mean and variance–covariance matrix which are robust (that is they are not unduly influenced by the atypical values). These estimators give full weight to observations that are assumed to come from the main body of the data but reduce the weight given to observations identified as aberrant. A measure called the Mahalanobis distance D is used to identify deviant members of the sample. Its square is defined by:

$$D^2 = (\mathbf{x_m} - \bar{\mathbf{x}})'\mathbf{S}^{-1}(\mathbf{x_m} - \bar{\mathbf{x}})$$

where m is the index counting the elements of the sample, x_m is the mth sample value (pixel vector). The sample mean vector is $\bar{\mathbf{x}}$ and $\mathbf{S}$ is the sample variance–covariance matrix. The transpose of vector $\mathbf{x}$ is written as $\mathbf{x}'$. The Mahalanobis distance, or some function of that distance, can be plotted against the normal probabilities and outlying elements of the sample can be visually identified (Healy, 1968; Sparks, 1985). Robust estimates (that is estimates that are less affected by outliers) of $\bar{\mathbf{x}}$ and $\mathbf{S}$ are computed using weights which are functions of the Mahalanobis distance. The effect is to downgrade pixel values with high Mahalanobis distances (i.e. low weights) that are associated with pixels that are relatively far from (dissimilar to) the mean of the training class taking into account the shape of the probability distribution of training-class members. For uncontaminated data these robust estimates are close to those obtained from the usual estimators. The procedure for obtaining the weights is described and illustrated by Campbell (1980); it is summarized here for completeness.

$$\bar{x}_k = \frac{\sum_{i=1}^{n} w_i x_{ki}}{\sum_{i=1}^{n} w_i} \qquad (k = 1, 2, \ldots, p)$$

$$s_{jk} = \sum_{i=1}^{n} w_i^2 (x_{ji} - \bar{x}_j)(x_{ki} - \bar{x}_k) \qquad (j = 1, 2, \ldots, p)$$

$$(k = j, j+1, \ldots, p)$$

where

n = number of pixels in the training sample,
p = number of features
w_i = weight for pixel i
x_{ki} = value for pixel i on feature k
$\bar{x}_j$ = mean of jth feature for this class
s_{jk} = element j, k of the variance-covariance matrix for this class.

The weights are found from:

$$\mathrm{w}_i = \mathrm{F}(d_i)/\mathrm{d}_i$$

given

$$F(d_i) = \begin{cases} d_i & d_i \le d_0 \\ d_0 \exp[-0.5(d_i - d_0)^2/b_2^2] & \text{otherwise} \end{cases}$$

and d_i Mahalanobis distance of pixel i for this class, $d_0 = \sqrt{p} + b_1/\sqrt{2}$, $b_1 = 2$ and $b_2 = 1.25$.

The weights w_i are initially computed from the Mahalanobis distances which in turn are computed from $\bar{\mathbf{x}}_j$ and $\mathbf{S_j}$ derived from the above formulae but using unit weights. The Mahalanobis distances and the weights are recalculated iteratively until successive weight vectors converge within an acceptable limit, when any aberrant pixel vectors should have been given very low weights, and will therefore contribute only negligibly to the final (robust) estimates of $\bar{\mathbf{x}}_j$ and S_j which are required in the ML classification scheme, which is described later. Kavzoglu (2009) also considers the make-up of training datasets, using visualization techniques.

The reason for going to such apparently great lengths to obtain robust estimates of the mean and variance–covariance matrix for each of the training samples for use in ML classification is that the probabilities of class membership of the individual pixels depend on these estimates. The performance of both statistical and neural classifiers depends to a considerable extent on the reliability and representativeness of the sample. It is easy to use an image-processing system to extract ‘training samples’ from an image, but it is a lot more difficult to ensure that these training samples are not contaminated either by spatial autocorrelation effects or by the inclusion in the training sample of pixels which are not ‘pure’ but ‘mixed’ and therefore atypical of the class which they are supposed to represent. Horne (2003) describes an alternative method of finding more robust estimates of the mean and variance-covariance matrix, whereas Kavzoglu (2009) uses visualization methods to remove outlying pixels from training datasets. This is useful in cases where the training data are being used in conjunction with a statistical classifier but may remove key information as far as SVM are concerned.

The use of unsupervised classification techniques applied to the training classes has already been described as a method of ensuring that the information classes have been well chosen to represent a single spectral class (that is one with a single mode or peak in their frequency distributions). An alternative way to provide a visual check of the distribution of the training sample values is to employ an ordination method. Ordination is the expression of a multivariate dataset in terms of a few dimensions (preferably two) with the minimum loss of information. The Nonlinear Mapping procedure of Sammon (1969) projects the data in a p-dimensional space onto an m-dimensional subspace (m being less than p) whilst minimizing the error introduced into the inter-point Euclidean distances. That is to say, the m-dimensional representation of the distances between data points is the best possible for that number of dimensions in terms of the maintenance of interpoint Euclidean relationships. If $m = 2$ or 3 the results can be presented graphically and the presence of multiple modes or outlying members of the sample can be picked out by eye. Figure 8.10 shows a training sample projected onto two dimensions using the MIPS Nonlinear Mapping module. The training data coordinates were collected using the MIPS **Classify|Collect Training Data** option and the pixel data were extracted (cut) from the log file and pasted into a new data file, which was then edited using Windows Notepad. The 'point and click' facility of the Nonlinear Mapping module allows the identification of extreme points, perhaps representing aberrant pixels.

The performance of a classifier is usually evaluated using measures of classification accuracy (Section 8.10). These accuracy measures use a test set of known data that is collected using the same principles as those described above for the training dataset. One could thus think of training data being used to calibrate the classifier and test data being used for validation, a point that is explored by Muchoney and Strahler (2002).

One often-omitted consideration is that of scale, in terms of the relationship between the number of classes selected, the complexity of the land surface features to be classified and the pixel size of the imagery to be used. The scale of a study is determined by its objectives, so that MODIS or AVHRR data are used for large-scale vegetation inventories (Townshend, DeFries and Zhan, 2002) whereas IKONOS 4 m resolution data would be completely inappropriate. Questions of scale are considered by Dell'Acqua, Gamba and Trianni (2006)

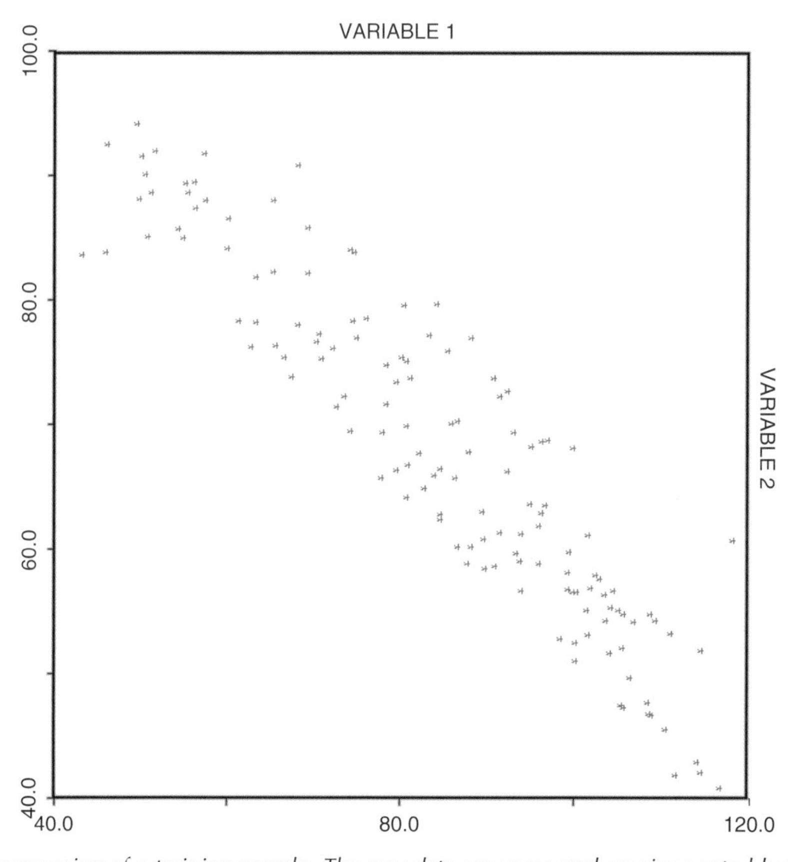

Figure 8.10 *Nonlinear mapping of a training sample. The raw data are measured on six spectral bands. Nonlinear mapping has been used to project the sample data onto two dimensions. The MIPS module* **Plot|NonlinearMapping** *allows the user to 'query' the identity of any aberrant or outlying pixels.*

and Ju, Gopal and Kolaczyk (2005) and are revisited in Section 8.5. Usually there is an inverse relationship between the detail of the classification and the spatial extent of the study area, so one can think of global scales (Meteosat or AVHRR datasets), regional scales (Landsat ETM+ or SPOT HRG data) and local (QuickBird, IKONOS, WorldView and other high resolution imagery). This produces a hierarchical view of a particular set of classes at high resolution melding into a smaller set of more generalized classes at a lower resolution.

Not all supervised classifiers require large training samples. In the preceding discussion it has been implicitly assumed that the training data were to be used in conjunction with a statistical classifier that requires the estimation of parameters such as the mean, variance and covariance matrix. SVMs (Section 8.6.1) attempt to separate classes by maximizing the minimum difference between a training sample from class a and a training sample from class b. There is no need for any other training data, as Foody and Mathur (2004a) point out. Foody and Mathur (2006) present results to show that the SVM classifier can work well with small training datasets collected from areas close to the boundaries of two classes – precisely the ones that would be weeded out using the methods described above for contaminated sample treatment. Foody *et al.* (2006) consider training class size when only one class is of interest, while Foody, McCullagh and Yates (1995) discuss training set size for ANNs. If such networks – which are the topic of Section 8.4.3 – are overtrained to recognize only the specific training samples to which they have been exposed then classification accuracy will be reduced as the network loses its ability to generalize (Kavzoglu and Mather, 2003).

The question of training data requirements is seen to be a complex one with no simple answer. Much depends on the classification algorithm that is used. Statistical methods such as ML need a representative sample that gives an unbiased estimate of the mean and variance–covariance matrix, whereas a SVM will function adequately with a much smaller sample size because it uses only those pixels that lie near class boundaries. For other classification methods a larger, more representative sample is needed. This is one of the key benefits of SVM (though, as we shall see, there are also drawbacks). Kavzoglu (2009) provides some ideas for the improvement of ANN performance by using refined training data. The scale of the study also bears on the question of sample size, as does the nature of the landscape (heterogeneous or uniform). Pal and Mather (2006, p. 2895) summarize the predilection for new classification algorithms by remarking that

> ... greater attention should be given to the collection of training and test data that represent the range of land surface variability at the spatial scale of the image.

Ultimately, it is the distribution of the training data in feature space that determines the positions of the decision boundaries, and it is the spatial scale of the land surface characteristics that are being classified relative to the image scale that affects the accuracy of the classifier.

8.4.2 Statistical Classifiers

Three algorithms are described in this section. All require that the number of categories (classes) be specified in advance, and that certain statistical characteristics of each class are known. The first method is called the parallelepiped or box classifier. A parallelepiped is simply a geometrical shape consisting of a body whose opposite sides are straight and parallel. A parallelogram is a two-dimensional parallelepiped. To define such a body all that is required is an estimate for each class of the values of the lowest and highest pixel values in each band or feature used in the analysis. Pixels are labelled by determining the identifier of the box into which they fall (Figure 8.11, Section 8.4.2.1).

The second method, which is analogous to the k-means unsupervised technique, uses information about the location of each class in the p-dimensional Cartesian space defined by the p bands (features) to be used as the basis of the classification. The location of each class in the p-space is given by the class mean or centroid (Figure 8.12, Section 8.4.2.2).

This third method also uses the mean as a measure of the location of the centre of each class in the p-space and, in addition, makes use of a measure summarizing the disposition or spread of values around the mean along each of the p axes of the feature space. The third method is that of ML (Section 8.4.2.3). All three methods

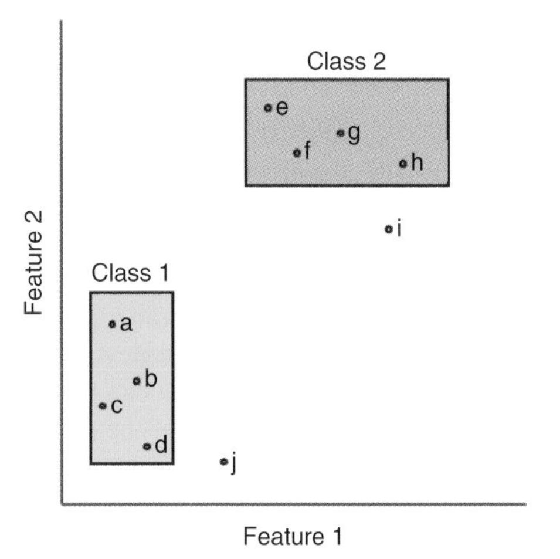

Figure 8.11 *Parallelepiped classifier in two dimensions. Points a, b, c, d lie in the region bounded by parallelepiped 1 and are therefore assigned to class 1. Points e, f, g, h are similarly labelled '2'. Points i and j are unclassified.*

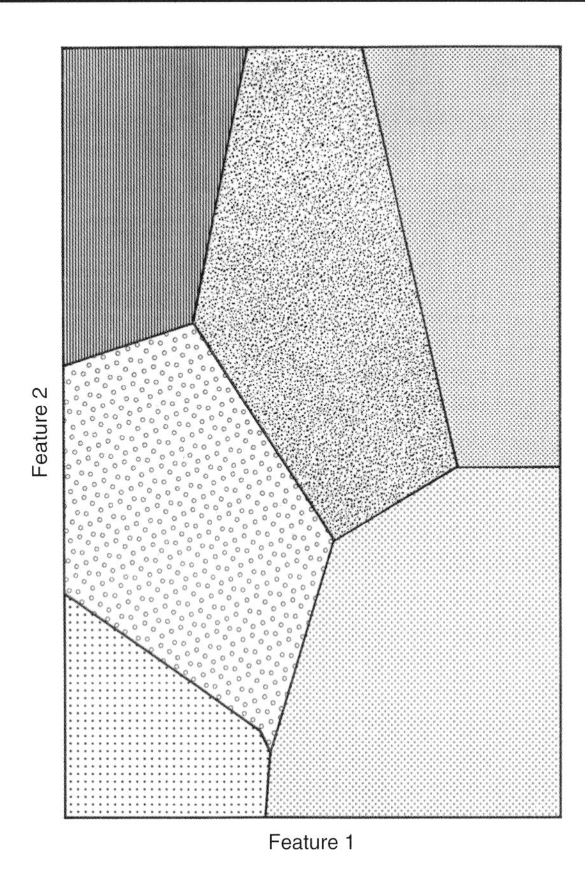

Figure 8.12 *Two-dimensional feature space partitioned according to distance from the centroid of the nearest group. There are six classes.*

require estimates of certain statistical characteristics of the classes to which the pixels are to be allocated. These estimates are derived from samples of pixels, called training samples, which are extracted from the image to be classified (Section 8.4.1).

8.4.2.1 Parallelepiped Classifier

The parallelepiped classifier requires the least information from the user of the statistical supervised classification methods described in this chapter. For each of the k classes specified, the user provides an estimate of the minimum and maximum pixel values on each of the p bands or features. Alternatively, a range, expressed in terms of a given number of standard deviation units on either side of the mean of each feature, can be used. These extreme values allow the estimation of the position of the boundaries of the parallelepipeds, which define regions of the p-dimensional feature space that are identified with particular land cover types (or information classes). Regions of the p-space lying outside the boundaries of the set of parallelepipeds form a *terra incognita* and pixels lying in these regions are usually assigned the label zero. The decision rule employed in the parallelepiped classifier is simple. Each pixel to be classified is taken in turn and its values on the p features are checked to see whether they lie inside any of the parallelepipeds. Two extreme cases might occur. In the first, the point in p-space representing a particular pixel does not lie inside any of the regions defined by the parallelepipeds. Such pixels are of an unknown type. In the second extreme case the point lies inside just one of the parallelepipeds, and the corresponding pixel is therefore labelled as a member of the class represented by that parallelepiped. However, there is the possibility that a point may lie inside two or more overlapping parallelepipeds, and the decision then becomes more complicated. The easiest way around the problem is to allocate the pixel to the first (or some other arbitrarily selected) parallelepiped inside whose boundaries it falls. The order of evaluation of the parallelepipeds then becomes of crucial importance, and there is often no sensible rule that can be employed to determine the best order.

The method can therefore be described as 'quick and dirty'. If the data are well structured (that is, there is no overlap between the classes) then the quick-and-dirty method might generate only a very few conflicts but, unfortunately, many image datasets are not well structured. A more complicated rule for the resolution of conflicts might be to calculate the Euclidean distance between the doubtful pixel and the centre point of each parallelepiped and use a 'minimum distance' rule to decide on the best classification. In effect, a boundary is drawn in the area of overlap between the parallelepipeds concerned. This boundary is equidistant from the centre points of the parallelepipeds, and pixels can be allocated on the basis of their position relative to the boundary line. On the other hand, a combination of the parallelepiped and some other, more powerful, decision rule could be used. If a pixel falls inside one single parallelepiped then it is allocated to the class that is represented by the parallelepiped. If the pixels falls inside two or more parallelepipeds, or is outside all of the parallelepiped areas, then a more sophisticated decision rule could be invoked to resolve the conflict.

Figure 8.11 shows a geometric representation of a simple case illustrating the parallelepiped classifier in action. Points a, b, c and d are allocated to class 1 and points e, f, g and h are allocated to class 2. Points i and j are not identified and are labelled as 'unknown'. The technique is easy to program and is relatively fast in operation. Since, however, the technique makes use only of the minimum and maximum values of each feature for each training set it should be realized that (i) these values may be unrepresentative of the actual spectral classes that they purport to represent and (ii) no information is garnered from those pixels in the training set other than the largest and the smallest in value on each band.

Furthermore it is assumed that the shape of the region in p-space occupied by a particular spectral class can be enclosed by a box. This is not necessarily so. Consequently the parallelepiped method should be considered as a cheap and rapid but not particularly accurate method of associating image pixels with information classes.

8.4.2.2 Centroid (k-Means) Classifier

The centroid or k-means method does make use of all the data in each training class, for it is based upon the 'nearest centre' decision rule that is described in Section 8.3. The centroid (mean centre) of each training class is computed – it is simply the vector comprising the mean of each of the p features used in the analysis, perhaps weighted to diminish the influence of extreme values as discussed in Section 8.4.1. The Euclidean distance from each unknown pixel is then calculated for each centre in turn and the pixel is given the label of the centre to which its Euclidean distance is smallest. In effect, the p-space is divided up into regions by a set of rectilinear boundary lines, each boundary being equidistant from two or more centres (Figure 8.12). Every pixel is classified by this method, for each point in p-space must be closer to one of the k centres than to the rest, excluding the case in which a pixel is equidistant from two or more centres. A modification to the 'closest distance' rule could be adopted to prevent freak or outlying pixel values from being attached to one or other of the classes. This modification could take the form of a distance threshold, which could vary for each class depending upon the expected degree of compactness of that class. Compactness might be estimated from the standard deviation for each feature of the pixels making up the training sample for a given class. Any pixel that is further away from the nearest centre than the threshold distance is left unclassified. This modified rule is actually changing the geometry of the decision boundaries from that shown in Figure 8.12 to that shown in Figure 8.13. In the latter, the p-space is subdivided into k hyperspherical regions each centred on a class mean.

In the same way that the parallelepiped method gets into difficulties with overlapping boxes and has to adopt a nearest-centre rule to break the deadlock, so the k-means method can be adapted to utilize additional information in order to make it intuitively more efficient. The alteration to the decision rule involving a threshold distance is effectively acknowledging that the shape of the region in p-space that is occupied by pixels belonging to a particular class is important. The third classification technique described in this section (the ML method) begins with this assumption and uses a rather more refined method of describing the shapes of the regions in p-space that are occupied by the members of each class.

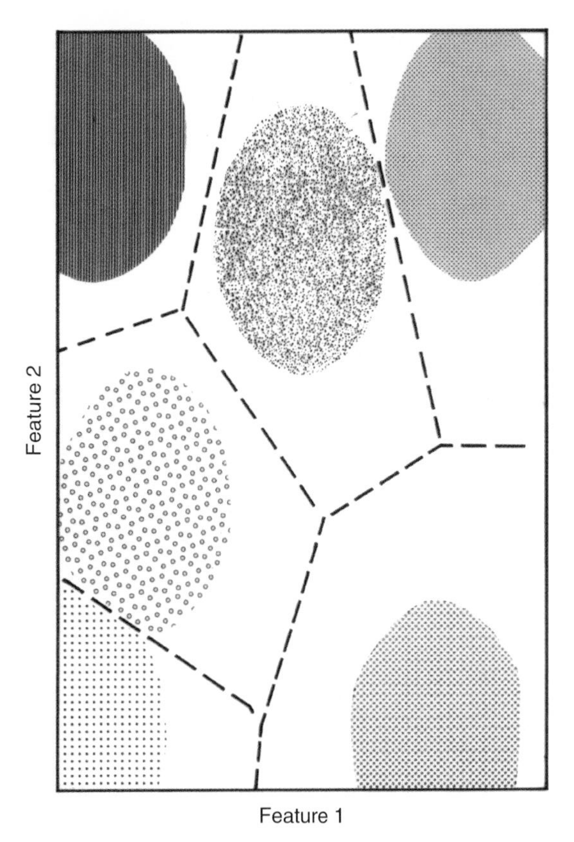

Figure 8.13 Group centroids are the same as in Figure 8.12 but a threshold has been used to limit the extent of each group. Blank areas represent regions of feature space that are not associated with a spectral class. Pixels located in these regions are not labelled.

8.4.2.3 Maximum Likelihood Method

The geometrical shape of a cloud of points representing a set of image pixels belonging to a class or category of interest can often be described by an ellipsoid (see Figure 8.10). This knowledge is used in Chapter 6 in the discussion of the principal components technique. In that chapter it is shown that the orientation and the relative dimensions of the enclosing ellipsoid (strictly speaking, a hyperellipsoid if p is greater than three) depends on the degree of covariance among the p features defining the pattern space.

Examples of two-dimensional ellipses are shown in Figure 8.14. A shape such as that of ellipse A (oriented with the longer axis sloping upwards to the right) implies high positive covariance between the two features. If the longer axis sloped upwards to the left the direction of covariance would be negative. The more circular shape of ellipse B implies lower covariances between the features represented by x and y. The lengths of the major and minor axes of the two ellipses projected onto the x- or y-axes are proportional to the variances of the two variables. The location, shape and size of the ellipse

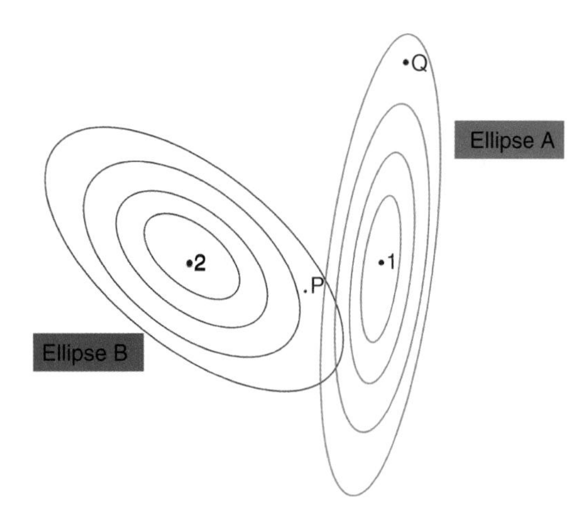

Figure 8.14 *Showing the equiprobability contours for two bivariate-normal distributions with means located at points 1 and 2. Point P is closer to the mean centre of distribution 1 than it is to the centre of distribution 2 yet, because of the shapes of the two ellipses, P is more likely to be a member of class 2 even though point P is closer to the centre of distribution 1 than is point Q.*

therefore reflects the means, variances and covariances of the two features, and the idea can easily be extended to three or more dimensions. The ellipses in Figure 8.14 do not enclose all the points that fall into a particular class; indeed, we could think of a family of concentric ellipses centred on the p-variate mean of a class, such as points '1' and '2' in Figure 8.14. A small ellipse centred on this mean point might enclose only a small percentage of the pixels which are members of the class, and progressively larger ellipses will enclose an increasingly larger proportion of the class members. These concentric ellipses represent contours of probability of membership of the class, with the probability of membership declining away from the mean centre. Thus, membership probability declines more rapidly along the direction of the shorter axis than along the longer axis.

Distance from the centre of the training data is not now the only criterion for deciding whether a point belongs to one class or another, for the shape of the probability contours depends on the relative dimensions of the axes of the ellipse as well as on its orientation. In Figure 8.14 point P is closer than point Q to the centre of class 1 yet, because of the shape of the probability contours, point Q is seen to be more likely to be a member of class 1 while point P is more likely to be a member of class 2.

If equiprobability contours can be defined for all k classes of interest then the probability that a pixel shown by a point in the p-dimensional feature belongs to class i $(i = 1, 2, \ldots, k)$ can be measured for each class in turn, and that pixel assigned to the class for which the probability of membership is highest. The resulting classification might be expected to be more accurate than those produced by either the parallelepiped or the k-means classifiers because the training sample data are being used to provide estimates of the shapes of the distribution of the membership of each class in the p-dimensional feature space as well as of the location of the centre point of each class. The coordinates of the centre point of each class are the mean values on each of the p features, while the shape of the frequency distribution of the class membership is defined by the covariances among the p features for that particular class, as we saw earlier.

It is important to realize that the ML method is based on the assumption that the frequency distribution of the class membership can be approximated by the multivariate normal probability distribution. This might appear to be an undue restriction for, as an eminent statistician once remarked, there is no such thing as a normal distribution. In practice, however, it is generally accepted that the assumption of normality holds reasonably well, and that the procedure described above is not too sensitive to small departures from the assumption provided that the actual frequency distribution of each class is unimodal (i.e. has one peak frequency). A clustering procedure (unsupervised classification) could be used to check the training sample data for each class to see if that class is multimodal, for the clustering method is really a technique for finding multiple modes.

The probability P($\mathbf{x}$) that a pixel vector $\mathbf{x}$ of p elements (a pattern defined in terms of p features) is a member of class k is given by the multivariate normal density:

$$P(\mathbf{x}) = 2\pi^{-0.5p}\,|\mathbf{S}_i|^{-0.5}\exp[-0.5(\mathbf{y}'\mathbf{S}_i^{-1}\mathbf{y})]$$

where |.| denotes the determinant of the specified matrix, $\mathbf{S}_i$ is the sample variance-covariance matrix for class i, $\mathbf{y} = (\mathbf{x} - \bar{\mathbf{x}}_j)$ and $\bar{\mathbf{x}}_i$ is the multivariate mean of class i. Note that the term $\mathbf{y}'\mathbf{S}^{-1}\mathbf{y}$ is the Mahalanobis distance, used in Section 8.4.1 to measure the distance of an observation from the class mean, corrected for the variance and covariance of class i.

Understanding of the relationship between equiprobability ellipses, the algebraic formula for class probability, and the placing of decision boundaries in feature space will be enhanced by a simple example. Figure 8.15 shows the bivariate frequency distribution of two samples, drawn respectively from (i) bands 1 and 2 and (ii) bands 5 and 7 of the TM images shown in Figures 1.10 and 1.11. The contours delimiting the equiprobability ellipses are projected onto the base of each of the diagrams. It is clear that the area of the two-dimensional feature space that is occupied by the band 5–7 combination (Figure 8.15b) is greater than that occupied by the band 1–2 combination. The orientation of the probability ellipses is similar, and the two ellipses are located at approximately the same point in the feature space. These observations can be related to the elements of the

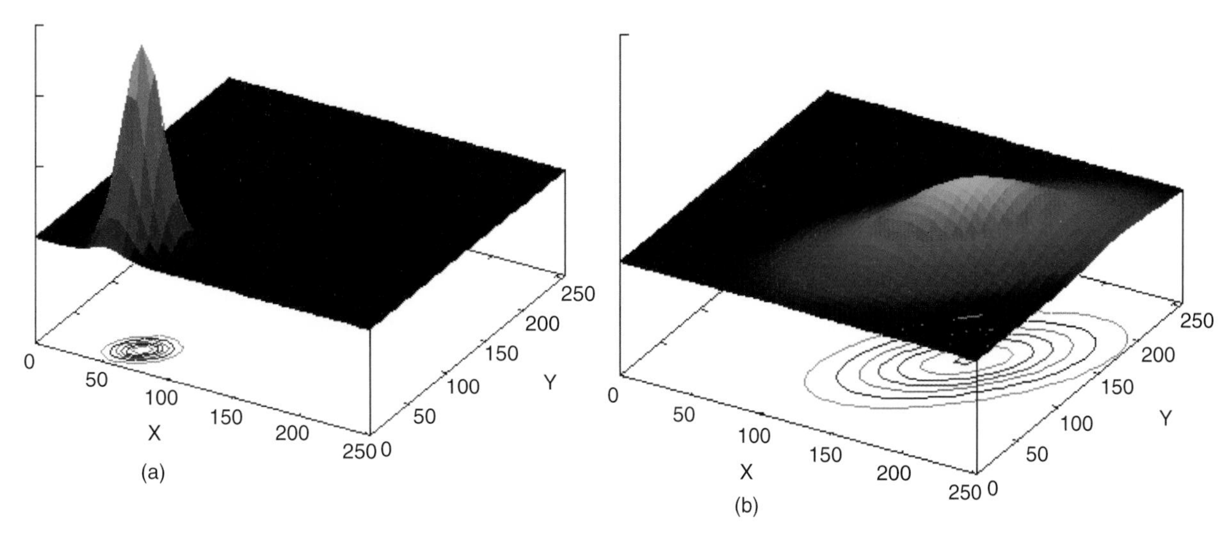

Figure 8.15 *Bivariate (xy) normal surface plots: (a) mean of x = 65, mean of y = 28, standard deviation of x = 11, standard deviation of y = 11, correlation (xy) = 0.4. (b) Mean of x = 175, mean of y = 125, standard deviation of x = 35, standard deviation of y = 40, correlation (xy) = 0.6.*

variance–covariance matrices on which the two plots are based. The following matrices were used in the derivation of the bivariate probability distributions Figure 8.15:

$$\bar{\mathbf{x}}_{12} = \begin{bmatrix} 65.812 \\ 28.033 \end{bmatrix} \quad \mathbf{S}_{12} = \begin{bmatrix} 78.669 & 45.904 \\ 45.904 & 31.945 \end{bmatrix}$$

$$\mathbf{S}_{12}^{-1} = \begin{bmatrix} 0.079 & -0.113 \\ -0.113 & 0.194 \end{bmatrix} \quad |\mathbf{S}_{12}| = 405.904$$

$$\bar{\mathbf{x}}_{57} = \begin{bmatrix} 64.258 \\ 23.895 \end{bmatrix} \quad \mathbf{S}_{57} = \begin{bmatrix} 329.336 & 181.563 \\ 181.563 & 128.299 \end{bmatrix}$$

$$\mathbf{S}_{57}^{-1} = \begin{bmatrix} 0.014 & -0.020 \\ -0.020 & 0.035 \end{bmatrix} \quad |\mathbf{S}_{57}| = 9288.356$$

The variances are the diagonal elements of the matrix **S**, and it is clear that the variances of bands 5 and 7 (329.336 and 128.299) are much larger than the variances of bands 1 and 2 (78.669 and 31.945). Thus, the 'spread' of the two ellipses in the x and y directions is substantially different. The covariance of bands 5 and 7 (181.563) and bands 1 and 2 (45.904) are both positive, so the ellipses are oriented upwards towards the $+x$ and $+y$ axes (if the plot were a two-dimensional one, we could say that the ellipses sloped upwards to the right, indicating a positive correlation between x and y). The covariance of bands 5 and 7 is larger than that of bands 1 and 2, so the degree of scatter is less for bands 5 and 7, relative to the magnitude of the variances. This example illustrates the fact that the mean controls the location of the ellipse in feature space, while the variance-covariance matrix controls the 'spread' and orientation of the ellipse. It is not possible to illustrate these principles in higher-dimensional spaces, though if it were then the same conclusions would be drawn.

The function $P(\mathbf{x})$ can be used to evaluate the probability that an unknown pattern $\mathbf{x}$ is a member of class i $i = 1, 2, \ldots, k$). The maximum value in this set can be chosen and $\mathbf{x}$ allocated to the corresponding class. However, the cost of carrying out these computations can be reduced by simplifying the expression. Savings can be made by first noting that we are only interested in the rank order of the values of $P(\mathbf{x})$. Since the logarithm to base e of a function has the same rank order as the function, the evaluation of the exponential term can be avoided by evaluating

$$\ln P(x) = -0.5p\ln(2\pi) - 0.5\ln|\mathbf{S}| - 0.5(\mathbf{y}'\mathbf{S}_i^{-1}\mathbf{y})$$

The rank order is unaffected if the right-hand side of this expression is multiplied by -2 and if the constant term $p\ln(2\pi)$ is dropped. The expression also looks tidier if it is multiplied by -1 and the smallest value for all k classes chosen, rather than the largest. These modifications reduce the expression to

$$-\ln P(\mathbf{x}) = \ln(|\mathbf{S}|) + \mathbf{y}'\mathbf{S}_{\mathbf{i}}^{-1}\mathbf{y}$$

Further savings can be made if the inverse and determinant of each $\mathbf{S}_i$ (the variance–covariance matrix for class i) are computed in advance and read from a file when required, rather than calculated when required. The computations then reduce to the derivation of the Mahalanobis distance, the addition of the logarithm of the determinant of the estimated variance-covariance matrix for each of the k classes in turn, and the selection of the minimum value from among the results. Note that, because we have multiplied the original expression by -0.5 we minimize $\{-\ln P(\mathbf{x})\}$ so as to achieve the same result as maximizing $P(\mathbf{x})$.

The ML equations given above are based upon the presumption that each of the k classes is equally likely. This may be the safest assumption if we have little knowledge of the extent of each land cover type in the area covered by the image. Sometimes the proportion of the area covered by each class can be estimated from reports, land-use maps, aerial photographs or previous classified images. An unsupervised classification of the area would also provide some guide to the areal extent of each cover type. An advantage of the ML approach to image classification is that this prior knowledge can be taken into account. *A priori* knowledge of the proportion of the area to be classified that is covered by each class can be expressed as a vector of *prior probabilities*. The probabilities are proportional to the area covered by each class, and can be thought of as weights. A high prior probability for class i in comparison with class j means that any pixel selected at random is more likely to be placed in class i than class j because class i is given greater weight. These weights are incorporated into the ML algorithm by subtracting twice the logarithm of the prior probability for class i from the log likelihood of the class as given by the equation above. Strahler (1980) provides further details and shows how different sets of prior probabilities can be used in cases where the image area can be stratified into regions according to an external variable such as elevation (for instance, regions described as high, intermediate or low elevation might have separate sets of prior probabilities as described in Section 8.7.2). Maselli *et al.* (1995b) discuss a non-parametric method of estimating prior probabilities for incorporation into a ML classifier.

In the same way that the parallelepiped classifier (Section 8.4.1) allows for the occurrence of pixels that are unlike any of the training patterns by consigning such pixels to a 'reject' class, so the probability of class membership can be used in the ML classifier to permit the rejection of pixel vectors for which the probability of membership of any of the k classes is considered to be too low. The Mahalanobis distance is distributed as chi-square, and the probability of obtaining a Mahalanobis distance as high as that observed for a given pixel can be found from tables, with degrees of freedom equal to p, the number of feature vectors used (Meyers, Gamst and Guarino, 2006, p. 67). For $p = 4$ the tabled chi-square values are 0.3 (99%), 1.06 (90%), 3.65 (50%), 7.78 (10%) and 15.08 (1%). The figures in brackets are probabilities, expressed in percentage form. They can be interpreted as follows: a Mahalanobis distance as high as (or higher than) 15.08 would, on average, be met in only 1% of cases in a long sequence of observations of four-band pixel vectors drawn from a multivariate normal population whose true mean and variance–covariance matrix are estimated by the mean and variance–covariance matrix on which the calculation of the Mahalanobis distance is based. A Mahalanobis distance of 1.06 or more would be observed in 90% of all such observations. It is self-evident that 100% of all observations drawn from the given population will have Mahalanobis distances of 0 or more. A suitable threshold probability (which need not be the same for each class) can be specified. Once the pixel has been tentatively allocated to a class using the ML decision rule, the value of the Mahalanobis distance (which is used in the ML calculation) can be tested against a threshold chi-square value. If this chi-square value is exceeded then the corresponding pixel is placed in the 'reject' class, conventionally labelled '0'. The use of the threshold probability helps to weed out atypical pixel vectors. It can also serve another function – to indicate the existence of spectral classes that may not have been recognized by the analyst and which, in the absence of the probability threshold, would have been allocated to the most similar (but incorrect) spectral class.

8.4.3 Neural Classifiers

The best image-interpretation system that we possess is the combination of our eyes and our brain (Gregory, 1998). Signals received by two sensors (our eyes) are converted into electrical impulses and transmitted to the brain which interprets them in real time, producing labelled (in the sense of recognized) three-dimensional images of our field of view (Section 5.2). Operationally speaking, the brain is thought to be composed of a very large number of simple processing units called neurons. Greenfield (1997, p. 79) estimates the number of neurons as being of the order of a 100 billion, a number that is of the same order of magnitude as the number of trees in the Amazon rain forest. Each neuron is connected to perhaps 10 000 other neurons (Beale and Jackson, 1990). Of course, brain size varies – for example men have – on average – larger brains than women (though a lot of women wonder why they do not use them). Many neurons are dedicated to image processing, which takes place in a parallel fashion (Gregory, 1999; Greenfield, 1997, p. 50). The brain's neurons are connected together in complex ways so that each neuron receives as input the results produced by other neurons, and it in turn outputs its signals to other neurons. It is not possible at the moment to specify how the brain actually works, or even whether the connectionist model really represents what is going on in the brain. It has been suggested that if there were fewer neurons in the brain then we might have a chance of understanding how they interact but, unfortunately, if our brains possessed fewer neurons we would probably be too stupid to understand the implications of the present discussion.

One model of the brain is that it is composed of sets of neural networks that perform specific functions such

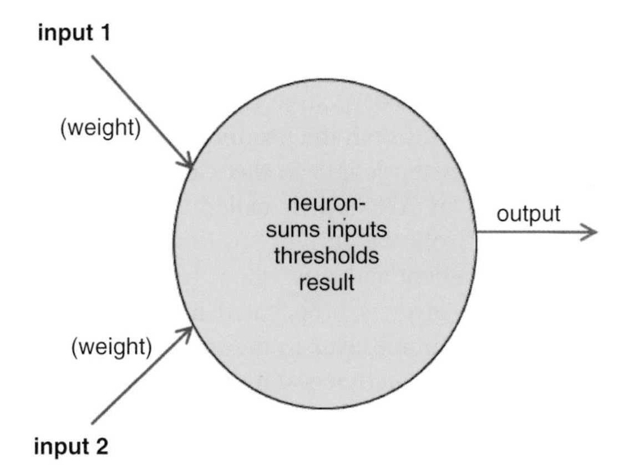

Figure 8.16 A basic neuron receives two weighted inputs, sums them, applies a threshold and outputs the result.

as vision or hearing. Artificial neural networks (ANNs) attempt, in a very simple way, to use this model of the brain by building sets of linked processing units (by analogy with the neurons of the brain) and using these to solve problems. Each neuron is a simple processing unit which receives weighted inputs from other neurons, sums these weighted inputs, performs a simple calculation on this sum such as thresholding, and then sends this output to other neurons (Figure 8.16).

The two functions of the artificial neuron are to sum the weighted inputs and to apply a thresholding function to this sum. The summation procedure can be expressed by:

$$S = \sum_{i=1}^{n} w_i x_i$$

where S represents the sum of the n weighted inputs, w_i is the weight associated with the ith input and x_i is the value of the ith input (which is an output from some other neuron). The thresholding procedure, at its simplest, a comparison between S and some pre-set value, say T. If S is greater than T then the neuron responds by sending an output to other neurons to which it is connected further 'down the line'. The term *feed-forward* is used to describe this kind of neural network model because information progresses from the initial inputs to the final outputs.

A simple ANN such as the one presented above lacks a vital component – the ability to learn. Some training is necessary before the connected set of neurons can perform a useful task. Learning is accomplished by providing training samples and comparing the actual output of the ANN with the expected output. If there is a difference between the two then the weights associated with the connections between the neurons forming the ANN are adjusted so as to improve the chances of a correct decision and diminish the chances of the wrong choice being made, and the training step is repeated. The weights are initially set to random values. This 'supervised learning' procedure is followed until the ANN gets the correct answer. To a parent, this is perhaps reminiscent of teaching a child to read; repeated correction of mistakes in identifying the letters of the alphabet and the sounds associated with them eventually results in the development of an ability to read. The method is called *Hebbian learning* after its developer, D.O. Hebb.

This simple model is called the *single layer perceptron* and it can solve only those classification problems in which the classes can be separated by a straight line (in other words, the decision boundary between classes is a straight line as in the simple example given in Section 8.2 and shown in Figure 8.1). Such problems are relatively trivial and the inability of the perceptron to solve more difficult problems led to a lack of interest in ANN on the part of computer scientists until the 1980s when a more complex model, the *multilayer perceptron*, was proposed. First, this model uses a more complex thresholding function rather than a step function, in which the output from the neuron is 1 if the threshold is exceeded and 0 otherwise. A sigmoid function is often used and the output from the neuron is a value somewhere between 0 and 1. Second, the neurons forming the ANN are arranged in layers as shown in Figure 8.17. There is a layer of input neurons which provide the link between the ANN and the input data, and a layer of output neurons that provide information on the category to which the input pixel vector belongs (for example if output neuron number 1 has a value near to 1 and the remaining output neurons have values near zero then the input pixel will be allocated to class 1).

The multilayer perceptron shown in Figure 8.17 could, for example be used to classify an image obtained from the SPOT HRV sensor. The three neurons on the input

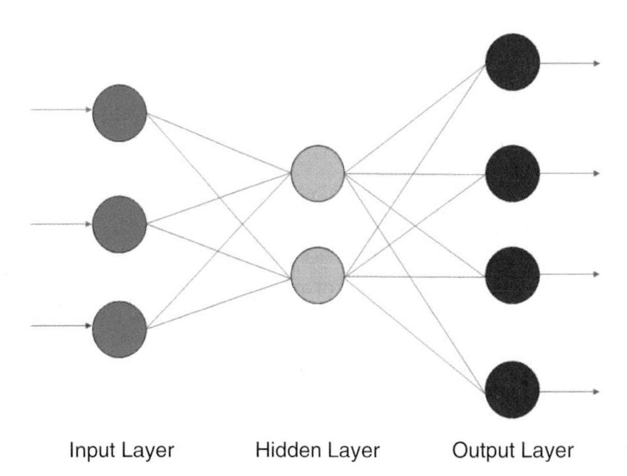

Figure 8.17 Multilayer perceptron. The input layer (left side, dark shading) connects to the hidden layer (centre, light shading) which in turn connects to the output layer (right, hatched).

layer represent the values for the three HRV spectral bands for a specific pixel. These inputs might be normalized, for example to the 0–1 range, as this procedure has been shown to improve the network's performance. The four neurons in the output layer provide outputs (or *activations*) which form the basis for the decision concerning the pixel's class membership, and the two neurons in the centre (hidden) layer perform summation and thresholding, possibly using a sigmoidal thresholding function. All the internal links (input to hidden and hidden to output) have associated weights. The input neurons do not perform any summation or thresholding – they simply pass on the pixel's value to the hidden layer neurons (so that input layer neuron 1 transmits the pixel value in band HRV-1, neuron 2 in the input layer passes the pixel value for band HRV-2 to the hidden layer neurons, and the third input neuron provides the pixel value for band HRV-3). The multilayer perceptron is trained using a slightly more complex learning rule than the one described above; the new rule is called the *generalized delta function rule* or the *back-propagation rule*. Hence the network is a feed-forward multilayer perceptron using the back-propagation learning rule, which is described by Paola and Schowengerdt (1995a).

Assume that a certain training data pixel vector, called a *training pattern*, is fed into the network and it is known that this training pattern is a member of class i. The output from the network consists of one value for each neuron in the output layer. If there are k possible classes then the expected output vector $\mathbf{o}$ should have elements equal to zero except for the ith element, which should be equal to one. The actual output vector, $\mathbf{a}$, differs from $\mathbf{o}$ by an amount called the error, e. Thus,

$$e = \frac{1}{2}\sum_{j=1}^{k}(o_j - a_j)^2$$

Multiplication by ½ is performed for arcane reasons of mathematics. The error is used to adjust the weights by a procedure that (in effect) maps the isolines or contours of the distribution of the error against the values of the weights, and then uses these isolines to determine the direction to move in order to find the minimum point in this map. If there are only two weights then a contour map of the value of the error term at point (w^1, w^2) can be easily visualized – it would look like a DEM. The aim is to find the coordinates of the point on this 'error surface' for which the value of the error term is a minimum. One way of doing this is to use the method of *steepest descent*. Imagine that you have reached the summit of a hill when a thick mist descends. To get down safely you need to move downhill, but you may not be confident enough to take long strides downhill for fear of falling over a cliff. Instead, you may prefer to take short steps when the slope is steep and longer steps on gentle slopes. The steepest descent method uses the same approach. The gradient is measured by the first derivative of the error in terms of the weight; this gives both the magnitude and direction of the gradient. The step length in this case is fixed and, in the terminology of ANNs, it is called the *learning rate*. A step is taken from the current position in the direction of maximum gradient and new values for the weights are determined. The error is propagated backwards through the net from the output layer to the input data, hence the term back-propagation. Steepest descent algorithms using variable step length (large where the surface is relatively flat and small whenever the slope is steep).

In your descent down a fog-bound hill you may come to an enclosed hollow. No matter which way you move, you will go uphill. You might think that you were at the bottom of the hill. In mathematical terms this is equivalent to finding a local minimum of the function relating weights to error, rather than a global minimum. The method of steepest descent thus may not give you the right answer; it may converge at a point that is far-removed from the real (global) minimum. Another problem may arise if you take steps of a fixed length. During the descent you may encounter a small valley. As you reach the bottom of this valley you take a step across the valley floor and immediately start to go uphill. So you step back, then forward, and so on in an endless dance rhythm. But you never get across the valley and continue your downhill march. This problem is similar to that of the local minimum and both may result in oscillations. Some ANN software allows you to do the equivalent of taking a long jump across the surface when such oscillations occur, and continuing the downhill search from the landing point. There are other, more powerful methods of locating the minimum of a function that can be used in ANN training, but these are beyond the scope of this book.

The advantages of the feed-forward multilayer ANN using the back-propagation training method are:

- It can accept all kinds of numerical inputs whether or not these conform to a statistical distribution or not. So non-remotely-sensed data can be added as additional inputs, and the user need not be concerned about multivariate normal distributions or multimodality. This feature is useful when remote sensing data are used within a GIS, for different types of spatial data can be easily registered and used together in order to improve classifier performance.
- ANNs can generalize. That is, they can recognize inputs that are similar to those which have been used to train them. They can generalize more successfully when the unknown patterns are intermediate between two known patterns, but they are less good at extending their ability to new patterns that exist beyond the

realms of the training patterns. Bigger networks (with more neurons) tend to have a poorer generalization capability than small networks.

- Because ANNs consist of a number of layers of neurons, connected by weighted links, they are tolerant to noise present in the training patterns. The overall result may not be significantly affected by the loss of one or two neurons as a result of noisy training data.

Disadvantages associated with the use of ANN in pattern recognition are:

- Problems in designing the network. Usually one or two hidden layers suffice for most problems, but how many hidden layers should be used in a given case? How many neurons are required on each hidden layer? (A generally used but purely empirical rule is that the number of hidden layer neurons should equal twice the number of input neurons.) Are all of the interneuron connections required? What is the best value for the learning rate parameter? Many authors do not say why they select a particular network architecture. For example, Kanellopoulos *et al.* (1992) use a four-layer network (one input, one output and two hidden layers). The two hidden layers contain, respectively, 18 and 54 neurons. Foody (1996a), on the other hand, uses a three-layer architecture with the hidden layer containing three neurons. Ardö, Pilesjö and Skidmore (1997) conclude that '... no significant difference was found between networks with different numbers of hidden nodes, or between networks with different numbers of hidden layers' (a conclusion also reached by Gong, Pu and Chen (1996) and Paola and Schowengerdt (1997)), though it has already been noted that network size and generalization capability appear to be inversely related. This factor is the motivation for network pruning (see below).
- Training times may be long, possibly of the order of several hours. In comparison, the ML statistical classifier requires collection of training data and calculation of the mean vector and variance–covariance matrix for each training class. *The* calculations are then straightforward rather than iterative. The ML algorithm can classify a 512×512 image on a relatively slow PC in less than a minute, while an ANN may require several hours of training in order to achieve the same level of accuracy.
- The steepest-descent algorithm may reach a local rather than a global minimum, or may oscillate.
- Results achieved by an ANN depend on the initial values given to the inter-neuron weights, which are usually set to small, random values. Differences in the initial weights may cause the network to converge on a different (local) minimum and thus different classification accuracies can be expected. Ardö, Pilesjö and Skidmore (1997), Blamire (1996) note differences in classification accuracy resulting from different initial (but still randomized) choice of weights (around 6% in Blamire's experiments, but up to 11% in the results reported by Ardö, Pilesjö and Skidmore (1997)). Skidmore *et al.* (1997, p. 511) remark that '... the oft-quoted advantages of neural networks ... were negated by the variable and unpredictable results generated'. Paola and Schowengerdt (1997) also find that where the number of neurons in the hidden layer is low then the effects of changes in the initial weights may be considerable.
- The generalizing ability of ANNs is dependent on a complex fashion on the numbers of neurons included in the hidden layers and on the number of iterations achieved during training. 'Overtraining' may result in the ANN becoming too closely adjusted to the characteristics of the training data and losing its ability to identify patterns that are not present in the training data. Pruning methods, which aim to remove interneuron links without reducing the classifier's performance, have not been widely used in remote sensing image classification, but appear to have some potential in producing smaller networks that can generalize better and run more quickly.

Further details of applications of ANNs in pattern recognition and spatial analysis are described by Aitkenhead and Aalders (2008), Aleksander and Morton (1990), Bishof, Schneider and Pinz (1992), Cappellini, Chiuderi and Fini (1995), Gopal and Woodcock (1996), Kanevski and Maignan (2004) and Mas and Flores (2008). Witten and Frank (2005) use ANNs in data mining, and also provide the Weika software which includes procedures for multilayer perceptrons. Austin *et al.* (1997) and Kanellopoulos *et al.* (1997) provide an excellent summary of research problems in the use of ANNs in remote sensing. The topic of pruning ANNs in order to improve their generalization capability is discussed by Kavzoglu and Mather (1999, 2003), Le Cun, Denker and Solla (1990) and Tidemann and Nielsen (1997). Jarvis and Stuart (1996) summarize the factors that affect the sensitivity of neural nets for classifying remotely-sensed data. A good general textbook is Bishop (1995). de Castro (2006) is a fascinating book on natural computing, with one chapter devoted to ANN. This book contains a lot of ideas but is written for mathematicians rather than the average geographer. Qiu and Jensen (2004) show how ANNs can be used in fuzzy (soft) classification schemes, which are described in Section 8.5.3.

Other useful reading includes Hepner *et al.* (1990) who compare the performance of an ANN classifier with that of ML, and find that (with a small training dataset) the ANN gives superior results, a finding that is repeated by Foody, McCullagh and Yates (1995). Kanellopoulos *et al.* (1992) compare the ML classifier's performance with that of a neural network with two hidden layers (with 18 and 54 neurons, respectively), and report that the classification accuracy (Section 8.10) rises from 51% (ML) to 81% (ANN), though this improvement is very much greater than that reported by other researchers. Paola and Schowengerdt (1995b) report on a detailed comparison of the performance of a standard ANN and the ML statistical technique for classifying urban areas. Their paper includes a careful analysis of decision boundary positions in feature space. Although the ANN slightly out-performed the ML classifier in terms of percentage correctly classified test pixels, only 62% or so of the pixels in the two classified images were in agreement, thus emphasizing the important point that measures of classification accuracy based upon error matrices (Section 8.10) do not take the spatial distribution of the classified pixels into account.

The feed-forward multilayer perceptron is not the only form of artificial neural net that has been used in remote sensing nor, indeed, is it necessarily true that ANNs are always used in supervised mode. Chiuderi and Cappellini (1996) describe a quite different network architecture, the Kohonen Self-Organizing Map (SOM). This is an unsupervised form of ANN. It is first trained to learn to distinguish between patterns in the input data ('clusters') rather than to allocate pixels to predefined categories. The clusters identified by the SOM are grouped on the basis of their mutual similarities, and then identified by reference to training data. The architecture differs from the conventional perceptron in that there are only two layers. The first is the input layer, which – as in the case of the perceptron – has one input neuron per feature. The input neurons are connected to all of the neurons in the output layer, and input pixels are allocated to a neighbourhood in the output layer, which is arranged in the form of a grid. Chiuderi and Cappellini (1996) use a 6 by 6 output layer and report values classification accuracy (Section 8.10) in excess of 85% in an application to land cover classification using airborne thematic mapper data. Schaale and Furrer (1995) successfully use a SOM network, also to classify land cover, while Hung (1993) gives a description of the learning mechanism used in SOM. Carpenter *et al.* (1997) report on the use of another type of neural network, the ART network, to classify vegetation. An interesting application is Bue and Stepinski (2006), who use an unsupervised SOM to classify Martian landforms based on topographic data.

A special issue of *International Journal of Remote Sensing* (volume 18, number 4, 1997) is devoted to 'Neural Networks in Remote Sensing'. A more recent text is authored by Sivanandam, Sumathi and Deepa (2006). This book provides coverage of all the main types of ANN and has examples in MATLAB 6.0 code.

Example 8.2 (next page) illustrates the use of ML and ANN classifiers in the remote sensing of agricultural crops.

8.5 Subpixel Classification Techniques

The techniques described in the first part of Chapter 8 are concerned with 'hard' pixel labelling. All of the different classification schemes require that each individual pixel is given a single, unambiguous, label. This objective is a justifiable one whenever regions of relatively homogeneous land cover occur in the image area. These regions should be large relative to the instantaneous field of view of the sensor, and may consist of fields of agricultural crops or deep, clear water bodies that are tens of pixels in each dimension. In other instances, though, the instantaneous field of view of the sensor may be too large for it to be safely assumed that a single pixel contains just a single land cover type. In many cases, a 1 km × 1 km pixel of an AVHRR or ATSR image is unlikely to contain just one single cover type. In areas covered by semi-natural vegetation, natural variability will be such as to ensure that, even in a 20 or 30 m^2 pixel, there will be a range of different cover types such as herbs, bare soil, bushes, trees and water. The question of scale is one that bedevils all spatial analyses.

The resolution of the sensor is not the only factor that relates to homogeneity, for much depends what is being sought. If generalized classes such as wheat, barley or rice are the targets then a resolution of 30 m rather than 1 km may be appropriate. A 30 m resolution would be quite inappropriate, however, if the investigator wished to classify individual trees. Fuzziness and hardness, heterogeneity and homogeneity are properties of the landscape at a particular geographical scale of observation, that is related to the aims of the investigator, a point that has been noted several times already in this chapter. Questions of scale are considered by de Cola (1994), Hengl (2006), Levin (1991) and Ustin *et al.* (1996). The use of geostatistics in estimating spatial scales of variation is summarized by Atkinson and Curran (1995) and Curran and Atkinson (1998). Other references relevant to the use of geostatistics are Hyppänen (1996), Jupp, Strahler and Woodcock (1988, 1989), Kanevski and Maignan (2004) (whose book includes a CD with geostatistical software), van Gardingen, Foody and Curran (1997), Woodcock, Strahler and Jupp (1988a, 1988b) and Woodcock and Strahler (1987). Selecting a suitable spatial scale and spatial resolution is the subject of papers by Atkinson and Aplin (2004), Atkinson and Curran (1997), Aplin (2006) and Barnsley, Barr and Tsang (1997). Baccini *et al.* (2007)

Example 8.2: Supervised Classification[1]

The study area selected is an agricultural area located near the town of Littleport in Cambridgeshire, in the eastern part of England. Landsat ETM+ data acquired on 19 June 2000 are used. The classification problem involves the identification of seven land cover types, namely, wheat, potato, sugar beet, onion, peas, lettuce and beans that cover the bulk of the area of interest (Example 8.2 Figure 1).

Example 8.2 Figure 1. *Ground reference data for the Littleport study area. See text for details.*

The ERDAS Imagine image processing software package (version 8.4) was used to register the images to the Ordnance Survey of Great Britain's National Grid by applying a linear least squares transformation (Section 4.3.2). The root mean squared error (RMSE) values estimated for image transformations were less than 1 pixel. An area of 307-pixel (columns) by 330-pixel (rows) covering the area of interest was then extracted for further analysis (Example 8.2 Figure 2).

Example 8.2 Figure 2. *Contrast-enhanced and geometrically corrected Landsat 7 ETM+ image of the Littleport study area. Image date: 19 June 2000.*

(Continues on next page)

[1] I am grateful to Dr Mahesh Pal for providing the material used in this example.

Field data for the relevant crops were collected from farmers and their representative agencies, and other areas were surveyed on the ground. The field boundaries visible on the multispectral image were then digitized using Arc Info software. A polygon file was created by applying a buffering operation of 1 pixel width to remove the boundary pixels during the classification process and each polygon is assigned a label corresponding to the crop it contained. Finally, a ground reference image was generated by using the polygon file (Example 8.2 Figure 1).

Random sampling methods were used to collect training and test datasets using ground reference data. The pixels collected by random sampling were divided into two subsets, one of which was used for training and the second for testing the classifiers, so as to remove any bias resulting from the use of the same set of pixels for both training and testing. Also, because the same test and training datasets are used for all of the classifiers, any differences resulting from sampling variations are avoided.

Four classification algorithms, the ML, univariate DT, back-propagation ANN and SVM were tested. A total of 2700 training and 2037 test pixels were used with all four classification algorithms. The resulting classification accuracies and kappa values are shown in Example 8.2 Table 1. The classified images are shown in Example 8.2 Figures 3–6.

Example 8.2 Table 1 *Percentage accuracy and corresponding kappa values for the classified images shown in Example 8.2 Figures 3–6.*

	Accuracy (%)	Kappa value
Decision tree	84.24	0.82
Maximum likelihood	82.90	0.80
Neural network	85.10	0.83
Support vector machines	87.90	0.87

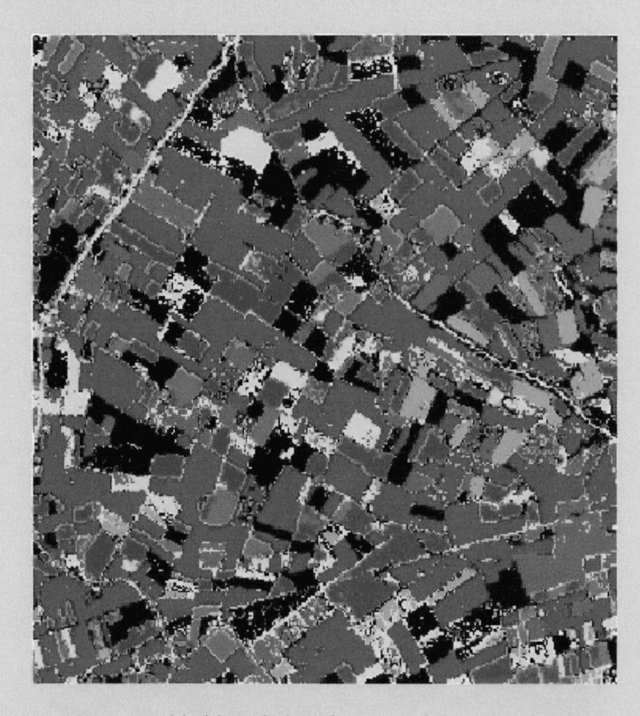

Example 8.2 Figure 3. *Maximum likelihood classification of the area shown in Example 8.2 Figure 2.*

The DT classifier employed error based pruning and used the gain ratio as an attribute selection measure. The standard back-propagation ANN classifier had one hidden layer with 26 neurons. In the SVM approach, the concept of the kernel was introduced to extend the capability of the SVM to deal with nonlinear decision surfaces. There is little guidance in the literature on the criteria to be used in selecting a kernel and the kernel-specific parameters,

Example 8.2 Figure 4. *Decision tree classification of the area shown in Example 8.2 Figure 2.*

Example 8.2 Figure 5. *ANN classification of the area shown in Example 8.2 Figure 2.*

so a number of trials were carried out using five different kernels with different kernel specific parameters, using classification accuracy as the measure of quality. A radial basis kernel function with parameters c = 2 and C = 5000 gave the highest overall classification accuracy. A 'one against one' strategy was used to deal with the multiple classes. Example 8.2 Table 1 provide the results with provided by different classification algorithms by using the same number of training and test dataset. Example 8.2 Figures 3–6 provide the classified images of the study

(Continues on next page)

Example 8.2 Figure 6. *Support vector machine classification of the area shown in Example 8.2 Figure 2.*

area by maximum likelihood, decision tree, and neural network and support vector machines classifiers respectively. It is seen that all the results exceed 80% classification accuracy with the SVM classifier performing best on both percent accuracy and kappa value. The ANN, DT and ML classifiers are each a few percent worse than the SVM method.

focus on the reverse problem – that of upscaling field data for use in calibration and validation of moderate resolution data.

To the investigator whose concern is to label each pixel unambiguously, the presence of large heterogeneous pixels or smaller pixels containing several cover types is a problem, since they do not fall clearly within one or other of the available classes. If a conventional 'hard' classifier is used, the result will be low classification accuracy. 'Mixed pixels' represent a significant problem in the description of the Earth's terrestrial surface where that surface is imaged by an instrument with a large (1 km or more) instantaneous field of view, when natural variability occurs over a small area, or where the scale of variability of the target of interest is less than the size of the observation unit, the pixel.

Several alternatives to the standard 'hard' classifier have been proposed. The method of mixture modelling starts from the explicit assumption that the characteristics of the observed pixels constitute mixtures of the characteristics of a small number of basic cover types, or end members. Alternatively, the investigator can use a 'soft' or 'fuzzy' classifier, which does not reach a definite conclusion in favour of one class or another. Instead, these soft classifiers present the user with a measure of the degree (termed membership grade) to which the given pixel belongs to some or all of the candidate classes, and leaves to the investigator the decision as to the category into which category the pixel should be placed. In this section, the use of linear mixture modelling is described, and the use of the ML and artificial neural net classifiers to provide 'soft' output is considered.

In this section we consider the main approaches to 'soft' classification, in which the pixels in the image are allowed to be 'mixed' (i.e. their value consists of the sum of two or more land cover types). In Section 8.5.1 the linear mixture model is introduced. This method has been particularly widely used by geologists. Section 5.2.2 describes the technique of ICA, which has very similar aims to the linear mixture model. Section 8.5.3 describes some approaches to soft or fuzzy classification.

8.5.1 The Linear Mixture Model

If it can be assumed that a single photon impinging upon a target on the Earth's surface is reflected into the field of view of the sensor without interacting with any other ground surface object, then the total number of photons reflected from a single pixel area on the ground and intercepted by a sensor can be described in terms of a simple linear model, as follows:

$$r_i = \sum_{j=1}^{n} a_{ij} f_j + e_i$$

in which r_i is the reflectance of a given pixel in the ith of m spectral bands. The number of mixture or fractional components is n, f_j is the value of the jth fractional component (proportion of end member j) in the makeup of r_i, and a_{ij} is the reflectance of end member j in spectral band i. The term e_i is the error term, which expresses the difference between the observed pixel reflectance r_i and the reflectance for that pixel computed from the model. In order for the components of $\mathbf{r}(= r_i)$ to be computable, the number of end members n must be less than the number of spectral bands, m. This model is simply expressing the fact that if there are n land cover types present in the area on the land surface, that is covered by a single pixel, and if each photon reflected from the pixel area interacts with only one of these n cover types, then the integrated signal received at the sensor in a given band (r_I) will be the linear sum of the n individual interactions. This model is quite well known; for example Figure 8.18 shows how an object (P) can be described in terms of three components (A, B and C). The object may be a soil sample, and A, B and C could be the percentage of sand, silt and clay in the sample.

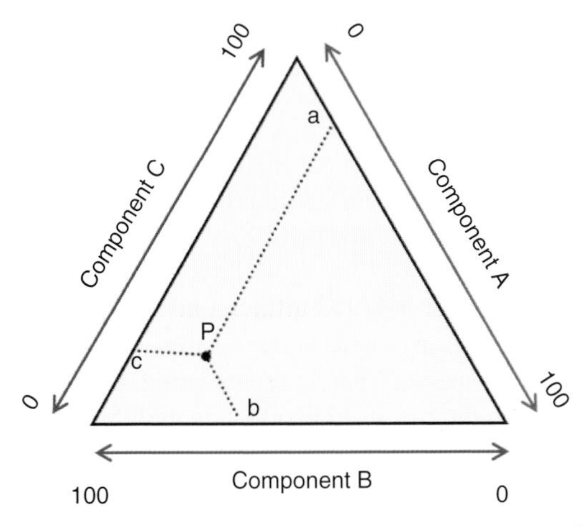

Figure 8.18 *Point P represents an object, such as a soil sample, which is composed of a mixture of three components, A, B and C. The proportions of the components (in percent) are a, b and c. The components could, for example represent the proportions of sand, silt and clay making up the soil sample. These proportions could be determined by sieving. In a remote sensing context, components A, B and C represent three land cover types. The objective of linear spectral unmixing is to determine statistically the proportions of the different land cover types within a single pixel, given the pixel's spectral reflectance curve and the spectral reflectance curves for components A, B and C.*

A simple example shows how the process works. Assume that we have a pixel of which 60% is covered by material with a spectral reflectance curve given by the lower curve in Figure 8.19 and 40% is covered by material with a spectral reflectance curve like the upper curve in Figure 8.19. The values 0.6 and 0.4 are the proportions of these two end members contributing to the pixel reflectance. The values of these mixture components are shown in the first two columns of Table 8.2, labelled C1 and C2. A 60 : 40 ratio mixture of the two mixture components is shown as the middle (dashed) curve in Figure 8.19 and in column M of Table 8.2. The rows of the table (b1, b2 and b3) represent three spectral bands in the green, red and near infrared respectively, so C1 may be turbid water and C2 may be vigorous vegetation of some kind. The data in Table 8.2 therefore describe two end members, with contributions shown by the proportions in columns C1 and C2, from which a mixed pixel vector, M, is derived. We will now try to recover the values of the mixture proportions f_1 and f_2 from these data, knowing in advance that the correct answer is 0.6 and 0.4.

First, define a matrix $\mathbf{A}$ with three rows (the spectral bands) and two columns (the end member proportions, represented by columns C1 and C2 in Table 8.2). Vector $\mathbf{b}$ holds the measurements for each spectral band of the mixed pixel (column M of Table 8.2). Finally $\mathbf{f}$ is an unknown vector which will contain the proportions f_1 and f_2 as the elements of its two rows. Assume that the relationship between $\mathbf{A}$ and $\mathbf{b}$ is of the form $\mathbf{Af} = \mathbf{b}$,

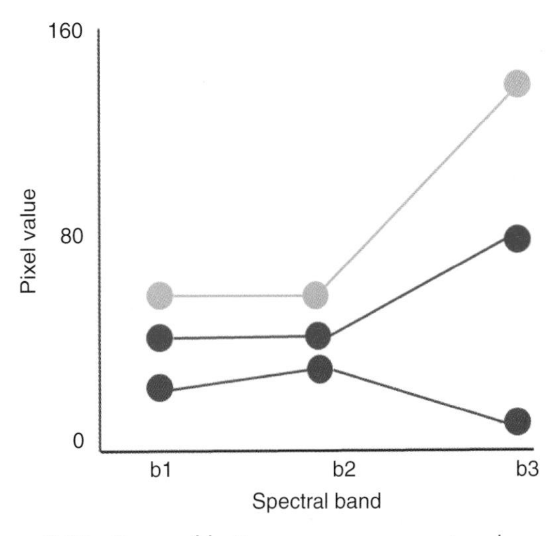

Figure 8.19 *Top and bottom curves represent end member spectra. The centre curve is the reflectance spectrum of an observed pixel, and is formed by taking 60% of the value of the lower curve and 40% of the value of the upper (dashed line).*

Table 8.2 *Columns C1 and C2 show the reflectance spectra for two pure types. Column M shows a 60 : 40 ratio mixture of C1 and C2. See text for discussion.*

	C1	C2	M
b1	46	62	52.4
b2	31	42	35.1
b3	12	160	68.8

which is equivalent to the following set of simultaneous linear equations:

$$46f_1 + 62f_2 = 54.2$$

$$31f_1 + 42f_2 = 35.4$$

$$12f_1 + 160f_2 = 71.2$$

so that $\mathbf{A} = \begin{matrix} 46 & 62 \\ 31 & 42 \\ 12 & 160 \end{matrix}$ $\mathbf{b} = \begin{matrix} 52.4 \\ 35.4 \\ 71.2 \end{matrix}$ and $\mathbf{f} = \begin{matrix} f_1 \\ f_2 \end{matrix}$.

Because we have made up the columns of **A** and **b** we know that f_1 and f_2 must equal 0.6 and 0.4 respectively, but readers should check that this is in fact the case by, for example entering the values of **A** and **b** into a spreadsheet and using the Multiple Regression option, remembering to set the intercept to 'zero' rather than 'computed'. This should give values of f_1 and f_2 equal to 0.6 and 0.4 respectively. If linear mixture modelling were being applied to an image then the values f_1 and f_2 for all pixels would be scaled onto the range 0–255 and written to file as the two output fraction images, thus generating an output fraction image for each mixture component.

The example given above assumes that the spectral reflectance curve, that is derived as a mixture of two other spectral reflectance curves is unique and is different from other spectral reflectance curves. In reality, this may not be the case, as Price (1994), shows. For instance, the spectral reflectance curve for corn is intermediate between the spectral reflectance curves of soybeans and winter wheat. Hence, the procedure outlined in the example above would be incapable of distinguishing a pixel that was split between two agricultural crops (soybeans and winter wheat) and a pure pixel covered by corn. Applications of spectral unmixing with multispectral data use general classes such as 'green vegetation' and 'soil', so this confusion may not be too serious. However, hyperspectral data may be used to identify specific mineral/rock types, and Price's point is relevant to such studies. Sohn and McCoy (1997) also consider problems in end member selection.

In a real application, rather than a fictitious example, it is necessary to go through a few more logical steps before determining the values of the fractional (end member) components, otherwise serious errors could result. First of all, in our example we knew that the mixed pixel values are composed of two components, C1 and C2 in Table 8.2. In reality, the number of end members is not known. Second, the values of the fractions (the proportions of the mixture components used to derive column M from columns C1 and C2) were known in advance to satisfy two logical requirements, namely:

$$0.0 \leq f_i \leq 1.0$$

$$\sum_{j=1}^{n} f_j \leq 1.0$$

These constraints specify that the individual fractions f_i must take values between 0 and 100%, and that the fractions for any given mixed pixel must sum to 100% or less. These two statements are a logical part of the specification of the mixture model which, in effect, assumes that the reflectance of a mixed pixel is composed of a linear weighted sum of a set of end member reflectances, with no individual proportion exceeding the range 0–1 and the sum of the proportions being 1.0 at most. This statement specifies the linear mixture model.

There are several possible ways of proceeding from this point. One approach, called 'unconstrained', is to solve the mixture model equation without considering the constraints at all. This could result in values of f_i that lie outside the 0–1 range (such illogical values are called undershoots or overshoots, depending whether they are less than 0 or greater than 1). However, the value of the unconstrained approach is that it allows the user to estimate how well the linear mixture model describes the data. The following criteria may be used to evaluate the goodness of fit of the model:

1. **The sizes of the residual terms, e_i in the mixture equation.** There is one residual value per pixel in each spectral band, representing the difference between the observed pixel value and the value computed from the linear mixture model equation. In the simple example given above, the three residuals are all zero. It is normal to take the square root sum of squares of all the residuals for a given pixel divided by the number of spectral bands, m, to give the Root Mean Squared Error (RMSE) for that pixel:

$$RMSE = \sqrt{\frac{\sum_{b=1}^{m} e_b^2}{m}}$$

The RMSE is calculated for all image pixels and scaled to the 0–255 range in order to create an RMSE image. The larger the RMSE, the worse the fit of the model. Since the residuals are assumed to be random, then any spatial pattern that is visible in the RMS images can be taken as evidence that the model has not fully accounted for the systematic variation in the image data, which in turn implies that potential end members have been omitted from the model, or that the selected end members are deficient, or that the assumed linear relationship $\mathbf{Af} = \mathbf{b}$ does not describe the relationship between the pixel spectrum and the end member spectra.

2. **The number of pixels that have proportions f_i that lie outside the logical range of 0–1.** Undershoots (f_i less than 0.0) and overshoots (f_i greater than 1.0) indicate that the model does not fit. If there are only a small percentage of pixels showing under- and overshoots then the result can be accepted, but if large numbers (say greater than 5%) of pixels under- or overshoot then the model does not fit well. Duran and Petrou (2009) discuss what they call 'negative and superunity abundancies'.

Undershoots and overshoots can be coded by a method that maps the legitimate range of the fractions f_i (i.e. 0–1) to the range 100–200 by multiplying each f_i by 100 and adding 100 to the result, which is constrained to lie in the range 0–255. This ensures that undershoots and overshoots are mapped onto the range 0–99 and 201–255 respectively. Any very large under- and overshoots map to 0 and 255, respectively. Each fraction image can then be inspected by, for example applying a pseudocolour transform to the range 100–200. All of the legitimate fractions will then appear in colour and the out of range fractions will appear as shades of grey. Alternatively, the ranges 0–99 and 201–255 can be pseudocoloured (Section 5.4).

If unconstrained mixture modelling is used then the number of under- and overshoots, and the overall RMSE, should be tabulated in order to allow others to evaluate the fit of the model. Some software packages simply map the range of the fractions for a given mixture component, including under- and overshoots, onto a 0–255 scale without reporting the presence of these under- and overshoots. Whilst this approach may produce an interpretable image for a naive user, it fails to recognize a basic scientific principle, which holds that sufficient information should be provided to permit others independently to replicate your experiments and test your results.

The model may not fit well for one or more of several reasons. First, the pixels representing specific mixture components may be badly chosen. They should represent 'pure' pixels, composed of a single land-cover type, that determine the characteristics of all of other image pixels. It is easy to think of simple situations where mixture component selection is straightforward; for example a flat, shadow-free landscape composed of dense forest, bare soil and deep, clear water. These three types can be thought of as the vertices of a triangle within which all image pixels are located, by analogy with the triangular sand–silt–clay diagram used in sedimentology (Figure 8.18). If the forest end member is badly chosen and, in reality, represents a ground area that is only 80% forest then any pixel with a forest cover of more than 80% will be an over-shoot. There may be cases in which no single pixel is completely forest-covered, so the forest end member will not represent the pure case.

Second, a mixture component may have been omitted. This is a problem where the landscape is complex and the number of image bands is small because, as noted earlier, the number of mixture components cannot exceed the number of bands. Third, the data distribution may not be amenable to description in terms of the given number of mixture components. Consider the triangle example again. The data distribution may be circular or elliptical, and thus all of the pixels in the image may not fit within the confines of the triangle defined by three mixture components. Fourth, the assumption that each photon reaching the sensor has interacted with only a single object on the ground may not be satisfied, and a non-linear mixing model may be required (Ray and Murray, 1996). The topic of choice of mixture components is considered further in the following paragraphs.

Some software uses a simple procedure to make the model appear to fit. Negative fractions (under-shoots) are set to zero, and the remaining fractions are scaled so that they lie in the range 0–1 and add to 1. This might seem like cheating, and probably is. An alternative is to use an algorithm that allows the solution of the linear mixture model equation subject to the constraints. This is equivalent to searching for a solution within a specified range, and is no different from the use of the square root key on a calculator. Generally speaking, when we enter a number and press the square root key we want to determine the positive square root, and so we ignore the fact that a negative square root also exists. If anyone were asked 'What is the square root of four?' he or she would be unlikely to answer 'Minus two', though this is a correct answer. Similarly, the solution of a quadratic equation may result in an answer that lies in the domain of complex numbers. That solution, in some instances, would be unacceptable and the alternative solution is therefore selected.

Lawson and Hansen (1995) provide a Fortran-90 subroutine, *BVLS*, which solves the equations $\mathbf{Af} = \mathbf{b}$ subject to constraints on the elements of $\mathbf{f}$. This routine, which is available on the Internet via the Netlib library, does not allow for a constraint on the sum of the f_i. An alternative is a Fortran-90 routine from the IMSL mathematical library provided with Microsoft Powerstation Fortran v4.0 Professional Edition. This routine, *LCLSQ*, allows the individual f_i to be constrained and it also allows the sum of the f_i to be specified.

The solution of the mixture model equation in the unconstrained mixture model requires some thought. The standard solution of the matrix equation $\mathbf{Af} = \mathbf{b}$ generally involves the calculation of the inverse of $\mathbf{A}$ (represented as $\mathbf{A}^{-1}$) from which $\mathbf{f}$ is found from the expression $\mathbf{f} = \mathbf{A}^{-1}\mathbf{b}$. A number of pitfalls are likely to be encountered in evaluating this innocent-looking expression. If $\mathbf{A}$ is an orthogonal matrix (that is its columns are uncorrelated) then the derivation of the inverse of $\mathbf{A}$ is easy (it is the transpose of $\mathbf{A}$). If the columns of $\mathbf{A}$ (which contain the reflectances of the end members) are correlated then error enters the calculations. The greater the degree of interdependence between the columns the more likely it is that error may become significant. When the columns of $\mathbf{A}$ are highly correlated (linearly dependent) then matrix $\mathbf{A}$ is said to be *near-singular*. If one column of $\mathbf{A}$ can be calculated from the values in the other columns then $\mathbf{A}$ is *singular*, which means that it does not possess an inverse (just has the value 0.0 does not possess a reciprocal). Consequently, the least-squares equations cannot be solved. However, it is the problem of near-singularity of $\mathbf{A}$ that should concern users of linear mixture modelling because it is likely that the end member spectra are similar. The solution (the elements of vector $\mathbf{f}$) may, in such cases, be significantly in error, especially in the case of multi-spectral (as opposed to hyperspectral) data when the least-squares equations are being solved with relatively small numbers of observations.

Boardman (1989) suggests that a procedure based on the singular value decomposition (SVD) handles the problem of near-singularity of $\mathbf{A}$ more effectively. The inverse of matrix $\mathbf{A}$ (i.e. $\mathbf{A}^{-1}$) is found using the SVD:

$$\mathbf{A} = \mathbf{UWV}'$$

where $\mathbf{U}$ is an $m \times n$ column-orthogonal matrix, $\mathbf{W}$ is an $n \times n$ matrix of singular values and $\mathbf{V}$ is an $n \times n$ matrix of orthogonal columns. $\mathbf{V}'$ is the matrix transpose of $\mathbf{V}$. The inverse of $\mathbf{A}$ is found from the following expression:

$$\mathbf{A}^{-1} = \mathbf{VW}^{-1}\mathbf{U}'$$

The singular values, contained in the diagonal elements of $\mathbf{W}$, give an indication of the dimensionality of the space containing the spectral end members. They can be thought of as analogous to the principal components of a correlation or covariance matrix. If the spectral end members are completely independent then the information in each dimension of the space defined by the spectral bands is equal and the singular values $\mathbf{W}_{ii}$ are all equal. Where one spectral end member is a linearly combination of the remaining end members then one of the diagonal elements of $\mathbf{W}$ will be zero. Usually neither of these extreme cases is met with in practice, and the singular values of $\mathbf{A}$ (like the eigenvalues of a correlation matrix, used in principal components analysis (PCA)) take different magnitudes. An estimate of the true number of end members can therefore be obtained by observing the magnitudes of the singular values, and eliminating any singular values that are close to zero. If this is done then the inverse of $\mathbf{A}$ can still be found, whereas if matrix inversion methods are used then the numerical procedure either fails or produces a result which is incorrect.

If the landscape is composed of a continuously varying mixture of idealized or pure types, it might appear to be illogical to search within that landscape for instances of these pure types. Hence, in a number of studies, laboratory spectra have been used to characterize the mixture components. Adams, Smith and Gillespie (1993) term these spectra *reference end members*. They are most frequently used in geological studies of arid and semi-arid areas. Since laboratory spectra are recorded in reflectance or radiance units, the image data must be calibrated and atmospherically corrected before use, as described in Chapter 4.

In other cases, empirical methods are used to determine which of the pixels present in the image set can be considered to represent the mixture components. Murphy and Wadge (1994) use PCA (Chapter 6) to identify candidate image end members. A graphical display of the first two principal components of the image set is inspected visually and

> ... the pixels representing the end member for each cover type should be located at the vertices of the polygon that bounds the data space of the principal components which contain information (Murphy and Wadge, 1994, p. 73).

This approach makes the assumption that the m dimensional space defined by the spectral bands can be collapsed onto two dimensions without significant loss of proximity information; in other words, it is assumed that pixels that lie close together in the PC1 – PC2 plot are actually close together in the m dimensional space, and *vice-versa*. PCA is discussed in Section 6.4, where it is shown that the technique is based on partitioning the total variance of the image dataset in such a way that the first principal component accounts for the maximum variance of a linear combination of the spectral bands, principal component two accounts for a maximum of the remaining variance, and so on with the restriction that the principal components are orthogonal. Since PCA has the aim of identifying dimensions of variance (in that the principal components are ordered on the basis of their variance), principal components 1 and 2 will, inevitably, contain much of the information present in the image dataset. However, significant variability may remain in the lower-order principal components. Furthermore, if the PCA is based on covariances, then the spectral bands may contribute unequally to the total variance analysed. Tompkins *et al.* (1997) present an elaborate methodology for end member selection; their procedure estimates both the end member proportions and the end member reflectances simultaneously. Zortea and Plaza (2009) use spatial as well as spectral information to identify end-members. Other references to the use of the linear mixture model are de Asis *et al.* (2008), Chang *et al.* (2006), Dennison and Roberts (2003), Dennison, Halligan and Roberts (2004), Eckmann, Roberts and Still (2008), Ngigi, Tateishi and Gachari (2009) and Plourde *et al.* (2007). Feng, Rivard and Sánchez-Azofeifa (2003) consider the impact of topographic normalization (Section 4.7) on the choice of end members. Users of the ENVI image processing system will be familiar with the pixel purity index (PPI), which is used for end-member selection. When mentioned in the literature its exact nature is shrouded in mystery. Chang and Plaza (2006) claim to have worked out how PPI is calculated and present their results.

An empirical method for selecting mixture components is Sammon's (1969) Nonlinear Mapping. The method uses a measure of goodness of fit between the inter-pixel distances measured in the original m-dimensional space and those measured in the reduced-dimensionality space, usually of two or three dimensions (Figure 8.10). The two- or three-dimensional representation is updated so as to reduce the error in the inter-pixel distances measured in the two- or three-dimensional space compared to the equivalent distance measured in the full space of m spectral bands. A minimum of the error function is sought using the method of steepest descent, similar to that used in the back-propagating artificial neural net. Study of the plots of pairs of dimensions provides some interesting insights into the nature of the materials in the image. Nonlinear Mapping analysis appears to be a better way of analysing image information content than the principal components approach, as the analysis is based on inter-pixel distances. Bateson and Curtiss (1996) discuss another multidimensional visualization method of projecting points (representing pixels) onto two dimensions. Their method is based on a procedure termed parallel coordinate representation (Wegman, 1990), which allows the derivation of 'synthetic' end member spectra that do not coincide with image spectra. The method is described in some detail by Bateson and Curtiss (1996).

Applications of linear mixture modelling in remote sensing are numerous, though some authors do not give enough details for their readers to discover whether or not the linear mixture model is a good fit. If the data do not fit the model then, at best, the status of the results is questionable. Good surveys of the method are provided by Adams, Smith and Gillespie (1989, 1993), Ichoku and Karnieli (1996), Mustard and Sunshine (1999), Settle and Drake (1993) and Settle and Campbell (1998). A selection of typical applications is: Cross *et al.* (1991), who use AVHRR data to provide subpixel estimates of tropical forest cover, and Gamon *et al.* (1993), who use mixture modelling to relate AVIRIS (imaging spectrometer) data to ground measurements of vegetation distribution. Shipman and Adams (1987) assess the detectability of minerals on desert alluvial fans. Smith *et al.* (1990) provide a critical appraisal of the use of mixture modelling in determining vegetation abundance in semiarid areas. Other useful references are Bateson and Curtiss (1996), Bryant (1996), Garcia-Haro, Gilabert and Melia (1996), Hecker *et al.* (2008), Hill and Horstert (1996), Kerdiles and Grondona (1995), Peddle and Smith (2005), Roberts, Adams and Smith (1993), Rogge *et al.* (2007), Shimabukuro and Smith (1991), Thomas, Hobbs and Dufour (1996), Ustin *et al.* (1996) and Wu and Schowengerdt (1993). Roberts, Adams and Smith (1993) consider an interesting variant on the standard mixture modelling technique which, in a sense, attempts a global fit in that each pixel in the image is assumed to be the sum or mixture of the same set of end members. Roberts, Adams and Smith (1993) suggest that the number and nature of the end members may vary over the image. This opens the possibility of using an approach similar to stepwise multiple regression (Grossman *et al.*, 1996) in which the best k from a pool of m possible end members are selected for each pixel in turn. Roberts *et al.* (1998) also consider a multiple end-member model approach to mapping chaparral. Foody *et al.* (1996, 1997) describe an approach to mixture modelling using ANNs (Section 8.4.3).

Spectral angle mapping (SAM) and ICA have similar aims to mixture modelling, and are described below in Sections 8.5.2 and 8.5.3.

8.5.2 Spectral Angle Mapping

A simple but effective alternative to linear spectral unmixing that does not have any statistical overtones is provided by the method of SAM, which is based on the well-known coefficient of proportional similarity, or cosine theta ($\cos\theta$). This coefficient measures the difference in the shapes of the spectral curves (Imbrie, 1963; Weinand, 1974). It is insensitive to the magnitudes of the spectral curves, so two curves of the same shape are considered to be identical. If two spectral reflectance curves have the same shape but differ in magnitude, one might conclude that the difference is due to changes in illumination conditions. This is a reasonable claim, but the method does not *correct* for illumination variations.

In the context of image processing, a small number of pixels is selected as the reference set $\mathbf{r}$ and the remaining image pixels (represented by vectors $\mathbf{t}$) are compared to these reference pixels. Each pixel is considered as a

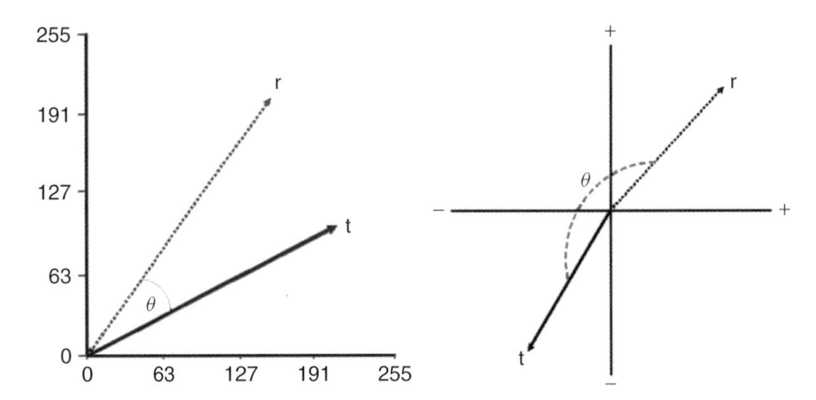

Figure 8.20 *The cosine of angle θ measures the degree of dissimilarity between the shapes of reflectance spectra. Here, an image pixel spectrum* **t** *is being compared to a reference spectrum* **r** *in a two-dimensional feature space. In (a) the 2 pixel vectors have components that are measured on a scale of 0–255, and so the angle θ lies within the range 0–90°, thus 0 ≤ cos(θ) ≤ 1.0. In (b) the elements of the two vectors* **t** *and* **r** *are unrestricted in range, and so θ lies in the range 0–360° (2π radians) and −1.0 ≤ cos(θ) ≤ 1.0.*

geometric vector. For simplicity, let us use a feature space defined by two spectral bands, as shown in Figure 8.20a. If the pixel values stored in **r** and **t** are measured on an 8-, 16- or 32-bit unsigned integer scale then the points representing pixels will lie in the first quadrant, as they are all non-negative. Figure 8.20a shows a reference vector **r** and an image pixel vector **t**. Both vectors are measured on a scale from 0 to 255, and $0 \leq \theta \leq 90^\circ$.

If the line (vector) joining the pixel of interest **t** to the origin is coincident with the reference pixel vector **r** then the angle between the two vectors (pixel and reference) is zero. The cosine of zero is 1, so $\cos(\theta) = 1$ means complete similarity. The maximum possible angle between a reference pixel vector and an image pixel vector is 90°, and $\cos(90^\circ) = 0$, which implies complete dissimilarity.

For image data measured on a 0–255 scale, the output image can either be the cosine of the angle θ (which lies on a 0–1 scale with 1 meaning similar and 0 meaning dissimilar, so that similar pixels will appear in light shades of grey) or the value of the angle θ (on a scale of 0–90° or 0–$\pi/2$ radians). Both of these representations would require that the output image be written in 32-bit real format (Section 3.2). If the angular representation is used then you must remember that pixels with a value of θ equal to 0° (i.e. complete similarity) will appear as black, while dissimilar pixels ($\theta = 90^\circ$ or $\pi/2$ radians) will appear in white. Of course, the real number scales 0.0–1.0, 0.0–90.0 or 0.0–$\pi/2$) can be mapped onto a 0–255 scale using one of the methods described in Section 3.2.1.

If the input data are represented in 32-bit real form, then it is possible that some image pixel values will be negative (for example principal component images can have negative pixel values). In this case, additional possibilities exist. A reference vector in two-dimensional feature space may take the values (90, 0) so that its corresponding vector joins the origin to the point (90, 0). Imagine a pixel vector with the values (−90, 0). This is equivalent to a vector joining the origin to the point (−90, 0), so that the angle between the reference and pixel vectors is °. The cosine of 180° is −1.0. This value does not mean that the two vectors have no relationship – they do, but it is an inverse one, so that the pixel vector takes the same values as the reference vector but with the opposite sign. The output pixel values will thus lie within the range (−1.0, 1.0). When the output image is displayed then pixel vectors that are identical to a reference vector will appear white (1.0), whereas pixels that are not related to a reference vector will take the value 127 (if a linear mapping is performed from (−1.0, 1.0) to (0, 255)). Unrelated pixels will appear as mid-grey, whereas pixels whose vectors are the reciprocals of the reference vector will appear black (−1.0). Figure 8.20b shows a reference pixel **r** and an image pixel **t**. Both **r** and **t** are 32-bit real quantities, and the angle θ between them is almost 180°, implying that **r** is the mirror image of **t** (or vice versa).

If the ith element of the reference pixel vector is represented by r_i and any other image pixel vector is written as t_i then the value of $\cos(\theta)$ is calculated from:

$$\cos(\theta) = \frac{\sum_{i=1}^{N} r_i t_i}{\left(\sum_{i=1}^{N} r_i^2\right)^{0.5} \left(\sum_{i=1}^{N} t_i^2\right)^{0.5}}$$

given N bands of image data. One image of $\cos(\theta)$ values is produced for each reference vector **r**. There are no statistical assumptions or other limitations on the number or nature of the reference vectors used in these calculations. Kruse *et al.* (1993), Ben-Dor and Kruse (1995) and van der Meer (1996c) provide examples of the use of the method, in comparison with the linear spectral unmixing technique. Yonezawa (2007) combines a ML classifier with Spectral Angle Mapper to classify a QuickBird multsipectral image.

8.5.3 ICA

Independent Components Analysis (ICA) is a well-known technique in signal processing, where it is used as a method of 'blind source separation', that is finding sources (end members) that, when mixed in specific proportions, produce the observed pixel reflectance values. It has found less than widespread use in the analysis of remotely-sensed images. The basic equation of ICA is similar to that of the linear mixture model, namely, $\mathbf{x} = \mathbf{A}\mathbf{s}$ where $\mathbf{x}$ is a vector of observations, $\mathbf{A}$ is the matrix of mixture proportions or abundances, and $\mathbf{s}$ is the vector of sources. Several texts on ICA use the cocktail party story to illustrate this model. If there are two people speaking in a room, say s_1 and s_2, and their voices are picked up by two microphones, say x_1 and x_2, then the two microphones will pick up mixtures of the two voices. The matrix $\mathbf{A}$ contains those mixture proportions, which are the weights in a weighted sum of the sounds made by the two speakers.

What is interesting about ICA is that it is not a statistical technique so there is no error term, as there is with linear mixture modelling. Furthermore, the technique assumes the exact opposite of what most statistical techniques assume. Standard statistical procedures require that the data are distributed in a Normal or Gaussian way, and many users of the ML classifier worry inordinately because their data are non-normal. The reason that the technique of ICA assumes non-normality is that normal or gaussian distributions are '... the most random, least structured, of all distributions' (Hyvärinen and Oja, 2000, p. 418). What we want are structured, systematic and independent mixing components that represent the data adequately. The measures used to ascertain the 'gaussianicity' of a variable are kurtosis and entropy. The former is a measure of the peakedness of the frequency distribution and is equal to zero for a Gaussian random variable whereas the latter measures information content. The ICA algorithm uses an iterative scheme to maximize the non-gaussianicity using one or other of the measures of degree of conformity to a Gaussian distribution. Hyvärinen and Oja (2000) and Hyvärinen, Karhunen and Oja (2001) provide a good discussion of these points.

If kurtosis is used as the basis for the decomposition of the observations (pixel values) into sources of variability (end members in mixture modelling parlance) then ICA can demonstrate one other potential advantage. Other methods use the mean and the variance/covariance matrix alone; these are called first and second order statistics, whereas the kurtosis is a fourth order statistic. It is claimed that the use of higher order statistics provides ICA with the ability to pick out smaller, more detailed, sources in the image data.

There is still a lot to learn about ICA and its potential in remote sensing data analysis. Standard references include Bayliss, Gualtieri and Cromp (1997), Fiori (2003), Gao *et al.* (in press), Hyvärinen and Oja (2000), Hyvärinen, Karhunen and Oja (2001), Lee (1998), Nascimento and Dias (2005), Roberts and Everson (2001), Shah, Varshney and Arora (2007), Stone (2004) and Wang and Chang (2006). MATLAB code is given by Gopi (2007). Hyvärinen, Karhunen and Oja (2001) and Stone (2004) give Internet references to sources of code.

8.5.4 Fuzzy Classifiers

The distinction between 'hard' and 'soft' classifiers is discussed in the opening paragraphs of the introduction to Section 8.5. The 'soft' or fuzzy classifier does not assign each image pixel to a single class in an unambiguous fashion. Instead, each pixel is given a 'membership grade' for each class. Membership grades range in value from 0 to 1, and provide a measure of the degree to which the pixel belongs to or resembles the specified class, just as the fractions or proportions used in linear mixture modelling (Section 8.5.1) represent the composition of the pixel in terms of a set of end members. It might appear that membership grade is equivalent to probability, and the use of probabilities in the ML classification rule might lend support to this view. Bezdek (1993) differentiates between precise and fuzzy data, vague rules and imprecise information. Crisp sets contain objects that satisfy unambiguous membership requirements. He notes that $H = \{r \in \Re | 6 \leq 8\}$ precisely represents the crisp or hard set of real numbers H from 6 to 8. Either a number is a member of the set H or it is not. If a set F is defined by a rule such as 'numbers that are close to 7' then a given number such as 7.2 does not have a precise membership grade for F, whereas its membership grade for H is 1.0. Bezdek (1993) also notes that '... the modeller must decide, based on the potential applications and properties desired for F, what $\mathrm{m_F}$ should be' (p. 1). The membership function m_F for set F can take any value between 0 and 1, so that numbers far away from 7 still have a membership grade. This, as Bezdek notes, is simultaneously both a strength and a weakness. Bezdek (1994) points out that fuzzy membership represents the similarity of an object to imprecisely defined properties, while probabilities convey information about relative frequencies.

Fuzzy information may not be of much value when decisions need to be taken. For example, a jury must decide, often in the basis of fuzzy information, whether or not the defendant is guilty. A fuzzy process may thus have a crisp outcome. The process of arriving at a crisp or hard conclusion from a fuzzy process is called defuzzification. In some cases, and the mixture model discussed in the previous section is an example, we may wish to retain some degree of flexibility in presenting our results. The membership grades themselves may be of interest as they may relate to the proportions of the pixel area that is represented by each of the end members. In seminatural landscapes one may accept that different land cover types merge in a transition zone. For instance, a forested area may merge gradually into grassland, and grassland in turn may merge imperceptibly into desert. In such cases, trying to draw hard and fast boundaries around land cover types may be a meaningless operation, akin to Kimble's (1951) definition of regional geography: putting '... boundaries that do not exist around areas that do not matter' (Kimble, 1951, p. 159). On the other hand, woodland patches, lakes and agricultural fields have sharp boundaries, though of course the meaning of 'sharp' depends upon the spatial scale of observation, as noted previously. A lake may shrink or dry up during a drought, or the transition between land cover types may be insignificant at the scale of the study. At a generalized scale it may be perfectly acceptable to draw sharp boundaries between land cover types. Wang (1990) discusses these points in the context of the ML decision rule, and proposes a fuzzy version of ML classification using weighted means and variance–covariance matrices similar to those described above (Section 8.4.1) in the context of deriving robust estimates from training samples. Gong, Pu and Chen (1996) use the idea of membership functions to estimate classification uncertainty. If the highest membership function value for a given pixel is considerably greater than the runner-up, then the classification output is reasonably certain, but where two or more membership function values are close together then the output is less certain.

One of the most widely used unsupervised fuzzy classifiers is the fuzzy c-means clustering algorithm. Bezdek *et al.* (1984) describe a Fortran program. Clustering is based on the distance (dissimilarity) between a set of cluster centres and each pixel. Either the Euclidean or the Mahalanobis distance can be used. These distances are weighted by a factor m which the user must select. A value of m equal to 1 indicates that cluster membership is 'hard' while all membership function values approach equality as m gets very large. Bezdek *et al.* (1984) suggest that m should lie in the range 1–30, though other users appear to choose a value of m of less than 1.5. Applications of this method are reported by Bastin (1997), Cannon *et al.* (1986), Du and Lee (1996), Foody (1996b) and Key, Maslanik and Barry (1989). Other uses of fuzzy classification procedures are reported by Blonda and Pasquariello (1991), Maselli *et al.* (1995a) and Maselli, Rodolfi and Conese (1996).

The basic ideas of the fuzzy c-means classification algorithm can be expressed as follows. **U** is the *membership grade matrix* with n columns (one per pixel) and p rows (one per cluster):

$$\mathbf{U} = \begin{bmatrix} u_{11} & \cdots & u_{1n} \\ \vdots & & \vdots \\ u_{p1} & \cdots & u_{pn} \end{bmatrix}$$

The sum of the membership grades for a given class (row of **U**) must be non-zero. The sum of the membership grades for a given pixel (column of **U**) must add to one, and the individual elements of the matrix, u_{ij}, must lie in the range 0–1 inclusive. The number of clusters p is specified by the user, and the initial locations of the cluster centres are either generated randomly or supplied by the user. The Euclidean distance from pixel i to cluster centre j is calculated as usual, that is

$$d_{ij} = \sqrt{\sum_{l=1}^{k} (x_{il} - c_{jl})^2}$$

and the centroids c_j are computed from:

$$c_{jl} = \sum_{i=1}^{n} u_{ji}^{m} x_{il} \Big/ \sum_{i=1}^{n} u_{ji}^{m}$$

This is simply a weighted average (with the elements u being the weights) of all pixels with respect to centre $j(1 \leq j \leq p)$. The term x_{il} is the measurement of the ith pixel $(1 \leq i \leq n)$ on the lth spectral band or feature. The exponent m is discussed above.

Each of the membership grade values u_{ij} is updated according to its Euclidean distance from each of the cluster centres:

$$u_{ij} = \frac{1}{\sum_{c=1}^{p} \left(\frac{d_{ij}}{d_{cj}} \right)^{\frac{2}{(m-1)}}}$$

where $1 \leq i \leq p$ and $1 \leq j \leq n$ (Bezdek *et al.*, 1984). The procedure converges when the elements of **U** differ by no more than a small amount between iterations.

The columns of **U** represent the membership grades for the pixels on the fuzzy clusters (rows of **U**). A process of 'defuzzification' can be used to determine the cluster membership for pixel i by choosing the element in column i of **U** that contains the largest value. Alternatively, and perhaps more informatively, a set of classified images could be produced, one per class, with the class membership grades (measured on a scale from 0 to 1) scaled onto the 0–255 range. If this were done, then the pixels with membership grades close to 1.0 would appear white, and pixels whose membership grade was close to zero would appear black. Colour composites of the classified images, taken three at a time, would be interpretable in terms of class membership. Thus, if class 1 were shown in red and class 2 in green, then a yellow pixel would be equally likely to belong to class 1 as to class 2.

The activations of the output neurons of an ANN (Foody, 2002a) or the class membership probabilities derived from a standard ML classifier, can also be used as approximations of the values of the membership function. In an ideal world, the output from a neural network for a given pixel will be represented by a single activation value of 1.0 in the output layer of the network, with all other activation levels being set to 0.0. Similarly, an ideal classification using the ML method would be achieved when the class membership probability for class i is 1.0 and the remaining class membership probabilities are zero. These ideal states are never achieved in practice. The use of a 'winner takes all' rule is normally used, so that the pixel to be classified is allocated to the class associated with the highest activation in the output layer of the network, or the highest class-membership probability, irrespective of the activation level or the magnitude of the maximum probability. It is, however, possible to generate one output image per class, rather than a single image representing all the classes of a 'hard' representation. The first image will show the activation for output neuron number one (or the class membership probability for that class, if the ML method is used) scaled to a 0–255 range. Pixels with membership function values close to 1.0 are bright, while pixels with a low membership function value are dark. This approach allows the evaluation of the degree to which each pixel belongs to a given class. Such information will be useful in determining the level of confidence that the user can place in the results of his or her classification procedure.

8.6 More Advanced Approaches to Image Classification

8.6.1 Support Vector Machines

SVMs have only recently been introduced into remote sensing as an effective and efficient means of classification, though they were originally developed in the 1970s. One of their first reported uses in remote sensing is Huang, Davis and Townshend (2002). SVM are effective because they produce classification accuracy values (Section 8.10) as high, if not higher, than other classification methods and they are efficient because they need only small amounts of training data that is located in those areas of feature space that lie near to interclass boundaries. In the two-class case shown in Figure 8.21 the two classes to be separated are shown as blue circles or green triangles. The margin is the distance between

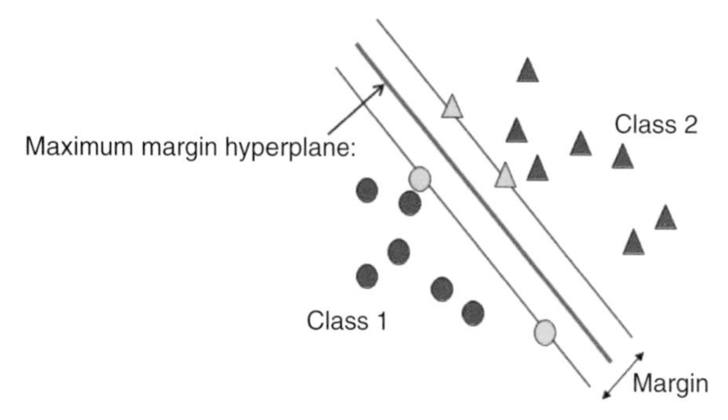

Figure 8.21 *Illustrating the concept of the support vector. The maximum margin hyperplane is shown in red, and the margin between the support vectors is shown by the parallel light blue lines. The two classes do not overlap. The support vectors (patterns that are on the margin) are shown as yellow shapes – circles for class 1, triangles for class 2.*

the two light blue parallel lines. There is one position of the margins that gives the maximum distance between the closest members of the two classes. Two blue samples and two green samples in Figure 8.21 are the closest pairs of data points in the feature space that are the furthest apart and they are shown in yellow. They are called support vectors. There must be a minimum of two support vectors but there can be more than two. The red line running down the centre of the margin is the maximum margin hyperplane (a hyperplane is a plane defined in more than three dimensions). Any new point can be identified by reference to the maximum margin hyperplane. If it is above and to the right it is a member of the green class. If it is lower than the maximum margin hyperplane and to the left then it is a member of the blue class. This example illustrates that only the support vectors are needed – the rest of the training data is irrelevant. Thus, the SVM classification remains stable if the training data are altered provided that the same support vectors emerge. The choice of training data for SVM is discussed by Foody and Mathur (2004a, 2006).

If the two classes are not linearly separable, as in Figure 8.21, then a more complicated approach needs to be adopted. A term is added to the SVM decision rule (it is called a slack variable) and it penalizes errors, that is the assignment of a training pixel to the wrong class. In essence, the original decision rule was to maximize the margin. The modified rule adds a further requirement, that of penalizing incorrect decisions.

So far, the idea underlying SVM appears to be simple. There are instances, however, where the members of each of the two classes in the training dataset can be separated in a nonlinear way (e.g. by a curved line). The data are then mapped into a higher-dimensional space that has the property of increasing inter-class separability. The mapping function is called a kernel, hence SVM are kernel-based methods. The idea of a kernel is surprisingly difficult to explain, and readers wishing to get involved in the mathematics can refer to Shawe-Taylor and Cristianini (2004), who deal with kernel-based methods of pattern recognition. Their book includes some software. Another book on the subject is Abe (2005), while yet another book by Kanevski and Maignan (2004) (which has been cited several times in this chapter) includes a chapter on SVM and some software. Tso and Mather (2009) provide a somewhat dense account of the non-linear mapping procedure. Burges (1998) provides a tutorial introduction but the level of mathematical sophistication required of the reader is considerable. Software to help readers experiment with SVM is available from Chang and Lin (2001). Readers who do not wish to pursue the mathematical details can consider kernel-based methods as analogous to the use of logarithms in manual calculation. Multiplication of large numbers can be reduced to addition of smaller numbers, resulting in a saving of time and effort. The same savings accrue to the nonlinear mapping provided by the kernel function. This analogy is, however, not strictly true for, as noted by Tso and Mather (2009, p. 132)

> ... the purpose of a kernel function is to enable operations to be performed in the current domain ... rather than the potentially high dimensional feature space [of the kernel] ... This provides a smart means of resolving the computational issue caused by high dimensionality.

The user of an SVM has to make a selection from a range of available kernel functions, which include: homogeneous and inhomogeneous polynomials, radial basis function and Gaussian radial basis function, and the sigmoid kernel. There is no clear evidence to support the use of one rather than another, though the radial basis function is widely used.

The equation of the maximum margin hyperplane is needed to label the members of the test dataset (and, ultimately, the image pixels). Optimization methods are used, similar to the steepest descent approach described in Section 8.4.3 in connection with training ANN classifiers. An alternative to the steepest descent algorithm when only two parameters are required is the grid search method, in which the function to be minimized is evaluated over the nodes of a grid and the accuracy of the allocation of test data pixels to their classes is measured. The highest value of classification accuracy (Section 8.10) is then located and the parameter values read from the axes (see Tso and Mather, 2009, p. 141).

Until this point it has been assumed that the problem facing the classifier is to allocate the image pixels to one of two classes. This is not a common situation in remote sensing, where there are usually multiple classes. Three possibilities present themselves: one could carry out an SVM classification for one class against the rest of the classes bundled together as a single class, for example class i might be tested against the rest of the k classes excluding class i. This is called 'one against the rest', and it is carried out k times with i running from 1 to k. A second possibility is to do a pairwise comparison of classes i and j, giving a total of $k(k-1)/2$ runs of the SVM program. This is called 'one against one'. A voting system is needed to assign a unique label to a pixel. Usually the majority vote is used. In the case of a tie then the label of a neighbouring pixel could be used. Another solution to the problem of multiple classes is to use a multiclass SVM rather than generate a multiclass SVM by the use of 'one against one' or 'one against all' strategies. However, the details of this technique are beyond the range of this book and interested readers can refer to Crammer and Singer (2002), Hsu and Lin (2002), Mathur and Foody (2008), Tso and Mather (2009) and Vapnik (1998).

There are a number of useful articles on SVM in remote sensing. Foody and Mathur (2004b) compare SVM with other classifiers. Pal and Mather (2005) compare SVM with ANN and ML and conclude that SVM can give higher classification accuracy with high-dimensional data and small training samples. Dixon and Candade (2008) compare ML, SVM and ANN classifiers, and conclude that SVM and ANN give higher classification accuracies than ML, but that training times for SVM are much lower than for ANN. Su *et al.* (2007) used SVMs for recognition of semiarid vegetation types using MISR multiangle imagery. Melgani and Bruzzoni (2004) and Plaza *et al.* (2009) consider the problem of classifying hyperspectral remote sensing images (Section 9.3) using SVMs. Pal and Mather (2004) also address this topic. Camps-Valls and Bruzzone (2005) use SVM and boosting (Section 8.6.3.3). Zhang and Ma (2008) present an improved SVM method (P-SVM) for classification of remotely-sensed data. Lizarazo (2008) gives details of an SVM-based segmentation and classification of remotely-sensed data. Zhu and Blumberg (2002) give a case study of the use of SVM and ASTER data. Marçal *et al.* (2005) use eCognition object-oriented software (Section 8.6.3.2) to segment an ASTER image for land cover classification. They then compare four classifiers, including SVM, but find no clear difference between them. Dixon and Candade (2008) claim that SVM behaviour is independent of dimensionality and thus this method does not require as much training data as, say, a ML classifier, which has a need for exponentially increasing amounts of training data as dimensionality increases. These authors also conclude that ANN and SVM give more accurate results than the ML classifier in a study of land cover classification. Smirnoff, Boisvert and Paradis (2008) give a geologist's perspective of SVM.

In comparison with Decision Trees (DTs), ML methods, and even ANNs, SVMs appear to be complicated and even, to the mathematically challenged, mysterious and magical. They do, however, have some very significant attractions, particularly their relatively little need for fully-representative training data. In comparison with ML, which needs a large training sample size especially if the data lie in high dimensional feature space, SVM requires only one sample per class provided, of course, that sample lies close to the boundary of the class. As training data are expensive and time consuming to collect, this is a major advantage. Disadvantages include the need to specify a kernel function and the relatively slow development of multiclass SVMs. The use of one against one and one against all strategies with binary SVMs is a little unaesthetic.

8.6.2 Decision Trees

It is sometimes the case that different patterns (or objects) can be distinguished on the basis of one, or a few, features. The one (or few) features may not be the same for each pair of patterns. On a casual basis, one might separate a vulture from a dog on the basis that one has wings and the other does not, while a dog and a sheep would be distinguished by the fact that the dog barks and the sheep says 'baa'. The feature 'possession of wings' is thus quite irrelevant when the decision to be made is 'is this a dog or a sheep?'. The DT classifier is an attempt to use this 'stratified' or 'layered' approach to the problem of discriminating between spectral classes. The approach is top–down rather than bottom–up, or pixel-based. The dataset is progressively subdivided into categories, on the basis of a splitting mechanism which chooses the 'best' feature on which to split the dataset. There is a range of measures available to calculate the information content of each feature, and the feature for which information is maximized is chosen.

Given p classes and k features, the DT classifier first selects a single feature on which to split the dataset into two parts. There are a number of measures that can be used to determine which of the k features should be used to split the dataset. For example, the information gain, measured by entropy, can be used. The concept of entropy is used in Chapter 6 in the discussion of image fusion or pan-sharpening techniques. Other measures including the Gini impurity index, the chi-square test are discussed in detail by Tso and Mather (2009) and Witten and Frank (2005). One notable feature of the DT approach is that feature selection methods (Section 8.9) are not required as the rules defining the tree select the 'best' feature at each split. A tree will normally have separate branches, and the feature used to cut the data need not be the same for each branch at each level. A simple DT is shown in Figure 8.22.

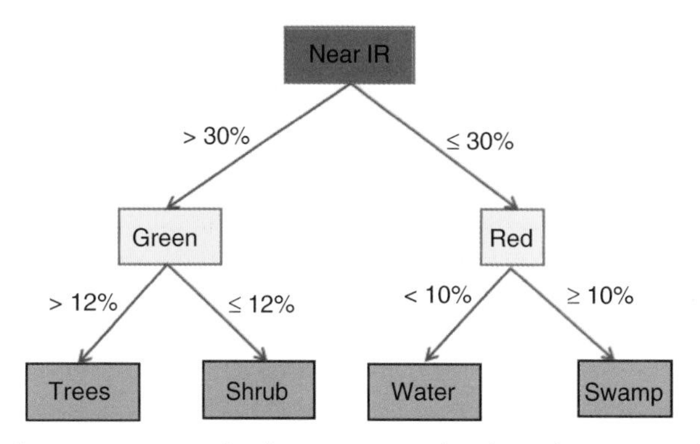

Figure 8.22 *Simple decision tree. The three features used in the classification (near-IR, red and green reflectance) are outlined in red. The root node is coloured green, the non-leaf nodes are yellow and the leaf nodes are light blue. The percentage values refer to reflectance.*

Descriptions of applications of DT techniques to the classification of remotely sensed data are provided by Friedl, Brodley and Strahler (1999) and Muchoney *et al.* (2000). Pal and Mather (2003) provide a comparison of the effectiveness of DT, ANN and ML methods in land cover classification. Tso and Mather (2009) provide a full description of the method. Other references include Laliberte, Fredrickson and Rango (2007), who compare object-oriented methods and DTs at different spatial scales, Liu, Gopal and Woodcock (2004) use an ARTMAP neural network and a DT to classify land cover. Rogan *et al.* (2003) assess land-cover change monitoring with classification trees using Landsat TM and ancillary data. Sesnie *et al.* (2008) attempt to integrate Landsat TM and variables computed from an interferometric DEM derived from the SRTM project (Section 9.2) with DTs for habitat classification and change detection in complex neotropical environments.

DTs are widely used in data mining (Witten and Frank, 2005). This text contains details of the Weka software that the authors have developed, which includes DT and random forest procedures. Quinlan (1993) also includes DT software (C4.5). Procedures such as 'pruning' DTs and building random forests are described by Witten and Frank (2005) and by Tso and Mather (2009). Random forests are collections of DTs built from random samples drawn from the training dataset. Each tree provides a label for every pixel, so if there are k trees then each pixel will have k labels. A voting procedure such as majority vote is then used to produce the final labelling scheme. Breiman (2001) suggests that the use of random forests is better than boosting because it appears to produce higher accuracy, it is robust to outliers and noise, it is faster computationally, and it gives some idea of the labelling error at each pixel. Esposito, Malerba and Semeraro (1997) conduct experiments on pruning methods. Gislason, Benediktsson and Sveinsson (2006) explore the use of random forests for land cover classification. Other worthwhile reading is Breiman (2001) and Chan and Paelinckx (2008) (who also use boosting methods described in Section 8.6.3.3), Waske and Braun (2009), Pal and Mather (2003) and Pal (2005).

Example 8.2 illustrates the results of four different classification methods – ML, DT, ANN and SVM.[2]

8.6.3 Other Methods of Classification

In this section, some less-widely used approaches to image classification are considered. The use of rule-based classifiers is described in Section 8.6.3.1, together with the simplification of the rule base using a genetic algorithm (GA). The newly developing methods of object-oriented classification are considered next. Most object-oriented classification exercises reported in the literature use the Definiens eCognition software, which is briefly described. Section 8.6.3.3 includes brief descriptions of a number of methods that are not widely applied, but which have potential for the future, including the use of hybrid and multiple classifiers, evidential reasoning and the procedures of bagging and boosting.

8.6.3.1 Rule-Based Classifiers and Genetic Algorithm

Rule-based classifications are also related to the DT approach, and represent an attempt to apply artificial intelligence methods to the problem of determining the class to which a pixel belongs (Srinivasana and Richards, 1990). The number of rules can become large, and may require the use of advanced search algorithms to perform an exhaustive analysis. The *GA*, which uses an iterative randomized approach based on the processes of mutation and crossover, has become popular in recent years as a robust search procedure. Its use in feature selection is described in Section 8.9. A brief introduction is provided by Tso and Mather (2009) while de Castro (2006) has a chapter on the topic. Seftor and Larch (1995) illustrate its use in optimizing a rule-based classifier, and Clark and Cañas (1995) compare the performance of a neural network and a GA for matching reflectance spectra. The GA is also described by Zhou and Civco (1996) as a tool for spatial decision-making. Gopi (2007) provides MATLAB code for the GA (and for several other methods mentioned in this chapter, including ICA (Section 8.5.2) and ANNs (Section 8.4.3)). Press, Teukolsky and Vetterling (2007) also make available some C++ code for GAs.

[2]I am grateful to Dr Mahesh Pal' assistance with this example.

8.6.3.2 Object-Oriented Methods

The image classification techniques discussed thus far are based on the labelling of individual pixels in the expectation that groups of neighbouring pixels will form regions or patches with some spatial coherence. This method is usually called the *per-pixel* approach to classification. An alternative approach is to use a process termed *segmentation*, which – as its name implies – involves the search for homogeneous areas in an image set and the identification of these homogeneous areas with information classes. A hierarchical approach to segmentation, using the concept of geographical scale, is used by Woodcock and Harward (1992) to delineate forest stands. This paper also contains an interesting discussion of the pros and cons of segmentation versus the per-pixel approach. Shandley, Franklin and White (1996) test the Woodcock–Harward image segmentation algorithm in a study of chaparral and woodland vegetation in southern California. Zhang *et al.* (2005) describe another semiautomatic approach to image segmentation.

Another alternative to the per-pixel approach requires *a priori* information about the boundaries of objects in the image, for example agricultural fields or forest stands. If the boundaries of these fields or stands are digitized and registered to the image, then some property or properties of the pixels lying within the boundaries of the field can be used to characterize that field. For instance, the means and standard deviations (or other statistical properties) of the six non-thermal Landsat TM bands of pixels lying within agricultural fields could be used as features defining the properties of the fields. The fields, rather than the pixels, are then classified. This method thus uses a *per-field* approach. Another technique might classify all the pixels within a field, perhaps excluding those in a possible buffer zone, as detailed in the next paragraph. The class memberships for the pixels in the field of interest are counted up and a voting procedure used to allocate the field to a class. One easy voting rule is 'winner takes all', but other rules (for example the winner must have more than 50% of the vote) can be considered.

Normally, the use of map and image data would take place within a GIS (Chapter 10), which provides facilities for manipulating digitized boundary lines (for example checking the set of lines to eliminate duplicated boundaries, ensuring that lines 'snap on' to nodes, and identifying illogical lines that end unexpectedly). One useful feature of most GIS is their ability to create buffer zones on either side of a boundary line (Gomboši and Žalik, 2005, Section 10.3.1). If a per-field approach is to be used then it would be sensible to create a buffer around the boundaries of the objects (e.g. agricultural fields) to be classified in order to remove pixels which are likely to be mixed. Such pixels may represent tractor-turning zones (headlands) as well as field boundary vegetation (hedges). The per-field approach is often used with SAR imagery because the individual pixels contain speckle noise, which leads to an unsatisfactory per-pixel classification. Averaging of the pixels within a defined geographical area such as a field generally gives to better results (Schotten, van Rooy and Janssen 1995; Wooding, Zmuda and Griffiths, 1993). Alberga (2007) and Alberga, Satalino and Staykova (2008) evaluate a window-based approach to classification of polarimetric SAR data. Lobo, Chic and Casterad (1996) discuss the per-pixel and per-field approaches in the context of Mediterranean agriculture. Tarabalka, Benediktsson and Chanussot (2009) use a combined unsupervised clustering method to aid segmentation and a per-pixel classifier, then fuse the two representations.

The use of the per-field approach in which segmentation precedes classification is an example of the object-oriented approach, in which the fields (objects) are classified rather than the pixels making up the image. This approach is described as object oriented (as opposed to pixel oriented). Object-oriented methods of image classification have become more popular in recent years due to the availability of software (eCognition) developed by the German company Definiens Imaging. This software uses a segmentation approach at different scale levels from coarse to fine, using both spectral properties and geometric attributes of the regions, such as scale, colour, smoothness and shape. A hierarchy of regions at different scale levels is thus developed. Esch *et al.* (2003) provide references to the technical details, and compare the performance of the eCognition approach with that of the standard ML method. The former proves to be about 8% more accurate than the latter, though the overall accuracy (Section 8.10) of the ML result is surprisingly high at 82%. See also Baatz *et al.* (2004), Chubey, Franklin and Wulder (2006), Ivits *et al.* (2005) and Yu *et al.* (2006) for examples of the use of object-based classification. Antonarakis, Richards and Brasington (2008) use airborne lidar data in an object-based land cover classification. Bork and Su (2007) also use lidar data in conjunction with multispectral imagery to classify rangeland data. Yu *et al.* (2006) report on a successful use of object-based classification using DAIS hyperspectral images. A full description is provided by Navullur (2006) who provides details of the operational use of eCognition software. Yet another example is provided by Marçal *et al.* (2005), who compare four methods of supervised classification of land cover, based on an initial segmentation using eCognition. Marçal and Rodrigues (2009) discuss methods of evaluating the quality of image segmentations. A comprehensive source of information and a good starting point for appreciating the object-oriented paradigm are the contributions to the book edited by Blaschke, Lang and Hay (2008).

8.6.3.3 Other Methods

8.6.3.3.1 Evidential Reasoning If a single image set is to be classified using spectral data alone then the absolute or relative 'reliability' of the features is not usually taken into consideration. All features are considered to be equally reliable, or important. Where multisource data are used, the question of reliability needs to be considered. For example, one might use data derived from maps, such as elevation, and may even use derivatives of these features, such as slope and aspect, as well as datasets acquired by different sensor systems (Section 8.7). The Dempster–Shafer theory of evidence (Shafer, 1979) has been used by researchers to develop the method of *evidential reasoning*, which is a formal procedure which weights individual data sources according to their reliability or importance. The method uses the concepts of belief and plausibility to create decision rules via which the image pixels are labelled. Papers by Cohen and Shoshan (2005), Duguay and Peddle (1996), Lein (2003), Lu and Weng (2007), Peddle (1995), Peddle and Ferguson (2002), Mertikas and Zervakis (2001) and Tso and Mather (2009) illustrate the uses of the evidential reasoning approach to classification, and compare its effectiveness with that of other approaches.

8.6.3.3.2 Bagging, Boosting and Ensembles of Classifiers Procedures known as 'bagging' and 'boosting' can be used to try to improve the performance of a classifier. Bagging operates by generating a series of training data subsets by random sampling with replacement from the training dataset, and using each of these subsets to produce a classification. A voting procedure is then used to determine the class of a given pixel. The simplest such method is majority voting; if the majority of the p classifiers place pixel j into class k then pixel j is given the label k (Breiman, 1996). Boosting involves the repeated presentation of the training dataset to a classifier, with each element of the training set being given a weight that is proportional to the difficulty of classifying that element correctly. The initial weights given to the pixels are unity. The Adaboost method of Freund and Schapire (1996) is probably the best known of these methods. Chan, Huang and DeFries (2001) provide a short example of the use of these procedures in land cover classification. They find that bagging and boosting do not always result in higher classification accuracy, and that behaviour varies with the classification algorithm used. Thus, they note that both bagging and boosting increased the accuracy of the DT classifier from 89.3 to 94.0% (bagging) and 95.8% (boosting). The classification accuracy for a SOM-type ANN increased slightly for bagging (from 86.9 to 87.6) but fell quite sharply for bagging (to 80.3%). An SVM classifier was unaffected by boosting and bagging.

Other interesting papers on these topics are Drucker, Schapire and Simard (1993) and Bauer and Kohavi (1999). Dietterich (2000) reports on an experimental comparison of bagging, boosting and randomisation for constructing ensembles of DTs. Pal (2008) looks at the use of bagging and boosting in the context of SVMs, and disagrees with authors who suggest that boosting (specifically AdaBoost) improves SVM performance. However, he concludes that bagging worked well in some instances. Sharkey (2000) considers using a number of ANNs in a multinet system, each of which classifies the image. Decision fusion then follows, usually a majority vote (with the possibility of classifying the pixel as 'other' if there is only a weak majority). The component ANNs may differ in terms of initial weights, or topology, or in input data (bagging again). Other references are Kittler *et al.* (1998) on combining classifiers, and Doan and Foody (2007) who study ways of increasing soft classification accuracy through the use of an ensemble of classifiers. Foody, Boyd and Sanchez-Hernandez (2007) examine the problem of mapping a specific class with an ensemble of classifiers. Ensembles of SVMs are used by Pal (2008) in land cover classification. Lee and Ersoy (2007) use ensembles of SOM (Section 8.4.3) in a hierarchical approach to classification. Liu, Gopal and Woodcock (2004) conclude by saying that

> ...The hybrid approach seems suitable to tackle a variety of classification problems in remote sensing and may ultimately aid map users in making more informed decisions (p. 963).

8.7 Incorporation of Non-spectral Features

Two types of feature in addition to spectral values can be included in a classification procedure. The first kind are measures of the texture of the neighbourhood of a pixel, while the second kind represent external (i.e. non-remotely-sensed) information such as terrain elevation values or information derived from soil or geology maps. Use of textural information has been limited in passive remote sensing, largely because of two difficulties. The first is the operational definition of texture in terms of its derivation from the image data, and the second is the computational cost of carrying out the texture calculations relative to the increase in classification accuracy, if any. External data have not been widely used either, though digital cartographic data have become much more readily available in recent years. A brief review of both these topics is provided in this section.

8.7.1 Texture

Getting a good definition of texture is almost as difficult as measuring it. While the grey level of a single pixel or a group of pixels in a greyscale image can be said to represent 'tone', the texture of the neighbourhood in which that pixel lies is a more elusive property, for several reasons. At a simple level, texture can be thought of as the variability in tone within a neighbourhood, or the pattern of spatial relationships among the grey levels of neighbouring pixels, and which is usually described in terms such as 'rough' or 'smooth'. Variability is a variable property, however; it is not necessarily random – indeed, it may be structured with respect to direction as, for instance, a sub-parallel drainage pattern on an area underlain by dipping beds of sandstone. The observation of texture depends on two factors. One is the scale of the variation that we are willing to call 'texture' – it might be local or regional. The second is the scale of observation. Microscale textures that might be detected by the panchromatic band of the SPOT HRV would not be detected by the NOAA AVHRR due to the different spatial resolutions of the two sensor systems (10 m and 1.1 km respectively). We must also be careful to distinguish between the real-world texture present, for example in a field of potatoes (which are generally planted in parallel rows) and the texture that is measurable from an image of that field at a given spatial resolution (Ferro and Warner, 2002; Foody and Curran, 1994).

The fact that texture is difficult to measure is no reason to ignore it. It has been found to be an important contributor to the ability to discriminate between targets of interest where the spatial resolution of the image is sufficient to make the concept a meaningful and useful one, for example in manual photo interpretation. The simplest method of estimating the texture property of an image is to measure the image variance for each of a number of moving windows. This approach could be especially useful in cases where a coregistered panchromatic image is supplied with the multispectral image (e.g. SPOT HRV, Landsat ETM+, IKONOS, QuickBird, etc.). The variance of the panchromatic channel would then provide an estimate of texture (Emerson, Siu-Ngan Lam and Quattrochi, 2005).

The earliest application of texture measurements to digital remotely-sensed image data was published by Haralick, Shanmugam and Dinstein (1973). These authors proposed what has become known as the Grey Level Co-occurrence Matrix (GLCM), which represents the distance and angular spatial relationships over an image subregion of specified size. Each element of the GLCM is a measure of the probability of occurrence of two grey scale values separated by a given distance in a given direction. The concept is more easily appreciated via a simple numerical example. Table 8.3a shows a small segment of a digital image quantized to four grey levels (0–3). The number of adjacent pixels with grey levels i and j is counted and placed in element (i, j) of the GLCM **P**. Four definitions of adjacency are used; horizontal (0°), vertical (90°), diagonal (bottom left to top right −45°) and diagonal (top left to bottom right −135°). The inter-pixel distance used in these calculations is 1 pixel. Thus, four GLCM are calculated, denoted $\mathbf{P}_0$, $\mathbf{P}_{90}$, $\mathbf{P}_{45}$ and $\mathbf{P}_{135}$ respectively. For example, the element $\mathbf{P}_0$ (0,0) is the number of times a pixel with grey scale value 0 is horizontally adjacent to a pixel which also has the grey scale value 0, scanning from left to right as well as right to left. Element $\mathbf{P}_0$ (1, 0) is the number of pixels with value 1 that are followed by pixels with value 0, while $\mathbf{P}_0$ (0, 1) is the number of pixels with value 0 that are followed by pixels with value 1, again looking in both the left-right and right-left directions. The four GLCM are shown in Table 8.3b–e.

Haralick, Shanmugam and Dinstein (1973) originally proposed 32 textural features to be derived from each of the four GLCM. Few instances of the use of all these features can be cited; Jensen and Toll (1982) use only one, derived from the Landsat 1–3 MSS band 5 image. The first two of the Haralick measures will be described here to illustrate the general approach. The first measure (f_1) is termed the *angular second moment*, and is a measure of homogeneity. It effectively measures the number of transitions from one grey level to another and is high for few transitions. Thus, low values indicate heterogeneity. The second Haralick texture feature, *contrast* (f_2), gives non-linearly increasing weight to transitions from low to high greyscale values. The weight is the square of the difference in grey level. Its value is a function of the number of high/low or low/high transitions in grey level. The two features are formally defined by:

$$f_1 = \sum_{i=1}^{N} \sum_{j=1}^{N} \left\{ \frac{P(i, j)}{R} \right\}^2$$

$$f_2 = \sum_{n=0}^{N-1} \sum_{i=1}^{N} \sum_{j=1}^{N} \frac{P(i, j)}{R} \qquad |i - j| = n$$

Table 8.3 *Example data and derived grey-tone spatial dependency matrices. (a) Test dataset. (b–e) Grey-tone spatial dependency matrices for angles of 0, 45, 90 and 135°, respectively.*

(a)

0	0	0	2	1
0	1	1	2	2
0	1	2	2	3
1	1	2	3	3

(b)

	0	**1**	**2**	**3**
0	4	2	1	0
1	2	4	4	0
2	1	4	4	2
3	0	0	2	2

(c)

	0	**1**	**2**	**3**
0	2	2	0	0
1	1	4	3	0
2	0	3	6	0
3	0	0	0	2

(d)

	0	**1**	**2**	**3**
0	4	3	0	0
1	3	4	2	0
2	0	2	6	2
3	0	0	2	2

(e)

	0	**1**	**2**	**3**
0	0	4	1	0
1	4	0	3	0
2	1	3	2	3
3	0	0	3	0

where N is the number of grey levels, $\mathbf{P}(i, j)$ is an element of one of the four GLCM listed above, and R is the number of pairs of pixels used in the computation of the corresponding $\mathbf{P}$. For the horizontal and vertical directions R is equal to $2N^2$ while in the diagonal direction R equals $2(N - 1)^2$. Haralick, Shanmugam and Dinstein (1973) and Haralick and Shanmugam (1974) give examples of the images and corresponding values of f_1 and f_2. A grassland area gave low (0.064–0.128) values of f_1 indicating low homogeneity and high contrast whereas a predominantly water area has values of f_1 ranging from 0.0741 to 0.1016 and of f_2 between 2.153 and 3.129 (higher homogeneity, lower contrast). These values are averages of the values of f_1 and f_2 for all four angular grey-tone spatial dependency matrices. The values of f_1 for the example data in Table 8.3 are (for angles of 0, 45, 90 and 135°): 0.074, 0.247, 0.104 and 0.216 while the values of f_2 for the same data are 0.688, 0.444, 0.438 and 1.555.

Rather than compute the values of these texture features for a sequence of moving windows surrounding a central pixel, Haralick and Shanmugam (1974) derive them for 64 × 64 pixel subimages. This shortcut is unnecessary nowadays, as sufficient computer power is available to compute local texture measures for individual pixels. They also use 16 rather than 256 quantization levels in order to reduce the size of the matrices $\mathbf{P}$. If all 256 quantization levels of the Landsat TM were to be used, for example then $\mathbf{P}$ would become very large. The reduction in the number of levels from 256 to 16 or 32 might be seen as an unacceptable price to pay, though if the levels are chosen after a histogram equalization enhancement (Section 5.3.2) to ensure equal probability for each level then a reduction from 256 to 64 grey levels will give acceptable results (Tso, 1997; Tso and Mather, 2009). Clausi (2002) presents some results on the effects of quantization levels on the GLCM texture statistics.

The rapid improvements in computing power in recent years have led to an increased interest in the use of texture measures. Paola and Schowengerdt (1997) incorporate texture features into a neural network-based classification by including as network inputs the greyscale values of the eight neighbours of the pixel to be classified. The central pixel is thus classified on the basis of its spectral reflectance properties plus the spectral reflectance properties of the neighbourhood. Although the number of input features is considerably greater than would be the case if the central pixel alone were to be input, Paola and Schowengerdt (1997) report that the extra size of the network is compensated by faster convergence during training. Clausi and Zhao (2003) present a computationally fast method of obtaining GLCM texture statistics, and provide computer code for their technique. Other interesting papers on GLCM based methods are Dutra (1999), Franklin, Maudie and Lavigne (2001), Puissant, Hirsch and Weber (2005), Ouma, Tetuko and Tateishi (2008) and Wang *et al.* (2004).

A second approach to texture quantization uses bandpass filters in the frequency domain (Section 7.5) to measure the relative proportion of high frequency information in each of a number of moving windows over the image for which texture is measured. Although this seems to be intuitively simple and attractive in theory, there have not been a large number of applications cited in the literature (Couteron, Barbier and Gautier, 2006; Mather, Tso and Koch, 1998; Proisy, Couteron and Fromard, 2007; Riou and Seyler, 1997).

A third approach is based on the calculation of the fractal or multifractal dimension of a moving window. There are several ways of calculating fractal dimension, which is used here as a measure of the roughness of a surface. Tso and Mather (2009, pp. 224–231) consider several ways of estimating the fractal dimension, including wavelet-based methods, the use of a fractal Brownian motion model and box counting methods. Emerson, Siu-Ngan Lam and Quattrochi (2005) also uses fractal dimension as a measure of texture.

Yet another method of quantifying texture is described by Ouma, Tetuko and Tateishi (2008). They determine an appropriate window size from the calculation of semi-variograms, then use eight texture methods derived from the GLCM to differentiate forest and non-forest land cover. It is difficult to imagine that a single window size could satisfy all the requirements for texture quantization of all the land cover types present in the image. In this study, the GLCM approach is contrasted with the use of discrete wavelet texture measures (Section 6.7). The authors obtained higher classification accuracies using GLCM measures and spectral information, with an overall accuracy (Section 8.10) of 74% compared with an average accuracy of 64% for the wavelet-based methods. Dell'Acqua *et al.*, 2006, use a multi-scale co-occurrence matrix approach to the classification of SAR images.

Bian (2003) describes a texture measure (entropy) derived from multiresolution wavelet decomposition. Entropy is a measure of order, and has been met with at several points in this book. It is defined for each central position of a moving window of size $n \times n$ by the relationship:

$$entropy = -\sum_{i=1}^{k} p(i) \log_2 (p(i))$$

Where k is the number of grey levels and $p(i)$ is the probability of grey level i, computed from $f(i)/n^2$, the frequency of the ith histogram class divided by the number of pixels in the window. The entropy measure can be calculated for one or all of the H_1, D_1 and V_1 quadrants of the wavelet decomposition (Figure 6.37) or for the corresponding quadrants at the second or higher decomposition levels. This ability to measure texture at different scales is a characteristic that might be developed further.

Further reading on alternative approaches to and applications of texture measures in image classification is provided by Bharati, Liu and MacGregor (2004), Bruzzone *et al.* (1997), Buddenbaum *et al.* (2002), Carlson and Ebell (1995), Chica-Olmo and Abarca-Hernandez (2000), Coburn and Roberts (2004), Dekker (2003), de Jong and Burrough (1995), Dikshit (1996), Lark (1996), Kuplich, Curran and Atkinson (2005), Maillard (2003), Ouma, Tetuko and Tateishi (2008), Soares *et al.* (1997), Tso (1997) and Tso and Mather (2009). Lloyd *et al.* (2004) consider per-field measures of texture.

Given the considerable effort that has gone into the study of texture measures, two things are surprising: first, the oldest of these measures, based on the GLCM, is still widely used and is still one of the top performers in comparative studies and, secondly, no single measure has proved to be superior. It is nevertheless the case that the inclusion of texture features enhances classifier performance. This is particularly appropriate to cases in which a high-resolution panchromatic band is supplied with the multispectral data (e.g. Landsat ETM+). The texture features can be based on the panchromatic band rather than on one or more of the multispectral bands. Tso and Mather (2009) find that the wavelet-based method, the multiplicative autoregressive random (MAR) field and the GLCM perform best in terms of classification accuracy (p. 251). Texture features are particularly badly affected by high signal-to-noise ratios (SNRs), as high-frequency noise can be confused with texture.

8.7.2 Use of External Data

The term external (or ancillary) is used to describe any data other than the original image data or measures derived from these data. Examples include elevation and soil type information or the results of a classification of another image of the same spatial area. Some such data are not measured on a continuous (ratio or interval) scale and it is therefore difficult to justify their inclusion as additional feature vectors. Soil type or previous classification results are examples, both being categorical variables. Where a continuous variable, such as elevation, is used difficulties are encountered in deriving training samples. Some classes (such as water) may have little relationship with land surface height and the incorporation of elevation information in the training class may well reduce rather than enhance the efficiency of the classifier in recognizing those categories.

An external variable may be used to stratify the image data into a number of categories. If the external variable is land elevation then, for example the image may be stratified in terms of land below 500 m, between 500 and 800 m and above 800 m. For each stratum of the data an estimate of the frequency of occurrence of each class must be provided by the user. This estimate might be derived from field observation, sampling of a previously classified image or from sample estimates obtained from air photographs or maps. The relative frequencies of each class are then used as estimates of the prior probabilities of a pixel belonging to each of the k classes and the ML algorithm used to take account of these prior probabilities (Section 8.4.2.3). The category or level of the external variable is used to point to a set of prior probabilities, which are then used in the estimation of probabilities of class membership. This would assist in the distinction between classes which are spectrally similar but which have different relationships with the external variable. Strahler, Logan and Bryant (1978) used elevation and aspect as external variables; both were separated into three categories and used as pointers to sets of prior probabilities. They found that the elevation information contributed considerably to the improvement in the accuracy of forest cover classification. Whereas the spectral features alone produced a classification with an accuracy estimated as 57%, the addition of terrain information and the introduction of prior probability estimates raised this accuracy level to a more acceptable 71%. The use of elevation and aspect to point to a set of prior probabilities raised the accuracy further to 77%. Strahler (1980) provides an excellent review of the use of external categorical variables and associated sets of prior probabilities in ML classification. He concludes that the method 'can be a powerful and effective aid to improving classification accuracy'. Another accessible reference is Hutchinson (1982), while Maselli *et al.* (1995b) discuss integration of ancillary data using a non-parametric method of estimating prior probabilities.

Elevation datasets are now available for many parts of the world, generally at a scale of 1 : 50 000 or coarser. DEMs can also be derived from stereo SPOT and ASTER images as well as from interferometric data from SAR sensors such as TerraSAR-X. The widespread availability of GIS means that many users of remotely sensed data can now derive DEM by digitizing published maps, or by using photogrammetric software to generate a DEM from stereoscopic images such as SPOT HRV, ASTER or IRS-1 LISS. Care should be taken to ensure that the scale of the DEM matches that of the image. GIS technology allows the user to alter the scale of a dataset, and if this operation is performed thoughtlessly then error is inevitable. Users of DEM derived from digitized contours should refer to one of the many GIS textbooks now available (for example Bonham-Carter, 1994; Longley *et al.*, 2005) to ensure that correct procedures are followed.

Rather than use an external variable to stratify the image data for improved classifier performance, users may prefer to use what has become known as the *stacked vector* approach, in which each feature (spectral, textural, external) is presented to the classifier as an independent input. Where a statistical classifier, such as ML, is used then this approach may well not be satisfactory. Some external variables, such as elevation, may be measured on a continuous scale but may not be normally distributed or even unimodal for a given class. Other variables, such as lithology or soil type, may be represented by a categorical label that the ML classifier cannot handle. The value of ANN and DT classifiers is that they are non-parametric, meaning that the frequency distribution and scale of measurement of the individual input feature is not restricted. Thus, the ANN-based classifier can accept all kinds of input features without any assumption concerning the normality or otherwise of the associated frequency distribution and without consideration of whether the feature is measured on a continuous, ordinal or categorical scale. One problem with an indiscriminate approach, however, is that all features may not have equal influence on the outcome of the classification process. If one is trying to distinguish between vultures and dogs, then 'possession of wings' is a more significant discriminating feature than 'colour of eyes', though the latter may have some value. Evidential reasoning (Section 8.6) offers a more satisfactory approach.

See Treitz and Howarth (2000) and references therein for a study of forest classifications involving hyperspectral reflectance, spectral–spatial, textural and geomorphometric variables. They conclude that, in a low to moderate relief environment, the use of external data together with remotely-sensed data leads to improved discrimination of forest ecosystem classes.

8.8 Contextual Information

Geographical phenomena generally display order or structure, as shown by the observation that landscapes are not, in general, randomly organized. Thus, trees grow together in forests and groups of buildings form towns and villages. The relationship between one element of a landscape and the whole defines the context of that element. So too the relationship between one pixel and the pixels in the remainder of the image is the context of that pixel. Contextual information is often taken into account after a preliminary classification has been produced, though at the research level investigations are proceeding into algorithms which can incorporate both contextual and spectral information simultaneously (Kittler and Föglein, 1984). The simplest methods are those which are applied following the classification of the pixels in an image using one of the methods described in Section 8.4. These methods are similar in operation to the spatial filtering techniques described in Chapter 7 for they use a moving window algorithm.

The first of these methods is called a 'majority filter'. It is a logical rather than numerical filter since a classified image consists of labels rather than quantized counts. The simplest form of the majority filter involves the use of a filter window, usually measuring three rows by three columns, is centred on the pixel of interest. The number of pixels allocated to each of the k classes is counted. If the centre pixel is not a member of the majority class (containing five or more pixels within the window) it is given the label of the majority class. A threshold other than five (the absolute majority) can be applied – for example if the centre pixel has fewer than n neighbours (in the window) that are not of the same class then relabel that pixel as a member of the majority class. The effect of this algorithm is to smooth the classified image by weeding-out isolated pixels which were initially given labels that were dissimilar to the labels assigned to the surrounding pixels. These initial dissimilar labels might be thought of as noise or they may be realistic. If the latter is the case then the effect of the majority filter is to treat them as detail of no interest at the scale of the study, just as contours on a 1 : 50 000-scale map are generalized (smoothed) in comparison with those on a map of the same area at a 1 : 25 000 scale. A modification of the algorithm just described is to disallow changes in pixel labelling if the centre pixel in the window is adjacent to a pixel with an identical label. In this context adjacent can mean having a common boundary (i.e. to the left or right, above or below) or having a common corner. The former definition allows four pixels to be adjacent to the centre pixel, the latter eight.

Harris (1981, 1985) describes a method of post-classification processing which uses a probabilistic relaxation model. An estimate of the probability that a given pixel will be labelled $l_i (i = 1, 2, \ldots, k)$ is required. Examination of the pixels surrounding the pixel under consideration is then undertaken to attempt to reduce the uncertainty in the pixel labelling by ensuring that pixel labels are locally consistent. The procedure is both iterative and rather complicated. The results reported by Harris (1985) show the ability of the technique to clean up a classified image by eliminating improbable occurrences (such as isolated urban pixels in a desert area) while at the same time avoiding smoothing-out significant and probably correct classifications. However, the computer time requirements are considerable. Further discussion of the probabilistic relaxation model is given in Rosenfeld (1976), Peleg (1980), Kittler (1983) and Kontoes and Rokos (1996), while an alternative approach to the problem of specifying an efficient spectral–spatial classifier is discussed by Landgrebe (2003). An excellent general survey is Gurney and Townshend (1983).

More recently, attention has been given to the use of geostatistical methods of characterizing the spatial context of a pixel that is to be classified. Geostatistical methods are summarized in Section 8.5. Image data are used to characterize the spectral properties of the candidate pixel, and geostatistical methods provide a summary of its spatial context, so that both are simultaneously considered in the decision-making process. See Lark (1998) and van der Meer (1994, 1996a, 1996b) for a fuller exposition. Flygare (1997) gives a review of advanced statistical methods of characterizing context. Wilson (1992) uses a modified ML approach to include neighbourhood information by the use of a penalty function which increases the 'cost' of labelling a pixel as being different from its neighbours. Sharma and Sarkar (1998) review a number of approaches to the inclusion of contextual information in image classification. Finally, recall the opening paragraph of this section, where it is noted that the Earth's surface is generally ordered, with similarity increasing with closeness. One could consider modelling this property using Markovian methods – what is the probability of the correct label for pixel (i, j) given the labels of the surrounding pixels? The use of Markov Random Fields is considered by Tso and Mather (2009, Chapter 8). MRFs have been widely used for 30 years for characterizing contextual information in image segmentation and image restoration applications. The level of mathematics required to fully comprehend the use of MRF is high, however.

8.9 Feature Selection

Developments in remote sensing instruments over the last 10 years have resulted in image data of increasingly higher resolution becoming available in more spectral channels. Thus, the volume and dimensionality of datasets being used in image classification is exceeding the ability of both available software systems and computer hardware to deal with it. However, as shown in the discussion of the DT approach to classification (Section 8.4.6), it is possible to base a classification on the consideration of the values measured on one spectral band at a time. In this section the idea will be extended, so that we will ask: can the dimensions of the data set be reduced (in order to save computer time) without losing too much of the information present in the data? If a subset of the available spectral bands (and other features such as textural and ancillary data) will provide almost as good a classification as the full set then there are very strong arguments for using the subset. We will consider what 'almost as good' means in this context in the following paragraphs.

Reduction of the dimensionality of a dataset is the aim of PCA (Section 6.4). An obvious way of performing the feature selection procedure would be to use the first m principal components in place of the original p features (m being smaller than p). This does not, however, provide a measure of the relative performance of the two classifications – one based on all p features, the other on m principal components. Methods of accuracy assessment (Section 8.10) might be used on training and test sites to evaluate the performance directly. Information, in terms of principal components, is directly related to variance or scatter and is not necessarily a function of inter-class differences. Thus, the information contained in the last $(p - m)$ components might represent the vital piece of information needed to discriminate between class x and class y, as shown in the example of PCA in Section 6.4. PCA might therefore be seen as a crude method of feature selection if it is employed without due care. It could be used in conjunction with a suitable method for determining which of the possible p components should be selected in order to maximize inter-class differences, as discussed below. Jia and Richards (1999) propose a method based on a modification of PCA.

Two widely used methods of feature selection are discussed in this section. The first is based on the derivation of a measure of the difference between all pairs from the k groups. It is called *divergence*. The second is more empirical. It evaluates the performance of a classifier in terms of a set of test data for which the correct class assignments have been established by ground observations or by the study of air photographs or maps. The classifier is applied to subsets of the p features and classification accuracy measured for each subset using the techniques described in Section 8.10. A subset is selected that gives a sufficiently high accuracy for a specific problem.

The technique based on the divergence measure requires that the measurements on the members of the k classes are distributed in multivariate normal form. The effect of departures from this assumption is not known, but one can be certain that the results of the analysis would be less reliable as the departures from normality increased. If the departures are severe then the results could well be misleading. Hence, the divergence method is only to be recommended for use in conjunction with statistical (rather than neural) classifiers. The divergence measure J based on a subset m of the p features is computed for classes i and j as follows (Singh, 1984) with a zero value indicating that the classes are identical. The greater the value of $J(i, j)$ the greater is the class separability based on the m selected features.

$$J(i, j) = 0.5\mathrm{tr}\left\{(\mathbf{S}_i - \mathbf{S}_j)(\mathbf{S}_j^{-1} - \mathbf{S}_i^{-1})\right\} + 0.5\mathrm{tr}\left\{(\mathbf{S}_i^{-1} + \mathbf{S}_j^{-1})(\bar{\mathbf{x}}_i - \bar{\mathbf{x}}_j)(\bar{\mathbf{x}}_i - \bar{\mathbf{x}}_j)'\right\}$$

The symbol tr(.) means the trace or the sum of the diagonal elements of the indicated matrix. $\mathbf{S}_i$ and $\mathbf{S}_j$ are the $m \times m$ sample variance–covariance matrices for classes i and j, computed for the m selected features, and $\bar{\mathbf{x}}_i$ and $\bar{\mathbf{x}}_j$ are the corresponding sample mean vectors. For $m = 1$ (a single feature) the divergence measure for classes i and j is:

$$J(i, j) = 0.5\left(\frac{s_i^2}{s_j^2} + \frac{s_j^2}{s_i^2} - 2\right)$$

where s_i^2 and s_j^2 are the variances of the single feature calculated separately for classes i and j. Since the divergence measure takes into account both the mean vectors and the variance–covariance matrices for the two classes being compared, it is clear that the interclass difference is being assessed in terms of (i) the shape of the frequency distribution and (ii) the location of the centre of the distribution. The divergence will therefore be zero only when the variance–covariance matrices and the mean vectors of the two classes being compared are identical.

The distribution of $J(i, j)$ is not well known so a measure called the *transformed divergence* is used instead. This has the effect of reducing the range of the statistic, the effect increasing with the magnitude of the divergence. Thus, when averages are taken, the influence of one or more pairs of widely separated classes will be reduced. The transformed divergence is obtained from:

$$J_T(i, j) = c(1 - \exp[-J(i, j)/8])$$

with c being a constant used to scale the values of J_T onto a desired range. Sometimes the value 2000 is used as a scaling factor, but a value of 100 seems to be equally reasonable as the values of J_T can then be interpreted in the same way as percentages. A value of J_T of 80 or more indicates good separability of the corresponding classes i and j. The values of $J_T(i, j)$ are averaged for all possible mutually exclusive pairs of classes i and j and the average pairwise divergence is denoted by J_{Tav}.

$$J_{Tav} = \frac{2}{k(k-1)} \sum_{i=1}^{k-1} \sum_{j=1}^{i} J_T(i, j)$$

Study of the individual $J_T(i, j)$ might show that some pairs of classes are not statistically separable on the basis of any subset of the available features. The feature selection process might then also include a class amalgamation component. It might be worth following another line of thought. If the aim of feature selection is to produce the subset of m features that best combines classification accuracy and computational economy then, instead of considering the average separability of all pairs of classes, why not try to find that set of m features that maximizes the minimum pairwise divergence? In effect, this is trying to find the subset of m features that best performs the most difficult classification task. The minimum pairwise divergence is:

$$J_{min}(i, j) = \min J(i, j) \qquad i < j$$

A measure called the *Bhattacharyya distance* is sometimes used in place of the divergence to measure the statistical separability (or, more correctly, the probability of correct classification) of a pair of spectral classes. It is computed from the expression:

$$B_{12} = \frac{1}{8}(\bar{\mathbf{x}}_1 - \bar{\mathbf{x}}_2)' \frac{\mathbf{S}_1 + \mathbf{S}_2}{2} (\bar{\mathbf{x}}_1 + \bar{\mathbf{x}}_2) + \frac{1}{2} \ln \frac{\frac{\mathbf{S}_1 - \mathbf{S}_2}{2}}{|\mathbf{S}_1|^{0.5} |\mathbf{S}_2|^{0.5}}$$

(Haralick and Fu, 1983). The quantity B_{ij} is computed for every pair of classes given m features. The sum of B_{ij} for all $k(k-1)/2$ classes is obtained and is a measure of the overall separability of the k classes using m features. All possible combinations of m out of p features are used to decide the best combination. Again, selection algorithms such as those described above for the transformed divergence could be used to improve the efficiency of the method. Like the divergence measure the Bhattacharyya distance is based on the assumption of multivariate normality.

Given that the *raison d'être* of feature selection is the availability of several (more than four) features, the selection of combinations of m from p features is a problem. The number of subsets of size m that can be drawn from a set with p elements is

$$\binom{p}{m} = \frac{p!}{m!(p-m)!}$$

The symbol '!' indicates 'factorial'; for example 3! is $3 \times 2 \times 1 = 6$. If p is large then the number of subsets soon becomes very considerable. Take the Daedalus airborne scanner as an example. This instrument generates 12 channels of spectral data. If we assume that no texture features or ancillary data are added, then the number of subsets of size $m = 4$ is 495. If subsets of size $m = 12$ are to be drawn from a dataset with $p = 24$ features then the number of subsets is 2 704 156. Clearly any brute-force method involving the computation of the average pairwise divergence for such a large number of subsets is out of the question. The problem of selection of optimal subsets is not dissimilar to the problem of determining the best subset of independent variables in multiple linear regression. Any of three main approaches can be used – these are the forward selection, backward elimination and stepwise procedures. The forward selection method starts with the best subset of size $m = 1$.

Call this feature f_1. Now find the best subset of size $m = 2$ including f_1, that is f_1 plus one other feature. The best subset at the end of the second cycle will be $\{f_1, f_2\}$. The procedure continues to determine subsets $\{f_1, f_2, f_3\}$, and so on until all features are included. The user can then evaluate the list of features included and corresponding divergence value, and must weigh up the advantages of using fewer features against the cost of lower classification accuracy.

The backward elimination method works the opposite way round. Starting with the complete set $\{f_1, f_2, \ldots, f_p\}$, remove that feature which contributes least to the average pairwise divergence. This is done by computing the average pairwise divergence for all subsets of size $p - 1$. Repeat until $m = 1$. Neither procedure is guaranteed to produce the optimal subset; indeed, both may produce differing results unless the dataset is so clearly structured that no selection procedure is needed.

Stepwise methods incorporate both the addition of features to the selected set, as in forward selection, and their removal, as in backward elimination. The single best feature is selected first, with 'best' being defined as 'generating the largest classification accuracy'. Call this feature f_1. Now add that feature drawn from the set of remaining features that, together with f_1, produces the highest classification accuracy for all pairs of features that include f_1 So now the best subset is $\{f_1, f_2\}$. The increase in classification accuracy resulting from the addition of f_2 to the best subset can be tested statistically; if the increase is not statistically significant then f_2 is eliminated and the procedure terminates. If the increase is acceptably large, then a third feature is added, and the testing procedure is applied again. A second statistical test is also used. It is concerned with the question of whether any of the features included in the best subset can be eliminated without any significant loss of classification accuracy. Features that are included at an early stage in the selection process can be eliminated later. Interaction (shown by high correlations) between variables is responsible for these apparent anomalies. The process terminates when no excluded features can be added and no included features can be eliminated.

Other methods include the use of the GA as a search procedure. The use of the GA in this role is noted in Section 8.6. It is also applicable to the feature selection problem. A good introduction to the workings of the GA is provided by Holland (1992). Other references are Coley (1999), Man, Tang and Kwong (1999), Mitchell (1996), van Coillie, Verbeke and De Wulf (2007) and Yu, De Backer and Scheunders (2002). The last of these papers uses an application based on the use of AVIRIS hyperspectral data. Univariate DT classifiers (Section 8.4.5) effectively select one feature at each level, using any one of a range of criteria (Gomez-Chova *et al.*, 2003; Tso and Mather, 2009). Bazi and Melgani (2006) also focus on hyperspectral data, but they use a SVM (Section 8.4.4) to extract features and classify data. They also conduct a comparative analysis of feature extraction techniques.

Kumar (1979) describes an experiment in which the exhaustive search, forward and backward selection algorithms were employed. He found that the forward selection method produced results that were almost as good as exhaustive search and which were better than those produced by the backward elimination method. Mao (2004) also uses forward and backwards selection in an orthogonal space.

Other studies of feature selection are provided by Aha and Bankert (1996), Baofeng, Gunn and Damper (2006), Kavzoglu and Mather (2002), Muasher and Landgrebe (1984) and Ormsby (1992). Yool *et al.* (1986) compare the use of transformed divergence and empirical approaches to the assessment of classification accuracy (Section 8.10). They found no clear agreement between the results from the two alternative approaches, and attributed the differences – which in some instances were considerable – to departures from normality and conclude that '...a divergence algorithm requiring normally-distributed data may not be a reliable indicator of performance' (Yool *et al.*, 1986, p. 689). However, if the empirical classification accuracy approach is used then a classification analysis must be carried out on test samples for each subset of m features. Kavzoglu and Mather (2000) use feature selection to reduce the size of an artificial neural net (Section 8.4.3) and thereby improve its generalization capabilities.

The availability of high-dimensional multispectral image data is thus seen to be a mixed blessing. Additional spectral channels provide more detailed or more extensive information on the spectral response of the ground-cover targets, though their use requires additional computer time. Classification accuracy is dependent on feature-set size, yet no clear and recommendable algorithm is available to determine the subset that will produce the best compromise between accuracy and cost. Factors other than dimensionality will affect the choice of subset; the number of classes and their relative separability will have some influence on the number and choice of features needed to discriminate between them. Studies that have been carried out to date indicate that statistical methods (based on the assumption of normal distributions) should be used with caution.

Non-parametric feature selection methods do not rely on assumptions concerning the frequency distribution of the features. One such method, which has not been widely used, is proposed by Lee and Landgrebe (1993). Benediktsson and Sveinsson (1997) demonstrate its application. Pal (2006) uses SVMs (Section 8.6.1) for feature selection.

8.10 Classification Accuracy

The methods discussed in Section 8.9 have as their aim the establishment of the degree of separability of the k spectral classes to which the image pixels are to be allocated (though the Bhattacharyya distance is more like a measure of the probability of mis-classification). Once a classification exercise has been carried out there is a need to determine the degree of error in the end product. These errors could be thought of as being due to incorrect labelling of the pixels. Conversely, the degree of accuracy could be sought. First of all, if a method allowing a 'reject' class has been used then the number of pixels assigned to this class (which is conventionally labelled '0') will be an indication of the overall representativeness of the training classes. If large numbers of pixels are labelled '0' then the representativeness of the training datasets is called into question – do they adequately sample the feature space? The most commonly used method of representing the degree of accuracy of a classification is to build a $k \times k$ *confusion* (or *error*) *matrix*. The elements of the rows i of this matrix give the number of pixels which the operator has identified as being members of class i that have been allocated to classes 1 to k by the classification procedure (see Table 8.4). Element i of row i (the ith diagonal element) contains the number of pixels identified by the operator as belonging to class i that have been correctly labelled by the classifier. The other elements of row i give the number and distribution of pixels that have been incorrectly labelled. The classification accuracy for class i is therefore the number of pixels in cell i divided by the total number of pixels identified by the operator from ground data as being class i pixels. The overall classification accuracy is the average of the individual class accuracies, which are usually expressed in percentage terms.

Some analysts use a statistical measure, the kappa coefficient, to summarize the information provided by the contingency matrix (Bishop, Fienberg and Holland, 1975). Kappa is computed from:

$$\hat{K} = \frac{N\sum_{i}^{k} x - \sum_{i=1}^{k} x_i + x + i}{N - \sum_{i=1}^{k} n_i + x + i}$$

where k is the number of classes and N the number of test (reference) data samples. The $k \times k$ matrix $\mathbf{X}$ is the confusion matrix.

The x_{ii} are the diagonal entries of the confusion matrix. The notation x_{i+} and x_{+i} indicates, respectively, the sum of row i and the sum of column i of the confusion matrix $\mathbf{X}$. Row totals (x_{i+}) for the confusion matrix shown in Table 8.4 are listed in the column headed (i) and column totals are given in the last row. The sum of the diagonal elements (x_{ii}) is 350 ($\sum_{i=1}^{k} x_{ii}$ for $k = 6$), and the sum of the products of the row and column marginal totals ($\sum_{i=1}^{k} x_{i+}x_{+i}$) is 28 739, and the value of kappa estimated from the test data is:

$$\hat{\kappa} = \frac{410 \times 350 - 28729}{168100 - 28739} = \frac{114761}{139361} = 0.823$$

A value of zero indicates no agreement between the classification and the test data, while a value of 1.0 shows perfect agreement between the classifier output and the reference data. Montserud and Leamans (1992) suggest that a value of kappa of 0.75 or greater shows a 'very good to excellent' classifier performance, while a value of less than 0.4 is 'poor'. However, these guidelines are only valid when the assumption that the data are randomly sampled from a multinomial distribution, with a large sample size, is met.

Table 8.4 *Confusion or error matrix for six classes. The row labels are those given by an operator using ground reference data. The column labels are those generated by the classification procedure. See text for explanation. (i) Number of pixels in class from ground reference data. (ii) Estimated classification accuracy (percent). (iii) Class* i *pixels in reference data but not given label by classifier. (iv) Pixels given label* i *by classifier but not class* i *in reference data. The sum of the diagonal elements of the confusion matrix is 350, and the overall accuracy is therefore (350/410) × 100 = 85.4%.*

Class	1	2	3	4	5	6	(i)	(ii)	(iii)	(iv)
Ref										
1|	50	3	0	0	2	5	60	83.3	10	21
2	4	62	3	0	0	1	70	88.5	8	10
3	4	4	70	0	8	3	89	81.4	19	6
4	0	0	0	64	0	0	64	100.0	0	3
5	3	0	2	0	71	1	77	92.2	6	10
6	10	3	1	3	0	33	50	66.0	17	10
Col. sums	71	72	76	67	81	43	410	–	60	60

The advantages of using kappa (or an estimate of kappa, to be more precise) is that it takes into account the probability that some of the agreement between test data and classified image is due to chance. Also the variance of the sample estimate of kappa, written $\hat{\text{var}}$ ($\hat{\kappa}$) can be computed. The hats over var and κ indicate that these quantities are sample estimates of the true but unknown population values computed from a large random sample. The estimated variance is normally distributed for a large sample and so, in this case, a z value or standard normal deviate can be computed. First, set up the null hypothesis that kappa is actually equal to zero. Next, compute the test statistic, z. The probability that a value of z as high as the computed value could occur if the null hypothesis were true can be found from standard statistical packages that provide a probability value, α, corresponding to the z value. Some examples are: $z = 1.96$, $\alpha = 0.025$ or 2.5%; $z = 2.58$, $\alpha = 0.01$ or 1%; $z = 1.64$, $\alpha = 0.05$ or 5%. To understand what these figures mean, imagine that you have collected a very large number of test and training datasets and have used each training dataset to carry out a classification. For each classification, calculate the estimates of kappa, the variance of kappa and the z value. If the true value of kappa really is equal to zero then a z value as high as 1.96 will be observed in only 2.5% of a large number of tests using different test and training data. Given a sample value of kappa equal to 2.58 most people would consider it reasonable to conclude that kappa is not equal to zero at a significance level of 0.05. For the data in Table 8.4 the estimated value of kappa is 0.823 (for comparison, the classification accuracy is 85.4%), the estimated variance of kappa is 0.000043 and z is 39.9. The α value corresponding to this high value of z is extremely small (it is so small that my computer program tells me it is zero). So we can confidently state that the classification on which Table 8.4 is based is extremely unlikely to have arisen by chance. The words 'by chance' mean that at each pixel position in the classified image a random label is chosen from the range $1 - k$.

Another advantage of kappa is that it can be used to test the null hypothesis that the difference between two classifications of the same dataset are statistically equivalent. That is to say, the differences between the two classification results could have occurred by chance.

The equation for estimated kappa is given above. The estimated variance of estimated kappa is given by

$$\text{var}(\hat{\kappa}) = \frac{1}{N}\left(\frac{\theta_1(1-\theta_1)}{(1-[\theta_2])^2} + \frac{2(1-\theta_1)(2\theta_1\theta_2 - \theta_3)}{(1-\theta_2)^3} + \frac{(1-\theta_1)^2\left(\theta_4 - 4\theta_2^2\right)}{(1-\theta_2)^4}\right)$$

where

$$\theta_1 = \frac{1}{N}\sum_{i=1}^{k} x_{ii}$$

$$\theta_2 = \frac{1}{N}\sum_{i=1}^{k} x_{i+}x_{+i}$$

$$\theta_3 = \frac{1}{N^2}\sum_{i=1}^{k} x_{ii}(x_{i+} + x_{+i})$$

and

$$\theta_4 = \frac{1}{N^3}\sum_{i=1}^{k}\sum_{j=1}^{k} x_{ij}\,(x_{i+}x_{+i})$$

The z values for the null hypothesis that the true kappa value is zero and for the null hypothesis that two classifications are equivalent are given by:

$$z = \frac{\hat{\kappa}}{var\,(\hat{\kappa})}$$

$$z = \frac{|\hat{\kappa}_1 - \hat{\kappa}_2|}{\sqrt{var(\hat{\kappa}_1) + var(\hat{\kappa}_2)}}$$

See Congalton and Green (2008), Kalkhan, Reich and Czaplewski (1997), Stehman (1997), Stehman and Wickham (2006), Stehman, Sohl and Loveland (2005) and Schott (2007) for more details. Koukoulas and Blackburn (2001) suggest new accuracy measures for classification of semi-natural woodlands. Stehman (2004) discusses the value of normalizing the confusion matrix, and comes to the conclusion that normalization generates more problems than it solves. Jenness and Wynne (2007) describe an ArcView 3× plugin that computes all of the details given above, and a lot more besides. Foody (2004) reviews the evaluation of the statistical significance of differences in classification accuracy. Foody (2005) shows that the overall classification accuracy can hide significant local spatial variations in accuracy. In one case he showed that while the overall accuracy of a classifier was 84.0% but that local accuracies ranged from 53.3 to 100.0%. In a later paper, Foody (2008) holds the view that the remote sensing approach to classification accuracy is unduly harsh, and considers some of the key issues. It is clear that (i) tests based on kappa are parametric, that is they require the data to follow a specific distribution and (ii) statistical comparisons of classifications say nothing about spatial patterns. It is conceivable that two classifications may produce the same kappa values (in terms of statistical significance) but the pattern of labels in the two images may be different.

The confusion matrix procedure stands or falls by the availability of a test sample of pixels for each of the k classes. The use of training-class pixels for this purpose is dubious and is not recommended – one cannot logically calibrate and evaluate a procedure using the same dataset. A separate set of test pixels should be used for the calculation of classification accuracy. Users of the method should be cautious in interpreting the results if the ground data from which the test pixels were identified was not collected on the same date as the remotely sensed image, for crops can be harvested or forests cleared. Other problems may arise as a result of differences in scale between test and training data and the image pixels being classified. Baccini *et al.* (2007) discuss the scaling of field data to calibrate and validate moderate resolution remote sensing models using an aggregation procedure.

The confusion matrix can be used to assess the nature of erroneous labels besides allowing the calculation of classification accuracy. Errors of omission are committed when patterns that are really class i become labelled as members of some other class, whereas errors of commission occur when pixels that are really members of some other class are labelled as members of class i. Table 8.4 shows how these error rates are calculated. From these error rates the user may be able to identify the main sources of classification inaccuracy and alter his or her strategy appropriately, for example by combining two or more classes that are not separable on the basis of spectral data alone, or by adding new discriminating features such as texture.

How to calculate the accuracy of a fuzzy classification might appear to be a difficult topic; refer to Foody and Arora (1996), Gómez, Biging and Montero (2008), Gopal and Woodcock (1994) and Silván-Cárdenas and Wang (2008). Burrough and Frank (1996) consider the more general problem of fuzzy geographical boundaries. The question of estimating area from classified remotely sensed images is discussed by Canters (1997) with reference to fuzzy methods. Dymond (1992) provides a formula to calculate the root mean square error of this area estimate for 'hard' classifications (see also Lawrence and Ripple, 1996). Czaplewski (1992) discusses the effect of misclassification on areal estimates derived from remotely sensed data.

The use of single summary statistics to describe the degree of association between the spatial distribution of class labels generated by a classification algorithm and the corresponding distribution of the true (but unknown) ground cover types is rather simplistic. First, these statistics tell us nothing about the spatial pattern of agreement or disagreement. An accuracy level of 50% for a particular class would be achieved if all the test pixels in the upper half of the image were correctly classified and those in the lower half of the image were incorrectly classified, assuming an equal number of test pixels in both halves of the image. The same degree of accuracy would be computed if the pixels in agreement (and disagreement) were randomly distributed over the image area. Second, statements of 'overall accuracy' levels can hide a multitude of sins. For example, a small number of generalized classes will usually be identified more accurately than would a larger number of more specific classes, especially if one of the general classes is 'water' (Atkinson and Aplin, 2004; Aplin, 2006; Rahman *et al.*, 2003, Hengl, 2006). More thought should perhaps be given to the use of measures of confidence in pixel labelling. It is more useful and interesting to state that the analyst assigns label x to a pixel, with the probability of correct labelling being y, especially if this information can be presented in quasi-map form. A possible measure might be the relationship between the first and second highest membership grades output by a soft classifier. The use of ground data to test the output from a classifier is, of course, necessary. It is not always sufficient, however, as a description or summary of the value or validity of the classification output.

Foody (2002b) provides a useful summary of the state of the art in land cover classification accuracy assessment. He describes a number of problems associated with procedures based on the confusion matrix, including misregistration, sampling issues and accuracy of the reference data, and concludes that 'it is unlikely that a single standardized method of accuracy assessment and reporting can be identified'. Vieira and Mather (2000) consider in more detail one of the points raised by Foody (2002b), namely, the spatial distribution of error and its visualization. Foody (2004) analyses the statistical significance of measures of accuracy. He evaluates measures of local accuracy in a later paper (Foody, 2005). A further analysis of accuracy assessment is well worth reading (Foody, 2008). Brown, Foody and Atkinson (2009) considers per-pixel thematic uncertainty rather than overall classification accuracy. Yu *et al.* (2008) discuss classification uncertainty in an image object-based vegetation mapping, while Hale and Rock (2003) evaluate the impact of topographic normalization (Section 4.7) on classification accuracy. Huang *et al.* (2008) also investigate the impact of atmospheric and topographic correction on accuracy.

The question of classification accuracy also involves the number of features used in the classification. Hughes (1968) noted that, once a certain point had been passed, then classification accuracy declined as the number of features increased. This characteristic is now known as the *Hughes Phenomenon*. If we consider a statistical classifier such as ML (Section 8.4.2) then we need to provide estimates of the mean vector $\mathbf{x}$ (with p elements), and the $p(p-1)$ elements of the variance–covariance matrix. Given a fixed sample size, then as the number of features p increases so the number of independent sample elements supporting these $p + p(p1)$ estimates declines, and they become less stable (in a statistical sense – see, for example Tadjudin and Landgrebe (2000) in the context of mixture modelling). Hence, the effectiveness of the sample in characterizing the class to which it relates will inevitably decline.

Classification accuracy thus depends not solely on the method used to measure it but on the properties of the training data (Section 8.4.1) such as degree of representativeness (which in turn relates to the heterogeneity of the pixels falling into the specific class) as well as sample size. There is some evidence to show that the Hughes Phenomenon starts to show itself at different numbers of features/sample size depending on the classifier used. For example, the SVM may perform well with small training samples for the same number of features compared to the ML statistical classifier; for example Lennon, Mercier and Hubert-Moy (2002) claim that SVM do not suffer from the effects of increasing dimensionality of the pattern space. Pal and Mather (2004) make the same claim. ANNs may also perform better (in the sense of producing higher classification accuracy via increased generalisability) with smaller training samples (Section 8.4.3). Foody, Boyd and Sanchez-Hernandez (2007) discuss the problem of sample size in the context of an SVM-based classification in which interest is focused on a single class. Dobbertin and Biging (1996), Blamire (1996), Foody, McCullagh and Yates (1995) and Ju, Gopal and Kolaczyk (2005) all contribute to the debate concerning the relationship between sample size, sampling method, spatial scale and spatial autocorrelation on classification accuracy. This relationship is worthy of further investigation, as the cost of collecting training data is high and windows of opportunity (i.e. the coincidence of a cloudless day and a satellite overpass) are rare. van Niel, McVicar and Datt (2005) remark that the general rule mentioned earlier in this chapter governing training class size (i.e. 10–30 times the number of features) does not take into account the nature of the problem, and they conduct simulation experiments and that in some instances the rule produces a figure that is needlessly high. These authors state that the size of the training dataset should reflect the complexity of the discrimination problem.

One of the conclusions to arise from the literature mentioned in the preceding section is that classification accuracy is a function of the classification algorithm being used, the complexity of the discrimination problem, the adequacy of sampling (in terms of spatial coverage as well as sample size), and the nature of the terrain (for instance, mountainous terrain poses greater problems than do areas of low relative relief). Pal and Mather (2006) conclude that small increases in classification accuracy can be obtained by using more sophisticated techniques, but that greater attention should be paid to the collection of both test and training data that represent the range of land surface variability at the spatial scale of interest. Plourde and Congalton (2003) focus on the effects of sample size and placement on classification accuracy. Good reviews are provided by Congalton and Green (2008), Liu, Frazier and Kumar (2007) and Lunetta and Lyon (2004). An alternative approach to classification accuracy estimation, using a Bayesian rather than a frequency-based method, is described by Denham, Mengersen and Witte (2009).

Classification accuracy must be considered carefully when land cover information, derived from remote sensing sources, is used as input to models of atmosphere/biosphere interactions. The use of remotely-sensed data in modelling within a GIS is discussed in Section 10.6. DeFries and Los (1999) and Bounoua, Masek and Tourre (2006) demonstrate the errors that can be generated when land cover classification maps produced from remotely-sensed data are introduced into biospheric models.

8.11 Summary

A number of developments in classification methodology have occurred in recent years. However, the level of mathematical and statistical sophistication required to understand and implement the newer methods is well beyond the scope of this book. Readers requiring more detail should refer to advanced texts such as Landgrebe (2003) or Tso and Mather (2009). The use of pattern recognition techniques is not exclusive to remote sensing, and useful information can be found from books and articles targeted at, for instance, medical imaging (Meyer-Bäse, 2004) or mathematics and statistics (Bishop, 2006; Theodoridis and Koutraumbas, 2006). Bishop (2006) includes MATLAB code for major algorithms.

Other questions that are important in an operational as well as an academic context have not been described in any detail, such as the effects of misregistration of image sets or the need for atmospheric correction when changes in classification labelling over time are used in change detection studies (this latter point is considered by Song *et al.* (2001)).

The use of ANNs, DTs, SVMs, multiple classifiers, fuzzy methods, new techniques for computing texture features, and new models of spatial context, which were introduced into remote sensing during the 2000s, are now better known. This chapter has hardly scratched the surface, and readers are encouraged to follow up the references provided at various points. I have deliberately avoided providing potted summaries of each paper or book to which reference is made in order to encourage readers to spend some of their time in the library (either real or electronic) because reading requires thought and discipline. However, 'learning by doing' is always to be encouraged. It is important, however, to acquire familiarity with the established methods of image classification before becoming involved in advanced methods and applications.

Despite the efforts of geographers following in the footsteps of Alexander von Humboldt over the last 150 years, we are still a long way from being able to state with any acceptable degree of accuracy the proportion of the Earth's land surface that is occupied by different cover types. At a regional scale, there is a continuing need to observe deforestation and other types of land cover change, and to monitor the extent and productivity of agricultural crops. More reliable, automatic, methods of image classification are needed if answers to these problems are to be provided in an efficient manner. New sources of data, at both coarse and fine resolution, are becoming available. The early years of the new millennium have seen a very considerable increase in the volumes of Earth observation data being collected from space platforms, and much greater computer power (with intelligent software) will be needed if the maximum value is to be obtained from these data. An integrated approach to geographical data analysis is now being adopted, and this having a significant effect on the way image classification is performed. The use of non-remotely-sensed data in image classification process is providing the possibility of greater accuracy, while – in turn – the greater reliability of image-based products is improving the capabilities of environmental GIS, particularly in respect to studies of temporal change.

All of these factors present challenges to the remote sensing and GIS communities. The focus of research will move away from specialized algorithm development to the search for methods that satisfy user needs, and which are broader in scope than the statistically based methods of the 1980s, which are still widely used in commercial GIS and image processing packages. If progress is to be made then high-quality interdisciplinary work is needed, involving mathematicians, statisticians, computer scientists and engineers as well as Earth scientists and geographers. See, for example a review by Gautama *et al.* (undated) of computer vision techniques for remote sensing. The future has never looked brighter for researchers in this fascinating and challenging area.

9 Advanced Topics

9.1 Introduction

This chapter deals with three topics (interferometric synthetic aperture radar (InSAR), imaging spectroscopy and lidar) that are unlikely to be covered in an undergraduate course but which will be of interest to students following Masters courses, or undertaking preliminary reading for postgraduate research degrees. In Section 9.2 the topic of synthetic aperture radar interferometry (SAR interferometry or InSAR) is introduced. InSAR is primarily used to acquire data that can be processed and calibrated to produce digital elevation models (DEMs) of a target area. differential interferometric synthetic aperture radar (DInSAR) uses a time-sequence of interferometric observations and can detect movements of the order of centimetres. It has been used in fields such as glaciology, volcanology and tectonics to measure small changes in the height of the Earth's surface or to measure movement of, for example glaciers or ice sheets (Rott, 2009).

Imaging spectroscopy has been a topic that has attracted interest from researchers over the past 15 years or so. The launch of Earth Observer-1 (EO-1), which carried an imaging spectrometer (Hyperion) into orbit for the first time, plus the more widespread availability of data from airborne sensors such as 'Hyperspectral Mapper' (Hymap), DAIS and Airborne Visible/Infrared Imaging Spectrometer (AVIRIS), means that access to such data is becoming easier. Section 9.3 provides an introductory description of methods of processing high-dimensional data in both one and two dimensions (i.e., spectrally and spatially). Methods of analysing spectra used in chemometric analysis are described in the context of remote sensing, and examples are provided to lead the reader through some of the more difficult material.

The third topic considered in this chapter is that of the interpretation and use of lidar data. Ice, Cloud and land Elevation SATellite (ICESat), launched in early 2003, carried the first space-borne lidar. It stopped functioning in late 2009. As in the case of imaging spectroscopy, lidar data collected by aircraft-mounted sensors are becoming more widely available, and so these data are becoming more familiar to students and researchers. Lidar, like InSAR, is a ranging technique that measures distance from the instrument to a point on or above the ground surface, such as the roof of a building or a tree canopy. Thus, lidar data can be used to generate a digital surface model (DSM) of ground and above-ground elevations. Lidar can penetrate vegetation canopies to a greater or lesser extent, and more modern lidar sensors can collect two or more returns from each ground point, making possible the study of the relationships between lidar penetration distances and the biophysical properties of vegetation canopies.

9.2 SAR Interferometry

9.2.1 Basic Principles

A SAR image is generated by processing of millions of pulses of microwave energy that are transmitted and received by airborne or satellite-borne antennae (Section 2.4). The transmitted pulses are scattered by a target, and the same antenna receives the return pulse of energy that is backscattered by the target on the ground. Distance to the target (in the range direction) can be computed from the time taken between transmission and reception of the pulse, as the microwave radiation travels at the speed of light, while the direction of movement of the platform defines the azimuth direction. When the magnitudes of the processed pulses, which relate to the strength of the backscattered signal, are displayed in both azimuth and range, then we see a radar image.

SAR instruments are described as *coherent* because they record information about the phase as well as the magnitude of the return pulses. Phase is measured as an angle. Figure 9.1 shows two curves. One is a plot of sin (x) and the second is a plot of $\sin(x + \pi/2)$. The first curve has a phase of zero (i.e. the value of the sine wave at $x = 0$ is zero). The second curve lags the first by $\pi/2$ radians. This is seen clearly around the point $x = 6.28$ radians. Imagine that the curve of $\sin(x)$ is typical of the microwave energy transmitted by a SAR,

Computer Processing of Remotely-Sensed Images: An Introduction, Fourth Edition Paul M. Mather and Magaly Koch

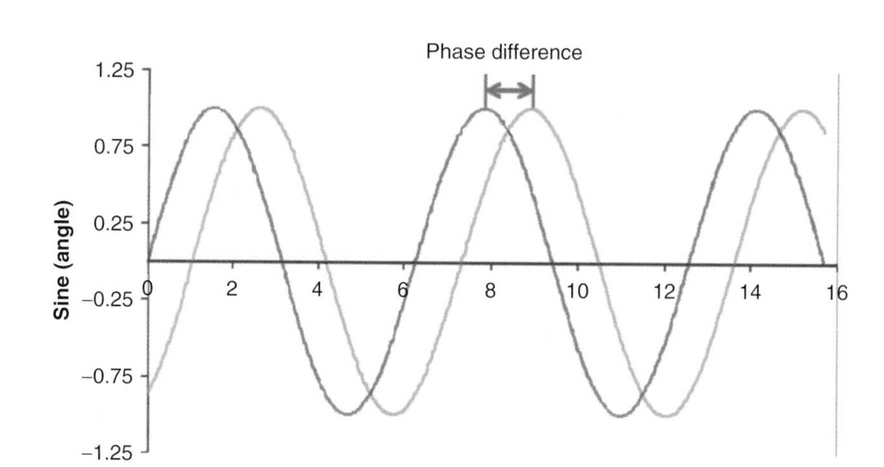

Figure 9.1 Phase difference between two sine waves. The x-axis is graduated in radians (2π radians = 360°). The shift of the peak from the first sine wave to the second is equal to $\pi/3$ radians, and this shift is the phase angle. Note that phase angle cannot be equal to or greater than 2π radians.

and $\sin(x + \pi/2)$ represents the return signal. The difference in phase between the transmitted and returned pulse is $\pi/2$ radians. The phase difference between transmitted and received energy as well as the magnitude of the return signal is recorded for each pixel in a SAR image, and datasets consisting of phase and magnitude information can be acquired in the form of single-look complex (SLC) images. The phase information for a single, independent image is of no practical value. However, the technique of interferometry makes use of two or more SLC images of the same area taken at the same or different times in order to recover information about the phase differences between them. The elevations of all pixels above some geodetic datum, such as WGS84, can be computed from this phase difference information.

The mathematical and algorithmic details of InSAR processing are well beyond the scope of this book. However, the underlying principles are reasonably straightforward, and these are presented in the following paragraphs. Readers requiring a more technical account should refer to Armour *et al.* (1998), Evans *et al.* (1992), Gens and van Genderen (1996a), Hanssen (2001), Massonet (2000), Massonet and Feigl (1998), Rosen *et al.* (2000) and Zebker and Goldstein (1986). InSAR uses the differences in phase between the signals received by two separate SAR antennae to construct a pixel-by-pixel map of ground surface elevations. The height h in Figure 9.2 is calculated from the difference in the phase (or path difference) of the signals received by antennae A1 and A2, the length of the baseline B that separates the antennae, and the look

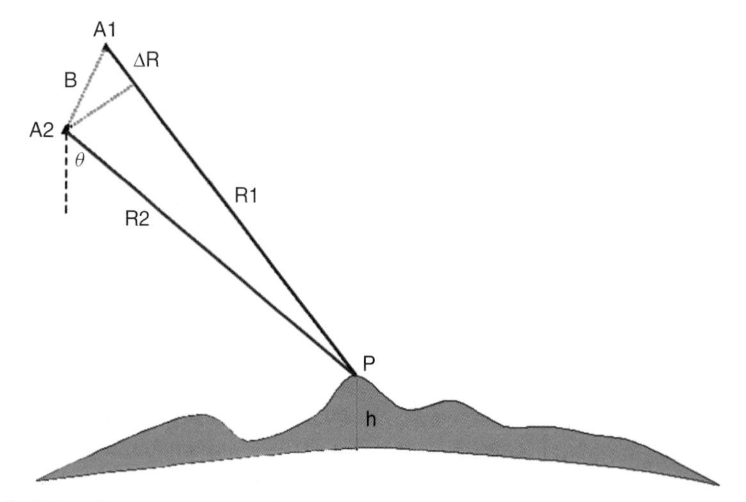

Figure 9.2 Geometry of SAR interferometry. A1 and A2 are the positions of two SAR antennae. B is the baseline (the distance between A1 and A2). R1 and R2 are the distances from A1 and A2 to the point P that has an elevation of h above a specified datum. Angle θ is the look angle of the radar.

angle of the radar. Figure 9.2 is idealized in the sense that any vegetation cover that is present may intercept the microwave radiation before it reaches the ground. The measurement h may therefore include the height of forest trees, buildings and other man-made structures and may therefore be described as a DSM rather than as a DEM.

The configuration of the two antennae A1 and A2 shown in Figure 9.2 can be achieved using one of two strategies. In *single-pass interferometry* the two antennae are carried by a single platform whereas in *repeat-pass interferometry* the signal is measured at position A1 on one orbit and at A2 on a later orbit. The US Shuttle Radar Topography Mission (SRTM), which took place during February 2000, used the single-pass approach. Two radars were carried during the mission. NASA reused the C-band system from its SIR-C experiment of 1994, and the German Space Agency (DLR) contributed an X-band radar (Rabus *et al.*, 2003; Rodriguez *et al.*, 2006; Slater *et al.*, 2006). For each instrument, one antenna was placed in the Shuttle's cargo hold and the other was located at the end of a 60 m mast that was deployed once the Shuttle reached its orbital altitude of 233 km. Single-pass systems need only one active antenna, so the backscatter from the signal transmitted from antenna A1 in Figure 9.2 is received by both antenna A1 and A2. Figure 9.3 shows a visualization of the Cape of Good Hope, South Africa, produced from C-band interferometric data collected during the SRTM with Landsat *ETM+* data as an overlay. Figure 9.4 is a relief map of Ireland generated from SRTM data. Quality assessment of SRTM data and products is provided by Smith and Sandwell (2003) and Walker, Kellndorfer and Pierce (2007). The availability of a constellation of radar satellites such as COSMO/Skymed (Section 2.4) opens up opportunities for the use of two satellites a few seconds apart in their orbit giving effectively single-pass interferometry or alternatively the orbit separation could be measured in days, giving repeat-pass interferometry. Both COSMO/Skymed and TerraSAR-X can produce SAR imagery with sub-metre spatial resolution; Eineder (2009) illustrates the use of high-resolution TerraSAR-X data in mapping buildings and other structures. The paper is worth looking at for the interferometrically generated image of the Eiffel Tower, Paris. Moreira *et al.* (2004) and Krieger *et al.* (2007) give details of the planned TanDEM approach to interferometry. The existing TerraSAR-X system is to be augmented by a second similar system in close orbit. The horizontal resolution will be of the order of 12 m with 10 m vertical accuracy. TeraSAR-X is described in Section 2.4. The TanDEM home page is at http://www.dlr.de/hr/en/desktopdefault.aspx/tabid-2317/3669_read-5488/, and a presentation to the 2009 IGARSS meeting is linked from the home page. To go to the presentation directly, link to http://www.dlr.de/Portaldata/32/

Figure 9.3 *Three-dimensional view of Cape Town and the Cape of Good Hope, South Africa. The relief data was obtained from interferometric processing of C-band InSAR data from the SRTM (February 2000) and the colour overlay from Landsat ETM bands 3, 2 and 1 in R, G and B (June, 2000). From JPL's Photojournal (http://photojournal.jpl.nasa.gov/catalog/PIA04961, accessed 31 July 2009). Courtesy of NASA/JPL/NIMA.*

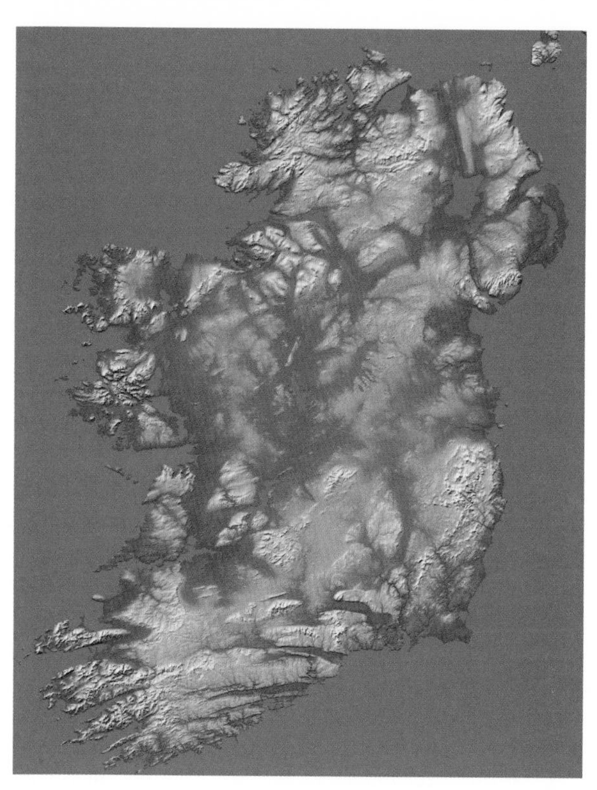

Figure 9.4 *Relief map of Ireland from C-band SRTM data. Illumination is from the north-west, so south-east facing slopes are darker. Colour coding (low–high elevations): green–yellow–brown–white. From JPL's Photojournal (http://photojournal.jpl.nasa.gov/catalog/PIA06672, accessed 31 July 2009). Courtesy of NASA/JPL/NGA.*

Resources/dokumente/tdmx/TanDEM-X_PreLaunch-Sci-breakenceMeeting_Agenda_Nov08.pdf.

Repeat-pass interferometric data are collected, as the name implies, at different times. SAR data acquired by the ERS, ASAR, JERS-1 and Radarsat radars have been used to derive interferometric DEMs. The first data take occurs when the satellite is at position A1 in Figure 9.2. At a later date, when the orbital track is different, a second set of data is acquired from antenna position A2. The length of time between successive and suitable ERS orbits varied from one day (when both ERS-1 and ERS-2 were operating in tandem mode) to as long as 35 days.

A single-pass interferometric configuration such as that used in the SRTM (and by aircraft-borne interferometric systems) has a number of advantages over a repeat-pass system. First, the target area is imaged under virtually identical conditions, so that backscattering from the target to the two antennae is effectively the same. The SAR images collected by the antennae A1 and A2 (Figure 9.2) are thus highly correlated. The correlation between two complex-valued (SLC) SAR images is termed coherence (Wegmüller, 1997; Zebker and Chen, 2005; also see below). High coherence is necessary for successful interferometry in that the assumption is made that the characteristics of the backscattered energy from each pair of corresponding points in the two images is the same. If a repeat-pass system is used then the backscattering characteristics of the target may have changed between the dates that the two SAR images were collected, and the degree of correlation between the two images would therefore be reduced, causing what is known as temporal decorrelation. Temporal decorrelation leads to a reduced level of accuracy in height determination and in extreme cases makes interferometry impossible. Some targets decorrelate very quickly. For example, Askne *et al.* (1997) show that decorrelation can occur within a few minutes for forest targets, as the individual leaf orientations change quickly relative to the SAR illumination direction as a result of wind action. Water bodies also decorrelate very rapidly, whereas agricultural areas show a moderate reduction of coherence as the time between the two successive SAR image collection dates increases. Urban areas show the lowest temporal decorrelation. In vegetated areas, the impact of temporal decorrelation depends on wavelength to some extent, as shorter wavelengths are scattered by the outer leaves whereas longer wavelengths penetrate more deeply into the canopy, as noted above. The rate at which decorrelation occurs can be used to distinguish between static and dynamic targets, for example urban areas and growing crops. Llu *et al.* (2004) use coherence to detect areas of rapid erosion in south-east Spain.

A second advantage of the single-pass approach is that atmospheric conditions are similar for the SAR images collected at antennae positions A1 and A2 in Figure 9.2. If the dates of image acquisition differ then atmospheric effects may result in errors in phase angle determination. The propagation of microwave energy through the atmosphere is affected both by the presence of water vapour and by tropospheric effects that are not well understood, but which can cause significant errors in phase determination from SAR images.

A disadvantage of single-pass interferometry is the limited baseline length that can be achieved (B in Figure 9.2). The SRTM used a 60 m mast to serve as the mounting point for the second antenna, whereas a repeat-pass configuration could involve a separation between the two orbits of several hundred metres. Baseline length is important as it has an effect on the sensitivity of the relationship between height and phase. The rate of change of height with phase difference (i.e. height sensitivity) is directly proportional to baseline length. Thus, if the baseline is short then – all other things being constant – small changes in phase angle produce relatively large changes in computed ground elevations. The opposite is the case for longer baselines. However, if the baseline becomes too long then the two SAR 'views' of the target become decorrelated and the phase differences that are used to calculate terrain elevation cannot be measured. This critical baseline length for repeat-pass interferometry using ERS-1 and -2 is of the order of 1 km, but best results for DEM generation are obtained with a baseline length of around 200–300 m. The baseline length must be known accurately. Reigber *et al.* (1996) state that, for ERS interferometry to be successful, the baseline length must be known to an accuracy of less than 5 cm. During the SRTM the length of the mast on which the C- and X-band receive-only antennae were placed was monitored closely, for – as Rabus *et al.* (2003) note – an error of 1 mm in measuring its length would lead to an elevation error of 0.5 m on the ground. Bending of the mast tip was also a problem. A star-tracking system and an electronic distance-measuring device were employed to ensure that the position of the mast tip was known to a sufficient degree of accuracy. More details of the SRTM mission are provided by Farr and Kobrick (2000).

Thus far, the question 'how do we know exactly where the antennae are?' has not been considered, yet it is a fundamental one. The orbits of the ERS-1 and ERS-2 satellites are determined using the Precise Range and Range Rate Equipment (PRARE) instrument, from laser retroreflector readings, and from the use of orbital models, which are described in Section 4.3. The use of lasers to determine the locations of satellites is called satellite laser ranging. Laser beams are directed from a ground station towards the satellite of interest, which carries efficient reflectors called retroreflectors that return the pulse back to the ground. The round-trip time is measured to a high accuracy and the distance is calculated from knowledge of the speed of light. The position of the satellite is then computed from this distance plus the angle of beam projection. Using this procedure, the position of the satellite can be fixed to within 1 m. Both ERS and Envisat are equipped with retroreflectors. Reigber *et al.* (1996) discuss the effect of the accuracy of orbit determination on interferometry.

For airborne InSAR, aircraft platform altitude and attitude are monitored using GPS and an inertial navigation system (mentioned in Section 9.4.2 in relation to lidar data collection from aircraft). During the SRTM the Shuttle's location in space was determined to an accuracy of about 1 m by GPS. An inertial navigation package also provided data on the Shuttle's attitude parameters (pitch, roll and yaw).

Both Radarsat and the Envisat ASAR can operate in a mode known as ScanSAR mode (also known as wide-swath or global monitoring modes in the case of ASAR). The radar antenna is capable of scanning several sub-swaths simultaneously, as illustrated in Figure 9.5. The penalty is a reduction in spatial resolution. For example, the ASAR onboard Envisat produces SAR imagery with a spatial resolution of 30 m for any single subswath in 'image' mode, whereas in 'global monitoring' mode all the subswaths are scanned, but at a resolution of 1 km. The derivation of interferograms from ScanSAR imagery presents a number of additional problems (Hellwich, 1999a, 1999b; Monte Guarnieri *et al.*, 1998). First, the sub swaths have to be scanned in an identical fashion on the two orbits required for repeat-pass interferometry and, second, the critical baseline length becomes shorter, at around 400 m. However, the advantage is a greater frequency of coverage and the possibility of building up a global database of interferometric measurements that could be of value in studies of surface motion. The spatial resolution of the InSAR is important – compare the 1 km resolution of ScanSAR with the submetre spatial resolution of the COSMO/Skymed SAR. Scale, as has been noted elsewhere in this book, is related to the characteristics of the proposed application and a conscious choice should be made depending on the nature of the use to which the InSAR data are to be put.

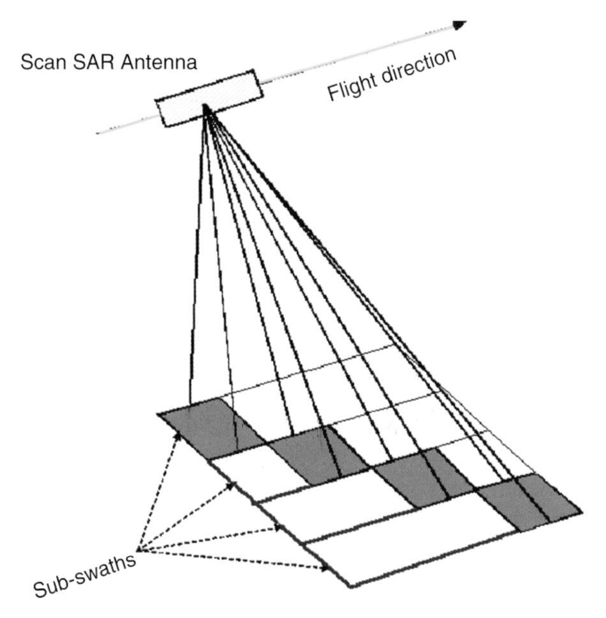

Figure 9.5 *Illustrating the principle of ScanSAR. The antenna directs separate bursts towards each sub-swath to build up a large image. The ASAR onboard Envisat can produce a swath of 400 km in ScanSAR mode, but the penalty is decreased spatial resolution of either 150 or 1000 m, depending on requirements.*

A further possibility is the use of polarimetric SAR in interferometry. The background to polarimetric SAR is provided in Section 2.4. The SIR-C experiment in 1994 provided polarimetric SAR from space for the first time. Today, the Envisat ASAR can provide polarimetric SAR imagery from which both medium (30 m) and coarse (150/1000 m) resolution interferograms can be derived. Hellwich (1999a, 1999b) notes that interferograms computed from different polarization modes can be used in a number of applications, including:

- combining the polarimetric interferograms in order to create datasets for land cover classification (Chapter 8),
- improving interferometric DEM accuracy by inter-comparison between coherence maps produced by polarimetric SAR and
- deriving information on vegetation height and structure from the differential interactions between differently polarized microwaves and the components of the surface vegetation cover.

Multiple-waveband SAR can also be used to infer the characteristics of the vegetation canopy. See Cloude and Papathanassiou (1998) for more details of polarimetric SAR interferometry. Applications of interferometry are summarized in Section 9.2.4.

DInSAR is the estimation of differences in surface location (both plan position and height) from interferometry. The principle is the same as the use of ground surveying to collect data from which contour maps are made. Assume that we have a raster (digital) contour map of a survey conducted 5 years ago and a second raster contour map that was completed last week. (*Raster* means made up of pixels; in this case, the pixel values represent elevation.) If the two maps are exactly the same then subtracting one from the other on a pixel-by-pixel basis will produce an array of zeros. If any pixel in the difference map has a non-zero value then change has occurred, and the magnitude of that change is proportional to the pixel value in the difference image (Section 6.8). In effect, we have used the first (old) raster contour map to 'remove the topography' from the newer map. The same could be done with two interferometric DEMs derived from SAR images collected before and after a significant event such as an earthquake. The difference between the two interferometric DEMs shows the change in surface geometry resulting from the earthquake. Four SLC SAR images would be required to produce two DEMs. Since the first interferometric DEM is intended to provide a good approximation to the land surface elevation, then it should be generated from a long-baseline interferometric pair, while the second interferometric DEM should show as much surface detail as possible and ideally would be derived from a short baseline interferometric pair. Where the target (such as a glacier or an ice-stream) is in motion when the second pair of SLC SAR images is acquired, the baseline should be even shorter. Critical baseline lengths should be of the order of 300, 20 and 5 m for DEM generation, ground displacement and motion analysis applications.

For DInSAR to be successful, the degree of decorrelation (as measured by loss of coherence) between the two interferograms should be as small as possible. Over water areas, or areas covered by forest, decorrelation occurs rapidly and it soon becomes impossible to separate the effects of ground deformation, land subsidence or ice-stream motion from the effects of decorrelation. Long wavelengths tend to decorrelate less rapidly than short wavelengths, as they penetrate vegetation canopies more deeply and are less likely to be affected by the geometry of the canopy surface. The ERS and ASAR radars both operate in C band, which makes them less useful for DInSAR in vegetated areas. A second disadvantage is that displacements are measured only along the line of sight of the SAR. Atmospheric effects may also produce spurious fringes in the interferograms, especially if the gap between the dates of the two acquisitions is relatively long. However, DInSAR is capable, in theory, of measuring displacements at the millimetre scale. Gabriel, Goldstein and Zebker (1989) and Massonet and Feigl (1998) are the definitive references on DInSAR. Applications of DInSAR are discussed in Section 9.2.4.

9.2.2 Interferometric Processing

The requirements for successful SAR interferometry are discussed in Section 9.2.1. Given a suitable pair of SLC images, with sufficiently high coherence, known and accurate orbital parameters and suitable baseline length, then processing can proceed. The main processing steps are: coregistering the two SLC images, complex multiplication of the two registered SLC images to generate the interferogram, computation of the coherence map, removal of the fringe pattern caused by Earth curvature, phase unwrapping and georeferencing (Figure 9.6).

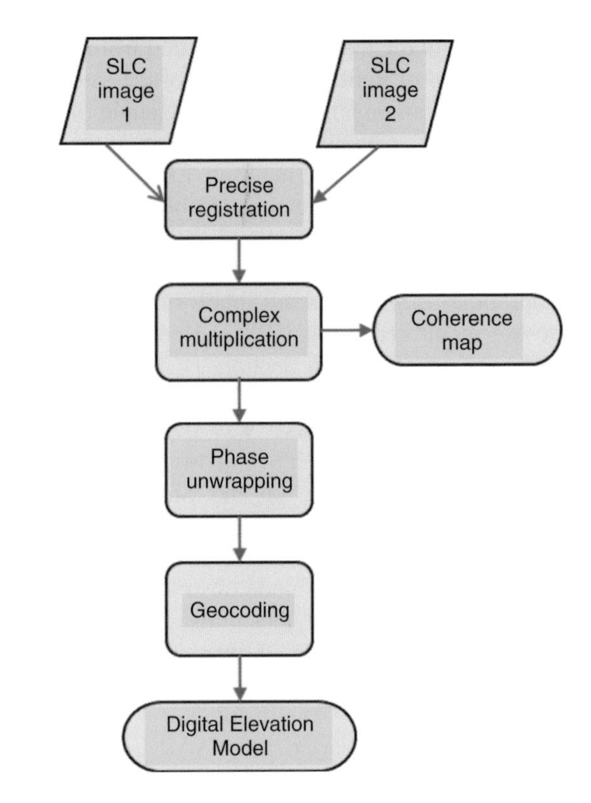

Figure 9.6 *Processing steps in SAR interferogram generation.*

In mathematical terms, multiplying one SLC SAR image by the complex conjugate of the other SLC SAR image generates an interferogram. Recall that in SLC images the backscatter characteristics of each pixel are represented by a pair of 32-bit real numbers, which form a single complex number. A complex number has two components, termed the real (a) and imaginary (b) parts in the expression $a + ib$. For example, the two elements of the complex number (4.5, 2.7) represent the terms a and b. This complex number could be written as $4.5 + i \times 2.7$, where $i = \sqrt{-1}$. The *complex conjugate* of a complex number $a + ib$ is simply $a - ib$. The SAR amplitude image is computed on a pixel-by-pixel basis from the expression $amplitude = \sqrt{a^2 + b^2}$ while the phase value is given by the expression $phase = \tan^{-1} a/b$. If the complex number (4.5, 2.7) were the value recorded for a given pixel in an SLC image then the corresponding amplitude value could be calculated as $\sqrt{4.5^2 + 2.7^2} = \sqrt{20.25 + 7.29} = \sqrt{27.54} = 5.25$, while the phase value for the same pixel would be $\tan^{-1} 2.7/4.5 = \tan^{-1}(0.6) = 0.54$ radians, or about 31°.

The complex multiplication operation is a fairly simple one, but it assumes that matching pixels in the two SLC images are precisely identifiable in the sense that if we are given the row and column coordinates of a pixel in the first SLC image then we can find the pixel covering exactly the same ground area in the second SLC image. The process of aligning the coordinate systems of two images is called *image registration*, and it is described in Section 4.3.4. As noted in the preceding section, the orbital characteristics of the platform (for single-pass interferometry) or platforms (for repeat-pass interferometry) must be known accurately, and so the principles of orbital geometry can be used to register Image 1 and Image 2. Correlation methods (Section 4.3) are also used to coregister the SLC images. The co-registration must be accurate to within 0.1 of a pixel. Once the two SLC images are co-registered the complex multiplication procedure is carried out to derive an interferogram and a coherence image.

The raw interferogram is represented as an image containing a repeating set of fringe patterns (Figure 9.7). Each fringe represents a single phase-difference cycle of 2π radians. The elevation range corresponding to a single phase difference cycle can be calculated. Conventionally, each individual fringe is displayed as a complete colour cycle from blue (0 radians) to red (2π radians) via cyan, green and yellow. The raw interferogram must be corrected further before surface elevations can be derived. These corrections are discussed below. The coherence image measures the correlation between the two complex SAR images over a number of small, overlapping rectangular windows (rather like the filter windows described in Section 7.2.1). The window is placed over the top left $n \times m$ area of the two registered SLC images, and a coherence (correlation) value is calculated for the $n \times m$ pixels in each SLC image lying beneath the window. This value is placed in the output image at the point corresponding to the centre of the window. The window is then moved right by 1 pixel, and the process repeated. When the window abuts against the right edge of the two images it is moved down by one line and back to the left hand side of the images, and the process repeated. The result is an output image that contains the coherence values for all possible positions of the $n \times m$ window. The first position of the rectangular moving window is shown in Figure 9.8. One problem is that the choice of window size will have an effect on the resulting coherence values. If the window is too small then the coherence values will vary considerably from pixel to pixel, while an over-large window will produce a generalized result. The coherence map indicates the degree of correspondence between the backscattered signals from equivalent pixels in the two SLC images on which the interferogram is based. High values (close to 1.0) indicate close correspondence and thus high reliability. In the case of repeat-pass interferometry, low values show that the one or both of the phase and amplitude of the backscatter from the two pixels has changed, and that temporal decorrelation has occurred (see above). Other factors causing loss of coherence are considered towards the end of this section.

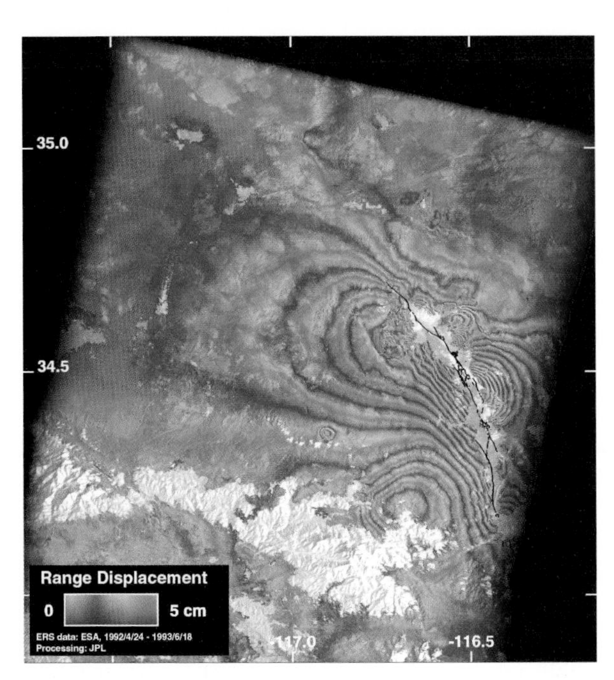

Figure 9.7 *Interferogram of the Landers area following the 1992 earthquake (see Massonet et al., 1993). One complete colour cycle corresponds to a range displacement of 5 cm. Grey areas are regions of low coherence and are masked. Surface rupture is shown by solid lines. The interferogram was calculated using two ERS-2 images, from 24 April 1992 and 18 June 1992. From http://www-radar.jpl.nasa.gov/insar4crust/LandersCo.html (accessed 31 July 2009). Courtesy NASA Jet Propulsion Laboratory. ERS-2 Data © ESA.*

A complication arises at this point because the phase differences shown in the raw interferogram are measured not in terms of the total number of full wavelength cycles but only in terms of an angular range of 2π radians (360°). Each full cycle of $0-360°$ or $0-2\pi$ radians represents one interferometric fringe. The phase differences must be 'unwrapped' by the addition of appropriate multiples of 2π before elevation can be extracted. This step is called phase unwrapping, and its implementation is difficult. A second difficulty also presents itself – an interferogram of a completely flat area has a fringe pattern that is parallel to the flight direction. This pattern is caused by Earth curvature. This 'flat Earth' fringe pattern must be removed before the fringes are calibrated in terms of elevation.

The 'flat Earth' correction can be accomplished by locating an area of flat terrain on the image and assuming that the fringe pattern shown in the corresponding region of the interferogram represents the desired 'flat Earth' pattern. This pattern is then removed from the

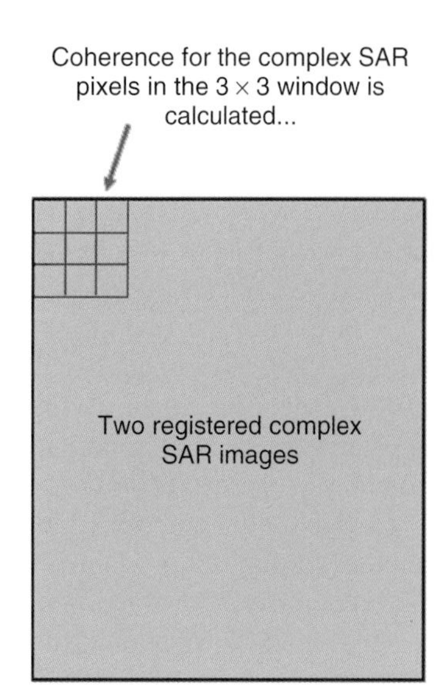

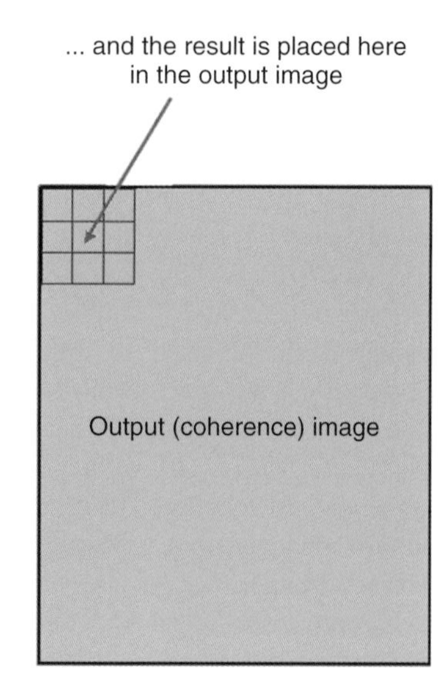

Figure 9.8 *Calculation of the coherence image. The initial position of the moving window (dimension 3 × 3 in this example) is shown on the left, where it is placed over the co-registered SLC images. The coherence for the two sets of nine pixels underlying the window is computed and the result placed in the output coherence image (right) at the point corresponding to the central window pixel. The moving window steps right by one pixel and the process is repeated. When the moving window reaches the right side of the image it drops down by one scan line and returns to the left hand side of the registered SAR images.*

interferogram as a whole. Recovering the unwrapped phase is a rather more difficult problem, and a number of algorithms have been developed to accomplish this task. None of them is completely satisfactory, and all are rather too complicated to be discussed here. Gens and van Genderen (1996a) provide a useful summary of available methods. The problem that is faced is that the phase difference between the signals received at the two antennae can be measured only on a range of 2π radians (360°). An appropriate integer multiple of 2π must be added to the calculated phase difference in order to estimate the true phase difference (Figure 9.9). The unwrapped phase image may also contain empty areas or holes, representing pixels at which the phase coherence value is too low for a phase difference to be computed, or they may represent areas of radar shadow. The wrapped phase image may also be rather noisy, so smoothing may be performed by an adaptive median filter (Premelatha, 2001), or by wavelet denoising (Section 9.3.2.2.3; Braunich, Wu and Kong, 2000) before phase unwrapping is begun. For the mathematically inclined reader, Gens (2003) and Ghiglia and Pritt (1998) provide reviews of phase unwrapping techniques.

At this point the interferogram represents the Earth surface elevation variations in the form of a DEM but is not calibrated in terms of height above a specific datum, nor does it fit a map projection. It is also expressed in 'slant range' form (the oblique view resulting from the fact that SAR is a sideways-looking instrument). The interferogram is first converted to 'ground range' form (meaning the vertical view from above). The next two steps are known as geocoding, which involves the warping of the interferometric DEM and interpolating or resampling the elevation values on to a regular grid. Both warping and resampling are discussed in Section 4.3. Unless the platform position is known very accurately, ground control points (defined in Section 4.3.2) will be needed in order to correct for global height offsets (i.e. the DEM may over- or under-estimate the surface elevation) and for other errors, described by Armour *et al.* (1998).

9.2.3 Problems in SAR Interferometry

The quality of an InSAR DEM is affected by a number of factors (Premelatha, 2001). These are: system characteristics, baseline length, terrain characteristics and processing parameters. System characteristics include wavelength and incidence angle. The property of height sensitivity is discussed in the preceding section, where it is related to baseline length. It is also inversely proportional to wavelength, incidence angle and slant range distance. These properties of an imaging radar system are examined in Section 2.4. Variations in slant range distance are not very significant for satellite-borne systems but may be significant for airborne InSAR. Height sensitivity increases as wavelength decreases, so it may be expected that X-band systems are capable of producing more detailed DEMs than are L-band systems. This is not a deterministic statement, but rather an indication of a tendency, for other factors such as temporal decorrelation may be more significant at shorter wavelengths. To some extent this is dependent on surface vegetation type, for longer

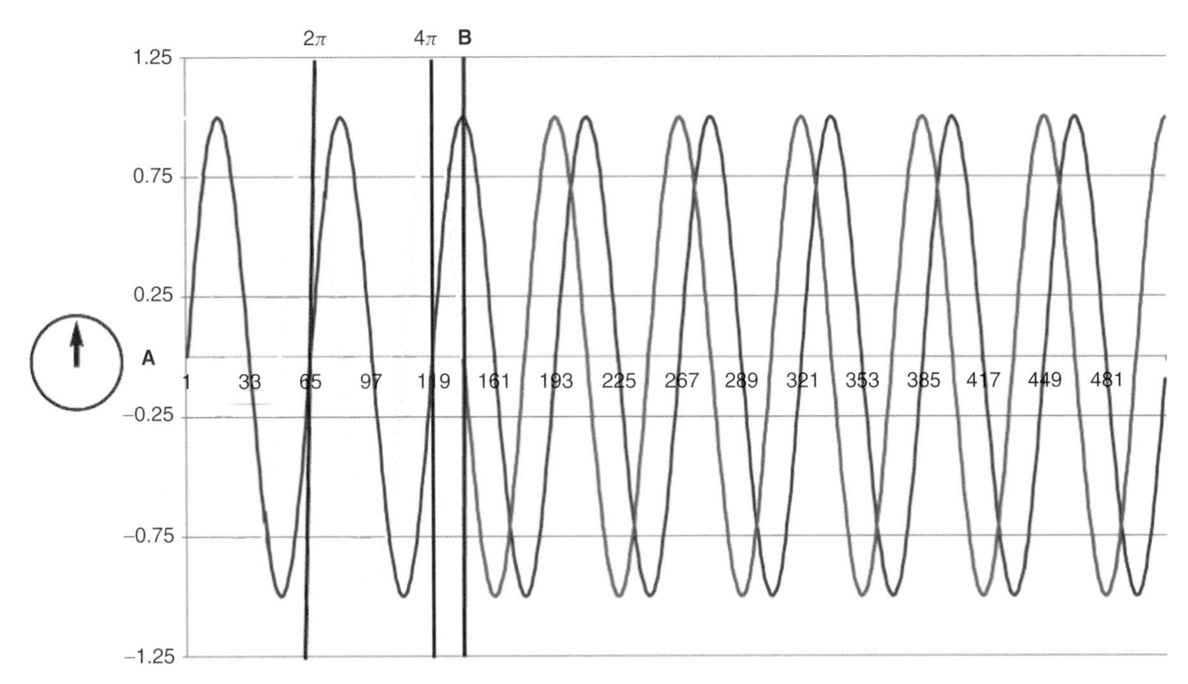

Figure 9.9 The need for phase unwrapping is shown by two signals, represented by solid and dashed lines. Imagine that these two curves are moving leftwards. The finger of the 'clock' at the origin, A, makes a complete revolution of 360° or 2π radians as each full sine wave passes. As point B passes the origin, the clock will show 2.25 revolutions, equivalent to an angular distance of 4.5π radians, which is the phase difference between the two curves. SAR interferometry measures only the phase difference in the range $0 - 2\pi$ and so the difference in the case shown here would be recorded as 0.5π. The process of phase unwrapping attempts to determine the number of integer multiples of 2π to add to the recorded phase difference. In this example, two complete cycles representing 4π radians must be added to produce the correct phase difference of 4.5π radians.

wavelengths penetrate more deeply into the canopy than do short wavelengths, and so longer wavelengths are less influenced by short-term changes in the uppermost layers of the canopy. Also, the critical baseline for longer wavelengths is greater than for shorter wavelengths.

The effect of vegetation on temporal decorrelation in repeat-pass interferometry has been mentioned already. Terrain relief is also an influential factor. In areas of rugged terrain, the effect of radar shadow is to create gaps in the spatial coverage of the radar, while foreshortening and overlay (Section 2.4) can also provide additional problems. Where slopes are steep, spatial decorrelation occurs at lower baseline lengths than is the case on flat terrain, because the local incidence angle changes with surface slope (Figure 2.17). Finally, the choice of method of SLC image registration, phase unwrapping and other processes such as filtering that are not mentioned here can produce results that differ substantially from each other. Refer to Gens and van Genderen (1996a, 1996b) for more detailed discussion of the geometric factors that influence SAR interferometry.

9.2.4 Applications of SAR Interferometry

The most common application of InSAR is to derive DSMs of the Earth's terrain. This was the primary objective of the SRTM. DInSAR applications are generally found in accurate measurement of Earth surface movements, including land subsidence, landslides, ground movement associated with volcanic activity, movement of ice-streams and glaciers, and of ocean currents. A comprehensive survey of DInSAR applications is provided by Massonet and Feigl (1998) and Massonet (2000). Smith (2002) is a useful source of reference on InSAR applications in geomorphology and hydrology. Applications of SAR interferometry in Earth system science are comprehensively covered by Rott (2009).

Elevation modelling, using repeat-pass interferometry or single-pass interferometry (represented by ERS, JERS-1 and Radarsat for the former and SRTM and aircraft systems for the latter) is the most common InSAR application. Planimetric (x, y) accuracies of 10 m and elevation (z) accuracies of 10–15 m over swaths of up to 100 km are claimed for many applications. Papers reporting the use of InSAR in elevation modelling include Albertini and Ponte (1996), Evans *et al.* (1992), Gabriel, Goldstein and Zebker (1989), Garvin *et al.* (1998), Madsen and Zebker (1993) and Zebker and Goldstein (1986). Sties *et al.* (2000) provide a comparative analysis of results achieved by InSAR and lidar (Section 9.4) in elevation modelling. Hodgson *et al.*

(2003) and Lin, Vesesky and Zebker (1994) compare elevation models generated from InSAR with DEMs generated from maps. Riedman and Haynes (2007) discuss the use of SAR interferometry in the study of geohazards. Colesanti and Wasowski (2006) and Rott and Nagler (2006) investigate landslides using InSAR, while Fischer, Rott and Björnsson (2003) study glacial surges in Iceland.

Studies of ocean currents using DInSAR techniques are described by Goldstein and Zebker (1987) and Shemer, Marom and Markman (1993). Mouginis-Mark (1995) and Rosen *et al.* (1996) use differential interferometry to monitor ground movements related to volcanic activity, and Perski and Jura (1999) use similar methods to study land subsidence patterns. Bindschlander (1998), Goldstein *et al.* (1993), Joughin, Winebrenner and Fahnestock (1995), Rabus and Fatland (2000) and Rignot, Forster and Isacks (1996) examine the potential uses of DInSAR in monitoring movements of ice-sheets and glaciers. The classic study of surface displacements following the Landers earthquake is Massonet *et al.* (1993), while Hooper, Bursik and Webb (2003) examine the potential of InSAR in mapping fault scarps for geomorphological purposes. Xu, Dvorkin and Nur (2001) use ERS SLC image pairs with a 105-day temporal separation to study subsidence in an oil field in southern California. Each fringe represents 30.4 mm vertical displacement. They were able to measure a vertical displacement of 25 cm over this time period. A DInSAR DEM of Antarctica derived from ERS-1/2 SAR images is compared to other sources of elevation information by Drews *et al.* (2009).

Repeat-pass InSAR has also been used in vegetation classification (Chapter 8; Engdahl and Hyyppa, 2003; Askne *et al.*, 2003) and in studies of canopy characteristics. The coherence image (described above) provides an indication of the rate of temporal change of surface conditions, as can differences in SAR intensity between the two dates of image acquisition. Some researchers create false colour composite images (Chapter 3) by displaying coherence, average SAR intensity, and difference between SAR intensities in red, green and blue respectively. See Askne *et al.* (1997), Dobson *et al.* (1995), Strozzi *et al.* (1999), Wegmüller and Werner (1997) and Wegmüller *et al.* (1995) for examples and further details.

9.3 Imaging Spectroscopy

9.3.1 Introduction

Imaging spectroscopy is the collection of measurements in a large number of contiguous and narrow spectral bands. The term *hyperspectral sensor* is in widespread use. Since the word *hyper* actually means 'beyond', it seems to be an inappropriate modifier to the term 'spectral', so *imaging spectrometer* is preferable. A good source of information to supplement the material contained in this section is van der Meer (2000). Many of the techniques used in the analysis of reflectance spectra were developed in the field of analytical chemistry. Huguenin and Jones (1986) provide an accessible account of developments in that subject. See also Borengasser, Hungate and Watkins (2007) for a short, introductory text. Recent developments are summarized by Plaza *et al.* (2009). Chang (2003) is another useful source, while Schaepman *et al.* (2009) review applications of imaging spectroscopy in Earth system science. In many applications of hyperspectral data (e.g., Choe *et al.*, 2008) it is important to carry out ground measurements for calibration purposes using field spectroscopy (Milton *et al.*, 2009).

It is noted in Chapter 1 that not all of the electromagnetic spectrum is available for remote sensing, due to the presence of absorption bands resulting from interactions between incoming radiation and molecules of gases such as water vapour, ozone and carbon dioxide. Regions known as *atmospheric windows* in which remote sensing is possible separate the absorption bands. The concept of the spectral reflectance curve that characterizes the reflectance distribution of a specific material is introduced in Section 1.3 and typical examples of such curves for materials such as the leaves of plants, rock surfaces and water are given in that section. The measurements made by an imaging spectrometer can be acquired only in regions of low atmospheric absorbance (atmospheric windows). The relationship between the positions of regions of atmospheric absorbance and the location of the wavebands used in a typical imaging spectrometer, the DAIS 7915, is illustrated in Figure 9.10 and Table 9.1. It is apparent that this instrument's spectral bands are located in regions of the spectrum with high transmittance.

The DAIS 7915 sensor is an airborne imaging spectrometer, built by the Geophysical Environmental Research Corporation and funded by the European Union and the German Space Agency, DLR. It has been used since spring 1995 for experimental remote sensing applications such as monitoring of land and marine ecosystems, vegetation status and stress studies, agriculture and forestry resource mapping, geological mapping and mineral exploration. The instrument is mounted on an aircraft, and upwelling radiance is directed onto detectors using a four-faced mirror (each face producing four scan lines, a procedure that can result in banding or striping – see Example 9.1) and a beam splitter. Geometric, radiometric and atmospheric corrections (Sections 4.3–4.7) are performed by DLR, and the data supplied to users are in the form of reflectance values.

One of the differences between multispectral data sets, with fewer than 10 bands of data, and imaging spectrometer data, collecting data in possibly more than

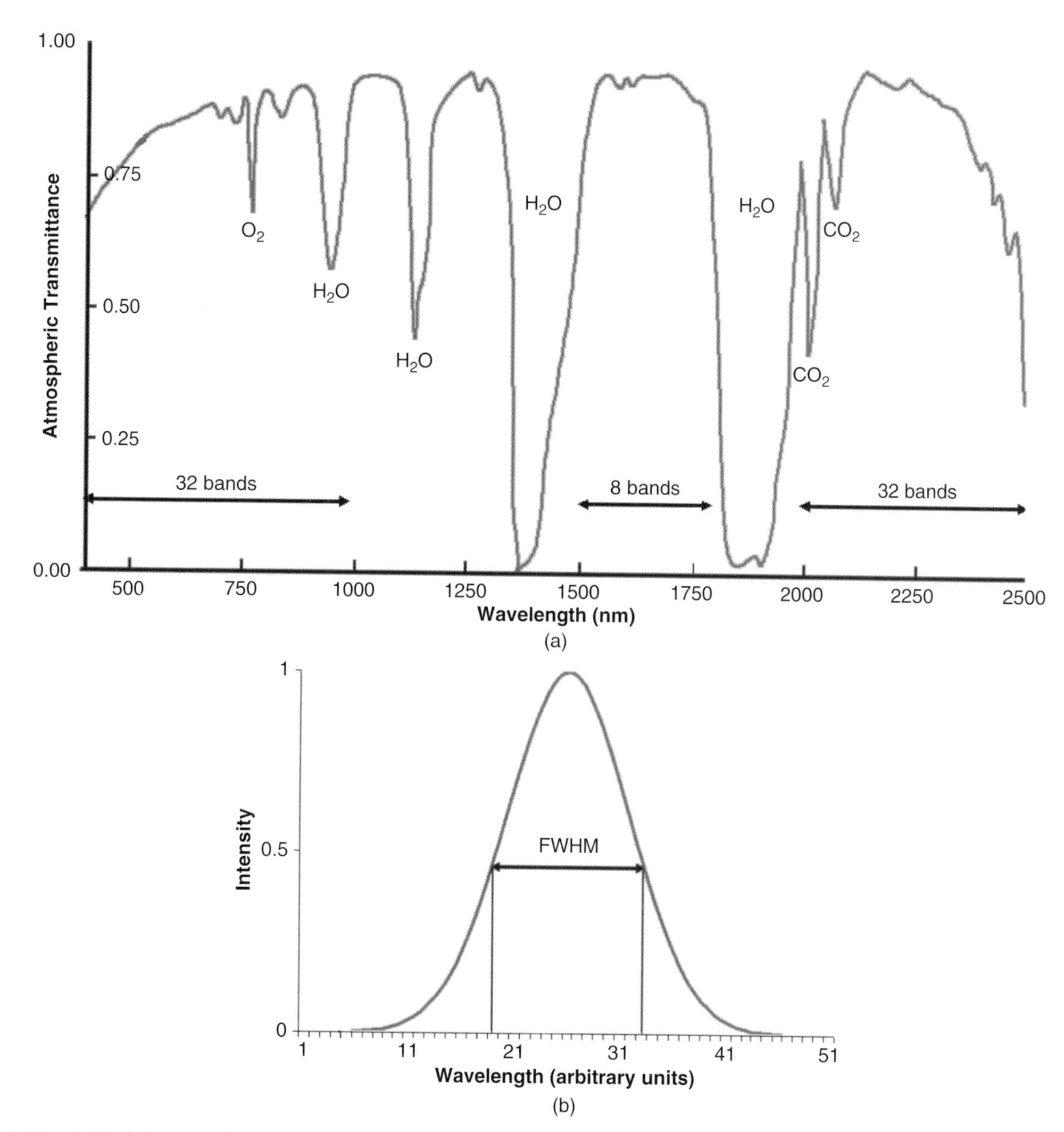

Figure 9.10 *(a) Relationship between regions of high atmospheric transmittance and the location of DAIS wavebands. DAIS is an airborne spectrometer with 79 bands, 32 of which cover the range 400–1000 nm (the visible and near infrared), with another 32 bands located in the 2000–2500 nm region. Reflectance in the narrow 'window' around 1650 nm is detected by a further 8 bands. Other bands not shown on this diagram are a single broad band covering the wavelengths between 3000 and 5000 nm, and six narrower bands in the thermal infrared (8000–12 600 nm). The width of the visible, near infrared and shortwave infrared bands varies from 15 to 45 nm. The bandwidth in the thermal infrared is 900 nm. The DAIS data were recorded by DLR in the framework of the EC funded 'Access to Research Infrastructures' project 'DAIS/ROSIS – Imaging Spectrometers at DLR'; Contract No. HPRI-CT-1999-00075 and were kindly made available by DLR (German Aerospace Agency). (b) The full width half maximum (FWHM) is the wavelength range defined by the two points at which the intensity level is 50% of its peak value. In the example, the FWHM is (33.5–19.5)= 14 units. The bandpass profile for a remote sensing instrument tends to be Gaussian in shape.*

200 narrow spectral bands, is the fact that the focus of interest is not limited to the analysis of spatial patterns. With a Landsat ETM+ image, for example we have $x = 6000$ pixels per line and $y = 6000$ scan lines, plus $z = 7$ bands. A DAIS image covers a smaller spatial area (up to $x = 512$ and $y = 3000$ pixels) but the z dimension is 79 bands. Hence, we have two ways of looking at image spectrometer data. One is to consider spatial patterns in the $x - y$ plane. The other is to consider the properties of the z dimension at specific

Example 9.1: DAIS 7915 Imaging Spectrometer Data

The three images shown in this example are extracts from a hyperspectral image set collected in June 2000 over a test site in the La Mancha region of central Spain, using the DAIS 7915 imaging spectrometer. The data were collected by the German Space Agency on behalf of a team led by Prof. J. Gumuzzio of the Autonomous University of Madrid.

Example 9.1 Figure 1. *Extract from DAIS 7915 band 7 image of the La Mancha area of central Spain. Band 7 is centred on a wavelength of 607 nm, in the red region of the electromagnetic spectrum. Vegetation absorbs electromagnetic energy in this region of the spectrum, and so photosynthetically active vegetation appears dark.*

Example 9.1 Figure 2. *This DAIS 7915 image extract covers the same geographical area as the image shown in Example 9.1 Figure 1. The centre waveband of this image is 2395 nm, on the edge of the optical region of the spectrum. The magnitude of incident radiation is low, and so the effects of striping are much more apparent than in Example 9.1 Figure 1.*

Example 9.1 Figure 1 measures reflectance in the green/red region of the electromagnetic spectrum (band 7, centre wavelength 607 nm). At this wavelength, the chlorophyll pigments in the leaves of photosynthetically-active vegetation absorb incident energy, and so appear dark.

The centre waveband in Example 9.1 Figure 2 lies in the short wave infrared region, with a centre wavelength of 2395 nm (band 72). Solar irradiance in this region of the spectrum is very low relative to irradiance in the visible bands (Figure 1.17) so the SNR of the image is much lower than the SNR of the green/red image (top) or the thermal infrared image (bottom). The mechanism used by the scanner also results in horizontal banding of the image, which is prominent in the centre image.

Example 9.1 Figure 3 shows thermal emission in a waveband centred at 10.941 μm (band 77). This image shows emitted rather than reflected radiation. DAIS band 77 is close to the Earth's emittance peak of approximately 10.5 μm. The relationship between land cover type and thermal emittance is evident from visual inspection of the patterns of light (higher thermal emission) and dark (lower thermal emission).

Example 9.1 Figure 3. *Whereas Example 7.2 Figures 1 and 2 show reflected solar radiation at wavelengths of 607 and 2395 nm respectively, this DAIS 7915 image was captured in the thermal infrared region of the spectrum. The centre wavelength is 10.941 μm, close to the Earth's emittance peak (see Figure 1.16).*

To assist in the process of visual interpretation, the images have been sharpened using the 'image minus Laplacian' procedure (Section 7.3.2) and then contrast stretched using a Gaussian stretch (Section 5.3.3).

'Q-mode' approaches in multivariate statistical analysis. R-mode analysis considers relationships such as correlations among the variables of interest that are measured on a sample of objects, while Q-mode analysis focuses on the relationships between the objects, each of which is characterized by a vector of measurements on a set of variables or features.

Thus, the analysis of imaging spectrometer data can take place in the 'spatial domain' (the $x - y$ plane) as illustrated in Example 9.1. Because measurements are made in a large number of narrow and contiguous wavebands, it is also possible to analyse variability in the z direction (across wavebands) at one or more points. Figure 9.12 shows the spectral reflectance curves derived from two pixels selected from the DAIS image data set shown in Example 9.1. Methods of processing imaging spectrometer data are considered below (Section 9.3.2). These methods have been derived from procedures used in analytical chemistry. In addition, some of the methods described elsewhere in this book in the context of multispectral data analysis can also be used to process imaging spectroscopy data; for example spatial and frequency domain filtering (Chapter 7), image classification (Chapter 8) and linear spectral unmixing (Section 8.5.1) as well as procedures used in filtering and denoising.

A second example of an airborne imaging spectrometer is Hymap, produced by Integrated Spectronics

Table 9.1 Bands 1–32 of the DAIS 7915 Imaging Spectrometer. The table shows the centre wavelength of each band together with the full width half maximum (FWHM) in nanometres (nm). The FWHM is related to the width of the band. See Figure 9.10.

Band number	Centre wavelength (nm)	FWHM (nm)	Band number	Centre wavelength (nm)	FWHM (nm)
1	502	23	17	783	29
2	517	21	18	802	27
3	535	20	19	819	30
4	554	18	20	837	27
5	571	22	21	854	28
6	589	20	22	873	28
7	607	20	23	890	27
8	625	22	24	906	31
9	641	22	25	923	28
10	659	24	26	939	30
11	678	25	27	955	28
12	695	25	28	972	28
13	711	28	29	990	38
14	729	27	30	1005	38
15	747	28	31	1020	36
16	766	29	32	1033	32

Figure 9.11 Hyperspectral data cube. Each horizontal slice (in the $x - y$ plane) represents a spectral band. The z direction (front to back) represents the spectra of individual pixels, as shown in Figure 9.12, while the rows (left–right) and columns (up–down) are the spatial coordinates. This data cube shows on its front face a false colour image of part of the Flinders Range in South Australia, captured on 8 December 2008 by the Hymap instrument using bands 102 (2.083 μm), 32 (0.895 μm) and 3 (0.483 μm), emulating bands 7, 4 and 1 of the Landsat ETM. There are 126 bands in all (see Table 9.2). The spatial resolution is 5 m at nadir. Reproduced with permission from HYVISTA.

Pty. Ltd., an Australian company. Like the DAIS 7915 sensor, Hymap can collect data in wavebands ranging from the visible to the thermal infrared. The version operated by Hyvista Ltd. collects data in the optical region only in 126 spectral bands (Table 9.2). The Hymap sensor uses the optomechanical principle described in Chapter 2 to direct upwelling radiance on to a beam splitter and thence to the detector elements. Note that Hymap has more spectral bands available in the 0.4 – 2.5 μm region than has DAIS 7915 (126 against 72) and that the bands are more closely spaced (spectral sampling interval 13–17 nm, compared with full width half maximum (FWHM) values of 18–40 nm for DAIS 7915; see Figure 9.12). Also note that Hymap bands are collected by four spectrometers (rows of Table 9.2) and that there is an overlap between the data collected by spectrometers 1 and 2. The bands in the overlap region are not presented in wavelength order, and this can lead to problems with some software. Specifically, band 31 has a longer wavelength than band 32.

The AVIRIS was first deployed by NASA in the late 1980s. Since then, the instrument has been continuously updated. AVIRIS has 224 spectral bands covering the region 0.4 – 2.45 μm. Each image is 614 pixels wide. Pixel size depends upon aircraft altitude; at a flying height of 20 km the pixel size is 20 × 20m. AVIRIS is described in detail in Vane (1987).

NASA's experimental EO-1 satellite was launched on November 21 2000. It is in the same orbit as Landsat-7 with an equatorial crossing time of 1 minute later than Landsat-7. It carries two Earth-observing sensors. The Advanced Land Imager (ALI) is a prototype for a Landsat-7 ETM+ replacement instrument, while the Hyperion Imaging Spectrometer is the first civilian high spatial resolution imaging spectrometer to be carried in orbit. The third instrument carried by EO-1 is a spectrometer that measures atmospheric water vapour content. The data produced by this instrument are used in the process of atmospheric correction (Section 4.4). ALI and Hyperion data are now available free of charge from the US Geological Survey.

One of the problems in handling hyperspectral data is the fact that dimensionality is high yet the colour composite image contains only three components (R, G and B). Principal components analysis of the hyperspectral data to reduce dimensionality does not completely solve the problem. Du *et al.* (2008) contains an interesting

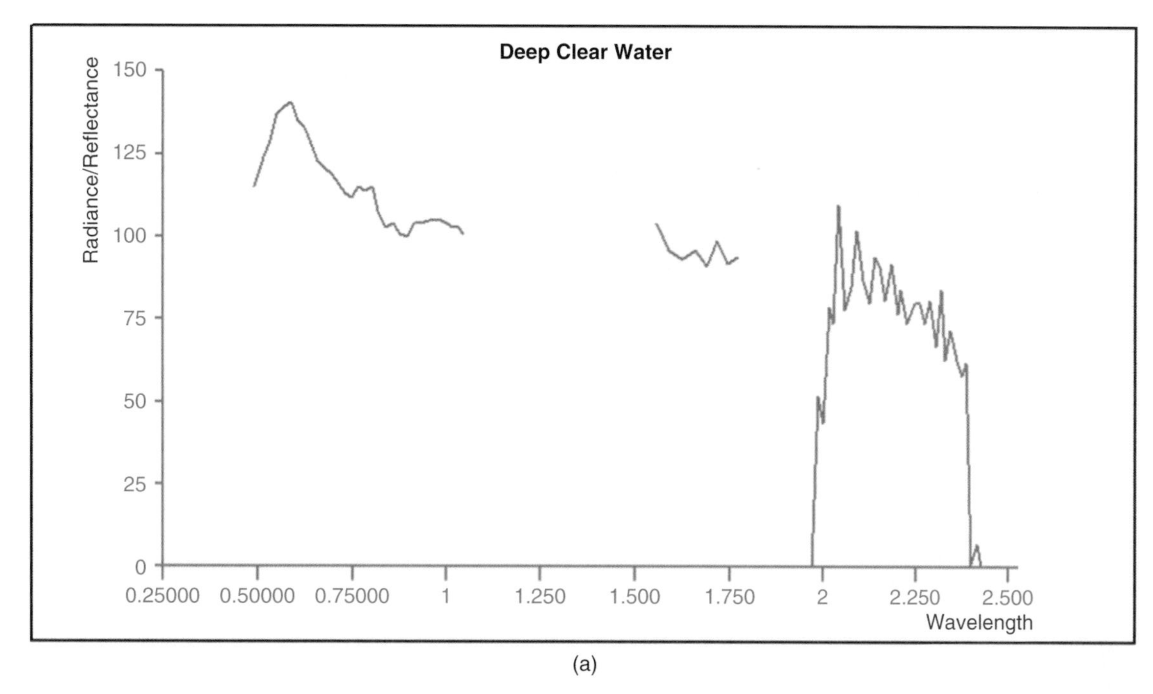

(a)

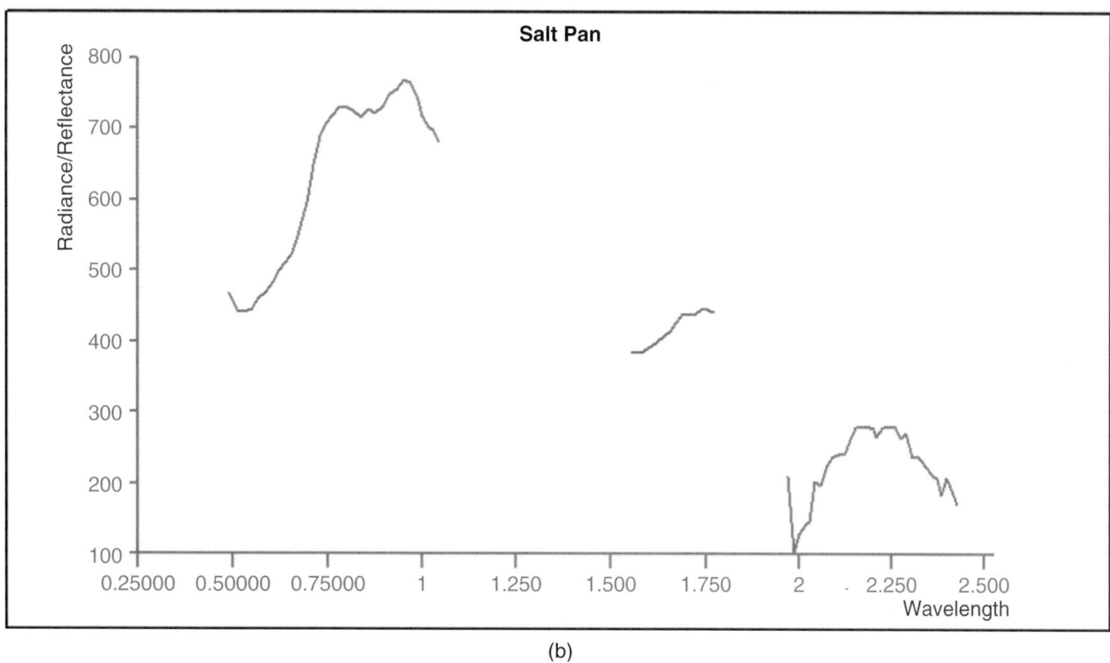

(b)

Figure 9.12 *(a) Plot of DAIS 72 band image pixel from deep clear water. (b) Plot of DAIS 72 band image pixel from salt pan area. The location of the DAIS bands in the electromagnetic spectrum is shown in Figure 9.10. The bandpass values of bands 1–32 are listed in Table 9.1.The DAIS data were recorded by DLR in the framework of the EC funded 'Access to Research Infrastructures' project 'DAIS/ROSIS – Imaging Spectrometers at DLR'; Contract No. HPRI-CT-1999-00075 and were kindly made available by DLR (German Aerospace Agency).*

Table 9.2 *Summary of Hymap imaging spectrometer wavebands, bandwidths and sampling intervals.*

Spectral region	Wavelength range (nm)	Bandwidth (nm)	Average spectral sampling interval (nm)
Visible	450–890	15–16	15
Near infrared	890–1350	15–16	15
Short-wave infrared 1	1400–1800	15–16	13
Short-wave infrared 2	1950–2480	18–20	17

Based on information from Cocks *et al.* (1998).

discussion of the problem. Feature selection is also a problem as the number of bands is considerable. Pu and Gong (2004) compare three methods – band selection, principal components analysis (Section 6.4) and wavelets (Section 6.7) and conclude that wavelet analysis produces the best subset of features for a subsequent regression analysis. Prasad and Bruce (2008) also consider dimensionality reduction and study the use of principal components analysis to select a reduced set of features prior to linear discriminant analysis (Esbensen, 2002; Mather, 1976; Rencher, 2002; Timm, 2002).

9.3.2 Processing Imaging Spectroscopy Data

9.3.2.1 *Derivative Analysis*

A derivative measures the rate of change of the variable being differentiated (x) with respect to some other variable (y), and is written as $\delta x/\delta y$. For example, given a moving object, one could compute the derivative of its velocity with respect to distance. The result would show whether velocity was increasing as distance travelled increased (acceleration, positive derivative), decreasing with distance (deceleration, negative derivative) or remaining constant (zero derivative). Acceleration is independent of velocity, so that two objects having quite different velocities could have the same acceleration (Section 7.3.2).

Derivatives can be computed only for continuous and single-valued functions. A function is simply an expression that returns a value when provided with an input. For instance, if x is an angle then the function $\cos(x)$will return a value between $+1$and -1 for any finite value of x, such as $-125.985°$ or $319276.24135°$. A continuous function returns a value for any permissible input value. In the case of the cos function, the only restriction on the argument x is that it is finite. A single-valued function returns only one answer, in contrast to a function such as *sqrt* which returns two values, except when the argument is zero. For example, *sqrt(4)* returns two values ($+2$ and -2) for a single value of the argument, and so the square root function cannot be differentiated.

A digital image can be considered to be an example of a two-dimensional function that returns the pixel value (grey level) at a given point defined by the image row r and column c, so that we could write *pixel value* $= f(r, c)$. However, it is not possible to differentiate $f(r, c)$ because it is not a continuous function that is defined for every possible value of r and c. The row and column indices must be integer values. Instead of the method of differentiation being used to calculate the rate of change at a particular point, a procedure called the method of differences is used instead.

The spectral reflectance curve of a target, as collected by a field radiometer or an imaging spectrometer, is drawn by interpolating between measured, discrete points which are spaced apart at intervals such as 15 nm. The measurements on which the curve is based are discrete or separate, and so the derivatives are estimated using the method of differences. If y_i and y_j represent adjacent, discrete, reflectance values on a spectral reflectance curve at wavelengths x_i and x_j then the first difference value is given by the expression $\frac{\Delta y}{\Delta x} = \frac{y_i - y_j}{x_i - x_j}$. The terms Δx and Δy are pronounced 'delta x' and 'delta y', and the left hand side of this equation is pronounced 'delta-x by delta-y'. The second difference (i.e. the difference of the first difference) is calculated in a similar way from the formula $\frac{\Delta^2 y}{\Delta x^2} = \frac{\Delta y_i - \Delta y_j}{\Delta x_i - \Delta x_j}$. The first difference gives the rate of change of the function y with distance along the x axis, which is the same as the slope of the graph representing the function. The second difference is the rate of change of slope with distance along the x-axis. If the curve is flat then both first and second derivatives are zero. If the curve slopes upwards to the right then the slope and the first differences are positive, and increase in magnitude as the curve becomes steeper until, if the curve becomes vertical, the first difference is infinite in magnitude. Conversely, as the slope decreases the first difference reduces in magnitude. When a turning point

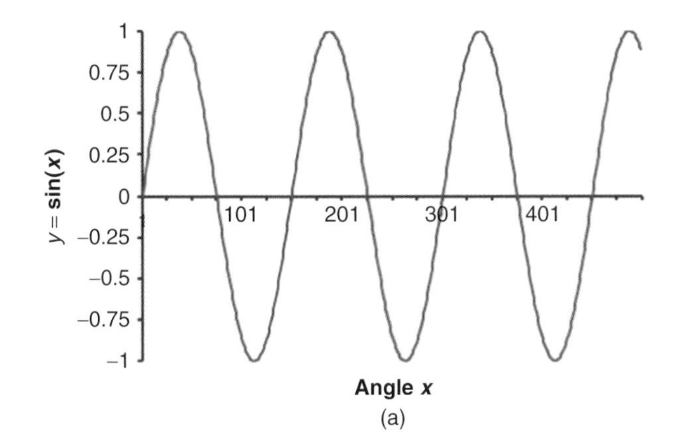

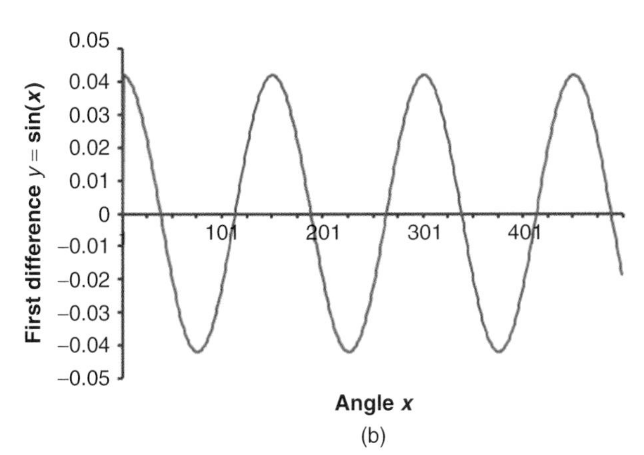

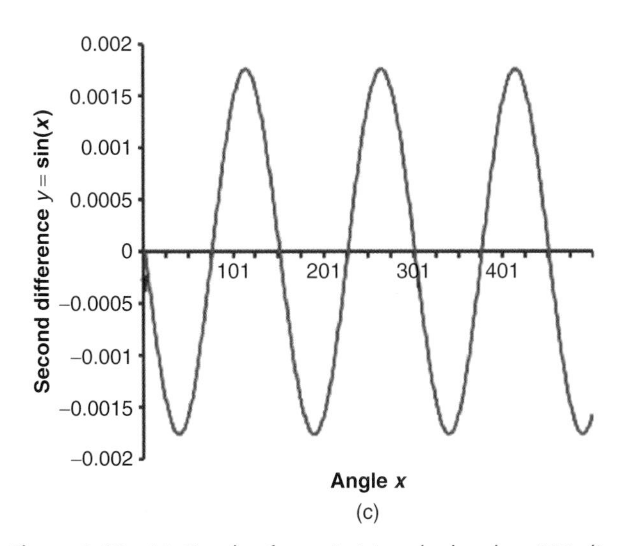

Figure 9.13 *(a) Graph of* $y = \sin(x)$ *calculated at 500 discrete points. (b) First difference of* $y = \sin(x)$ *and (c) Graph of second differences of* $y = \sin(x)$. *Note differences in the scale of the y-axis. The slope of* $y = \sin(x)$ *is initially positive, then decreases to zero at a turning point at* $x = 39$, $y = 1$. *The slope then becomes negative, increases towards* $y = 0$ *and reaches a minimum at* $x = 114$, $y = -1$. *The same pattern then repeats for each cycle of the sine wave. Figure 9.13b, c show the first and second differences, which you should attempt to interpret.*

is reached, for example at a maximum or minimum of the curve, the first difference is zero. If the graph slopes down to the right, the slope is negative, and it decreases in value until the curve reaches a minimum. The second difference shows how rapidly the slope is changing. Where slope is constant, such as at a maximum or minimum, then the second difference is zero. Where the slope gets steeper, the second difference becomes larger and, conversely, where the slope gets less steep so the second difference becomes smaller in magnitude. These ideas are illustrated in Figure 9.13.

First and second differences calculated for one-dimensional spectra or two-dimensional images are often described as 'derivatives', though it is clear that they provide a means to approximate the derivatives of a discrete function that cannot be calculated. Nevertheless, the term 'derivative' is used in the remainder of this section in order to ensure compatibility with the literature. The first derivative measures a rate of change. It is not dependent on the magnitude of the function. For example, if $x_1 = 6$ and $x_2 = 12$ then the difference is 6. The difference is also 6 if $x_1 = 106$ and $x_2 = 112$. If $y_1 = 1$ and $y_2 = 3$ then value of the first difference is $6/2 = 3$ in both cases. This means spectral reflectance curves with the same shape will have the same first derivative curve, irrespective of their measured reflectance values. This property can be useful because it means that objects with similar spectral reflectance properties located in shadow and in direct sunlight, as shown in Figure 6.5, have the same derivative though their apparent (at-sensor) radiances are different.

The first and second derivatives of a single image band are used in Chapter 7 without particular reference to the concepts outlined above. For instance, the Roberts Gradient edge detection operator is an example of a first derivative, while the Laplacian operator is an example of a second derivative function (Section 7.3.2). Other examples of the use of the derivative in image processing include the analysis of the position and magnitude of absorption bands in the pixel spectrum (Blackburn, 1998; Demetriades-Shah, Steven and Clark, 1990; Philpott, 1991; Tsai and Philpott, 1998). Gong, Pu and Yu (1997) use first derivatives as inputs to a neural network for classifying coniferous species. Laba *et al.* (2005) also use first derivatives in an attempt to find the optimum dates for the discrimination of invasive species. Bruce and Li (2001) consider the role of wavelets in smoothing the pixel reflectance spectrum prior to derivative calculation. Adams, Philpot and Norvell (1999) use the second derivative to generate a yellowness index to identify stressed vegetation. Beckera, Luschb and Qi (2005) identify optimum spectral bands from second derivative analysis.

9.3.2.2 *Smoothing and Denoising the Reflectance Spectrum*

One of the characteristics of derivative-based methods is that they amplify any noise that is present in the data. Even the presence of moderate noise can make the derivative spectrum unusable. Various methods of noise removal, ranging from simple filtering to more complex wavelet-based methods, have been applied to remote sensing data. In Section 9.3.2.2.1 a method based on the fitting of local polynomials that have special properties is described. It was first described by Savitzky and Golay (1964) in the context of analytical chemistry. Not surprisingly, it is known as the Savitzky–Golay (SG) method. An alternative procedure using the one-dimensional discrete wavelet transform (DWT, Section 6.7) is described in Section 9.3.2.2.2. The DWT decomposes a data series into a set of scale components. The lower-order scale components (levels $1, 2, \ldots$) are referred to as 'detail coefficients', and the basic wavelet-based denoising method involves the thresholding of these detail coefficients in order to separate noise and high frequency information. In contrast, simple filtering methods, such as the moving average, make no distinction between high frequency information and noise, and are best described as smoothing rather than denoising functions.

9.3.2.2.1 *Savitzky–Golay Polynomial Smoothing*

Savitzky and Golay (1964) introduced a technique that combines smoothing (i.e. low-pass filtering, Section 7.2) and calculation of derivatives in an elegant and computationally effective fashion. Smoothing is performed by approximating the data series by a low-order local polynomial, using a moving window technique. One might reasonably distinguish between smoothing and denoising. Smoothing is a purely mathematical operation that is designed to remove some or all of the high-frequency components of the data series, either to reduce the level of detail or to eliminate noise. Denoising has the aim of characterizing the statistical nature of the noise and of using estimates of these characteristics to reduce or remove the effects of that noise.

The user of the SG method must specify *a priori* (i) the order of the polynomial and (ii) the size of the moving window. The larger the window the greater the smoothing effect. Calculation of a least-squares polynomial at every moving window position in a data series might seem to be a lengthy task, but the SG method is a clever one. Only one set of coefficients is calculated, and this is applied to the data in every window simply by multiplying the value at each data point in the window by the corresponding coefficient value (Figure 9.14). Usually, a polynomial of order 2 is selected for data smoothing, and an order of 4 is recommended for derivative calculation. The method is easy to program but more difficult to describe verbally in the abstract, so a simple example is used.

Assume that we specify a window size of 5, as shown in Figure 9.14. The centre point (where the coefficient is labelled '0') has two coefficients to the left (N_L, labelled -1 and -2) and two to the right (N_R, labelled 1 and 2). A second-order polynomial ($M = 2$) is selected, as the example relates to data smoothing. Recall from Section 4.3.2 that the classical least-squares equation is:

$$\mathbf{c} = (\mathbf{A}'\mathbf{A})^{-1}\mathbf{A}'\mathbf{g} \qquad (9.1)$$

where $\mathbf{A}$ is the design matrix, $\mathbf{c}$ is the vector of least-squares coefficients and $\mathbf{g}$ is the data vector. The elements a_{ij} of the SG design matrix $\mathbf{A}$ are given by $a_{i0} = 1$ and $a_{ij} = i^j$ for $i = -N_L$ to N_R and j from 1 to M, the order of the polynomial. We are using a window size of 5 so $N_L = N_R = 2$ and we can define $\mathrm{N_{TOT}} = \mathrm{N_R} + \mathrm{N_L} + 1 = 5$. For this example, $\mathbf{A}$ has N_{TOT} rows and $M + 1$ columns and its contents are shown in Table 9.3a.

Equation 9.1 shows that we need to compute the matrix product $\mathbf{A}'\mathbf{A}$ and then find its inverse $(\mathbf{A}'\mathbf{A})^{-1}$. The final calculation involves the premultiplication of $\mathbf{A}'$ by $(\mathbf{A}'\mathbf{A})^{-1}$. These matrices are shown in tables Table 9.3b–d. The first row of Table 9.3d contains the required coefficients for data smoothing. None of the real data is required in the computation of

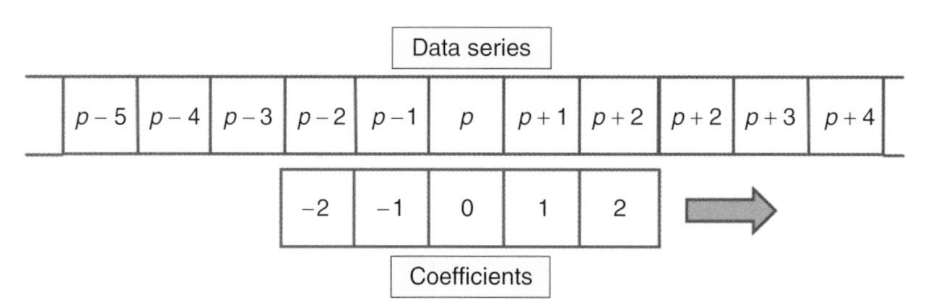

Figure 9.14 *One-dimensional moving window. The window starts on the left of the data series and the filter coefficients are multiplied by the corresponding data value. The products are added to give the filter output. In this example, the output is {data value at* (p − 2)x *coefficient* {(−2)} + {*data value at*(p − 1)x *coefficient* (−1)} + {*data value at* (p) x *coefficient* (0)} { *data value at* (p + 1)x *coefficient* (1)}{ *data value at* (p + 2)x *coefficient* (2)}.

Table 9.3 *Matrices and vectors used in Savitzky–Golay example. (a) The design matrix,* **A***. (b) Matrix product* **A′A***. (c) Inverse matrix* $(\mathbf{A'A})^{-1}$ *and (d) the matrix product* $(\mathbf{A'A})^{-1}\mathbf{A'}$*.*

(a) Design matrix **A**		
1.000	−2.000	4.000
1.000	−1.000	1.000
1.000	0.000	0.000
1.000	1.000	1.000
1.000	2.000	4.000

(b) Matrix product **A′A**		
5.000	0.000	10.000
0.000	10.000	0.000
10.000	0.000	34.000

(c) Inverse of (**A′A**)		
0.486	0.000	−0.143
0.000	0.100	0.000
−0.143	0.000	0.071

(d) Matrix $(\mathbf{A'A})^{-1}\mathbf{A'}$				
−0.086	0.343	0.486	0.343	−0.086
−0.200	−0.100	0.000	0.100	0.200
0.143	−0.071	−0.143	−0.071	0.143

these coefficients – that is the advantage of the SG method. All we need do to calculate the smoothed value at data point i is to perform an element-by-element multiplication of the raw spectrum values $\{r_{i-2}r_{i-1}\, r_i r_{i+1}\, r_{i+2}\}$ for $i = N_L + 1, N - N_R$ with the coefficients $\{-0.086\ 0.343\ 0.486\ 0.343\ \ -0.086\}$, as shown in Figure 9.15. Figure 9.16 shows the result of this operation carried out on a small sample of arbitrary data. The smoothed curve was calculated using the coefficients vector shown above. Note that smoothed values for points 1 and 2 at the beginning and points 14 and 15 at the end of the data series cannot be calculated as $N_L = 2$ and $N_R = 2$. Remember too that one assumption of the method is that the data points are equally spaced.

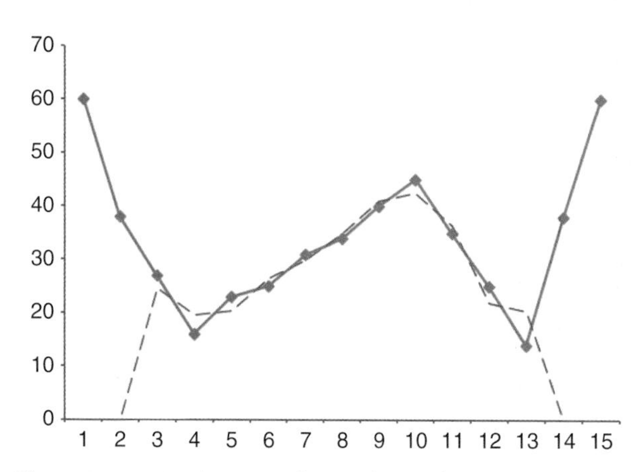

Figure 9.15 *Application of Savitzky–Golay smoothing to an arbitrary set of set of 15 data points. The raw data points are indicated by the diamond symbols, which are joined by a solid line. The smoothed curve (values of which can only be computed for data points 3–13 inclusive) is shown by the dashed green line.*

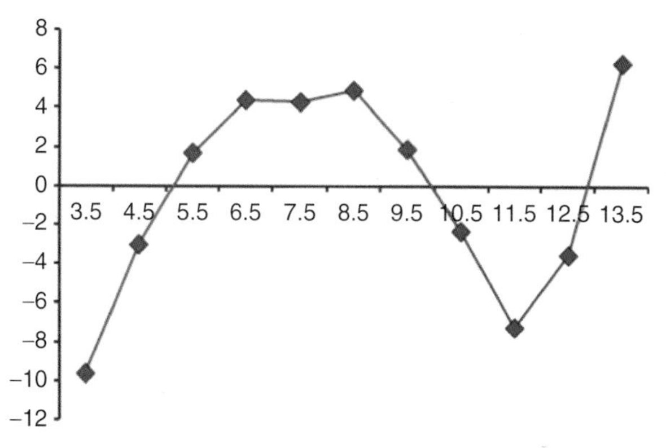

Figure 9.16 *First derivative of raw data series shown in Figure 9.25 (solid line) computed using Savitzky–Golay smoothing polynomial procedure. Note that the data points are shifted right along the x-axis by half a unit because the derivative is estimated using differences between adjacent raw data points.*

The values making up the second row of Table 9.3d are, perhaps surprisingly, the coefficients required to calculate the first derivative of the SG smoothed polynomial. To illustrate this, the result of applying the second row of coefficients (Table 9.3d) to the set of arbitrary data shown by the solid line in Figure 9.15 is given in Figure 9.16. The third row of coefficients is used to compute the second derivative, and so on. Usually, however, one would use a polynomial order of 4 rather than 2 for derivative calculation (Press *et al.*, 1992).

In summary, the one-dimensional SG procedure provides a computationally efficient means of smoothing a one-dimensional data series such as a reflectance spectrum. It is, however, a smoothing procedure and, as such, differs from denoising techniques based on the DWT, as noted earlier in this section. The following points should be remembered:

- The data points are assumed to be equally spaced. Press *et al.* (1992) suggest that moderate departures from this assumption are not likely to have any considerable effect on the result.
- The degree of smoothing depends on the order of the polynomial and on the number of points in the moving window.
- A polynomial order of 2 for smoothing and 4 for derivative calculation is generally used.
- The number of points to the left and to the right of the point of interest (i.e. N_L and N_R in the discussion above) should normally be equal for smoothing reflectance spectra.

The SG method can also be applied to two-dimensional data such as images, in which case a local smoothing polynomial surface is fitted. The computational procedure is described by Krumm (2001). The moving window is now rectangular, and the design matrix is defined in terms of powers of x and y. Again, it easier to explain by example. The moving window is size $x = 5$ and $y = 5$, but the x and y values are counted from -2 to $+2$ rather than 1 to 5. The 25 cells in the 5×5 moving window are labelled 0–24, counting sequentially left to right along rows in a zig-zag fashion starting from the top row (Table 9.4). The leftmost cell in row 1 is labelled '0'.

First, think of the 25 cells in Table 9.4 as being stored as a column vector, with p_0 at the top and p_{24} at the bottom. The design matrix **A** is formed by calculating, for each of these 25 rows, a vector of powers and cross-products of the (x, y) values associated with that cell. The arrangement of the terms in this vector is exactly the same as that used in Section 4.3.1 in the context of geometric correction. For example, a first order polynomial has terms $\{1, x, y\}$. At position p_0, the values $x = -2$ and $y = -2$ would be substituted into this expression, so that the vector of powers and cross-products at this position for a first-order polynomial is $\{1, -2, -2\}$.

Table 9.4 *Two-dimensional moving window. The cell values are referenced by the* x *and* y *coordinates in the usual way.*

	X				
Y	−2	−1	0	1	2
−2	p_0	p_1	p_2	p_3	p_4
−1	p_5	p_6	p_7	p_8	p_9
0	p_{10}	p_{11}	p_{12}	p_{13}	p_{14}
1	p_{15}	p_{16}	p_{17}	p_{18}	p_{19}
2	p_{20}	p_{21}	p_{22}	p_{23}	p_{24}

At position 17 (p_{17} in Table 9.4), $x = 0$ and $y =$, so the vector of powers and cross products for a first-order polynomial is $\{1, 0, 1\}$. The vectors of powers and cross products for low-order polynomials are as follows:

Order 1: $\{1, x, y\}$
Order 2: $\{1, x, y, x^2, xy, y^2\}$
Order 3: $\{1, x, y, x^2, xy, y^2, x^3, x^2y, xy^2, y^3\}$
Order 4: $\{1, x, y, x^2, xy, y^2, x^3, x^2y, xy^2, y^3, x^4, x^3y, x^2y^2, xy^3, y^4\}$

For example, the vector of powers and cross-products at position p_{11} (which has coordinates $x = -1$ and $y = 0$) for a third order polynomial would be $\{1, -1, 0, 1, 0, 0, 1, 0, 0, 0\}$, a total of 10 coefficients. This vector is calculated simply by substituting $x = -1$ and $y = 0$ into the definition of the order 3 powers and cross products given above, that is $\{1, x, y, x^2, xy, y^2, x^3, x^2y, xy^2, y^3\}$.

Thus, the SG design matrix **A**, in the case of a third-order polynomial, has 25 rows corresponding to the 25 elements of the 5×5 moving window and 10 columns, corresponding to the 10 power and cross-product terms for a third-order polynomial. We now compute $\mathbf{C} = (\mathbf{A}'\mathbf{A})^{-1}\mathbf{A}'$as before and find that the 25 smoothing polynomial coefficients are contained in row 1 of **C**. Rows 2 and 3 of **C** contain the coefficients for the first partial derivatives with respect to x and y, respectively. Again, note that the values of the elements of matrix **C** do not depend on the real data values in the image. These coefficients are the same for all images. The 25 values underlying the 5×5 moving window are multiplied by the corresponding coefficient values in the first row of **C**, using the labelling scheme shown in Table 9.4. The 25 products are summed to give the smoothed value at point $(x = 0, y = 0)$ in the output image. The moving window, like the classical finger, moves on and $(x = 0, y = 0)$ now overlies the next pixel to the left. The concept of the moving window is illustrated in Figure 9.8.

Example 9.2 illustrates the use of the SG technique in image smoothing.

9.3.2.2.2 Denoising Using the Discrete Wavelet Transform The DWT is introduced in Section 6.7.1. It is shown there that a one-dimensional data series can be decomposed into a collection of subsets of detail coefficients, with $n/2$ first-level detail coefficients, $n/4$ second-level detail coefficients, and so on, as illustrated in Figures 6.22 and 6.23. Donoho and Johnstone (1995) show that the variance of Gaussian-distributed white noise with zero mean can be estimated from the higher-order wavelet detail coefficients, and that a threshold value, which they call the universal threshold

Example 9.2: Savitzky–Golay Smoothing

This example shows the result of applying a SG smoothing filter to a single band (band 30) of the DAIS 7915 imaging spectrometer data set discussed in Example 9.1. The moving average window size is $x = 5$ and $y = 5$. A second-order polynomial is used for image smoothing (Figure 9.15) and a fourth-order polynomial is used to calculate the first derivative images. Example 9.2 Figure 1a shows the original DAIS 7915 band 30 image. Next, Example 9.2 Figure 1b shows the image after smoothing using the SG procedure. Three first derivative images are displayed in Example 9.2 Figure 1c–e. The first, in Example 9.2 Figure 1a, estimates the horizontal grey level gradient along the scan lines (rows) of the image. The second (Example 9.2 Figure 1b) is the vertical grey level gradient, measured down the columns of the image and the third (Example 9.2 Figure 1c) is the spatial derivative (calculated with respect to both x and y). You can use the `Filter|Savitzky-Golay` item on the MIPS main menu to carry out a similar exercise. Compare Example 9.2 Figure 1c–e. Can you say that the filter separates the horizontal (c) and vertical (d) edges? Why are the edges shown in the spatial derivative image (e) more marked than those in Example 9.2 Figure 1c,d?

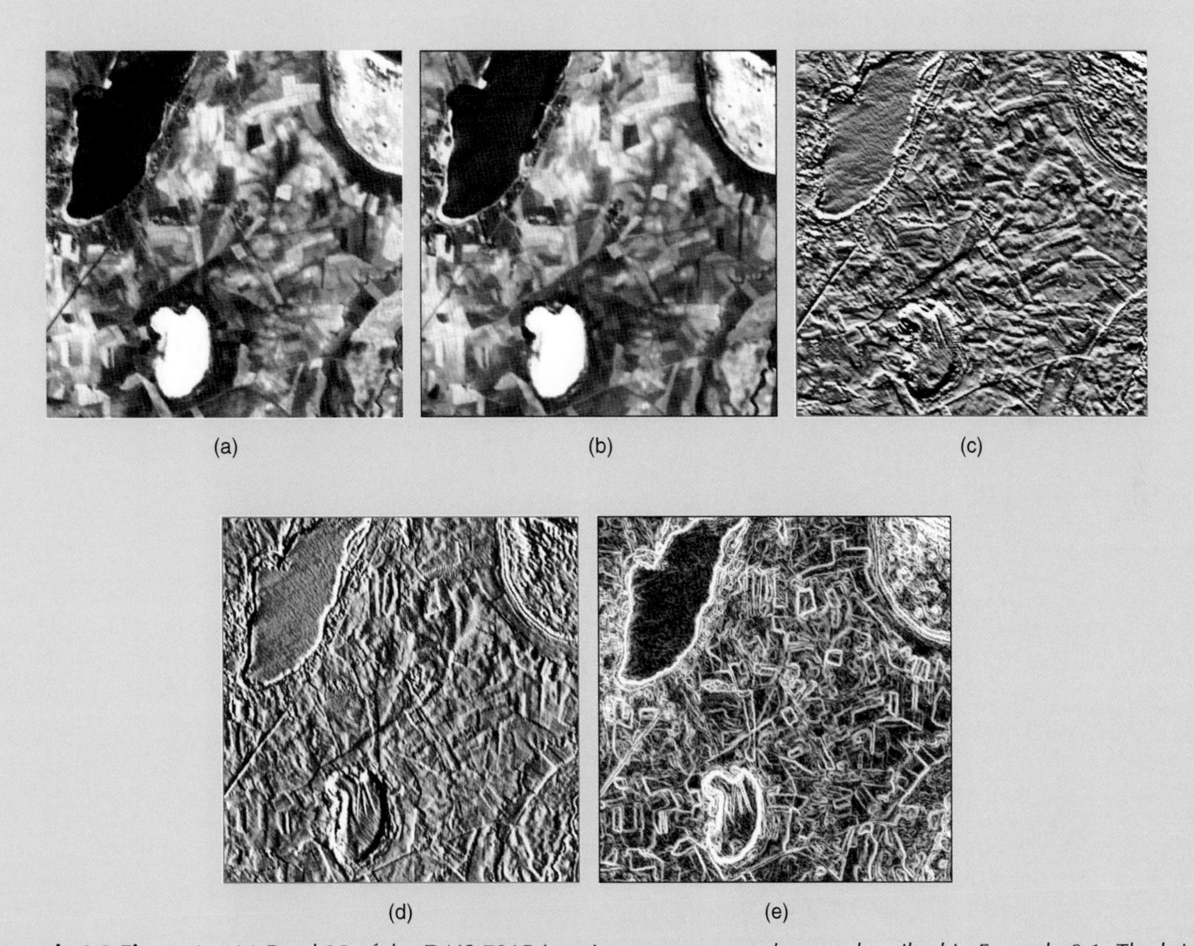

Example 9.2 Figure 1. *(a) Band 30 of the DAIS 7915 imaging spectrometry data set described in Example 9.1. The bright area in the lower centre is a dry salt lake, and the dark area in the upper left corner is a water-filled lake. (b) Image shown in (a) after smoothing using a Savitzky–Golay polynomial filter (second-order polynomial, window size 5 × 5). (c) First partial derivative (horizontal) of the image shown in (a) using a Savitzky–Golay polynomial filter. (d) First partial derivative (vertical) of the image shown in (a) using a Savitzky–Golay polynomial filter. (e) First spatial derivative (with respect to x and y) of the image shown in (a) using a Savitzky–Golay polynomial filter.*

(UT), can be computed from this noise variance. Other studies, such as those reported by Cai and Harrington (1998), Zervakis, Sundararajan and Parhi (2001) and Horgan (1999) discuss the use of the Donoho–Johnstone method, and suggest that the UT may overestimate the noise level in the data. Other problems include the choice of wavelet function (the 'mother wavelet'), the selection of the number of levels of detail coefficients, and the fact that the procedure ideally requires equi-spaced data points that ideally number a power of 2. The assumption is also made that the data series is circular in the sense that we can use points from the end of the series in to precede the first data point. This may work reasonably well with functions such as sine waves measured over a range that is a multiple of 180°, in which the data values are similar in magnitude at both ends of the series. If there is a discrepancy in data magnitude at the start and end of the series then some instability is present in the wavelet coefficients. Shafri (2003) gives a detailed analysis of these problems in relation to both one- and two-dimensional data series, and Taswell (2000) gives a tutorial guide to wavelet shrinkage denoising. See also Shafri and Mather (2005) for further details.

The steps involved in the Donoho and Johnstone (1995) procedure are:

1. If the data series length, n, is not a power of 2 then add sufficient zero values to the series so that $n = 2^j$.
2. Select an appropriate mother wavelet (Section 6.7) and decompose the data series into a set of j detail coefficients using the forward DWT.
3. Determine the noise variance from the detail coefficients at levels $1 - p(p < j)$.
4. Evaluate the UT and determine an appropriate multiple of the threshold for this data (usually between 0.4 and 1.0).
5. Use either hard or soft thresholding (see below) to modify all wavelet coefficients (levels $1 - j$).
6. Perform an inverse DWT to reconstruct the denoised data.

The results of each stage are shown graphically in Figure 9.17. Although it is difficult to see any difference between the graphs shown in Figure 9.17a, there does appear to be a small amount of noise present when the two series are differenced (Figure 9.17e). Because the derivative is very sensitive to noise in the raw data, the use of the wavelet denoising procedure can be justified on the grounds that the data series is essentially unaffected if no noise is present, yet the procedure will remove noise when it is present. Roger and Arnold (1996) discuss the estimation of noise levels in hyperspectral (AVIRIS) data.

The degree or severity of denoising is related to the choice of mother wavelet. The Daubechies wavelet (Section 6.7), for example is implemented in MIPS in three different forms, with 4, 12 and 20 coefficients.

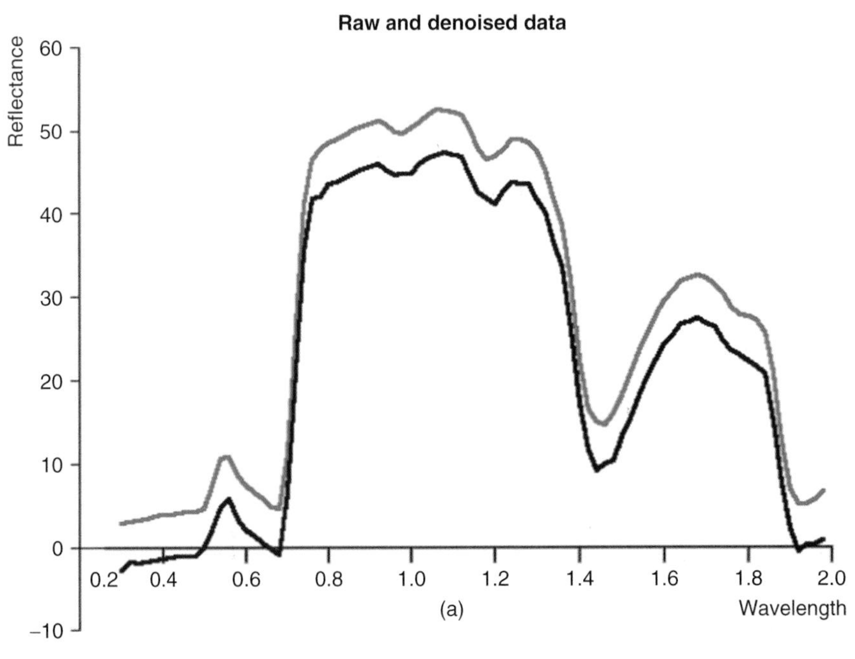

Figure 9.17 Stages in denoising a one-dimensional spectrum of a deciduous leaf. (a) Raw and denoised data. The denoised data is offset vertically downwards for clarity. (b) First derivative calculated from raw data. (c) Wavelet coefficients. (d) First derivative calculated from denoised data. (e) Difference between (b) and (d). Data from the ASTER Spectral Library through the courtesy of the Jet Propulsion Laboratory, California Institute of Technology, Pasadena, California. © 1999, California Institute of Technology. All rights reserved.

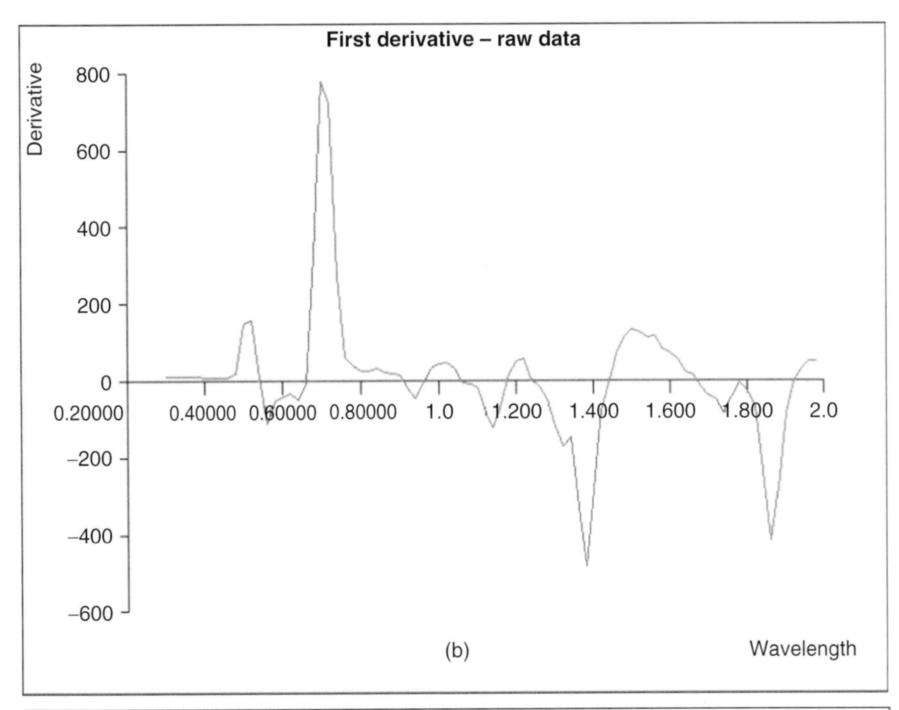

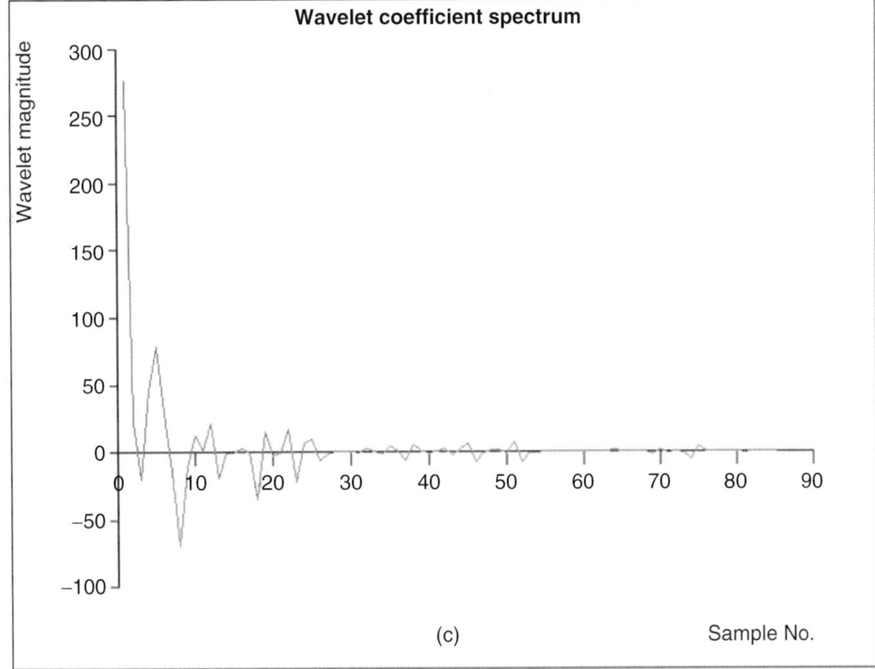

Figure 9.17 (continued)

The greater the number of coefficients, the greater the degree of smoothing. In the example illustrated in Figure 9.17, a Daubechies-20 wavelet was employed. The noise variance is computed (following Donoho and Johnstone, 1995) by firstly selecting the number of levels of detail coefficients to be used in noise estimation. In the example above, in which the series length is 512 (after zero padding), a total of five out of a possible nine levels of detail coefficients were selected ($512 = 2^9$; see also Figure 6.22). The median of the selected detail coefficients is computed first, then the absolute deviations of the detail coefficients from this median are calculated. The median of the absolute deviations (MADs) is multiplied by $\sqrt{2n}$, where n is the number of detail coefficients, and divided by the constant 0.6435 to give the value of the Donoho and Johnstone (1995) UT.

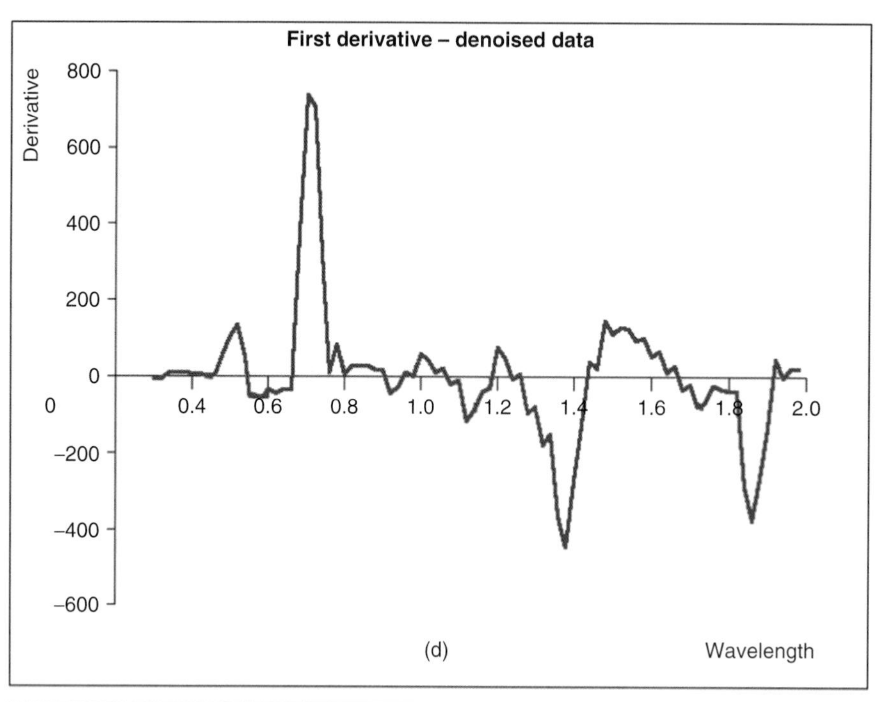

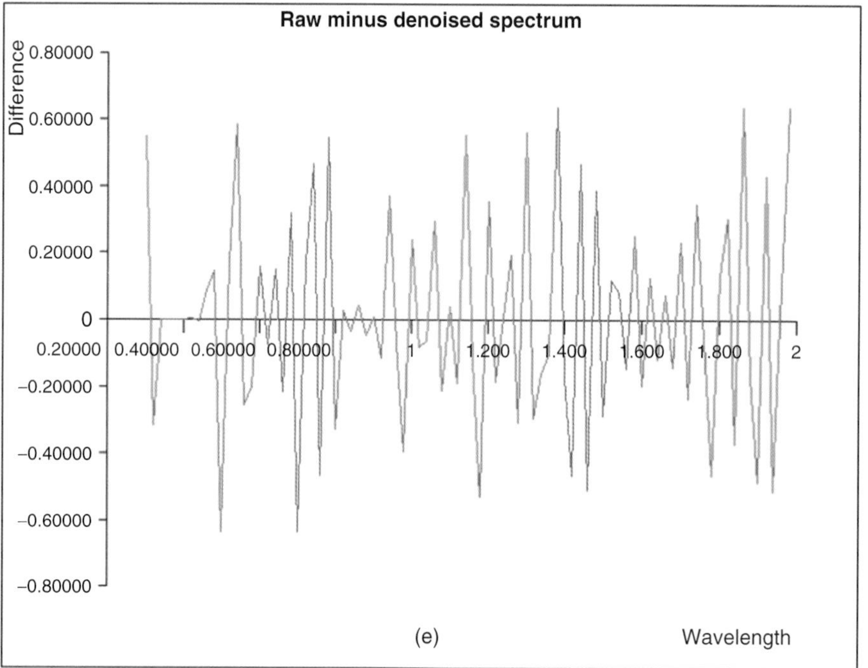

Figure 9.17 (continued)

The UT can be multiplied or divided by a scaling factor to increase or decrease its value (Horgan, 1999, provides some examples using synthetic data). If hard thresholding is used, all wavelet detail coefficients in the chosen levels (five in the example above) that are smaller in magnitude than the scaled UT are set to zero. Other detail coefficients are left unaltered. If soft thresholding is used then hard thresholding is followed by the subtraction of the scaled UT from all the remaining detail coefficients. Again, Horgan (1999) provides some examples. The inverse DWT is then used to transform the thresholded detail coefficients back to the spatial domain; this operation generates the denoised data series.

Two-dimensional denoising follows a similar pattern. Figure 6.36 shows a three-level wavelet decomposition of an image. The detail coefficients at each level are divided into horizontal, vertical and diagonal components. Any or all of these can be selected at a chosen number of levels for the computation of the UT. Often the diagonal detail coefficients at level one are used. The computation of the noise variance and the UT follow the same steps as described above, and hard or soft thresholding is applied before the inverse wavelet transform is computed. See Bruce, Mathur and Byrd (2006) for an example of the use of wavelets in denoising vegetation signatures from MODIS data.

9.3.2.3 Determination of 'Red Edge' Characteristics of Vegetation

The 'red edge' in the reflectance spectrum of active vegetation has become more widely used as a diagnostic feature as the volume of data collected by imaging spectrometers has increased (Boochs and Kupfer, 1990; Clevers, Kooistra and Salas, 2004; Curran *et al.*, 1991; Horler, Dockray and Barber, 1983; Smith, Steven and Colls, 2004; Ustin *et al.*, 1999). An example of a typical vegetation spectrum is shown in Figure 1.21. The biophysical factors that give the spectrum of active vegetation its typical shape can be summarized as follows: leaf chemistry is responsible for the absorption characteristics of the leaf spectrum in the visible wavebands, while the high reflectivity in the near-infrared wavebands is explained by the internal leaf structure. Different vegetation types generally have different spectral reflectance curves, and these differences can be sufficient to allow these different vegetation types to be discriminated and mapped using classification techniques (Chapter 8). Also, the spectrum of a given plant (or group of plants, depending size of the individual ground element that is 'seen' by the sensor) will change during the day and from day to day, depending on season, moisture availability and other stress factors. A number of methods of characterizing vegetation in terms of biomass, leaf area index, vigour (or response to stress) using features of its spectral reflectance curve have been developed, such as vegetation indices (Section 6.2.4). Both the simple vegetation ratio (IR reflectance divided by red reflectance) and the Normalized Difference Vegetation Index or NDVI attempt to characterize the spectral reflectance curve of vegetation by contrasting the high reflectance in the near-infrared wavebands and the low reflection in the visible red wavebands. In effect, such ratios are measuring the steepness of the red-infrared region of the spectral reflectance curve, but not the position in the spectrum of the steep section. Using imaging spectrometer data it is now possible to attempt to characterize this steep rise in the reflectance curve in terms of a single wavelength (though in practice the accuracy of such a determination depends on the width and the spacing of the wavebands in which data are acquired by the imaging spectrometer). The position of this steep rise in reflectance can be characterized by the *red edge wavelength* and the *red edge magnitude*.

The most common definition of the red edge position is the point of inflection of the spectral reflectance curve of vegetation in the red/near-infrared region. A point of inflection is that point on an upward-sloping curve at which the gradient (steepness) of the curve stops rising and starts falling – that is. it is the point of maximum gradient and it is also the point at which the rate of change of gradient is zero.

This idea is illustrated in Figure 9.18, in which the uppermost curve represents an idealized vegetation reflectance spectrum for the wavelength range 650–800 nm. This curve is referred to as 'the function' in the following sentences. The central plot in Figure 9.18 shows the slope or first derivative of the function, defined as the rate of change of the function value (y-axis) per unit step along the x-axis. The first derivative increases in value, reaches a maximum and then declines again. The rate of change of slope (bottom plot in Figure 9.18) per unit step along the x-axis is the second derivative. The second derivative curve crosses the x-axis at the same wavelength as the first derivative reaches a maximum. This is the point of inflection – defined as a point at which the first derivative reaches a maximum and the value of the second derivative changes from negative to positive (or vice versa). The point at which the value of a function changes from positive to negative (or vice versa) is known as a 'zero crossing'.

If there is random noise in a data series then derivative analysis will amplify it. Some researchers, for example Clevers and Jongschaap (2001), whose ideas on red edge determination are summarized below, suggest that derivative-based methods are not robust. However, noise removal ('denoising') using methods such as the one-dimensional DWT (Section 6.7) are effective in removing additive random noise from both one and two-dimensional datasets. Thus, it is sensible to use denoising procedures before carrying out derivative-based red edge determinations.

An alternative and simpler method of computing the red edge wavelength is given by Guyot and Baret (1988) (an accessible account is provided by Clevers and Jongschaap (2001)). Their method requires only four reflectance measurements in the red/near-infrared region of the spectrum. It is therefore suitable for use with image data that are measured in relatively broad wavebands, such as the data collected by the MODIS sensor (Section 2.3.3). Call these four wavebands R_1, R_2, R_3 and R_4 and assume that they are measured at points 1, 2, 3 and 4 on the spectrum, with point 1 being close to 670 nm and point 4 being near 780 nm. The red edge radiance R_e is simply the average of R1 and R4. The red edge wavelength λ_e is found by linear interpolation: $\lambda_e = R_2 + WI\,[(R_e - R_2)/(R_3 - R_2)]$ The term WI is the wavelength interval. In Figure 9.19 the WI from 700 to 780 nm is 40 nm. Points R_1, R_2, R_3 and R_4 are measured at wavelengths of 670, 700, 740 and 780 nm (shown in Figure 9.19 as R_{670}, R_{700}, R_{740} and R_{780}). Clevers and Jongschaap (2001) suggest that the Guyot–Baret procedure is more robust than the derivative analysis described above, and that it produces results that are comparable with those achieved by more complicated methods.

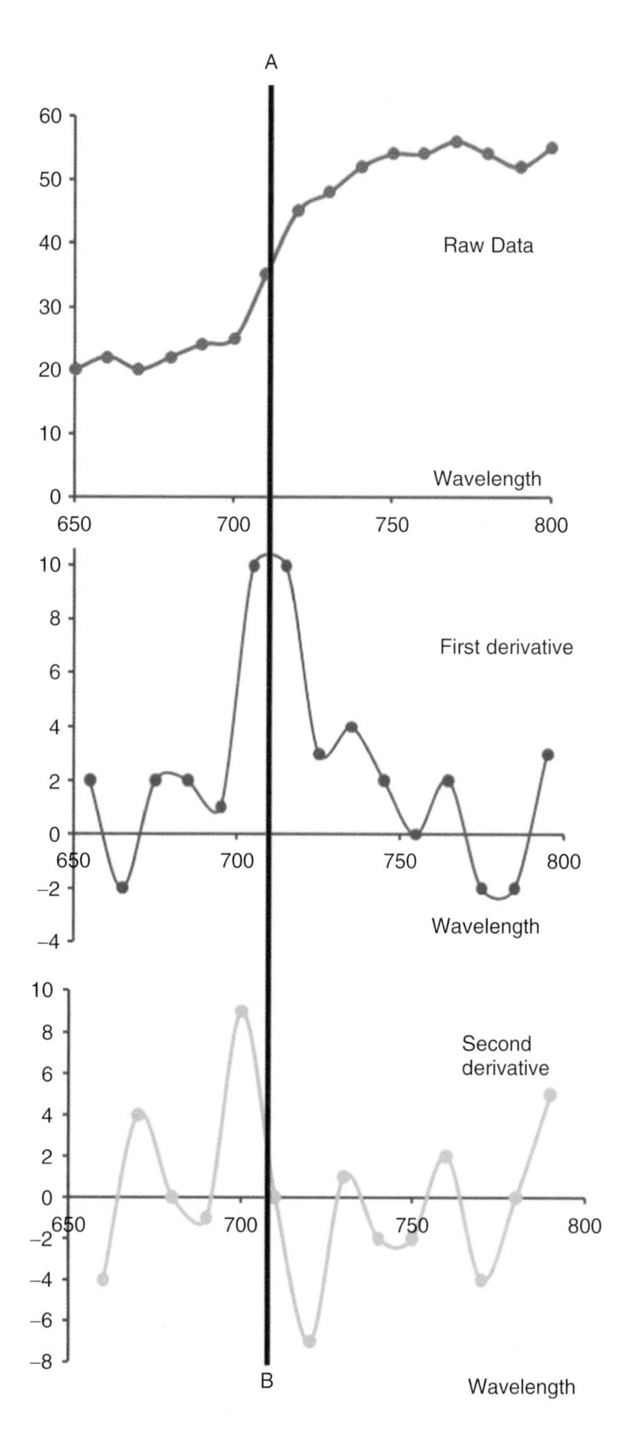

Figure 9.18 *The top curve is a plot of a reflectance spectrum covering the wavelengths 600–800 nm. Note the sudden increase in the gradient of the curve between 680 and 750 nm. The middle graph shows the gradient of the reflectance spectrum, calculated by the method of first differences, which approximates the first derivative. The gradient increases from zero in the green–red wavelengths (600 nm), reaches a maximum, and declines back to zero in the near-infrared wavelengths (800 nm). The rate of change of this gradient (the second difference/derivative of the reflectance spectrum) is shown in the bottom graph. The point at which the second difference (derivative) curve crosses the x-axis (i.e. changes from positive to negative, or vice versa) is called a zero crossing. A point of inflection on a curve is indicated by the correspondence of (i) a zero crossing in the second derivative and (ii) a maximum in the first derivative. In this example, such a point is indicated by the vertical line AB. This point of inflection is often used as an estimate of the red edge wavelength (approximately 710 nm in this case).*

Bonham-Carter (1988) describes another method, based in fitting an inverted Gaussian model. He also provides a Fortran program to implement the method. Cho and Skidmore (2006) suggest that there is a bimodal distribution of red edge position, at around 700 and 725 nm, and they suggest a method based on linear interpolation. This measure is compared to existing methods of estimating the red edge position. Baranoski and Rokne (2005) propose another method using what they describe as a practical approach. Clevers *et al.* (2002) describe a procedure to compute the red edge position for MERIS data.

Two procedures are available in MIPS to compute the red edge position. The first uses the derivative-based approach as follows:

1. Denoise the individual pixel spectra using a DWT (for example based on the Daubechies-4 wavelet).
2. For each pixel, compute the ratio between the reflectance values at the wavebands closest to 800 and 660 nm. If the magnitude of this ratio is less than a specified threshold (such as 2.0) then mark this pixel as 'not vegetation'.
3. For all 'vegetation' pixels, calculate the first and second derivatives of the spectrum.
4. Locate a zero crossing (a change from negative to positive values or vice-versa) in the second derivative in the 660–820 nm spectral region that corresponds to a maximum of the first derivative.
5. Use linear interpolation to estimate the wavelength of the zero crossing. For example, the zero crossing is indicated by a positive value in waveband i and a negative value in waveband $i + 1$. The magnitudes of the second derivative at points i and $i + 1$ are known, so the wavelength at which the value of the second derivative are zero can be interpolated.

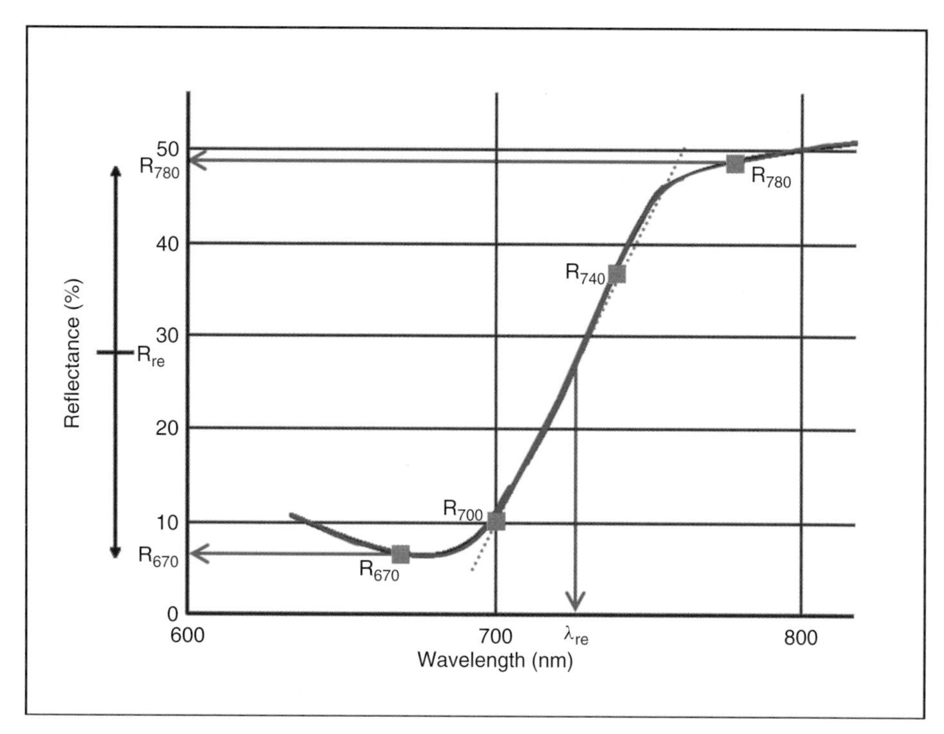

Figure 9.19 *Guyot and Baret's (1988) linear method of red edge determination uses the reflectance at four points on the spectrum (the values 670, 700, 740 and 780 nm are used here). These points are marked* R_{670}, R_{700}, R_{740} and R_{780}. *The reflectance* R_e *at the red edge is the average of the reflectance at 780 and 670 nm. The red edge wavelength is determined by a linear interpolation between the 700 and 780 nm points. Based on Figure 9.2 of Clevers and Jongschaap (2001), Imaging spectrometry for agriculture. In F. van der Meer and S.M. de Jong (eds),* Imaging Spectrometry: Principles and Applications. *Dordrecht: Kluwer Academic Publishers, pp. 157–199. Reproduced with permission from Springer SBM NL.*

6. Output the red edge wavelength, the magnitude of the first derivative at the red edge, and the area under the first derivative curve between (*red edge* −30) and (*red edge* +30) nm.

The output from the derivative approach using data collected over Thetford Forest in eastern England by the Hymap sensor is shown in Figure 9.20a. The same image data, showing spatial variations in the red edge wavelength, is shown in Figure 9.20b after the application of a 3 × 3 median filter. Longer wavelengths are displayed in lighter shades of grey. The black area is that which has been masked by the application of a vegetation index mask, as described above. The spatial variations in red edge wavelength and magnitude correlate well with information about tree species and age for each stand. Hansen and Schjoerring (2003) do not specifically use the red edge wavelength position but find, in a study of a variety of narrow-band vegetation ratios applied to imaging spectrometer data, that most of the ratios used bands that were located in the red edge region of the spectrum. Example 9.3 illustrates the application of wavelet denoising to two-dimensional images.

9.3.2.4 Continuum Removal

Workers in the field of analytical chemistry have found that removal of the local trend from a one-dimensional derivative spectrum can enhance its interpretability. Continuum removal emphasizes absorption bands that depart from their local trend line. The local trend line is usually defined as the upper surface of a convex hull surrounding the data points. Consider a plot of reflectance (y-axis) against centre wavelength for a number of spectral bands at a given pixel (x, y) position. The convex hull is a line that surrounds the scatter of points representing measurements of reflectance at each waveband centre (Figure 9.21). For present purposes the upper surface of the convex hull is required. The upper convex hull is defined by a series of unequally spaced points. Values on the hull at each waveband centre position are interpolated, and the ratio of the reflectance value at waveband centre i to the corresponding interpolated hull value is computed to give the continuum removed spectrum. An example is shown in Figure 9.22. The data used for this example were derived from a pixel representative of an area of deciduous woodland in the top right of the region shown in Figure 9.20 near the north–south river valley. Clark (1999) and Ustin *et al.* (1999) describe the use of

Example 9.3: Image Denoising Using the Wavelet Transform

In this example, a single-band greyscale image is decomposed using the DWT, then the noise variance is estimated from the detail coefficients, and the denoised image is reconstructed. The image used in this example is referenced by the dictionary file **etm_pan2.inf**. You should locate this file before starting the exercise.

Begin by displaying the image **etm_pan2.img**, which is a 1024 × 1024 Landsat ETM panchromatic band image of an agricultural area in eastern England. Display the image using **View|Display Image** and enhance it using **Enhance|Stretch|Use Percentage Limits**, selecting 5 and 95% as the lower and upper bounds. This step is not strictly needed, but it will be useful later to have the original image on-screen for comparative purposes.

Now follow these steps:

1. Choose **Transform|Wavelet (2D) Transform**.
2. Check the radio button for **Mode 1: Single Band, 3 output images**.
3. Identify the **INF** file to be used by double clicking on the entry **etm_pan2.inf** when the **File Selection** dialog box appears.
4. There is only a single band in this image set, so enter 1 in the next dialog box.
5. Choose the default number of decomposition levels, that is 2.
6. Check all three radio buttons to compute the transformed, noise and denoised images.
7. Check all three radio buttons to select the horizontal, vertical and diagonal detail coefficients at level 1 for noise estimation.
8. Do not select any of the three radio buttons for level 2 noise estimation.
9. After a short wait, select the Daubechies 4 mother wavelet.
10. Opt to save the specially stretched transformed output image (shown in Example 9.3 Figure 2) in which each quadrant is separately stretched (you can see why that is done at a later stage). Supply a data band sequential (**BSQ**), header (**HDR**) and dictionary (**INF**) file name for this special 8-bit image.
11. Provide the name of an output **BSQ** file and the corresponding header (**HDR**) file that will hold the three output images (in order: transformed, noise, denoised).
12. Wait a while as the DWT is applied to each of the 1024 rows and 1024 columns of the image.
13. Use a threshold multiplier value of 1.0 (the default) plus hard thresholding.
14. The image is now denoised and the inverse DWT is computed. Eventually, you will see the message **Finished**.
15. Select **File|INF File Operations|Create ENVI INF File** and follow the instructions to create an **INF** file that references the three output files created at step 6.
16. Finally, use **View|Display Image** to view each of the three output files separately.

Example 9.3 Figure 1 shows the original image. The DWT decomposition is shown in Example 9.3 Figure 2. We chose to do a two-level decomposition simply to discover what the resulting decomposed image would look like, and we see that the image at level 1 has been transformed into quadrants, as explained in the main text. The top left quadrant is decomposed into four subquadrants at level 2.

Noise is computed from the $3n/2$ detail coefficients in the top right, bottom left and bottom right quadrants. The noise image is shown in Example 9.3 Figure 3. The denoised image, after hard thresholding using the UT, is shown in Example 9.3 Figure 4.

There is no noticeable (visible) noise in the original image. You might like to investigate the validity of this statement by repeating the experiment and using a different mother wavelet. Other questions that you could investigate are:

- Is there any apparent difference between the results of hard and soft thresholding?
- What happens if you increase or decrease the threshold for wavelet shrinkage by changing the value of the multiplier at step 12? (The default is 1.0.)
- Wavelet shrinkage is designed for the removal of *additive* noise. Could you use the method to remove speckle noise in SAR images? If so, how?

- What happens if you base the noise threshold on the level 2 detail coefficients (steps 7 and 8) rather than on level 1? Or if you base the noise estimate on a single quadrant (one of horizontal, vertical and diagonal) at either level 1 or level 2? Or even level 3 (step 5)?

Example 9.3 Figure 1. *Landsat* ETM+ *panchromatic image of an agricultural area of eastern England.*

Example 9.3 Figure 2. *Two-level DWT of the image shown in Example 9.3 Figure 9.1, using the Daubechies-4 mother wavelet.*

(Continues on next page)

Note that the output **BSQ** file contains the transformed, noise and denoised images. However, you created a special 8-bit output image at step 10. Display this image, and compare it to the transformed image in the main output file (it is the first of three images, the others being the noise image and the denoised image). You will see that a single contrast stretch cannot accommodate the range of values present in the transformed image; that is why the special image is created. Each of the quadrants and subquadrants in the special image is stretched individually in order to achieve the optimum display.

Example 9.3 Figure 3. *Noise removed from Example 9.3 Figure 1 using wavelet shrinkage. The noise was estimated from the level 1 (horizontal, vertical and diagonal) detail coefficients of Example 9.3 Figure 2 using hard thresholding.*

Example 9.3 Figure 4. *Denoised image.*

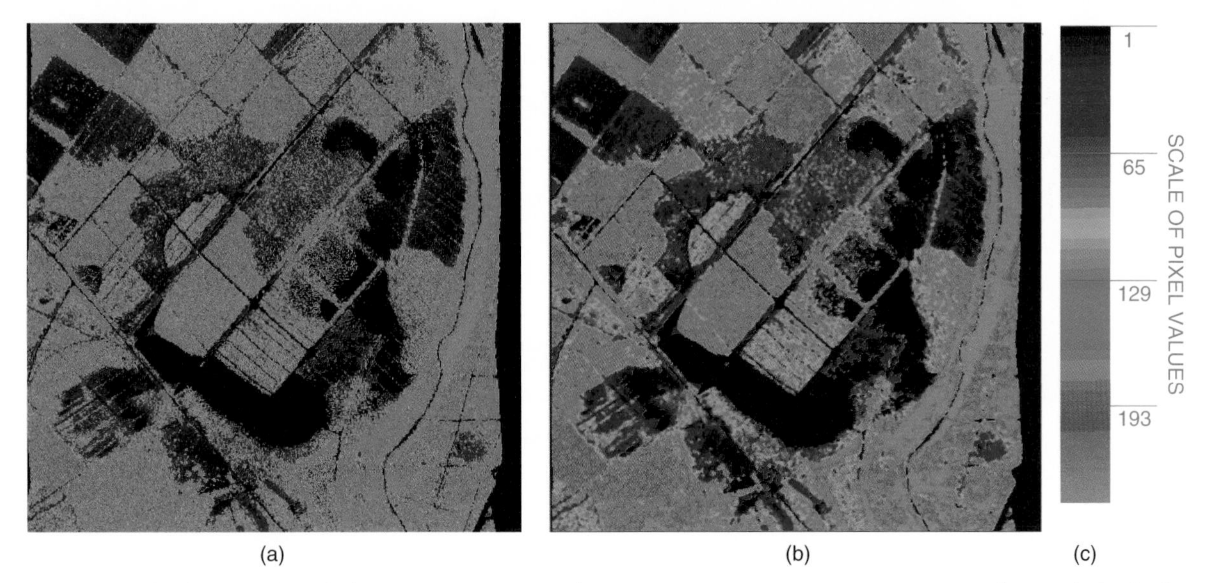

Figure 9.20 *Red edge wavelength for a 512 × 512 area of a Hymap imaging spectrometer image of part of Thetford Forest, eastern England. (a) Raw output from the MIPS derivative-based red edge wavelength procedure, (b) image shown in (a) after the application of a 3 × 3 median filter. (c) Colour table used in (a) and (b). Black areas are masked using a vegetation index threshold and represent bare soil and non-vegetated areas, plus left and right marginal areas resulting from geometric correction of the image. Data collected for the BNSC/NERC SHAC campaign, 2000.*

continuum removal in the context of rock and mineral identification and geobotany, respectively. Mutanga, Skidmore and Prins (2004) use continuum-removed absorption features to predict pasture quality.

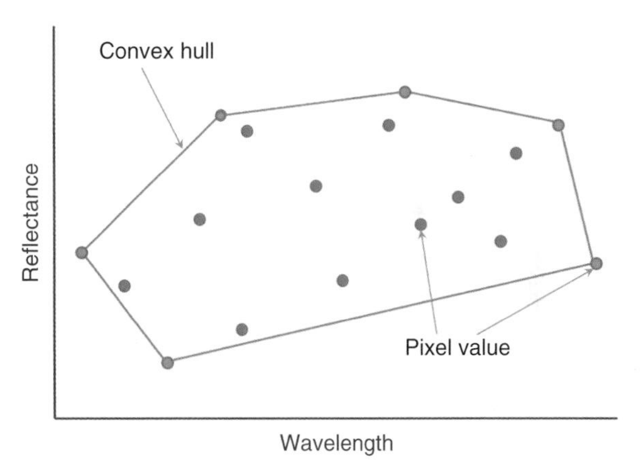

Figure 9.21 *The black circles show the reflectance (y-axis) plotted against waveband centre (x-axis) for 10 spectral wavebands. The solid line joining the extreme points is the convex hull. Only the upper surface of the hull between the first and last data points is required.*

9.4 Lidar

9.4.1 Introduction

The word 'lidar' is an acronym derived from light detection and ranging. The same system is also known as LASER detection and ranging (LADAR), and also as LASER altimetry. The word LASER is yet another acronym, of light amplification by stimulated emission of radiation. The lower-case word 'lidar' will be used here to be consistent with our use of the word 'radar', which is also an acronym. Like radar, a lidar sensor is an active sensor, the differences between lidar and radar being: (i) lidar uses electromagnetic energy in the visible and near-infrared (VNIR) wavelengths, whereas a radar sensor uses microwave energy, (ii) lidar is a nadir-looking, but radar is a side-looking instrument, (iii) a lidar records information at discrete points across the swath, which is not therefore formed of contiguous pixels, as is the case with a radar sensor and (iv) because lidar operates in the VNIR wavelengths, its signal is affected by atmospheric conditions, whereas at the wavelengths used in remote sensing, a radar sensor is weather-independent. Unlike imaging sensors operating in the VNIR wavebands, and which record upwelling electromagnetic energy that is emitted by or reflected from objects on the Earth's surface, a lidar instrument measures the time taken by an energy pulse to reach the ground, and for a part of the scattered radiation to return to the sensor. A lidar thus measures the distance from the sensor to the ground, because electromagnetic energy travels at the speed of light and so the time taken for the energy pulse to travel from the lidar instrument to the ground and back can easily be converted to a distance.

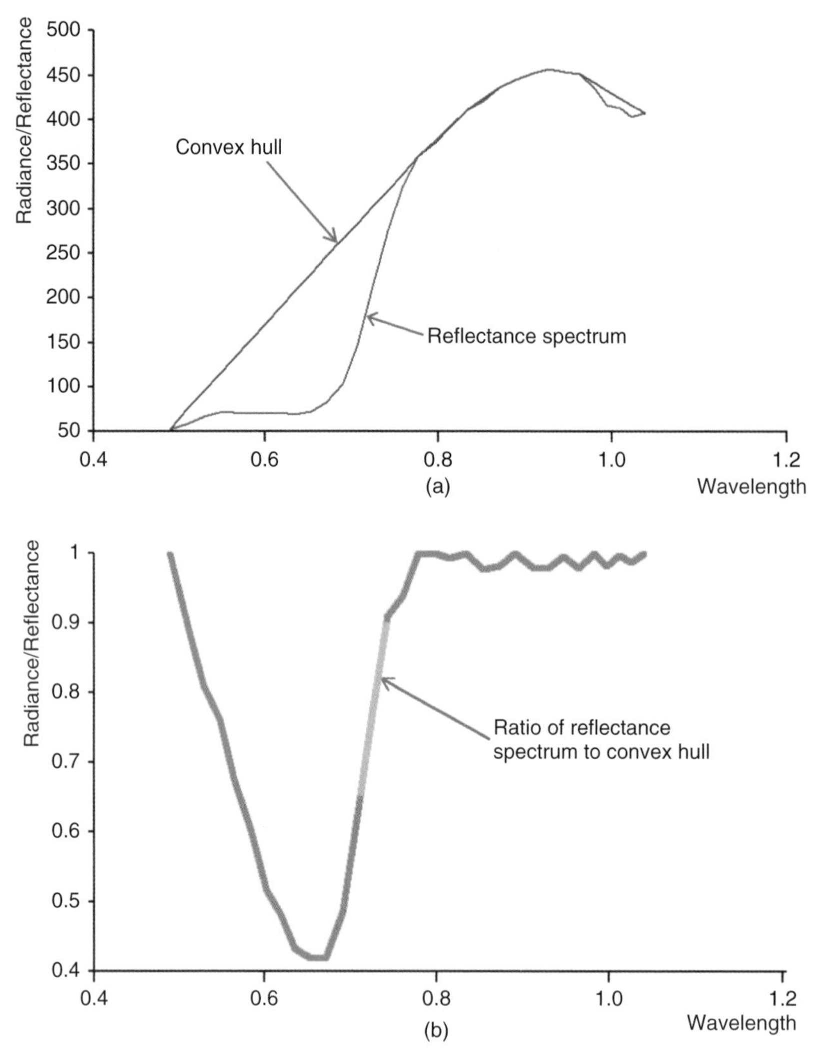

Figure 9.22 *Continuum removal. (a) Reflectance spectrum of the selected pixel and the upper segment of the convex hull. (b) Continuum-removed spectrum derived from the ratio of the reflectance spectrum and the corresponding convex hull value at each waveband centre. The positions and features of the absorption bands are more clearly perceived.*

If the position of the instrument is known to a sufficient level of accuracy then these distances can be converted to elevations above a specified geodetic datum, and a raster map of these elevations can be generated. The set of elevation values produced by a lidar does not necessarily define a DEM, because the lidar pulse is reflected back by the first object of sufficient size and density that it meets as it travels downwards from the instrument. This object may be a branch, a tree crown, the ground surface or the top of a building, depending on the area being viewed and the size (or footprint) of the lidar pulse. The set of point elevation values can be used to generate a DSM, which shows the elevation of the highest reflective object on the ground. Thus, data from lidar sensors can be used to map the highest point on a building, or of a tree. In order to generate a DEM from the DSM, the surface objects must be removed, or the height of the superimposed object must be estimated. The DSM may be useful in itself. For example, Figure 9.23 is a lidar image produced by National Atmospheric and Oceanic Administration (NOAA) showing the site of 'Ground Zero' in Manhattan, New York, in September 2001. Lidar images collected by aircraft can be used to generate three-dimensional urban models, or they can show objects such as forests that project above the ground surface. As well as recording range or distance, a lidar records the intensity of the return signal. Donoghue *et al.* (2007) attempt to differentiate between coniferous species using both lidar height and intensity data.

There is a considerable literature in both photogrammetric and remote sensing journals on the theme of lidar measurements of ground surface phenomena. Hodgson and Bresnahan (2004) report the results of an analysis of lidar-derived elevation measurements. The ISPRS

Figure 9.23 *Lidar image of 'Ground Zero', Manhattan, New York, taken on 17 September 2001. The data collection program was the result of a collaboration between the US National Atmospheric and Oceanic Administration (NOAA), the US Army Joint Precision Strike Demonstration (JSD) and the University of Florida, using an aircraft-mounted Optech lidar. The colours represent elevations between 0 and 200 m. The three-dimensional model helped to locate original support structures, stairwells, elevator shafts, basements, and so on. Credit: NOAA/U.S. Army JPSD. (From http://www.noaanews.noaa.gov/stories/s781.htm; accessed 18 August 2009.)*

Journal of Photogrammetry and Remote Sensing (2008) is a special issue on terrestrial laser scanning. Lichti, Pfeifer and Maas (2008) edited a theme issue of *ISPRS Journal of Photogrammetry and Remote Sensing* on the subject of terrestrial laser scanning. Fraser, Schroeder and Baudoin (2006) edited another special issue of the same journal on extraction of topographic information from high-resolution satellite imagery. Anderson, Thompson and Austin (2005) consider the influence of lidar point density and interpolation effects on elevation estimates. Antonarakis, Richards and Brasington (2008) use lidar data in a land cover classification exercise, as do Bork and Su (2007). Uses of lidar in forest classifications are described by Donoghue *et al.* (2007), Hill and Thomson (2005) and Lim *et al.* (2003). Liu (2008) gives an overview of issues surrounding the use of lidar in the generation of DEMs. Kobler *et al.* (2007) present an algorithm for extracting a DTM in forested terrain. Shan and Toth (2008) is an edited volume of papers relevant to various aspects of the use of laser ranging. Su and Bork (2006) also consider DEM generation, in this case studying the effects of vegetation cover, slope angle and lidar view angle on elevation estimates. Zhang and Whitman (2005) and Zhang *et al.* (2003) discuss the process of filtering of lidar data.

We saw in the preceding paragraphs that the conversion of a lidar DSM to a DEM requires that we estimate the height of the highest reflective object on the ground at a given point. The lidar instrument itself can in fact, perform this task. The account given above of lidar operation describes what is known as a 'first return' or 'first bounce' system, which – as the name implies – records the time taken between the emission of the pulse of light energy and the reception at the sensor of the first backscattered return. More sophisticated instruments (with more sophisticated signal processing software) can generate two signals for each pulse. One of these is the first return, as described earlier, while the second is the position of the last indication of backscatter. This second event is called the last return, so these systems give 'first return – last return' data for each grid cell of the raster. There is no difference between the first and last return if the target object does not transmit light; if it does, then the time difference between the first and last bounce is proportional to the height of the object above the ground. A concrete surface does not transmit light, but a forest canopy does. Even more sophisticated systems can record the backscatter events between the first and last return, and so provide a profile of this backscattering between the first and last bounce points (Figures 9.24 and 9.25). As noted already, solid objects like buildings are not penetrated by the electromagnetic energy emitted by a lidar. For these targets, and for the ground itself, only a single return is recorded. The 'first return–last return' and the profile data are returned by objects such as trees, forests and other types of vegetation that are capable of transmitting light energy. See Mallet and Bretar (2009) for a discussion of what is called full-waveform topographic lidar.

So far, the operation of a lidar system has been described in terms of the emission of energy pulses, and the timing of the returned (backscattered) energy. Not all lidars operate in this way. Some use the continuous wave (CW) principle, which is described in more detail in Section 9.2 in the context of SAR interferometry. Instead of generating discrete pulses of energy, a CW lidar emits energy in the form of a sinusoidal wave of known wavelength. Recall that the phase of a wave (Section 1.2) is the offset between the y-axis and the waveform crest (Figure 9.1). In effect, the number of waveforms that are required to cover the distance between the lidar sensor and the ground is calculated, with the fractional part being estimated by the phase difference between the original and the received wave.

An important distinction can be made between 'small footprint' (5–30 cm) and 'large footprint' (10–25 m) lidar sensors. A large-footprint system has a greater swath width than a small-footprint system. Small-footprint systems are used for detailed local mapping of surface elevations, as might be required for floodplain mapping. However, the spacing between the points at which the lidar pulse hits the target in a small-footprint system may be such that several points on the surface of a vegetation

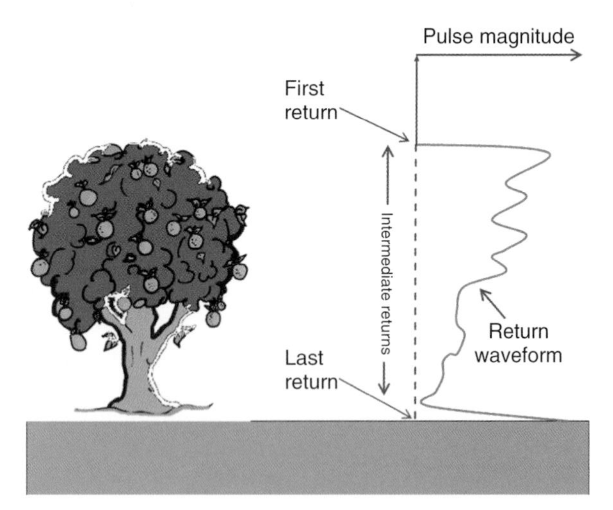

Figure 9.24 Profile of the return pulse magnitude of lidar interaction with a tree. The lidar sensor receives the returned (back-scattered) signal (solid line on right). Some systems record the time from pulse transmission to receipt of the first return, others record the time to the last return, while more sophisticated systems take a sample of the intermediate returns. The distance (range) from the sensor to the target is computed from these timings. Clipart tree from http://www.clipsahoy.com/webgraphics2/as3313.htm (accessed 4 January 2009).

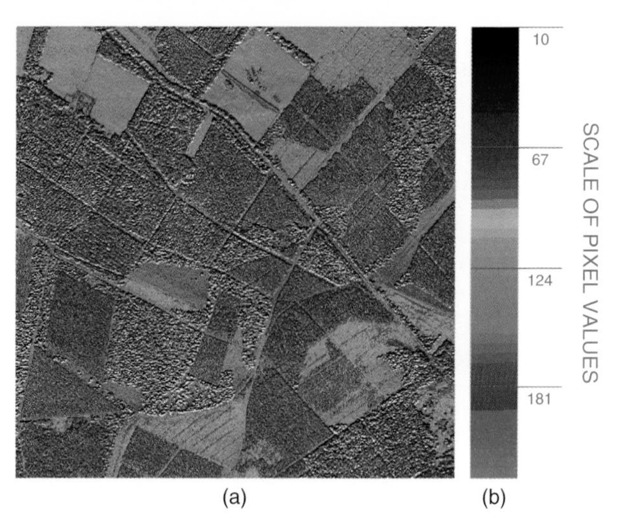

Figure 9.25 (a) Difference between first and last return lidar data for a 2 × 2 km area near Thetford, Norfolk, United Kingdom. The difference between the two returns is related to canopy height and to the vertical structure of the vegetation. The contrast between smooth and busy textures of the forested and non-forested areas is clear. The visual interpretability of the image has been be enhanced via the use of a pseudocolour transform. (b) Colour wedge for Figure 9.26a. Lidar data © Environment Agency Geomatics Group 2009.

canopy may be measured, giving a detailed representation of that canopy, whereas the large-footprint system will collect an average value for a greater area of the canopy surface. This latter measurement may be more useful for studies of forest canopy response.

Most lidar sensors are flown onboard aircraft. One experimental lidar sensor, the Lidar In-space Technology Experiment or LITE, was flown in September 1994 as part of the STS-64 mission. LITE is a three-wavelength profiling lidar developed by NASA Langley Research Center, and is primarily designed for measuring atmospheric rather than terrestrial phenomena. It takes simultaneous measurements in three harmonically related wavelengths of 1064 nm (infrared), 532 nm (visible green) and 355 nm (ultraviolet) along a profile measuring approximately 300 m wide at the Earth's surface.

The Shuttle Laser Altimeter (SLA) was carried on board two Space Shuttle missions, in January 1996 and August 1997, respectively. SLA incorporates a laser operating at a wavelength of 1068 nm, with a sampling rate of 10 Hz. It has a footprint radius of 100 m. The first SLA mission (SLA-1) was not as successful as expected, as the dynamic range of the backscattered echoes was greater than had been allowed for in the system design. SLA-2 was modified to provide more flexibility. Data from the SLA-2 mission can be downloaded from the Internet (search for 'SLA data'). Garvin *et al.* (1998) provide a review of the SLA program.

ICESat, was launched on 13 January 2003. It carries the Geoscience Laser Altimeter System (GLAS), which operates at two wavelengths – 1064 and 532 nm. The position of the ICESat platform is determined by GPS and by stellar navigation systems. The footprint of the GLAS is 75 m, and the spacing between points is 175 m. A comprehensive description of the GLAS instrument and its applications is provided by Zwally *et al.* (2002). Schutz *et al.* (2005) give an overview of the mission. Remote sensing of snow and ice is covered by Massom and Lubin (2006) discuss remote sensing of ice sheets.

9.4.2 Lidar Details

The material in this section summarizes two review papers, by Baltsavias (1999) and Wehr and Lohr (1999), to which readers should refer for more detailed accounts.

The basic principle of operation of a lidar sensor is described briefly above. A typical lidar instrument incorporates (i) a laser ranging unit, (ii) an optomechanical scanner and (iii) a control and processing unit. The laser ranging unit contains the laser transmitter and receiver. The transmitter is able to generate a narrow beam of electromagnetic energy, while the receiver 'looks' along the same path as the transmitter in order to capture the backscattered return. Most of the present generation of lidar sensors use the pulse principle, described above; Wehr and Lohr (1999) note that only one commercial airborne lidar employs the CW principle to calculate range. The principle of the pulsed lidar is quite straightforward; if the time between transmission and reception of the pulse is t and if R is the distance between the lidar transmitter and the target then $R = ct/2$, where c is the speed of light. A CW lidar transmits a continuous signal on which a sinusoidal wave of known period is superimposed. If the phase difference between the transmitted and received signals is computed then the range R is related to the number of full waveforms plus the phase difference. Since the phase difference could be more than 360° some ambiguity could be introduced, analogous to the 'phase unwrapping' problem in SAR interferometry (Section 9.2).

The maximum range of a pulsed lidar system depends on the maximum time interval that can be measured by the control unit and on the strength of the backscattered signal, which is to some extent dependent on the power of the transmitted pulse (it also depends on the reflectivity of the surface). An analogy can be made between a laser and a torch. The range of the torch depends on the battery power, the properties of the bulb and the focusing power of the lens. A 'high-power' torch can transmit a narrow beam of light over a considerable distance. If you take the torch to a dark and isolated location and direct the torchlight upwards, you will not see anything because there are no reflective objects within the maximum range of the torch. The accuracy of the measurements made by a lidar instrument depends upon the signal to noise ratio (SNR), which in turn is dependent on electronic noise in the components of the lidar sensor, as well as on the power of the transmitted signal.

Wehr and Lohr (1999) note that the most sensitive detectors for use in the receiver unit operate in the 800–1000 nm region (photographic infrared). In this wavelength range, eye safety is a consideration and so longer wavelengths (around 1500 nm) are employed, because higher power lasers can be used at these wavelengths without compromising safety. A further advantage that comes from the use of longer wavelengths is that the background level of solar radiation is lower than in the 800–1000 nm region (see Figure 1.7). Lidar systems that are used for measuring bathymetry, rather than the properties of terrestrial targets, use shorter wavelengths (of the order of 500–550 nm) because electromagnetic energy at longer wavelengths is absorbed by water bodies, rather than transmitted or reflected (Figure 1.23).

The laser ranging unit described above emits and receives a pulse of light energy, and the range (or distance to the target) is calculated from the time difference between transmission and reflection. A two-dimensional field of measurements is generated first by the forward movement of the platform and second by the employment of a side-to-side scanning system (Figure 9.26). The lidar footprint, that is the size of the small area on the ground that is viewed by the lidar, depends on the instantaneous field of view (IFOV) of the instrument, on the altitude of the platform, and on the angle of view. The radius of the footprint is greater at the edge of the scan than at the centre. The footprint points are collected at equal angle intervals across the scan, so that their ground spacing is spatially unequal. The number of points collected is related to the pulse rate of the lidar transmitter and to the height of the aircraft above the ground. Most lidar sensors employ the scanning mirror principle, as used by the Landsat ETM+ and NOAA AVHRR sensors. The distribution of the observed ground points when an oscillating mirror is used results in a zig-zag pattern, shown in Figure 9.25. One problem experienced with some oscillating mirror scanners is that the mirror has to slow down, stop and accelerate at the end of each scan. Other problems are caused by variations in the altitude and attitude (pitch, roll and yaw) of the aircraft. These variations result in displacements of the ground points from their theoretical positions.

The range or distance from an aircraft to a point on the ground is merely of local interest in that it must be more than zero at nadir, unless the aircraft has landed. To be of scientific use, range information must be placed in the context of a coordinate system, that is it must be converted to a height above an accepted datum such as WGS84. This transformation can only be achieved if the position of the sensor relative to some reference point is known to an acceptable degree of accuracy. Information

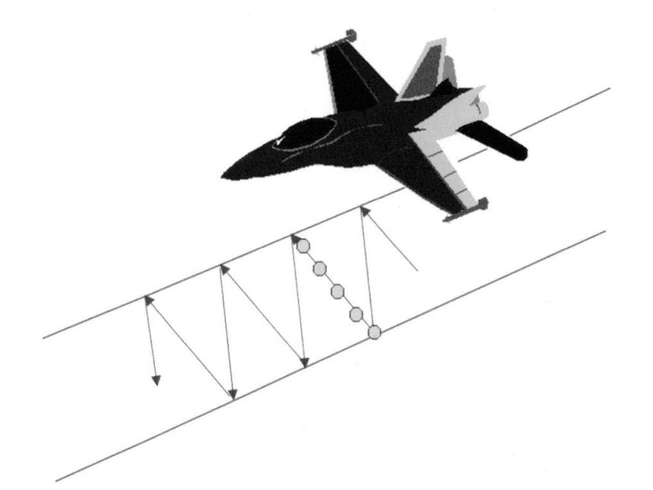

Figure 9.26 *Schematic illustration of airborne lidar scanner operation. The mirror oscillates from side to side in the vertical plane, and the forward motion of the platform results in a scan line that is oblique to the flight direction. Further disturbances result from the pitch, roll and yaw of the platform. The measurement points (yellow circles) are collected at equal-angle steps across the scan, so their ground spacing is not constant.*

Example 9.4: Lidar First/Last Return

Example 9.4 Figure 1 shows the difference between a first-return lidar dataset and a last-return lidar dataset for a 2 × 2 km area near Thetford, Norfolk, in eastern England. The data are processed by the provider, and are supplied in gridded form as first and last return measurements, with a nominal spatial resolution of 2 m. The accuracy of the surface elevation measurements is 0.15 m at 1200 m altitude (one sigma, that is. 66% of all points will be within 15 cm of the true elevation, and 95% will be within 30 cm). The horizontal accuracy is quoted as 0.002 × aircraft altitude.

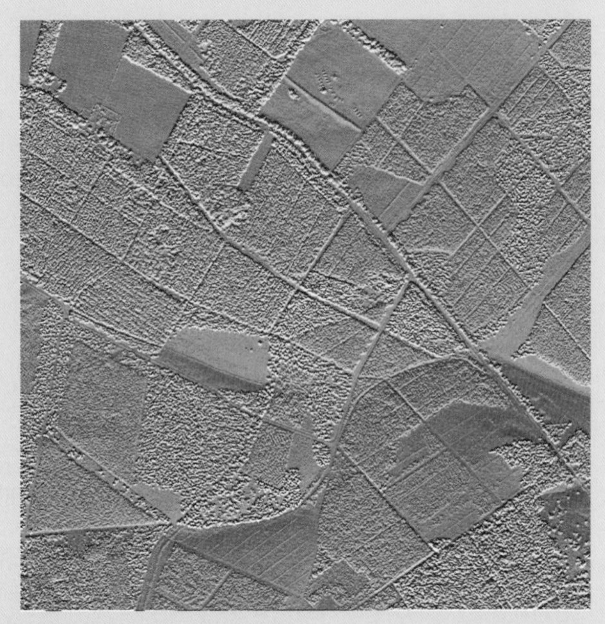

Example 9.4 Figure 1. *The difference between the first and the last return lidar data for a 2 × 2 km area near Thetford, Norfolk, United Kingdom. The difference between the two returns is related to canopy height and to the vertical structure of the vegetation. The contrast between smooth and busy textured areas is clear. (The visual interpretability (and impact) of this type of image can be enhanced via the use of a pseudocolour transform as shown in Example 9.4 Figure 2a).*

on the aircraft's attitude is also required if the positions of the points on the ground are to be calculated accurately. Data relating to position and accuracy are collected by the control and processing unit, which contains a GPS receiver and an inertial navigation system. The results of a lidar mission thus consist of two data sets. The first consists of the measurements made by the lidar sensor and the second contains the positional data collected by the GPS and inertial navigation unit. Both data sets are used at the processing stage, in which the lidar measurements are converted to a regular raster format. The range and position of each measured ground point are computed, and the resulting irregular spatial pattern is re-sampled on a regular grid to produce the output data set. Further processing is necessary if a 'bare earth' DEM is required, as the lidar range is measured between the sensor and the first reflector, in the case of a 'first-return' lidar. Maas (2002) provides a useful survey of methods of analysing errors in lidar data.

The preceding description of the *modus operandi* of airborne lidar sensors makes only one reference to the IFOV of the instrument, and implicitly assumes that the purpose of any investigation using lidar is to generate a DEM or a DSM. This is an oversimplification. One of the main areas of research using lidar is ecology. Here, some interest lies in the measurement of canopy heights, especially of forests, but there is an equal if not greater interest in the measurement of the three-dimensional characteristics of vegetation. Small-footprint lidars 'see' only a small area on the ground, and these small footprints are separated by 'unseen' areas (Figure 9.25). The spatial distribution of these small footprints over a forest canopy may be such that gaps between trees are missed, and it is also possible for the crown of a tree to be left unobserved by a small footprint system. The small footprint

Another application of connectivity analysis is to determine the connectivity of patches in the landscape that might, for instance, provide specific habitats for plants and animals (Nikolakaki, 2004; Schumaker, 1996). One everyday application of network connectivity is in in-car navigation, which is reliant on accurate road maps and an algorithm for finding the shortest/fastest/most economical route between two points. The shortest route problem is described by Dantzig (1960). A more elaborate version of the problem is known as the travelling salesman problem (Kruskal, 1956) and involves visiting all nodes in a network in the shortest time or in the shortest distance.

10.5 Spatial Analysis

Spatial analysis (or geospatial analysis) is usually defined as the application of numerical (statistical or mathematical) procedures to data that have location coordinates. The analysis may relate to the pattern of the spatial entities (points, lines or polygons) or to the values associated with those points, lines or polygons. In a raster model, spatial analysis would refer to patterns and covariations in pixel values of one or more data layer. Topics include network analysis, point pattern analysis, trend analysis, spatial interpolation, sampling and exploratory data analysis. For example, we may have a sample pattern of points representing wells and wish to ascertain the probability of the observed sample point pattern occurring by chance when the true but unknown population pattern is uncorrelated or random. The alternative hypothesis is that there is some systematic process that is influencing the point pattern, for example lithology. Another example is the compression of data layers measured over a set of polygons, using a technique such as principal components analysis (PCA) (Section 6.4). A third example is the derivation of a statistical relationship between a set of sample field measurements and corresponding values or combinations of values in a remotely-sensed dataset, so that this relationship might be applied to the dataset as a whole. This kind of analysis is often performed when the values of a geophysical or biophysical parameter is to be determined from remotely-sensed imagery. Examples of the latter might include the determination of soil moisture status from microwave images or the derivation of LAI values from combinations (such as ratios) of remotely-sensed data.

The material in this section is derived from a number of sources, including Atkinson and Lloyd (2009) on geostatistics and spatial interpolation, and the texts on geospatial analysis by de Smith, Goodchild and Longley (2007), Fortin and Dale (2005), Fotheringham, Brundson and Charlton (2000), Fotheringham and Rogerson (2009), Haining (2003) and O'Sullivan and Unwin (2003). Anselin, Syabri and Kho (2006) describe GeoDa, a free software package for spatial analysis.

10.5.1 Point Patterns and Interpolation

Spatial analysis of point patterns, the first of the three examples in the preceding paragraph, encompasses a number of applications in geography. The example above relates to the statistical hypothesis that a point pattern is random, that is there are no systematic processes at work. Two instances of the potential use of point pattern analysis in environmental GIS are: (i) the analysis of patterns of sample data points used in calibrating field measurements against remotely-sensed data and (ii) the analysis of locations of ground control points used in geometric correction of remotely-sensed images (Section 4.3). In both cases, a random pattern is desirable. Texts such as Ripley (1981) should be consulted to discover methods of determining the nature of point patterns, which are generally described by the terms clustered, uniform and random. Random sampling patterns are generally preferred, but – in cases where the study area is subdivided into separate regions – a stratified random sample may be preferable. The paper on two-dimensional systematic sampling of land use by Dunn and Harrison (1993) covers spatial aspects of sampling. Moisen, Edwards and Cutler (1994) deal with spatial sampling for assessing classification accuracy (Section 8.10).

If the points represent sampling locations where the value of a random variable such as soil moisture content or surface temperature has been measured then one may wish to interpolate estimated values for all the cells of a grid, so that the ground sample can be related to values estimated from remotely-sensed data. One way of achieving this aim is to fit a low-order least-squares polynomial (a trend surface) to the data, and read off the desired values. Computationally, this procedure is similar to the process of geometric correction of remotely-sensed imagery using least-squares methods, described in Chapter 4, whereas the derivation of the grid cell values is akin to resampling, also dealt with in Chapter 4. A low-order, two-dimensional polynomial is computed from the set of point locations (x, y) and the point value (z). Usually a polynomial order of 4 or less is used. The procedure attempts to differentiate between patterns of systematic spatial variation and patterns that are generated by random or local processes. Figure 10.7a shows a second-order trend surface for the average annual rainfall collected at 42 points (rain gauges) near Nottingham. Figure 10.7b shows the pattern of residuals or deviations of the measured values from the trend surface. Figure 10.7c shows a visualization of the trend surface.

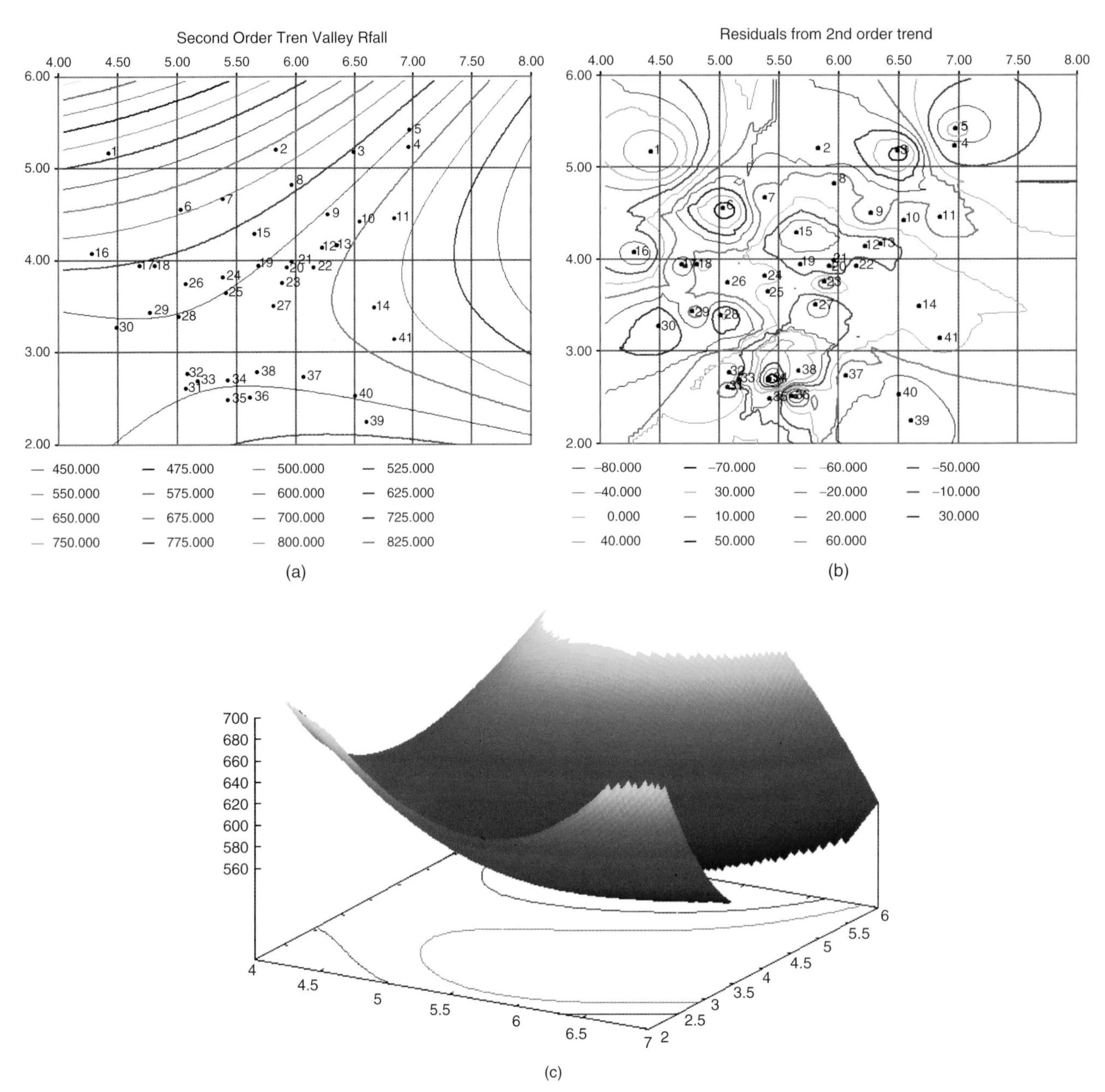

Figure 10.7 (a) Least squares estimation of a second order least squares polynomial surface from a set of points (shown by numbered black dots). The points actually represent rain gauges located to the north and south of the Trent valley around Nottingham. (b) Residuals from the trend surface. Estimated values are computed at each of the sample points and interpolated onto a grid using an inverse distance weighted procedure and (c) visualization of the trend surface. The x and y axes show geographic position. The z axis is mean annual rainfall (millimetres).

Other interpolation methods include inverse distance weighted interpolation and kriging. Inverse distance weighted interpolation takes a set of point values (e.g. the black dots representing rainfall stations in Figure 10.7a, b) and computes an estimate for each cell in the grid, with a weight that is inversely proportional to the distance from the point to be interpolated and the known point (Bonham-Carter, 1994; Lu and Wong, 2008). If $\hat{z}_i$ is the value to be interpolated, w_i is the weight for the ith known data point, n is the number of data points and z_i is any data point other than the interpolated point, then

$$\hat{z}_i = \frac{\sum_1^n w_i z_i}{\sum_1^n z_i}$$

The weights are usually set to be the reciprocal of some power, usually 2, of the distance from the interpolation point to its neighbours. A circular zone around the interpolation point can be specified so that only those points within the circle have an influence on the interpolated value. inverse distance weighting (IDW) is the simplest interpolator, and one that is incorporated into a number of GIS. However, the surface produced by IDW does not honour (i.e. pass through) all of the known data points. Also, the values of the weights are user-selected and different weights will give different surfaces.

An alternative is based on geostatistical methods known as kriging, in which the weights are proportional both to distance and the directional correlation between the data points (here, the words *directional correlation* are used to mean the correlation in a specific direction, such as north). Atkinson and Lloyd (2009) and Bonham-Carter (1994) consider the problem of interpolation from the point of view of geostatistics. The basic equation for the kriging procedure is very similar, superficially, to the IDW method:

$$\hat{z}_i = \sum_{i=1}^{n} w_i z_i$$

where the parameters have the same meaning as in the IDW example. The weights are defined by the reciprocals of the distances from the data points to the points being estimated, written in vector form as $\mathbf{d}\ (= d_i, i = 1, n)$ and by the matrix $\mathbf{C}$ of spatial correlations between n pairs of points, where n is the number of data points (Bonham-Carter, 1994). An autocorrelation factor, μ, is also involved. The use of spatial correlations ensures that the estimated value will be in the same range as the measured data points. Fisher (1998) uses geostatistics to estimate elevation error, Robertson (1987) describes geostatistical methods of interpolation in ecology, while other useful references are Oliver and Webster (1990), Goovaerts (1999) and Webster and Oliver (2007). Trauth *et al.* (2007) contains MATLAB code and discussion of geostatistics and other geoscience techniques. Pebesma and Wesseling (1998) give details of a geostatistical software library, GSTAT.

Interpolation from a point pattern onto a grid can also be accomplished by the use of triangulation methods or TINs (see Section 10.2.4). The Delaunay method is the most popular, as the triangles it generates are as near equilateral as possible. Values associated with points lying within, on the boundaries of, or at the nodes or vertices of individual triangles can be interpolated from the values at the three vertices. One advantage of the method is that the resulting surface passes through the data points exactly. The initial result can appear irregular, and so smoothing using a low-pass filter can be performed to blur or smear the result. Further details of the Delaunay triangulation method are given by Bonham-Carter (1994), Worboys and Duckham (2004) and de Smith, Goodchild and Longley (2007).

10.5.2 Relating Field and Remotely-Sensed Measurements: Statistical Analysis

One of the oldest and most widely used methods of spatial analysis is regression analysis, which is used to establish a relationship between selected spectral bands of a remotely-sensed image set and a geophysical or biophysical variable. An example of a geophysical variable is soil moisture content. An example of a biophysical variable is LAI. Neither soil moisture content nor LAI can be measured directly by the instruments carried by aircraft or satellites, so regression analysis is used to establish a relationship between the few and expensive ground measurements of soil moisture content or LAI and the pixel values in multispectral imagery in which the sample points lie. The relationship is of the form:

$$y = \beta_0 + \beta_1 x_1 + \beta_2 x_2 + \beta_3 x_3 + \varepsilon$$

where y is known as the dependent variable as its value depends on the values of the explanatory variables x_1, x_2 and x_3. The model can handle any number of explanatory variables; the three x values used here could be the values of pixels in a SPOT multispectral image. The beta values are coefficients or weights, the values of which are to be estimated from the data, and ε is the error. More specifically, y is a difficult or expensive to measure variable such as LAI, and the x variables are cheap or easily obtained measurements, such as SPOT HRV bands 1–3, that are correlated with y. If we obtain a sample of y values and read the corresponding SPOT pixel values from the relevant image then we can calculate sample estimates of the β_i using standard formulae (O'Sullivan and Unwin, 2003; Mather, 1976). Note that we calculate estimates of β_i not the β_i themselves. The quality of these estimates can be obtained from formulae involving the magnitude of ε, the error term, which is a measure of the goodness of fit of the regression line. Statistical tests are available to assess the goodness of fit of the line. These tests assume that (i) the y values follow a Normal or Gaussian distribution, (ii) that the x values are fixed values and do not have a probability distribution and (iii) the sample size is adequate – usually a value of 30 or more is considered large. If the sample size is inadequate then the error term can become large and the goodness of fit of the data to the model can be inadequate. Given field measurements of LAI or soil moisture content or any difficult to measure geo- or biogeophysical property, estimates of their can be obtained from regression analysis, and the result interpolated to every pixel in a raster image data layer using one of the methods described in the preceding section to produce a new, derived, data layer. Care should be taken to acknowledge the error present in the results, especially for small sample sizes. Estimates of bio- and geophysical properties are often derived in this way and provide inputs to geomorphological, hydrological and atmospheric models (Chuvieco, 2008; Steffen *et al.*, 2005).

10.5.3 Exploratory Data Analysis and Data Mining

Other widely used forms of mathematical and statistical analysis can be grouped together under the general heading of exploratory data analysis, also known as data mining. Data mining is a kind of search for meaningful patterns, trends and anomalies (Witten and Frank, 2005) in large datasets. Data mining includes clustering, neural nets and decision trees, which are mentioned in Chapter 8, under the heading 'classification', while PCA is described in Chapter 6. Other methods that can be used to discover patterns within large datasets are those which project an n-dimensional pattern onto a subspace of two or three dimensions, for visual analysis. These methods can be lumped together under the heading of ordination techniques, and they include principal coordinates analysis, and Sammon's non-linear mapping (NLM) method. The latter method is described in Section 8.5.1 in connection with the selection of end-members in linear mixture modelling.

All of these techniques can be considered to have the aim of extracting structure or pattern from data. As such they have found widespread application in a range of disciplines, such as chemoinformatics (Leach and Gillet, 2007) and biosciences (Fielding, 2007). Other references are: Lee and Verleysen (2007), Han and Kamber (2006), Mitra and Acharva (2003) and Abonyi and Feil (2007), as well as that old but still readable text, Mather (1976).

As Grey notes in his preface to Witten and Frank (2005), data mining has made 'stunning progress' through the coming together of statistics, machine learning, information theory and computing. Old friends like PCA and unsupervised classification (clustering) are present as well as relatively new ones like decision trees, artificial neural networks, fuzzy methods and Bayesian networks. Given the growth in digital observations of the Earth's environment, a data mining approach will be necessary in order to extract patterns from the contents of large and perhaps disparate databases. This is an area ripe for exploitation.

10.6 Environmental Modelling

In a scientific context, the word *model* means a simplified representation of reality. It represents the modeller's view of reality in terms of the objects that are interesting and the processes that are important. It can be the expression of a theory or hypothesis, or an attempt to explain how and why certain features of the Earth's surface come to exist. Models are necessarily selective and identify important objects, processes and relationships. One of the most interesting ways of looking at dynamic models is by treating them as models of open systems, which are made up of a related set of phenomena together with the relationships between them. Open systems exchange energy and material with other systems, and have inputs and outputs as well as internal connections, flows and storages of energy and matter.

Modelling requires a good understanding of the system that is being modelled, and it incorporates several steps (Barnsley, 2007; Figure 10.8). The first of these is the definition of the problem in precise terms. The problem should be amenable to solution at an appropriate spatial scale, and the detailed knowledge required to make the solution practicable should be available or, in the case of primary research, that knowledge should be attainable in a reasonable time. Any datasets that are needed should be accessible and should satisfy the spatial and physical resolutions of the project. Problems can be applications-driven, relating to a specific topic such as biomass estimation of forests, or conjectural, that is scientific. The problem definition should be limited to a specific domain in time and space. A fuzzy problem definition leads to ambiguities, gaps and possibly contradictory statements. Problem definition also includes setting of timescales, derivation of programmes of work and the determination of funding priorities.

The second step, and the one that requires intellectual ability and imagination as well as detailed subject-matter knowledge, is the development of a conceptual model that gives both an overview and a detailed description of the model being proposed as a solution to the problem identified at stage 1. The conceptual model defines the model parameters and the relationships between them in terms that can be translated into mathematical

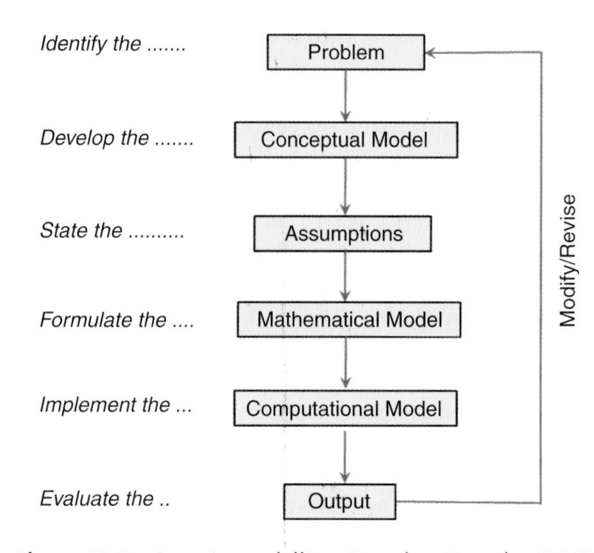

Figure 10.8 Steps in modelling. Based on Barnsley (2007).

procedures. These procedures can be converted into a set of mathematical statements, subject to assumptions that limit the domain of the model, for example to a particular climatic zone.

Once the model has been defined in conceptual terms it has to be set out in a mathematical and/or statistical form. There are several possible combinations: mathematical or statistical, and empirical or theoretical. A mathematical model could also be called 'deterministic' and a statistical model could be described as probabilistic or stochastic. Models can also be static or dynamic. A static model takes inputs that specify the parameters of the model and generate an output, whereas a dynamic model simulates the changes in model output over time. Whatever form the model takes, it is specified as a mathematical model by a series of equations that link model components (sometimes called variables such as annual rainfall) and convert from inputs to outputs. The definition of a mathematical model results in a specification that can be used to develop tractable computer code, using either a high-level or modelling language or a symbolic editor like ERDAS's Spatial Modeller. ArcGIS uses a similar interface, called ModelBuilder (McCoy, 2004; Goodchild, 2005). DeMers (2001) describes the use of a modelling language (PCRaster), as do Wainwright and Mulligan (2005).

The PROSPECT model is a good example as it is relatively simple. The model was developed by Jacquemoud and Baret (1990) and it considers the reflectance, transmittance and absorption of electromagnetic radiation in the 400–2500 nm range using a series of equations to control the model's behaviour. The output is in the form of a reflectance spectrum (Liang, 2004, Section 3.2.1). The computer code for the PROSPECT model is available from the CD that accompanies Liang's (2004) book, and it is also incorporated into MIPS. Other models use a statistical rather than a mathematical formulation of the relationships between model components, for example the regression model (Section 10.5.2) may be used to establish a relationship between input data and the results from the model. Regression analysis may also be used to predict or estimate a difficult-to-measure variable from an easily-measured variable. Thus, the slope of the land surface can be measured from a DEM relatively easily, but soil depth measurement requires fieldwork, which is time consuming and expensive. A regression of soil depth against DEM-derived slope could be used, if the regression relationship was a strong one, within a model of, for instance, landslide prediction.

Sometimes a statistical model is called a stochastic model, while a mathematical model is said to be deterministic. Stochastic models are used whenever there are a number of small, independent process acting; these components cannot be identified individually but collectively they can be accounted for by the use of statistical relationships (between processes and outcomes) or as random changes to model parameters. The variance of the random changes is frequently proportional to our ability to predict the value of the parameter concerned. Stochastic models of hydrological processes encompass equations that determine the values of infiltration, runoff, overland flow, water storage and other model parameters but they do have a certain level of randomness in the rainfall input that drives the model. This randomness could be considered to be accounting for local effects such as slope, aspect and wind direction on the amount of rain factually received at a given site. The distinction between a hydrological model and the PROSPECT leaf reflectance model is that the hydrological model is dynamic whereas the leaf reflectance model is static. A dynamic model contains feedbacks (either positive or negative) that influence the relationships between the processes and storages at each model cycle. Whereas the leaf reflectance model produces a single spectral reflectance graph, the dynamic model may run in time steps of minutes, hours, days or even years to produce a simulated sequence of events. The depth of knowledge required to specify such models is considerable. Whatever approach is adopted, the result is a mathematical–statistical specification of the model, with inputs, processes and outputs linked by equations.

Once a model is built it must be calibrated and tested using real data. Calibration refers to the selection of parameter values in equations and specifications, while testing refers to the running of the whole model or parts of it and comparing the results with reality. This is the calibration and validation stage, where model outputs (for static models) and scenarios (for dynamic or simulation models) are compared to real-world behaviour. It is often difficult to acquire adequate real-world data sets that can be used to test models; for example GCMs which are dynamic in nature may be tested on their ability to 'postdict' the past rather than predict the future. Simpler models like PROSPECT can be calibrated in the laboratory. The goodness of fit of the model outputs at the testing stage frequently is less than satisfactory, so the conceptual model and its implementation in computer code may require modification. These modifications may be the result of gross errors, for example adding rather than subtracting two numbers, or they may be due to the misspecification or overlooking of relationships between model components requiring the alteration of equations and the recoding of one or more routines. Another important attribute of a model is the sensitivity of the output to changes in the values of the mathematical and statistical parameters specified at the conceptual stage. Sensitivity can be estimated by changing each coefficient in turn by a small amount, and assessing the resultant change in model behaviour. A good model is stable rather than sensitive.

Once the model has been calibrated and tested, it can be evaluated by being run on new data in an operational context. If the result is acceptable, the model can be considered to be reliable within its specific domain of application. The greater the number of successful uses of a model, especially for prediction purposes, the greater its acceptability until it becomes the basis of policy-making or education. For example, the MIPS implementation of the PROSPECT model can be used to ask (and answer) the question: 'What happens if leaf water content falls by 50%?' The result is shown in Figure 10.9.

Modelling is an important activity in environmental RS, and inputs from RS are routinely incorporated into models on scales ranging from the global scale to the regional and local. Justice *et al.* (1998) describe the global data products produced by the two MODIS instruments currently in orbit, while Zhan *et al.* (2002) discuss the use of MODIS products in measuring land cover change over time. See also Chuvieco (2008) for a survey of RS contributions to global scale modelling.

At a more local level, remotely-sensed data has been used in modelling of the environment, for example *habitat suitability models* for identifying favourable habitat conditions for particular species (Krivoruchko and Gotway Crawford, 2005), *wildlife management models* to help save endangered raptor species (Scally, 2006); *landscape-scale habitat use* (Osborne, Alonso and Bryant, 2001); *transport models* for evaluating the environmental impacts of traffic as well as planning new infrastructures (Israelsen and Frederiksen, 2005); and *hydrological modelling* or a variety of water planning and management tasks (Maidment, Robayo and Merwade, 2005). The use of RS and GIS in forest fire management is considered by Chuvieco and Salas (1996), while Brivio *et al.* (2002) and Townsend and Walsh (1998) discuss the use of GIS and RS in flood extent mapping. RS and GIS contributions to monitoring natural disasters are spelled out by Tralli *et al.* (2005). Paylor, Evans and Tralli (2005) edit a special issue of *ISPRS Journal of Photogrammetry and Remote Sensing* on the theme of 'RS and geospatial information for natural hazards characterization'. Other useful references are Brimicombe (2010), Gillespie *et al.* (2007), Paegelow and Camacho Olmedo (2008), Schumann *et al.* (2007), Skidmore (2002) and Wainwright and Mulligan (2005). DeFries and Los (1999) and Bounoua, Masek and Tourre (2006) discuss the use of land cover classification maps produced from remotely-sensed data and demonstrate the substantial errors that occur when the maps are upscaled from 1 to 100 km^2.

10.7 Visualization

In a scientific context, depicting reality by visual methods is called visualization. If the representation is of a spatial scene then the technique is called geovisualization (Demšar, 2009; Dykes, MacEachren and Kraak, 2005; Hutchinson, 2008; Kraak and Ormeling, 2002). Worboys and Duckham (2004, p. 305) say that

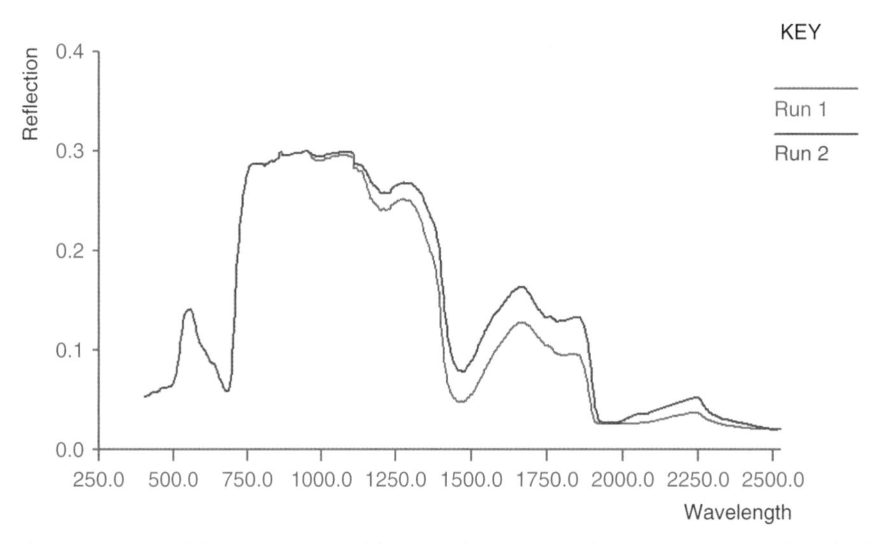

Figure 10.9 *Output from two runs of the Prospect model, using the MIPS implementation. Run 1 has the leaf equivalent water content (LEWC) set to 0.024 cm, while run 2 has a LEWC of 0.012. The 50% decrease in water content does not affect the visible and SWIR portion of the spectrum but from about 900 to 2500 nm the effect of the lower LEWC is to increase reflectance (Run 2 is shown in green). Reproduced with permission of A. Gaber and F. El-Baz.*

> Geovisualisation is the process of using computer systems to gain insights into and understanding of geospatial information.

Other definitions are similar in nature; for example Kraak and Ormeling (2002) see visualization as

> ... integrating approaches from image analysis, exploratory data analysis, and GIS to provide theory, methods and tools for visual exploration, analysis and presentation of geospatial data, that is any data having spatial referencing.

Visualization includes, obviously, a depiction of the phenomenon of interest – in other words we can visualize feature space (Chapter 8) as well as geographic space. Visualization can also include interactivity, animation, active querying of objects and entities in the visualization. One extreme form of visualization is virtual reality (VR), which is used, for example in training pilots on a flight simulator. The 'fly-through' modules on some image processing systems, like the cinematic experience, do not include interaction, which is a feature of VR. Information can be captured by a user far more quickly if it is visual in nature rather than text or tables. Of course, this facility can be misused to distort reality, in ways that Monmonier (1996) describes and illuminates in his book *How to Lie with Maps*. A good example of the use of visualization tools in both two and three dimensions is the *Atlas of Switzerland* (Atlas of Switzerland2, undated; Sieber and Huber, 2007).

Why visualize? The main reason is that the human eye and brain form the best object-oriented image processing system yet developed, and can make intuitive judgements on the basis of vision alone (e.g. is the old man in a good mood?) as well as acting as a guidance system for our interactions with the real world. Some examples will aid understanding. It is invariably assumed that decision-makers and managers have no time to read long reports or understand multivariate statistics. At the same time it is also assumed that these decision-makers and managers can take in facts from a graphic display of the same data at a glance. Several methods can be used as illustration. The first is called Chernoff faces, after Herman Chernoff who first introduced the technique in 1973. The characteristics of a face are each related to some statistical property of the data being visualized. For example, smiling/not smiling could mean negative/positive correlation, while distance between the eyes might be related to variance. More details are provided by Everett and Nicholls (1975) and some example faces are shown in Figure 10.10.

A second use of visualization was met in Example 8.1 in which a Landsat TM image of part of the Mississippi valley was subjected to an ISODATA classification. The relationships between the different classes was not easy to see on the screen, but a dendrogram derived from a hierarchical cluster analysis was used to simplify the transmission of information. Example 8.2 Figures 1 and 2 are reproduced here as Figure 10.11 for convenience.

Figure 10.10 *Examples of Chernoff faces. From http://people.cs.uchicago.edu/~wiseman/chernoff/.*

The classified image uses colour-coding to separate the classes. The relations between the class centroids are visualized using a dendrogram, which is derived from hierarchical clustering. The dendrogram indicates the similarity (or dissimilarity) between the classes, with dissimilarity increasing towards the right hand side of the tree. Visual analysis of the tree diagram indicates that there are two major sets of classes, that is {1, 7, 3, 11, 5, 8, 12, 9, 14} and {2, 15, 10, 4, 3, 6}. The identity of these classes can be read from the colour bar (which is not displayed here). The two sets of classes join at a dissimilarity value of almost 18. The two sets stay separate until the subclass {9, 14} splits off from the first of the two sets at a dissimilarity value of about 10.75. Individuals split off at decreasing dissimilarity levels until the individual class centroids remain. Centroids of the ISODATA classes 1 and 7 are least dissimilar. The visualization in the form of a dendrogram has therefore revealed a hierarchical structure in the ISODATA results. This hierarchical representation can be used to reclassify the image by combining the least dissimilar classes (as displayed on the dendrogram) and then assigning these groups of classes to the same colour. This action will reduce the displayed number of different classes and make the classification more general. The penultimate reclassification would give the two sets discussed above the names 'land' and 'water'.

The third example of visualization is the interactive scatter diagram as depicted in Figure 8.10 and reproduced here as Figure 10.12. In this case, visualization is linked to data mining to address the problem of finding outliers in a multi- or hyperspectral training dataset. The approach is described in Section 8.4.2. Basically, the approach described by Sammon (1969) is used to map the

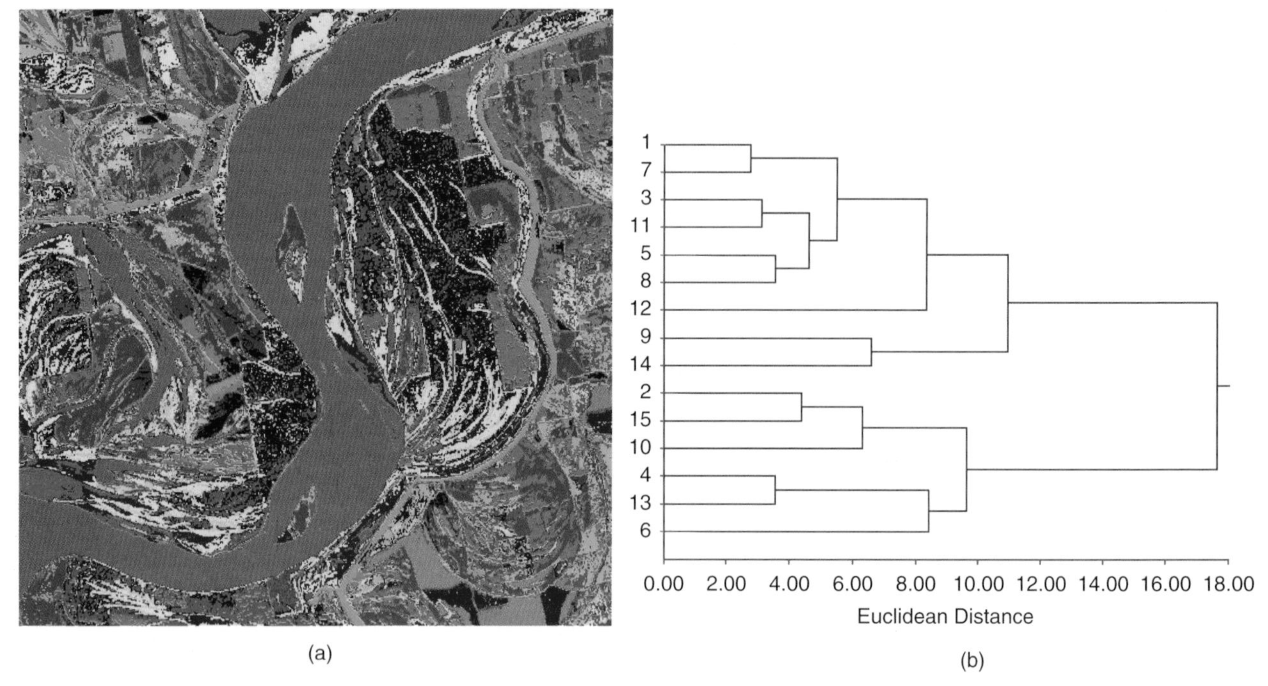

Figure 10.11 *(a) ISODATA classification of Mississippi Landsat TM image and (b) Dendrogram showing similarity between classes (horizontal axis).*

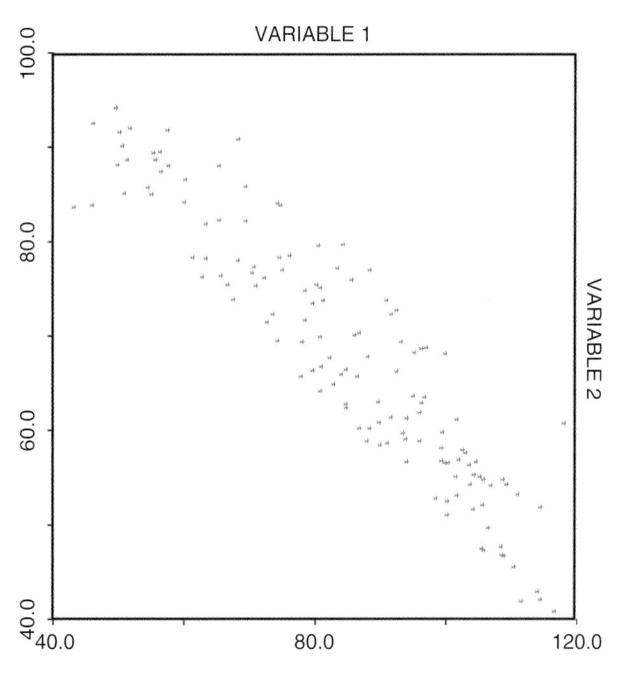

Figure 10.12 *Two-dimensional interactive visualization of the structure of a training sample dataset. The original number of dimensions is six. Sammon's (1969) nonlinear mapping was used to perform the dimensionality reduction. The visualization is interactive as the numeric ID of any point can be obtained via a mouse-click. The scaling on the two axes is arbitrary.*

point pattern from p to two dimensions by minimizing the sum of the interpoint distances. The program is listed in Mather (1976). A more recent account is by Lee and Verleysen (2007). The result can be shown as a scatter diagram in which each point represents a pixel. In the MIPS implementation, moving the mouse cursor close to a point will list the point number. A more sophisticated method would list the spectral band values of the selected point. This illustration shows how data mining methods, in particular dimensionality reduction techniques, can be allied to visualization to give considerable 'value added' to the result.

Sammon's (1969) Nonlinear Mapping (NLM) takes n points in a p-dimensional space, where n is the number of pixels in the sample and p is the number of spectral bands, and outputs a configuration of the n points in a k-dimensional space, where $k < p$. If d_{ij} is the Euclidean distance between points i and j in the original p-dimensional space and δ_{ij} is the corresponding distance in the k-dimensional space, then the function to be minimized is E:

$$E = \frac{1.0}{\sum_{i=1}^{n}\sum_{j=1}^{n} d_{ij}} \sum_{i=1}^{n}\sum_{j=1}^{n} \frac{(\delta_{ij} - d_{ij})^2}{\delta_{ij}^2}$$

The term E is minimized using a non-linear optimization procedure, based on the first and second partial derivatives of E. A scaling factor (called the 'magic factor' by

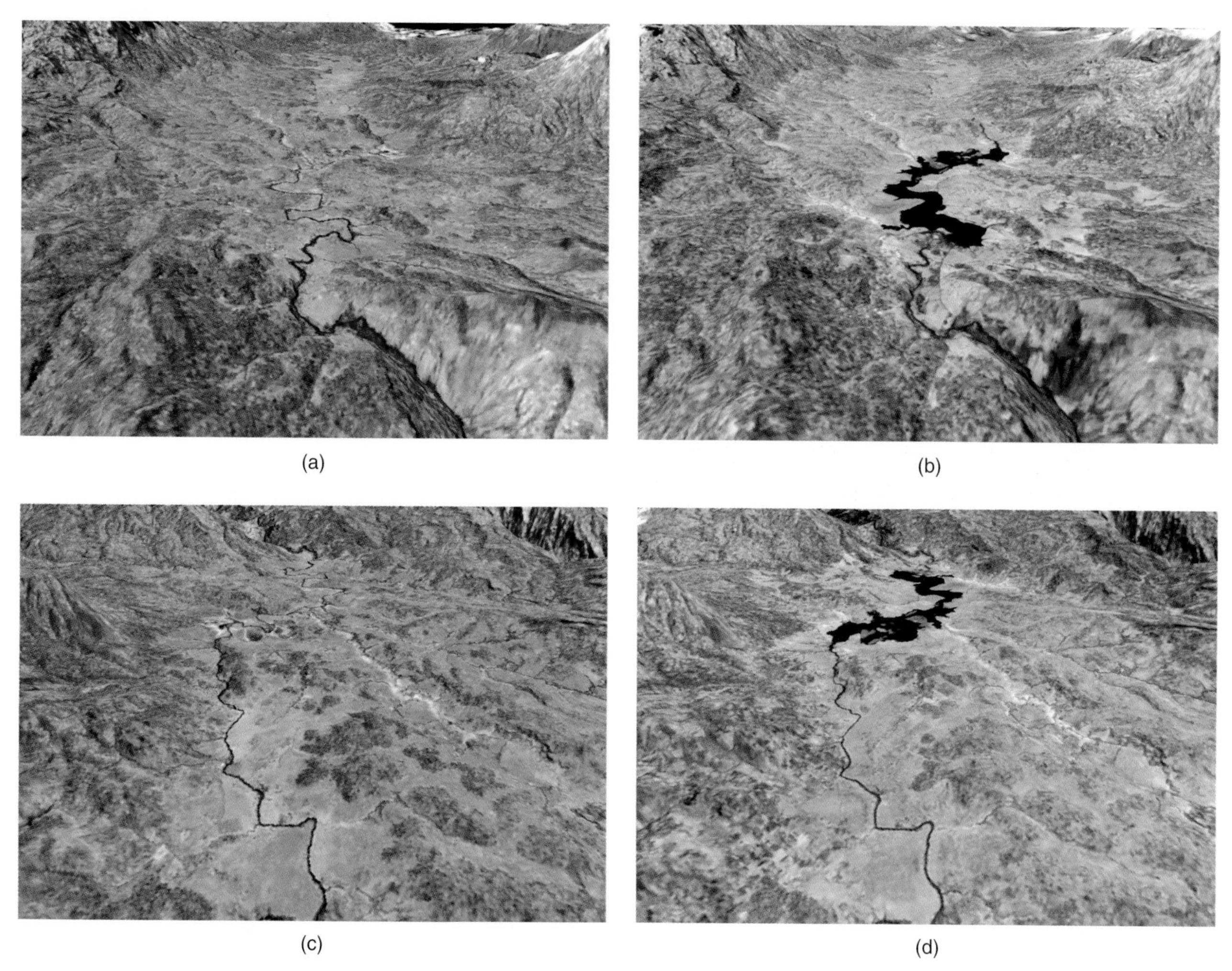

Figure 10.13 *(a) View towards the west down the Ghibe Valley in the SW Ethiopian highlands. The TM image was acquired in 1984. It is draped over an ASTER GDEM and the viewpoint is user-selected. (b) Visualization of the Ghibe Valley using an ASTER GDEM and a Landsat ETM image acquired in 2003. Changes between 1984 and 2003 include the construction of a large reservoir. The viewpoint is the same as in Figure 10.13a. (c) Looking eastwards up the valley – Landsat TM image (1984) overlain on a DEM. (d) Looking eastwards in 2003 from the same viewpoint as that used in Figure 10.13c but using a Landsat ETM image as overlay.*

Sammon) is required; it usually lies in the range 0.3–0.4. A value of 0.3 was used here. The process iterates until the output coordinates (x_{ij}) stabilize, which can take a while if the sample is large. This example of the visualization of multidimensional data reduction illustrates the incorporation of data mining methods in the process of geovisualization using interactive graphical displays.

Visualization can also be used to give a three-dimensional view of a landscape. In an interactive display the user can select the viewpoint in (x, y, z) coordinates or, in the case of a DEM, can move or fly through the virtual landscape. Figure 10.13a–d show the Ghibe Valley in the Ethiopian highlands near the city of Jimma. In each case a Landsat TM or ETM image is draped over a DEM (ASTER GDEM). These are freely available DEMs that can be downloaded from the following web sites: http://www.gdem.aster.ersdac.or.jp/ and http://asterweb.jpl.nasa.gov/gdem.asp. The images are separated by a 19-year gap (1984–2003) and significant differences can be picked out, the most obvious being a new reservoir. The view for Figure 10.13a, b is towards the west. Figure 10.13c, d are looking towards the east.

Overlaying satellite or aircraft imagery on a DEM is a widely used visualization technique. Some RS expertise is needed in order to maximize the information content of the three-dimensional view, for example the choice of spectral bands to display should be based on the characteristics of the target, or some transformation of the image data (such as Tasselled Cap or PCA) may be applied before the overlay operation. Contrast enhancements using single bands (such as linear contrast stretch or histogram equalization) or using all three selected bands together (such as decorrelation stretch or HSI) will improve the visual quality of the image. Some smoothing or sharpening may be required in addition to contrast enhancement. Finally, the image set and the DEM need to be geometrically registered and resampled to a common pixel size.

A second example of using three-dimensional visualization of a landscape shows how the combination of a DEM with a natural colour digital overlay can be used in landscape design.[3] The study area is the east-west Mosedale valley, through which the River Caldew flows, and which is located to the north-east of Keswick in the English Lake District. The view is towards the west. The valley floor has an elevation of about 225 m while the high ground to the north, west and south of the valley rises to around 700 m. The small lake in the south (left) of the area is Bowscale Tarn, and it is a glacial corrie lake. The landscape is smooth and rounded rather than craggy, as it is formed on the Palaeozoic Skiddaw Slates with some outcrops of gabbro. The degree of smoothness is related to the number of triangles used in approximating the ground surface, as discussed below.

Figure 10.14 was produced by first taking NEXTMap's 5 m resolution InSAR DEM and approximating the shape of the resulting surface by triangulation, with the degree of generalization being dependent on the number of triangles selected, as shown in Figure 10.15. Next, the GetMapping colour imagery is projected onto the triangles. Then those screen pixels that are visible from the chosen viewpoint are identified, and the corresponding photo texture is determined and sent to the screen. All these images and triangles can therefore be at different resolutions as what matters is projected screen pixel resolution required. Thus screen pixels which represent areas closer to the viewer will use an effectively higher resolution from the triangles and photo raster, and hence more detail, than those further away. (M.J. McCullagh, personal communication, 2009). The free Google SketchUp software (http://sketchup.google.com/) was used to perform the triangulation and projection. A total of 3000 triangles was used in the generation of Figure 10.14. For comparison, triangulations with 1000 and 30 000 vertices are illustrated in Figure 10.15.

[3]The data processing to produce the figures in this example was carried out by M.J. McCullagh who also commented helpfully on the text.

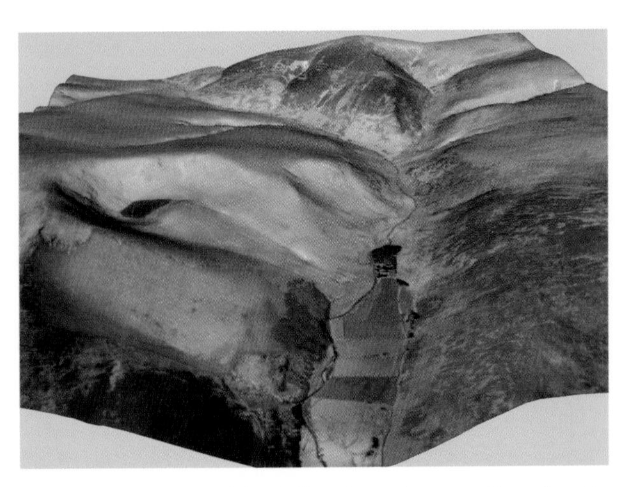

Figure 10.14 *Mosedale Valley from the east. The elevation model is derived from NEXTMap's airborne InSAR and has a spatial resolution of 5 m. A set of triangles is computed to approximate the elevation data, with the number of triangles determining the level of generalization. See Figure 10.15 for examples of triangulation. Finally, GetMapping's airborne colour imagery is projected onto the triangles, and visible pixels are drawn. See text for discussion. NEXTmap® is a registered trademark of Intermap Technologies. All rights reserved. Copyright Getmapping Plc.*

Now let us assume that a landscape planner has been employed by a water supply company to carry out an initial study and local consultation for a programme to build a dam in the upper Mosedale Valley, and to publish his thoughts on the Internet. He could use an Ordnance Survey map with overlays showing the position of the proposed dam and the extent of the impounded area. Only a relatively small proportion of the population can 'read' a map to the extent that they can visualize the shape of the land from the contours shown on the map. Our landscape inspector could, however, choose to use widely available data to build a scientific visualization which would be easy to understand and comprehend. It would be possible to interact with the visualization by rotating it in the horizontal plane or tilting in the vertical plane. More advanced interactions may use attributes of places visible on the diagram.

The planning process is aided by the fact that local people would have an informed opinion on the subject and thus engage in rational debate. The visual impact of a fictitious dam built in the valley can be estimated from Figure 10.16.

This example demonstrates (i) the combined use of an airborne InSAR DEM and of digital air photography, (ii) the value of scientific visualization of landscapes in the environmental planning process and (iii) the use of triangulation of gridded elevation data in providing data storage economies.

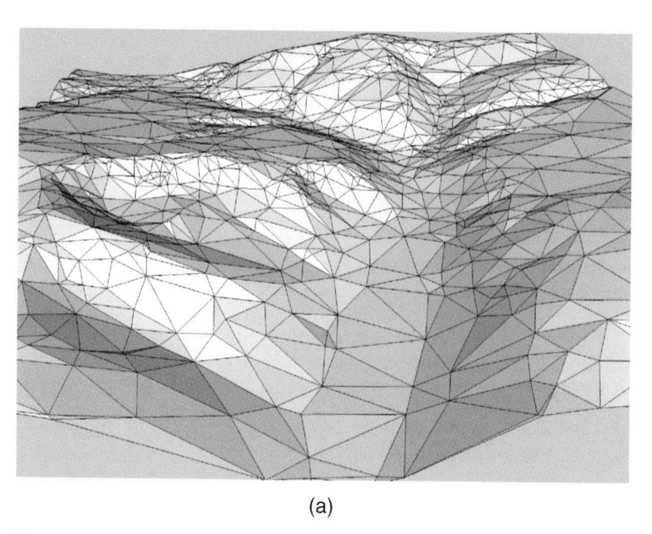
(a)

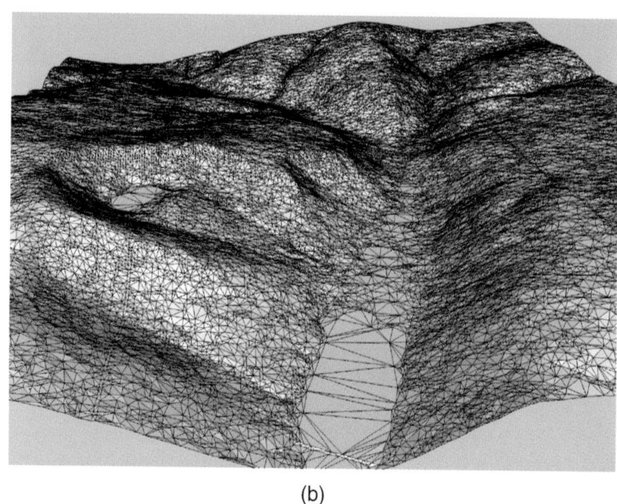
(b)

Figure 10.15 Approximate triangulation based on NEXTMap InSAR DEM using (a) 1000 and (b) 30 000 vertices. The savings in storage space compared to a raster grid are considerable, even when 30 000 vertices are used. NEXTmap® is a registered trademark of Intermap Technologies. All rights reserved. Copyright Getmapping Plc.

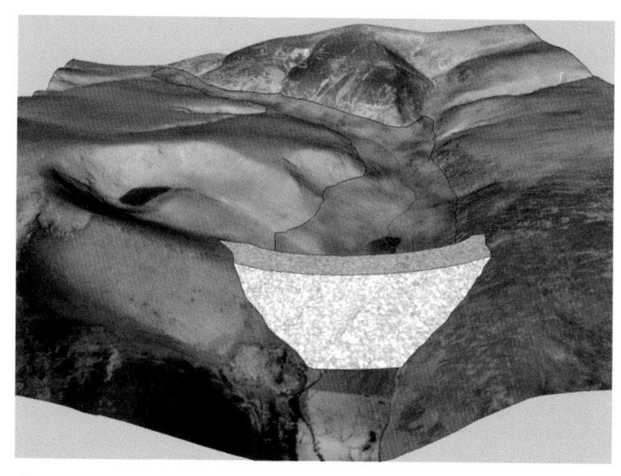

Figure 10.16 Visual impact of a dam in upper Mosedale. The extent of inundation by the impounded reservoir is shown in transparent light blue so as not to obscure the valley-floor and valley-side features. Depending on the nature and number of comments by members of the public, the colour and texture and perhaps the type of dam could be changed. InSAR elevation data and colour photography. NEXTmap® is a registered trademark of Intermap Technologies. All rights reserved. Copyright Getmapping Plc.

10.8 Multicriteria Decision Analysis of Groundwater Recharge Zones

10.8.1 Introduction

This case study shows how GIS and RS can be used in an integrated way for locating potential groundwater resources in arid environments using MCDA (Section 10.3.2). The identification of areas favourable to groundwater recharge is based on establishing the degree and extent of the spatial relationship between several factors conducive to surface and subsurface water accumulation, infiltration and storage. Among the primary factors affecting groundwater recharge in arid and semiarid regions are: precipitation, terrain elevation and slope, lithology and soil, drainage and fracture intersection as well as drainage and fracture densities. Their importance can be expressed in a GIS overlay analysis by ranking the individual thematic layers representing these factors and assigning relative weights to each of them. Such overlay analysis is also called multicriteria decision analysis (see Section 10.3.2) in which each of the layers is associated with a weight value that may be determined by various methods, including statistical analysis, published studies and/or professional experience (Ji and Ma, 2008; Murthy and Mamo, 2009). Figure 10.17 shows the location of the study area in northern United Arab Emirates (UAE) and Figure 10.18 illustrates schematically the steps followed in the MCDA which was used to establish favourable conditions for groundwater recharge in the arid mountain range of UAE. The thematic layers (step 1 in the flowchart of Figure 10.18) were produced by Boston University's Center for Remote Sensing as part of a study conducted in northern UAE[4] (El-Baz *et al.*, 2004) and are used in this case study as input layers for the recharge model that follows.

[4]This study was carried out by staff of the Center for Remote Sensing, Boston University, under contract to the Government of Sharjah, UAE. A project summary is available at the BU-CRS webpage: http://www.bu.edu/remotesensing/research/uae-groundwater/index.html.

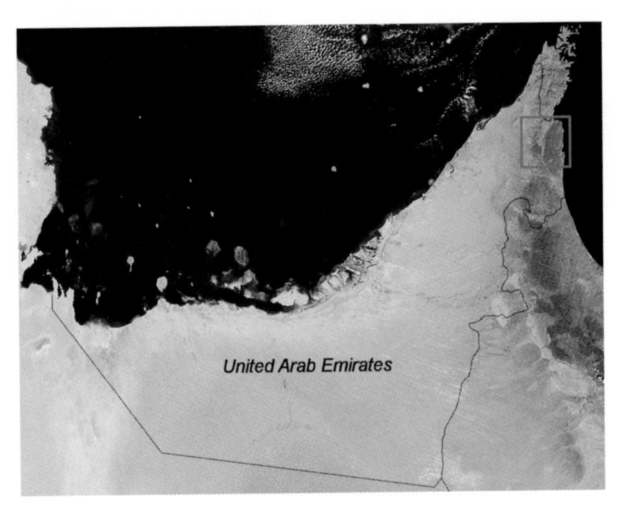

Figure 10.17 *MODIS image of UAE showing the study area utilized in the weighted overlay analysis outlined in red. Credit: Jacques Descloitres, MODIS Rapid Response Team, NASA/GSFC (http://visibleearth.nasa.gov/view_rec.php?id=6083.*

10.8.2 Data Characteristics

As in all decision models, the first step is to decide on the number of factors that may contribute to a particular environmental process or function, in this case surface water accumulation and infiltration to recharge the aquifer. Some processes of the hydrological system can be directly observed and measured (e.g. precipitation) while others can only be indirectly identified or inferred (e.g. subsurface flow accumulation). Satellite images and DEMs can provide very useful information on terrain features and characteristics that are indicative of groundwater occurrence. Terrain features that provide direct evidences of groundwater flow (e.g. vegetation and fresh water springs) have the highest likelihood of indicating groundwater, whereas indirect evidences (e.g. tensional or open faults, buried paleo-channels, or alluvial fans) are based on deductive conclusions of appropriate geological settings for trapping and accumulating groundwater. Many of these indicators can be identified and mapped using image processing techniques such as vegetation indices, band ratios, principal component analysis, classification algorithms and change detection procedures (see Chapters 6 and 8). Although image processing tools are invaluable in extracting information and producing useful thematic maps, GIS is the most appropriate tool to evaluate and model the resulting image processing products. In this case study thematic maps are derived from remotely-sensed data (including DEMs) and evaluated in terms of spatial and statistical coincidence of favourable surface and subsurface terrain characteristics for groundwater occurrence and accumulation in an arid environment.

Table 10.2 lists the thematic maps utilized in this study and the numerical rating scheme used to assign weights to the individual layers depending on their importance or influence in determining potential areas for groundwater recharge. A recharge index equation was developed based on expert knowledge and published studies on groundwater recharge processes in arid lands, particularly in fractured hard-rock aquifer systems (Krishnamurthy *et al.*, 1996; Saraf and Choudhury, 1998; Jaiswal *et al.*, 2003; Jasrotia, Kumar and Saraf, 2007; Dinesh Kumar, Gopinath and Seralathan, 2007; Murthy and Mamo, 2009). The index is calculated by establishing the sum of ranked or weighted scores of individual thematic layers.

In order to conduct such an analysis, thematic map layers need first to be generated and prepared to serve as model input parameters in the equation listed at the end of Table 10.2. In this example, the geological/geomorphological map was generated from classifying ASTER scenes (Section 2.3.8) using a combination of PCA (Section 6.4) and a decision tree algorithm (Section 8.6.2). PCA is used to reduce the number of original bands to produce composites that provide the most significant spectral separation between individual rock units. These significant principal components were used as inputs in the supervised decision tree classification (Section 8.6.2) procedure that was based on a decision tree algorithm called C4.5 (Quinlan, 1993). The algorithm was trained using spectral characteristics of different geological units that were previously obtained from training sites identified by an expert. Figure 10.19 shows that the resulting geological/geomorphologic classification map is composed of a few broad categories that were reclassified and weighted (Table 10.2) according to their hydrological significance. The reclassification was based on rock properties such as rock type, solubility, permeability and weathering/fracturing. Mapping efforts were mainly focused in the mountain area (crystalline basement rocks and limestone) and the gravel plains (alluvium) at the foot of the mountain range where flash floods occur that may recharge the underlying aquifer. The desert basin area, and its potential for storing and transmitting water, would need to be treated separately as it represents a different type of aquifer system (sedimentary aquifer) from the one encountered in the mountain area (hard-rock aquifer). Koch and Mather (1997) demonstrate the use of RS data in characterizing the hydrological function of crystalline basement aquifers in arid environments.

The precipitation map used in this analysis is the result of averaging 10 years of annual rainfall. It shows long-term rainfall distribution pattern rather than yearly patterns because in arid environments groundwater recharge occurs predominantly through the cumulative

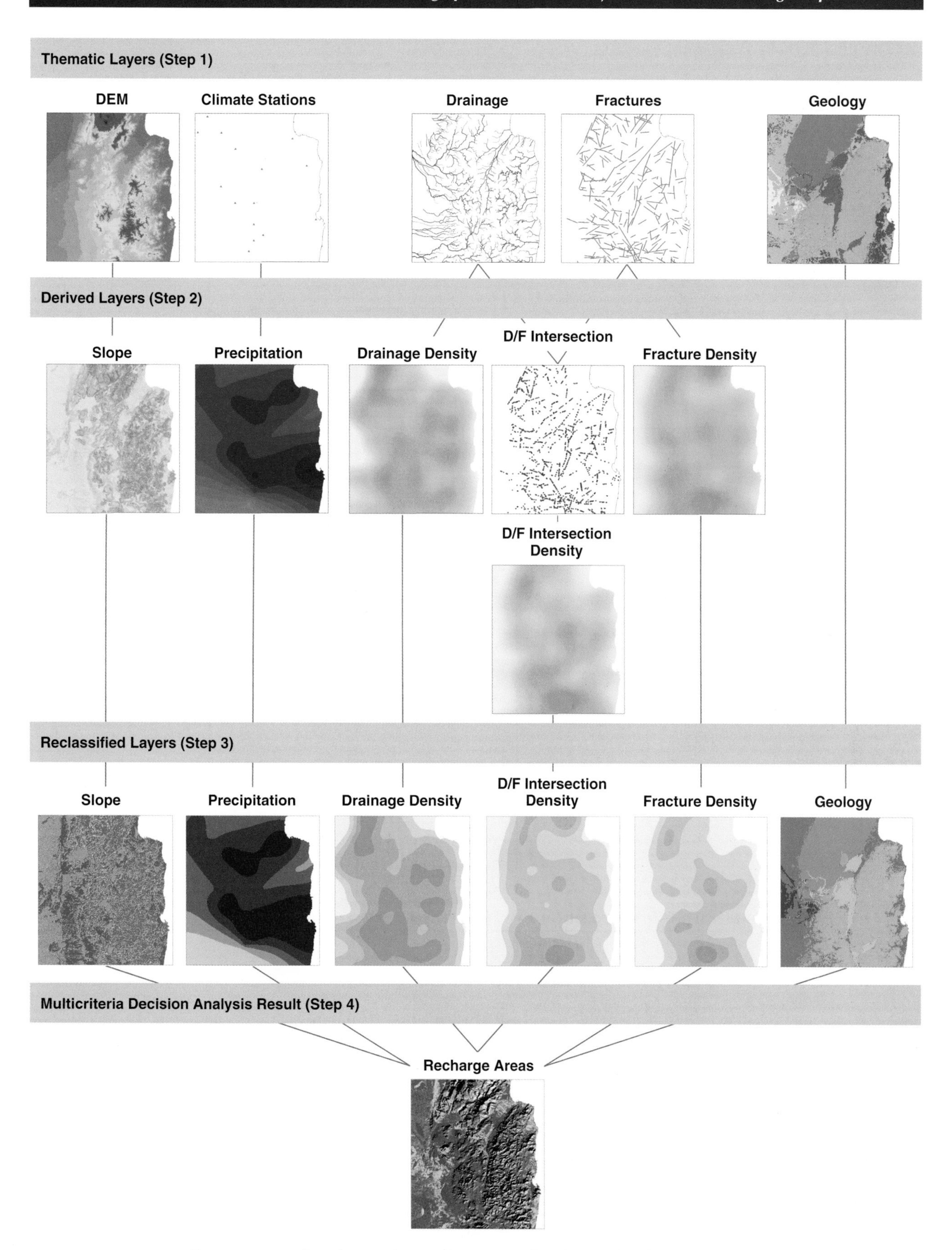

Figure 10.18 Flowchart outlining the steps used in the weighted overlay analysis.

Table 10.2 Weights and scores for thematic layers and their classes.

Thematic layers	Classes	Weights 0–100%	Scores 1–9
Geology/geomorphology (hydrogeol.)	Desert sand	30	7
	Alluvium		9
	Limestone		5
	Metamorphic rocks		3
	Igneous rocks		1
Intersection of drainage and fractures (D&F)	Low	20	1
	Medium-low		3
	Medium		5
	Medium-high		7
	High		9
Precipitation in mm 10-year average	<145	10	1
	145–160		3
	160–170		5
	170–180		7
	180–210		9
Terrain slope in percentage	0–1	20	3
	1–5		9
	5–10		7
	10–15		3
	>15		1
Fracture density (FD)	Low	10	1
	Medium-low		3
	Medium		5
	Medium-high		7
	High		9
Drainage density (DD)	Low	10	9
	Medium-low		7
	Medium		5
	Medium-high		3
	High		1
(Hydrogeology) + (D&F) + (precipitation) + (slope) + (FD) + (DD) = recharge potential			

Scores range from 1 (least suitable) to 9 (most suitable). Weights sum to 100%.

effect of rainfall. Precipitation maps can be generated either from the interpolation of rainfall amounts measured at climatic stations or from satellite radar sensors such as tropical rainfall mapping mission (TRMM, http://trmm.gsfc.nasa.gov/) or NEXRAD (NOAA's Next Generation Radar; http://www.roc.noaa.gov/WSR88D/). The latter method is especially attractive in regions that do not have a dense network of climatic stations. However, there are two problems attached to space-derived rainfall measurements in arid lands: one is the coarse spatial resolution of current radar sensors and the second is the lack of long-term archived records. Rainstorms are very much localized in arid environments and occur in large temporal scales. The precipitation map used in this study was generated from the interpolation (using the inverse distance weighted method in ArcGIS) of rainfall data obtained at 15 climatic station (Figure 10.20) between the years 1990–1991 and 1999–2000.

The slope map was generated from a DEM obtained from remotely-sensed data (Figure 10.21). In the past DEMs were created from digitizing or scanning contours on a topographic map. Nowadays,

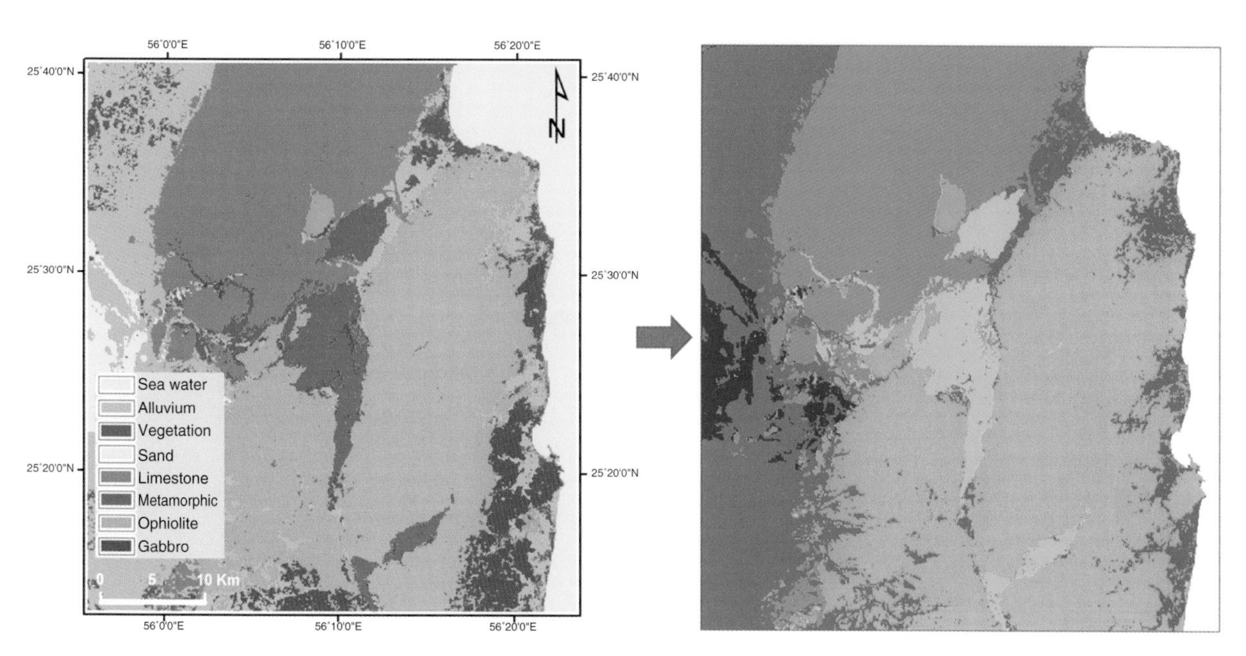

Figure 10.19 Geological map before and after reclassification. Bedrock units were generalized and the vegetation class was removed.

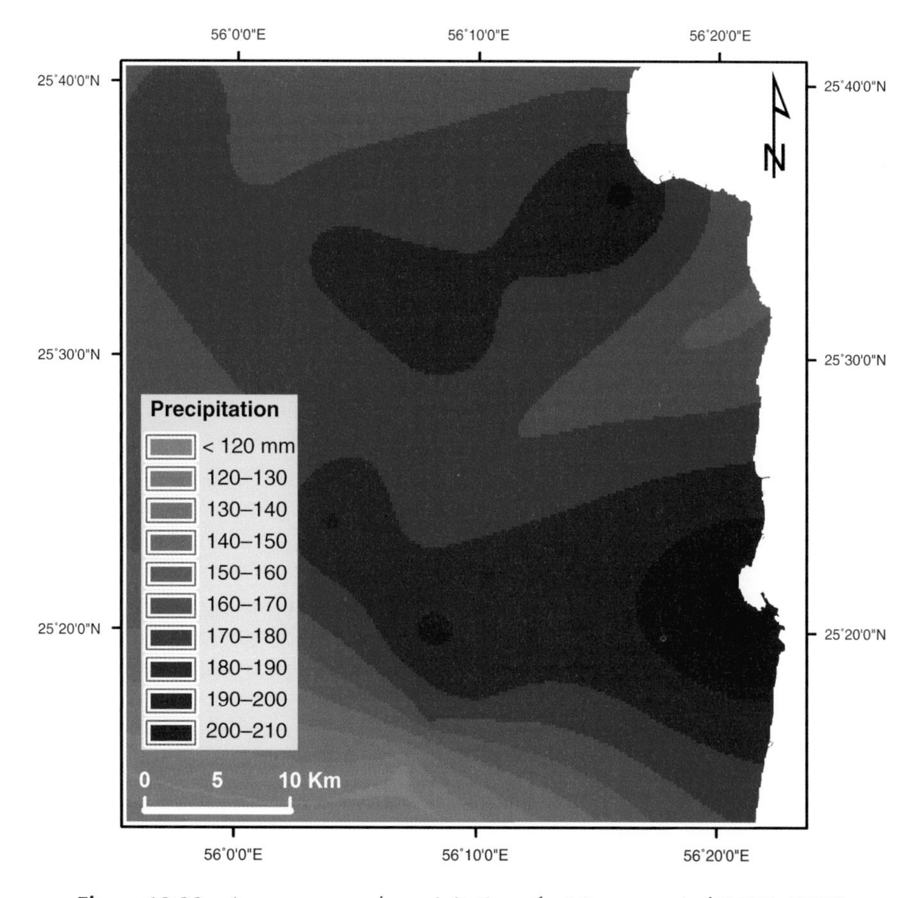

Figure 10.20 Average annual precipitation of a 10-year period (1990–2000).

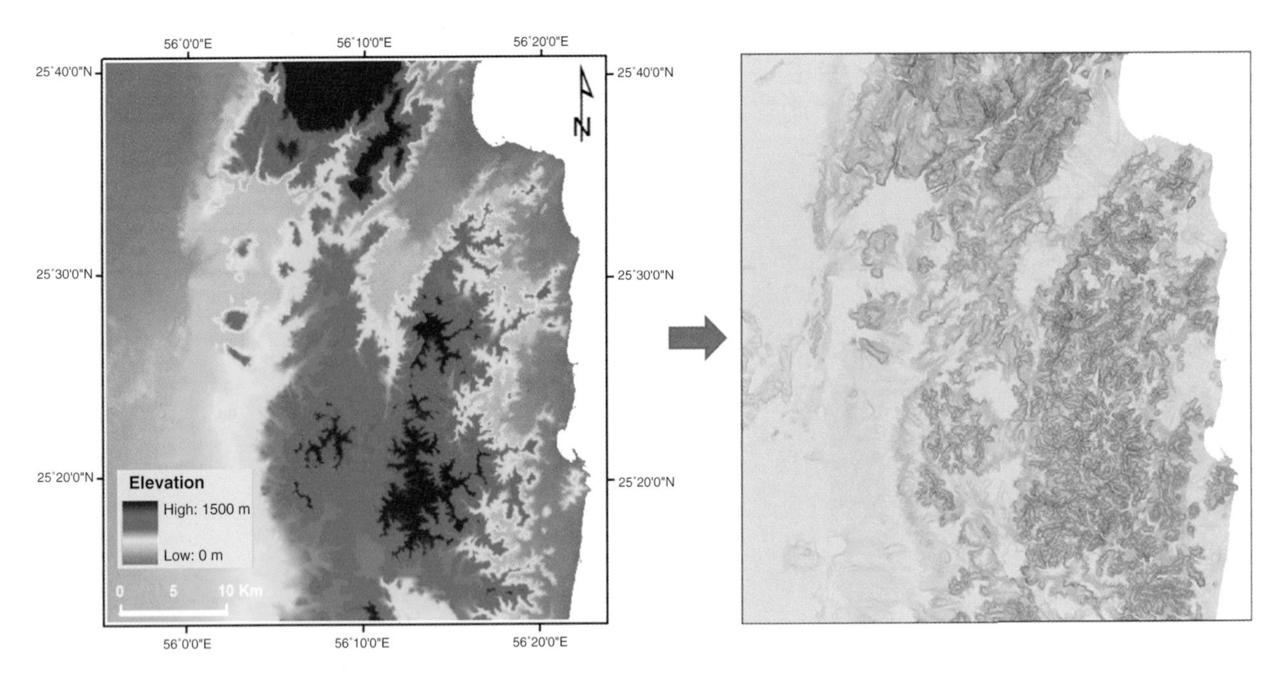

Figure 10.21 Slope map derived from a DEM generated by the SRTM.

many medium to coarse resolution (>20 m) DEMs are produced from satellite-borne (SRTM (Section 9.2) or ASTER (Section 2.3.8; Figure 2.22)) or airborne systems (lidar) (Section 9.4) and are increasingly being made available to the general public at no cost (http://www2.jpl.nasa.gov/srtm/and https://wist.echo.nasa.gov/api/). A slope map was derived from the SRTM DEM (Section 9.2), which has a resolution of 90 m. This DEM was reclassified into five elevation zones (i.e. 0–1, 1–5, 5–10, 10–1 5 and >15; values are in percentages), which represent categories of terrain in terms of suitability for rainwater to infiltrate. The scores are assigned according to terrain slope and water flow conditions. Gentle slopes (1–5%) show the best recharge potential for runoff, while building a certain hydraulic gradient for near surface water to continue flowing downhill, whereas areas having zero to very low (0–1%) and moderate to very high (10–15 and >15%) slopes are considered unfavourable because water may flow too slowly and stagnate and pound or simply flow too fast for infiltration to occur. Therefore, slopes between 1 and 10% receive the highest scores. This slope classification is also supported by groundwater recharge studies conducted in similar semiarid and arid environments (Saraf and Choudhury, 1998; Jaiswal *et al.*, 2003; Jasrotia, Kumar and Saraf, 2007).

DEMs are also useful for deriving drainage channel networks. However, they are prone to error, especially in flattish areas, so visual inspection, either on the ground or using high resolution images, is recommended. Fracture networks can be easily mapped on satellite images especially in areas with no or little vegetation cover. In this case study both drainage and fracture network maps were obtained from satellite data through visual interpretation and onscreen digitizing by an expert (Figure 10.22). Drainage channels and lineaments representing fractures and faults were mapped at a scale of 1:50 000 from an edge-enhanced ASTER colour composite using the first three bands with a pixel resolution of 15 m. The respective density maps were produced by counting the number or length of lines (either fractures or drainage channels) per unit area (1 km^2) and interpolating the resulting density values. The scores in each category reflect recharge potential with low values for high drainage density and high values for low drainage density areas. In the case of fracture density an inverse relationship applies, with high weights corresponding to high fracture density values and vice versa. Similarly, the drainage and fracture intersection density map was produced and scores were assigned in accordance to recharge suitability (i.e. high intersection density values = high recharge potential and vice versa).

The purpose of representing the fracture or drainage network in the form of density maps is to facilitate the comparison of their spatial distribution with other terrain characteristics (e.g. geology, geomorphology, etc.) and to identify the main trends in the data. For instance, long and possibly deep seated faults that cross an area with high fracture density values (brittle rocks) have

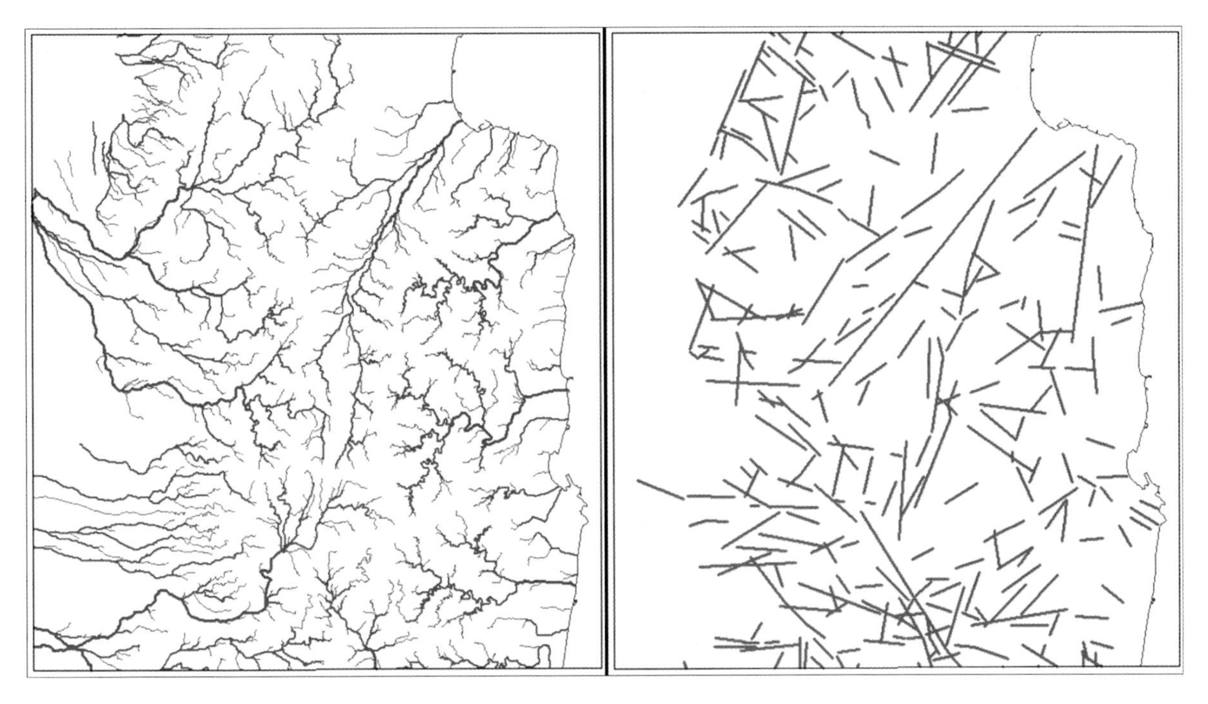

Figure 10.22 Drainage and fracture networks derived from the interpretation of ASTER satellite images.

the potential of being good groundwater collectors and transmitters if other terrain factors such as slope, rock type and tectonic forces are favourable. Furthermore, overlaying or intersecting the drainage and fracture network helps to identify which drainage systems may act as potential recharge zones, especially if they are prone to flash floods. High intersection density areas in the drainage/fracture intersection map denote good crosscutting relationships between wadis (ephemeral channels) and fractures and are often depicted as isolated clusters.

10.8.3 Multicriteria Decision Analysis

The thematic layers were finally ranked or weighted according to importance giving higher weights to those layers that have more influence to groundwater recharge occurrence. The total sum of the weights is always 1 or 100% (depending on whether floating point or discrete integer rasters are used), while the scores always range from one extreme of the scale to the other extreme. The evaluation scale may contain any number of increments; however, the same scale has to be applied to all thematic layers. In this case study we selected a range of nine values (from 1 to 9) and applied the scores to the individual layers according to expert knowledge (e.g. annual precipitations between 0 and 145 mm are less probable to generate sufficient runoff for subsurface infiltration to occur) or by simply dividing the whole range into equally space increments (e.g. density maps where the extremes clearly represent either less or highly suitable attributes but the values in-between are more difficult to assess). Weights are more difficult to assign as they determine the combined effect of all thematic layers in the model. Here again expert knowledge is essential in determining the degree of influence that each layer should have in the model. In this case study the geological/geomorphologic map (with a weight of 30%) is ranked as the most important factor followed by the drainage and fracture intersection map and slope map (with a weight of 20% respectively). All other thematic layers are given a 10% weight, meaning their influence is treated equally, but is less than the layers representing lithology, slope and the relationship between wadis (dry river channels) and fractures. The thematic layers were then multiplied by their corresponding scores and these products were then summed to obtain the groundwater recharge potential map (Figure 10.23). This was done by utilizing the arithmetic overlay approach (weighted overlay or weighted sum) built into ArcGIS *ModelBuilder*. Such an arithmetic overlay process accepts both continuous and discrete grid layers, and the derived recharge potential map is a discrete grid data layer with seven classes that were reclassified into three categories: no or little recharge, moderate, high to very high recharge. The first four classes were added to one class (low recharge potential), and the last three classes were kept and relabelled to moderate (fifth class) and high (sixth and seventh class) recharge potential. The reclassification was based on visual comparison of the resulting recharge map with the original thematic

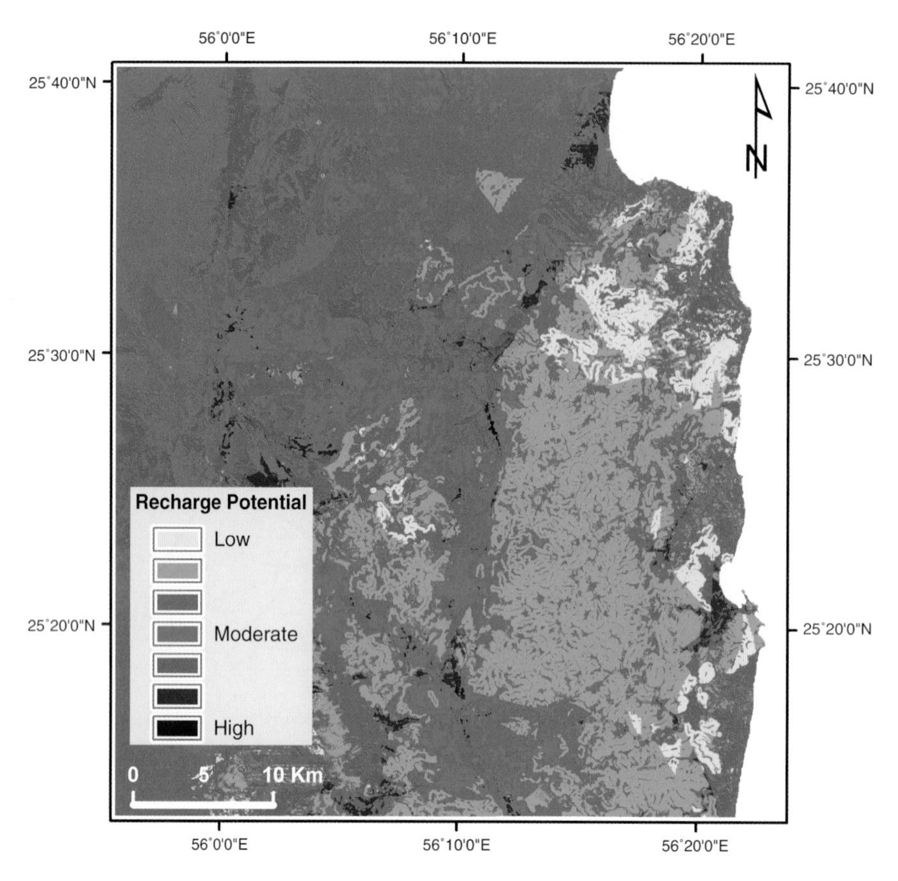

Figure 10.23 Output map of the recharge potential equation in Table 10.2.

layers used in the analysis, especially geology, fracture and drainage intersection and slope. Figure 10.24 shows the resulting recharge potential classes draped over a hill-shaded SRTM DEM of the study area and Figure 10.25 shows a three-dimensional view with major faults (representing structural trends) superimposed.

10.8.4 Evaluation

Inspection of the potential recharge areas (shown in light and dark blue colours in Figures 10.24 and 10.25) in relation to main fracture network, terrain characteristics and geology (Figure 10.26) reveals: (i) a close relationship between potential recharge areas in the alluvial gravel plain area with major structures (faults); existing wells located in the proximity of the highlighted areas confirm the high recharge potentiality (El-Baz *et al.*, 2004); (ii) the coastal plain area in the north-east shows high recharge potential as it receives both direct (rainfall) and indirect (flash flood) recharge; and (iii) the alluvial plain west of the mountain front shows a higher recharge potential than its eastern counterpart because it has more space to develop large alluvial fans and is not abruptly terminated by the sea. Also rainfall patterns are another factor contributing to its higher recharge potential, since the north-western region receives rain during the winter and summer whereas the eastern region receives rain mainly during the winter.

Correlation of the final recharge potential map in Figure 10.24 with the distribution pattern of vegetation as derived by applying a vegetation index (NDVI) to the ASTER images confirms the location of high recharge potential areas (Figure 10.27). In arid environments vegetation is an excellent indicator of surface or near surface water. The large patches of vegetated areas in Figure 10.27 are irrigated farms, mostly date palms and fruit trees. They are generally located where groundwater is accessible via shallow wells. The NDVI values showing the location of vegetated areas were not used in the MCDA and the distribution of NDVI values therefore represents an independent test of the recharge potential map. Comparison of the spatial distribution of NDVI with modelled recharge potential indicates a significant agreement between both data sets and thus validates the model used in this case study.

10.8.5 Conclusions

In summary, MCDA is a very attractive geospatial analysis technique because of its simplicity and straightforwardness. It uses a set of maps that provide evidence

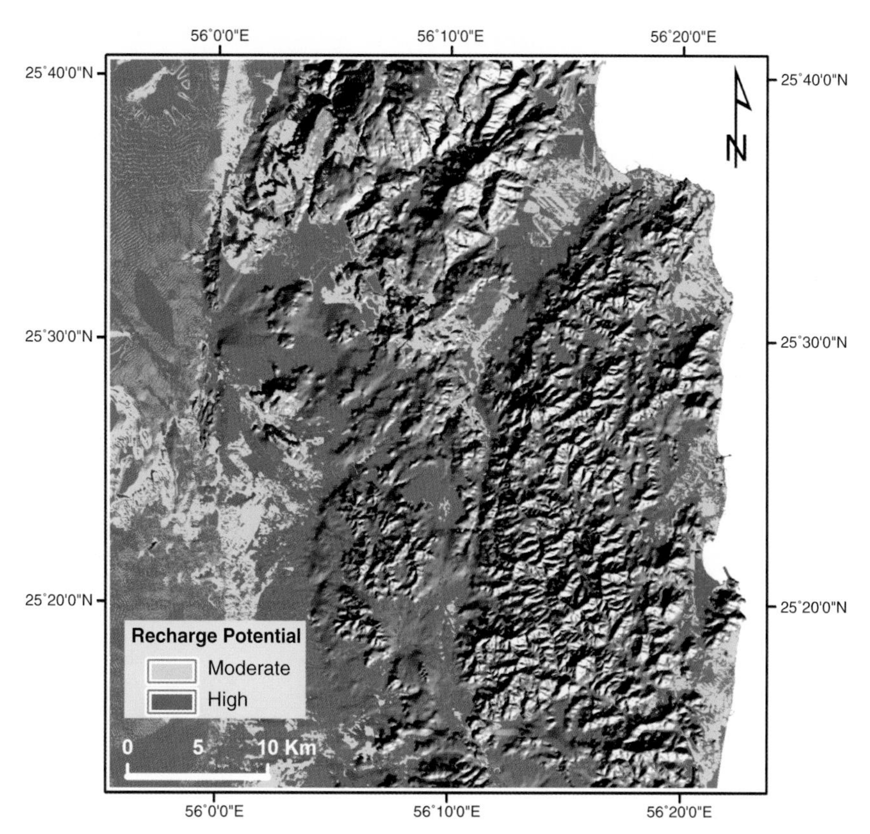

Figure 10.24 *Recharge potential map of the study area. Background layer is a hill-shaded SRTM DEM. See text for discussion of the derived recharge potentiality map.*

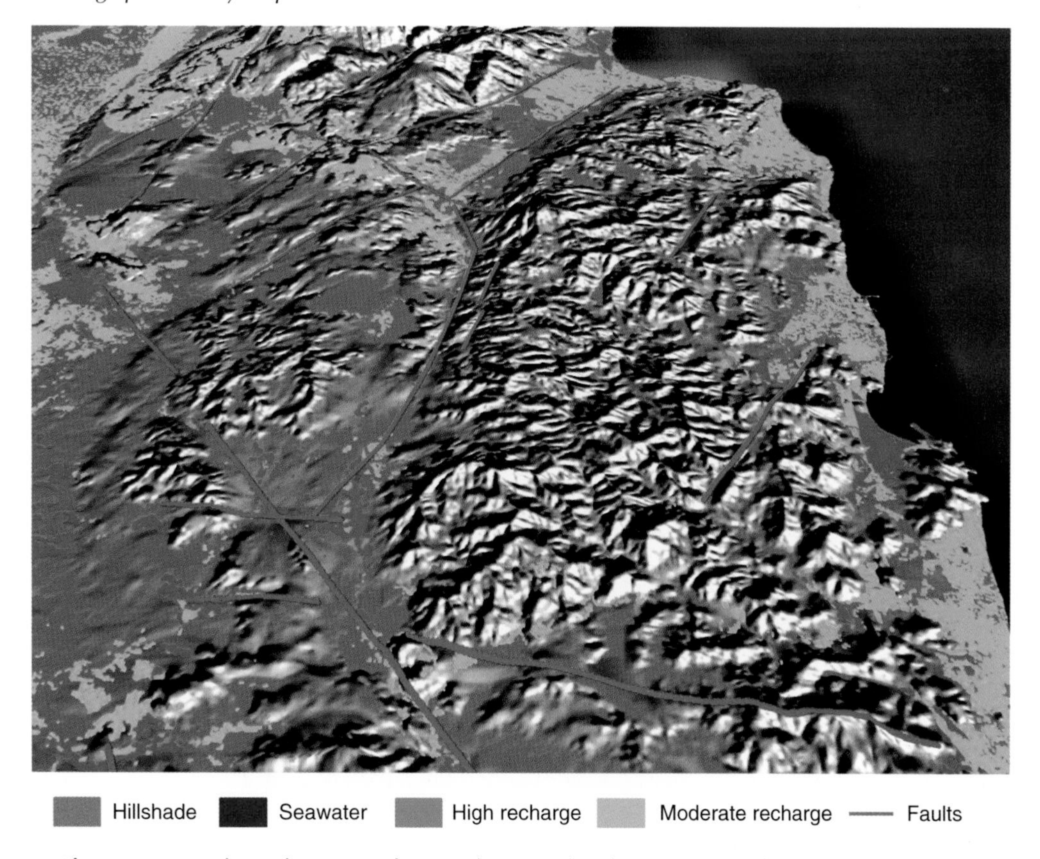

Figure 10.25 *Three-dimensional view of potential recharge areas with major faults overlaid.*

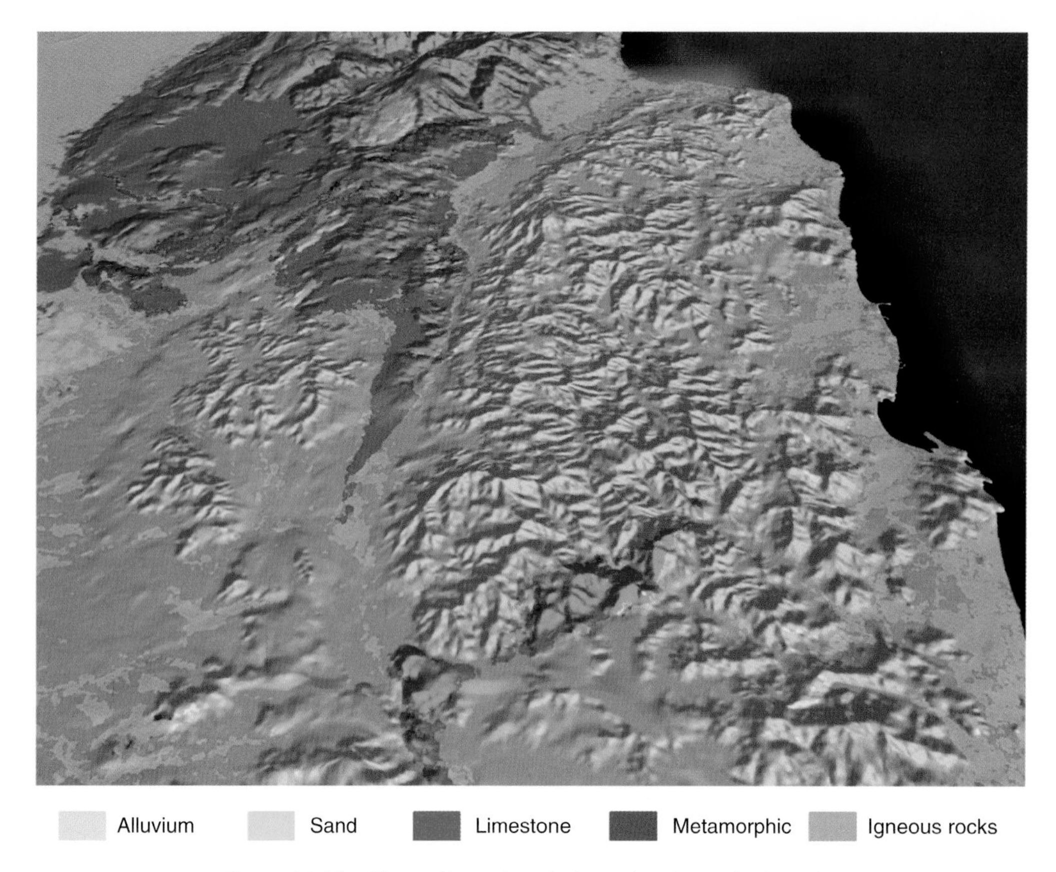

Figure 10.26 *Three-dimensional view of main geologic units.*

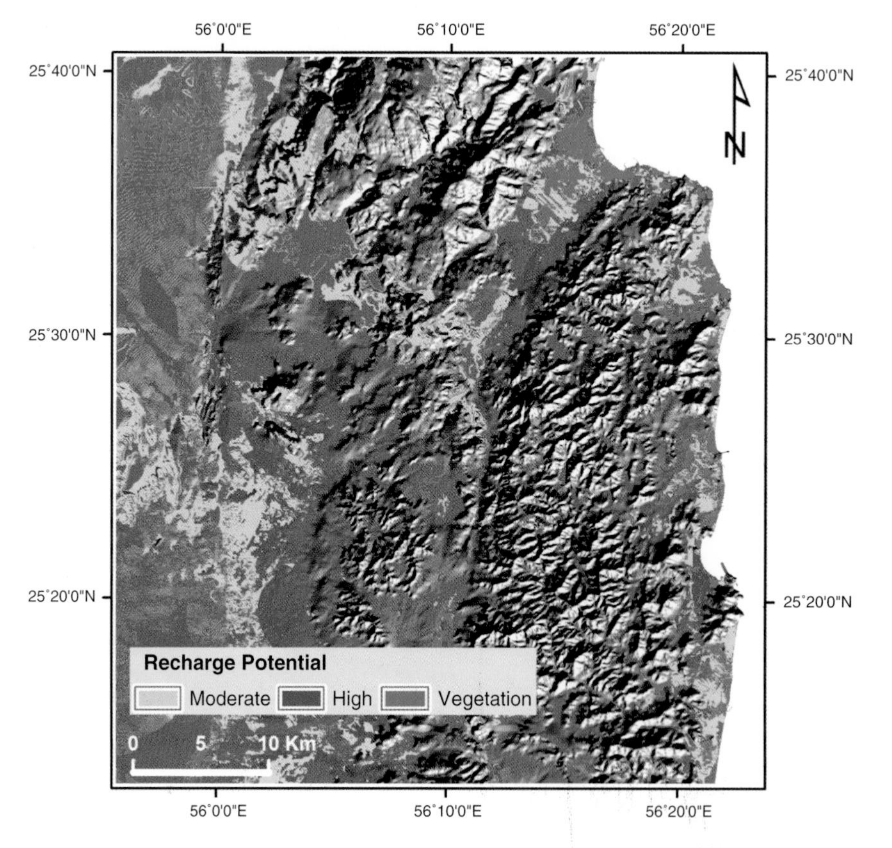

Figure 10.27 *Recharge potential map together with vegetation distribution derived from an ASTER NDVI image.*

of a certain process (in this case groundwater recharge) to occur if certain criteria are fulfilled. Of course the analysis has to be based on a sound conceptual model otherwise the predictive model will not work. Nevertheless, there are some shortcomings related to this method of which the user should be aware. First, the selection of weights is not an easy task even for an expert in the particular research field being investigated. Although professional judgement and/or statistical methods are often used for assigning weights, they may still be subject to errors. Second, the impact or sensitivity that slight changes in the weight values may have on the model needs to be assessed. Some factors (thematic maps) describing the phenomenon or process that is under investigation, are less sensitive to weight-value changes than others. Therefore, variations of the weights and their contrast should be tested before deciding which set of weights to use for each map layer. Third, possible dependencies between the data sets selected for predicting a phenomenon should also be checked. This can be done by adding/removing a map layer from the analysis and then checking the model result for any significant output variations. In general, input maps should represent independent variables otherwise the information they supply to the model will be redundant. By taking these and other possible sources of errors into account, MCDA can be a very viable and reliable prediction tool.

10.9 Assessing Flash Flood Hazards by Classifying Wadi Deposits in Arid Environments

10.9.1 Introduction

Surface water resources in arid regions are scarce due to low annual rainfall amounts and high evaporation rates. In the Middle East region, ephemeral watercourses (wadis) are of great importance for water supply. Following infrequent rainfall events, surface water accumulates in these wadis, producing two different effects, namely, flash floods and groundwater recharge. With rising water demands for population and agriculture in arid regions, proper surface water management through flood mitigation and harvesting schemes is increasingly needed. However, the lack of long-term rainfall and runoff/discharge records makes it difficult to predict which of the wadi systems are most active in carrying and discharging surface water in the event of a rainstorm. In this example we present a methodology for mapping and classifying drainage systems and identifying active wadis (i.e. flash flood prone wadis) in arid lands using satellite images. Optical multispectral (Landsat ETM+) images are used in conjunction with microwave (Radarsat-1 and PALSAR)[5] images to determine the spectral and textural properties of the alluvial wadi floor materials. Indicators of recent surface water flow, such as source rock composition of alluvial fills, presence of desert varnish and/or vegetation and textural characteristics of the bed load are derived from these images by a combination of data fusion and classification methodology. The results are used to classify the wadis in terms of their efficiency in carrying bed load sediments and, thus, surface water flow. A recent study by Gaber, Koch and El-Baz (2010) in Wadi Feiran basin of south-western Sinai Peninsula, Egypt, is used to illustrate this methodology.

10.9.2 Water Resources in Arid Lands

The availability of water resources in arid lands is still poorly understood because of the difficulty of data acquisition and inaccessibility of many such areas. Flash flooding in arid regions is a double-edged natural phenomenon: being a blessing and at the same time a curse. It is a blessing as it is a major source of groundwater recharge in arid regions. It is, however, a curse as the tremendous power of the water flow causes devastation along its pathway. Flash floods in desert regions are extremely difficult to predict due to the sporadic and spatially variable nature of rainfall-runoff events (Graef and Haigis, 2001; Gheith and Sultan, 2002; Masoud, 2009). Generally, desert rainstorms are highly localized, with only part of a drainage basin receiving rain and contributing directly to surface runoff along drainage channels or wadis (Sharon, 1972). The lack of sufficiently long rainfall and river discharge records limits the application of hydrological models of storm runoff generation. These models are often inappropriate because they were developed and calibrated in more humid environments (Ouattara, Gwyn and Dubois 2004; Masoud, 2009).

The ability to monitor and predict sites prone to flash flooding is also necessary to ensure the safety of people, as well as the protection of property and infrastructure. Moreover, the scarcity of fresh water in these regions presses the need for harvesting floods in order to take advantage of these water resources before reaching the coast and flowing into the sea or evaporating into the atmosphere. However, before these floodwaters can be managed, a more comprehensive understanding must be achieved to improve our ability to monitor and predict flood events. In the absence of detailed field measurements (a common problem in arid environments) an

[5]The Radarsat-1 image was supplied by the KACST facility in Riyadh, Saudi Arabia. PALSAR images were supplied by JAXA, Japan. Both datasets were obtained by BU-CRS through a data user agreement.

alternative approach is to assess flash flood generation potential by studying associated sedimentary transport and deposition behaviour in ephemeral streams. This can be achieved by using satellite images to identify and classify sediment types and transport in active wadis based on terrain properties (slope, channel surface roughness and moisture) and spectral properties of wadi fill materials (source-rock fragments, residuals from suspended sediments, desert varnish and vegetation). A study conducted by Laronne and Reid (1993) showed that ephemeral rivers in the Israeli drylands are up to 400 times more efficient at transporting coarse material than their perennial counterparts in humid zones.

10.9.3 Case Study from the Sinai Peninsula, Egypt

The example presented here is based on a study conducted by Gaber, Koch and El-Baz (2010) in Wadi Feiran drainage basin in southwest Sinai Peninsula, Egypt (Figure 10.28). The basin covers an area of approximately 1850 km^2, and is largely composed of a rugged mountain range of igneous and metamorphic rocks of Pre-Cambrian age. Wadi Feiran is intersected by numerous tributaries, all potentially contributing runoff to a large alluvial fan located at the basin outlet at the coast of the Gulf of Suez. This basin was chosen to illustrate the use of image processing techniques (data fusion and classification) supported by GIS analysis techniques to identify tributaries that are prone to flash flooding and erosion or may serve as groundwater recharge areas based on their bed load and terrain characteristics. High stream power generally produces more erosion and therefore larger quantities of large rock fragments. Low stream power carries smaller rock fragments often great distances from the source rock area. If wadi deposits can be characterized in terms of their fragment size and rock composition then this information could be used to identify their original source area and transportation route as well as their flash flood generation capacity. For an in-depth discussion on stream power index and other geomorphic indices to determine the ability of a stream to erode its bed and transport sediments, the reader is

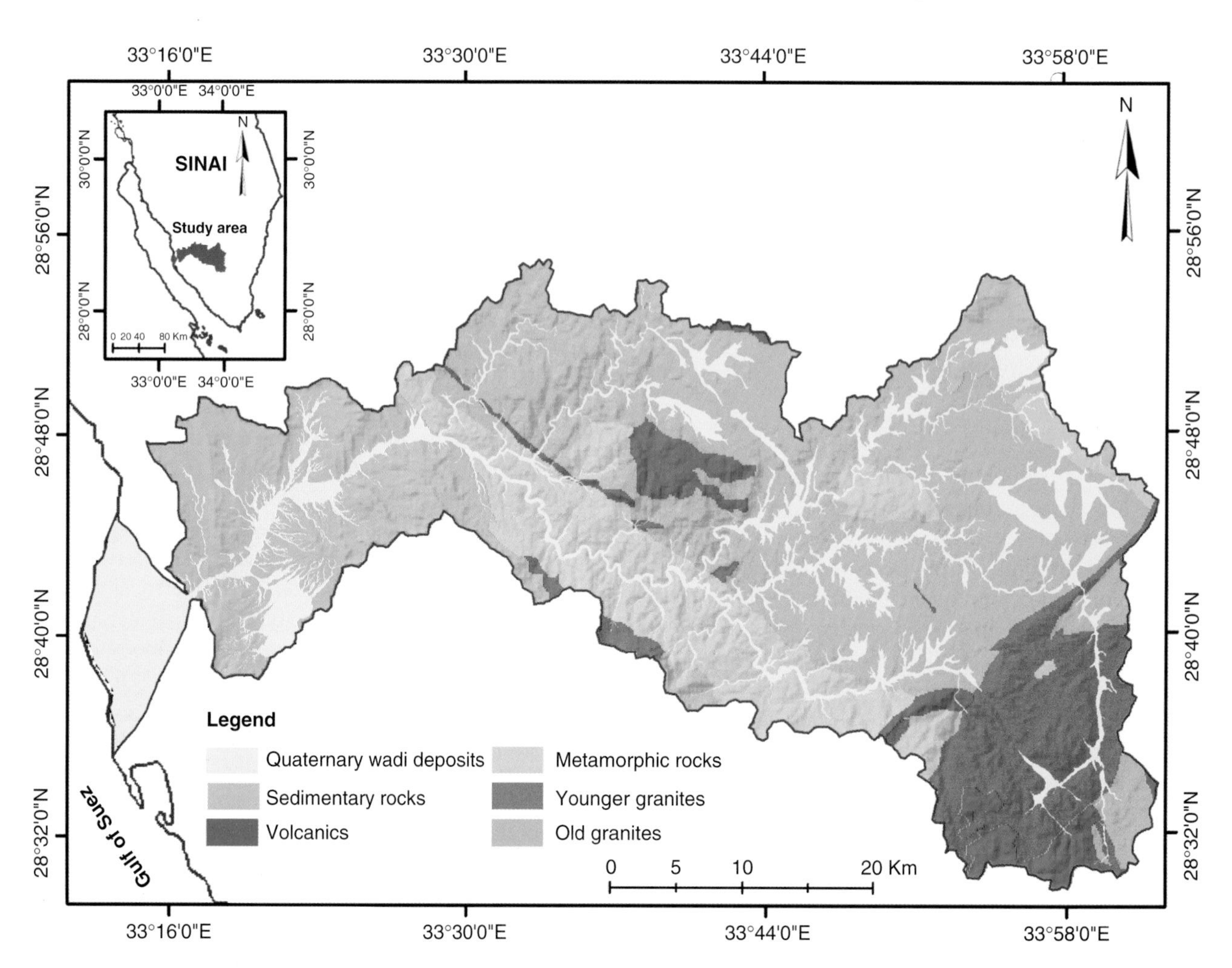

Figure 10.28 Simplified geological map of the Wadi Feiran basin, southwest Sinai. © Gaber, Koch and El-Baz (2010).

referred to the excellent work of Kelley and Pinter (2002) and more recently that of Masoud (2009) on estimating flash flood risks in southern Sinai Peninsula, Egypt.

10.9.4 Optical and Microwave Data Fusion

Radarsat-1 (C-band, HH-polarized) and PALSAR (L-band, HH-polarized) SAR images were used together with Landsat ETM+ images to analyse the drainage basin area covered by alluvium. Two sets of optical/microwave images were utilized in this study. One set comprises the ETM+/Radarsat-1 datasets for November–December 2000 and the second set represents the ETM+/PALSAR of January–March 2008 (Figure 10.29). The Landsat scenes were selected to match the acquisition dates of the microwave images as close as possible. For the first dataset only single Radarsat-1 and Landsat ETM+ scene were needed to cover the basin area. However, for the second dataset two PALSAR as well as two Landsat ETM+ scenes needed to be mosaicked to cover the entire study area. In addition, the Landsat ETM+ of 2008 shows numerous stripes with no data due to the malfunctioning of the scan line corrector (SLC). A list of image scenes and their main characteristics is shown in Table 10.3.

Data fusion was performed by applying a principal component transformation to the Landsat ETM+ scene and replacing PC1 with the SAR image as the high-resolution image followed by an inverse transformation (Section 6.9). The resulting fused image is a hybrid image in which the colour component of the multispectral image is added to an intensity component derived from the SAR image. In this study, the integration of ETM+ and SAR data introduces information that is correlated with the surface roughness of the wadi floors. Surface roughness in wadi floors and alluvial fans is a function of the distribution and density of sand, pebbles, boulders and vegetation (mainly shrubs) (Arkin, Ichoku and Karnieli, 1999). Different surface types show different backscattering coefficient values that relate to (i) the radar characteristics,

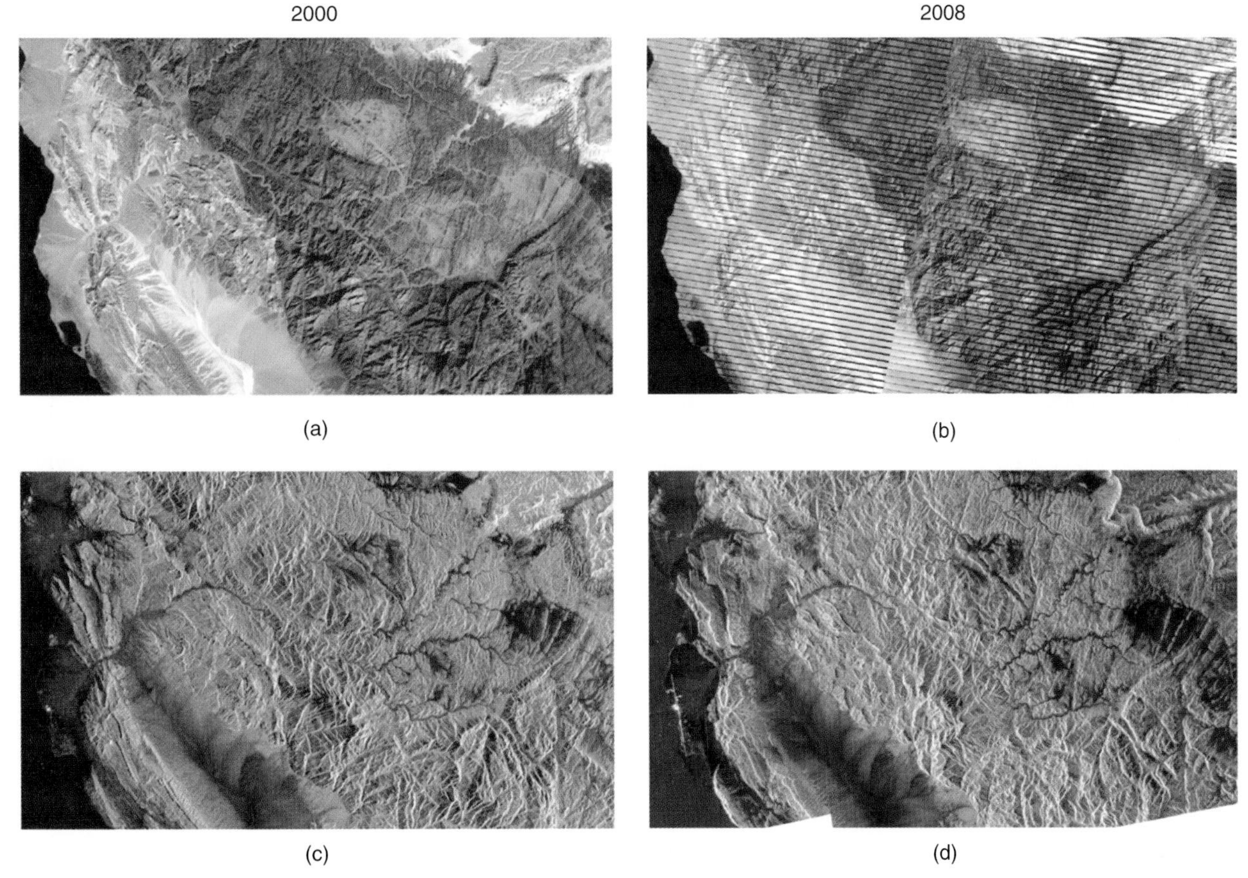

Figure 10.29 Optical/radar data sets used in this example: Landsat ETM+ of 2000 (a) and of 2008 (b) were respectively merged with Radarsat-1 of 2000 (c) and PALSAR of 2008 (d). The 2000 set represents single scenes whereas the 2008 represents mosaicked scenes. Credits: Radarsat-1 data © CSA 2000, recieved and processed by KACST, King Abdullah Aziz City for Science & Technology, Riyadh, Saudi Arabia. Distribyted under licence from MDA Geospatial Services Inc. PALSAR data was provided by JAXA/METI.

Table 10.3 Satellite dataset characteristics.

Sensor type	Radar		Optical	
	Radarsat-1	PALSAR	ETM + (SLC on)	ETM + (SLC off)
Acquisition date	18 November 2000	26 January and 24 February 2008	22 December 2000	21 February and 15 March 2008
Bands (polarization)	1 C-band (HH)	1 L-band (HH)	7 reflective bands	7 reflective bands
Incident angle	35.085°	34.3°	Nadir	Nadir
Resolution (m)	12.5	6.25	28.5/14.25	28.5/14.25
Swath width (km)	150	70	185	185

for example wavelength, polarization and look angle and (ii) the surface material properties, such as surface slope, roughness, orientation and dielectric constant (Henderson and Lewis, 1998). SAR images are more sensitive to surface roughness variations than optical spectral images. In relatively flat areas with little vegetation, subtle textural variations of the alluvial infill are mainly due to changing particle sizes of wadi deposits and soil moisture content. Therefore, these surface properties can be used together with the spectral information obtained from Landsat ETM+ data to improve the classification of bed load materials in wadi systems. The aim of classifying wadi beds is to distinguish fine to medium-grained sandy areas (indicating low stream power values) from surfaces completely covered by boulders (indicating high stream power values). The ability of SAR data to discriminate different surface textures is especially useful in areas that show similar spectral responses on ETM+ images, as is the case for example in the interior Wadi Feiran basin area, which is made up of low contrasting granites and metamorphic rocks (Figures 10.28 and 10.29).

Prior to the multisensor data fusion a Lee filter (Section 7.2.3) with a kernel size of 3 × 3 was applied to the raw Radarsat-1 and PALSAR images before converting from 16- to 8-bit representation for data fusion. This ensured that some noise (speckle) inherent in the radar images was removed without too much information loss (when filtering the smaller 8-bit data range) while improving the appearance of the image for subsequent ground control point selection (Pohl and van Genderen, 1998). Using the filtered 16-bit data, control points were selected only in the flat alluvial areas because no topographic distortion correction was performed to the SAR image to compensate for foreshortening, layover and shadow effects in mountainous areas. Since we are mainly interested in the textural information of low relief areas (i.e. wadi deposits), geometric distortions in mountain areas were considered to be irrelevant to the analysis. The SAR images were resampled to the same pixel size as the panchromatic band of the ETM+ image using 35 well-distributed control points in the case of the Radarsat-1 image and 48 in the case of the PALSAR image. A second order polynomial function was used to register the ETM+ and SAR images, with an RMS error of less than 1 pixel (14.25 m) (Section 4.3.2). Furthermore, all reflective ETM+ bands (1–5 and 7) were resampled to the same resolution as band 8 (14.25 m) before fusing them with the SAR data.

Figure 10.30 shows the two resulting hybrid ETM+/SAR images of Wadi Feiran basin. Figure 10.30a represents the PCA fused ETM+/Radarsat-1 images of November–December 2000 while Figure 10.30b represents the ETM+/PALSAR images of January–March 2008. Note that for generating the 2008 fused image (Figure 10.30b) pairs of ETM+ scenes (February and March 2008) and PALSAR scenes (January and February 2008) (Figure 10.29b, d) had to be mosaicked, resulting in a slightly different colour-contrast hybrid image compared to the one of 2000 (Figure 10.30a) which is produced by fusing single scenes only (Figure 10.29a, c). Also, the 2008 hybrid image shows strips with no data due to the SLC problem that affects all ETM+ scenes after May 2003. Therefore, the colour contrast between the limestone, granite and metamorphic units is slightly different (i.e. less pronounced) in the ETM+/Radarsat-1 image than in the ETM+/PALSAR image although the same band combinations were used. These colour differences are also due to the different backscatter properties of Radarsat-1 and PALSAR. Alluvial areas show, in general, good contrast with respect to the rugged rock outcrops due to the smoothness of their surfaces especially in wide wadi beds (dark areas in Figure 10.30). The next section explores in more detail the different backscatter properties of Radarsat-1 C-band and PALSAR L-band in relation to wadi bed deposits. It is postulated that the higher frequency of C-band radar is capable of differentiating more accurately surface deposits of various sizes (sand, pebbles, cobbles and boulders) than the lower frequency of L-band, even though the spatial resolution of the C-band image is lower (12.5 m) than the L-band image (6.25 m).

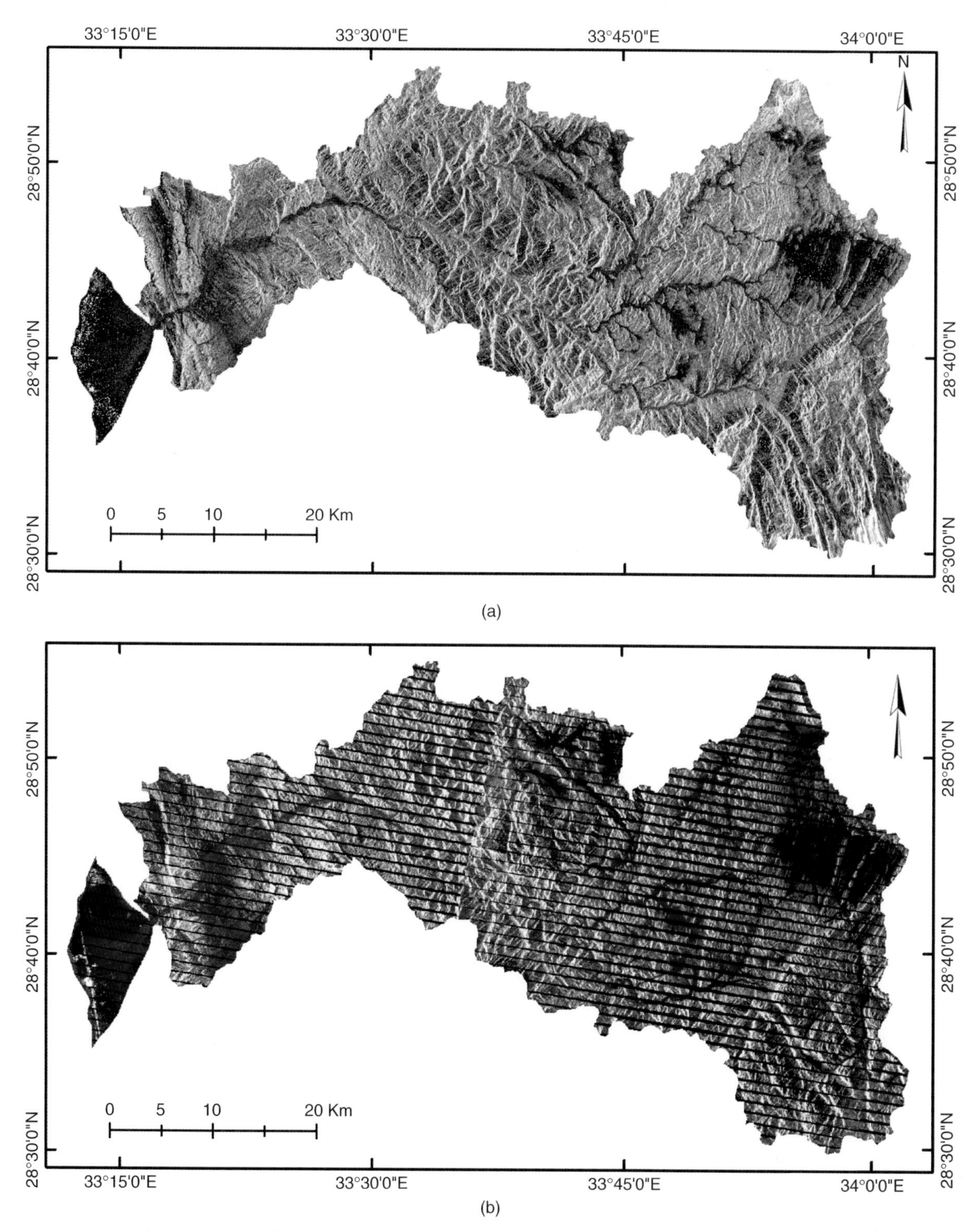

Figure 10.30 *Fused images of Landsat ETM+ (bands 7, 4, 2 as RGB) with (a) Radarsat-1 and (b) PALSAR data. © Gaber, Koch and El-Baz (2010). See Figure 10.29 for credits.*

10.9.5 Classification of Wadi Deposits

A classification using an unsupervised k-means clustering algorithm (with 10 initial classes and five iterations) was performed on each of the ETM+/SAR fused images. K-means clustering is described in Section 8.3.1 and Example 8.1. Each hybrid data set consists of six reflective ETM+ bands fused with the corresponding SAR image by means of PCA (see previous section). An unsupervised classification was chosen rather than a supervised classification because of the complexity of the spectral and textural response pattern of wadi deposits. Unsupervised classifiers are ideal for exploring the natural groupings (clusters) of spectral and textural pixel values contained in an image. Wadi deposits are by nature a mixture of rock types of different mineral compositions and sizes and, therefore, produce a whole range of mixed pixels in an image. Prior to running the k-means classifier a mask was applied to restrict the classification to alluvial areas. Restricting the image data to the area of interest helped to reduce the number of classes to five and improved the interpretability of the classification results. A majority filter with a 5×5 kernel size was applied to the resulting classifications in order to remove noise (spurious pixels) from the data and thus reducing the variance within the resulting clusters.

Figure 10.31a, b shows the respective classification results for the ETM+/Radarsat-1 (a) and ETM+/PALSAR (b) images. The colours of the five resulting classes were matched based on their spatial relationship to facilitate visual comparison of the class distribution in the two datasets. However, similarly coloured classes do not necessarily represent the exact same surface features. Two examples are illustrated in Figure 10.31 (boxes A and B), where slight differences in class distribution are due to the way in which C- and L-band SAR detect textural variations of fluvial surface deposits. These differences were further investigated by extracting the mean backscatter values of Radarsat-1 and PALSAR at 41 field sites using ArcGIS `Extract Values to Points` tool. This function extracts the cell values of a raster based on a set of points, in this case the field points. The field points also served to verify the surface roughness conditions of the five hybrid classes and were labelled accordingly. The scatter plot shown in Figure 10.32 shows the relationship between the mean backscatter values of Radarsat-1 and PALSAR at the 41 field sites. The points are colour-coded according to the surface roughness they represent in the field (with red being the smoothest class and cyan the roughest class). A clear trend is visible in the relationship between surface roughness and SAR backscatter values. Furthermore, the backscatter values derived from Radarsat-1 and PALSAR data show a positive correlation, with PALSAR showing consistently higher values than Radarsat-1. This is probably due to the fact that uncalibrated raw DN values are being used in this comparison.

Nevertheless, the scatter plot shown in Figure 10.32 shows that both datasets produced comparable classes. The red and green classes represent very smooth to intermediate smooth areas and are found in the alluvial fan area (west) as well as in the wider sections of the wadis in the upper basin area (east). The yellow and cyan colours represent intermediate to very rough surfaces and are found mainly in the smaller and narrower tributaries. The blue class is a textural class lying between both extremes. In order to interpret the meaning of the different classes, and evaluate their usefulness in mapping active wadi systems, a comparison of both classification results with the distribution of lithological units, terrain slope and field observations was performed using GIS techniques.

10.9.6 Correlation of Classification Results with Geology and Terrain Data

A geological map (CONOCO, 1987) at scale 1:500 000 and a digital terrain model (DEM) obtained from the SRTM (Section 9.2) with a 3-arcsec (90 m) horizontal resolution and a 16 m vertical accuracy (USGS, 2008) were used as reference data on which the interpretation of the classification results was based. The geological map was used as a reference for identifying the main host rock components of the bed load sediments. The DEM provided useful information on terrain properties such as slope and elevation that correlates with textural and spectral properties of main wadi infill and alluvial fan deposits. A field survey was conducted in February 2008 along the main channel of Wadi Feiran basin. A total of 41 field sites was investigated (Figure 10.31), and at each site wadi deposit type and size as well as their GPS locations were recorded and photographs were taken (Figure 10.33). The correlation between image derived information (e.g. classification results and radar backscatter coefficients), maps (geology and DEM) and field observation was carried out in a GIS environment.

The alluvial area of the Wadi Feiran drainage basin shows the following general surface sediment characteristics: the soils are typical of desert environments (aridosols and entisols) and important accumulations are found in few vegetated areas mainly in wadi gorges and terraces. The soils are light or yellowish brown in colour, and sandy to loamy sand in texture with low content of silt and clay (Abd El-Wahab *et al.*, 2006). Wadi sediment deposits become more gravelly and rocky as one moves away from the coast into the mountains, showing rocky surfaces with large boulders in the upper portions of the basin (Figure 10.33). According to pedological studies conducted in Wadi Feiran the soils here are of fluvial origin (torriorthents and torrifluvents) and of relatively

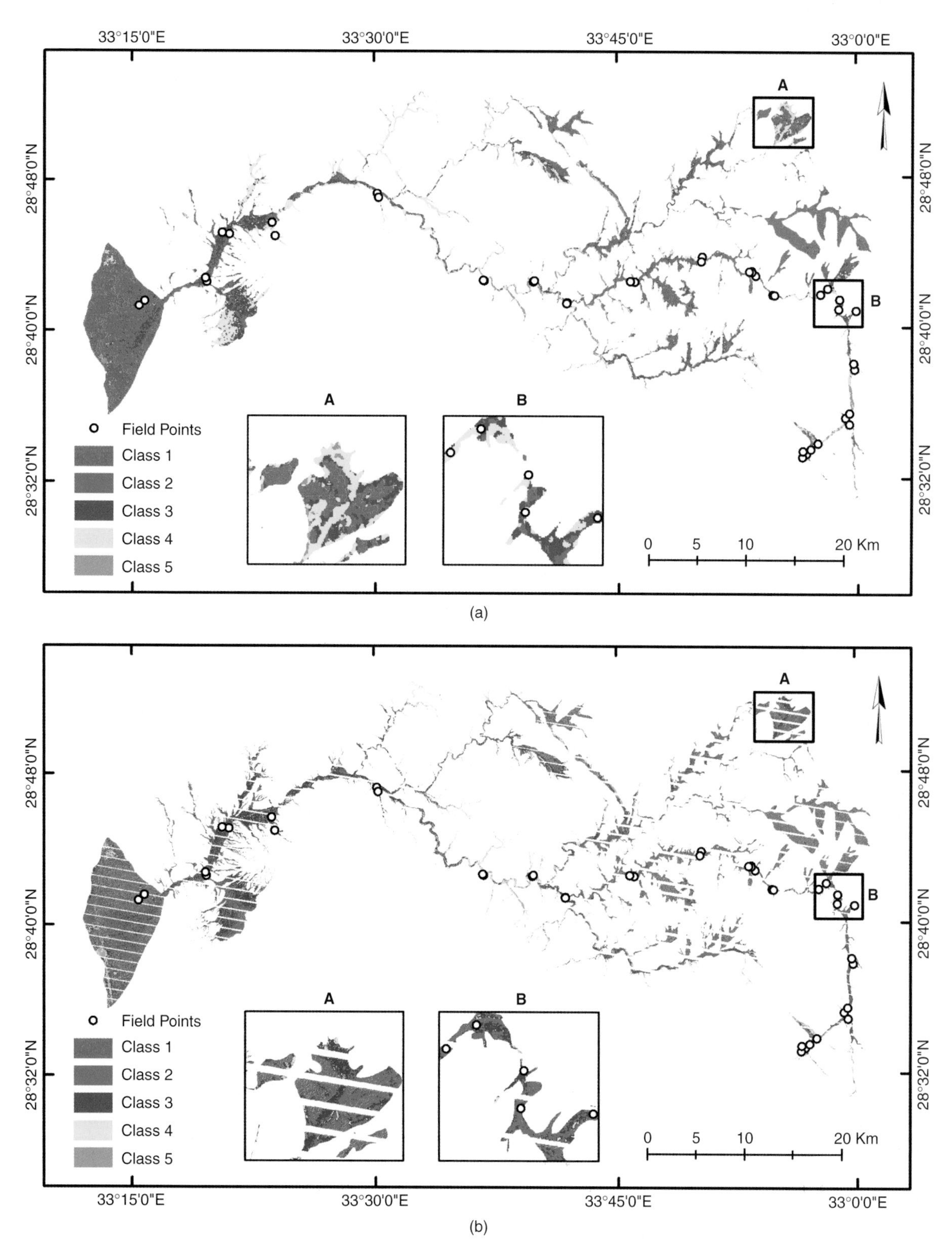

Figure 10.31 Unsupervised classification of wadi deposits using two sets of fused optical/microwave images: (a) ETM+/Radarsat-1 and (b) ETM+/PALSAR. Enlarged sections (A and B) highlight class distribution differences. © Gaber, Koch and El-Baz (2010). See Figure 10.29 for credits.

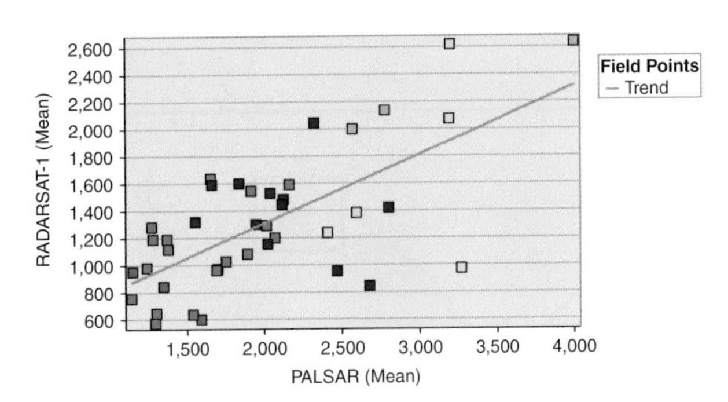

Figure 10.32 Correlation of mean backscatter DN values of Radarsat-1 and PALSAR at surveyed field locations along Wadi Feiran. Point colours match class colours in Figure 10.31.

good soil quality (El-Araby and El-Demerdashe, 1981; Abd El-Wahab *et al.*, 2006). Torriorthents and torrifluvents are very gravelly soils usually found on recently formed terraces and incized channels, and are associated to flooding events.

In order to investigate the nature of the unsupervised classification results of the ETM+/SAR hybrid images, individual classes were cross-checked with information obtained in the field (at 41 sites) as well as with map information on geology and topography. The main goal was to determine whether wadi deposits were successfully classified into similar spectral groups (i.e. the main lithological units, such as sedimentary, volcanic, granitic and metamorphic rocks) and textural groups (i.e. rough, medium, smooth surfaces). In addition, differences between the two ETM+/SAR classification results were examined, as C- and L-band show different sensitivities in detecting surface roughness. Correlation procedures were carried out in ArcGIS using its **Zonal Statistics Tools** which calculates statistics on values of a raster within the zones of another dataset. Zones can be rasterized points (e.g. field sites), polygons (e.g. lithological units) or classes from a classified image (e.g. hybrid classes). Statistics for each zone are extracted from another raster image or map (e.g. backscatter or slope values) and include pixel count, area, minimum and maximum values, range, mean, standard deviation and sum depending on the type of statistical analysis performed. For example the command **Zonal Statistics As Table** would have the following syntax in ArcGIS:

```
ZonalStatisticsAsTable<in_zone_data>
  <zone_field><in_value_raster><out_table>
  {DATA|NODATA}
```

where:

`in_zone_data` = dataset that defines the zone

`zone_field` = field that holds the values that define each zone

`in_value_raster` = raster that contains the values for which to calculate a statistic

`out_table` = output table that will contain the summary of the values in each zone

`DATA|NODATA` = denotes whether NoData values are ignored.

Zonal Statistics was one of the main GIS functions used in the Wadi Feiran study conducted by Gaber, Koch and El-Baz (2010). Here we illustrate some of the steps used in deriving the results published in that paper. However, for a more complete description of the study the reader is referred to the original publication.

In order to carry out the statistical correlation analysis, individual classes produced by the two sets of unsupervised classifications were used as zones, that is overlaid on the respective median-filtered 16-bit SAR data to determine the median backscatter coefficient (in DN values) corresponding to each class. The reason for using the original 16-bit data instead of the 8-bit data used for data fusion with ETM+ (which are in 8-bit) is that better class discrimination could be achieved by using the original unscaled image. The five classes resulting from the two hybrid classifications fall into well-defined data ranges that indicate an increase in surface roughness and thus rock fragment/sediment grain size with increasing backscatter median values (Table 10.4).

The results displayed in Table 10.4 show that class 1 (red in Figure 10.31) represents the smoothest wadi surface made up of sand and pebbles (Figure 10.33a) and classes 2 (green) and 3 (blue) are more gravelly surfaces (Figure 10.33b). Class 4 (yellow) corresponds to a mixture of sand, stones and cobbles (Figure 10.33c). Class 5 (cyan) has the highest backscatter value as it represents a mixture of small to large boulders (Figure 10.33d).

Second, a spatial correlation of individual hybrid classes with main lithological units (Figure 10.28) and slope was performed. The classified hybrid images were used as zone layers for which statistics were calculated for each zone (class) in terms of main lithological unit underlying each class and their mean slope value. A schematic representation of this procedure is shown in Figure 10.34. The results reveal that hybrid classes of both data sets (ETM/RSAT-1 and ETM/PALSAR) show similar trends with respect to underlying geology and slope (Table 10.5). Note that wadi sediments are a mixture of eroded bedrock fragments, some of which travel a long distance and their composition does not necessarily need to resemble that of the host rock flanking the wadi segment where they are deposited. However, landforms are a function of geological substratum, tectonic forces and surface processes acting upon them. Therefore, the interpretation of the results in Table 10.4 (surface roughness/fragment size) and Table 10.5 (geology/slope) may tell us something about the nature, transportation

Figure 10.33 *Field photographs displaying the typical grain/fragment sizes of wadi deposits corresponding to the following hybrid classes: (a) class 1 and 2: sand and pebbles, (b) class 3: gravels, (c) class 4: stones and cobbles and (d) class 5: small and large boulders. Photographs taken by A. Gaber.*

and deposition of wadi materials, that is the erosive power of streams carrying those materials and their resistance to weathering processes.

Classes 1 and 2 which represent the smoothest wadi surface deposits are mainly located in gently sloping terrains within the old weathered granite unit which makes up most of the central part of the basin. This unit is characterized by relatively gentle landforms and class 1 and 2 deposits are predominantly found in low relief plains where wadi beds become wider (box A in Figure 10.31a, b). Class 3 is mainly represented by the sedimentary rocks (sands, marls and carbonates) which are found at the lower and upper reaches of the basin (Figure 10.28). A combination of very wide and relatively flat wadi floors, with short, narrow tributaries draining into the flats characterizes this unit as less resistant to erosion than the older Pre-Cambrian units. However, some escarpments are found here along the more resistant layers depending on the sediment material.

Class 4 deposits are located predominantly in the younger granitic unit (Figure 10.28) which is characterized by high mountain relief, straight and narrow wadis (gorges) and very resistant (undeformed) granitic rocks. The average slopes of class 3 and 4 show increased steepness of the wadi floor surface. Class 5 has the highest average wadi floor surface slope values as well as surface roughness and is predominantly located in the metamorphic unit that underlies the main wadi trunk where it becomes narrower and starts meandering.

The correlation analysis described above highlights the distribution of predominant wadi sediment texture and size within the mountain area where bedrocks outcrop. It

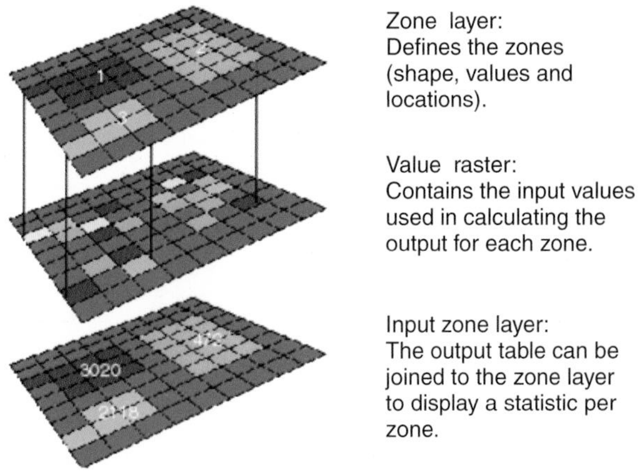

Figure 10.34 *Concept of zonal statistic analysis in ArcGIS. Copyright © ESRI. All Rights Reserved. Used by permission. www.esri.com.*

does not include the delta plain (alluvial fan area) at the mouth of the basin where finer particles (class 1) tend to concentrate. Also, the analysis does not conclusively reveal the mineral composition of the wadi sediments, as for this type of analysis sub-pixel classifications such as spectral unmixing (Section 8.5) are more appropriate techniques than a combination of data fusion and unsupervised classification. Indeed, spectral unmixing was the approached used by Gaber, Koch and El-Baz (2010) in Wadi Feiran and their results show a slightly different distribution of rock composition within each hybrid class. Their results are summarized in Table 10.6 and show that the predominant end member (rock type) making up each hybrid class does not necessarily match the predominant lithology where the hybrid classes are found. This means that the rock mineral composition of wadi bed materials differ from the host rocks where they are eventually deposited. It also shows the degree of compositional heterogeneity of wadi deposits which makes it so difficult to accurately classify them with medium spatial resolution images with a limited number of spectral bands.

Comparison of Tables 10.4 and 10.6 further reveals that as the surface roughness of wadi materials increases (from pebbles to cobbles and boulders) their composition seems to shift from rock units that are presently most prone to erosion (especially carbonates and young granites) to those rock units that show largely eroded and therefore less rugged surfaces (older weathered granites) or are more resistant to erosion (metamorphics and volcanics). This finding is also in accordance with the landforms and elevations where these rock-units outcrop, with older granites forming low relief areas in mid-altitude plateaus and younger granites forming high relief areas with pronounced peaks in the upper reaches of Wadi Feiran basin.

Table 10.4 *Median backscatter (DN) values of Radarsat-1 and PALSAR data for each of the five classes with corresponding roughness/grain size as observed in the field.*

Class (colour)	Backscatter (median)		Roughness and grain size
	Radarsat-1	PALSAR	
Class 1 (red)	862	1378	Smooth (sand, pebble)
Class 2 (green)	1060	1720	
Class 3 (blue)	1280	2193	To ↓
Class 4 (yellow)	1581	2761	Rough (cobble, boulder)
Class 5 (cyan)	1913	3698	

Modified after Gaber, Koch and El-Baz (2010).

Table 10.5 *Spatial correlation of hybrid classes (ETM+/Radarsat-1 and ETM+/PALSAR) with main underlying lithological units and mean slope values.*

Class (colour)	Geology		Slope (%)		
	ETM/Radarsat-1	ETM/PALSAR	ETM/RSAT-1		ETM/PALSAR
Class 1 (red)	Old granite	Old granite	2.50	Gentle	2.34
Class 2 (green)	Old granite	Old granite	3.09		3.02
Class 3 (blue)	Sedimentary	Sedimentary	4.25	to ↓	4.93
Class 4 (yellow)	Young granite	Young granite	6.40	Steep	9.55
Class 5 (cyan)	Metamorphics	Metamorphics	9.16		13.22

Table 10.6 Predominant rock composition (end members) within each class produced by unsupervised classification of the hybrid image.

Classes	Carbonate	Old granite	Young granite	Metamorphic	Volcanic	Predominant EM
1	XXX	XX	XXX	X	XX	Young granite and carbonate
2	XXXX	X	XXXX	XXX	X	Young granite, carbonate and metamorphic
3	XXXX	XX	XXX	XXX	X	Carbonate, young granite and metamorphic
4	XX	XXXX	XX	XX	XXX	Old granite and volcanic
5	X	XXX	X	XXXX	XXXX	Metamorphic and volcanic

X = Low, XX = Low to Medium, XXX = Medium to High, XXXX = High and EM = End member.

10.9.7 Conclusions

This case study demonstrates how GIS spatial correlation tools (**Value Extraction** and **Zonal Statistics Tools** in ArcGIS) can assist in the interpretation of unsupervised classification results of multisensor data, in this case optical and microwave images. The example used in this classification study is especially complex as it deals with very heterogeneous materials (wadi bed loads) that are transported over long distances where they mix and change in shape and size before finally being deposited as fine sediments in the coastal alluvial plain. Textural and compositional characteristics of wadi deposits were assessed using a combination of RS techniques supported by GIS statistical analysis to determine the source area of wadi sediments that are most actively being eroded.

10.10 Remote Sensing and GIS in Archaeological Studies

10.10.1 Introduction

In the preceding examples we demonstrated the use of RS and GIS applications in water resources studies. These types of studies are typically conducted by first analysing the regional context in which a particular natural resource occurs before selecting promising areas where detailed studies, conducted at a local scale, may reveal additional information about the status and availability of the resource. Thus, the typical approach in most geological exploration and resources assessment studies is firstly to study the natural system as a whole before focusing on specific target areas for detailed examination.

In archaeological studies, an opposite approach has been traditionally adopted, in that site-specific studies are usually conducted first before analysis is undertaken of the broader environmental context in which they occur. More recently, however, as geospatial technologies and high resolution images become more accessible, archaeologists are increasingly initiating their surveys with a regional exploration of archaeological sites. However, one reason for using a local approach first in archaeological studies is that by studying individual archaeological sites much knowledge can be gained about the cultural and natural environment that existed at the time when these ancient settlements were built and occupied. Once a number of archaeological sites are discovered and surveyed, their distribution in space and time may become a priority research goal as they may hold important clues in understanding how humans interacted with their environment and how the environment may have contributed to the rise and fall of ancient civilizations. Landscape archaeology is an important aspect in archaeological studies and includes the reconstruction of the paleoenvironment and human–environment interactions.

In this third case study we illustrate the use of RS and GIS technology in archaeology by presenting two archaeological field studies conducted in very different environments. The first example deals with the recent discovery and exploration of hidden Maya temples under the thick rainforest of the Holmul region in Guatemala, and is based on an investigation led by Estrada-Belli and Koch (2007). In this study, a combination of a radar derived DEM and multispectral images were used to pin-point promising areas for detailed surveys. The second example is located in northern Ethiopia where an important and powerful civilization (the ancient Aksumite kingdom) existed for many centuries. The Aksumites developed a sophisticated system of utilizing the natural resources (soils, rocks, water) which enabled them to become an important socioeconomic centre in the Horn of Africa (Michels, 1979, 1994, 2005). In the example presented here, RS and GIS tools were utilized by Sernicola (2009) to map and correlate the distribution of changing settlement patterns in relation to natural resources availability.

10.10.2 Homul (Guatemala) Case Study

Homul is an ancient Maya city located in the Petén region in northeast Guatemala near the Belize border

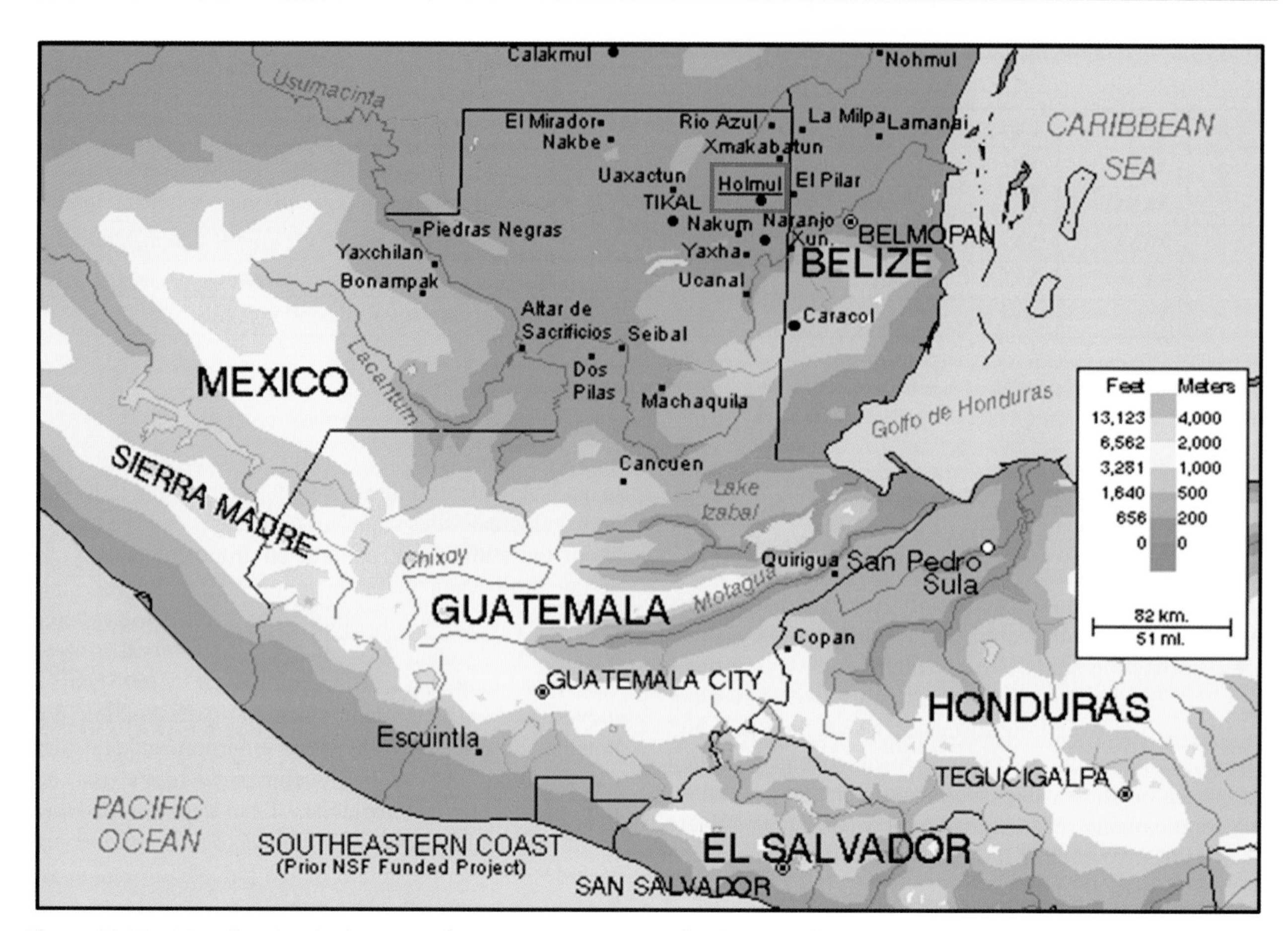

Figure 10.35 Map showing the location of ancient Maya cities. Holmul is one of them and is located in northeastern Guatemala near the border with Belize. Reproduced with permission from F. Estrada Belli, Archaeology Dept, Boston University.

(Figure 10.35). It lies in a wetland area known as the Maya Lowlands and existed between 1000 BC and 900 AD when it was finally abandoned. The causes for the collapse of this longest-occupied Maya settlement remains unclear to the present time, but possible explanations are destructive wars between rival Mayan cities, environmental change caused by deforestation and draining of the wetland areas as well as overpopulation leading to the depletion of natural resources (Culbert, 1988; Shaw, 2003; Beach *et al.*, 2009).

Detailed mapping of ancient Maya settlements in the Holmul area has been carried out between 2000 and 2009 by Estrada-Belli and his team utilizing a combination of RS, field work and test excavations of sample sites (Estrada-Belli, 2003, 2010; Estrada-Belli and Koch, 2007). The landscape in Holmul is characterized by limestone ridges and basins covered with a thick rainforest canopy (Figure 10.36). The thick vegetation makes it extremely difficult to locate buried structures from the ground as well as from above (Figures 10.37 and 10.38). Applying standard image processing techniques to optical imagery does not necessarily reveal the buried pyramids beneath the forest. However, there are some clues that may reveal these structures on satellite images. One clue is that archaeologists know that Mayans built their temples or pyramids as sacred sites mimicking mountains and hills, thus creating a topographic rise in an overall flat canopy surface. Another clue is that Maya sites show very characteristic vegetation types and densities that are well adapted to the type of rocks (limestone) used as building material for the temples and the thin soil cover found in these sites (Estrada-Belli and Koch, 2007). In essence, archaeologists use the anomalies in vegetation colours as proxies for the presence/absence of buried Maya architecture. These two clues can be used to identify hidden structures on remotely sensed images as illustrated in the following paragraphs.

One way archaeologists may discover new Maya sites is by mapping rainforest texture to locate textural anomalies caused by trees growing on top of pyramid structures (man-made hills) that would cause a bump in the overall flat canopy surface (Sever and Irwin, 2003). Textural mapping is best done by radar images because this type of imagery is sensitive to the relative roughness

Figure 10.36 *Three dimensional view of a Landsat ETM+ image showing a sequence of limestone ridges and seasonal wetlands (bajos shown in pink colours). The location names refer to discovered Maya sites. Deforestation activities are responsible for the bright colours in the lower right corner. Reproduced with permission from F. Estrada Belli, Archaeology Dept, Boston University.*

and smoothness of target surfaces (textural anomalies). In addition, radar images can be acquired regardless of cloud cover, an important aspect to consider in tropical environments.

High resolution DEMs are also very useful in identifying slight height variations in canopy surfaces. In this respect, DEMs generated from radar interferometry (InSAR, Section 9.2) are particularly helpful in rainforest environments as they are able to detect vegetation height variations with high precision (Sever and Irwin, 2003). A DEM generated from airborne STAR-3*i* radar imagery of the Holmul region and acquired through NASA's Scientific Data Purchase (SDP) programme, was used to explore the canopy surface for height anomalies (Figure 10.39). This InSAR DEM has a spatial resolution of 10 m with an absolute vertical and horizontal accuracy of respectively 3 and 2.5 m and was originally developed by NASA's Jet Propulsion Laboratory (JPL) and operated by Intermap Technologies (Sever and Irwin, 2003). A shaded relief image was produced from this high precision DEM and the vertical axis was exaggerated to enhance very subtle elevation variations (Figure 10.40). Figure 10.40 shows the rainforest canopy with groups or individual tree crowns forming a rugged surface. The most elevated areas in the canopy surface are promising areas for discovering new Maya pyramids.

In order to predict the location of new undiscovered Maya sites, a GIS analysis of least-cost pathways across the Holmul landscape was performed (Estrada-Belli and Koch, 2007). Least-cost or least-resistance pathway analysis, as the name implies, consists of calculating the least-cost path or paths in terms of one or more variables or quantities between two selected locations. To find the path, the user computes a friction map in which landscape features such as slope and aspect serve as factors that add cost to movement in any direction. In addition, a cumulative cost is generated detailing the pixel-by-pixel accumulation of cost outward from one location

Figure 10.37 Maya temples in Holmul are hardly visible from above (aerial photographs or high resolution images) as well as from the ground because they are overgrown by dense rainforest vegetation. Reproduced with permission from F. Estrada Belli, Archaeology Dept, Boston University.

Figure 10.38 Rainforest canopy of the Holmul region viewed from the top of a Maya pyramid. Elevated areas in the background are forested limestone ridges. Photo: M. Koch.

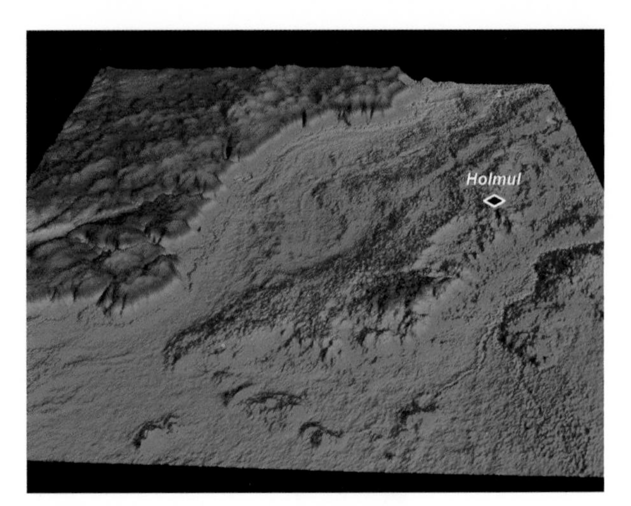

Figure 10.39 High resolution radar STAR-3i DEM of Holmul region. Lowlands (wetlands) are shown in green and highlands (limestone ridges) in blue and purple colours. Reproduced with permission from F. Estrada Belli, Archaeology Dept, Boston University.

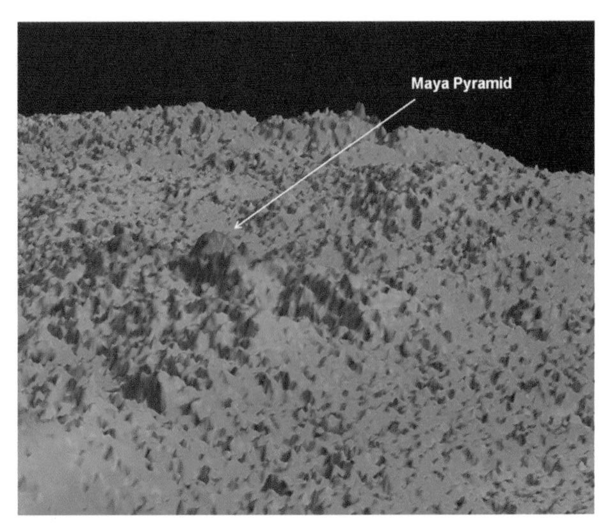

Figure 10.40 Vertically exaggerated shaded relief map of radar STAR-3i DEM highlighting elevated areas in the rainforest canopy in blue colours. These are possible sites of Mayan pyramids. Reproduced with permission from F. Estrada Belli, Archaeology Dept, Boston University.

to another. The cost feature uses the friction map as a basis for determining the set of pixels of lowest cumulative cost (or path of least resistance) from the starting to the stopping location to form a least-cost route.

In the Holmul case, two separate sets of friction maps were generated as input criteria for the least-cost route analysis. The first set included the simple terrain features of slope and aspect. The slope map was reclassified to reduce the map's original values of slope expressed in percentages from 0 to 90%, into friction factors 1–4 (1 = flat, 2 = low gradient, 3 = steep, 4 = very steep). The aspect map was reclassified to reduce the original values expressed in compass degrees (0–360) into friction factors 1–7. The factors were adjusted according to the direction of movement. For example, for an eastward route the west-facing side of hills received a factor value of 7 while the east-facing hillsides were given a factor value of 1. All other sides of hills were given a mild difficulty factor of 2. In this manner an aspect factor map was generated for each of four directions of movement. The second set of friction maps included the criteria of visibility. Areas visible from the destination site were given a neutral cost factor value of 1, while areas not visible from the same location received a friction-cost factor of 7.

The following step included generating a cost map which would map out the cost of traversing the various friction features located on the landscape. This was done with an algorithm that calculated the incremental costs of movement on a pixel-by-pixel basis as measured from the destination location outwards towards the edges of the map. In this manner, the cost of crossing any pixel was multiplied by the relative friction factor for that pixel and added the initial cost for the next pixel calculation.

The final step was to run an algorithm that searched the map for the lowest-cost pixels that connected the starting location to the stopping location. Once located, the lowest cost pixels were threaded together to form a continuous path. As an option, the user could request that the pixels for the generated path represent cumulative cost values as measured from the starting location.

Through the above described steps, Estrada-Belli and Koch (2007) generated optimal routes from a number of random locations along the edges of the map for all directions, all ending at the site of Holmul (Figure 10.41). Interesting patterns immediately emerged. First, the various paths from any particular random point along the edge of the map quickly converged and followed the same route into Holmul. As a result, there was really only one best route for approaching Holmul from each direction as one came within 5 km from the centre. A second interesting pattern was that the set of routes generated according to the visibility criterion also converged at certain distances from Holmul. In the east, south and west these location tended to coincide with the edge of the upland plateau where Holmul's peripheral centres of T'ot, Riverona and K'o were situated (Figure 10.42). Each site was located at approximately 5 km distance from Holmul. Because of these observed regularities in three areas (south, west and east) Estrada-Belli and Koch (2007) hypothesized that in the two remaining areas, which had not been surveyed, to the north and northeast of Holmul, as well, there might be an important ceremonial centre at the specified location along the optimal path and at the expected distance of roughly 5 km. In the following field season Estrada-Belli and his team were able to confirm these hypothesis by locating the two ceremonial centres of Hamontun, located to the northeast of Holmul at 5 km distance and along the visibility-predicted optimal path, and Hahakab, located to the north of Holmul at the expected location where the visibility and the non-visibility paths converged at 4 km from Holmul (Figure 10.41). The discovery of these two sites was significant not only because it completed the set of secondary centres around Holmul, but also because reified the initial hypothesis according to which secondary centres were located in locations that controlled land resources around Holmul and access to the centre from all directions. The latter function may have served ritual as well as protection purposes. Each of the centres included a large elite palace, the residence of lesser lords of the kingdom closely allied with the Holmul rulers.

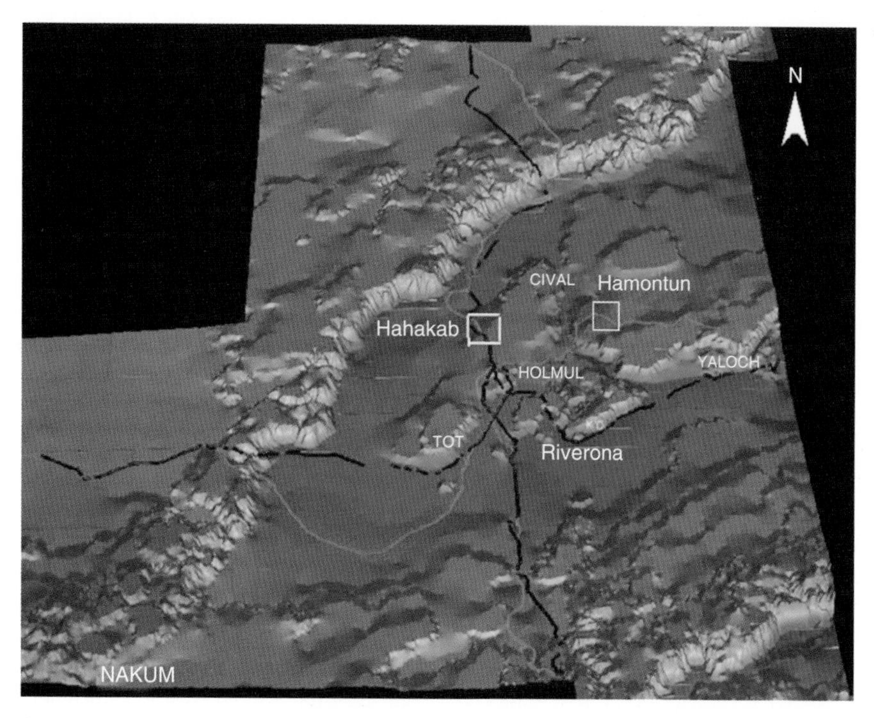

Figure 10.41 *Digital elevation model generated from a 1:50 000 topographic map showing paths (in red and black) of least resistance to Holmul (centre) and associated Maya sites. Squares indicate new Maya sites discovered by means of least-cost route analysis. Reproduced with permission from F. Estrada Belli, Archaeology Dept, Boston University.*

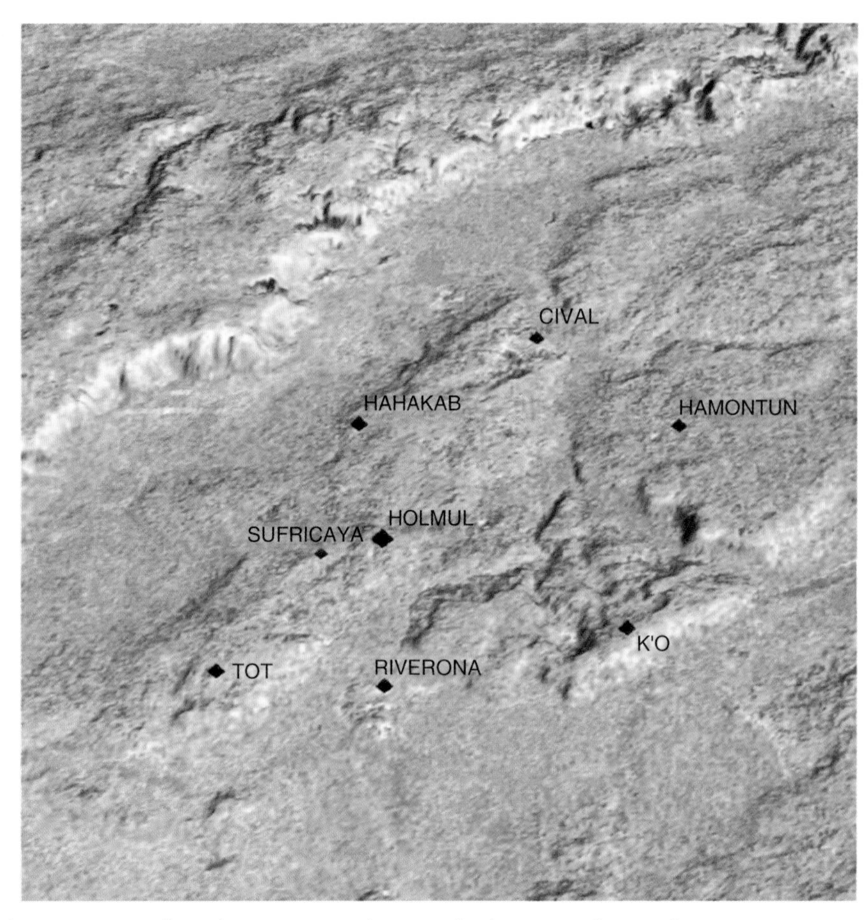

Figure 10.42 *Landsat TM image draped over a DEM showing the location of secondary Maya centres around Holmul, most of them situated along the edges of the upland plateau. Reproduced with permission from F. Estrada Belli, Archaeology Dept, Boston University.*

Aside from the least-cost path, another source of data led Estrada-Belli and Koch (2007) to locating the secondary centres of Holmul, a DEM with ground resolution of 10 m. This dataset was produced by interferometry on data obtained in 1999 for NASA by the STAR-3*i* AIRSAR radar instrument. The elevation data, when displayed in a three-dimensional tool such as the NVIZ visualization programme in the GRASS GIS software (Neteler and Mitasova, 2002) allows the user to identify anomalies in the ground cover of the Holmul region. Because the area is uniformly covered by 30 m tall trees, only the largest of Maya buildings can be detected using this procedure. These are typically pyramids and large platforms that range from 10 to 30 m in height and up to 100 m in width. In the radar imagery groups of tree crowns can be discriminated. Those trees that grow on pyramids and platforms appear as anomalous sharp peaks in the forest canopy (Figure 10.40). Clearly, in many cases, larger tree species can exist in the forest canopies and these also appear as peaks in the forest. These are typically rare, such as the mahogany or the ceiba trees. However, because these trees grow as individually and are surrounded by other smaller trees they can be easily separated from those anomalies reflecting Maya ruins. The pyramids lie under groups of several trees that appear to be taller than their surroundings. Using these expedient procedures Estrada-Belli and his team (Estrada-Belli, 2003, 2010; Estrada-Belli and Koch, 2007) were able to isolate with high precision the location of several of the ceremonial centres around Holmul that included large pyramids and platforms. These included Hahakab and Hamontun, whose taller structures were observed along the least-cost paths.

This case study demonstrates that GIS can be successfully used in archaeological studies, but (i) a high level of detailed subject-matter knowledge is required to define what can or cannot be seen from remotely-sensed imagery, (ii) a good knowledge of different types of remotely-sensed data and their properties is essential, leading to the intelligent use of GIS layers derived from remotely-sensed data, such as DEMs and their derivatives

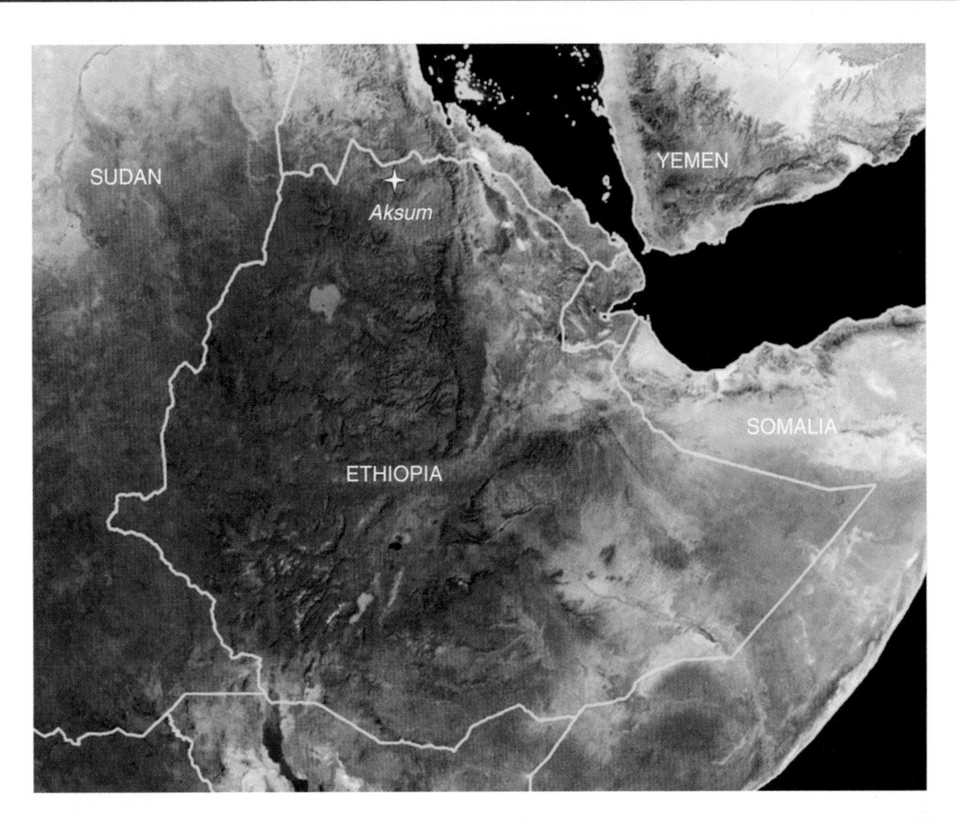

Figure 10.43 *Location of Aksum in Tigray region, northern Ethiopia. Data courtesy NASA World Wind.*

and (iii) the need to use data that have a spatial resolution that is appropriate for the phenomena of interest, such as the Maya temples in this case study.

10.10.3 Aksum (Ethiopia) Case Study

The Aksum region of Tigray province, in the northern highlands of Ethiopia (Figure 10.43) is of considerable importance in terms of the development of Holocene culture in the Horn of Africa. The earliest evidence of large-scale sedentary settlement in this region dates to the Pre-Aksumite period (about 800 to 300 BC) when the first towns and urban centres appeared (Michels, 1979, 1994, 2005). In the centuries that followed, the regional population grew and settlements proliferated as the Aksumite kingdom (about 300 BC to 800 AD) arose and eventually became a powerful empire with economic ties to the Graeco-Roman and Indian Ocean worlds (Bard *et al.*, 2000, 2003; Fattovich *et al.*, 2000; DiBlasi, 2005).

Today, however, Tigray is considered one of Ethiopia's most environmentally degraded regions (Feoli, Vuerich and Woldu, 2002a, 2002b). The ecosystem is very sensitive to climatic changes and the region as a whole is prone to severe erosion, depletion of the natural resource base, and frequent droughts (Feoli, Vuerich and Woldu, 2002a, 2002b; Machado, Pérez-González and Benito, 1998). Despite this region's important history and its potential as a study area for investigating the dynamics of environmental change and degradation, very few systematic regional studies have been conducted that integrate environmental and archaeological research to examine the history of human-environment interactions that contributed to Tigray's current environmental conditions (see Bard, 1997; Nyssen *et al.*, 2004; Sernicola, 2009).

Geospatial technology tools are opening new opportunities for archaeologists to integrate historical, cultural and environmental variables. Maps depicting geomorphology, hydrology, geology, vegetation and soil can be generated from satellite images and DEMs, and complemented by field visits. These maps can be used to identify environmental correlates of settlement and examine how these have changed through time. One example of such an approach is currently underway in the Aksum region in northern Ethiopia by the collaborative efforts of an international team of researchers from the United States, Italy, Spain, United Kingdom and Ethiopia (Bard *et al.*, 2000; Fattovich *et al.*, 2000; Schmid *et al.*, 2008; Ciampalini *et al.*, 2006; Sernicola, 2009; Sulas, Madella and French, 2009; French *et al.*, 2009).

The integration of GIS landscape analysis techniques with archaeological research enhances the ability to understand the history of landscape development and land use in the Aksum region. By integrating

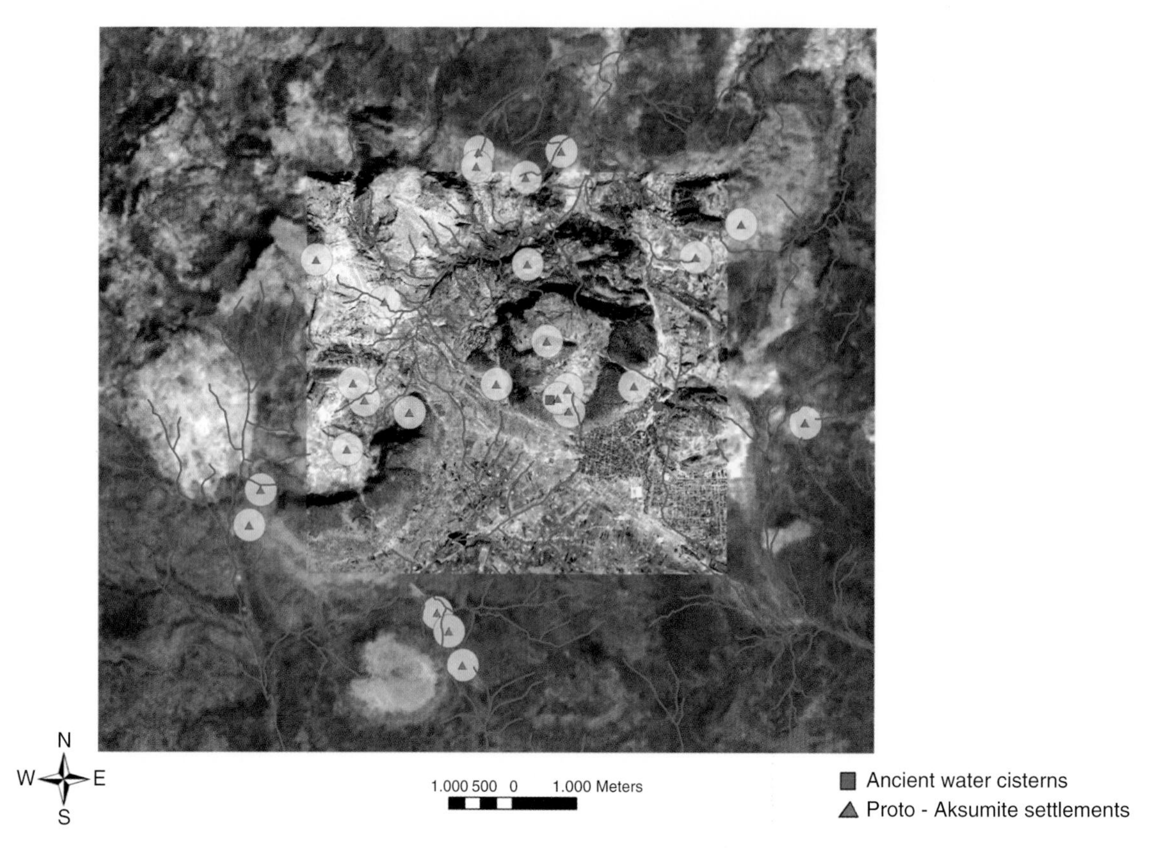

Figure 10.44 *Proximity analysis of Proto-Aksumite settlements (circa 400 to 50 BC) to water resources (streams and ancient cisterns) in Aksum, Ethiopia. Background images: Ikonos (inner square) and Landsat TM (outer square). Reproduced with permission from Luisa Sernicoli, University of Naples "L'Orientale", Dept of African and Arabian Studies, Pzza S. Domenico Maggiore 12, Naples, Italy.*

archaeological settlement distribution data with geoenvironmental maps in a GIS database, one can reconstruct and interpret how human populations – ancient and modern – interacted with the landscape and how those interactions changed over time. Using such an approach, archaeologists can model the dynamics of settlement growth and decline (Billari and Prskawetz, 2003; Dean *et al.*, 2000). Variables such as distance to a river, terrain type, accessibility, as well as proximity to other towns and coasts may play an important role in shaping settlement patterns. The following paragraphs describe the GIS correlation analyses carried out in Aksum mainly by the work of Sernicola (2009).

Aksum – located in the north–central sector of Tigray region, about 35 km south-west of Adwa (Figure 10.43) – flourished between the end of the first millennium BC and the first millennium AD as the capital city of the Aksumite kingdom. Even though the Aksum region has been the subject of archaeological study since the first decade of the twentieth century (Michels, 1979; Fattovich, 1992; Phillipson, 1998), the detailed reconstruction of ancient settlement dynamics in the area is a recent development. In fact, starting in 2000 an intensive and systematic archaeological survey was initiated with the aim of providing complete coverage of the entire Aksum territory (Bard *et al.*, 2000; Fattovich *et al.*, 2000; Ciampalini *et al.*, 2006; Sernicola, 2009). The main goal of the survey was to reconstruct the settlement history of the area in order to provide new insights for long-term analyses of human–environment interactions in this region. To this end, archaeological surveys and environmental studies have been supplemented by GISs analyses.

Quantitative and statistical analyses were conducted by querying archaeological records that were surveyed and stored in a GIS database. The database consisted of a vector file with the location of 700 surveyed archaeological sites and linked to several attribute tables with information on site location and description, typology and density of surface artefact assemblages and cultural and chronological attributions. This database was used to model the changes in settlement density, settlement dimension and location throughout the different cultural phases. For instance, the **`Nearest Neighbour`** spatial statistic tool available in the ArcGIS software was used to evaluate the type of settlement distribution (clustered or

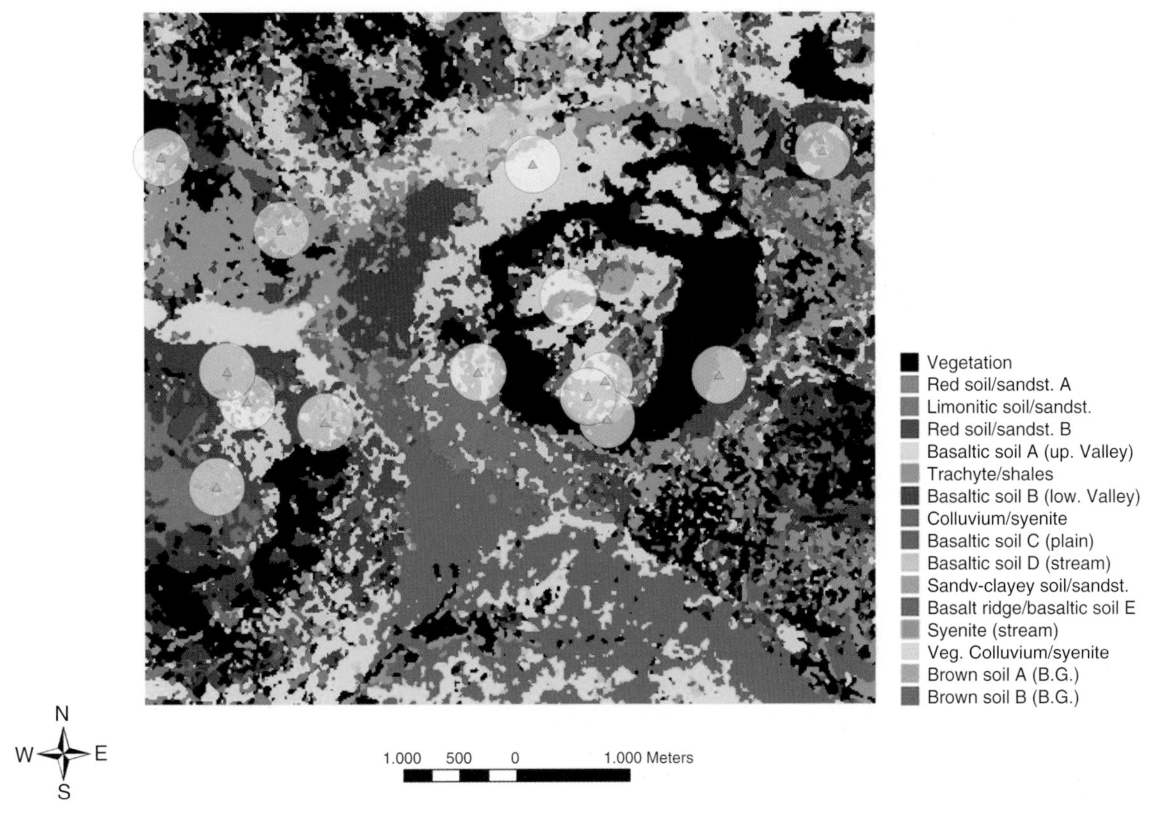

Figure 10.45 *Location of Proto-Aksumite settlements (circa 400 to 50* BC*) with respect to soils and rock types. More productive soils are generally those developed from syenite rocks (light violet class = vegetated colluvium/syenite) as they are rich in clay minerals. A supervised classification of ASTER image was used to obtain the surface lithology map (Schmid* et al.*, 2008). Reproduced with permission from Luisa Sernicoli, University of Naples "L'Orientale", Dept of African and Arabian Studies, Pzza S. Domenico Maggiore 12, Naples, Italy.*

dispersed) during the different cultural phases. This was done by measuring the distance between each site centroid and its nearest neighbour's centroid location, and by comparing the resulting average with the average of a hypothetical random distribution.

Factors that are commonly identified as being significant in the establishment of agricultural settlements in the Tygrean plateau of northern Ethiopia include water sources for domestic uses and animals, and soil fertility (Sulas, Madella and French, 2009; Sernicola and Sulas, in press). Therefore, proximity analyses like **`Buffer`** and **`Multiple Ring Buffer`** (in ArcGIS) were used to investigate whether settlement locations were selected according to water and productive soils availability. In the first case, buffer polygons were created around the sites of each phase (e.g. Proto-Aksumite sites shown in Figure 10.44) to a specified distance of 250 m (which is commonly adopted for catchment analyses conducted on sedentary rural settlements) in order to evaluate their proximity to rivers, streams or ancient water cisterns. In the second case the same type of buffer polygons was created around the settlements and analysed using as background the soil map elaborated by means of remotely sensed data coupled with ground surveys (Figure 10.45) (Schmid *et al.*, 2008).

Finally, viewshed analysis available in the **`3D Analyst`** extension of ArcGIS software was used to evaluate the visibility of a specific category of Middle Aksumite (about 350–550 AD) monuments traditionally called 'Mestah Worki' within the surrounding landscape. The results from the archaeological excavations conducted in one of the four Mestah Worki found in the Aksum territory suggest that this type of monuments represents a ritual structure, however, the occurrence of coins and tokens suggest that administrative activities were also performed at these sites (Fattovich, 2005). This hypothesis seems also to be confirmed by the local traditional name given to these monuments; that is Mestah Worki means 'the place where gold is spread'. The prominent position, the chronological classification, the occurrence of both ritual features and administrative devices, and the local name suggest that this type of structure, located in the northern sector of the plain of Aksum (Figure 10.46), played an important role as a landmark of the territory

Figure 10.46 *Viewshed analysis showing the visibility ranges (in beige) from four selected Middle Aksumite (circa 350–550 AD) monuments traditionally called 'Mestah Worki' and located along the four main valley systems leading to Aksum. Background images: Ikonos (inner square) and Landsat TM (outer square). Reproduced with permission from Luisa Sernicoli, University of Naples "L'Orientale", Dept of African and Arabian Studies, Pzza S. Domenico Maggiore 12, Naples, Italy.*

and perhaps as a place where foreigners or members of the Aksumite kingdom paid their tributes to the 'King of the Kings' of Aksum (Sernicola, 2009). Similar functions were probably carried out by all the other Mestah Worki found in the Aksum territory.

In order to determine the viewable areas from the four Mestah Worki monuments, Sernicola (2009) draped a Landsat TM satellite image over a DEM that was previously produced by digitizing and interpolating 20 m spaced contour lines from a 1:50 000 topographic map of the area. The `Viewshed` tool was then applied to identify the cells that could be seen from any one of the four Mestah Worki which were selected as the observer points with their respective heights as measured in the field with a hand-held GPS receiver (Figure 10.46). This type of GIS analysis seems to confirm the interpretation given to Mestah Worki monuments because of the prominent position and location of these structures right at the entrance of the four main valley systems leading to the capital city of Aksum from the west, north-west, north and north-east. Such data, together with the information resulted from the excavations of the Mestah Worki at Akeltegna (Fattovich, 2005), might give new insights into the interpretation of this still poorly known category of monuments (Michels, 2005; Sernicola, 2009).

10.10.4 Conclusions

Geospatial technologies are becoming an integral part in archaeological studies, primarily because of two reasons: (i) it is a non-destructive survey method which enables studying the sites while preserving them; and (ii) it is a tool that enables the integration of detailed excavation results with regional landscape studies. From the very beginning of GIS and RS development, archaeologists understood very well the great value of this mapping and analysis tool. In fact, archaeology was one of the first disciplines in exploring the use of aerial photography as a survey method in archaeological site investigations. Only recently has satellite imagery been adopted as a standard mapping and survey data source in archaeology, mainly because of the increased spatial resolution of recent satellite sensors that enables studying archaeological structures at submetre levels. Other characteristics of RS sensors, such as thermal and microwave sensing capabilities are also increasingly being adopted in archaeological studies because of their capability of detecting and mapping subsurface structures that are not necessarily visible from the ground. These types of sensors can play an important role in guiding and limiting ground excavation efforts, and by doing so are crucial in site preservation and maintenance.

From the practical examples illustrated in this chapter, it is clear that recent developments in RS and GIS technologies are ever more becoming intertwined as both offer complementary information sources as well as analysis tools. For additional application examples in RS the reader is referred to the MIPS exercises included in the publisher's web site, www.wiley.com/go/mather4.

Appendix A

Accessing MIPS

MIPS is a software package written by the senior author to implement many of the procedures described in this book. It is written mainly in Silverfrost Fortran 95, but some of the routines are compiled using Microsoft's Visual C++. Earlier versions were distributed on CD with the second and third editions. The latest edition of MIPS can now be downloaded by ftp from www.wiley.com/go/mather4. Simply download the file **`mips.zip`** into a temporary folder, say **`C:\mips_temp`**, and then use an unzipping program to unpack the files from **`mips.zip`**, so that folder **`C:\mips temp`** holds the zip file and its components. Arrange the files and subfolders according to type, and double click on the file **`install_mips.exe`**.

The MIPS installation program will then start. It was written when MIPS was distributed on CD so some references to CDs will be found – simply think of the folder **`C:\mips temp`** as holding the contents of the CD. You will then be asked to provide three pieces of information. The first is the install folder, or the name of the folder which is to contain all the MIPS executable files, plus the **`help`** and **`image`** subfolders. This install folder will be created for you. Unless you know what you are doing, enter **`C:\mips`** as the name of the install folder. Second, you must provide the name of the source folder. This is the folder into which **`mips.zip`** was unzipped, namely, **`C:\mips_temp`**. Finally, provide your name (ask someone if you don't know).

Press the Install button and you will be guided through a sequence of not-very-difficult questions. Quit the install program and read Appendices B and C of this book before proceeding.

MIPS has run successfully on Windows 2000, Windows Vista, Windows XP and Windows 7. However, the author cannot guarantee that it will work on any computer or operating system, and further states that the software comes without warranty, express or implied. It should only be used for teaching and research purposes.

Several exercises are contained in an Examples folder on the web site. These are pdf files prepared by Dr Koch and illustrating the use of MIPS in a series of case studies. The data for these case studies is also provided, as are the figures used in the examples. These figures are saved in TIFF and JPEG formats and can easily be converted to PowerPoint presentations.

Computer Processing of Remotely-Sensed Images: An Introduction, Fourth Edition Paul M. Mather and Magaly Koch

Appendix B

Getting Started with MIPS

The MIPS installation folder contains a text file called **readme.txt**. This file can be opened with Windows Notepad, and it contains all the information you need to install MIPS. In these worked examples, it is assumed that you have a directory structure as shown in Figure B.1. MIPS also creates a file called **mymips.ini** in your Windows directory (usually **c:\windows**). You should never edit or delete this file. File **readme.txt** also contains instructions on how to create a short-cut icon on your desktop to access MIPS. Assuming that you have created a short-cut icon on the Windows desktop, double-click on it and MIPS will start. First, accept the licence agreement. Next, note carefully the warning that MIPS allows a maximum of eight image windows to be open at any one time, that the maximum path plus filename length is 200 characters and that some modules refuse to accept paths/filenames with embedded blanks. Finally, use the **Open File Dialog Box**[1] to create a log file to hold details of your MIPS session. Remember that you can get help by clicking **Help|Help** on the main menu bar.

Before proceeding to display an image, we need to understand the way in which MIPS references and stores image data. The original MIPS format uses an image dictionary file to store the names (and other details) of one or more related image files (an image data set). Each of these image files is stored separately. In order to provide compatibility with the ENVI image processing software, a second storage format was added. This uses a single file to hold all of the image data (with band 1 preceding band 2, etc.). Such a file is said to be in band-sequential or BSQ format. Associated with the BSQ file is a header file with the suffix **.hdr**. Finally, a standard MIPS **INF** file provides access details. The two methods of storing and referencing image data sets are shown in Figure B.2.

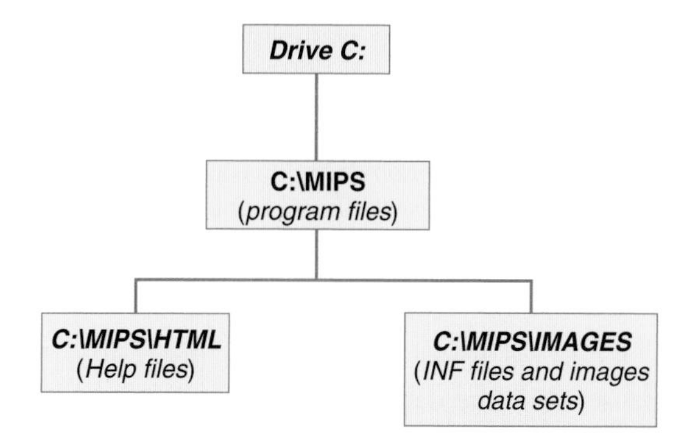

Figure B.1 *Folders in MIPS. The main folder,* `c:\mips`*, holds the executable programs (MIPS plus ancillary programs). The subfolder* `c:\mips\help` *stores the HTML files associated with the MIPS Help function while* `c:\mips\images` *holds a selection of images that are described in Appendix C.*

The MIPS toolbar is automatically switched on when you start MIPS for the first time. You can switch it on and off via the **Toolbar** menu item. MIPS remembers whether the toolbar is on or off at the end of the previous session. Some people like it; others don't, so you can customize it to suit your own tastes.

Either click the 'eye' icon on the toolbar (there is 'tooltip help' to identify the icons – just leave the cursor over the 'eye' for a second or two) or select **View|View Image**. Select **c:/mips/images** as your default folder. A list of MIPS **INF** files is provided in the **Open File Dialog Box**. Select **litcolorado.inf**. This **INF** file references a traditional MIPS image set. First, provide a title for the window (such as Little Colorado River). Then select **False Colour** from the drop-down list, and finally highlight and select the three image files to be combined to form a false colour image (this is done three times in RGB order, once for each of the three selected

[1] Some of the modules making up MIPS were compiled using Microsoft Visual C++ (VC). The version of VC used here does not allow spaces embedded in file or directory names so it is probably best to use underscores (_) to replace spaces.

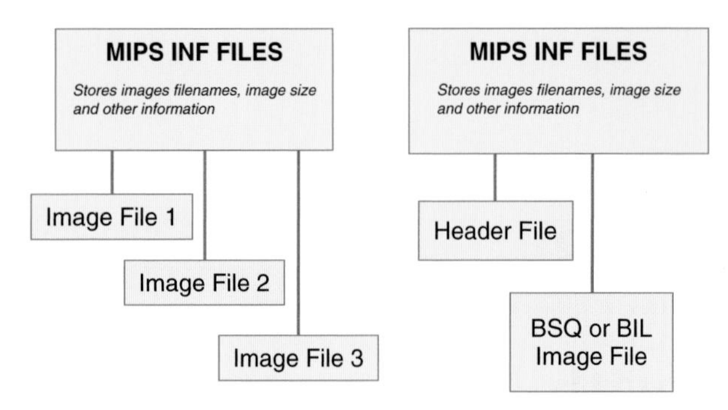

Figure B.2 *Showing the two types of INF file. The left-hand INF file holds details about the image dataset, and also stores the filenames of the original image bands. There would, for example be seven filenames (and paths) for a Landsat ETM+ image. The right-hand INF file stores the image data in one file (band sequential or band interleaved by line) and so holds one filename (plus image information). Also, there is a header file which is included for compatibility with the ENVI software package. ENVI will access the header and image files and so datasets can be swapped easily.*

files). I selected bands 7, 4 and 1 in that order. There is a pause while the data are read from disc, and a window then opens with the title 'Little Colorado River', and your false colour image is displayed. You may move the window around, or minimize it.

Now open **LaManchaETM.inf**. You will see an information message, telling you that this INF file references an ENVI 8-bit data file (8-bit means 256 levels). You can also see the size of the image file (512 lines × 512 pixels per line × 6 bands). Click **OK**. Now provide a window caption (for example 'La Mancha'), again select **False Colour** as the display type, and finally enter three file identification numbers in RGB order. These file identification numbers are the positions of the image files in the band sequential image data file. A value of 1 means the first image in the **bsq** file. Since there are six bands stored in the file **LaManchaETM bsq**, we cannot input a number greater than 6 or less than 1. I selected 4, 3, 2 (TM bands 4, 3 and 2) and, after a short pause while the data are read from disk, a window with the title 'La Mancha' appears, containing a false colour image.

Try to display a selection of images from the **C:\mips\images** directory. You should practise moving the windows around, closing them, minimizing them, and generally familiarizing yourself with starting the MIPS program and displaying both greyscale and false colour images.

Exit MIPS either by using the **EXIT** icon on the toolbar or clicking **File|Exit**. You do not need to close the windows containing the images before you exit, but it may sometimes be useful to close all of them at once. To do this, select **File|Close All Windows**.

Appendix C

Description of Sample Image Datasets

Acknowledgements

I am grateful to the following organizations and individuals for permission to use copyrighted material:

- NASA Observatorium (Joanie Straub, Curator) for permission to include the images marked 1. Some of these images were made available to the NASA Observatorium by Intermountain Digital Imaging, LC and CORE Software Technology, whose assistance is acknowledged.
- Images marked 2 are the copyright © ESA 1994, distributed by Eurimage Ltd.
- Mr. R. D. Freeman, DERA Radarsat Marketing Manager, kindly supplied the Radarsat image extracts which are indicated in the list below by the symbol 3. All of these images are © Canadian Space Agency/Agence spatiale canadienne, and were received by the Canada Centre for Remote Sensing (CCRS) and processed and distributed by Radarsat International. These images, plus a three full Radarsat scenes and over 30 other image extracts, are contained on the CD-ROM 'Radarsat Images', produced by Radarsat International in 1996.
- The SPOT images contained on this CD (and marked with the symbol 4) are copyright © CNES 1986/1998 – Spot Image Distribution 1982, and permission to use the data was kindly provided by Spot Image, 5 rue des Satellites, BP 4359, F-31030 Toulouse Cedex, France. I am grateful to Mme Isabelle Guidolin for her patience.
- Images marked 5 were obtained from the US Geological Survey, EROS Data Center, Sioux Falls, SD, USA.

Using the Image Datasets

The images listed below are contained in the subfolder **mips/images** in the installation file. When the installation starts, an opening page is displayed, and you can select **install MIPS** from the main menu. The MIPS installer gives you the option to transfer the example image files from the installation file to your hard disk. If you select this option then all of the image dictionary (**INF**) files, as described in Appendix B, are changed so that the new location of the image datasets is correctly recorded. This is the easiest option.

Alternatively, you can use Windows Explorer to copy the **/images** subfolder from the installation file to your hard disk (let's say you choose to move them to **C:/mips/images**). The image dictionary (**INF**) files will need to be updated, as they are set up to operate from my CD drive, which has the drive letter **G**. For example, the **rio.inf** file contains the following information:

```
IMAGE
G:/images/rio1.img
G:/mages/rio2.img
G:/images/rio3.img
G:/images/rio4.img
G:/images/rio5.img
G:/images/rio6.img
G:/images/rio7.img
-1 -1
255 255 255 255 255 255 255
5 8 1985
{Blank Line}
G:/images/rio.hst
{Blank Line}
```

Computer Processing of Remotely-Sensed Images: An Introduction, Fourth Edition Paul M. Mather and Magaly Koch

Note that the expression {Blank Line} means leave a blank record.

If I want to move this **INF** file manually to the **C:/mips/images** folder then I must firstly copy it from its location at **G:/images** to **C:/mips/images** using Windows Explorer then edit it so that it looks like this:

```
IMAGE
C:/mips/images/rio1.img
C:/mips/images/rio2.img
C:/mips/images/rio3.img
C:/mips/images/rio4.img
C:/mips/images/rio5.img
C:/mips/images/rio6.img
C:/mips/images/rio7.img
-1 -1
255 255 255 255 255 255 255
5 8 1985
{Blank Line}
C:/mips/images/rio.hst
{Blank Line}
```

There is a MIPS module under **File|INF File Operations** that will do these changes automatically. It will only work with 'classic' **INF** files, and not with the newer type of **INF** file. See the online **Help** in MIPS for more details.

Descriptions of Image Datasets Included on the CD

1. Mississippi River[1]
2. Little Colorado River[1]
3. London[1]
4. Paris[1]
5. San Joaquin Valley, California[1]
6. Morro Bay, California[1]
7. Candlewood Lake, Connecticut[1]
8. Rio de Janeiro, Brazil[1]
9. Nottingham[4]
10. Chott el Guettar, Tunisia[5]
11. Gregory Rift Valley, Kenya[5]
12. Littleport, Cambridgeshire[2]
13. Los Monegros, NE Spain[2]
14. Fry Canyon, Arizona/Utah[2]
15. Tanzanian Coast[5]
16. The Camargue, France[4]
17. Red Sea Hills, Sudan[5]
18. Radarsat images: East Anglia, The Netherlands, Indonesia and western Canada[3]
19. AVHRR browse data, Australia
20. Landsat TM, The Wash, eastern England[2]

1. **Mississippi River**: The centre latitude and longitude of this Landsat TM image set is 34°46′N, 90°27′W, a point to the south-west of Memphis, Tennessee. The date of acquisition is 13 January 1983. The image has 512 rows, 512 columns and 7 bands. The image set is referenced by the file **missis.inf**.
2. **Little Colorado River**: The centre latitude and longitude of this Landsat TM image set is 36°12′N, 111°47′W, to the east of the San Francisco peaks, Arizona. Here the Little Colorado River flows NNE across the Painted Desert. The date of acquisition is 24 August 1985. The image has 512 rows, 512 columns and 7 bands. The image set is referenced by the INF file **litcolorado.inf**.
3. **London**: This Landsat TM image set covers the centre of London (51°30′N, 0°20′W). The date of acquisition is 18 August 1984. The image has 512 rows, 512 columns and 7 bands. The image set is referenced by the INF file **london.inf**.
4. **Paris**: This Landsat TM image set covers the centre of Paris (48°50′N, 2°20′E), and was acquired on 9 May 1987. The size of the mages is 512 lines of 512 pixels. The image set is referenced by the INF file **paris.inf**.
5. **San Joaquin Valley, California**: The centre of this Landsat TM image set is at the point 35°11′N, 119°06′W, near Fresno, California. The date of acquisition is 15 September 1986. The image has 512 rows, 512 columns and 7 bands. The image set is referenced by the INF file **san-joaq.inf**.
6. **Morro Bay, California**: The centre of this Landsat TM image set is at the point 35°21′N, 120°49′W, to the north of San Luis Obispo, California. The date of acquisition is 19 November 1984. The image has 512 rows, 512 columns and 7 bands. The image set is referenced by the INF file **morrobay.inf**.
7. **Candlewood Lake, Connecticut**: The centre of this Landsat TM image set is at the point 41°30′N, 73°30′W, in the state of Connecticut. Two image sets are provided. The first was acquired in the summer, on 10 June 1984, and the second in autumn, on 9 October 1986. Both images have 512 rows, 512 columns and 7 bands. The summer image set is referenced by the INF file **cans.inf** and the autumn (fall) data are referenced in **canf.inf**.
8. **Rio de Janeiro, Brazil**: This Landsat TM image set shows the city of Rio de Janeiro, Brazil (23°S, 43°W), and its surroundings. The date of acquisition is 5 August 1985. The image has 512 rows, 512 columns and 7 bands. The image set is referenced by the file **rio.inf**.
9. **Nottingham**: This multispectral SPOT image covers the area around the city of Nottingham

(52°50′N, 1°10′W). The image has 1024 rows and 1024 columns. The date of acquisition is unknown. The name of the file that references this image set is **`notspot1.inf`**.

10. **Chott el Guettar, Tunisia**: The Chott el Guettar is a dry saline lake located near 33°50′N, 8°30′E in Tunisia. The area shown on this 512 × 512 pixel subimage covers part of the salt lake bed, with the southern slopes of the Djebel Ortaba to the north of the lake. The main road from Gabes to Gafsa runs along the northern side of the saline lake in a general NW-SE direction. Three Landsat TM bands (2, 3 and 4) are provided, and these are referenced in the file **`egmar.inf`**. The date of image acquisition is not known.
11. **Gregory Rift Valley, Kenya**: This TM image set contains threeLandsat TM bands (3, 5 and 7) and is 512 × 512 pixels in size. The Gregory rift valley is part of the East African Rift system. The area shown on this image lies to the north of Lake Baringo and south of Silali volcano, at approximately 1°N, 36°E. This area is sparsely vegetated and several lava flows of different ages and compositions are clearly apparent. The eroded area to the west of the small volcanic cone visible in the lower right of the image is a pyroclastic deposit. Alluvial areas extend on either side of the river that flows eastwards across the northern part of the image area. The image acquisition date is 30 July 1884. The three TM bands are referenced by the file **`kenya.inf`**.
12. **Littleport, Cambridgeshire**: The small town of Littleport is located about 5 km north of the city of Ely, Cambridgeshire, UK, at 52°25′N, 0°20′E. The main features of the area shown on the image are the River Ouse, running NE on the right-hand side of the image, and the parallel Old and New Bedford Rivers, also running in a NE direction. This area is low-lying (around sea level) and flat. The blue areas on the band 432 colour composite are either ploughed fields or towns and villages. The main crops in this fertile region of the Fens are wheat, barley and sugar beet. The image area contains a few small clouds and associated shadows. Six of the seven Landsat TM bands are included (band 6 is excluded). Image size is 512 × 512 pixels. The file for this image set is **`littlept.inf`**.
13. **Los Monegros, NE Spain**: The Los Monegros region lies in the Spanish province of Aragon. It is a semi-arid and sparsely populated upland region, with an average altitude of 350 m. Recent developments include the building of irrigation canals to bring water for agriculture from the Flumen river system to the north of the area shown on this Landsat TM image, acquired on 7 July 1997. The image covers the area between the village of Bujaraloz (in the top right corner of the image) and the northern slopes of the Ebro Valley. The approximate latitude and longitude of the image centre are 41°30′N and 0°15′W. The Laguna la Playa, which is generally dry in summer, lies to the south of Bujaraloz. The southern part of the area is mainly dryland farming (wheat and barley). The north-west part of the image covers a dissected upland area covered by low coniferous trees. Bands 1–7 of this Landsat TM image are referenced by the file **`LosMoneg97.inf`**. Bands 1–5 and 7 of a Landsat TM sub-image, acquired in 1991 and covering approximately the same area, is referenced in **`LosMoneg91.inf`**. Differences between the 1991 and 1997 images are due to changes in agriculture, including crop rotation and extension of the ploughed area, and differences in the weather conditions in the days preceding image acquisition. The period before the 1997 image was acquired was particularly wet.
14. **Fry Canyon, Arizona/Utah**: This four-band Landsat MSS image has been sub-sampled (taking every sixth pixel on every fourth line). It covers the area from the north of the Grand Canyon in Arizona (around the latitude of the city of Page) to the Canyonlands National Park, south of Moab, Utah (approximate centre position: 36°N, 111°W). Lake Powell is a prominent water feature in the south of the region. The main tributary of the Colorado River, the San Juan, joins near the northern end of Lake Powell. The upland areas to the west of the Colorado include the snow-covered Henry Mountains (well-known to readers of John Wesley Powell's famous monograph on their geology and geomorphology), the Waterpocket Fold, the Kaiparowitz Plateau and Smoky Mountain. Navajo Mountain lies to the south of the Colorado–San Juan confluence. The mountain ridges are vegetated, mainly by low trees, and the lowland areas are semi-arid. The file **`fry.inf`** references four Landsat images, the MSS bands labelled 4, 5, 6 and 7.
15. **Tanzanian Coast**: This three-band Landsat MSS image set covers an area of the Tanzanian coast south of Dar-es-Salaam and north of Mafia Island, at a latitude and longitude of 7°30′S, 39°25′W. The northern part of the coast is fringed with coral reefs, and a small coral island is apparent in the lower centre of the image. The reddish colours over the land indicate forest. Brownish and whitish areas have been cleared. The coastal waters are clear, and the colour variations here refer to water depth as the light in MSS bands 4 and 5 is reflected from

the sea bed. The edge of the continental shelf is clearly visible. This image set is 512 × 512 pixels in size and is referenced by **tanzcoas.inf**. Note that Landsat MSS bands 4, 5 and 7 are provided, and that the dynamic range of these bands is 7-bit for bands 4 and 5, and 6-bit for band 7.

16. **The Camargue, France**: Two SPOT multispectral images, each of 512 × 512 pixels, show the area to the north of the town of Tarascon in the Camargue area of the Rhône delta in the south of France (43°50′N, 4°45′E). The first image set (referenced by **camarg.inf**) was collected on 12 January 1987. The second (**camarg1.inf**) was collected on 17 January of the same year. The two image sets show the changes that occurred between these two dates, largely the result of a comprehensive snowfall. These images are extracts from a demonstration CD ('SPOT Scene') produced by SPOT Image.
17. **Red Sea Hills, Sudan**: The Red Sea Hills are located in eastern Sudan (between 18° and 19°N latitude and around 36–38°E longitude). The area is described by Koch and Mather (1997; see also Mather, Tso and Koch, 1998) as a 200 km wide range of mountains, rising steeply from the coastal plain to elevations of 1000 m and more. The sandy Nubian Desert lies to the west. The upland area is heavily dissected by a network of drainage channels called khors. The underlying rock is volcanic and granitic, with some limestone areas, and is heavily faulted with the main directions of faulting being N–S, with a subsidiary E–W trend. Two image sets are provided. One is a degraded version of a Landsat TM image from 1984, which has been resampled at a 6-pixel interval in both rows and columns and then padded with zeros to give an image size of 1024 × 1024 pixels. The second is a full resolution extract from the same image, and is also 1024 × 1024 pixels in size. The degraded image is referenced by the dictionary file **sudanlo.inf**, while the full resolution image is referenced by **sudanhi.inf**.
18. **Radarsat Images**: Four Radarsat image sets are included on this CD. Each of the images is © Canadian Space Agency/Agence spatiale canadienne, and was received by the CCRS and processed and distributed by RADARSAT International. The image dictionary files **hengelo.inf**, **indonesia.inf** and **okanagan.inf** each reference a single RADARSAT SAR image which is 512 × 512 pixels in size. The **hengelo** image covers an agricultural area near the town of Hengelo, in The Netherlands. The **indonesia** image shows a volcanic peak on the island of Java, while the **okanagan** image is of a forested area around Okanagan in British Columbia, Canada. The **east_anglia.inf** image dictionary file references four Radarsat images, each 512 × 512 pixels in size, covering an area of East Anglia (UK) to the south-west of Lakenheath. These four images were collected at different times during the crop-growing season. They have been geometrically corrected, and can therefore be overlaid. Colour differences then indicate differential growth, which affects the back-scattering properties of the surface. I am grateful to Mr. G. Gill for providing these images.
19. **AVHRR, Australia**: This is a browse image of a five-band NOAA-14 scene covering parts of New Guinea, Northern Territory and Queensland. The area of the image is approximately 2000 by 4500 km, giving a pixel size of around 5 km across the scan lines (columns) and 4 km in the along-track (row) direction. The date of image capture was 20 July 1996 at 04.31 GMT. I am grateful to Susan Campbell of the CSIRO Earth Observation Centre (EOC), Canberra, Australia, for her help in providing the image data. The data were collected by the Australian Institute of Marine Science (AIMS) in Townsville, Queensland. The scale of this image is roughly that of the Global Area Coverage (GAC) data, and shows the problems involved in obtaining cloud-free NDVI composites from daily AVHRR data. The image dictionary file for this image set is **avhrr.inf**.
20. **Landsat TM, The Wash, eastern England**: Four of the seven TM bands are included in this image extract – bands 2, 3, 4 and 5. The image size is 707 scan lines with 800 pixels per line. The image, which has been geometrically registered to the UK National Grid, covers the southern end of The Wash, a shallow area of sea on the east coast of England. The Great Ouse River passes through the town of Kings Lynn before discharging into The Wash. To the north-west of the river the coastal lands are salt marshes and reclaimed marshland, which provide excellent agricultural land. King John of England lost his country's treasury in these marshes in the early thirteenth century. It has never been recovered. The area to the east of the river is more undulating with forested low hills, within which Sandringham House, one of HM Queen Elizabeth II's residences, is located. The four TM bands for this 1990 image are referenced by the INF file **wash90.inf**.

Appendix D

Acronyms and Abbreviations

ADEOS	Advanced Earth Observing Satellite (Japan)
ALI	Advanced Land Imager (NASA)
ALOS	Advanced Land Observing Satellite (Japan)
ANN	Artificial neural network
ASAR	Advanced SAR (Envisat)
ASTER GDEM	ASTER Global DEM
ASTER	Advanced Spaceborne Thermal Emission and ReflectiveSpectrometer
ATREM	ATmospheric REMoval (model)
AU	Astronomical Unit
AVHRR	Advanced Very High Resolution Radiometer
AVIRIS	Airborne Visible and Infrared Spectrometer (JPL)
BDRF	Bidirectional Reflectance Distribution Function
BIL	Band Interleaved by Line
BIP	Band Interleaved by Pixel
BNSC	British National Space Centre (now UK Space Agency)
BSQ	Band Sequential
BU-CRS	Boston University Center for Remote Sensing
CCD	Charge Coupled Device
CEOS	Committee on Earth Observation Satellites
CHRIS	Compact High Resolution Imaging Spectrometer
CNES	Centre National d'Etudes Spatiales (French Space Agency)
CO_2	carbon dioxide
COSMO/Skymed	COnstellation of small Satellites for Mediterranean basin Observation
CW	Continuous Wave
CZCS	Coastal Zone Colour Scanner
DAIS	Digital Airborne Imaging Spectrometer
DEM	Digital Elevation Model
DFT	Discrete Fourier Transform
DInSAR	Differential INSAR
DIODE	Détermination Immédiate d'Orbite par DORIS Embarqué (Instantaneous determination of orbit using onboard DORIS)
DLG	Digital Line Graph
DLR	German Space Agency
DORIS	Doppler Orbitography and Radiopositioning Integrated by Satellite
DSM	Digital Surface Model

Computer Processing of Remotely-Sensed Images: An Introduction, Fourth Edition Paul M. Mather and Magaly Koch

DT	decision tree
DWT	Discrete Wavelet Transform
DXF	Drawing Exchange Format
EIFOV	Effective Instantaneous Field Of View
EME	ElectroMagnetic Energy
EMR	ElectroMagnetic Radiation
EO-1	Earth Observing Mission 1 (USA)
ERE	Effective Resolution Element
ERS	European Remote sensing Satellite
ESA	European Space Agency
ESRI	Environmental Systems Research Institute
ETM+	Enhanced Thematic Mapper Plus (Landsat-7, USA)
FFT	Fast Fourier Transform
FOV	Field of View
FWHM	Full Width Half Maximum
GCM	global circulation model
GCP	Ground Control Point
GDEM	Global DEM (ASTER)
GIS	Geographical Information System
GLAS	Geoscience Laser Altimeter System
GLCM	Grey Level Co-occurrence Matrix
GPS	Global Positioning System
GRASS	Geographic Resources Analysis Support System
H_2O	water/water vapour
HRG	High Resolution Geometric (SPOT)
HRS	High Resolution Stereoscopy (SPOT)
HRV	High Resolution Visible (SPOT)
HRV-IR	High Resolution Visible and InfraRed (SPOT)
HSI	Hue-Saturation-Intensity
HSV	Hue-Saturation-Value (same as HSI)
HyMap	Hyperspectral Mapper
ICESat	Ice, Cloud and land Elevation SATellite
ICA	Independent Components Analysis
ID	IDentifier
IEEE	Institute of Electrical and Electronic Engineers
IFOV	Instantaneous FOV
IGARSS	International Geoscience and Remote Sensing Symposium (IEEE)
InSAR	Interferometric SAR
IR	Infrared
IRS	Indian Remote Sensing Satellite
ISPRS	International Society for Photogrammetry and Remote Sensing
ISS	International Space Station
JERS	Japanese Earth Resources Satellite
JPL	NASA Jet Propulsion Laboratory, Pasadena, California
LAI	Leaf Area Index
LDCM	Landsat Data Continuity Mission
Lidar	Light Detection and Ranging
LISS	Linear Self Scanning Sensor (India)
m	metre
MAD	Median of Absolute Deviations
MCDA	Multiple Criteria Decision Analysis
MISR	Multi-angle Imaging SpectroRadiometer (US)
MODIS	Moderate Resolution Imaging Spectrometer
MOS	Marine Observation Satellite (Japan)

MSS	Multispectral Scanner
MTF	Modulation Transfer Function
NASA	National Aeronautics and Space Administration
NPOESS	National Polar-orbiting Operational Environmental Satellite System
NDVI	Normalised Difference Vegetation Index
NEST	Next European Space Agency Synthetic Aperture Radar Toolbox
NEXRAD	Next Generation Radar (NOAA)
NIR	Near Infrared
NLM	NonLinear Mapping
nm	nanometre
NOAA	National Oceanic and Atmospheric Administration (USA)
O_2	oxygen
OCTS	Ocean Colour and Temperature Sensor
OO	Object Oriented
OSAVI	Optimized Soil-Adjusted Vegetation Index
OSTM	Ocean Surface Topography Mission
PALSAR	Phased Array L-band SAR (ALOS)
PCA	Principal Components Analysis
PPI	Pixel Purity Index
PRARE	Precise Range and Range Rate Equipment
PSF	Point Spread Function
PV	Pixel Value
PVI	Perpendicular Vegetation Index
Radar	Radio Detection and Ranging
RAM	Random Access Memory
RGB	Red-Green-Blue
RMSE	Root Mean Square Error
RS	Remote Sensing
SAM	Spectral Angle Mapping
SAR	Synthetic Aperture Radar
SAVI	Soil Adjusted Vegetation Index
SeaWIFS	Sea-viewing Wide Field of View Sensor (USA)
SG	Savitzky–Golay
SHOALS	Scanning Hydrographic Operational Airborne Lidar Survey
SIR	Shuttle Imaging Radar
SLAR	Side-Looking Airborne Radar
SLC	(i) scan line corrector (Landsat ETM+) (ii) Single-Look Complex (SAR image)
SLIM-6	Surrey Linear Imager-6
SOM	Self Organising Map
SPOT	Satellite Pour l'Observation de la Terre (Earth Observation Satellite) (France)
SRTM	Shuttle Radar Topographic Mission
SSTL	Surrey Space Technology Ltd.
SVD	Singular Value Decomposition
SVM	Support Vector Machine
SWIR	ShortWave InfraRed
TDRS	Tracking and Data Relay Satellite System
TIFF	Tagged Image File Format
TIN	Triangulated Irregular Network
TIR	Thermal Infrared
TM	Thematic Mapper (Landsat-5, USA)
TSAVI	Transformed Soil-Adjusted Vegetation Index
TRMM	Tropical Rainfall Mapping Mission
UAE	United Arab Emirates
UAV	Unmanned Aerial Vehicle

USGS United States Geological Survey
UT Universal Threshold
VI Vegetation Index
VIIRS Visible/Infrared Imager Radiometer Suite
VIS VISible
VNIR Visible and Near Infrared
Wm^{-2} sr^{-1} Watts per square metre per steradian
Mm micrometre

References

Abd El-Wahab, R.H., Zayed, A.M.Z., Moustafa, A.A. *et al.* (2006) Landforms, vegetation, and soil quality in south Sinai, Egypt. *Catrina*, **1**, 127–138.

Abe, S. (2005) *Support Vector Machines for Pattern Classification*, Springer-Verlag, London.

Abonyi, J. and Feil, B. (2007) *Cluster Analysis for Data Mining and System Identification*, Springer-Verlag, Berlin.

Abramowitz, M. and Stegun, I.A. (eds) (1972) *Handbook of Mathematical Functions*, Dover Books, New York.

Abrams, M. (2000) The advanced spaceborne thermal emission and reflection radiometer (ASTER): data products for the high spatial resolution imager on NASA's terra platform. *International Journal of Remote Sensing*, **21**, 847–859.

Ackermann, F. (1984) Digital image correlation: performance and potential application in photogrammetry. *Photogrammetric Record*, **64**, 429–439.

Adams, J.B. and Gillespie, A.R. (2006) *Remote Sensing of Landscapes with Spectral Images*, Cambridge University Press, Cambridge.

Adams, J.B., Smith, M.O. and Gillespie, A.R. (1989) Simple models for complex natural surfaces: a strategy for the hyperspectral era of remote sensing. *Proceedings of the IEEE International Geoscience and Remote Sensing Symposium, IGARSS'89, 10–14 July 1989, Vancouver, British Columbia*, vol. 1, IEEE Press, Piscataway.

Adams, J.B., Smith, M.O. and Gillespie, A.R. (1993) Imaging spectroscopy: interpretation based on spectral mixture analysis, in *Remote Geochemical Analysis: Elemental and Mineralogical Composition* (eds C.M. Pieters, and P.A.J. Englert), Cambridge University Press, Cambridge, pp. 145–166.

Adams, M.L., Philpot, W.D. and Norvell, W.A. (1999) Yellowness index: an application of spectral second derivatives to estimate chlorosis of leaves in stressed vegetation. *International Journal of Remote Sensing*, **20**, 3663–3675.

Addison, P.S. (2002) *The Illustrated Wavelet Transform Handbook: Introductory Theory and Applications in Science, Engineering, Medicine and Finance*, Institute of Physics Publishing, Bristol.

Aguilar, M.A., Aguera, F., Aguilar, F.J. and Carvajal, F. (2008) Geometric accuracy assessment of the orthorectification process from very high resolution satellite imagery for Common Agricultural Policy purposes. *International Journal of Remote Sensing*, **29**, 7181–7197.

Aha, D.W. and Bankert, R.L. (1996) A comparative evaluation of sequential feature selection algorithms, in *Learning from Data: Artificial Intelligence and Statistics V* (eds D. Fisher, and J.-H. Lenz), Springer-Verlag, New York, pp. 199–206.

Aiazzi, B., Alparone, L., Baronti, S. *et al.* (2007) Context-sensitive pan-sharpening of multispectral images, in *Proceeedings of Semantic Multimedia: Second International Conference on Semantic and Digital Media Technologies, SAMT 2007, Genoa, Italy, December 5–7, 2007* (eds B. Falcidieno, M. Spagnuolo, Y. Avrithis, *et al.*), Springer-Verlag, Berlin, pp. 121–125.

Aitkenhead, M.J. and Aalders, I.H. (2008) Classification of landsat thematic mapper imagery for land cover using neural networks. *International Journal of Remote Sensing*, **29**(7), 2075–2084.

Alberga, V. (2007) A study of land cover classification using polarimetric SAR parameters. *International Journal of Remote Sensing*, **28**, 3851–3870.

Alberga, V., Satalino, G. and Staykova, D.K. (2008) Comparison of polarimetric SAR observables in terms of classification performance. *International Journal of Remote Sensing*, **29**, 4129–4150.

Albertini, G. and Ponte, S. (1996) Three dimensional digital elevation model of Mt Vesuvius from NASA/JPL TOPSAR. *International Journal of Remote Sensing*, **17**, 1797–1801.

Albrecht, J. (2007) *Key Concepts and Techniques in GIS*, Sage Publications, Ltd, London.

Aleksander, I. and Morton, J. (1990) *An Introduction to Neural Computing*, Chapman & Hall, London.

Al-Hinai, K.G., Khan, M.A. and Canaas, A.A. (1991) Enhancement of sand dune texture from Landsat imagery using difference of Gaussian filter. *International Journal of Remote Sensing*, **12**, 1063–1069.

Alley, R.E. (1995) Algorithm Theoretical Basis Document, Version 2.0, March 1, 1995, Jet Propulsion Laboratory, Pasadena.

Alparone, L., Baronti, S., Carla, R. and Pugilisi, C. (1996) An adaptive order-statistics filter for SAR images. *International Journal of Remote Sensing*, **17**, 1357–1365.

Al-Rousan, N. and Petrie, G. (1998) System calibration, geometric accuracy testing and validation of DEM and orthoimage data extracted from SPOT stereopairs using commercially available image processing systems. *International Archives of Photogrammetry and Remote Sensing*, **32**, 8–15.

Al-Rousan, N., Cheng, P., Petrie, G. *et al.* (1997) Automated DEM extraction and orthoimage generation from SPOT Level 1B imagery. *Photogrammetric Engineering and Remote Sensing*, **63**, 965–974.

Amarsaikhan, D. and Douglas, T. (2004) Data fusion and multisource image classification. *International Journal of Remote Sensing*, **25**, 3529–3539.

Amolins, K., Zhang, Y. and Dare, P. (2007) Wavelet based image fusion techniques – An introduction, review and comparison. *ISPRS Journal of Photogrammetry and Remote Sensing*, **62**, 249–263.

Anderson, E.S., Thompson, J.A. and Austin, R.E. (2005) LIDAR density and linear interpolator effects on elevation estimates. *International Journal of Remote Sensing*, **26**, 3889–3900.

Andreadis, I., Glavas, E. and Tsalides, Ph. (1995) Image enhancement using colour information. *International Journal of Remote Sensing*, **16**, 2285–2289.

Anon (undated) Spectral Response for DigitalGlobe WorldView 1 and WorldView 2 Earth Imaging Instruments. http://digitalglobe.com/downloads/WV1_WV2_SpectralResponse.pdf (accessed 23 January 2009).

Anselin, L., Syabri, I. and Kho, Y. (2006) GeoDa: An introduction to spatial data analysis. *Geographical Analysis*, **38**, 5–22.

Antonarakis, A.S., Richards, K.S. and Brasington, J. (2008) Object-based land cover classification using airborne LiDAR. *Remote Sensing of Environment*, **112**, 2988–2998.

Anuta, P.E. (1970) Spatial registration of multispectral and multitemporal digital imagery using fast fourier transform techniques. *IEEE Transactions on Geoscience Electronics*, **8**, 353–368.

Aplin, P. (2004) Remote sensing as a means of ecological investigation. Proceedings of the XXth Congress of the International Society for Photogrammetry and Remote Sensing (ISPRS), Istanbul, Turkey, Vol. 29(1) (ed. M.O. Altan), p. 104.

Aplin, P. (2005) Progress report – remote sensing: ecology. *Progress in Physical Geography*, **29**, 104–113.

Aplin, P. (2006) On scales and dynamics in observing the environment. *International Journal of Remote Sensing*, **27**, 2123–2140.

Arbia, G., Griffith, D.A. and Haining, R.P. (2003) Spatial error propagation when computing linear combinations of spectral bands: the case of vegetation indices. *Environmental and Ecological Statistics*, **10**, 375–396.

Arce, G.R. (2004) *Nonlinear Signal Processing: A Statistical Approach*, John Wiley & Sons, Inc., New York.

Ardö, J., Pilesjö, P. and Skidmore, A. (1997) Neural networks, multitemporal Landsat Thematic Mapper data and topographic data to classify forest damage in the Czech Republic. *Canadian Journal of Remote Sensing*, **23**, 217–219.

Arévalo, V. and González, J. (2008) An experimental evaluation of non-rigid registration techniques on Quickbird satellite imagery. *International Journal of Remote Sensing*, **29**, 513–527.

Arkin, Y., Ichoku, C. and Karnieli, A. (1999) Fault traces in the arid Arava valley floor, Israel, revealed by RADARSAT surface roughness classification. *Canadian Journal of Remote Sensing*, **25**, 302–310.

Armour, B., Tanaka, A., Ohkura, H. and Saito, G. (1998) Radar interferometry for environmental change detection, in *Remote Sensing Change Detection: Environmental Monitoring Methods and Applications* (eds R.S. Lunetta, and C.D. Elvidge), Ann Arbor Press, Chelsea, Michigan, pp. 245–279.

Arnaud, M. (1994) The SPOT programme, in *TERRA 2: Understanding the Terrestrial Environment – Remote Sensing Data Systems and Networks* (ed. P.M. Mather), John Wiley & Sons, Ltd, Chichester, pp. 29–39.

Aronoff, S. (2005) *Remote Sensing for GIS Managers*, ESRI Press, Redlands.

Askne, J., Dammert, P.B.G., Ulander, L.M.H. and Smith, G. (1997) C-band repeat-pass interferometric SAR observations of the forest. *IEEE Transactions on Geoscience and Remote Sensing*, **35**, 25–35.

Askne, J., Santoro, M., Smith, G. and Fransson, J.E.S. (2003) Multitemporal repeat-pass SAR interferometry of boreal forests. *IEEE Transactions on Geoscience and Remote Sensing*, **41**, 1540–1550.

Aspinall, R.J., Marcus, W.A. and Boardman, J.W. (2002) Considerations in collecting, processing, and analysing high spectral resolution hyperspectral data for environmental investigations. *Journal of Geographical Systems*, **4**, 15–29.

Asrar, G. (ed.) (1989) *Theory and Applications of Optical Remote Sensing*, John Wiley & Sons, Inc., New York.

Assali, S. and Menenti, M. (2000) Mapping vegetation–soil–climate complexes in southern Africa using temporal Fourier analysis of NOAA-AVHRR NDVI data. *International Journal of Remote Sensing*, **21**, 973–996.

Atkinson, P.M. and Aplin, P. (2004) Spatial variation in land cover and choice of spatial resolution for remote sensing. *International Journal of Remote Sensing*, **25**, 3687–3702.

Atkinson, P.M. and Curran, P.J. (1995) Defining an optimal size of support for remote sensing investigation. *IEEE Transactions on Geoscience and Remote Sensing*, **33**, 768–776.

Atkinson, P.M. and Curran, P.J. (1997) Choosing an appropriate spatial resolution. *Photogrammetric Engineering and Remote Sensing*, **63**, 1345–1351.

Atkinson, P.M. and Lloyd, C.D. (2009) Geostatistics and spatial interpolation, in *The SAGE Handbook of Spatial Analysis* (eds A.S. Fotheringham, and P.A. Rogerson), SAGE Publications, London, pp. 159–182.

Atkinson, P.M., Sargent, I.M., Foody, G.M. and Williams, J. (2005) Interpreting image-based methods for estimating the signal-to-noise ratio. *International Journal of Remote Sensing*, **26**, 5099–5115.

Atkinson, P.M., Sargent, I.M., Foody, G.M. and Williams, J. (2007) Exploring the geostatistical method for estimating the signal-to-noise ratio of images. *Photogrammetric Engineering and Remote Sensing*, **73**, 841–850.

Attema, E., Edwards, P., Levrini, G. *et al.* (2007) Sentinel-1, the radar mission for GMES operational land and sea services. *ESA Bulletin*, **131**, 11–17.

Austin, J., Harding, S., Kanellopoulos, I. *et al.* (1997) Connectionist Computation in Earth Observation, Report EUR 17314 EN, Joint Research Centre, European Commission, Brussels.

Avena, G.C., Ricotta, C. and Volpe, F. (1999) The influence of principal component analysis on the spatial structure of a multispectral dataset. *International Journal of Remote Sensing*, **20**, 3367–3376.

Baatz, M., Heynen, M., Hofmann, P. *et al.* (2004) eCognition User Guide 4.0: Object Oriented Image Analysis, Definiens Imaging GmbH, Munich.

Baccini, A., Friedl, M.A., Woodcock, C.E. and Zhu, Z. (2007) Scaling field data to calibrate and validate moderate spatial resolution remote sensing models. *Photogrammetric Engineering and Remote Sensing*, **73**, 945–954.

Baghdadi, N., Holah, N. and Zribi, M. (2006) Soil moisture estimation using multi-incidence and multi-polarization ASAR data. *International Journal of Remote Sensing*, **27**, 1907–1920.

Bakker, W.H. (2000) Satellite and sensor systems for environmental monitoring, in *Encyclopaedia of Analytical Chemistry* (ed. R.A. Meyers), John Wiley & Sons, Ltd, Chichester, pp. 8693–8746.

Baldridge, A.M., Hook, S.J., Grove, C.I. and Rivera, G. (2009) The ASTER spectral library version 2.0. *Remote Sensing of Environment*, **113**, 711–715.

Baltsavias, E.P. (1999) Airborne laser scanning: basic relations and formulas. *ISPRS Journal of Photogrammetry and Remote Sensing*, **54**, 199–214.

Bannari, A., Morin, D., Bénié, G.B. and Bonn, F.J. (1995a) A theoretical review of different mathematical models of geometric corrections applied to remote sensing images. *Remote Sensing Reviews*, **13**, 27–47.

Bannari, A., Morin, D., Bonn, F. and Huete, A. (1995b) A review of vegetation indices. *Remote Sensing Reviews*, **13**, 95–120.

Baofeng, G., Gunn, S.R. and Damper, R.I. (2006) Band selection for hyperspectral image classification using mutual information. *IEEE Geoscience and Remote Sensing Letters*, **3**, 522–526.

Baranoski, G.V.G. and Rokne, J.G. (2005) A practical approach for estimating the red edge position of plant leaf reflectance. *International Journal of Remote Sensing*, **26**, 503–521.

Bard, K.A. (ed.) (1997) *The Environmental History and Human Ecology of Northern Ethiopia in the Late Holocene*, Istituto Universitario Orientale, Naples.

Bard, K.A., Coltorti, M., DiBlasi, M. *et al.* (2000) The environmental history of Tigray (Northern Ethiopia) in the Middle and Late Holocene: a preliminary outline. *African Archaeological Review*, **17**, 65–86.

Bard, K.A., DiBlasi, M., Koch, M. *et al.* (2003) The joint archaeological project at Bieta Giyorgis (Aksum, Ethiopia) of the Istituto Universitario Orientale, Naples (Italy), and Boston University, Boston (USA): Results, research procedures and preliminary computer applications, in *The Reconstruction of Archaeological Landscapes Through Digital Technologies*, BAR International Series, Vol. **1151** (eds M. Forte, and P.R. Williams), Archaeopress, Oxford, pp. 1–13.

Baret, F. and Buis, S. (2008) Estimating canopy characteristics from remote sensing observations: Review of methods and associated problems, in *Advances in Land Remote Sensing* (ed. S. Liang), Springer, Berlin, pp. 173–201.

Baret, F. and Guyot, G. (1991) Potential and limits of vegetation indices for LAI and PAR assessment. *Remote Sensing of Environment*, **35**, 161–173.

Baret, F., Jacquemond, S. and Hanocq, J.F. (1993) The soil line concept in remote sensing. *Remote Sensing Reviews*, **7**, 65–82.

Barnea, D.I. and Silverman, H.F. (1972) A class of algorithm for fast digital image registration. *IEEE Transactions on Computers*, **21**, 179–186.

Barnsley, M.J. (1983) The implications of view angle effects on the use of multispectral data for vegetation studies. Proceedings of the International Conference on Remote Sensing for Rangeland Monitoring and Management, Silsoe, Bedfordshire, England, The Remote Sensing Society, Nottingham, pp. 173–177.

Barnsley, M.J. (2007) *Environmental Modelling: A Practical Introduction*, CRC Press, Boca Raton.

Barnsley, M.J. and Kay, S.A.W. (1990) The relationship between sensor geometry, vegetation-canopy geometry and image variance. *International Journal of Remote Sensing*, **11**, 1075–1083.

Barnsley, M.J., Barr, S.L. and Tsang, T. (1997) Scaling and generalization in land cover mapping from satellite sensors, in *Scaling-Up, From Cell to Landscape* (eds P.R. van Gardingen, G.M. Foody, and P.J. Curran), Cambridge University Press, Cambridge, pp. 173–199.

Barnsley, M.J., Settle, J.J., Cutter, M.A. *et al.* (2004) The PROBA/CHRIS mission: a low-cost smallsat for hyperspectral multiangle observations of the Earth surface and atmosphere. *IEEE Transactions on Geoscience and Remote Sensing*, **42**, 1512–1520.

Bastin, L. (1997) Comparison of fuzzy c-mean classification, linear mixture modelling and MLC probabilities as tools for unmixing coarse pixels. *International Journal of Remote Sensing*, **18**, 3629–3648.

Basu, J.P. and Odell, P.L. (1974) Effects of intraclass correlation among training samples on the misclassification probabilities of Bayes' procedure. *Pattern Recognition*, **6**, 13–16.

Bateson, A. and Curtiss, B. (1996) A method for manual endmember selection and spectral unmixing. *Remote Sensing of Environment*, **55**, 229–243.

Bauer, E. and Kohavi, R. (1999) An empirical comparison of voting classification algorithms: bagging, boosting and variants. *Machine Learning*, **36**, 105–142.

Baugh, W.M. and Groeneveld, D.P. (2008) Empirical proof of the empirical line. *International Journal of Remote Sensing*, **29**, 665–672.

Bayliss, J., Gualtieri, A. and Cromp, R.F. (1997) Analyzing hyperspectral data with independent component analysis. *Proceedings of SPIE (International Society for Optical Engineering)*, **3240**, 133–143.

Bazi, Y. and Melgani, F. (2006) Toward an optimal SVM classification system for hyperspectral remote sensing images. *IEEE Transactions on Geoscience and Remote Sensing*, **44**, 3374–3385.

Beach, T., Luzzadder-Beach, S., Dunning, N. *et al.* (2009) A review of human and natural changes in Maya lowland wetlands over the Holocene. *Quaternary Science Reviews*, **28**, 1710–1724.

Beale, R. and Jackson, T. (1990) *Neural Computing: An Introduction*, Adam Hilger, Bristol.

Beauchemin, M., Thomson, K.B.P. and Edwards, G. (1996) Edge detection and speckle adaptive filtering based on a second-order textural measure. *International Journal of Remote Sensing*, **17**, 1751–1759.

Beckera, B.L., Luschb, D.P. and Qi, J. (2005) Identifying optimal spectral bands from in situ measurements of Great Lakes coastal wetlands using second-derivative analysis. *Remote Sensing of Environment*, **97**, 238–248.

Begni, G. (1988) Absolute calibration of SPOT data. *SPOT Newsletter*, **10**, 2–3.

Begni, G., Dinguirard, M.C., Jackson, R.D. and Slater, P.N. (1988) Absolute calibration of the SPOT-1 HRV cameras. *SPIE (Society of Photo-Optical Instrumentation Engineers)*, **660**, 66–76.

Beinat, E. and Nijkamp, P. (eds) (1998) *Multicriteria Analysis for Land-use Management*, Kluwer Academic Publishers, Dordrecht.

Ben-Dor, E. and Kruse, F.A. (1995) Surface mineral mapping of the Makhtesh Ramon Negev, Israel, using GER 63 channel scanner data. *International Journal of Remote Sensing*, **16**, 3529–3553.

Benediktsson, J.A. and Sveinsson, J.R. (1997) Feature extraction for neural network classifiers, in *Neurocomputation in Remote Sensing Data Analysis* (eds I. Kanellopoulos, G. Wilkinson, F. Roli, and J. Austin), Springer, Heidelberg, pp. 97–104.

Benny, A.H. (1981) Automatic relocation of ground control points in Landsat imagery, *Proceedings of the International Conference Matching Remote Sensing Technologies and their Applications*, The Remote Sensing Society, Nottingham, pp. 307–315.

Bergland, G.D. (1969) A guided tour of the fast fourier transform. *IEEE Spectrum*, **6**, 41–45.

Bergland, G.D. and Dolan, M.T. (1979) Fast Fourier transform algorithms, in *Programs for Digital Signal Processing*, Section 1–2 pp. 1–18, IEEE Acoustics, Speech and Signal Processing Society, IEEE Press/John Wiley & Sons, Inc., New York.

Berk, A.L., Anderson, G.P., Bernstein, L.S. *et al.* (1999) MODTRAN4: Radiative transfer modelling for atmospheric correction. Society of Photo-Optical Instrumentation Engineers (SPIE) Proceeding, Optical Spectroscopic Techniques and Instrumentation for Atmospheric and Space Research III, Vol. 3756, Available from http://www.spectral.com/pdf/sr116.pdf (accessed 22 January 2009).

Bernstein, R., Lotspiech, J.B., Myers, J. *et al.* (1984) Analysis and processing of Landsat-4 sensor data using advanced image processing techniques and technologies. *IEEE Transactions on Geoscience and Remote Sensing*, **22**, 192–221.

Bezdek, J.C. (1993) Editorial: Fuzzy models – what are they, and why. *IEEE Transactions on Fuzzy Systems*, **1**, 1–6.

Bezdek, J.C. (1994) The thirsty traveler visits Gamont: A rejoinder to 'Comments on fuzzy sets – what are they and why?' *IEEE Transactions on Fuzzy Systems*, **2**, 43–45.

Bezdek, J.C., Ehrlich, R. and Full, W., (1984) FCM: The fuzzy *c*-means clustering algorithm http://www.sciencedirect.com/science/journal/00983004 *Computers and Geosciences*, **10**, 191–203.

Bharati, M.H., Liu, J.J. and MacGregor, J.F. (2004) Image texture analysis: methods and comparisons. *Chemometrics and Intelligent Laboratory Systems*, **72**, 57–71.

Bian, L. (2003) Retrieving urban objects using a wavelet transform approach. *Photogrammetric Engineering and Remote Sensing*, **69**, 133–141.

Billari, F.G. and Prskawetz, A. (2003) *Agent-Based Computational Demography: Using Simulation to Improve our Understanding of Demographic Behavior*, Physica-Verlag, Heidelberg.

Billingsley, F.C. (ed.) (1983) Data processing and reprocessing, in *Manual of Remote Sensing*, 2 vols (ed. R.N. Colwell), American Society of Photogrammetry, Falls Church, pp. 719–792.

Bindschlander, R. (1998) Monitoring ice sheet behaviour from space. *Reviews of Geophysics*, **36**, 79–104.

Bird, A.C. (1991a) Principles of remote sensing: electromagnetic radiation, reflectance and emissivity, in Belward, A. and Valenzuela, C.R. (eds.) (1991) *Remote Sensing and Geographical Information Systems for Resource Management in Developing Countries*. Euro-Courses: Remote Sensing, Volume 1. Dordrecht: Kluwer Academic Publishers, 1–15.

Bird, A.C. (1991b) Principles of remote sensing: interaction of electromagnetic radiation with the atmosphere and the Earth, in Belward, A. and Valenzuela, C.R. (eds.) (1991) *Remote Sensing and Geographical Information Systems for Resource Management in Developing Countries*. Euro-Courses: Remote Sensing, Volume 1. Dordrecht: Kluwer Academic Publishers, 17–30.

Bishof, H., Schneider, W. and Pinz, A.J. (1992) Multispectral classification of Landsat images using neural networks. *IEEE Transactions on Geoscience and Remote Sensing*, **30**, 482–490.

Bishop, C.M. (1995) *Neural Networks for Pattern Recognition*, Clarendon Press, Oxford.

Bishop, C.M. (2006) *Pattern Recognition and Machine Learning*, Springer-Verlag, Berlin.

Bishop, M.P. and Colby, J.D. (2002) Anisotropic reflectance correction of SPOT-3 HRV imagery. *International Journal of Remote Sensing*, **23**, 2125–2131.

Bishop, Y.M., Fienberg, S.E. and Holland, P.W. (1975) *Discrete Multivariate Analysis: Theory and Practice*, MIT Press, Cambridge.

Blackburn, G.A. (1998) Quantifying chlorophyll and carotenoids at leaf and canopy scales: an evaluation of some hyperspectral approaches. *Remote Sensing of Environment*, **66**, 273–285.

Blackburn, G.A. (2007a) Wavelet decomposition of hyperspectral data: a novel approach to quantifying pigment concentrations in vegetation. *International Journal of Remote Sensing*, **28**, 2831–2855.

Blackburn, G.A. (2007b) Hyperspectral remote sensing of plant pigments. *Journal of Experimental Botany*, **58**, 855–867. Available at http://jxb.oxfordjournals.org/cgi/content/abstract/58/4/855 (accessed 22 June, 2009).

Blackburn, G.A. and Milton, E.J. (1997) An ecological survey of deciduous woodlands using airborne remote sensing and geographical information systems (GIS). *International Journal of Remote Sensing*, **18**, 1919–1935.

Blamire, P. (1996) The influence of relative sample size in training artificial neural networks. *International Journal of Remote Sensing*, **17**, 223–230.

Blaschke, T., Lang, S. and Hay, G. (eds) (2008) *Object-Based Image Analysis: Spatial Concepts for Knowledge-Driven Remote Sensing Applications (Lecture Notes in Geoinformation and Cartography)*, Springer-Verlag, Berlin.

Blaser, T.J. and Caloz, R. (1991) Digital ortho-image registration from a SPOT panchromatic image using a digital elevation model. *IEEE Transactions on Geoscience and Remote Sensing*, **29**, 2431–2434.

Blesius, L. and Weirich, F. (2005) The use of the Minnaert correction for land-cover classification in mountainous terrain. *International Journal of Remote Sensing*, **26**, 3831–3851.

Blom, R.G. (1988) Effects of variations in look angle and wavelength in radar images of volcanic and aeolian terrains, or now you see it, now you don't. *International Journal of Remote Sensing*, **9**, 945–965.

Blom, R.G. and Daily, M. (1982) Radar image processing for rock-type discrimination. *IEEE Transactions on Geoscience Electronics*, **20**, 343–351.

Blonda, P.N. and Pasquariello, G. (1991) An experiment for the interpretation of multitemporal remotely sensed images based on a fuzzy logic approach. *International Journal of Remote Sensing*, **12**, 463–476.

Boardman, J. (1989) Inversion of imaging spectrometer data using singular value decomposition. *Proceedings of the IEEE International Geoscience and Remote Sensing Symposium, IGARSS'89, 10-14 July 1989, Vancouver, British Columbia*, vol. 4, IEEE Press, Piscataway, pp. 2069–2072.

Bolstad, P.V. and Lillesand, T.M. (1992) Semi-automated training approaches for spectral class definition. *International Journal of Remote Sensing*, **13**, 3157–3166.

Bolstad, P.V. and Stowe, T. (1994) An evaluation of DEM accuracy: elevation, slope, and aspect. *Photogrammetric Engineering and Remote Sensing*, **60**, 1327–1332.

Bolstad, P.V., Gessler, P. and Thomas, M.L. (1990) Positional uncertainty in manually-digitised map data. *International Journal of Geographical Information Systems*, **4**, 39–42.

Bonham-Carter, G.F. (1988) Numerical procedures and computer program for fitting an inverted Gaussian model to vegetation reflectance data. *Computers and Geosciences*, **14**, 339–356.

Bonham-Carter, G.F. (1994) *Geographic Information Systems for Geoscientists: Modelling with GIS*, Pergamon/Elsevier Science Publications, Oxford.

Bonhomme, R. (1993) The solar radiation: characteristics and distribution in the canopy, in *Crop Structures and Light Microclimate, Characteristics and Applications* (eds C. Varlet-Grancher, R. Bonhomme, and H. Sinoquet), INRA Editions, Paris, pp. 17–28.

Boochs, F. and Kupfer, G. (1990) Shape of the red edge as vitality indicator for plants. *International Journal of Remote Sensing*, **11**, 1741–1753.

Borengasser, M., Hungate, W.S. and Watkins, R. (2007) *Hyperspectral Remote Sensing – Principles and Applications*, CRC Press, Boca Raton.

Bork, E.W. and Su, J.G. (2007) Integrating LIDAR data and multispectral imagery for enhanced classification of rangeland vegetation: A meta analysis. *Remote Sensing of Environment*, **111**, 11–24.

Bounoua, L., Masek, J. and Tourre, Y.M. (2006) Sensitivity of surface climate to land surface parameters: A case study using the simple biosphere model SiB2. *Journal of Geophysical Research*, **111**, D22S09. doi:10.1029/2006JD007309.

Bow, S.T. (ed.) (2002) *Pattern Recognition and Image Preprocessing*, 2nd edn, Marcel Dekker, New York.

Boyd, D. and Petitcolin, F. (2004) Remote sensing of the terrestrial environment using middle infrared radiation. *International Journal of Remote Sensing*, **25**, 3343–3368.

Brady, M. (1982) Computational approaches to image understanding. *Association of Computer Manufacturers' (ACM) Computing Surveys*, **14**, 3–71.

Braunich, H., Wu, B.-I. and Kong, J.A. (2000) Phase unwrapping of SAR interferograms after wavelet denoising, *Proceedings of the IEEE International Geoscience and Remote Sensing Symposium, IGARSS 2000, Honolulu*, IEEE Press, Piscataway, pp. 752–754.

Brown, K.M., Foody, G.M. and Atkinson, P.M. (2007) Modelling geometric and mis-registration error in airborne sensor data to enhance change detection. *International Journal of Remote Sensing*, **28**, 2857–2879.

Breiman, L. (1996) Bagging predictors. *Machine Learning*, **24**, 123–140.

Breiman, L. (2001) Random forests. *Machine Learning*, **45**, 5–32.

Brierley, G., Fryirs, K. and Jain, V. (2006) Landscape connectivity: the geographic basis of geomorphic applications. *Area*, **38**, 165–174.

Brimicombe, A. (2003) *GIS, Environmental Modelling, and Engineering*, Taylor and Francis Ltd, London.

Brimicombe, A. (2010) *GIS, Environmental Modelling, and Engineering*, 2nd edn, Taylor & Francis Ltd, London

Brivio, P.A., Colombo, R., Maggi, M. and Tomasoni, R. (2002) Integration of remote sensing data and GIS for accurate mapping of flooded areas. *International ournal of Remote Sensing*, **23**, 429–441.

Brown, K.M., Foody, G.M. and Atkinson, P.M. (2009) Estimating per-pixel thematic uncertainty in remote sensing classifications. *International Journal of Remote Sensing*, **30**, 209–229.

Brown, S.R. and Scholz, C.H. (1985) Broad bandwidth study of the topography of natural rock surfaces. *Journal of Geophysical Research*, **90**, 12575–12582.

Brownrigg, D.R.K. (1984) The weighted median filter. *Communications of the Association of Computer Manufacturers (CACM)*, **27**, 807–818.

Bruce, L.M. and Li, J. (2001) Wavelets for computationally efficient hyperspectral derivative analysis. *IEEE Transactions on Geoscience and Remote Sensing*, **39**, 1540–1546.

Bruce, L.M., Mathur, A. and Byrd, J.D. Jr. (2006) Denoising and wavelet based feature extraction of MODIS multi-temporal vegetation signatures. *IEEE Transactions on Geoscience and Remote Sensing*, **43**, 67–77.

Bruce, V., Green, P.R. and Georgeson, M.A. (2003) *Visual Perception: Physiology, Psychology and Ecology*, 2nd edn, Psychology Press, London.

Brush, R.J.H. (1985) A method for real-time navigation of AVHRR imagery. *IEEE Transactions on Geoscience and Remote Sensing*, **23**, 876–887.

Bruzzone, L., Conese, C., Maselli, F. and Roli, F. (1997) Multisource classification of complex rural areas using statistical and neural-network approaches. *Photogrammetric Engineering and Remote Sensing*, **63**, 523–533.

Bryant, R.G. (1996) Validated linear mixture modelling of Landsat TM data for mapping evaporite minerals on a playa surface: methods and applications. *International Journal of Remote Sensing*, **17**, 315–330.

Bubenzer, O. and Bolten, A. (2008) The use of new elevation data (SRTM/ASTER) for the detection and morphometric quantification of Pleistocene megadunes (draa) in the eastern Sahara and the southern Namib. *Geomorphology*, **102**, 221–231.

Buckingham, W.F. and Sommer, S.E. (1983) Mineralogical characterization of rock surfaces formed by hydrothermal alteration and weathering: application to remote sensing. *Economic Geology*, **78**, 664–674.

Budkewitsch, P., Newton, G. and Hynes, A.J. (1994) Characterisation and extraction of linear features from digital images. *Canadian Journal of Remote Sensing*, **20**, 268–279.

Bue, B.D. and Stepinski, T.F. (2006) Automated classification of landforms on Mars. *Computers and Geosciences*, **32**, 604–614.

Burges, C.J.C. (1998) A tutorial on support vector machines for pattern recognition. *Data Mining and Knowledge Discovery*, **2**, 121–167.

Burrough, P.A. and Frank, A. (1996) *Geographic Objects with Indeterminate Boundaries*, Taylor & Francis, London.

Burrough, P.A. and McDonnell, R.A. (1998) *Principles of Geographical Information Systems for Land Resources Assessment*, Oxford University Press, Oxford.

Cai, C. and Harrington, P.D.B. (1998) Different discrete wavelet transforms applied to denoising analytical data. *Journal of Chemical Information and Computer Sciences*, **38**, 1161–1170.

Calder, N. (1991) *Spaceship Earth*, Viking Books and Channel Four Television, London.

Campbell, J.B. (1981) Spatial correlation effects upon accuracy of supervised classification of land cover. *Photogrammetric Engineering and Remote Sensing*, **47**, 355–357.

Campbell, J.B. (2006) *Introduction to Remote Sensing*, 4th edn, Taylor and Francis, London.

Campbell, N.A. (1980) Robust procedure in multivariate analysis. I: robust covariance estimation. *Applied Statistics*, **29**, 231–237.

Campbell, N.A. (1996) The decorrelation stretch transform. *International Journal of Remote Sensing*, **17**, 1939–1949.

Camps-Valls, G. and Bruzzone, L. (2005) Kernel-based methods for hyperspectral image classification. *IEEE Transactions on Geoscience and Remote Sensing*, **43**, 1351–1362.

Canadian Journal of Remote Sensing (2004) Special issue on Radarsat-2. *Canadian Journal of Remote Sensing*, **30**(3), 221–571.

Canisius, F., Turral, H. and Molden, D. (2007) Fourier analysis of historical NOAA time series data to estimate bimodal agriculture. *International Journal of Remote Sensing*, **28**, 5503–5522.

Cannon, R.J., Dave, J.A., Bezdek, J.C. and Trivedi, M.M. (1986) Segmentation of a Thematic Mapper image using the fuzzy c-means clustering algorithm. *IEEE Transactions on Geoscience and Remote Sensing*, **24**, 400–408.

Canters, F. (1997) Evaluating the uncertainty of area estimates derived from fuzzy land cover classification. *Photogrammetric Engineering and Remote Sensing*, **63**, 403–414.

Canty, M.J. (2007) *Image Analysis, Classification and Change Detection in Remote Sensing (with Algorithms for ENVI/IDL).*, CRC Press, Boca Raton.

Cappellini, V., Chiuderi, A. and Fini, S. (1995) Neural networks in remote sensing multisensor data processing, in *Sensors and Environmental Applications of Remote Sensing*, Proceedings of the 14th EARSeL Symposium, Goteborg, Sweden, 6–8 June, 1994 (ed. J. Askne), A.A. Balkema, Rotterdam, pp. 457–462.

Carara, A., Bitelli, G. and Carla, T. (1997) Comparison of techniques for generating digital terrain models from contour lines. *International Journal of Geographical Information Science*, **11**, 451–473.

Carlson, G.E. and Ebell, W.J. (1995) Co-occurrence matrices for small region texture measurements and comparison. *International Journal of Remote Sensing*, **16**, 1417–1423.

Carlson, T.N. and Ripley, D.A. (1997) On the relation between NDVI, fractional vegetation cover, and leaf area index. *Remote Sensing of Environment*, **62**, 241–252.

Carpenter, G.A., Gjaja, M.N., Gopal, S. and Woodcock, C.E. (1997) ART neural networks for remote sensing: vegetation classification from Landsat TM and terrain data. *IEEE Transactions on Geoscience and Remote Sensing*, **35**, 308–325.

CEOS (2008) CEOS EO Handbook – Earth Observation Satellite Capabilities and Plans. http://www.eohandbook.com/eohb2008/earthobservation.htm (accessed 25 January 2009).

Cetin, M. and Musaoglu, N. (2009) Merging hyperspectral and panchromatic image data: qualitative and quantitative analysis. *International Journal of Remote Sensing*, **30**, 1779–1804.

Chan, J.C.-W. and Paelinckx, D. (2008) Evaluation of Random Forest and Adaboost tree-based ensemble classification and spectral band selection for ecotope mapping using airborne hyperspectral imagery. *Remote Sensing of Environment*, **112**, 2999–3011.

Chan, J.C.-W., Huang, C. and DeFries, R. (2001) Enhanced algorithm performance for land cover classification from remotely sensed data using bagging and boosting. *IEEE Transactions on Geoscience and Remote Sensing*, **39**, 693–695.

Chan, R.H., Ho, C.W. and Nikolova, M. (2005) Salt-and-pepper noise removal by median-type noise detectors and detail-preserving regularization. *IEEE Transactions on Image Processing*, **14**, 1479–1485.

Chander, G., Markham, B.L. and Helder, D.L. (2009a) Summary of current radiometric calibration coefficients for Landsat MSS, TM, ETM+, and EO-1 ALI sensors. *Remote Sensing of Environment*, **113**, 893–903.

Chander, G., Saunier, S., Choate, M.J. and Scaramuzza, P.L. (2009b) SSTL UK-DMC SLIM-6 data quality assessment. *IEEE Transactions on Geoscience and Remote Sensing*, **47**, 2380–2391.

Chang, C. and Plaza, A. (2006) A fast iterative algorithm for implementation of pixel purity index. *IEEE Geoscience and Remote Sensing Letters*, **3**, 63–67.

Chang, C.C. and Lin, C.J. (2001) LIBSVM: A Library for Support Vector Machines. Department of Computer Science and Information Engineering, National Taiwan University, Taiwan. Available online at: http://www.csie.ntu.edu.tw/~cjlin/libsvm (accessed 5 September 2008).

Chang, C.-I. (2003) *Hyperspectral Imaging: Techniques for Spectral Detection and Classification*, Kluwer Academic Publishers, Norwell.

Chang, C.-I., Wu, C.-C., Liu, W. and Ouyang, Y.-C. (2006) A new growing method for simplex based endmember extraction algorithm. *IEEE Transactions on Geoscience and Remote Sensing*, **44**, 2804–2819.

Chapman, R.E. (1995) *Physics for Geologists*, UCL Press, London.

Chavez, P.S. Jr. (1988) An improved dark-object subtraction technique for atmospheric scattering correction of multispectral data. *Remote Sensing of Environment*, **24**, 459–479.

Chavez, P.S. Jr. (1996) Image-based atmospheric corrections – revisited and improved. *Photogrammetric Engineering and Remote Sensing*, **62**, 1025–1036.

Che, N. and Price, J.C. (1992) Survey of radiometric calibration results and methods for visible and near-infrared channels of NOAA-7, -9, and -11 AVHRRs. *Remote Sensing of Environment*, **41**, 19–27.

Chen, C.H. (ed.) (2007) *Image Processing for Remote Sensing*, CRC Press, Boca Raton.

Chen, J., Gong, P., He, C. *et al.* (2003) Land-use/land-cover change detection using improved change-vector analysis. *Photogrammetric Engineering and Remote Sensing*, **69**, 369–379.

Chen, L., Teo, T. and Liu, C. (2006) The geometrical comparisons of RSM and RFM for FORMOSAT-2 satellite images. *Photogrammetric Engineering and Remote Sensing*, **72**, 573–579.

Chen, T., Ma, K.-K. and Li-Hui Chen, L.-H. (1999) Tri-state median filter for image denoising. *IEEE Transactions on Image Processing*, **8**, 1834–1838.

Cheng, H.D., Jiang, X.H., Sun, Y. and Wang, J. (2001) Colour image segmentation: advances and prospects. *Pattern Recognition*, **34**, 2259–2281.

Cheng, P. and Sustera, J. (2009) Automated high-speed high-accuracy orthorectification and mosaicing. *GeoInformatics*, **12**, 36–40.

Chica-Olmo, M. and Abarca-Hernandez, F. (2000) Computing geostatistical image texture for remotely sensed data classification. *Computers and Geosciences*, **26**, 373–383.

Chitroub, S. (2005) Neural network model for standard PCA and its variants applied to remote sensing. *International Journal of Remote Sensing*, **26**, 2197–2218.

Chiuderi, A. and Cappellini, V. (1996) A Kohonen's self organising map for land cover classification, in *Progress in Environmental Remote Sensing Research and Applications*, Proceedings of the 15th EARSeL Symposium, Basle, Switzerland, 4–6 September 1996 (ed. E. Parlow), A.A. Balkema, Rotterdam, pp. 107–112.

Cho, M.A. and Skidmore, A.K. (2006) A new technique for extracting the red edge position from hyperspectral data: The linear extrapolation method. *Remote Sensing of Environment*, **101**, 181–193.

Choe, E., van der Meer, F., van Ruitenbeek, F. *et al.* (2008) Mapping of heavy metal pollution in stream sediments using combined geochemistry, field spectroscopy, and hyperspectral remote sensing: a case study of the Rodalquilar mining area, S.E. Spain. *Remote Sensing of Environment*, **112**, 3222–3233.

Chopping, M.J., Rango, A., Havstad, K.M. *et al.* (2003) Canopy attributes of desert grassland and transition communities derived from multiangular airborne imagery. *Remote Sensing of Environment*, **85**, 339–354.

Chubey, M.S., Franklin, S.E. and Wulder, M.A. (2006) Object-based analysis of Ikonos-2 imagery for extraction of forest inventory parameters. *Photogrammetric Engineering and Remote Sensing*, **72**, 383–394.

Chuviceo, E. and Martin, M.P. (1994) Global fire mapping and fire danger estimation using AVHRR images. *Photogrammetric Engineering and Remote Sensing*, **60**, 563–570.

Chuvieco, E. (ed.) (2008) *Earth Observation of Global Change: The Role of Satellite Remote Sensing in Monitoring the Global Environment*, Springer-Verlag, New York.

Chuvieco, E. and Salas, J. (1996) Mapping the spatial distribution of forest fire danger using GIS. *International Journal of Geographic Information Systems*, **10**, 333–345.

Chuvieco, E., Cocero, D., Riaño, D. *et al.* (2004) Combining NDVI and surface temperature for the estimation of live fuel moisture content in forest fire danger rating. *Remote Sensing of Environment*, **92**, 322–331.

Ciampalini, R., Manzo, A., Perlingieri, C. and Sernicola, L. (2006) Landscape archaeology and GIS for the eco-cultural heritage management of the Aksum region, Ethiopia, in *From Space to Place: Second International Conference on Remote Sensing in Archaeology*, BAR International Series, Vol. 1568, Proceedings of the Second International Workshop, CNR, December 4–7 2006, Rome (eds M. Forte, and S. Campana), Archaeopress, Oxford, pp. 219–226.

Clark, C. and Cañas, A. (1995) Spectral identification by artificial neural network and genetic algorithm. *International Journal of Remote Sensing*, **16**, 2255–2275.

Clark, M.L., Clark, D.B. and Roberts, D.A. (2004) Small-footprint lidar estimation of sub-canopy elevation and tree height in a tropical rain forest landscape. *Remote Sensing of Environment*, **91**, 68–89.

Clark, R.N. (1999) Spectroscopy of rocks and minerals and principles of spectroscopy, in *Manual of Remote Sensing*, Remote Sensing for the Earth Sciences, Vol. 3, 3rd edn (ed. A.N. Rencz), John Wiley & Sons, Inc., New York, pp. 3–58.

Clausi, D.A. (2002) An analysis of co-occurrence texture statistics as a function of grey level quantization. *Canadian Journal of Remote Sensing*, **28**, 45–62.

Clausi, D.A. and Zhao, Y. (2003) Grey level co-occurrence integrated algorithm (GLCIA): a superior computational method to rapidly determine co-occurrence probability texture features. *Computers and Geosciences*, **29**, 837–850. Code at http://www.iamg.org/CGEditor/index.htm (accessed 26 September 2008).

Clavet, D., Lasserre, M. and Pouliot, J. (1993) GPS control for 1 : 50 000-scale topographic mapping from satellite images. *Photogrammetric Engineering and Remote Sensing*, **59**, 107–111.

Clevers, J.P.G.W. and Jongschamp, R. (2001) Imaging spectrometry for agricultural applications, in *Imaging Spectrometry: Basic Principles and Applications* (eds F.D. van der Meer, and S.M. de Jong), Kluwer Academic Publishers, Dordrecht, pp. 157–199.

Clevers, J.G.P.W., Kooistra, L. and Salas, E.A.L. (2004) Study of heavy metal contamination in river floodplains using the red-edge position in spectroscopic data. *International Journal of Remote Sensing*, **25**, 3883–3895.

Clevers, J.P.G.W., de Jong, S.M., Epema, G.F. *et al.* (2002) Derivation of the red edge index using MERIS standard band setting. *International Journal of Remote Sensing*, **23**, 3169–3184.

Cliff, A.D. and Ord, J.K. (1973) *Spatial Autocorrelation*, Pion Press, London.

Cloude, S.R. and Papathanassiou, K.P. (1998) Polarimetric SAR interferometry. *IEEE Transactions on Geoscience and Remote Sensing*, **36**, 1551–1565.

Cobby, D.M., Mason, D.C. and Davenport, I.J. (2001) Image processing of airborne scanning laser altimetry data for improved river flood modelling. *ISPRS Journal of Photogrammetry and Remote Sensing*, **56**, 121–138.

Coburn, C.A. and Roberts, A.C.B. (2004) A multiscale texture analysis procedure for improved forest stand classification. *International Journal of Remote Sensing*, **25**, 4287–4308.

Cocks, T., Jenssen, R., Stewart, A. *et al.* (1998) The Hymap airborne hyperspectral sensor: the system, calibration and performance. First EARSeL Workshop on Imaging Spectroscopy. Zurich, Switzerland, European Association of Remote Sensing Laboratories (EARSeL), Paris.

Cohen, Y. and Shoshan, Y. (2005) Analysis of convergent evidence in an evidential reasoning knowledge-based classification. *Remote Sensing of Environment*, **96**, 518–528.

Coley, D.A. (1999) *An Introduction to Genetic Algorithms for Scientists and Engineers*, World Scientific, Singapore.

Coll, C., Galve, J.M., Sánchez, J.M. and Caselles, V. (2010) Validation of Landsat-7/ETM+ thermal band calibration and atmospheric correction with ground-based measurements. *IEEE Transactions on Geoscience and Remote Sensing,* **48**, 547–555).

Collins, J.B. and Woodcock, C.E. (1996) Assessment of several linear change detection techniques for mapping forest mortality using multitemporal Landsat TM data. *Remote Sensing of Environment*, **56**, 66–77.

Combal, B. and Isaka, H. (2002) The effect of small topographic variations on reflectance. *IEEE Transactions on Geoscience and Remote Sensing*, **40**, 663–670.

Committee on Earth Studies, Space Studies Board. Commission on Physical Sciences, Mathematics, and Applications, National Research Council (2000) *The Role of Small Satellites in NASA and NOAA Earth Observation Programs*, National Academy Press, Washington, DC.

Conese, C., Gilabert, M.A., Maselli, F. and Bottai, L. (1993) Topographic normalisation of TM scenes through the use of an atmospheric correction method and digital terrain models. *Photogrammetric Engineering and Remote Sensing*, **59**, 1745–1753.

Congalton, R.G. and Green, K. (2008) *Assessing the Accuracy of Remotely Sensed Data: Principles and Practices*, 2nd edn, CRC Press, Boca Raton.

CONOCO, The Egyptian General Petroleum Corporation (1987) Geological Map of South Sinai, Egypt, Scale 1 : 500 000.

Cook, A.E. and Pinder, J.E. III (1996) Relative accuracy of rectifications of coordinates determined from maps and the global positioning system. *Photogrammetric Engineering and Remote Sensing*, **62**, 73–77.

Cooley, W.W. and Lohnes, P.R. (1962) *Multivariate Procedures for the Behavioural Sciences*, John Wiley & Sons, Inc., New York.

Coppin, P., Jonckeere, I., Nackaerts, K. *et al.* (2004) Digital change detection methods in ecosystem monitoring: a review. *International Journal of Remote Sensing*, **25**, 1565–1596.

Cormen, T.H., Leiserson, C.E., Rivest, R.L. and Stein, C. (2001) *Introduction to Algorithms*, 2nd edn, The MIT Press, Cambridge.

Costa-Posada, C.R. and Devereux, B.J. (1995) Reduction of the topographic effect in SPOT imagery: an examination of the Minnaert model. *Proceedings of the Series SPIE*, **2579**, 137–149.

Couteron, P., Barbier, N. and Gautier, D. (2006) Textural ordination based on fourier spectral decomposition: a method to analyze and compare landscape patterns. *Landscape Ecology*, **21**, 555–567.

Cracknell, A.P. (1997) *The Advanced Very High Resolution Radiometer*, Taylor and Francis, London.

Cracknell, A.P. (1998) Synergy in remote sensing – what's in a pixel? *International Journal of Remote Sensing*, **19**, 2025–2047.

Craig, R.G. (1979) Autocorrelation in Landsat data. Proceedings of the 13th International Symposium on Remote Sensing of Environment, Environmental Research Institute of Michigan, Ann Arbor, pp. 1517–1524.

Crammer, K. and Singer, Y. (2002) On the learnability and design of output codes for multi-class problems. *Machine Learning*, **47**, 201–233.

Crawford, P.S., Brooks, A.R. and Brush, R.J.H. (1996) Fast navigation of AVHRR images using complex orbital models. *International Journal of Remote Sensing*, **17**, 197–212.

Crippen, R.W. (1989) A simple spatial filtering routine for the cosmetic removal of scan line noise from Landsat TM P-tape imagery. *Photogrammetric Engineering and Remote Sensing*, **55**, 327–331.

Crist, E.P. (1983) The TM tasseled cap – a preliminary formulation. Proceedings of the Symposium on Machine Processing of Remotely-Sensed Data 1983, Purdue University, West Lafayette, pp. 357–364.

Crist, E.P. and Cicone, R.C. (1984a) A physically-based transformation of thematic mapper data – the TM Tasseled Cap. *IEEE Transactions on Geoscience and Remote Sensing*, **22**, 256–263.

Crist, E.P. and Cicone, R.C. (1984b) Comparison of the dimensionality and features of simulated Landsat-4 MSS and TM data. *Remote Sensing of Environment*, **14**, 235–246.

Crist, E.P. and Kauth, R.J. (1986) The Tasselled Cap demystified. *Photogrammetric Engineering and Remote Sensing*, **52**, 81–86.

Cross, A.M., Settle, J.J., Drake, N.A. and Paivinen, R.T.M. (1991) Subpixel measurement of tropical forest cover using AVHRR data. *International Journal of Remote Sensing*, **12**, 1119–1129.

Crósta, A.P., De Souza Filho, C.R., Azevedo, F. and Brodie, C. (2003) Targeting key alteration minerals in epithermal deposits in Patagonia, Argentina, using ASTER imagery and principal component analysis. *International Journal of Remote Sensing*, **24**, 4233–4240.

Cuartero, A., Felicisimo, A.M. and Ariza, F.J. (2005) Accuracy, reliability and depuration of SPOT HRV and Terra ASTER digital elevation models. *IEEE Transactions on Geoscience and Remote Sensing*, **43**, 404–407.

Culbert, T.P. (1988) The collapse of classic Maya civilization, in *The Collapse of Ancient States and Civilizations* (eds N. Yoffee, and G.L. Cowgill), University of Arizona Press, Tucson, pp. 69–101.

Curran, P.J. and Atkinson, P.M. (1998) Geostatistics and remote sensing. *Progress in Physical Geography*, **22**, 61–78.

Curran, P.J., Dungan, J.I., Maesler, B.A. and Plummer, S.E. (1991) The effect of a red leaf pigment on the relationship between red edge and chlorophyll concentration. *Remote Sensing of Environment*, **35**, 69–76.

Czaplewski, R.L. (1992) Misclassification bias in areal estimates. *Photogrammetric Engineering and Remote Sensing*, **58**, 189–192.

Daily, M. (1983) Huesaturationintensity split-spectrum processing of Seasat radar images. *Photogrammetric Engineering and Remote Sensing*, **49**, 349–355.

Dale, P.E.R., Chandica, A.L. and Evans, M. (1996) Using image subtraction and classification to evaluate change in sub-tropical inter-tidal wetlands. *International Journal of Remote Sensing*, **17**, 703–719.

Danaher, T.J. (2002) An empirical BRDF correction for Landsat TM and ETM+ imagery. Proceedings of the 11th Australasian Remote Sensing and Photogrammetry Conference, Brisbane, pp. 966–977.

Danaher, T.J., Xiolaing, W. and Campbell, N.A. (2001) Bi-directional reflectance distribution function approaches to radiometric calibration of Landsat ETM+ imagery, *Proceedings of the IEEE International Geoscience and Remote Sensing Symposium, IGARSS 2001. Sydney, Australia*, IEEE Press, Piscataway, pp. 2654–2657.

Danielsson, P.E. (1981) Getting the median faster. *Computer Graphics and Image Processing*, **15**, 71–78.

Dantzig, G.B. (1960) On the shortest route through a network. *Management Science*, **6**, 187–190.

Davis, J.C. (1973) *Statistics and Data Analysis in Geology*. John Wiley & Sons, Inc., New York.

de Asis, A.M., Omassa, K., Oki, K. and Shimuzu, Y. (2008) Accuracy and applicability of linear spectral unmixing in delineating potential erosion areas in tropical watersheds. *International Journal of Remote Sensing*, **29**, 4151–4171.

de Castro, L.N. (2006) *Fundamentals of Natural Computing: Basic Concepts, Algorithms, and Applications*, Chapman & Hall and CRC Press, Boca Raton.

de Cola, L. (1994) Simulating and mapping spatial complexity using multi-scale techniques. *International Journal of Geographical Information Systems*, **8**, 411–427.

de Jong, S.M. and Burrough, P.A. (1995) A fractal approach to the classification of Mediterranean vegetation types using remotely sensed data. *Photogrammetric Engineering and Remote Sensing*, **61**, 1041–1063.

de Smith, M.J., Goodchild, M.F. and Longley, P.A. (2007) *Geospatial Analysis: A Comprehensive Guide to Principles, Techniques and Software Tools*, 2nd edn, Troubador Publishing, Leicester.

de Souza Filho, C.R., Drury, S.A., Denniss, A.M. *et al.* (1996) Restoration of corrupted optical Fuyo-1 (JERS-1) data using frequency domain techniques. *Photogrammetric Engineering and Remote Sensing*, **62**, 1037–1047.

Dean, J.S., Gumerman, G.J., Epstein, J.M. *et al.* (2000) *Understanding Anasazi Culture Change Through Agent-Based Modeling*, Oxford University Press, Oxford.

DeFries, R.S. and Los, S.O. (1999) Implications of land cover misclassification for parameter estimates in global land-surface models: an example from the Simple Biosphere Model (SiB2). *Photogrammetric Engineering and Remote Sensing*, **65**, 1083–1088.

Dekker, R.J. (2003) Texture analysis and classification of ERS SAR images for map updating of urban areas in The Netherlands. *IEEE Tranactions on Geoscience and Remote Sensing*, **41**, 1950–1958.

Dell'Acqua, F., Gamba, P. and Trianni, G. (2006) Semi-automatic choice of scale-dependent features for satellite SAR image classification. *Pattern Recognition Letters*, **27**, 244–251.

DeMers, M.N. (2001) *GIS Modelling in Raster*. John Wiley & Sons, Inc., New York.

Demetriades-Shah, T.H., Steven, M.D. and Clark, A.C. (1990) High resolution derivative spectra in remote sensing. *Remote Sensing of Environment*, **33**, 55–64.

Demšar, U. (2009) Geovisualisation and geovisual analysis, in *The SAGE Handbook of Spatial Analysis* (eds A.S. Fotheringham, and P.A. Rogerson), SAGE Publications, London, pp. 41–62.

Denham, R., Mengersen, K. and Witte, C. (2009) Bayesian analysis of thematic map accuracy data. *Remote Sensing of Environment*, **113**, 371–379.

Dennison, P.E. and Roberts, D.A. (2003) Endmember selection for multiple endmember spectral mixture analysis using endmember average RMSE. *Remote Sensing of Environment*, **87**, 123–135.

Dennison, P.E., Halligan, K.Q. and Roberts, D.A. (2004) A comparison of error metrics and constraints for multiple end member spectral mixture analysis and spectral angle mapper. *Remote Sensing of Environment*, **93**, 359–367.

Deschamps, P.Y., Herman, M. and Tanré, D. (1983) Definitions of atmospheric radiance and transmittances in remote sensing. *Remote Sensing of Environment*, **13**, 89–92.

Desnos, Y.-L. and Matteini, V. (1993) Review on structural detection and speckle filtering on ERS-1 images. *EARSeL Advances in Remote Sensing*, **2**, 52–65.

Desnos, Y.-L., Buck, C., Guijarro, J. *et al.* (2000a) The ENVISAT advanced synthetic aperture radar system. Proceedings of the IEEE International Geoscience and Remote Sensing Symposium, 2000. IGARSS 2000 24–28 July, vol. 3, Honolulu, Hawaii, pp. III-1171–III-1173.

Desnos, Y.-L., Buck, C., Guijarro, J., *et al.* (2000b) ASAR – Envisat's advanced synthetic aperture radar: building on ERS achievements towards future earth watch missions. *ESA Bulletin*, **102**, 91–100. Available at http://esapub.esrin.esa.it/bulletin/bullet102/Desnos102.pdf (accessed 4 September, 2009).

Di, K., Ma, R. and Li, R.X. (2004) Rational functions and potential for rigorous sensor model recovery. *Photogrammetric Engineering and Remote Sensing*, **69**, 33–41.

Dial, G., Bowen, H., Gerlach, F. *et al.* (2003) IKONOS satellite, imagery, and products. *Remote Sensing of Environment*, **88**, 23–36.

DiBlasi, M. (2005) Foreword, in *Changing Settlement Patterns in the Aksum-Yeha Region of Ethiopia: 700 BC – 850 AD*, Cambridge Monographs in African Archaeology, British

Archaeological Reports International Series (ed. J.W. Michels), Archaeopress, Oxford, pp. xv–xxviii.

Dietterich, T.G. (2000) An experimental comparison of three methods for constructing ensembles of decision trees: Bagging, boosting, and randomization. *Machine Learning*, **40**, 139–158.

Dikshit, O. (1996) Textural classification for ecological research using ATM images. *International Journal of Remote Sensing*, **17**, 887–915.

Diner, D.J., Bruegge, C.J., Martonchik, J.V. *et al.* (1991) A Multi-angle Imaging SpectroRadiometer for terrestrial remote sensing from the Earth Observing System. *International Journal of Imaging Systems and Technology*, **3**, 92–107.

Dinesh Kumar, P.K., Gopinath, G. and Seralathan, P. (2007) Application of remote sensing and GIS for the demarcation of groundwater potential zones of a river basin in Kerala, southwest coast of India. *International Journal of Remote Sensing*, **28**, 5583–5601.

Dixon, B. and Candade, N. (2008) Multispectral land-use classification using neural networks and support vector machines: one or the other, or both? *International Journal of Remote Sensing*, **29**, 1185–1206.

Doan, H.T.X. and Foody, G.M. (2007) Increasing soft classification accuracy through the use of an ensemble of classifiers. *International Journal of Remote Sensing*, **28**, 4609–4623.

Dobbertin, M. and Biging, G.S. (1996) A simulation study of the effect of scene autocorrelation, training size and sampling method on classification accuracy. *Canadian Journal of Remote Sensing*, **22**, 360–367.

Dobson, M.C., Ulaby, F.T., Pierce, L.E. *et al.* (1995) Estimation of forest biophysical characteristics in Northern Michigan with SIR-C/X-SAR. *IEEE Transactions on Geoscience and Remote Sensing*, **33**, 877–895.

Dong, J., Kaufmann, R.K., Myneni, R.B. *et al.* (2003) Remote sensing estimates of boreal and temperate forest woody biomass: carbon pools, sources, and sinks. *Remote Sensing of Environment*, **84**, 393–410.

Dong, Y., Forster, C., Milne, K. and Morgan, G.A. (1998) Speckle suppression using recursive wavelet transforms. *International Journal of Remote Sensing*, **19**, 317–330.

Dong, Y., Milne, A.K. and Forster, B.C. (2001) Toward edge sharpening: a SAR speckle filtering algorithm. *IEEE Transactions on Geoscience and Remote Sensing*, **39**, 851–863.

Donoghue, D.N.M., Watt, P.J., Cox, N.J. and Wilson, J. (2007) Remote sensing of species mixtures in conifer plantations using LiDAR height and intensity data. *Remote Sensing of Environment*, **110**, 509–522.

Donoho, D.L. and Johnstone, L.M. (1995) Adapting to unknown smoothness via wavelet shrinkage. *Journal of the American Statistical Association*, **90**, 1200–1224.

Dowman, I. (1992) The geometry of SAR images for geocoding and stereo applications. *International Journal of Remote Sensing*, **13**, 1609–1617.

Dowman, I. and Dare, P. (1999) Automated procedures for multisensor registration and orthorectification of satellite images. *International Archives of Photogrammetry and Remote Sensing*, **32** (Part 7-4-3 W6) (unpaginated; available from http://www.data-fusion.org/ps/sig/meeting/Spain99ps/dowman.pdf (accessed 17 June 2009).

Dowman, I. and Dolloff, J. (2000) An evaluation of rational functions for photogrammetric restitution. *International Archives of Photogrammetry and Remote Sensing*, **33** (B3), 254–266.

Dowman, I. and Neto, F. (1994) The accuracy of along track stereoscopic data for mapping: Results from simulations and JERS OPS. *International Archives of Photogrammetry and Remote Sensing*, **30**, 216–221.

Dowman, I., Laycock, J. and Whalley, J. (1993) Geocoding in the UK, in *SAR Geocoding: Data and Systems* (ed. G. Schreier), Herbert Wichmann, Karlsruhe, pp. 373–387.

Drews, R., Rack, W., Wesche, C. and Helm, V. (2009) A spatially adjusted elevation model in Dronning Maud Land, Antarctica, based on differential SAR interferometry. *IEEE Transactions on Geoscience and Remote Sensing*, **47**, 2501–2509.

Drucker, H., Schapire, R. and Simard, P.Y. (1993) Boosting performance in neural networks. *International Journal of Pattern recognition and Artificial Intelligence*, **7**, 705–719.

Drury, S.A. (2004) *Image Interpretation in Geology*, 3rd edn, Nelson Thornes, Cheltenham.

Du, L. and Lee, J.S. (1996) Fuzzy classification of earth terrain covers using complex polarimetric SAR data. *International Journal of Remote Sensing*, **17**, 809–826.

Du, Q., Raksuntorn, N., Cai, S. and Moorhead, R.J. II (2008) Colour display for hyperspectral imagery. *IEEE Transactions on Geoscience and Remote Sensing*, **46**, 1858–1866.

Du, Q., Younan, N.H., King, R. and Shah, V.P. (2007) On the performance evaluation of pan-sharpening techniques. *IEEE Geoscience and Remote Sensing Letters*, **4**, 518–522.

Du, Y., Guindon, B. and Cihlar, J. (2002) Haze detection and removal in high resolution satellite image with wavelet analysis. *IEEE Transactions on Geoscience and Remote Sensing*, **40**, 210–217.

Dubayah, R. and Drake, J. (2000) Lidar remote sensing for forestry applications. *Journal of Forestry*, **98**, 44–46.

Dubayah, R., Knox, R.G., Hofton, M.A. *et al.* (2000) Land surface characteristics using lidar remote sensing, in *Spatial Information for Land Use Management* (eds M. Hill, and R. Aspinall), International Publishers Direct, Singapore, pp. 238–255.

Duda, R.O., Hart, P.E. and Stork, D.G. (2000) *Pattern Classification*, 2nd edn, John Wiley & Sons, Inc., New York.

Dueker, K.J. (1979) Land resource information systems: a review of fifteen years experience. *Geo-Processing*, **1**, 105–128.

Duggin, M.J. (1985) Factors influencing the discrimination and quantification of terrestrial features using remotely-sensed radiance. *International Journal of Remote Sensing*, **6**, 3–28.

Duguay, C.R. and Peddle, D.R. (1996) Comparison of evidential reasoning and neural network approaches in a multi-source classification of alpine tundra vegetation. *Canadian Journal of Remote Sensing*, **22**, 433–440.

Dunn, R. and Harrison, A.R. (1993) Two-dimensional systematic sampling of land use *Journal of the Royal Statistical Society. Series C (Applied Statistics)*, **42** (4), 585–601.

Duran, O. and Petrou, M. (2009) Spectral unmixing with negative and superunity abundances for subpixel anomaly detection. *IEEE Geoscience and Remote Sensing Letters*, **6**, 152–156.

Dutra, L.V. (1999) Feature extraction and selection for ERS-1/2 InSAR classification. *International Journal of Remote Sensing*, **20**, 993–1016.

Dykes, J., MacEachren, A.M. and Kraak, M.-J. (eds) (2005) *Exploring Geovisualization*, Elsevier, Amsterdam.

Dymond, C.C., Mladenoff, D.J. and Radeloff, V.C. (2002) Phenological differences in Tasseled Cap indices improve deciduous forest classification. *Remote Sensing of Environment*, **80**, 460–472.

Dymond, J.R. (1992) How accurately do image classifiers estimate area. *International Journal of Remote Sensing*, **13**, 1735–1742.

Eckmann, T.C., Roberts, D.A. and Still, C.J. (2008) Using multiple end-member spectral mixture analysis to retrieve sub-pixel fire properties from MODIS. *Remote Sensing of Environment*, **112**, 3773–3783.

Egbert, D.D. and Ulaby, F.T. (1972) Effect of angles on reflectivity. *Photogrammetric Engineering*, **29**, 556–564.

Eghbali, H.J. (1979) A K-S test for detecting changes from Landsat imagery data. *IEEE Transactions on Systems, Man and Cybernetics*, **9**, 17–23.

Ehlers, M. (1990) Remote sensing and GIS: towards integrated spatial information processing. *IEEE Transactions on Geoscience and Remote Sensing*, **28**, 763–766.

Elachi, C. (1988) *Spaceborne Remote Sensing: Applications and Techniques.*, IEEE Press, New York.

Elachi, C. and van Zyl, J. (2006) *Introduction to the Physics and Techniques of Remote Sensing*, 2nd edn, John Wiley & Sons, Inc., New York.

El-Araby, A.M. and El-Demerdash, S. (1981) Pedology of Wadi Feiran, South Sinai with special reference to mode and nature of depositional environments. *The Desert Institute Bulletin (Egypt)*, **31**, 97–106.

El-Baz, F., Koch, M., Robinson, C. *et al.* (2004) Use of Space Images for Groundwater Exploration in the Northern United Arab Emirates. Third Annual Report submitted to SEWA, Sharjah, United Arab Emirates by the Center for Remote Sensing, Boston University, Boston, MA.

Eineder, M., Adam, N., Bamler, R., Yague-Martinez, N. and Breit, H. (2009) Spaceborne spotlight SAR interferometry with TerraSAR-X. *IEEE Transactions on Geoscience and Remote Sensing*, **47**, 1524–1535.

Eliason, E.M. and McEwen, A.S. (1990) Adaptive box filters for the removal of random noise from digital images. *Photogrammetric Engineering and Remote Sensing*, **56**, 453–458.

Elvidge, C.D., Yuan, D., Weerackoon, R.D. and Lunetta, R.S. (1995) Relative radiometric normalisation of Landsat Multispectral Scanner (MSS) data using an automatic scattergram-controlled regression. *Photogrammetric Engineering and Remote Sensing*, **61**, 1255–1260.

Emerson, C.W., Siu-Ngan Lam, N. and Quattrochi, D.A. (2005) A comparison of local variance, fractal dimension, and Moran's *I* as aids to multispectral image classification. *International Journal of Remote Sensing*, **26**, 1575–1588.

Emery, W.J., Baldwin, D. and Matthews, D. (2003) Maximum cross correlation automatic satellite image navigation and attitude corrections for open-ocean image navigation. *IEEE Transactions on Geoscience and Remote Sensing*, **41**, 33–42.

Engdahl, M.E. and Hyyppa, J.M. (2003) Land-cover classification using multitemporal ERS-1/2 InSAR data. *IEEE Transactions on Geoscience and Remote Sensing*, **41**, 1629–1637.

Esaias, W.E., Abbot, M.R., Barton, I. *et al.* (1998) An overview of MODIS capabilities for ocean science observations. *IEEE Transactions on Geoscience and Remote Sensing*, **36**, 1250–1265.

Esch, T., Roth, A., Strunz, G., and Dech, S. (2003) Object-oriented classification of Landsat-7 data for regional purposes. In: *Proceedings of ISPRS WG VII Workshop on Remote Sensing of Urban Areas*, Regensburg, Germany, 27–29 July, 2003, ed. Carstens, J. pp. 50–55.

Esposito, F., Malerba, D. and Semeraro, G. (1997) A comparative analysis of methods for pruning decision trees. *IEEE Transactions on Pattern Analysis and Machine Intelligence*, **19**, 476–491.

Estrada-Belli, F. (2003) Archaeological Investigations at Holmul, Petén, Guatemala, http://www.famsi.org/reports/01009/index.html (accessed 31 December 2009).

Estrada-Belli, F. (2010) *The First Maya Civilization. Ritual and Power before the Classic Period*, Routledge, London.

Estrada-Belli, F. and Koch, M. (2007) Remote sensing and GIS analysis of a Maya city and its landscape: Holmul, Guatemala, in *Remote Sensing in Archaeology* (eds J.R. Wiseman, and F. El-Baz), Springer, New York, pp. 263–281.

Evans, D.L., Farr, T.G., Zebker, H.A. *et al.* (1992) Radar interferometry studies of the Earth's topography. *Eos: Transactions of the American Geophysical Union*, **73**, 553–558.

Everett, B.S. and Nicholls, P. (1975) Visual techniques for representing multivariate data. *The Statistician*, **24**, 37–49.

Everitt, B. (1993) *Cluster Analysis*, Hodder Arnold, London.

Fairchild, M.D. (2005) *Color Appearance Models*, John Wiley & Sons, Ltd, Chichester.

Farag, A.A. (1992) Edge-based image segmentation. *Remote Sensing Reviews*, **6**, 95–122.

Farr, T.G. and Kobrick, M. (2000) Shuttle Radar Topography Mission produces a wealth of data. *American Geophysical Union Eos*, **81**, 583–585.

Fattovich, R. (1992) *Lineamenti di storia dell'archeologia dell'Etiopia e della Somalia*, Supplemento n. 71 agli Annali dell'Istituto Universitario Orientale di Napoli, Vol. 52, Istituto Universitario Orientale, Napoli.

Fattovich, R. (2005) The Archaeological Area of Aksum (Ethiopia): Remote Sensing and GIS for a Reconstruction of the Ancient Landscape and the Archaeological Heritage Management, technical report of the archaeological investigations conducted at Aksum by the University of Naples 'LOrientale', Naples (Italy) between 2003 and 2005 submitted to Regione Campania, Naples.

Fattovich, R., Bard, K.A., Petrassi, L. and Pisano, V. (2000) The Aksum Archaeological Area: A Preliminary Assessment, Working Paper 1, Istituto Universitario Orientale, Naples.

Faugeras, O. (1993) *Three-dimensional Computer Vision*, MIT Press, Cambridge.

Favey, E., Pateraki, M., Baltsavias, E.P. *et al.* (2000) Surface modelling for Alpine glacier monitoring by airborne laser scanning and digital photogrammetry. XIXth Congress of the ISPRS, Amsterdam, pp. 269–277.

Feng, J., Rivard, B. and Sánchez-Azofeifa, A. (2003) The topographic normalization of hyperspectral data: implications for the selection of spectral end members and lithologic mapping. *Remote Sensing of Environment*, **85**, 221–231.

Fenna, D. (2006) *Cartographic Science: A Compendium of Map Projections, with Derivations*, CRC Press, Boca Raton.

Feoli, E., Vuerich, L.G. and Woldu, Z. (2002a) Evaluation of environmental degradation in northern Ethiopia using GIS to integrate vegetation, geomorphological, erosion and socio-economic factors. *Agriculture, Ecosystems and Environment*, **91**, 313–325.

Feoli, E., Vuerich, L.G. and Woldu, Z. (2002b) Processes of environmental degradation and opportunities for rehabilitation in Adwa, Northern Ethiopia. *Landscape Ecology*, **17**, 315–325.

Ferrari, M.C. (1992) Improved decorrelation stretching of TM data for geological applications: first results in Northern Somalia. *International Journal of Remote Sensing*, **13**, 841–851.

Ferro, C.J. and Warner, T.A. (2002) Scale and texture in digital image classification. *Photogrammetric Engineering and Remote Sensing*, **68**, 51–63.

Feynman, R. (1985) *QED: The Strange Theory of Light and Matter*, Princeton University Press, Princeton. Also published by Penguin Books, London (1990).

Fielding, A.H. (2007) *Cluster and Classification Techniques for the Biosciences*, Cambridge University Press, Cambridge.

Fiori, S. (2003) Overview of independent component analysis technique with an application to synthetic aperture radar (SAR) imagery processing. *Neural Networks*, **16**, 453–467.

Fischer, A., Rott, H. and Björnsson, H. (2003) Observation of recent surges of Vatnajökull, Iceland, by means of ERS SAR interferometry. *Annals of Glaciology*, **37**, 69–76.

Fisher, P.F. (1997) The pixel: a snare and a delusion. *International Journal of Remote Sensing*, **18**, 679–685.

Fisher, P.F. (1998) Improved modelling of elevation error with geostatistics. *Geoinformatica*, **2**, 215–233.

Flygare, A.-M. (1997) A comparison of contextual classification methods using Landsat TM. *International Journal of Remote Sensing*, **18**, 3835–3842.

Fogel, D.N. and Tinney, L.R. (1996) Image Registration using Multiquadric Functions, the Finite Element Method, Bivariate Mapping Polynomials and Thin Plate Spline. Technical Report 96-1. National Center for Geographic Information and Analysis (NCGIA), Simonett Center for Spatial Analysis (University of California), the State University of New York, and the University of Maine.

Foley, J.D., van Dam, A., Feiner, S.K. and Hughes, H.F. (1997) *Computer Graphics: Principles and Practice in C* (2nd edition). Addison-Wesley, Boston.

Foody, G.M. (1996a) Relating the land-cover composition of mixed pixels to artificial neural network classification output. *Photogrammetric Engineering & Remote Sensing*, **62**, 491–499.

Foody, G.M. (1996b) Approaches for the production and evaluation of fuzzy land cover classification from remotely-sensed data. *International Journal of Remote Sensing*, **17**, 1317–1340.

Foody, G.M. (2002a) Hard and soft classifications by a neural network with a non-exhaustively defined set of classes. *International Journal of Remote Sensing*, **23**, 3853–3864.

Foody, G.M. (2002b) Status of land cover classification accuracy assessment. *Remote Sensing of Environment*, **80**, 185–201.

Foody, G.M. (2004) Thematic map comparison: evaluating the statistical significance of differences in classification accuracy. *Photogrammetric Engineering and Remote Sensing*, **70**, 627–633.

Foody, G.M. (2005) Local characterization of thematic classification accuracy through spatially constrained confusion matrices. *International Journal of Remote Sensing*, **26**, 1217–1228.

Foody, G.M. (2008) Harshness in image classification accuracy assessment. *International Journal of Remote Sensing*, **29**, 3137–3158.

Foody, G.M. and Arora, M.K. (1996) Fuzzy thematic mapping: incorporating mixed pixels in the training, allocation and testing stages of supervised image classification, in *Proceedings of the 14th International Workshop on Soft Computing in Remote Sensing Data Analysis, Milan, 4–5 December 1995* (eds E. Binaghi, P.A. Brivio, and A. Rampini), World Scientific, Singapore, pp. 43–52.

Foody, G.M. and Curran, P.J. (1994) Scale and environmental remote sensing, in *Environmental Remote Sensing from Regional to Global Scales* (eds G.M. Foody, and P.J. Curran), John Wiley & Sons, Ltd, Chichester, pp. 223–232.

Foody, G.M. and Mathur, A. (2004a) Toward intelligent training of supervised image classifications: directing training data acquisition for SVM classification. *Remote Sensing of Environment*, **93**, 107–117.

Foody, G.M. and Mathur, A. (2004b) A relative evaluation of multiclass image classification by support vector machine. *IEEE Transactions on Geoscience and Remote Sensing*, **42**, 1335–1343.

Foody, G.M. and Mathur, A. (2006) The use of small training sets containing mixed pixels for accurate hard image classification: training on mixed spectral responses for classification by a SVM. *Remote Sensing of Environment*, **103**, 179–189.

Foody, G.M., Boyd, D.S. and Sanchez-Hernandez, C. (2007) Mapping a specific class with an ensemble of classifiers. *International Journal of Remote Sensing*, **28**, 1733–1746.

Foody, G.M., Lucas, R.M., Curran, P.J. and Honzak, M. (1996) Estimation of the areal extent of land-cover classes that only occur at the sub-pixel level. *Canadian Journal of Remote Sensing*, **22**, 428–432.

Foody, G.M., Lucas, R.M., Curran, P.J. and Honzak, M. (1997) Non-linear mixture modelling without end members using an artificial neural network. *International Journal of Remote Sensing*, **18**, 937–953.

Foody, G.M., Mathur, A., Sanchez-Hernandez, C. and Boyd, D.S. (2006) Training set size requirements for the classification of a specific class. *Remote Sensing of Environment*, **104**, 1–14.

Foody, G.M., McCullagh, M.B. and Yates, W.B. (1995) The effect of training set size and composition on artificial neural net classification. *International Journal of Remote Sensing*, **16**, 1707–1723.

Forshaw, M.R.B., Haskell, A., Miller, P.F. *et al.* (1983) Spatial resolution of remotely-sensed imagery: a review paper. *International Journal of Remote Sensing*, **4**, 497–520.

Fortin, M.J. and Dale, M.R.T. (2005) *Spatial Analysis: A Guide for Ecologists*, Cambridge University Press, Cambridge.

Fotheringham, A.S. and Rogerson, P.A. (eds) (2009) *The SAGE Handbook of Spatial Analysis*, SAGE Publications, London.

Fotheringham, A.S., Brundson, C. and Charlton, M. (2000) *Quantitative Geography: Perspectives on Spatial Data Analysis*, SAGE Publications, London.

Fourty, T. and Baret, F. (1998) On spectral estimates of fresh leaf biochemistry. *International Journal of Remote Sensing*, **19**, 1283–1297.

Franceschetti, G. and Lanari, R. (1999) *Synthetic Aperture Radar Processing*, CRC Press, Boca Raton.

Frank, T.D. (1985) Differentiating semiarid environments using Landsat reflectance data. *Professional Geographer*, **37**, 36–46.

Franklin, S.E. and Giles, P.T. (1995) Radiometric processing of aerial and satellite remote-sensing imagery. *Computers and Geosciences*, **21**, 413–425.

Franklin, S.E., Maudie, A.J. and Lavigne, M.B. (2001) Using spatial co-occurrence texture to increase forest structure and species composition classification accuracy. *Photogrammetric Engineering and Remote Sensing*, **67**, 849–855.

Fraser, C.S. and Yamakawa, T. (2004) Insights into the affine model for high-resolution satellite sensor orientation. *ISPRS Journal of Photogrammetry and Remote Sensing*, **58**, 275–288.

Fraser, C.S., Schroeder, M. and Baudoin, A. (eds) (2006), Special issue on extraction of topographic information from high-resolution satellite imagery. *ISPRS Journal of Photogrammetry and Remote Sensing*, **60**(3), 131–244.

Freeman, A., Villasenor, J., Klein, J.D. *et al.* (1994) On the use of multi-frequency and polarimetric radar backscatter features for classification of agricultural crops. *International Journal of Remote Sensing*, **15**, 1799–1812.

Frei, U., Graf, K.Chr. and Meier, E. (1993) Cartographic reference systems, in *SAR Geocoding: Data and Systems* (ed. G. Schreier), Herbert Wichmann, Karlsruhe, pp. 213–234.

French, C., Sulas, F., and Madella, M. (2009) New geoarchaeological investigations of the valley systems in the Aksum area of northern Ethiopia. *Catena*, **78**, 218–233.

Freund, Y. and Schapire, R.E. (1996) Experiments with a new boosting algorithm. Proceedings of the 13thInternational Conference on Machine Learning, Bari, Italy, 3–6 July, 1996, pp. 148–156.

Friedl, M.A., Brodley, C.E. and Strahler, A.H. (1999) Maximising land cover classification accuracies produced by decision trees at continental to global scales. *IEEE Transactions on Geoscience and Remote Sensing*, **37**, 969–977.

Frost, V.S., Stiles, J.A., Shanmugan, K.S. and Holtzman, J.C. (1982) A model for radar images and its application to adaptive digital filtering of multiplicative noise. *IEEE Transactions on Pattern Analysis and Machine Intelligence*, **4**, 157–166.

Frulla, L.A., Milovich, J.A. and Gagliardini, D.A. (1995) Illumination and observational geometry for NOAA-AVHRR imagery. *International Journal of Remote Sensing*, **16**, 2233–2253.

Fu, L.-L. and Cazenava, A. (eds) (2000) *Satellite Altimetry and Earth Sciences: A Handbook of Techniques and Applications*, Academic Press, San Diego.

Fusco, L. and Trevese, D. (1985) On the reconstruction of lost data in images of more than one band. *International Journal of Remote Sensing*, **6**, 1535–1544.

Gaber, A., Koch, M. and El-Baz, F. (2010) Textural and compositional characterization of Wadi Feiran deposits, Sinai Peninsula, Egypt, using Radarsat-1, PALSAR, SRTM and ETM+ Data. *Remote Sensing*, **2**(1), 52–75. Available from http://www.mdpi.com/2072-4292/2/1/52 (accessed 7 June 2010).

Gabriel, A., Goldstein, R. and Zebker, H.A. (1989) Mapping small elevation changes over large areas: differential radar interferometry. *Journal of Geophysical Research*, **94**, 9183–9191.

Gadallah, F.L., Csillag, F. and Smith, E.J.M. (2000) Destriping multisensor imagery with moment matching. *International Journal of Remote Sensing*, **21**, 2505–2511.

Gallego, F.J. (2004) Remote sensing and land cover area estimation. *International Journal of Remote Sensing*, **25**, 3019–3047.

Gamon, J.A., Field, C.B., Roberts, D.A. *et al.* (1993) Functional patterns in an annual grassland during and AVIRIS overflight. *Remote Sensing of Environment*, **44**, 239–253.

Gao, B.-C., Davis, C. and Goetz, A. (2006) A review of atmospheric correction techniques for hyperspectral remote sensing of land surfaces and ocean colour, *Proceeedings of the International Geoscience and Remote Sensing Symposium, IGARSS 2006, Denver, Colorado, July 31–4 August 2006*, IEEE Press, Piscataway, pp. 1979–1981.

Gao, B.-C., Heidebrecht, K.B. and Goetz, A.F.H. (1993) Derivation of scaled surface reflectances from AVIRIS data. *Remote Sensing of Environment*, **44**, 145–163.

Gao, J. (2009) *Digital Analysis of Remotely Sensed Imagery*, McGraw-Hill, New York.

Gao, J. and Liu, Y. (2008) Mapping of land degradation from space: a comparative study of Landsat ETM+ and ASTER data. *International Journal of Remote Sensing*, **29**, 4029–4043.

Gao, Q., Zhang, L., Zhang, D. and Xu, H. (2010) Independent components extraction from image matrix. *Pattern Recognition Letters* **31**, 171–178.

Garcia-Haro, F.J., Gilabert, M.A. and Melia, J. (1996) Linear spectral mixture modelling to estimate vegetation amount from optical spectral data. *International Journal of Remote Sensing*, **17**, 3373–3400.

Garguet-Duport, B., Girel, J., Chassery, J. and Pautou, G. (1996) The use of multiresolution analysis and wavelets transform for merging SPOT panchromatic and multispectral image data. *Photogrammetric Engineering and Remote Sensing*, **62**, 1057–1066.

Garvin, J., Bufton, J., Blair, J. *et al.* (1998) Observations of the earth's topography from the Shuttle Laser Altimeter (SLA): laser-pulse echo-recovery measurements of terrestrial surfaces. *Physics and Chemistry of the Earth*, **23**, 1053–1068.

Gautama, S., Heen, G., Pires, R. *et al.* (Undated) Computer Vision Techniques for Remote Sensing, http://telin.ugent.be/~sid/papers/gautama00dwtc.pdf (acessed 29 July 2009).

Gauthier, Y., Bernier and Fortin J.-P. (1998) Aspect and incidence angle sensitivity in ERS-1 SAR data. *International Journal of Remote Sensing*, 19, 2001–2006.

Gens, R. (2003) Two-dimensional phase unwrapping for radar interferometry: developments and new challenges. *International Journal of Remote Sensing*, **24**, 703–710.

Gens, R. and van Genderen, J.L. (1996a) Analysis of the geometric parameters of SAR interferometry for spaceborne systems. *International Archives of Photogrammetry and Remote Sensing*, **XXXI** (B2), 107–110.

Gens, R. and van Genderen, J.L. (1996b) SAR interferometry: issues, techniques, applications. *International Journal of Remote Sensing*, **17**, 1803–1835.

Gheith, H. and Sultan, M. (2002) Construction of a hydrologic model for estimating Wadi runoff and groundwater recharge in the Eastern Desert, Egypt. *Journal of Hydrology*, **263**, 36–55.

Ghiglia, D.C. and Pritt, M.D. (1998) *Two-Dimensional Phase Unwrapping: Theory, Algorithms and Software*, John Wiley & Sons, Inc., New York.

Ghosh, A. and Pal, S.K. (eds) (2002) *Soft Computing Approach to Pattern Recognition and Image Processing*, World Scientific Publishing Company, Singapore.

Gilabert, M.A., Gonzales-Piqueras, J., Garcia-Haro, F.J. and Melia, J. (2002) A generalised soil-adjusted vegetation index. *Remote Sensing of Environment*, **82**, 303–310.

Giles, P.T. and Franklin, S.E. (1996) Comparison of derivative topographic surfaces of a DEM generated from stereoscopic SPOT images with field measurements. *Photogrammetric Engineering and Remote Sensing*, **62**, 1165–1171.

Gillespie, A.R., Kahle, A.B. and Walker, R.E. (1986) Colour enhancement of highly correlated images. I: decorrelation and HSI contrast stretches. *Remote Sensing of Environment*, **20**, 209–235.

Gillespie, T.W., Chu, J., Frankenberg, E. and Thomas, D. (2007) Assessment and prediction of natural hazards from satellite imagery. *Progress in Physical Geography*, **31**, 459–470.

Gillies, R.R., Carlson, T.N., Cui, J. *et al.* (1997) Verification of the 'triangle' method for obtaining surface soil water content and energy fluxes from remote measurements of the Normalized Difference Vegetation Index NDVI and surface radiant temperature. *International Journal of Remote Sensing*, **18**, 3145–3166.

Gislason, P.O., Benediktsson, J.A. and Sveinsson, J.R. (2006) Random forests for land cover classification. *Pattern Recognition Letters*, **27**, 294–300.

Gitas, I.Z. and Devereux, B.J. (2006) The role of topographic correction in mapping recently burned Mediterranean forest areas from LANDSAT TM images. *International Journal of Remote Sensing*, **27**, 41–54.

Gitelson, A.A., Kaufman, Y.J., Stark, R. and Rundquist, D. (2002) Novel algorithms for remote estimation of vegetation fraction. *Remote Sensing of Environment*, **80**, 76–87.

Goetz, A.F.H. (1989) Spectral remote sensing in geology, in *Theory and Applications of Optical Remote Sensing* (ed. G. Asrar), John Wiley & Sons, Inc., New York, pp. 491–526.

Goetz, A.F.H., Rock, B.N. and Rowan, L.C. (1983) Remote sensing for exploration: an overview. *Economic Geology*, **78**, 573–589.

Goldstein, R.M., Engelhardt, H., Kamb, B. and Frolich, R.M. (1993) Satellite radar interferometry for monitoring ice-sheet motion: application to Antarctic ice stream. *Science*, **262**, 1525.

Gomboši, M. and Žalik, B. (2005) Point-in-polygon tests for geometric buffers. *Computers and Geosciences*, **31**, 1201–1212.

Gómez, D., Biging, G. and Montero, J. (2008) Accuracy statistics for judging soft classification. *International Journal of Remote Sensing*, **29**, 693–709.

Gomez-Chova, L., Calpe, J., Soria, E. *et al.* (2003) CART-based feature selection of hyperspectral images for crop cover classification. Proceedings of the 2003 IEEE International Conference on Image Processing (ICIP 2003), Barcelona, Spain, September 14–18, vol. 3, pp. 589–592.

Gong, P., Ledrew, E.F. and Miller, J.R. (1992) Registration noise reduction in difference images for change detection. *International Journal of Remote Sensing*, **13**, 773–739.

Gong, P., Pu, R. and Chen, J. (1996) Mapping ecological land systems and classification uncertainties from digital elevation and forest-cover data using neural networks. *Photogrammetric Engineering and Remote Sensing*, **62**, 1249–1260.

Gong, P., Pu, R. and Yu, B. (1997) Conifer species recognition: an exploratory analysis of *in situ* hyperspectral data. *Remote Sensing of Environment*, **62**, 189–200.

Gong, P., Pu, R., Biging, G.S. and Larrieu, M.R. (2003) Estimation of forest leaf area index using vegetation indices derived from Hyperion hyperspectral data. *IEEE Transactions on Geoscience and Remote Sensing*, **41**, 1355–1362.

Gonzales, R.C. and Woods, R.E. (2007) *Digital Image Processing*, 2nd edn, Prentice-Hall, Englewood Cliffs.

Goodchild, M.F. (2005) GIS, spatial analysis, and modelling overview, in *GIS, Spatial Analysis, and Modelling*, (eds D.J. Maguire, M. Batty and M.F. Goodchild), ESRI Press, Redlands, pp. 1–17.

Goovaerts, P. (1999) Geostatistics in soil science: state-of-the-art and perspectives. *Geoderma*, **89**, 1–45.

Gopal, S. and Woodcock, C. (1994) Theory and methods for accuracy assessment of thematic maps using fuzzy sets. *Photogrammetric Engineering and Remote Sensing*, **60**, 181–188.

Gopal, S. and Woodcock, C. (1996) Remote sensing of forest change using artificial neural networks. *IEEE Transactions on Geoscience and Remote Sensing*, **34**, 398–403.

Gopi, E.S. (2007) *Algorithm Collections for Digital Signal Processing Applications Using Matlab*, Springer-Verlag, Berlin.

Goshtasby, A.A. (2005) *2-D and 3-D Image Registration for Medical, Remote Sensing, and Industrial Applications*, John Wiley & Sons, Inc., New York.

Gowe, J.F.R. (ed.) (2006) *Manual of Remote Sensing*, Remote Sensing of the Marine Environment, Vol. 6, American Society for Photogrammetry and Remote Sensing, Bethesda.

Graef, F. and Haigis, J. (2001) Spatial and temporal rainfall variability in the Sahel and its effects on farmers' management strategies. *Journal of Arid Environments*, **48**, 221–231.

Grafarend, E.W. and Krumm, F.W. (2006) *Map Projections: Cartographic Information Systems*, Springer-Verlag, Berlin.

Grafarend, E.W. and Krumm, F.W. (2006) *Map Projections: Cartographic Information Systems*, Springer-Verlag, Berlin.

Green, A.A., Berman, M., Switzer, P. and Craig, M.D. (1988) A transformation for ordering multispectral data in terms of image quality with implications for noise removal. *IEEE Transactions on Geoscience and Remote Sensing*, **26**, 65–64.

Greenfield, S. (1997) *The Human Brain: A Guided Tour*, Weidenfeld and Nicholson, London.

Gregory, R.L. (1998) *Eye and Brain: The Psychology of Seeing*, 5th edn, Oxford University Press, Oxford.

Grey, W.M.F., Luckman, A.J. and Holland, D. (2003) Mapping urban change in the UK using satellite radar interferometry. *Remote Sensing of Environment*, **87**, 16–22.

Gribben, J. (1984) *In Search of Schrödinger's Cat*, Wildwood House, Ltd, London.

Grings, F.M., Ferrazzoli, P., Jacobo-Berlles, J.C. *et al.* (2006) Monitoring flood condition in marshes using EM models and Envisat ASAR observations. *IEEE Transactions on Geoscience and Remote Sensing*, **44**, 936–942.

Grossman, Y.L., Ustin, S.L., Jacquemond, S. *et al.* (1996) Critique of stepwise multiple linear regression for the extraction of leaf biochemistry information from leaf reflectance data. *Remote Sensing of Environment*, **56**, 182–193.

Gruen, A. and Li, H. (1995) Road extraction from aerial and satellite images by dynamic programming. *ISPRS Journal of Photogrammetry and Remote Sensing*, **50**, 11–20.

Gu, D., Gillespie, A.R., Adams, J.B. and Weeks, R. (1999) A statistical approach for topographic correction of satellite images using spatial context information. *IEEE Transactions on Geoscience and Remote Sensing*, **37**, 236–246.

Guanter, L., Alonso, L. and Moreno, J. (2005) First results from the PROBA/CHRIS hyperspectral/multi-angular satellite system over land and water targets. *Geoscience and Remote Sensing Letters*, **2**, 250–254.

Guanter, L., Richter, R. and Kaufmann, H. (2009) On the application of the MODTRAN4 atmospheric radiative transfer code to optical remote sensing. *International Journal of Remote Sensing*, **30**, 1407–1424.

Guo, L.J. and Moore, J.Mc.M. (1996) Direct decorrelation stretch technique for RGB colour composition. *International Journal of Remote Sensing*, **17**, 1005–1018.

Gupta, R.P. (2003) *Remote Sensing in Geology*, 2nd edn, Springer-Verlag, Berlin.

Gurney, C.M. and Townshend, J.R.G. (1983) The use of contextual information in the classification of remotely sensed data. *Photogrammetric Engineering and Remote Sensing*, **49**, 55–64.

Gutman, G. and Ignatov, A. (1995) Global land monitoring from AVHRR: potential and limitations. *International Journal of Remote Sensing*, **16**, 2301–2309.

Guyot, G. and Baret, F. (1988) Utilisation de la haute resolution spectrale pour suivre l'etat des couverts vegetaux. Fourth International Colloquium on Spectral Signatures of Objects in Remote Sensing, European Space Agency, Aussois, pp. 279–286.

Guyot, G. and Gu, X.-F. (1994) Effect of radiometric corrections on NDVI determined from SPOT-HRV and Landsat-TM data. *Remote Sensing of Environment*, **49**, 169–180.

Haboudane, D., Miller, J.R., Pattey, E. *et al.* (2004) Hyperspectral vegetation indices and novel algorithms for predicting green LAI of crop canopies: modeling and validation in the context of precision agriculture. *Remote Sensing of Environment*, **90**, 337–352.

Haboudane, D., Miller, J.R., Tremblay, N. *et al.* (2002) Integrated narrow-band vegetation indices for prediction of crop chlorophyll content for application to precision agriculture. *Remote Sensing of Environment*, **81**, 416–426.

Haining, R.P. (2003) *Spatial Data Analysis: Theory and Practice*, Cambridge University Press, Cambridge.

Hale, S.R. and Rock, B.N. (2003) Impact of topographic normalization on land-cover classification accuracy. *Photogrammetric Engineering and Remote Sensing*, **69**, 785–791.

Hall, F.G., Strebel, D.E., Nickeson, J.R. and Goetz, S.J. (1991) Radiometric rectification: toward a common radiometric response among multidate, multisensor images. *Remote Sensing of Environment*, **35**, 11–27.

Han, J. and Kamber, M. (2006) *Data Mining: Concepts and Techniques*, Elsevier and Morgan Kaufmann, Amsterdam.

Hansen, M., Stehman, S., Potapov, P. *et al.* (2008) Humid tropical forest clearing from 2000 to 2005 quantified using multi-temporal and multi-resolution remotely sensed data. *Proceedings National Academy of Sciences*, **10** (27), 9439–9444.

Hansen, P.M. and Schjoerring, J.K. (2003) Reflectance measurement of canopy biomass and nitrogen status in wheat crops using normalized difference vegetation indices and partial least squares regression. *Remote Sensing of Environment*, **86**, 542–553.

Hanssen, R. (2001) *Radar Interferometry*, Kluwer Academic Publishing, Dordrecht.

Haralick, R.M. and Fu, K.-S. (1983) Pattern recognition and classification, in *Manual of Remote Sensing*, vol. 2 (ed. R.N. Colwell), American Society of Photogrammetry, Falls Church, pp. 793–805.

Haralick, R.M. and Shanmugam, K.S. (1974) Combined spectral and spatial processing of ERTS imagery data. *Remote Sensing of Environment*, **3**, 3–13.

Haralick, R.M., Shanmugam, K.S. and Dinstein, I. (1973) Textural features for image classification. *IEEE Transactions on Systems, Man and Cybernetics*, **3**, 610–622.

Harding, D.J., Lefsky, M.A., Parker, G.G. and Blair, J.B. (2001) Laser altimetry waveform measurement of vegetation canopy structures: measurement and validation for closed-canopy, broadleaved forests. *Remote Sensing of Environment*, **76**, 283–297.

Harries, J., Llewellyn-Jones, D.T., Mutlow, C. *et al.* (1994) The ATSR programme: Instruments, data and science, in *TERRA 2: Understanding the Terrestrial Environment – Remote Sensing Data Systems and Networks* (ed. P.M. Mather), John Wiley & Sons, Ltd, Chichester, pp. 20–28.

Harris, R. (1981) An experiment in probabilistic relaxation for terrain cover classification of Kuwait from Landsat imagery. Proceedings of the 15th International Conference on Remote Sensing of Environment, Environmental Research Institute of Michigan, Ann Arbor, Michigan, pp. 1245–1252.

Harris, R. (1985) Contextual classification post-processing of Landsat data using a probabilistic relaxation model. *International Journal of Remote Sensing*, **6**, 847–866.

Hay, J.C. and Mackay, D.C. (1985) Estimating solar irradiances on inclined surfaces: a review and assessment of methodologies. *International Journal of Solar Energy*, **3**, 203–240.

Healy, M.J.R. (1968) Multivariate normal plotting. *Applied Statistics*, **17**, 157–161.

Hearn, D. and Baker, M.P. (1994), *Computer Graphics*, 2nd edn, Prentice-Hall International, London.

Hearn, D. and Baker, M.P. (1997) *Computer Graphics, C Version*. Prentice-Hall, Upper Saddle River, NJ.

Hecht, E. (2001) *Optics*, 4th edn, Addison-Wesley, Reading.

Hecker, C., van der Meijde, M., van der Werff, H. and van der Meer, F.D. (2008) Assessing the influence of reference spectra on synthetic SAM classification results. *IEEE Transactions on Geoscience and Remote Sensing*, **47**, 4162–4172.

Held, A., Ticehurst, C., Lymburner, L. and Williams, N. (2003) High resolution mapping of tropical mangrove ecosystems using hyperspectral and radar remote sensing. *International Journal of Remote Sensing*, **24**, 2739–2759.

Helder, D.L. and Ruggles, T.A. (2004) Landsat thematic mapper reflective-band radiometric artifacts. *IEEE Transactions on Geoscience and Remote Sensing*, **42**, 2704–2716.

Helder, D.L., Markham, B.L., Thome, K.J. *et al.* (2008) Updated radiometric calibration for the Landsat-5 Thematic Mapper reflective bands. *IEEE Transactions on Geoscience and Remote Sensing*, **46**, 3309–3325.

Hellwich, O. (1999a) Basic principles and current issues of SAR interferometry. ISPRS Joint Workshop Sensors and Mapping from Space 1999, Hannover, Institute for Photogrammetry and Engineering Surveying, University of Hannover.

Hellwich, O. (1999b) SAR interferometry: principles, processing and perspectives, in *Festschrift für Prof. Dr.-Ing. Heinrich Ebner* (eds C. Heipke, and H. Mayer), Lehrstuhl für Photogrammetrie und Fernerkundung, Technische Universität München, Munich, pp. 109–120.

Henderson, F.M., and Lewis, A.J. (eds) (1998) Principles and applications of imaging radar, in *Manual of Remote Sensing*, 3rd edn, vol. 2, John Wiley & Sons (in cooperation with ASPRS), New York.

Henebry, G.M. (1997) Advantages of principal components analysis for land cover segmentation from SAR image series. *ERS Symposium on Space at the Service of our Environment*, 14–21 March, 1997, Florence, Italy, pp. 175–178.

Hengl, T. (2006) Finding the right pixel size. *Computers and Geosciences*, **32**, 1283–1298.

Hepner, G.F., Logan, T., Ritter, N. and Bryant, N. (1990) Artificial neural net classification using a minimal training set: comparison to conventional supervised classification. *Photogrammetric Engineering and Remote Sensing*, **56**, 469–473.

Heurich, M. (2008) Automatic recognition and measurement of single trees based on data from airborne laser scanning over the richly structured natural forests of the Bavarian Forest National Park. *Forest Ecology and Management*, **255**, 2416–2433.

Heywood, I., Cornelius, S. and Carver, S. (2002) *An Introduction to Geographical Information Systems*, 2nd edn, Pearson Education Ltd, Harlow.

Higham, N.J. (2002) *Accuracy and Stability of Numerical Algorithms*, 2nd edn, SIAM, Philadelphia.

Hill, J. (1991) A quantitative approach to remote sensing: sensor calibration and comparison, in *Remote Sensing and Geographical Information Systems for Resource Management in Developing Countries*, Euro-Courses: Remote Sensing, Vol. 1 (eds A.S. Belward, and C.R. Valenzuela), Kluwer Academic Publishers, Dordrecht, pp. 97–110.

Hill, J. and Aifadopoulou, D. (1990) Comparative analysis of Landsat-5 TM and SPOT HRV-1 data for use in multiple sensor approaches. *Remote Sensing of Environment*, **34**, 55–70.

Hill, J. and Horstert, P.O. (1996) Monitoring the growth of a Mediterranean metropolis based on the analysis of spectral mixtures – a case study on Athens, in *Progress in Environmental Remote Sensing Research and Applications*, Proceedings of the 15th EARSeL Symposium, Basle, Switzerland, 4-6 September 1996 (ed. E. Parlow), A.A. Balkema, Rotterdam, pp. 21–31.

Hill, J., Mehl, W. and Radeloff, V. (1995) Improved terrain mapping by combining correction of atmospheric and topographic effects in Landsat TM imagery, in *Sensors and Environmental Applications of Remote Sensing*, Proceedings of the 14th EARSeL Symposium, Goteborg, Sweden, 6–8 June, 1994 (ed. J. Askne), A.A. Balkema, Rotterdam, pp. 143–151.

Hill, M.J., Braaten, R., Veitch, S.M. *et al.* (2005) Multi-criteria decision analysis in spatial decision support: the ASSESS analytic hierarchy process and the role of quantitative methods and spatially explicit analysis. *Environmental Modelling and Software*, **20**, 955–976.

Hill, R.A. and Thomson, A.G. (2005) Mapping woodland species composition and structure using airborne spectral and LIDAR data. *International Journal of Remote Sensing*, **26**, 3763–3779.

Hirano, A., Welch, R. and Lang, H. (2003) Mapping from ASTER stereo image data: DEM validation and accuracy assessment. *ISPRS Journal of Photogrammetry and Remote Sensing*, **57**, 356–370.

Hird, J.N. and McDermid, G.J. (2009) Noise reduction of NDVI time series: an empirical comparison of selected techniques. *Remote Sensing of Environment*, **113**, 248–258.

Hobbs, R.J. and Mooney, H.A. (eds) (1990) *Remote Sensing of Biospheric Functioning*, Springer-Verlag, New York.

Hodgson, M.E. and Bresnahan, P. (2004) Accuracy of airborne lidar-derived elevation: empirical assessment and error budget. *Photogrammetric Engineering and Remote Sensing*, **70**, 331–339.

Hodgson, M.E., Jensen, J.R., Schmidt, L. *et al.* (2003) An evaluation of LiDAR- and IFSAR-derived digital elevation models in leaf-on conditions with USGS Level 1 and Level 2 DEMs. *Remote Sensing of Environment*, **84**, 295–308.

Holben, B.N. and Fraser, R.S. (1984) Red and near-infrared sensor response to off-nadir viewing. *International Journal of Remote Sensing*, **5**, 145–160.

Holben, B.N. and Kimes, D. (1986) Directional reflectance response in AVHRR red and near-infrared bands for three cover types and varying atmospheric conditions. *Remote Sensing of Environment*, **19**, 213–226.

Holland, D.A., Boyd, D.S. and Marshall, P. (2006) Updating topographic mapping in Great Britain using imagery from high-resolution satellite sensors. *ISPRS Journal of Photogrammetry and Remote Sensing*, **60**, 212–223.

Holland, J.H. (1992) Genetic algorithms. *Scientific American*, **267**, 44–50.

Holm, R.G., Moran, M.S., Jackson, R.D. *et al.* (1989) Surface reflectance factor retrieval from Thematic Mapper data. *Remote Sensing of Environment*, **27**, 47–57.

Hong, G. and Zhang, Y. (2008a) A comparative study on radiometric normalization using high resolution satellite images. *International Journal of Remote Sensing*, **29**, 425–438.

Hong, G. and Zhang, Y. (2008b) Comparison and improvement of wavelet-based image fusion. *International Journal of Remote Sensing*, **29**, 673–691.

Hong, T.D. and Schowengerdt, R.A. (2005) A robust technique for precise registration of radar and optical satellite images. *Photogrammetric Engineering and Remote Sensing*, **71**, 585–593.

Hooper, D.M., Bursik, M.I. and Webb, F.E. (2003) Application of high-resolution, interferometric DEMs to geomorphic studies of fault scarps, Fish Lake Valley, Nevada-California, USA. *Remote Sensing of Environment*, **84**, 255–267.

Horgan, G.W. (1999) Using wavelets for data smoothing: a simulation study. *Journal of Applied Statistics*, **26**, 923–932.

Horler, D.N.H., Dockray, M. and Barber, J. (1983) The red edge of plant leaf reflectance. *International Journal of Remote Sensing*, **4**, 273–288.

Horn, B.K.P. and Woodham, R.J. (1979) Destriping Landsat MSS images by histogram modification. *Computer Graphics and Image Processing*, **10**, 69–83.

Horne, J.H. (2003) A Tasseled Cap transformation for Ikonos images. ASPRS 2003 Annual Conference Proceedings, May 2003, Anchorage, Alaska. Available from: http://www2.erdas.com/SupportSite/downloads/model_files/IKONOStcap.pdf (accessed 4 April 2009).

Hsieh, W.W. (2008) Nonlinear principal components analysis, in *Artificial Intelligence Methods in the Environmental Sciences* (eds S.E. Haupt, A. Pasini, and C. Marzban), Springer-Verlag, Berlin, pp. 173–190.

Hsu, C.-W. and Lin, C.-J. (2002) A comparison of methods for multi-class support vector machines. *IEEE Transaction on Neural Networks*, **13**, 415–425.

Huang, C., Davis, L.S. and Townshend, J.R.G. (2002a) An assessment of support vector machines for land cover classification. *International Journal of Remote Sensing*, **23**, 725–749.

Huang, C., Townshend, J.R.G., Liang, S. *et al.* (2002b) Impact of a sensor's point spread function on land cover characterization: assessment and deconvolution. *Remote Sensing of Environment*, **80**, 203–212.

Huang, C., Wylie, B., Yang, L. *et al.* (2002c) Derivation of a tasseled cap transformation based on Landsat 7 at-satellite reflectance. *International Journal of Remote Sensing*, **23**, 1741–1748.

Huang, H., Gong, P., Clinton, N. and Hui, F. (2008) Reduction of atmospheric and topographic effect on Landsat TM data for forest classification. *International Journal of Remote Sensing*, **29**, 5623–5642.

Huang, H.-L. and Antonelli, P. (2001) Application of principal component analysis to high-resolution infrared measurement compression and retrieval. *Journal of Applied Meteorology*, **40**, 365–388.

Huang, T.S., Yang, G.J. and Tang, G.Y. (1979) A fast two-dimensional median filtering algorithm. *IEEE Transactions on Acoustics, Speech and Signal Processing*, **27**, 13–18.

Huete, A. (1989) Soil influences in remotely sensed vegetation-canopy spectra, in *Theory and Applications of Optical Remote Sensing* (ed. G. Asrar), John Wiley & Sons, Inc., New York, pp. 107–141.

Huete, A. (2004) Remote sensing of soils and soil processes, in *Remote Sensing for Natural Resources Management and Environmental Monitoring.*, Manual of Remote Sensing, Vol. 4, 3rd edn (ed. S. Ustin), John Wiley & Sons, Inc., New York, pp. 1–48.

Huete, A., Didan, K., Miura, T. *et al.* (2002) Overview of the radiometric and biophysical performance of the MODIS vegetation indices. *Remote Sensing of Environment*, **83**, 195–213.

Hughes, G.F. (1968) On the mean accuracy of statistical pattern recognizers. *IEEE Transactions on Information Theory*, **14**, 55–63.

Huguenin, R.L. and Jones, J.L. (1986) Intelligent information extraction from reflectance spectra: absorption band positions. *Journal of Geophysical Research*, **19** (B9), 9585–9598.

Hung, C.C. (1993) Competitive learning networks for unsupervised training. *International Journal of Remote Sensing*, **14**, 2411–2415.

Hunt, G.R. (1977) Spectral signatures of particulate minerals in the visible and near-infrared. *Geophysics*, **42**, 501–513.

Hunt, G.R. (1979) Near-infrared (1.3–2.4 μm) spectra of alteration minerals: potential for use in remote sensing. *Geophysics*, **44**, 1974–1986.

Hunt, G.R. and Ashley, R.P. (1979) Spectra of altered rocks in the visible and near infrared. *Economic Geology*, **74**, 1613–1629.

Hunt, G.R. and Salisbury, J.W. (1970) Visible and near-infrared spectra of minerals and rocks: I silicate minerals. *Modern Geology*, **1**, 283–300.

Hunt, G.R. and Salisbury, J.W. (1971) Visible and near-infrared spectra of minerals and rocks: II carbonates. *Modern Geology*, **2**, 23–30.

Hunt, G.R., Salisbury, J.W. and Lenhoff, D.J. (1971) Visible and near-infrared spectra of minerals and rocks: III oxides and hydroxides. *Modern Geology*, **2**, 195–205.

Hunter, G.J. and Goodchild, M.F. (1997) Modeling the uncertainty of slope and aspect estimates derived from spatial databases. *Geographical Analysis*, **29**, 35–49.

Hussein, H.A., Rabie, S.I. and Abdel Nabie, S. (1996) Structural interpretation of aeromagnetic data, Gabel Zubeir area, North Eastern Desert, Egypt. *International Journal of Remote Sensing*, **17**, 1997–2012.

Hutchinson, C.F. (1982) Techniques for combining Landsat and ancillary data for digital classification improvement. *Photogrammetric Engineering and Remote Sensing*, **48**, 123–130.

Hutchinson, M.F. (2008) Adding the Z dimension, in *The Handbook of Geographic Information Science* (eds J.P. Wilson, and A.S. Fotheringham), Blackwell Publishing, Ltd, Oxford, pp. 144–168.

Hutchison, K.D. and Cracknell, A.P. (2005) *Visible Infrared Imager Radiometer Suite: A New Operational Cloud Imager*, CRC Press, Boca Raton.

Hyder, A.K., Shabazian, E. and Waltz, E. (eds) (2002) *Multisensor Fusion*, Kluwer Academic Publishers, Dordrecht.

Hyman, A.H. and Barnsley, M.J. (1997) On the potential for land cover mapping from multiple-view-angle (MVA) remotely sensed images, in *Proceedings of the 23rd Annual Conference and Exhibition of the Remote Sensing Society, University of Reading, 2–4 September 1997*, The Remote Sensing Society, Nottingham, pp. 135–140.

Hyppänen, H. (1996) Spatial autocorrelation and optimum spatial resolution of optical remote sensing data in boreal forest environment. *International Journal of Remote Sensing*, **17**, 3441–3452.

Hyvärinen, A. and Oja, E. (2000) Independent component analysis: algorithms and applications. *Neural Networks*,

13, 411–430. Also available at http://www.cs.helsinki.fi/u/ahyvarin/papers/NN00new.pdf (accessed 29 September 2008).

Hyvärinen, A., Karhunen, J. and Oja, E. (2001) *Independent Component Analysis*, John Wiley & Sons, Inc., New York.

Hyyppä, J., Hyyppä, H., Leckie, D. *et al.* (2008) Review of methods of small-footprint airborne laser scanning for extracting forest inventory data in boreal forests. *International Journal of Remote Sensing*, **29**, 1339–1366.

Ichoku, C. and Karnieli, A. (1996) A review of mixture modelling techniques for sub-pixel land cover estimation. *Remote Sensing Reviews*, **13**, 161–186.

Ichoku, C., Karnieli, A., Meisels, A. and Chorowicz, J. (1996) Detection of drainage channel networks on satellite imagery. *International Journal of Remote Sensing*, **17**, 1659–1678.

IEEE Transactions on Geoscience and Remote Sensing (2005) Special issue on the ASTER instrument. *IEEE Transactions on Geoscience and Remote Sensing*, **43**(12).

IEEE Transactions on Geoscience and Remote Sensing (2007) Special issue on Pattern Recognition in Remote Sensing. *IEEE Transactions on Geoscience and Remote Sensing*, **45**(12).

IEEE Transactions on Geoscience and Remote Sensing (2008) Special issue on data fusion. *IEEE Transactions on Geoscience and Remote Sensing*, **46**(5).

International Journal of Remote Sensing (2006) Special issue on scales and dynamics in observing the environment. *International Journal of Remote Sensing*, **27** (11).

Imbrie, J. (1963) Factor and Vector Analysis Programs for Analyzing Geologic Data. Technical Report 6, Office of Naval Research Geography Branch, Task No. 389-135, Contract No. 1228(26), Northwestern University, Evanston, October 1963.

Irish, J.L. and Lillycrop, W.J. (1999) Scanning laser mapping of the coastal zone: the SHOALS system. *ISPRS Journal of Photogrammetry and Remote Sensing*, **54**, 123–129.

Irish, R. (2008) *Landsat 7 Science Data Users Handbook*, NASA Goddard Spaceflight Centre, Maryland, URL: http://landsathandbook.gsfc.nasa.gov/handbook.html (accessed 23 January 2009).

Irons, J.R., Weismiller, R.A. and Petersen, G.W. (1989) Soil reflectance, in *Theory and Applications of Optical Remote Sensing* (ed. G. Asrar), John Wiley & Sons, Inc., New York, pp. 66–106.

ISPRS Journal of Photogrammetry and Remote Sensing (2006) Special issue on extraction of topographic information from high-resolution satellite imagery, ed. Fraser, C., Schroeder, M. and Baudoin, A. *ISPRS Journal of Photogrammetry and Remote Sensing*, **60**(3), 131–224.

ISPRS Journal of Photogrammetry and Remote Sensing (2008) Special issue on terrestrial laser scanning. *ISPRS Journal of Photogrammetry and Remote Sensing*, **63**(1), pp. 1–154.

Israelsen, T. and Frederiksen, R.D. (2005) The use of GIS in transport modelling, in *GIS, Spatial Analysis, and Modelling* (eds D.J. Maguire, M. Batty, and M.F. Goodchild), ESRI Press, Redlands, pp. 265–288.

Itten, K.I. and Meyer, P. (1993) Geometric and radiometric correction of TM data of mountainous forested areas. *IEEE Transactions on Geoscience and Remote Sensing*, **31**, 764–770.

Ivits, E., Koch, B., Blaschke, T. *et al.* (2005) Landscape structure assessment with image grey-values and object-based classification. *International Journal of Remote Sensing*, **26**, 2975–2993.

Jackson, J.E. (1991) *A User's Guide to Principal Components*, John Wiley & Sons, Inc., New York.

Jackson, R.D. (1983) Spectral indices in n-space. *Remote Sensing of Environment*, **13**, 409–421.

Jackson, R.D., Slater, P.N. and Pinter, P.J. Jr. (1983) Discrimination of growth and water stress in wheat by various vegetation indices through clear and turbid atmospheres. *Remote Sensing of Environment*, **13**, 187–208.

Jacquemoud, S. and Baret, F. (1990) PROSPECT: a model of leaf optical properties spectra. *Remote Sensing of Environment*, **34**, 75–91.

Jacquez, G.M., Maruca, S. and Fortin, M.-J. (2000) From fields to objects: A review of geographic boundary analysis. *Journal of Geographical Systems*, **2**, 221–241.

Jaiswal, R.K., Mukherjee, S., Krishnamurthy, J. and Saxena, R. (2003) Role of remote sensing and GIS techniques for generation of groundwater prospect zones towards rural development – an approach. *International Journal of Remote Sensing*, **24**, 993–1008.

Jakubauskas, M.E., Legates, D.R. and Kastens, J.H. (2001) Harmonic analysis of time-series AVHRR NDVI data. *Photogrammetric Engineering and Remote Sensing*, **67**, 461–470.

Jarvis, C.H. and Stuart, N. (1996) The sensitivity of a neural net for classifying remotely sensed imagery. *Computers and Geosciences*, **22**, 959–967.

Jasrotia, A.S., Kumar, R. and Saraf, A.K. (2007) Delineation of groundwater recharge sites using integrated remote sensing and GIS in Jammu district, India. *International Journal of Remote Sensing*, **28**, 5019–5036.

Jenness, J. and Wynne, J.J. (2007) Cohen's Kappa and classification table metrics 2.1a: an ArcView 3x extension for accuracy assessment of spatially-explicit models. http://www.jennessent.com/arcview/kappa_stats.htm (accessed 29 July 2009).

Jensen, J.R. (1996) *Introductory Digital Image Processing – A Remote Sensing Perspective*, 2nd edn, Prentice-Hall, Englewood Cliffs.

Jensen, J.R. and Toll, D.L. (1982) Detecting residential land-use development at the urban fringe. *Photogrammetric Engineering and Remote Sensing*, **48**, 629–643.

Ji, W. and Ma, J. (2008) Geospatial decision models for assessing the vulnerability of wetlands to potential human impacts, in *Wetland and Water Resource Modeling and Assessment* (ed. W. Ji), CRC Press, Boca Raton, pp. 215–230.

Jia, X. and Richards, J.A. (1999) Segmented principal components transformation for efficient hyperspectral remote-sensing image display and classification. *IEEE Transactions on Geoscience and Remote Sensing*, **37**, 538–542.

Johnsen, H., Lauknes, L. and Guneriussen, T. (1995) Geocoding of fast-delivery SAR image mode products using DEM data. *International Journal of Remote Sensing*, **16**, 1957–1968.

Johnson, R.D. and Kasischke, E.S. (1998) Change vector analysis: a technique for the multispectral monitoring of land cover and condition. *International Journal of Remote Sensing*, **19**, 411–426.

Jolliffe, I.T. (2002) *Principal Components Analysis*, 2nd edn, Springer-Verlag, Berlin.

Jones, A.F., Brewer, P.J.and Johnstone, E. (2007) High-resolution interpretative geomorphological mapping of river valley environments using airborne LiDAR data. *Earth Surface Processes and Landforms*, **32**, 1574–1592.

Jones, A.R., Settle, J.J. and Wyatt, B.K. (1988) Use of digital terrain data in the interpretation of SPOT-1 HRV data. *International Journal of Remote Sensing*, **9**, 669–682.

Jones, K.H. (1998) A comparison of algorithms used to compute hill slope as a property of the DEM. *Computers and Geosciences*, **24**, 315–323.

Jönsson, P. and Eklundh, L. (2004) TIMESAT – a program for analyzing time-series of satellite sensor data. *Computers and Geosciences*, **30**, 833–845. Code at http://www.iamg.org/CGEditor/index.htm (accessed 1 |June 2010).

Joseph, G. (1996) Imaging sensors for remote sensing. *Remote Sensing Reviews*, **13**, 257–342.

Joseph, G. (2000) How well do we understand Earth observation electro-optical sensor parameters? *Remote Sensing of Environment*, **55**, 9–12.

Joughin, I.R., Winebrenner, D.P. and Fahnestock, M.A. (1995) Observations of ice-sheet motion in Greenland using satellite radar interferometry. *Geophysics Research Letters*, **22**, 571–574.

Ju, J., Gopal, S. and Kolaczyk, E.D. (2005) On the choice of spatial and categorical scale in remote sensing land cover classification. *Remote Sensing of Environment*, **96**, 62–77.

Jupp, D.L.B. and Mayo, K.K. (1982) The use of residual images in Landsat image analysis. *Photogrammetric Engineering and Remote Sensing*, **48**, 595–604.

Jupp, D.L.B., Strahler, A.H. and Woodcock, C.E. (1988) Autocorrelation and regularisation in digital images. I: basic theory. *IEEE Transactions on Geoscience and Remote Sensing*, **26**, 463–473.

Jupp, D.L.B., Strahler, A.H. and Woodcock, C.E. (1989) Autocorrelation and regularisation in digital images. II: simple image models. *IEEE Transactions on Geoscience and Remote Sensing*, **27**, 247–258.

Justice, C., Giglio, L., Boschetti, L. *et al.* (2006) Algorithm Technical Background Document MODIS Fire Products, Version 2.3, EOS ID# 2741. 1 October 2006. http://modis.gsfc.nasa.gov/data/atbd/atbd_mod14.pdf (accessed 20 January 2009).

Justice, C.O., Vermote, E., Townshend, J.R.G. *et al.* (1998) The Moderate Resolution Imaging Spectroradiometer (MODIS): land remote sensing for global change research. *IEEE Transactions on Geoscience and Remote Sensing*, **36**, 1228–1249.

Jutz, S.L. and Chorowicz, J. (1993) Geological mapping and detection of oblique extensional structures in the Kenyan Rift Valley with a SPOT/Landsat data merge. *International Journal of Remote Sensing*, **14**, 1677–1688.

Kaewpijit, S., Le Moigne, J. and El-Ghazawi, T. (2003) Automatic reduction of hyperspectral imagery using wavelet spectral analysis. *IEEE Transactions on Geoscience and Remote Sensing*, **41**, 863–871.

Kahle, A.B. and Rowan, L.C. (1980) Evaluation of multispectral infrared aircraft images for lithological mapping in the East Tintic Mountains, Utah. *Geology*, **8**, 234–239.

Kalkhan, M.A., Reich, R.M. and Czaplewski, R.L. (1997) Variance estimates and confidence intervals for the kappa measure of classification accuracy. *Canadian Journal of Remote Sensing*, **23**, 210–216.

Kalpoma, K.A. and Kudoh, J. (2007) Image fusion processing for IKONOS 1-m color imagery. *IEEE Transactions on Geoscience and Remote Sensing*, **45**, 3075–3086.

Kaneko, T. (1976) Evaluation of Landsat image registration accuracy. *Photogrammetric Engineering and Remote Sensing*, **42**, 1285–1299.

Kanellopoulos, I., Varas, A., Wilkinson, G.G. and Mégier, J. (1992) Land-cover discrimination in SPOT HRV imagery using an artificial neural network – a 20-class experiment. *International Journal of Remote Sensing*, **13**, 917–924.

Kanellopoulos, I., Wilkinson, G.G., Roli, F. and Austin, J. (eds) (1997) *Neurocomputation in Remote Sensing Data Analysis*, Springer, Heidelberg.

Kanevski, M. and Maignan, M. (2004) *Analysis and Modelling of Spatial Environmental Data*, EPFL Press, Lausanne.

Karathanassi, V., Kolokousis, P. and Ioannidou, S. (2007) A comparison study on fusion methods using evaluation indicators. *International Journal of Remote Sensing*, **28**, 2309–2341.

Kardoulas, N.G., Bird, A.C. and Lawan, A.I. (1996) Geometric correction of SPOT and Landsat imagery: a comparison of map and GPS-derived control points. *Photogrammetric Engineering and Remote Sensing*, **62**, 1173–1177.

Karpouzli, E. and Malthus, T. (2003) The empirical line method for atmospheric correction of IKONOS imagery. *International Journal of Remote Sensing*, **24**, 1143–1150.

Katawa, Y., Ueno, S. and Kusaka, T. (1986) Radiometric correction for atmospheric and topographic effects on Landsat MSS images. *International Journal of Remote Sensing*, **9**, 729–748.

Kaufman, L. and Rousseeuw, P.J. (2005) *Finding Groups in Data: An Introduction to Cluster Analysis*, John Wiley & Sons, Inc., New York.

Kaufman, Y.J., Herring, D.D., Ranson, K.R. and Collatz, G.J. (1998) Earth Observing System AM1 mission to Earth. *IEEE Transactions on Geoscience and Remote Sensing*, **36**, 1045–1055.

Kaufmann, H., Berger, M., Meissner, M. and Süssenguth, G. (1996) MOMS-02/STS-55 design and validation of spectral and panchromatic modules. Proceedings of the MOMS Symposium, Cologne, Germany, 5–7 July, 1995, European Association of Remote Sensing Laboratories (EARSeL), Paris, pp. 26–38.

Kauth, R.J. and Thomas, G. (1976) The tasselled cap – a graphic description of the spectral-temporal development of agricultural crops as seen by Landsat. Proceedings of the Symposium on Machine Processing of Remotely-Sensed Data 1976, Purdue University, West Lafayette, vol. 4B, pp. 41–51.

Kavzoglu, T. (2004) Simulating Landsat ETM+ imagery using DAIS 7915 hyperspectral scanner data. *International Journal of Remote Sensing*, **25**, 5049–5067.

Kavzoglu, T. (2009) Increasing the accuracy of neural network classification using refined training data. *Environmental Modelling and Software*, **24**, 850–858.

Kavzoglu, T. and Mather, P.M. (1999) Pruning artificial neural networks: an example using land cover classification of multisensor images. *International Journal of Remote Sensing*, **20**, 2787–2803.

Kavzoglu, T. and Mather, P.M. (2000) Using feature selection techniques to produce smaller neural networks with better generalisation capabilities, *Proceedings of the IEEE Geoscience and Remote Sensing Symposium, 2000 (IGARSS 2000), 24–28 July, 2000, Honolulu*, vol. 7, IEEE Press, Piscataway, pp. 3069–3071.

Kavzoglu, T. and Mather, P.M. (2002) The role of feature selection in artificial neural network applications. *International Journal of Remote Sensing*, **23**, 2919–2937.

Kavzoglu, T. and Mather, P.M. (2003) The use of backpropagating artificial neural networks in land cover classification. *International Journal of Remote Sensing*, **24**, 4907–4938.

Kelley, E.A. and Pinter, N. (2002) *Active Tectonics: Earthquakes, Uplift, and Landscape*, 2nd edn, Prentice Hall, Upper Saddle River.

Kennedy, W.J. and Gentle, J.E. (1980) *Statistical Computing*, CRC Press, Boca Raton.

Kennet, M. and Eiken, T. (1997) Airborne measurements of glacier surface elevation by scanning laser altimetry. *Annals of Glaciology*, **24**, 293–296.

Kerdiles, H. and Grondona, M.O. (1995) NOAA-AVHRR NDVI decomposition and sub-pixel classification using linear mixing in the Argentine Pampa. *International Journal of Remote Sensing*, **16**, 1303–1325.

Kess, B.L., Steinwand, D.R. and Reichenbach, S.E. (1996) Compression of Global Land 1-km AVHRR dataset. *International Journal of Remote Sensing*, **17**, 2955–2969.

Key, J.R., Maslanik, J.A. and Barry, R.G. (1989) Cloud classification from satellite data using a fuzzy sets algorithm: a polar example. *International Journal of Remote Sensing*, **10**, 1823–1842.

Khan, B., Hayes, L.W.B. and Cracknell, A.P. (1995) The effects of higher-order resampling on AVHRR data. *International Journal of Remote Sensing*, **16**, 147–163.

Khorram, S.K., Biging, G.S., Chrisman, N.R. *et al.* (1999) *Accuracy Assessment of Remote Sensing-derived Change Detection*, Monograph Series, American Society for Photogrammetry and Remote Sensing, Bethesda.

Kimble, G. (1951) The inadequacy of the regional concept, in *London Essays in Geography: Rodwell Jones Memorial Volume* (eds L. Dudley Stamp, and S.W. Wooldridge), Longmans Green and Company, London, pp. 151–174.

Kingsley, S. and Quegan, S. (1992) *Understanding Radar Systems*, McGraw-Hill, London.

Kirkland, K. (2007) *Light and Optics*, Facts On File Inc., New York.

Kittler, J. (1983) Image processing for remote sensing. *Philosophical Transactions of the Royal Society of London, A*, **309**, 323–335.

Kittler, J. and Föglein, J. (1984) Contextual classification of multispectral pixel data. *Image and Vision Computing*, **2**, 13–29.

Kittler, J., Hatef, M., Duin, R.P.W. and Matas, J. (1998) On combining classifiers. *IEEE Transactions on Pattern Analysis and Machine Intelligence*, **20**, 226–239.

Kneizys, F.X., Shettle, E.P., Abreu, L.W. *et al.* (1988) Users Guide to LOWTRAN 7, U.S. Air Force Geophysics Laboratory, Hanscomb Air Force Base, 137 p.

Knuth, D.E. (1998) *The Art of Computer Programming*, Seminumerical Algorithms, Vol. 2, 3rd edn, Addison-Wesley, Boston.

Kobler, A., Pfeifer, N., Ogrinc, P. *et al.* (2007) Repetitive interpolation: a robust algorithm for DTM generation from Aerial Laser Scanner Data in forested terrain. *Remote Sensing of Environment*, **108**, 9–23.

Koch, B., Heyder, U. and Welnacker, H. (2006) Detection of individual tree crowns in airborne lidar data. *Photogrammetric Engineering and Remote Sensing*, **72**, 357–363.

Koch, B., Straub, C., Dees, M. *et al.* (2009) Airborne laser data for stand delineation and information extraction. *International Journal of Remote Sensing*, **30**, 935–963.

Koch, M. (2000) Geological controls of land degradation as detected by remote sensing: a case study in Los Monegros, north-east Spain. *International Journal of Remote Sensing*, **21**, 457–473.

Koch, M. and Mather, P.M. (1997) Lineament mapping for groundwater resource assessment: a comparison of digital Synthetic Aperture Radar imagery and stereoscopic Large Format Camera photographs in the Red Sea Hills, Sudan. *International Journal of Remote Sensing*, **18**, 1465–1482.

Kohl, H.G. and Hill, J. (1988) Geometric registration of multitemporal TM data over mountainous areas by use of a low resolution digital elevation model. Proceedings of the 8th EARSeL Symposium on Alpine and Mediterranean Areas: A Challenge for Remote Sensing, Capri, 17–20 May, 1998, European Association of Remote Sensing Laboratories (EARSeL), Paris, pp. 323–355.

Konecny, G. (2003) *Geoinformation: Remote Sensing, Photogrammetry and Geographic Information Systems*, Taylor and Francis, London.

Kontoes, C.C. (2008) Operational land cover change detection using change vector analysis. *International Journal of Remote Sensing*, **29**, 4757–4779.

Kontoes, C.C. and Rokos, D. (1996) The integration of spatial context information in an experimental knowledge-based system and the supervised relaxation algorithm – two successful approaches to improving SPOT-XS classification. *International Journal of Remote Sensing*, **17**, 3093–3106.

Koukoulas, S. and Blackburn, G.A. (2001) Introducing new indices for accuracy evaluation of classified images representing semi-natural woodland environments. *Photogrammetric Engineering and Remote Sensing*, **67**, 499–510.

Kowalik, W.S., Lyon, R.J.P. and Switzer, P. (1983) The effects of additive radiance terms on ratios of Landsat data. *Photogrammetric Engineering and Remote Sensing*, **49**, 659–669.

Kraak, M-J. and Ormeling, F.J. (2002) *Cartography: Visualization of Geospatial Data*, 2nd edn, Pearson Education, Harlow.

Kramer, H. and Cracknell, A.P. (2008) An overview of small satellites in remote sensing. *International Journal of Remote Sensing*, **29**, 4285–4337.

Kramer, H.J. (2002) *Observation of the Earth and its Environment – Survey of Missions and Sensors*, 4th edn, Springer-Verlag, Berlin.

Krause, G., Bock, M., Weiers, S. and Braun, G. (2004) Mapping land-cover and mangrove structures with remote sensing techniques: a contribution to a synoptic GIS in support of coastal management in north Brazil. *Environmental Management*, **34**, 429–440.

Krieger, G., Moreira, A., Fiedler, H. *et al.* (2007) TanDEM-X: a satellite formation for high-resolution SAR interferometry.

IEEE Transactions on Geoscience and Remote Sensing, **45**, 3317–3341.

Krishnamurthy, J., Manalavan, P. and Saivasan, V. (1992) Application of digital enhancement techniques for groundwater exploration in hard rock terrains. *International Journal of Remote Sensing*, **13**, 2925–2942.

Krishnamurthy, J., Venkatesa Kumar, N., Jayaraman, V. and Manivel, M. (1996) An approach to demarcate groundwater potential zones through remote sensing and a geographical information system. *International Journal of Remote Sensing*, **17**, 1867–1884.

Krivoruchko, K. and Gotway Crawford, C. (2005) Assessing the uncertainty resulting from geoprocessing operations, in *GIS, Spatial Analysis, and Modelling* (eds D.J. Maguire, M. Batty, and M.F. Goodchild), ESRI Press, Redlands, pp. 67–92.

Kropatsch, W.G. and Strobl, D. (1990) The generation of SAR layover and shadow maps from digital elevation models. *IEEE Transactions on Geoscience and Remote Sensing*, **28**, 90–107.

Krumm, J. (2001) Savitzky-Golay Filters for 2D Images. http://www.research.microsoft.com/users/jckrumm/savgol/savgol.htm (accessed 1 August 2009).

Kruse, F.A., Lefkoff, A.B., Boardman, J.W. *et al.* (1993) The Spectral Image Processing System (SIPS) – interactive visualization and analysis of imaging spectrometer data. *Remote Sensing of Environment*, **44**, 145–163.

Kruskal, J.B. Jr. (1956) On the shortest spanning subtree of a graph and the travelling salesman problem. *Proceedings of the American Mathematical Society*, **7**, 48–50.

Kuemmerle, T., Damm, A. and Hostert, P. (2008) A method to detect and correct single-band missing pixels in Landsat TM and ETM+ data. *Computers and Geosciences*, **34**, 445–455.

Kumar, R. (1979) Comparison of feature selection techniques for Earth resources data, in *Proceedings of an International Conference on Applications of Machine-Aided Image Analysis, Oxford, 1979*, Conference Series, Vol. 44 (ed. W.E. Gardner), Institute of Physics, Bristol, pp. 238–243.

Kuplich, T.M., Curran, P.J. and Atkinson, P.M. (2005) Relating SAR image texture to the biomass of regenerating tropical forests. *International Journal of Remote Sensing*, **26**, 4829–4854.

Kwok, R., Curlander, J.C. and Pang, S.S. (1987) Rectification of terrain-induced distortions in radar images. *Photogrammetric Engineering and Remote Sensing*, **53**, 507–513.

Laba, M., Tsai, F., Ogurcak, D. *et al.* (2005) Field determination of optimal dates for the discrimination of invasive wetland plant species using derivative spectral analysis. *Photogrammetric Engineering and Remote Sensing*, **71**, 603–611.

Labovitz, M.L. and Marvin, J.W. (1986) Precision in geodetic correction of TM data as a function of the number, spatial distribution, and success in matching control points: a simulation. *Remote Sensing of Environment*, **20**, 237–252.

Labovitz, M.L. and Matsuoko, E.J. (1984) The influence of autocorrelation on signature extraction – an example from a geobotanical investigation of Cotter Basin, Montana. *International Journal of Remote Sensing*, **5**, 315–332.

Labovitz, M.L., Toll, D.L. and Kennard, R.E. (1982) Preliminary evidence for the influence of physiography and scale upon the autocorrelation function of remotely sensed data. *International Journal of Remote Sensing*, **3**, 13–30.

Laliberte, A.S., Fredrickson, E.L. and Rango, A. (2007) Combining decision trees with hierarchical object-oriented image analysis for mapping arid rangelands. *Photogrammetric Engineering and Remote Sensing*, **73**, 197–207.

Lambin, E. and Strahler, A. (1994b) Change-vector analysis in multitemporal space: a tool to detect and categorize land cover change processes using high temporal resolution satellite data. *Remote Sensing of Environment*, **48**, 231–244.

Lambin, E.F. (1996) Change detection at multiple temporal scales: seasonal and annual variations in landscape variables. *Photogrammetric Engineering and Remote Sensing*, **62**, 931–938.

Lambin, E.F. and Geist, H. (2006) *Land-Use and Land-Cover Change: Local Processes and Global Impacts*, Springer-Verlag, Berlin, Heidelberg.

Lambin, E.F. and Strahler, A.H. (1994a) Indicators for land cover change for change-vector analysis in multitemporal space at coarse spatial scales. *International Journal of Remote Sensing*, **15**, 2099–2119.

Landgrebe, D.A. (2003) *Signal Theory Methods in Multispectral Remote Sensing*, John Wiley & Sons, Inc., New York.

Landgrebe, D.A. and staff (1974) *A Study of the Utilisation of ERTS-1 Data from the Wabash River Basin*. Laboratory for the Applications of Remote Sensing (LARS), Purdue University, West Lafayette, Indiana, LARS Information Note 052375.

Lang, H. and Welch, R. (1996) Algorithm Theoretical Basis Document for ASTER Digital Elevation Models (Standard Product AST14), Version 3, ATBD-AST-08, NASA Jet Propulsion Laboratory, Pasadena.

Lark, R.M. (1995) A reappraisal of unsupervised classification, I: correspondence between spectral and conceptual classes. *International Journal of Remote Sensing*, **16**, 1425–1443.

Lark, R.M. (1996) Geostatistical description of texture on an aerial photograph for discriminating classes of land cover. *International Journal of Remote Sensing*, **17**, 2115–2133.

Lark, R.M. (1998) Forming spatially-coherent regions by classification of multi-variate data: an example from the analysis of maps of crop yield. *International Journal of Remote Sensing*, **19**, 83–98.

Laronne, J.B. and Reid, I. (1993) Very high rates of bedload sediment transport by ephemeral desert rivers. *Nature*, **366**, 148–150.

Laur, H., Bally, P., Meadows, P. *et al.* (2002) ERS SAR Calibration: Derivation of the Backscattering Coefficient Sigma-0 in ESA ERS SAR PRI Products, Document No. ES-TN-RS-PM-HL09, Issue 2, Rev. 5d, European Space Agency, Paris.

Lawrence, R.L. and Ripple, W.J. (1996) Determining patch perimeters in raster image processing and geographic information systems. *International Journal of Remote Sensing*, **17**, 1255–1259.

Lawson, C.L. and Hansen, R.J. (eds) (1995) *Solving Least Squares Problems*, Prentice Hall, Englewood Cliffs.

Le Cun, Y., Denker, J.S. and Solla, S.A. (1990) Optimal brain damage, in *Advances in Neural Information* (ed. D.S. Touretsky), Morgan Kaufmann, San Mateo, pp. 598–605.

Leach, A.R. and Gillet, V.J. (2007) *An Introduction to Chemoinformatics*, Springer-Verlag, Berlin.

Leachtenauer, J.C. (1977) Optical power spectrum analysis: scale and resolution effects. *Photogrammetric Engineering and Remote Sensing*, **43**, 1117–1125.

Leberl, F.W. (1990) *Radargrammetric Image Processing*, Artech House, Dedham.

Lee, C. and Landgrebe, P.A. (1993) Decision boundary feature extraction for non-parametric classifiers. *IEEE Transactions on Systems, Man, and Cybernetics*, **23**, 433–444.

Lee, J. and Ersoy, O.K. (2007) Consensual and hierarchical classification of remotely sensed multispectral images. *IEEE Transactions on Geoscience and Remote Sensing*, **45**, 2953–2963.

Lee, J.A. and Verleysen, M. (2007) *Nonlinear Dimensionality Reduction*, Springer-Verlag, Berlin.

Lee, J.S. (1983a) Digital image smoothing and the sigma filter. *Computer Graphics, Vision and Image Processing*, **17**, 24–32.

Lee, J.S. (1983b) A simple speckle smoothing algorithm for synthetic aperture radar images. *IEEE Transactions on Systems, Man and Cybernetics*, **13**, 85–89.

Lee, J.-S. and Pottier, E. (2009) *Polarimetric Radar Imaging: From Basics to Applications*, CRC Press, Boca Raton.

Lee, J.S., Jurkevich, I., Dewaele, P. *et al.* (1994) Speckle filtering of synthetic aperture radar images: a review. *Remote Sensing Reviews*, **8**, 313–340.

Lee, J.-S., Wen, J.H., Ainsworth, T.L. *et al.* (2009) Improved sigma filter for speckle filtering of SAR imagery. *IEEE Transactions on Geoscience and Remote Sensing*, **47**, 202–213.

Lee, T.-W. (1998) *Independent Component Analysis: Theory and Applications*, Kluwer Academic Publishers, Dordrecht.

Lefsky, M.A., Cohen, W.B., Parker, G.G. and Harding, D.J. (2002) Lidar remote sensing for ecological studies. *BioScience*, **52**, 19–30.

Lei, Q., Henkel, J., Frei, M. *et al.* (1996) Radiometric noise correction of panchromatic high resolution data of MOMS-02. Proceedings of the MOMS Symposium, Cologne, Germany, 5–7 July, 1995, European Association of Remote Sensing Laboratories (EARSeL), Paris, pp. 303–313.

Lein, J.K. (2003) Applying evidential reasoning methods to agricultural land cover classification. *International Journal of Remote Sensing*, **24**, 4161–4180.

Lennon, M., Mercier, G. and Hubert-Moy, L. (2002) Classification of hyperspectral images with nonlinear filtering and support vector machines, *Proceedings of the IEEE International Geoscience and Remote Sensing Symposium, 2002 (IGARSS '02), 24–28 June 2002, Toronto*, vol. 3, IEEE Press, Piscataway, pp. 1670–1672.

Leprieur, C., Kerr, Y.H. and Pichon, J.M. (1996) Critical assessment of vegetation indices from AVHRR in a semi-arid environment. *International Journal of Remote Sensing*, **17**, 2549–2563.

Leprince, S., Barbot, S., Ayoub, F. and Avouac, J.-P. (2007) Automatic and precise orthorectification, coregistration, and subpixel correlation of satellite images: application to ground deformation measurements. *IEEE Transactions on Geoscience and Remote Sensing*, **45**, 1529–1558.

Levin, S.A. (1991) Concepts of scale at the local level, in (eds J.R. Ehleringer, and C.B. Field), *Scaling Physiological Processes – Leaf to Globe*. Academic Press, San Diego, pp. 7–19.

Lewis, A.J., Henderson, F.M. and Holcomb, D.W. (1998) Radar fundamentals: the geoscience perspective, in *Principles and Applications of Imaging Radar* (eds F.M. Henderson, and A.J. Lewis), John Wiley & Sons, Inc., New York, pp. 131–181.

Li, M., Daels, L. and Antrop, M. (1996) Lambertian and Minnaert relation simulation for topographic normalization. Proceedings of the 11th Thematic Conference and Workshops on Applied Geologic Remote Sensing, vol. 2, Las Vegas, Nevada, 27–29 February 1996, Environmental Research Institute of Michigan (ERIM), Ann Arbor, pp. 133–141.

Li, S., Kwok, J.T. and Wang, Y. (2002) Using the discrete wavelet frame transform to merge Landsat TM and SPOT panchromatic images. *Information Fusion*, **3**, 17–23.

Li, X. and Yeh, A.G.O. (1998) Principal component analysis of stacked multi-temporal images for the monitoring of rapid urban expansion in the Pearl River Delta. *International Journal of Remote Sensing*, **19**, 1501–1518.

Liang, S. (2004) *Quantitative Remote Sensing of Land Surfaces*, John Wiley & Sons, Inc., New York.

Liang, S. (2007) Recent developments in estimating land surface biogeophysical variables from optical remote sensing. *Progress in Physical Geography*, **31**, 501–516.

Liang, S., Schaepman, M., Jackson, T. *et al.* (2008) Emerging issues in land remote sensing, in *Advances in Land Remote Sensing: System, Modeling, Inversion and Application* (ed. S. Liang), Springer-Verlag, Berlin, pp. 485–494.

Lichti, D., Pfeifer, N. and Maas, H.-G. (eds) (2008) Theme issue: terrestrial laser scanning. *ISPRS Journal of Photogrammetry and Remote Sensing*, **63**, 1–154.

Lillesand, T.M., Kiefer, R.W. and Chipman, J.W. (2008) *Remote Sensing and Image Interpretation*, 6th edn, John Wiley & Sons, Inc., New York.

Lim, K., Treitz, P., Wulder, M. *et al.* (2003) LiDAR remote sensing of forest structure. *Progress in Physical Geography*, **27**, 88–106.

Lin, Q., Vesesky, J.F. and Zebker, H.A. (1994) Comparison of elevation derived from InSAR data with DEM over large relief terrain. *International Journal of Remote Sensing*, **15**, 1775–1790.

Ling, Y., Ehlers, M., Usery, E.L. and Madden, M. (2007) FFT-enhanced IHS transform method for fusing high-resolution satellite images. *ISPRS Journal of Photogrammetry and Remote Sensing*, **61**, 381–392.

Liu, C., Frazier, P. and Kumar, L. (2007) Comparative assessment of the measures of thematic classification accuracy. *Remote Sensing of Environment*, **107**, 606–616.

Liu, H. and Jezek, K.C. (2004) A complete high-resolution coastline of Antarctica extracted from orthorectified Radarsat SAR imagery. *Photogrammetric Engineering and Remote Sensing*, **70**, 605–616.

Liu, J. (2000) Smoothing filter-based intensity modulation: a spectral preserve image fusion technique for improving spatial details. *International Journal of Remote Sensing*, **21**, 3461–3472.

Liu, W.G., Gopal, S. and Woodcock, C.E. (2004) Uncertainty and confidence in land cover classification using a hybrid classifier approach. *Photogrammetric Engineering and Remote Sensing*, **70**, 963–971.

Liu, X. (2008) Airborne LiDAR for DEM generation: some critical issues. *Progress in Physical Geography*, **32**, 31–49.

Lizarazo, I. (2008) SVM-based segmentation and classification of remotely sensed data. *International Journal of Remote Sensing*, **29**, 7277–7283.

Lloyd, C.D. (2006) *Local Models for Spatial Analysis*, CRC Press, Boca Raton.

Lloyd, C.D., Berberoglu, S., Curran, P.J. and Atkinson, P.M. (2004) A comparison of texture measures for the per-field classification of Mediterranean land cover. *International Journal of Remote Sensing*, **25**, 3943–3965.

Llu, J.G., Mason, P., Hilton, F. and Lee, H. (2004) Detection of rapid erosion in SE Spain: a GIS approach based on ERS SAR coherence imagery: InSAR application. *Photogrammetric Engineering and Remote Sensing*, **70**, 1179–1185.

Lo, C.P. and Yeung, A.K.W. (2007) *Concepts and Techniques of Geographic Information Systems*, 2nd edn, Pearson Prentice Hall, Upper Saddle River.

Lobo, A., Chic, O. and Casterad, A. (1996) Classification of Mediterranean crops with multi-sensor data: per-pixel versus per-object statistics and image segmentation.

Lobser, S.E. and Cohen, W.B. (2007) MODIS tasselled cap: land cover characteristics expressed through transformed MODIS data. *International Journal of Remote Sensing*, **28**, 5079–5101.

Lobser, S.E. and Cohen, W.B. (2007) MODIS tasselled cap: land cover characteristics expressed through transformed MODIS data. *International Journal of Remote Sensing*, **28**, 5079–5101.

Loew, A. and Mauser, W. (2007) Generation of geometrically and radiometrically terrain corrected SAR image products. *Remote Sensing of Environment*, **106**, 337–349.

Loizou, C.P. and Pattichis, C.S. (2008) *Despeckle Filtering Algorithms and Software for Ultrasound Imaging*, Morgan and Claypool, San Rafael.

Longley, P., Goodchild, M.F., Maguire, D.J. and Rhind, D.W. (2005) *Geographic Information Systems and Science*, 2nd edn, John Wiley & Sons, Ltd, Chichester.

Lopes, A., Nezry, E., Touzi, R. and Laur, H. (1993) Structure detection and statistical adaptive speckle filtering in SAR images. *International Journal of Remote Sensing*, **14**, 1735–1758.

Lopes, A., Touzi, R. and Nezry, E. (1990) Adaptive speckle filters and scene heterogeneity. *IEEE Transactions on Geoscience and Remote Sensing*, **28**, 992–1000.

Lowman, P.D. Jr. (1994) Radar geology of the Canadian Shield: a 10-year review. *Canadian Journal of Remote Sensing*, **20**, 198–209.

Lu, D. (2006) The potential and challenge of remote sensing-based biomass estimation. *International Journal of Remote Sensing*, **27**, 1297–1328.

Lu, D. and Weng, Q. (2007) A survey of image classification methods and techniques for improving classification performance. *International Journal of Remote Sensing*, **28**, 823–870.

Lu, D., Mausel, P., Brondízio, E. and Moran, E. (2004) Change detection techniques. *International Journal of Remote Sensing*, **25**, 2365–2401.

Lu, G.Y. and Wong, D.W. (2008) An adaptive inverse-distance weighting spatial interpolation technique. *Computers and Geosciences*, **34**, 1044–1055.

Lu, X., Liu, R., Liu, J. and Liang, S. (2007) Removal of noise by wavelet method to generate high quality temporal data of terrestrial MODIS products. *Photogrammetric Engineering and Remote Sensing*, **73**, 1129–1139.

Lu, Z., Kwoun, O. and Rhykus, R. (2007) Interferometric Synthetic Aperture Radar (InSAR): its past, present and future. *Photogrammetric Engineering and Remote Sensing*, **73**, 217–221.

Lunetta, R.S. and Elvidge, C.D. (eds) (1998) *Remote Sensing Change Detection: Environmental Monitoring, Methods and Applications*, Ann Arbor Press, Chelsea.

Lunetta, R.S. and Lyon, J.G. (eds) (2004) *Remote Sensing and GIS Accuracy Assessment*, CRC Press, Boca Raton.

Lunetta, R.S., Knight, J.F., Ediriwickrema, J. *et al.* (2006) Land-cover change detection using multi-temporal MODIS NDVI data. *Remote Sensing of Environment*, **105**, 142–154.

Lynn, P.A. (1982) *An Introduction to the Analysis and Processing of Signals*, 2nd edn, Macmillan, London.

Maas, H.-G. (2002) Methods for measuring height and planimetry discrepancies in airborne laser scanner data. *Photogrammetric Engineering and Remote Sensing*, **68**, 933–940.

Machado, M.J., Pérez-González, A. and Benito, G. (1998) Paleoenvironmental changes during the last 4000 years in the Tigray, Northern Ethiopia. *Quaternary Research*, **49**, 312–321.

Madden, M. (ed.) (2009) *Manual of Geographic Information Systems*, American Society for Photogrammetry and Remote Sensing, Bethesda.

Madsen, S.N. and Zebker, H.A. (1998) Synthetic aperture radar interferometry: principles and applications, in *Manual of Remote Sensing, Principles and Applications of Imaging Radar*, vol. 2 (eds F.M. Henderson, and A.J. Lewis), John Wiley & Sons, Inc., New York, pp. 359–379.

Maguire, D.J., Batty, M. and Goodchild, M.F. (2005) *GIS, Spatial Analysis, and Modelling*, ESRI Press, Redlands.

Maidment, D.R. (2002) *Arc Hydro: GIS for Water Resources*, ESRI Press, Redlands.

Maidment, D.R., Robayo, O., and Merwade, V. (2005) Hydrologic modeling, in *GIS, Spatial Analysis, and Modelling* (eds D.J. Maguire, M. Batty, and M.F. Goodchild), ESRI Press, Redlands, pp. 319–332.

Maillard, P. (2003) Comparing texture analysis methods through classification. *Photogrammetric Engineering and Remote Sensing*, **69**, 357–367.

Malacara, D. (2002) *Color Vision and Colorimetry: Theory and Applications*, Society of Photo-Optical Instrumentation Engineers (SPIE) Press, Bellingham, Washington, DC.

Malczewski, J. (1999) *GIS and Multicriteria Decision Analysis*, John Wiley & Sons, Inc., New York.

Mallat, S. (1998) *A Wavelet Tour of Signal Processing*, Academic Press, San Diego.

Mallet, C. and Bretar, F. (2009) Full-waveform topographic lidar: state-of-the-art. *ISPRS Journal of Photogrammetry and Remote Sensing*, **64**, 1–16.

Malpica, J.A. (2007) Hue adjustment to IHS pan-sharpened IKONOS imagery for vegetation enhancement. *IEEE Geoscience and Remote Sensing Letters*, **4**, 27–31.

Man, K.F., Tang, K.S. and Kwong, S. (1999), *Genetic Algorithms: Concepts and Designs*, Springer-Verlag, London.

Manninen, T., Stenberg, P., Rautiainen, M. *et al.* (2005) Leaf area index estimation of boreal forest using ENVISAT ASAR. *IEEE Transactions on Geoscience and Remote Sensing*, **43**, 2627–2635.

Mao, K.Z. (2004) Orthogonal forward selection and backward elimination algorithms for feature subset selection. *IEEE Transactions on Systems, Man, and Cybernetics, Part B: Cybernetics*, **34**, 629–634.

Marçal, A.R.S. and Rodrigues, A.S. (2009) A method for multispectral image segmentation evaluation based on synthetic images. *Computers and Geosciences*, **35**, 1574–1581.

Marçal, A.R.S., Borges, J.S., Gomes, J.A. and Costa, J.F.P.D. (2005) Land cover update by supervised classification of segmented ASTER images. *International Journal of Remote Sensing*, **26**, 1347–1362.

Markham, B.L. and Barker, J.L. (1987) Thematic Mapper bandpass solar exoatmospheric irradiances. *International Journal of Remote Sensing*, **8**, 517–523.

Markham, B.L., Halthore, R.N. and Goetz, S.J. (1992) Surface reflectance retrieval from satellite and aircraft sensors: result of sensor and algorithm comparisons during FIFE. *Journal of Geophysical Research*, **97**, 18785–18795.

Marr, D. and Hildreth, E. (1980) Theory of edge detection. *Proceedings of the Royal Society of London, Series B*, **209**, 187–217.

Martimor, P., Arino, O., Berger, M. *et al.* (2007) Sentinel-2 optical high resolution mission for GMES operational services. IEEE International Geoscience and Remote Sensing Symposium, 2007 (IGARSS 2007), 23-28 July, 2007, Barcelona, pp. 2677–2680.

Martin, F.J. and Turner, R.W. (1993) SAR speckle reduction by weighted filtering. *International Journal of Remote Sensing*, **14**, 1759–1774.

Martin, S. (2004) *An Introduction to Ocean Remote Sensing*, Cambridge University Press, Cambridge.

Martonchik, J.V., Bruegge, C.J. and Strahler, A. (2000) A review of reflectance nomenclature used in remote sensing. *Remote Sensing Reviews*, **19**, 9–20.

Mas, J.-F. (1999) Monitoring land-cover changes: a comparison of change detection techniques. *International Journal of Remote Sensing*, **20**, 139–152.

Mas, J.-F. and Flores, J.J. (2008) The application of artificial neural networks to the analysis of remotely sensed data. *International Journal of Remote Sensing*, **29**, 617–663.

Masek, J.G., Honzak, M., Goward, S.N. *et al.* (2001) Landsat-7 ETM+ as an observatory for land cover: Initial radiometric and geometric comparisons with Landsat-5 Thematic Mapper. *Remote Sensing of Environment*, **78**, 118–130.

Masek, J.G., Vermote, E.F., Saleous, N.E. *et al.* (2006) A Landsat surface reflectance dataset for North America, 1990–2000. *IEEE Geoscience and Remote Sensing Letters*, **3**, 68–72.

Maselli, F., Conese, C., de Filippis, T. and Norcini, S. (1995a) Estimation of forest parameters through fuzzy classification. *IEEE Transactions on Geoscience and Remote Sensing*, **33**, 77–84.

Maselli, F., Conese, C., de Filippis, T. and Romani, M. (1995b) Integration of ancillary data into a maximum likelihood classification with nonparametric priors. *ISPRS Journal of Photogrammetry and Remote Sensing*, **50**, 2–11.

Maselli, F., Rodolfi, A. and Conese, C. (1996) Fuzzy classification of spatially degraded TM data for the estimation of sub-pixel components. *International Journal of Remote Sensing*, **17**, 537–551.

Masoud, A. (2009) Runoff modelling of the wadi systems for estimating flash flood and groundwater recharge potential in Southern Sinai, Egypt. *Arabian Journal of Geosciences.* doi: 10.1007/s12517-009-0090-9 (published online) See http://www.doi.org/ (accessed 25 September 2009).

Massom, R. and Lubin, D. (2006) *Polar Remote Sensing: Ice Sheets*, vol. 2, CRC Press, Boca Raton.

Massonet, D. (1993) Geoscientific applications at CNES, in *SAR Geocoding: Data and Systems* (ed. G. Schreier), Herbert Wichmann, Karlsruhe, pp. 397–415.

Massonet, D. (2000) Elevation modelling and displacement mapping using radar interferometry, in *Encyclopaedia of Analytical Chemistry* (ed. R.A., Meyers), John Wiley & Sons, Ltd, Chichester, pp. 8533–8543.

Massonet, D. and Feigl, K. (1998) Radar interferometry and its application to changes in the Earth's surface. *Reviews of Geophysics*, **36**, 441–500.

Massonnet, D., Rossi, M., Carmona, C. *et al.* (1993) The displacement field of the Landers earthquake mapped by radar interferometry. *Nature*, **364**, 138–142.

Masuoka, E., Fleig, A., Wolfe, R.E. and Patt, F. (1998) Key characteristics of MODIS data products. *IEEE Transactions on Geoscience and Remote Sensing*, **36**, 1313–1323.

Mather, P.M. (1976) *Computational Methods of Multivariate Analysis in Physical Geography*, John Wiley & Sons, Ltd, Chichester.

Mather, P.M. (1985) A computationally-efficient maximum likelihood classifier employing prior probabilities for remotely-sensed data. *International Journal of Remote Sensing*, **6**, 369–376.

Mather, P.M. (1991) *Computer Applications in Geography*, John Wiley & Sons, Ltd, Chichester.

Mather, P.M. (1995) Map–image registration using least-squares polynomials. *International Journal of Geographical Information Systems*, **9**, 543–554.

Mather, P.M., Tso, B. and Koch, M. (1998) An evaluation of Landsat TM spectral data and SAR-derived textural information for lithological discrimination in the Red Sea Hills, Sudan. *International Journal of Remote Sensing*, **19**, 587–604.

Mathur, A. and Foody, G.M. (2008) Multiclass and binary SVM classification: Implications for training and classification users. *IEEE Geoscience and Remote Sensing Letters*, **5**, 241–245.

Maune, D.F. (2001) *Digital Elevation Model Technologies and Applications: The DEM Users Manual*, The American Society for Photogrammetry and Remote Sensing, Besthesda.

McCloy, K.R. (2006) *Resource Management Information Systems: Remote Sensing, GIS and Modelling*, 2nd edn, CRC Press, Boca Raton.

McCoy, J. (2004) *ArcGIS 9: Geoprocessing in ArcGIS*, ESRI Press, Redlands.

McCullagh, M.J. (1988) Terrain and surface modelling systems: theory and practice. *The PhotogrammetricRecord*, **12**, 747–779.

McCullagh, M.J. and Davis J.C. (1972) Optical analysis of two-dimensional patterns. *Annals of the Association of American Geographers*, **62**, 561–577.

McCullough, B.D. (1998) Assessing the reliability of statistical software: part I. *The American Statistician*, **52**, 358–366.

McCullough, B.D. (1999) Assessing the reliability of statistical software: part II. *The American Statistician*, **53**, 149–159.

McGuire, D.J., Goodchild, M.F. and Rhind, D.W. (eds.) (1991) *Geographical Information Systems – Principles and Applications*. Harlow, Essex: Longmans Scientific and Technical.

McMaster, R.B. and Usery, E.L. (eds) (2004) *A Research Agenda for Geographic Information Science*, CRC Press, Boca Raton.

McMorrow, J.M., Cutler, M.E.J., Evans, M.G. and Al-Roichdi, A. (2004) Hyperspectral indices for characterizing upland peat composition. *International Journal of Remote Sensing*, **25**, 313–325.

Meadows, P. (1995) The calibration of ERS-1 synthetic aperture radar images using UK-PAF imagery, in *Sensors and Environmental Applications of Remote Sensing*, Proceedings of the 14th EARSeL Symposium, Goteborg, 6–8 June, 1994 (ed. J. Askne), A.A. Balkema, Rotterdam, pp. 423–430.

Melgani, F. and Bruzzoni, L. (2004) Classification of hyperspectral remote sensing images with support vector machines. *IEEE Transactions on Geoscience and Remote Sensing*, **42**, 1778–1790.

Memarsadeghi, N., Mount, D.M., Netanyahu, N.S. and Le Moigne, J. (2007) A fast implementation of the ISODATA clustering algorithm. *International Journal of Computational Geometry and Applications*, **17**, 71–103.

Mena, J.B. (2003) State of the art on automatic road extraction for GIS update: a novel classification. *Pattern Recognition Letters*, **24**, 3037–3058.

Mendoza, G.A. and Martins, H. (2006) Multi-criteria decision analysis in natural resource management: a critical review of methods and new modelling paradigms. *Forest Ecology and Management*, **230**, 1–22.

Menenti, M., Azzali, S., Verhoef, W. and van Swol, R. (1993) Mapping agro-ecological zones and time-lag in vegetation growth by means of Fourier analysis of time series of NDVI images. *Advances in Space Research*, **13**, 233–237.

Merchant, J.W. and Narumalani, S. (2009) Integrating remote sensing and GIS, in *The SAGE Handbook of Remote Sensing* (eds T.A. Warner, M.D. Nellis, and G.M. Foody), Sage Publications, London, pp. 257–268.

Mertikas, P. and Zervakis, M.E. (2001) Exemplifying the theory of evidence in remote sensing image classification. *International Journal of Remote Sensing*, **22**, 1081–1095.

Mesev, V. (ed.) (2007) *Integration of GIS and Remote Sensing*, John Wiley & Sons, Ltd, Chichester.

Meyer-Bäse, A. (2004) *Pattern Recognition for Medical Imaging*, Elsevier and Academic Press, Amsterdam.

Meyers, L.S., Gamst, G. and Guarino, A.J. (2006) *Applied Multivariate Research: Design and Interpretation*, Sage Publications, London.

Michels, J.W. (1979) Axumite archaeology: an introductory essay, in *Axum* (ed. Y.M. Kobishchanov), University of Pennsylvania Press, Philadelphia, pp. 1–34.

Michels, J.W. (1994) Regional political organization in the Axum–Yeha area during the pre-axumite and axumite eras, in *Etudes Ethiopiennes*, vol 1 (ed. C. Lapage), Societe Francaise des Etudes Ethiopiennes, Paris, pp. 61–80.

Michels, J.W. (2005) *Changing Settlement Patterns in the Aksum-Yeha Region of Ethiopia: 700 BC–850 AD*, Cambridge Monographs in African Archaeology, British Archaeological Reports International Series, Archaeopress, Oxford.

Miller, R.L. and McKee, B.A. (2004) Using MODIS Terra 250 m imagery to map concentrations of total suspended matter in coastal waters. *Remote Sensing of Environment*, **93**, 259–266.

Millington, A.C., Walsh, S.J. and Osborne, P.E. (eds) (2004) *GIS and Remote Sensing Applications in Biogeography and Ecology*, Kluwer Academic Publishers, Dordrecht.

Milman, A.S. (1999) *Mathematical Principles of Remote Sensing: Making Sense of Noisy Data*, Sleeping Bear Press, Chelsea.

Milovich, J.A., Frulla, L.A. and Gagliardini, D.A. (1995) Environmental contribution to the atmospheric correction for Landsat-MSS images. *International Journal of Remote Sensing*, **16**, 2515–2537.

Milton, E.A., Schaepman, M.E., Anderson, K. *et al.* (2009) Progress in field spectroscopy. *Remote Sensing of Environment*, **113** (Suppl. 1), S92–S109.

Mitchell, M. (1996) *An Introduction to Genetic Algorithms*, The MIT Press, Cambridge.

Mitra, S. and Acharva, T. (2003) *Data Mining: Multimedia, Soft Computing, and Bioinformatics*, John Wiley & Sons, Inc., New York.

Mitri, G.H. and Gitas, I.Z. (2004) A performance evaluation of a burned area object-based classification model when applied to topographically and non-topographically corrected TM imagery. *International Journal of Remote Sensing*, **25**, 2863–2870.

Mohr, J.J. and Madsen, S.N. (2001) Geometric calibration of ERS satellite SAR images. *IEEE Transactions on Geoscience and Remote Sensing*, **39**, 842–850.

Moik, J.G. (1980) *Digital Processing of Remotely Sensed Images*. NASA Special Paper 432. US Govt. Printing Office, Washington DC.

Moisen, G.G., Edwards, T.C. Jr. and Cutler, D.R. (1994) Spatial sampling to assess classification accuracy of remotely-sensed data, in *Environmental Information Management and Analysis: Ecosystem to Global Scales* (eds W.K. Michener., J.W. Brunt, and S.G. Stafford), Taylor & Francis, London, pp. 159–176.

Monmonier, M. (1994) *How to Lie with Maps*, 2nd edn, University of Chicago Press, Chicago.

Monte Guarnieri, A., Prati, C., Rocca, F. and Desnos, Y.-L. (1998) Wide baseline interferometry with very low resolution SAR systems, in *EUSAR'98: European Conference on Synthetic Aperture Radar*, VDE-Verlag GmbH, Friedrichshafen, Berlin, pp. 361–364.

Montserud, R.A. and Leamans, R. (1992) Comparing global vegetation maps with the kappa statistic. *Ecological Modelling*, **62**, 275–293.

Moore, G.K. and Waltz, F.A. (1983) Objective procedures for lineament enhancement and extraction. *Photogrammetric Engineering and Remote Sensing*, **49**, 641–647.

Morad, M., Chalmers, A.I. and O'Regan, P.R. (1996) The role of root-mean-square error in the geo-transformation of images in GIS. *International Journal of Geographical Information Systems*, **10**, 347–353.

Morain, S. and Budge, A.M. (2004) *Post-launch Calibration of Satellite Sensors*, Proceedings of the International Workshop on Radiometric and Geometric Calibration, December, 2003, Mississippi, CRC Press, Boca Raton.

Moran, M.S., Bryant, R., Thome, K. *et al.* (2001) A refined empirical line approach for reflectance factor retrieval from Landsat-5 TM and Landsat-7 ETM+. *Remote Sensing of Environment*, **78**, 71–82.

Moran, M.S., Jackson, R.D., Clarke, T.R. *et al.* (1995) Reflectance factor retrieval from Landsat TM and SPOT HRV data for bright and dark targets. *Remote Sensing of Environment*, **52**, 218–230.

Moran, M.S., Jackson, R.D., Hart, G.F. *et al.* (1990) Obtaining surface reflectance factors from atmospheric and view angle corrected SPOT-1 HRV data. *Remote Sensing of Environment*, **32**, 203–214.

Moran, M.S., Jackson, R.D., Slater, P.N. and Teillet, P.M. (1992) Evaluation of simplified procedures for retrieval of land surface reflectance factors from satellite sensor output. *Remote Sensing of Environment*, **41**, 169–184.

Moreira, A., Krieger, G., Hajnsek, I. *et al.* (2004) TanDEM-X: a TerraSAR-X add-on satellite for single-pass SAR interferometry, *Proceedings, IEEE International Symposium on Geoscience and Remote Sensing Symposium (IGARSS '04), 20–24 September 2004, Anchorage, Alaska*, vol. 2, IEEE Press, Piscataway, pp. 1000–1003.

Moreno, J.E. and Melia, J. (1993) A method for accurate geometric correction of NOAA AVHRR HRPT data. *IEEE Transactions on Geoscience and Remote Sensing*, **31**, 204–226.

Motrena, P. and Rebordão, J.M. (1998) Invariant models for ground control points in high-resolution images. *International Journal of Remote Sensing*, **19**, 1359–1375.

Mott, H. (2007) *Remote Sensing with Polarimetric Radar*, John Wiley & Sons, Inc., New York.

Mouginis-Mark, P.J. (1995) Analysis of volcanic hazards using radar interferometry. *Earth Observation Quarterly*, **47**, 6–10.

Muasher, M.J. and Landgrebe, D.A. (1984) A binary tree feature selection technique for limited training sample size. *Remote Sensing of Environment*, **16**, 183–194.

Muchoney, D. and Strahler, A.H. (2002) Pixel- and site-based calibration and validation methods for evaluating supervised classification of remotely sensed data. *Remote Sensing of Environment*, **81**, 290–299.

Muchoney, D., Borak, J., Chi, H. *et al.* (2000) Application of MODIS global supervised classification model to vegetation and land cover mapping in Central America. *International Journal of Remote Sensing*, **21**, 1115–1138.

Mulder, N.J. (1980) A view on digital image processing. *ITC Journal*, **3**, 452–476.

Muller, E. (1993) Evaluation and correction of angular anisotropic effects in multidate SPOT and Thematic Mapper data. *Remote Sensing of Environment*, **45**, 295–309.

Murphy, R.E., Ardanuy, P., Deluccia, F. *et al.* (2006) The visible infrared imaging radiometer suite, in *Earth Science Satellite Remote Sensing: Science and Instruments*, vol. 1, Chapter 1 (eds J.J. Qu, W. Gao, M. Kafatos, *et al.*), Springer-Verlag, New York, pp. 199–223 (in association with Tsingua University Press, Beijing, China).

Murphy, R.J. and Wadge, G. (1994) The effects of vegetation on the ability to map soils using imaging spectrometer data. *International Journal of Remote Sensing*, **15**, 63–86.

Murthy, K.S.R. and Mamo, A.G. (2009) Multi-criteria decision evaluation in groundwater zones identification in Moyale-Teltele subbasin, South Ethiopia. *International Journal of Remote Sensing*, **30**, 2729–2740.

Mustard, J.F. and Sunshine, J.M. (1999), Spectral analysis for Earth science: investigations using remote sensing data, in *Manual of Remote Sensing, Remote Sensing for the Earth Sciences*, 3rd edn, vol. 3 (ed. A.N. Rencz), John Wiley & Sons, Inc., New York, pp. 251–306.

Mutanga, O., Skidmore, A.K. and Prins, H.H.T. (2004) Predicting in situ pasture quality in the Kruger National Park, South Africa, using continuum-removed absorption features. *Remote Sensing of Environment*, **89**, 393–408.

Myneni, R.B., Hall, F.G., Sellers, P.J. and Marshak, A.L. (1995) The interpretation of spectral vegetation indices. *IEEE Transactions on Geoscience and Remote Sensing*, **33**, 481–486.

Nachtegael, M., Weken, D. van der, Kerre, E.E. and Philips, W. (eds.) *Soft Computing in Image Processing*. Studies in Fuzziness and Soft Computing, vol. 210. Springer-Verlag, New York.

Nagao, M. and Matsuyama, T. (1979) Edge preserving smoothing. *Computer Graphics and Image Processing*, **9**, 394–407.

Naik, S.K. and Murthy, C.A. (2003) Hue-preserving colour image enhancement without gamut problem. *IEEE Transactions on Image Processing*, **12**, 1591–1598.

Nalbant, S.S. and Alptekin, O. (1995) The use of Landsat Thematic Mapper imagery for analysing lithology and structure of Korucu-Dugla area in Western Turkey. *International Journal of Remote Sensing*, **16**, 2357–2374.

Napieralski, J., Li, B. and Harbor, J. (2006) Comparing predicted and observed spatial boundaries of geologic phenomena: automated proximity and conformity analysis applied to ice sheet reconstructions. *Computers and Geosciences*, **32**, 124–134.

NASA (1988) *Earth Observing System Instrument Panel Report, Synthetic Aperture Radar*, vol. 2f, National Aeronautics and Space Administration, Washington, DC.

Nascimento, J.M.P. and Dias, J.M.B. (2005) Does independent component analysis play a role in unmixing hyperspectral data? *IEEE Transactions on Geoscience and Remote Sensing*, **43**, 175–187.

Nataraj, P.S.V. (undated) Reliable Computing. www.sc.iitb.ac.in/~nataraj/invited_talks_lects/icresh2005/Reliable_Computing.pdf (accessed 5 February 2009).

Navulur, K. (2006) *Multispectral Image Analysis using the Object-Oriented Paradigm*, CRC Press, Boca Raton.

Neteler, M. and Mitasova, H. (2002) *Open Source GIS: A GRASS GIS Approach*, 2nd edn, Springer-Verlag, New York.

Neuenschwander, A.L., Crawford, M.M. and Ringrose, S. (2005) Results from the EO-1 experiment – a comparative study of Earth Observing-1 Advanced Land Imager (ALI) and Landsat ETM+ data for land cover mapping in the Okavango Delta, Botswana. *International Journal of Remote Sensing*, **26**, 4321–4337.

Ngigi, T.G., Tateishi, R. and Gachari, M. (2009) Global mean values in linear spectral unmixing: double fallacy. *International Journal of Remote Sensing*, **30**, 1109–1125.

Nicodemus, F.E., Richmond, J.C., Hsia, J.J. *et al.* (1977) Geometrical Considerations and Nomenclature for Reflectance, National Bureau of Standards, US Department of Commerce, Washington, DC. URL: http://physics.nist.gov/Divisions/Div844/facilities/specphoto/pdf/geoConsid.pdf (accessed 24 May 2010).

Nielsen, A.A. (1994) Analysis of regularly and irregularly sampled spatial, multivariate and multitemporal data. PhD thesis. Technical University of Denmark, Lyngsby. http://www2.imm.dtu.dk/~aa/phd/, (click on phd_no_figs.pdf. The individual figures can be accessed individually; accessed 2 April 2009).

Nikolakaki, P. (2004) A GIS site-selection process for habitat creation: estimating connectivity of habitat patches. *Landscape and Urban Planning*, **68**, 77–94.

Nishida, K., Nemani, R., Glassy, J. and Running, S. (2003) Development of an evapotranspiration index from Aqua/MODIS for monitoring surface moisture status. *IEEE Transactions on Geoscience and Remote Sensing*, **41**, 493–450.

Novak, K. (1992) Rectification of digital imagery. *Photogrammetric Engineering and Remote Sensing*, **58**, 339–344.

Nunes, A.S.L., Marcal, A.R.S. and Vaughan, R.A. (2008) Fast over-land atmospheric correction of visible and near infra-red satellite images. *International Journal of Remote Sensing*, **29**, 3523–3531.

Nyssen, J., Poesen, J., Moeyersons, J. *et al.* (2004) Human impact on the environment in the Ethiopian and Eritrean highlands – a state of the art. *Earth-Science Reviews*, **64**, 273–320.

O'Leary, D.W., Friedmann, J.D. and Pohn, H.A. (1976) Lineament, linear, lineation: some proposed new standards for old terms. *Bulletin of the Geological Society of America*, **87**, 1463–1469.

O'Sullivan, D.O. and Unwin, D.J. (2003) *Geographic Information Analysis*, John Wiley & Sons, Inc, New York.

Oksanen, J. and Sarjakoski, T. (2005) Error propagation of DEM-based surface derivatives. *Computers and Geosciences*, **31**, 1015–1027.

Oliveira, S. and Stewart, D.E. (2006) *Writing Scientific Software: A Guide for Good Style*. Cambridge University Press, Cambridge.

Oliver, C. and Quegan, S. (2004) *Understanding Synthetic Aperture Radar Images*, CRC Press, Boca Raton.

Oliver, M.A. and Webster, R. (1990) Kriging: A method of interpolation for geographical information systems. *International Journal of Geographical Information Systems*, **4**, 313–332.

Olsen, R.C. (2007) *Remote Sensing from Air and Space*, SPIE Press, Bellingham.

Olsson, H. (1993) Regression functions for multitemporal relative calibrations of Thematic Mapper data over boreal forests. *Remote Sensing of Environment*, **46**, 89–102.

Olsson, H. (1995) Radiometric calibration of Thematic Mapper data for forest change detection. *International Journal of Remote Sensing*, **16**, 81–96.

Olsson, L. and Ekhlund, L. (1994) Fourier series for analysis of temporal sequences of satellite sensor imagery. *International Journal of Remote Sensing*, **15**, 3735–3741.

Ormsby, J. (1992) Evaluation of natural and man-made features using Landsat TM data. *International Journal of Remote Sensing*, **13**, 303–318.

Osborne, P.E., Alonso, J.C. and Bryant, R.G. (2001) Modelling landscape-scale habitat use with GIS and remote sensing: a case study with Great Bustards. *Journal of Applied Ecology*, **38**, 458–471.

Ouattara, T., Gwyn, Q.H.J. and Dubois, J.-M.M. (2004) Evaluation of the runoff potential in high relief semi-arid regions using remote sensing data: application to Bolivia. *International Journal of Remote Sensing*, **25**(2), 423–435.

Ouma, Y., Tetuko, J. and Tateishi, R. (2008) Analysis of co-occurrence and wavelet transform textures for differentiation of forest and non-forest vegetation in very high resolution optical sensor imagery. *International Journal of Remote Sensing*, **29**, 3417–3456.

Paegelow, M. and Camacho Olmedo, M.T. (eds) (2008) *Modelling Environmental Dynamics: Advances in Geomatic Solutions*, Springer-Verlag, Berlin.

Pajares, G. and de la Cruz, J.M. (2004) A wavelet-based image fusion tutorial. *Pattern Recognition*, **37**, 1855–1872.

Pal, M. (2005) Random forest classifier for remote sensing classification. *International Journal of Remote Sensing*, **26**, 217–222.

Pal, M. (2006) Support vector machine-based feature selection for land cover classification: a case study with DAIS hyperspectral data. *International Journal of Remote Sensing*, **27**, 2877–2894.

Pal, M. (2008) Ensemble of support vector machines for land cover classification. *International Journal of Remote Sensing*, **29**, 3043–3049.

Pal, M. and Mather, P.M. (2003) An assessment of the effectiveness of decision tree methods for land cover classification. *Remote Sensing of Environment*, **86**, 554–565.

Pal, M. and Mather, P.M. (2004) Assessment of the effectiveness of support vector machines for hyperspectral data. *Future Generation Computer Systems*, **20**, 1215–1225.

Pal, M. and Mather, P.M. (2005) Support vector machines for classification in remote sensing. *International Journal of Remote Sensing*, **26**, 1007–1011.

Pal, M. and Mather, P.M. (2006) Some issues in the classification of DAIS hyperspectral data. *International Journal of Remote Sensing*, **27**, 2895–2916.

Pal, N.R. and Pal, S.K. (1993) A review on image segmentation techniques. *Pattern Recognition*, **26**, 1277–1294.

Palü, V. and Pons, X. (1995) Incorporation of relief in polynomial-based geometric corrections. *Photogrammetric Engineering and Remote Sensing*, **61**, 935–944.

Palubinskas, G., Müller, R., Reinartz, P. and Schroeder, M. (2007) Radiometric normalization of sensor scan angle effects in optical remote sensing imagery. *International Journal of Remote Sensing*, **28**, 4453–4469.

Pan, J.-J. and Chang, C.-I. (1992) Destriping of Landsat MSS images by filtering techniques. *Photogrammetric Engineering and Remote Sensing*, **58**, 1417–1423.

Paola, J. and Schowengerdt, R.A. (1995a) A review and analysis of backpropagation neural networks for classification of remotely-sensed multi-spectral imagery. *International Journal of Remote Sensing*, **16**, 3033–3058.

Paola, J. and Schowengerdt, R.A. (1995b) A detailed comparison of backpropagation neural networks and maximum-likelihood classifiers for urban land use classification. *IEEE Transactions on Geoscience and Remote Sensing*, **33**, 981–996.

Paola, J. and Schowengerdt, R.A. (1997) The effect of neural network structure on multispectral land-use/land-cover classification. *Photogrammetric Engineering and Remote Sensing*, **63**, 535–544.

Paolini, L., Grings, F., Sobrino, J.A. *et al.* (2006) Radiometric correction effects in Landsat multi-date/multi-sensor change detection studies. *International Journal of Remote Sensing*, **27**, 685–704.

Park, J.-M., Song, W.J. and Pearlman, W.A. (1999) Speckle filtering of SAR images based on adaptive windowing. *IEEE Proceedings on Vision, Image and Signal Processing*, **146**, 191–197.

Parlow, E. (ed.) (1996) *Progress in Environmental Remote Sensing Research and Applications*, Proceedings of the 15th EARSeL Symposium, Basle, 4–6 September 1996, A.A. Balkema, Rotterdam.

Pavlidis, T. (1982) *Algorithms for Graphics and Image Processing*, Springer-Verlag, Berlin.

Paylor, E.D. II, Evans, D.L. and Tralli, D.M. (2005) Special Issue of Remote sensing and geospatial information for natural hazards characterization. *ISPRS Journal of Photogrammetry and Remote Sensing*, **59**, 181–253.

Pebesma, E. and Wesseling, C.G. (1998) GSTAT: a program for geostatistical modelling, prediction and simulation. *Computers and Geosciences*, **24**, 17–31.

Peddle, D.R. (1995) Knowledge foundation for supervised evidential classification. *Photogrammetric Engineering and Remote Sensing*, **61**, 409–417.

Peddle, D.R. and Ferguson, D.T. (2002) Optimisation of multisource data analysis: an example using evidential reasoning for GIS data classification. *Computers and Geosciences*, **28**, 45–52.

Peddle, D.R. and Smith, A.M. (2005) Spectral mixture analysis of agricultural crops: end-member validation and biophysical estimation in potato plots. *International Journal of Remote Sensing*, **26**, 4959–4979.

Peleg, S. (1980) A new probabilistic relaxation scheme. *IEEE Transactions on Pattern Analysis and Machine Intelligence*, **2**, 362–369.

Perry, C.R. and Lautenschlager, L.F. (1984) Functional equivalence of spectral vegetation indices. *Remote Sensing of Environment*, **14**, 169–182.

Perski, Z. and Jura, D. (1999) ERS SAR interferometry for land subsidence detection in coal mining areas (amended version). *ESA Earth Observation Quarterly*, **64**. http://esapub.esrin.esa.it/eoq/eoq64/reprint63.pdf (accessed 1 August 2009).

Péteri, R. and Ranchin, T. (2007) Road networks from high spatial resolution remote sensing data, in *Remote Sensing of Impervious Surfaces* (ed. Q. Weng), CRC Press, Boca Raton, pp. 216–233.

Petrie, G. (2008) Spaceborne digital imaging and sensing systems, in *Advances in Photogrammetry, Remote Sensing and Spatial Information Science*, ISPRS Book Series (eds Z. Li, E. Baltsavias, and J. Chen), Taylor and Francis, London, pp. 29–44.

Pettit, C., Cartwright, W., Bishop, I. *et al.* (eds) (2008) *Landscape Analysis and Visualisation: Spatial Models for Natural Resource Management and Planning*, Springer-Verlag, Berlin.

Philipp, I. and Rath, T. (2002) Improving plant discrimination in image processing by use of different colour space transformations. *Computers and Electronics in Agriculture*, **35**, 1–15.

Phillipson, D.W. (1998) *Ancient Ethiopia*, British Museum, London.

Philpott, W.D. (1991) The derivative ratio algorithm: avoiding atmospheric effects in remote sensing. *IEEE Transactions on Geoscience and Remote Sensing*, **29**, 350–357.

Pinter, P.J., Jackson, R.D., Idso, S.B. and Reginato, R.J. (1983) Diurnal patterns of wheat spectral reflectance. *IEEE Transactions on Geoscience and Remote Sensing*, **21**, 156–163.

Pinty, B., Leprieur, C. and Verstraete, M.M. (1993) Towards a quantitative interpretation of vegetation indices. Part I: biophysical canopy properties and classical indices. *Remote Sensing Reviews*, **7**, 127–150.

Pitas, I. (1993) *Digital Image Processing Algorithms*, Prentice Hall, New York.

Pizurica, A., Philips, W., Limahieu, I. and Acheroi, M. (2001) Despeckling SAR images using wavelets and a new class of adaptive shrinkage regulators. Proceedings of the IEEE Conference on Image Processing, Thessaloniki, Greece, 7–10 October, 2001, pp. 233–236.

Plataniotis, K.N. and Venetsanopoulos, A.N. (2000) *Color Image Processing and Applications*, Springer-Verlag, Berlin.

Plaza, A., Benediktsson, J.A., Boardman, J.W. *et al.* (2009) Recent advances in techniques for hyperspectral image processing. *Remote Sensing of Environment*, **113** (Suppl. 1), S110–S122.

Plaza, A.J. and Chang, C.-I. (2008) *High Performance Computing in Remote Sensing*, Chapman & Hall and CRC Press, Boca Raton.

Plourde, L. and Congalton, R.G. (2003) Sampling method and sample placement: how do they affect the accuracy of remotely sensed maps? *Photogrammetric Engineering and Remote Sensing*, **69**, 289–297.

Plourde, L.C., Ollinger, S.V., Smith, M.-L. and Martin, M.E. (2007) Estimating species abundance in a northern temperate forest using spectral mixture analysis. *Photogrammetric Engineering and Remote Sensing*, **73**, 829–840.

Pohl, C. and van Genderen, J.L. (1998) Multisensor image fusion in remote sensing: concepts, methods and application. *International Journal of Remote Sensing*, **19**, 823–854.

Pohl, C. and van Genderen, J.L. (1998) Multisensor image fusion in remote sensing: concepts, methods and application. *International Journal of Remote Sensing*, **19**, 823–854.

Popp, T. (1995) Correcting atmospheric masking to retrieve the spectral albedo of land surfaces from satellite measurements. *International Journal of Remote Sensing*, **16**, 3843–3508.

Powell, S.L., Pfugmacher, D., Kirschbaum, A.A. *et al.* (2007) Moderate resolution remote sensing alternatives: a review of Landsat-like sensors and their applications. *Journal of Applied Remote Sensing*, **1**, 012506 (online journal). See http://spiedl.aip.org/dbt/dbt.jsp?KEY=JARSC4&Volume=1&Issue=1. (accessed 1 August 2009).

Prasad, L. and Iyengar, S.S. (1997) *Wavelet Analysis with Applications to Image Processing*, CRC Press, Boca Raton.

Prasad, S. and Bruce, L.M. (2008) Limitations of principal components analysis for hyperspectral target recognition. *IEEE Geoscience and Remote Sensing Letters*, **5**, 625–629.

Pratt, W.K. (1978) *Digital Image Processing*, John Wiley & Sons, Inc., New York.

Premelatha, M. (2001) Quality assessment of interferometrically derived digital elevation models. PhD thesis. University of Nottingham, Nottingham.

Press, W.A., Teukolsky, S.A., Vetterling, W.T. and Flannery, B.P. (1992) *Numerical Recipes in Fortran*, 2nd edn, Cambridge University Press, Cambridge.

Press, W.H., Teukolsky, S.A. and Vetterling, W.T. (2007) *Numerical Recipes: The Art of Scientific Computing*, 3rd edn, Cambridge University Press, Cambridge.

Price, J.C. (1987) Calibration of satellite radiometers and the comparison of vegetation indices. *Remote Sensing of Environment*, **21**, 15–27.

Price, J.C. (1988) An update on visible and near infrared calibration of satellite instruments. *Remote Sensing of Environment*, **24**, 419–422.

Price, J.C. (1994) How unique are spectral signatures? *Remote Sensing of Environment*, **49**, 181–186.

Proisy, C., Couteron, P. and Fromard, F. (2007) Predicting and mapping mangrove biomass from canopy grain analysis using Fourier-based textural ordination of IKONOS images. *Remote Sensing of Environment*, **109**, 379–392.

Prost, G.L. (2002) *Remote Sensing for Geologists: A Guide to Image Interpretation*, CRC Press, Boca Raton.

Proy, C., Tanré, D. and Deschamps, P.Y. (1989) Evaluation of topographic effects in remotely sensed data. *Remote Sensing of Environment*, **30**, 21–32.

Pu, R.L. and Gong, P. (2004) Wavelet transform applied to EO-1 hyperspectral data for forest LAI and crown closure mapping. *Remote Sensing of Environment*, **91**, 212–224.

Puissant, A., Hirsch, J. and Weber, C. (2005) The utility of texture analysis to improve per-pixel classification for high to very high spatial resolution imagery. *International Journal of Remote Sensing*, **26**, 733–745.

Qiu, F. and Jensen, J.R. (2004) Opening the black box of neural networks for remote sensing image classification. *International Journal of Remote Sensing*, **25**, 1749–1768.

Qu, J.J., Gao, W., Kafatos, M. *et al.* (eds) (2006) *Earth Science Satellite Remote Sensing: Science and Instruments*, vol. 1, Springer-Verlag, New York (in association with Tsingua University Press, Beijing, China).

Qu, J.J., Gao, W., Kafatos, M., Murphy, R.E. and Salomonson, V. (eds) (2007) *Earth Science Satellite Remote Sensing: Data, Computational Processing, and Tools*, vol. 2, Springer-Verlag, Berlin.

Quattrochi, D.A. and Goodchild, M.F. (eds) (1997) *Scale in Remote Sensing and GIS*, Lewis Publishers, New York.

Quegan, S. and Rhodes, I. (1994) Relating polarimetric SAR data to surface properties – the MAC-Europe experiment, in Mather, P.M. (ed.), *TERRA 2: Understanding the Terrestrial Environment – Remote Sensing Data Systems and Networks*. Chichester: John Wiley & Sons, Ltd., pp. 159–174.

Quegan, S. and Wright, A. (1984) Automatic segmentation techniques for satellite-borne synthetic aperture radar (SAR) images, *Proceedings of the Tenth Anniversary Conference of the Remote Sensing Society*, The Remote Sensing Society, Nottingham, pp. 161–167.

Quinlan, J.R. (1993) *C4.5: Algorithm for Machine Learning*, Morgan Kaufmann, San Mateo.

Rabus, B. and Fatland, D.R. (2000) Comparison of SAR-interferometric and surveyed velocities on a mountain glacier: black rapids glacier. *Journal of Glaciology*, **46**, 119–128.

Rabus, B., Eineder, M., Roth, A. and Bamler, R. (2003) The Shuttle Radar Topographic Mission – a new class of digital elevation models acquired by spaceborne radar. *ISPRS Journal of Photogrammetry and Remote Sensing*, **57**, 241–262.

Rahman, A.F., Gamon, J.A., Sims, D.A. and Schmidts, M. (2003) Optimum pixel size for hyperspectral studies of ecosystem function in southern California chaparral and grassland. *Remote Sensing of Environment*, **84**, 192–207.

Ralston, A. and Rabinowitz, P. (2000) *A First Course in Numerical Analysis*. Dover Publications, Mineola, NY.

Ramirez, R.W. (1985) *The Fast Fourier Transform: Fundamentals and Concepts*, Prentice Hall, Englewood Cliffs.

Ranchin, T. (2002a) Data fusion in remote sensing and improvement of the spatial resolution of satellite images, in *Multisensor Fusion* (eds A.K. Hyder, E. Shabazian, and E. Waltz), Kluwer Academic Publishers, Dordrecht, pp. 633–656.

Ranchin, T. (2002b) Wavelets for modelling and data fusion in remote sensing, in *Multisensor Fusion* (eds A.K. Hyder, E. Shabazian, and E. Waltz), Kluwer Academic Publishers, Dordrecht, pp. 361–363.

Ranchin, T. and Wald, L. (1993) The wavelet transform for the analysis of remotely sensed images. *International Journal of Remote Sensing*, **14**, 615–619.

Ranchin, T. and Wald, L. (2000) Fusion of high spatial and spectral resolution images: the ARSIS concept and its implementation. *Photogrammetric Engineering and Remote Sensing*, **66**, 49–61.

Ranson, K.J., Biehl, L.L. and Bauer, M.E. (1985) Variation in spectral response of soybeans with respect to illumination, view and canopy geometry. *International Journal of Remote Sensing*, **6**, 1827–1842.

Rao, C.R.N. and Chen, J. (1996) Post-launch calibration of the visible and near-infrared channels on the advanced very high resolution radiometer on the NOAA-14 spacecraft. *International Journal of Remote Sensing*, **17**, 2743–2747.

Ray, T.W. and Murray, B.C. (1996) Nonlinear spectral mixing in desert vegetation. *Remote Sensing of Environment*, **55**, 59–74.

Rayner, J.N. (1971) *An Introduction to Spectral Analysis*, Pion Press, London.

Rees, W.G. (2001) *Physical Principles of Remote Sensing*, 2nd edn, Cambridge University Press, Cambridge.

Rees, W.G. and Satchell, M.J.F. (1997) The effect of median filtering on synthetic aperture radar images. *International Journal of Remote Sensing*, **18**, 2887–2893.

Reigber, C., Xia, Y., Kaufmann, H. *et al.* (1996) Impact of precise orbits on SAR interferometry. FRINGE '96: ESA Workshop on Applications of ERS SAR Interferometry, European Space Agency, ESA SP-406, Zurich, Switzerland, Paris, pp. 223–232.

Remote Sensing of Environment (2005) Special issue: scientific results from ASTER. *Remote Sensing of Environment*, **99** (1–2) , pp. 1–220.

Rencher, A.C. (2002) *Methods of Multivariate Analysis*, John Wiley & Sons, Inc., New York.

Riaño, D., Chuvieco, E., Salas, J. and Aguado, I. (2003) Assessment of different topographic corrections on Landsat-TM data for mapping vegetation types. *IEEE Transactions on Geoscience and Remote Sensing*, **41**, 1056–1061.

Riazanoff, S., Cervelle, B. and Chorowicz, J. (1990) Parametrizable skeletonisation of binary and multi-level images. *Pattern Recognition Letters*, **11**, 25–33.

Richards, J.A. and Jia, X. (2005) *Remote Sensing Digital Image Analysis – An Introduction*, 4th edn, Springer-Verlag, Berlin.

Richardson, A.J. and Wiegand, C.L. (1977) Distinguishing vegetation from soil background information. *Photogrammetric Engineering and Remote Sensing*, **43**, 1541–1552.

Richter, R. (1996) A spatially-adaptive fast atmospheric correction algorithm. *International Journal of Remote Sensing*, **17**, 1201–1214.

Richter, R. (1997) On the in-flight absolute calibration of high spatial resolution spaceborne sensors using small ground targets. *International Journal of Remote Sensing*, **18**, 2827–2833.

Richter, R., Bachmann, M., Dorigo, W. and Muller, A. (2006) Influence of the adjacency effect on ground reflectance measurements. *IEEE Geoscience and Remote Sensing Letters*, **3**, 565–569.

Riedman, M. and Haynes, M. (2007) Developments in synthetic aperture radar interferometry for monitoring geohazards, in *Mapping Hazardous Terrain Using Remote Sensing*, Special Publication 283 (ed. R.W. Teeuw), Geological Society of London, London, pp. 45–52.

Rignot, E.J.M., Forster, R.R. and Isacks, B.L. (1996) Mapping of glacial motion and surface topography of Hielo Patagonico Norte, Chile, using satellite SAR L-band interferometry data. *Annals of Glaciology*, **23**, 209–216.

Rio, J.N.R. and Lozano-Garcia, D.F. (2000) Spatial filtering of radar data (RADARSAT) for wetlands (brackish marshes) classification. *Remote Sensing of Environment*, **73**, 143–151.

Riou, R. and Seyler, S. (1997) Texture analysis of tropical rain forest infrared satellite images. *Photogrammetric Engineering and Remote Sensing*, **63**, 515–521.

Ripley, B.D. (1981) *Spatial Statistics*, John Wiley & Sons, Inc., New York.

Roberts, D.A., Adams, J.B. and Smith, M.O. (1993) Discriminating green vegetation, non-photosynthetic vegetation and soils in AVIRIS data. *Remote Sensing of Environment*, **44**, 1–25.

Roberts, D.A., Gardner, M., Church, R. *et al.* (1998) Mapping chaparral in the Santa Monica Mountains using multiple endmember spectral mixture models. *Remote Sensing of Environment*, **65**, 267–279.

Roberts, S. and Everson, R. (eds) (2001) *Independent Component Analysis: Principles and Practice*, Cambridge University Press, Cambridge.

Robertson, B.C. (2003) Rigorous geometric modelling and correction of QuickBird imagery, *Proceedings of the IEEE International Geoscience and Remote Sensing Symposium, 2003 (IGARSS '03), Toulouse, 21–25 July, 2003*, vol. 2, IEEE Press, Piscastaway, pp. 797–802.

Robertson, G.P. (1987) Geostatistics in ecology: interpolating with known variance. *Ecology*, **68**, 744–774.

Robinson, I.S. (2004) *Measuring the Oceans from Space: The Principles and Methods of Satellite Oceanography*, Springer/Praxis, Berlin.

Rodriguez, E., Morris, C.S. and Belz J.E. (2006) A global assessment of the SRTM performance. *Photogrammetric. Enginering and Remote Sensing*, **72**, 249–260.

Roerink, G.J., Menenti, M., Soepboer, W. and Su, Z. (2003) Assessment of climate impact on vegetation dynamics by using remote sensing. *Physics and Chemistry of the Earth*, **28**, 103–109. (Special Issue on Applications of Quantitative Remote Sensing to Hydrology).

Rogan, J. and Chen, D.M. (2004) Remote sensing technology for mapping and monitoring land-cover and land-use change. *Progress in Planning*, **61**, 301–325.

Rogan, J., Miller, J., Stow, D. *et al.* (2003) Land-cover change monitoring with classification trees using Landsat TM and ancillary data. *Photogrammetric Engineering and Remote Sensing*, **69**, 793–804.

Roger, R.E. (1996) Principal components transform with simple, automatic noise adjustment. *International Journal of Remote Sensing*, **17**, 2719–2727.

Roger, R.E. and Arnold, J.F. (1996) Reliably estimating the noise in AVIRIS hyperspectral images. *International Journal of Remote Sensing*, **17**, 1951–1962.

Rogge, D.M., Rivard, B., Zhang, J. *et al.* (2007) Integration of spatial–spectral information for the improved extraction of end members. *Remote Sensing of Environment*, **110**, 287–303.

Rondeaux, G. (1995) Vegetation monitoring by remote sensing: a review of biophysical indices. *Photointerpretation*, **33**, 197–216.

Rondeaux, G., Steven, M.D. and Baret, F. (1996) Optimisation of soil-adjusted vegetation indices. *Remote Sensing of Environment*, **55**, 95–107.

Rosen, P.A., Hensley, S., Joughin, I. *et al.* (2000) Synthetic aperture radar interferometry. *Proceedings of the IEEE*, **88**, 333–382.

Rosen, P.A., Hensley, S., Zebker, H.A. *et al.* (1996) Surface deformation and coherence measurements of Kilauea Volcano, Hawaii, from SIR-C radar interferometry. *Journal of Geophysical Research*, **101**, 101–123.

Rosenfeld, A. (1976) Iterative methods in image analysis. *Pattern Recognition*, **10**, 181–187.

Rosenfeld, A. and Kak, A.C. (1982) *Digital Picture Processing*, 2nd edn, vol. 1, Academic Press, New York.

Rosenqvist, A., Shimada, M., Ito, N. and Watanabe, M. (2007) ALOS PALSAR: a pathfinder mission for global-scale monitoring of the environment. *IEEE Transactions on Geoscience and Remote Sensing*, **45**, 3307–3316.

Rothery, D.A. and Hunt, G.A. (1990) A simple way to perform decorrelation stretching and related techniques on menu-driven image processing systems. *International Journal of Remote Sensing*, **11**, 133–137.

Rott, H. (2009) Advances in interferometric syntetic aperture radar (SAR) in Earth system science. *Progress in Physical Geography*, **33**, 769–791.

Rott, H. and Nagler, T. (2006) The contribution of radar interferometry to the assessment of landslide hazards. *Advances in Space Research*, **37**, 710–719.

Rowan, L.C. and Mars, J.C. (2003) Lithologic mapping in the Mountain Pass, California area using Advanced Spaceborne Thermal Emission and Reflection Radiometer (ASTER) data. *Remote Sensing of Environment*, **84**, 350–366.

Sakamoto, T., Yokozawa, M., Toritani, H. *et al.* (2005) A crop phenology detection method using time-series MODIS data. *Remote Sensing of Environment*, **96**, 366–337.

Salomonson, V.V., Barnes, W., Xiong, J. *et al.* (2002) An overview of the earth observing system MODIS instrument and associated data systems performance, *Proceedings of the International Geoscience and Remote Sensing Symposium, IGARSS-02, Piscataway*, vol. 2, IEEE Press, pp. 1174–1176.

Samet, H. (1990) *The Design and Analysis of Spatial Data Structures*, Addison-Wesley, Reading.

Sammon, J.W. Jr. (1969) A nonlinear mapping algorithm for data structure analysis. *IEEE Transactions on Computers*, **18**, 401–409.

Sanchez, J. and Canton, M.P. (1998) *Space Image Processing*, CRC Press, Boca Raton.

Sandau, R., Röser, H.-P. and Valenzuela, A. (eds) (2008) *Small Satellites for Earth Observation*, Springer-Verlag, Berlin.

Sander, P. (2007) Lineaments in groundwater exploration: a review of applications and limitations. *Hydrogeology Journal*, **15**, 71–74.

Sansosti, E., Berardino, P., Manunta, M. *et al.* (2006) Geometrical SAR image registration. *IEEE Transactions on Geoscience and Remote Sensing*, **44**, 2861–2870.

Saraf, A.K. and Choudhury, P.R. (1998) Integrated remote sensing and GIS for groundwater exploration and identification of artificial recharge sites. *International Journal of Remote Sensing*, **19**, 1825–1841.

Savitsky, A. and Golay, M.J.E. (1964) Smoothing and differentiation of data by simplified least squares procedures. *Analytical Chemistry*, **36**, 1627–1639.

Sawter, R., Deuze, J.L., Devaux, G. *et al.* (1991) SPOT Calibration on the Test Site at La Crau, France, Cinquième Colloque International Mesures Physiques et Signatures en Telédétection, Courcheval, 14–18 January 1991. ESA-SP-319, European Space Agency, Paris, pp. 77–80.

Scally, R. (2006) *GIS for Environmental Management*, ESRI Press, Redlands.

Schaale, M. and Furrer, R. (1995) Land surface classification by neural networks. *International Journal of Remote Sensing*, **16**, 3003–3031.

Schaepman, M.E., Ustin, S.L., Plaza, A.J. *et al.* (2009) Earth system science related imaging spectroscopy – An assessment. *Remote Sensing of Environment*, **113** (Suppl. 1), S123–S137.

Schaepman-Strub, G., Schaepman, M.E., Painter, T.H. *et al.* (2006) Reflectance quantities in optical remote sensing-definitions and case studies. *Remote Sensing of Environment*, **103**, 27–42.

Schetselaar, E.M. (1998) Fusion by the IHS transform: should we use cylindrical or spherical coordinates? *International Journal of Remote Sensing*, **19**, 759–765.

Schläpfer, D. and Richter, R. (2002) Geo-atmospheric processing of airborne imaging spectrometry data. Part 1: parametric orthorectification. *International Journal of Remote Sensing*, **23**, 2609–2630.

Schmid, T., Koch, M., DiBlasi, M. and Hagos, M. (2008) Spatial and spectral analysis of land cover properties for an archaeological area in Aksum, Ethiopia: Applying high and medium resolution data. *Catena*, **75**, 93–101.

Schott, J.R. (2007) *Remote Sensing: The Image Chain Approach*, 2nd edn, Oxford University Press, Oxford.

Schott, J.R. and Volchock, W.J. (1985) Thematic Mapper thermal infrared calibration. *Photogrammetric Engineering and Remote Sensing*, **51**, 1351–1357.

Schotten, C.G.J., van Rooy, W.W.L. and Janssen, L.L.F. (1995) Assessment of the capabilities of multi-temporal ERS-1 SAR data to discriminate between agricultural crops. *International Journal of Remote Sensing*, **16**, 2619–2637.

Schowengerdt, R.E. (2006) *Remote Sensing: Models and Methods for Image Processing*, Academic Press and Elsevier, Amsterdam.

Schreier, G. (ed.) (1993a) *SAR Geocoding: Data and Systems*, Herbert Wichmann, Karlsruhe.

Schreier, G. (1993b) Geometrical properties of SAR images, in *SAR Geocoding: Data and Systems* (ed. G. Schreier), Herbert Wichmann, Karlsruhe, pp. 103–134.

Schreier, G. and Dech, S. (2005) High resolution earth observation satellites and services in the next decade – a European perspective. *Acta Astronautica*, **57**, 520–533.

Schumaker, N.H. (1996) Using landscape indices to predict habitat connectivity. *Ecology*, **77**, 1210–1225.

Schumann, G., Hostache, R., Puech, C. *et al.* (2007) High-resolution 3-D flood information from radar imagery for flood hazard management. *IEEE Transactions on Geoscience and Remote Sensing*, **45**, 1715–1725.

Schutz, B.E., Zwally, H.J., Shuman, C.A. *et al.* (2005) Overview of the ICESat mission. *Geophysical Research Letters*, **32**, L21S01. doi: 10.1029/2005GL024009.

Seftor, J.L. and Larch, D. (1995) The use of the genetic algorithm to optimise rule-based classifiers for land cover categorization. *Canadian Journal of Remote Sensing*, **21**, 412–420.

Sellers, P. (1989) Vegetation-canopy reflectance and biophysical properties, in *Theory and Applications of Optical Remote Sensing* (ed. G. Asrar), John Wiley & Sons, Inc., New York, pp. 297–335.

Serkan, M., Musaoglu, N., Kirkici, H. and Ormeci, C. (2008) Edge and fine detail preservation in SAR images through speckle reduction with an adaptive mean filter. *International Journal of Remote Sensing*, **29**, 6727–6738.

Sernicola, L. (2009) Il modello d'insediamento sull'altopiano tigrino (Etiopia settentrionale/Eritrea centrale) in epoca Pre-Aksumita e Aksumita (ca. 700 a.Cr. – 800 d.Cr.). Un contributo da Aksum. Unpublished PhD dissertation. Department of African and Arabic Studies, University of Naples 'L'Orientale', Naples.

Sernicola, L. and Sulas, F. (in press) Continuità e cambiamento nel paesaggio rurale di Aksum: dati archeologici, etnografici e ambientali, in *Studi in Onore di Yaqob Beyene*, *Studi Africanistici – Serie Etiopica*, Vol. 7 (eds C. Baffioni, A. Bausi, F. Ersilia, and A. Manzo), Università degli Studi di Napoli 'L'Orientale', Naples.

Serra, P., Pons, X. and Sauri, D. (2003) Post-classification change detection with data from different sensors: some accuracy considerations. *International Journal of Remote Sensing*, **24**, 3311–3340.

Serrano, L., Filella, I. and Peñuelas, J. (2000) Remote sensing of biomass and yield of winter wheat under different nitrogen supplies. *Crop Science*, **40**, 723–731.

Seshadri, K.S.V., Rao, M., Jayaraman, V. *et al.* (2005) Resourcesat-1: A global multi-observation mission for resources monitoring. *Acta Astronautica*, **57**, 534–539.

Sesnie, S.E., Gessler, P.E., Finegan, B. and Thessler, S. (2008) Integrating Landsat TM and SRTM-DEM derived variables with decision trees for habitat classification and change detection in complex neotropical environments. *Remote Sensing of Environment*, **112**, 2145–2159.

Settle, J.J. and Campbell, N. (1998) On the errors of two estimators of sub-pixel fractional cover when mixing is linear. *IEEE Transactions on Geoscience and Remote Sensing*, **36**, 163–170.

Settle, J.J. and Drake, N.A. (1993) Linear mixing and the estimation of ground cover proportions. *International Journal of Remote Sensing*, **14**, 1159–1177.

Sever, T.L. and Irwin, D.E. (2003) Landscape archaeology: Remote-sensing investigation of the ancient Maya in the Peten rainforest of northern Guatemala. *Ancient Mesoamerica*, **14**, 113–122.

Shafer, G. (1979) *A Mathematical Theory of Evidence*, Princeton University Press, Princeton.

Shafri, H.Z.M. (2003) An assessment of the potential of wavelet-based de-noising in the analysis of remotely sensed data. PhD thesis. School of Geography, The University of Nottingham.

Shafri, H.Z.M. and Mather, P.M. (2005) Wavelet shrinkage in noise removal of hyperspectral remote sensing data. *American Journal of Applied Sciences*, **2**, 1169–1173.

Shah, C.A., Varshney, P.K. and Arora, M.K. (2007) ICA mixture model algorithm for unsupervised classification of remote sensing imagery. *International Journal of Remote Sensing*, **28**, 1711–1731.

Shan, J. and Stilla, U. (ed.) (2008) Special Issue: Remote Sensing Data Fusion. *Photogrammetric Engineering and Remote Sensing*, **74** (2).

Shan, J. and Toth, C.K. (eds) (2008) *Topographic Laser Ranging and Scanning: Principles and Processing*, CRC Press, Boca Raton.

Shandley, J., Franklin, J. and White, T. (1996) Testing the Woodcock–Harward image segmentation algorithm in an area of southern California chaparral and woodland vegetation. *International Journal of Remote Sensing*, **17**, 983–1004.

Shapiro, L.G. and Stockman, G.C. (2001) *Computer Vision*, Prentice Hall, Englewood Cliffs.

Sharkey, A.J.C. (2000) Combining artificial neural networks, in *Encyclopedia of Computer Science and Technology*, vol. 42, Suppl. 27 (eds A. Kent, and J.G. Williams), CRC Press, Boca Raton, pp. 1–22.

Sharma, K.M.S. and Sarkar, A. (1998) A modified contextual classification technique for remote sensing data. *Photogrammetric Engineering and Remote Sensing,* **64**, 273–280.

Sharon, D. (1972) The spottiness of rainfall in a desert area. *Journal of Hydrology*, **17**, 161–175.

Shaw, J.M. (2003) Climate change and deforestation: implications for the Maya collapse. *Ancient Mesoamerica*, **14**, 157–167.

Shaw, R., Sowers, L. and Sanchez, E. (1982) A comparative study of linear and nonlinear edge finding techniques for Landsat multispectral data, in *Proceedings of the Pecora VII Symposium, Sioux Falls, South Dakota* (ed. B.F. Richason, Jr.), American Society of Photogrammetry and Remote Sensing, Falls Church, pp. 529–542.

Shawe-Taylor, J. and Cristianini, N. (2004) *Kernel Methods for Pattern Analysis*, Cambridge University Press, Cambridge.

Shemer, L., Marom, M. and Markman, D. (1993) Estimates of currents in the nearshore ocean region using interferometric synthetic aperture radar. *Journal of Geophysical Research*, **98**, 7001–7010.

Shi, W. and Zhu, C. (2002) The line segment match method for extracting road network from high-resolution satellite images. *IEEE Transactions on Geoscience and Remote Sensing*, **40**, 511–514.

Shimabukuro, Y.E. and Smith, J.A. (1991) The least-squares mixing models to generate fraction images derived from remote sensing multispectral data. *IEEE Transactions on Geoscience and Remote Sensing*, **29**, 16–21.

Shipman, H. and Adams, J.B. (1987) Detectability of minerals on desert alluvial fans using reflectance spectra. *Journal of Geophysical Research*, **92**, 10931–10402.

Shlien, S. (1979) Geometric correction, registration and resampling of Landsat imagery. *Canadian Journal of Remote Sensing*, **5**, 74–87.

Sieber, R. and Huber, S.M. (2007) Atlas of Switzerland 2 – a highly interactive thematic national atlas, in *Multimedia Cartography*, 2nd edn (eds W. Cartwright, M.P. Peterson, and G.F. Gartner), Springer-Verlag, Berlin, pp. 161–182.

Siljestrom, P.A. and Moreno, A. (1995) Monitoring burnt areas by principal components analysis of multi-temporal TM data. *International Journal of Remote Sensing*, **16**, 1577–1587.

Siljestrom, P.A., Moreno, A., Vikgren, G. and Cáceres, L.M. (1997) The application of selective principal components analysis (SPCA) to a Thematic Mapper (TM) image for the recognition of geomorphologic features configuration. *International Journal of Remote Sensing*, **18**, 3843–3852.

Silván-Cárdenas, J.L. and Wang, L. (2008) Sub-pixel confusion – uncertainty matrix for assessing soft classifications. *Remote Sensing of Environment*, **112**, 1081–1095.

Simonett, D.S. (1983) (ed.) The development and principles of remote sensing, in *Manual of Remote Sensing*, **2** vols. (ed. R.N. Colwell), American Society of Photogrammetry, Falls Church, pp. 1–36.

Singh, A. (1984) Some clarifications about the pairwise divergence method in remote sensing. *International Journal of Remote Sensing*, **5**, 623–627.

Singh, A. and Harrison, A. (1985) Standardized principal components. *International Journal of Remote Sensing*, **6**, 883–896.

Singleton, R.C. (1979a) Mixed radix fast Fourier transform, *Programs for Digital Signal Processing*, Section 1.4-1, IEEE Acoustics, Speech and Signal Processing Society, IEEE Press and John Wiley & Sons, Inc., New York.

Singleton, R.C. (1979b) Two-dimensional mixed radix mass storage Fourier transform, *Programs for Digital Signal Processing*, Section 1.9-1, IEEE Acoustics, Speech and Signal Processing Society, IEEE Press and John Wiley & Sons, Inc., New York, pp. 1.9-1–1.9-8.

Sithole, G. and Vosselman, G. (2004) Experimental comparison of filter algorithms for bare-earth extraction from airborne laser scanning point clouds. *ISPRS Journal of Photogrammetry and Remote Sensing*, **59**, 85–101.

Sivanandam, S.N., Sumathi, S. and Deepa, S.N. (2006) *Introduction to Neural Networks using Matlab 6.0*, Tata McGraw-Hill, New Delhi.

Skidmore, A. (2002) *Environmental Modelling with GIS and Remote Sensing*, Taylor & Francis, New York.

Slater, J.A., Garvey, G., Johnston, C. *et al.* (2006) The SRTM data "finishing" process and products. *Photogrammetric Engineering and Remote Sensing*, **72**, 237–247.

Slater, P.N. (1980) *Remote Sensing: Optics and Optical Systems*, Addison-Wesley, Reading.

Slater, P.N., Biggar, S.F., Holm, R.G. *et al.* (1987) Reflectance- and radiance-based methods for the in-flight calibration of multi-spectral sensors. *Remote Sensing of Environment*, **22**, 11–37.

Smirnoff, A., Boisvert, E. and Paradis, S.J. (2008) Support Vector Machine for 3D modelling from sparse geological information of various origins. *Computers and Geosciences*, **34**, 127–143.

Smith, B. and Sandwell, D. (2003) Accuracy and resolution of shuttle radar topography mission data. *Geophysical Research Letters*, **30**, 1467–1470.

Smith, C. (2001) *Environmental Physics*, Routledge, London.

Smith, D.M. (1996) Speckle reduction and segmentation of synthetic aperture radar images. *International Journal of Remote Sensing*, **17**, 2043–2057.

Smith, G.M. and Curran, P. (1996) The signal-to-noise ratio (SNR) required for the estimation of foliar biochemical concentrations. *International Journal of Remote Sensing*, **17**, 1031–1058.

Smith, G.M. and Curran, P.J. (1999) Methods for estimating image signal-to-noise ratio (SNR), in *Advances in Remote Sensing and GIS Analysis* (eds P.M. Atkinson, and N.J. Tate), John Wiley & Sons, Ltd, Chichester, pp. 61–74.

Smith, G.M. and Milton, E.J. (1999) The use of the empirical line method to calibrate remotely sensed data to reflectance. *International Journal of Remote Sensing*, **20**, 2653–2662.

Smith, J.A., Lin, T.L. and Ranson, K.J. (1980) The Lambertian assumption and Landsat data. *Photogrammetric Engineering and Remote Sensing*, **46**, 1183–1189.

Smith, K.L., Steven, M.D. and Colls, J.J. (2004) Use of hyperspectral derivative ratios in the red edge region to identify plant stress responses to gas leak. *Remote Sensing of Environment*, **92**, 207–217.

Smith, L.C. (2002) Emerging applications of Interferometric Synthetic Aperture Radar (InSAR) in geomorphology and hydrology. *Annals of the Association of American Geographers*, **93**, 385–398.

Smith, M.O., Ustin, S.L., Adams, J.B. and Gillespie, A.R. (1990) Vegetation in deserts: I. Regional measure of abundance from multispectral images. *Remote Sensing of Environment*, **31**, 1–26.

Snyder, J.P. (1982) Map Projections Used by the US Geological Survey. United States Geological Survey Bulletin 1532, US Government Printing Office, Washington, DC.

Soares Galvao, L., Pizarro, M.A. and Neves Epiphanio, J.C. (2001) Variations in reflectance of tropical soils: Spectral-chemical composition relationships from AVIRIS data. *Remote Sensing of Environment*, **75**, 245–255.

Soares, J.V., Rennó, C.D., Formaggio, A.R. *et al.* (1997) An investigation into the selection of texture features for crop discrimination using SAR imagery. *Remote Sensing of Environment*, **59**, 234–247.

Sohn, Y. and McCoy, R.M. (1997) Mapping desert shrub rangeland using spectral unmixing and modelling spectral mixtures with TM data. *Photogrammetric Engineering and Remote Sensing*, **63**, 707–716.

Solbø, S. and Eltoft, T. (2004) C-WMAP: a statistical speckle filter operating in the wavelet domain. *International Journal of Remote Sensing*, **25**, 1019–1036.

Song, C., Woodcock, C.E., Seto, K.C. *et al.* (2001) Classification and change detection using Landsat TM data: when and how to correct atmospheric effects? *Remote Sensing of Environment*, **75**, 230–244.

Sparks, D.N. (1985) Half-Normal plotting, in *Applied Statistical Algorithms* (eds P. Griffiths, and I.D. Hill), Ellis-Horwood, Chichester, pp. 65–69.

Srinivasama, A. and Richards, J.A. (1990) Knowledge-based techniques for multisource classification. *International Journal of Remote Sensing*, **11**, 505–525.

Srokosz, M.A. (2000) Biological oceanography by remote sensing, in *Encyclopaedia of Analytical Chemistry* (ed. R.A. Meyers), John Wiley & Sons, Ltd, Chichester, pp. 8506–8533.

Starck, J-L., Murtagh, F. and Bijaou, A. (1998) *Image Processing and Data Analysis: The Multiscale Approach*, Cambridge University Press, Cambridge.

Stathaki, T. (ed.) (2008) *Image Fusion Algorithms and Applications*, Academic Press, Boston.

Steffen, W., Sanderson, A., Tyson, P. *et al.* (2005) *Global Change and the Earth System: A Planet Under Pressure*, Springer-Verlag, New York.

Stehman, S.V. (1997) Selecting and interpreting measures of thematic classification accuracy. *Remote Sensing of Environment*, **62**, 77–89.

Stehman, S.V. (2004) A critical evaluation of the normalized error matrix in map accuracy assessment. *Photogrammetric Engineering and Remote Sensing*, **70**, 743–751.

Stehman, S.V. and Wickham, J.D. (2006) Assessing accuracy of net change derived from land cover maps. *Photogrammetric Engineering and Remote Sensing*, **72**, 175–185.

Stehman, S.V., Sohl, T.L. and Loveland, T.R. (2005) An evaluation of sampling strategies to improve precision of estimates of gross change in land use and land cover. *International Journal of Remote Sensing*, **26**, 4941–4957.

Steigler, S.E. (1978) *Dictionary of Earth Sciences*, Pan Books, London.

Steven, M.D. (1998) The sensitivity of the OSAVI vegetation index to observational parameters. *Remote Sensing of Environment*, **63**, 49–60.

Steven, M.D., Malthus, T.J., Baret, F. *et al.* (2003) Intercalibration of vegetation indices from different sensor systems. *Remote Sensing of Environment*, **88**, 412–422.

Sties, M., Kruger, S., Mercer, J.B. and Schnick, S. (2000) Comparison of digital elevation data from airborne laser and interferometric SAR systems. XIXth Congress of the ISPRS, Amsterdam, pp. 866–873.

Stone, J.V. (2004) *Independent Component Analysis: A Tutorial Introduction*, MIT Press, Cambridge.

Storvold, R. and Malnes, E. (2004) Snow covered area retrieval using Envisat ASAR wideswath in mountainous areas, *Proceedings, IEEE International Geoscience and Remote Sensing Symposium, 2004. IGARSS '04, Anchorage, Alaska, 20–24 September, 2004*, vol. 3, IEEE Press, Piscataway, pp. III-1845–III-1848.

Straatsma, M.W. and Middelkoop, A. (2006) Airborne laser scanning as a tool for lowland floodplain vegetation monitoring. *Hydrobiologia*, **565**, 87–103.

Strahler, A.H. (1980) The use of prior probabilities in maximum likelihood classification of remotely sensed data. *Remote Sensing of Environment*, **10**, 135–163.

Strahler, A.H., Logan, T.L. and Bryant, N.A. (1978) Improving forest classification from Landsat by incorporating topographic information. Proceedings of the 12th International Conference on Remote Sensing of Environment, Environmental Research Institute of Michigan, Ann Arbor, pp. 927–942.

Stramondo, S., Moro, M., Doumaz, F. and Cinti, F.R. (2005) The 26 December 2003, Bam, Iran earthquake: surface displacement from Envisat ASAR interferometry. *International Journal of Remote Sensing*, **26**, 1027–1034.

Strang, G. (1994) Wavelets. *American Scientist*, **82**, 250–255.

Strozzi, T., Dammert, P., Wegmüller, U. *et al.* (1999) Forest mapping with SAR interferometry. *ESA Earth Observation Quarterly*, **62**, 17–20.

Su, J.G. and Bork, E.W. (2006) Influence of vegetation, slope and LIDAR sampling angle on DEM accuracy. *Photogrammetric Engineering and Remote Sensing*, **72**, 1265–1274.

Su, L. (2009) Optimizing support vector machine learning for semi-arid vegetation mapping by using clustering analysis. *ISPRS Journal of Photogrammetry and Remote Sensing*, **64**, 407–413.

Su, L., Chopping, M.J., Rango, A. *et al.* (2007) Support vector machines for recognition of semi-arid vegetation types using MISR multi-angle imagery. *Remote Sensing of Environment*, **107**, 299–311.

Sui, H., Zhou, Q., Gong, J. and Ma, G. (2008) Processing of multitemporal data and change detection, in *Advances in Photogrammetry, Remote Sensing and Spatial Information Science*, ISPRS Book Series (eds Z. Li, E. Baltsavias, and J. Chen), Taylor and Francis, London, pp. 227–250.

Suits, G.H. (1983) The nature of electromagnetic radiation, in *Manual of Remote Sensing*, vol. I (ed. R.N. Colwell), American Society of Photogrammetry, Falls Church, pp. 37–60.

Sulas, F., Madella, M. and French, C. (2009) State formation and water resources management in the Horn of Africa: the Aksumite Kingdom of the northern Ethiopian highlands. *World Archaeology*, **41**, 2–15.

Switzer, P., Kowalik, W.S. and Lyon, R.J.P. (1981) Estimation of atmospheric path radiance by the covariance matrix method. *Photogrammetric Engineering and Remote Sensing*, **47**, 1469–1476.

Tadjudin, S. and Landgrebe, D.A. (2000) Robust parameter estimation for mixture model. *IEEE Transactions on Geoscience and Remote Sensing*, **34**, 439–444.

Tanré, D., Deroo, C., Duhaut, P. *et al.* (1986) Simulation of the Satellite Signal in the Solar Spectrum, Laboratoire d'Optique Atmospherique, Universite des Sciences et Techniques de Lille, 59655 Villeneuve D'Ascq Cedex, France/Centre Spatiale de Toulouse, 31055 Toulouse Cedex, France.

Tao, C.V. and Hu, Y. (2001) A comprehensive study on the rational function model for photogrammetric processing. *Photogrammetric Engineering and Remote Sensing*, **67**, 1347–1357.

Tarabalka, Y., Benediktsson, J.A. and Chanussot, J. (2009) Spectral–spatial classification of hyperspectral imagery based on partitional clustering techniques. *IEEE Transactions on Geoscience and Remote Sensing*, **47**, 2973–2987.

Taswell, C. (2000) The what, how and why of wavelet shrinkage denoising. *Computing in Science and Engineering*, **2**, 12–19.

Tatem, A.J., Nayar, A. and Hay, S.I. (2006) Scene selection and the use of NASA's global orthorectified Landsat dataset for land cover and land use change monitoring. *International Journal of Remote Sensing*, **27** (14), 3073–3078.

Teillet, P.M. and Fedosejevs, G. (1995) On the dark target approach to atmospheric correction of remotely-sensed data. *Canadian Journal of Remote Sensing*, **21**, 374–387.

Teillet, P.M., Barker, J.L., Markham, B.L. *et al.* (2001) Radiometric cross-calibration of the Landsat-7 ETM+ and Landsat-5 TM sensors based on tandem data sets. *Remote Sensing of Environment*, **78**, 39–54.

Teillet, P.M., Fedosejevs, G., Thome, K.J. and Barker, J.L. (2007) Impacts of spectral band difference effects on radiometric cross-calibration between satellite sensors in the solar-reflective spectral domain. *Remote Sensing of Environment*, **110**, 393–409.

Teillet, P.M., Guindon, B. and Goodenough, D.G. (1982) On the slope-aspect correction of multispectral data. *Canadian Journal of Remote Sensing*, **8**, 84–106.

Terhalle, U. and Bodechtel, J. (1986) Landsat TM data enhancement technique for mapping arid geomorphic features, Proceedings of the ISPRS/Remote Sensing Society Symposium, Mapping from Modern Imagery, Edinburgh, September 1986, The Remote Sensing Society, Nottingham, pp. 725–729.

Theodoridis, S. and Koutraumbas, K. (2006) *Pattern Recognition*, 3rd edn, Elsevier and Academic Press, Amsterdam.

Theodossiou, E.I. and Dowman, I.J. (1990) Heighting accuracy of SPOT. *Photogrammetric Engineering and Remote Sensing*, **56**, 1643–1649.

Thomas, G., Hobbs, S.E. and Dufour, M. (1996) Woodland area estimation by spectral mixing: applying a goodness of fit solution method. *International Journal of Remote Sensing*, **17**, 291–301.

Thomas, R., Krabill, W., Frederick, E. and Jezek, K. (1995) Thickening of Jacobshavns Isbrae, West Greenland, measured by airborne laser altimetry. *Annals of Glaciology*, **21**, 259–262.

Thome, K.J. (2001) Absolute radiometric calibration of Landsat 7 ETM+ using the reflectance-based method. *Remote Sensing of Environment*, **78**, 27–38.

Thome, K.J., Gellman, D.I., Parada, R.J. *et al.* (1993) Absolute radiometric calibration of Thematic Mapper. *Society of Photo-Optical Instrumentation Engineers (SPIE) Proceedings*, **600**, 2–8.

Thome, K.J., Markham, B., Barker, J. *et al.* (1997) Radiometric calibration of Landsat. *Photogrammetric Engineering and Remote Sensing*, **63**, 853–858.

Thomson, G.H. (2009) A note on spatial resolution measurement and its implications for image radiometry. *International Journal of Remote Sensing*, **30**, 1–8.

Tidemann, J. and Nielsen, A.A. (1997) A simple neural network contextual classifier, in *Neurocomputation in Remote Sensing Data Analysis* (eds I. Kanellopoulos, G.G. Wilkinson, F. Roli, and J. Austin), Springer, Heidelberg, pp. 186–193.

Timm, N.H. (2002) *Applied Multivariate Analysis*, Springer, New York.

Tokunaga, M. and Hara, S. (1996) DEM accuracy derived from ASTER data. Proceedings of the 17th Asian Conference on Remote Sensing, November 4-8, 1996, Colombo, pp. J-7-1–J-7-5.

Tompkins, S., Mustard, J.F., Pieters, C.M. and Forsythe, D.W. (1997) Optimization of end members for spectral mixture analysis. *Remote Sensing of Environment*, **59**, 472–489.

Torres, J. and Infante, S.O. (2001) Wavelet analysis for the elimination of striping noise in satellite images. *Optical Engineering*, **40**, 1309–1314.

Tou, J. and Gonzales, R. (1974) *Pattern Recognition Principles*, Addison-Wesley, Reading.

Toutin, T. (1995) Multisource data integration with integrated and unified geometric modelling, in *Sensors and Environmental Applications of Remote Sensing*, Proceedings of the 14th EARSeL Symposium, 6–8 June, Goteborg (ed. J. Askne), A.A. Balkema, Rotterdam, pp. 163–174.

Toutin, T. (2002) Three-dimensional topographic mapping with ASTER stereo data in rugged topography. *IEEE Transactions on Geoscience and Remote Sensing*, **40**, 2241–2247.

Toutin, T. (2004) Review paper: geometric processing of remote sensing images: models, algorithms and methods. *International Journal of Remote Sensing*, **25**, 1893–1924.

Toutin, T. (2008) ASTER DEMs for geomatic and geoscientific applications: a review. *International Journal of Remote Sensing*, **29**, 1855–1875.

Touzi, R. (2002) A review of speckle filtering in the context of estimation theory. *IEEE Transactions on Geoscience and Remote Sensing*, **40**, 2392–2404.

Tovée, M.J. (1996) *An Introduction to the Visual System*, Cambridge University Press, Cambridge.

Townsend, P.A. and Walsh, S.J. (1998) Modelling floodplain inundation using an integrated GIS with radar and optical remote sensing. *Geomorphology*, **21**, 295–312.

Townshend, J.R.G. (1980) The Spatial Resolving Power of Earth Resources Satellites: A Review. NASA Technical Memorandum 82020, Goddard Spaceflight Center, Greenbelt. See also: *Progress in Physical Geography*, **5**, 33–35.

Townshend, J.R.G. (1984) Agricultural land-cover discrimination using thematic mapper spectral bands. *International Journal of Remote Sensing*, **5**, 681–698.

Townshend, J.R.G. and Harrison, A. (1984) Estimation of the spatial resolving power of the Thematic Mapper of Landsat-4, *Proceedings of the Tenth Anniversary International Conference of the Remote Sensing Society*, The Remote Sensing Society, Nottingham, pp. 67–72.

Townshend, J.R.G. and Justice, C.O. (2002) Towards operational monitoring of terrestrial systems by moderate-resolution remote sensing. *Remote Sensing of Environment*, **83**, 351–359.

Townshend, J.R.G. and Skole, D.L. (1994) The global 1 km data set from the advanced very high resolution radiometer, in *TERRA 2: Understanding the Terrestrial Environment – Remote Sensing Data Systems and Networks* (ed. P.M. Mather), John Wiley & Sons, Ltd, Chichester, pp. 75–82.

Townshend, J.R.G., DeFries, R.S. and Zhan, X. (2002) MODIS 250m and 500m time series data for change detection and continuous representation of vegetation characteristics, in *Analysis of Multitemporal Remote Sensing Images* (eds L. Bruzzone, and P. Smits), World Scientific Publishing, Singapore, pp. 233–241.

Townshend, J.R.G., Huang, C., Kalluri, S.N.V. *et al.* (2000) Beware of per-pixel characterization of land cover. *International Journal of Remote Sensing*, **21**, 839–843.

Townshend, J.R.G., Justice, C., Gurney, C. and McManus, J. (1992) The impact of misregistration on change detection. *IEEE Transactions on Geoscience and Remote Sensing*, **30**, 1054–1060.

Tozawa, Y. (1983) Fast geometric correction of NOAA AVHRR. Proceedings of the Symposium on Machine Processing of Remotely-Sensed Data 1983, Purdue University, West Lafayette, pp. 46–53.

Tralli, D.M., Blom, R.G., Zlotnicki, V. *et al.* (2005) Satellite remote sensing of earthquake, volcano, flood, landslide and coastal inundation hazards. *ISPRS Journal of Photogrammetry and Remote Sensing*, **59**, 185–198.

Tran, T.N., Wehrens, R. and Buydens, L.M.C. (2005) Clustering multispectral images: a tutorial. *Chemometrics and Intelligent Laboratory Systems*, **77**, 3–17.

Trauth, M.H., Gebbers, R., Sillmann, E. and Marwan, N. (2007) *MATLAB Recipes for Earth Sciences*, Springer-Verlag, Berlin.

Treitz, P. and Howarth, P. (2000) Integrating spectral, spatial, and terrain variables for forest ecosystem classification. *Photogrammetric Engineering and Remote Sensing*, **66**, 305–317.

Tripathi, N.K., Gokhale, K.V.G.K. and Siddiqui, M.U. (2000) Directional morphological image transforms for lineament extraction from remotely sensed images. *International Journal of Remote Sensing*, **21**, 3281–3292.

Tsai, F. and Chen, W.W. (2008) Striping noise detection and correction of remote sensing images. *IEEE Transactions on Geoscience and Remote Sensing*, **46**, 4122–4131.

Tsai, F. and Philpott, W.D. (1998) Derivative analysis of hyperspectral data. *Remote Sensing of Environment*, **66**, 41–51.

Tso, B. (1997) The investigation of alternative strategies for incorporating spectral, textural and contextual information in remote sensing image classification. Ph.D. thesis, Department of Geography, The University of Nottingham.

Tso, B. and Mather, P.M. (2009) *Classification Methods for Remotely Sensed Data*, 2nd edn, CRC Press, Boca Raton.

Tu, T.-M., Ping, S., Huang, S. *et al.* (2004) A fast intensity–hue–saturation fusion technique with spectral adjustment for IKONOS imagery. *IEEE Geoscience and Remote Sensing Letters*, **1**, 309–312.

Tucker, C.J. (1979) Red and photographic infrared linear combinations for monitoring vegetation. *Remote Sensing of Environment*, **10**, 127–150.

Tucker, C.J., Grant, D.M. and Dykstra, J.J. (2004) NASA's global orthorectified Landsat data set. *Photogrammetric Engineering and Remote Sensing*, **70**, 313–322.

Tukey, J.W. (1977) *Exploratory Data Analysis*, Addison-Wesley, Reading.

Ulaby, F.T. and Elachi, C. (eds) (1990) *Radar Polarimetry for Geoscience Applications*, Artech House, Norwood.

Ulaby, F.T., Moore, R.K. and Fung, A.K. (1981–1986) *Microwave Remote Sensing: Active and Passive*, **3** vols., Artech House, Dedham.

Ungar, S.G., Pearlman, J., Mendenhall, J.A. and Reuter, D. (2003) Overview of the Earth Observing-1 (EO-1) Mission. *IEEE Transactions on Geoscience and Remote Sensing*, **41**, 1149–1159.

Unwin, D.J. and Wrigley, N. (1987) Towards a general theory of control point distribution effects in trend surface models. *Computers and Geosciences*, **13**, 351–355.

US Geological Survey (2003) Earth Observing-1 Extended Mission Fact Sheet 032-03 (March 2003). http://erg.usgs.gov/isb/pubs/factsheets/fs03203.html (accessed 26 January 2009).

US Geological Survey (2008) Seamless Shuttle Radar Topography Mission (SRTM) Finished 3 Arc Second (~90m resolution). http://seamless.usgs.gov/products/srtm3arc.php (accessed 21 November 2009).

Ustin, S.L., Hart, Q.J., Duan, L. and Scheer, G. (1996) Vegetation mapping on hardwood rangelands in California. *International Journal of Remote Sensing*, **17**, 3015–3036.

Ustin, S.L., Smith, M.O., Jacquemoud, S. *et al.* (1999) Geobotany: vegetation mapping for earth Sciences, in *Manual of Remote Sensing*, Remote Sensing for the Earth Sciences, Vol. 3, 3rd edn (ed. A.N. Rencz), John Wiley & Sons, Inc., New York, pp. 189–248.

van Asselen, S. and Seijmonsbergen, A.C. (2006) Expert-driven semi-automated geomorphological mapping for a mountainous area using a laser DTM. *Geomorphology*, **78**, 309–320.

van Coillie, F.M.B., Verbeke, L.P.C. and De Wulf, R.R. (2007) Feature selection by genetic algorithms in object-based classification of IKONOS imagery for forest mapping in Flanders, Belgium. *Remote Sensing of Environment*, **110**, 476–487.

van der Meer, F. (1994) Extraction of mineral absorption features from high spectral resolution data using nonparametric geostatistical techniques. *International Journal of Remote Sensing*, **15**, 2193–2214.

van der Meer, F. (1996a) Classification of remotely-sensed imagery using an indicator kriging approach: application to the problem of calcite-dolomite mineral mapping. *International Journal of Remote Sensing*, **17**, 1233–1249.

van der Meer, F. (1996b) Performance characteristics of the indicator classifier on simulated data. *International Journal of Remote Sensing*, **17**, 621–627.

van der Meer, F. (1996c) Metamorphic facies zonation in the Ronda peridotites: spectroscopic results from field and GER imaging spectrometer data. *International Journal of Remote Sensing*, **17**, 1633–1657.

van der Meer, F., van Dijk, P.M. and Westerhof, A.B. (1995) Digital classification of the contact metamorphic aureole along the Los Pedroches batholith, south-central Spain, using Landsat Thematic Mapper data. *International Journal of Remote Sensing*, **16**, 1043–1062.

van der Meer, F.D. (2000) Imaging spectrometry for geological applications, in *Encyclopedia of Analytical Chemistry* (ed. R.A. Meyers), John Wiley & Sons, Ltd, Chichester, pp. 8601–8638.

van Gardingen, P.R., Foody, G.M. and Curran, P.J. (eds) (1997) *Scaling-Up, From Cell to Landscape*, Cambridge University Press, Cambridge.

van Leeuwen, W.J., Orr, B.J., Marsh, S.E. and Herrmann, S.M. (2006) Multi-sensor NDVI data continuity: Uncertainties and implications for vegetation monitoring applications. *Remote Sensing of Environment*, **100**, 67–81.

van Niel, T.G., McVicar, T.R. and Datt, B. (2005) On the relationship between training sample size and data dimensionality: Monte Carlo analysis of broadband multi-temporal classification. *Remote Sensing of Environment*, **98**, 468–480.

van Niel, T.G., McVicar, T.R. and Datt, B. (2005) On the relationship between training sample size and data dimensionality: Monte Carlo analysis of broadband multi-temporal classification. *Remote Sensing of Environment*, **98**, 468–480.

Vane, G. (ed.) (1987) *Airborne Visible/Infrared Imaging Spectrometer (AVIRIS): A Description of the Sensor, Ground Data Processing Facility, Laboratory Calibration and First Results*, JPL Publication, NASA Jet Propulsion Laboratory, Pasadena, pp. 87–38.

Vapnik, V. (1998) *Statistical Learning Theory*, John Wiley & Sons, Inc., New York.

Varjo, J. (1996) Controlling continuously updated forest data by satellite remote sensing. *International Journal of Remote Sensing*, **17**, 43–67.

Vaudour, E., Moeys, J., Gilliot, J.M. and Coquet, Y. (2008) Spatial retrieval of soil reflectance from SPOT multispectral data using the empirical line method. *International Journal of Remote Sensing*, **29**, 5571–5584.

Vermote, E. and Kaufman, Y.J. (1995) Absolute calibration of AVHRR visible and near-infrared channels using ocean and cloud views. *International Journal of Remote Sensing*, **16**, 2317–2340.

Vermote, E.F., Tanré, D., Deuze, J.L. *et al.* (1997) Second simulation of the satellite signal in the solar spectrum, *6S*: an overview. *IEEE Transactions on Geoscience and Remote Sensing*, **35**, 675–686.

Verstraete, M.M. and Pinty, B. (1992) Extracting surface properties from satellite data in the visible and near-infrared wavelengths, in *TERRA-1: Understanding the Terrestrial Environment-The Role of Earth Observations from Space* (ed. P.M. Mather), Taylor and Francis, Ltd, London, pp. 203–209.

Verstraete, M.M., Pinty, B. and Myeni, R.B. (1996) Potential and limitations of information extraction on the terrestrial biosphere from satellite remote sensing. *Remote Sensing of Environment*, **58**, 201–214.

Vidal-Pantaleoni, A. and Martí, D. (2004) Comparison of different speckle-reduction techniques in SAR images using wavelet transform. *International Journal of Remote Sensing*, **25**, 4915–4932.

Vieira, C.A.O. and Mather, P.M. (2000) Visualisation of measures of classification reliability and error in remote sensing, in *Proceedings of the Fourth International Symposium on Spatial Accuracy Assessment in Natural Resources and Environmental Science* (eds G.B.M. Heuvelink, and M.J.P.M. Lemmens), Delft University Press, Delft, pp. 701–708.

Vincent, R.K. (1997) *Fundamentals of Geological Remote Sensing*, Prentice-Hall, Englewood Cliffs.

Vogelmann, J.E., Helder, D., Morfitt, R. *et al.* (2001) Effects of Landsat 5 Thematic Mapper and Landsat 7 Enhanced Thematic Mapper Plus radiometric and geometric calibrations and corrections on landscape characterisation. *Remote Sensing of Environment*, **78**, 55–70.

Vosselman, G. (2000) Slope based filtering of laser altimetry data, *International Archives of Photogrammetry and Remote Sensing*, Vol. 33, B3/2, ISPRS, Amsterdam, pp. 935–942.

Wainwright, J. and Mulligan, M. (eds) (2005) *Environmental Modelling: Finding Simplicity in Complexity*, John Wiley & Sons, Ltd, Chichester.

Wakabayeshi, H. and Arai, K. (1996) A new method for SAR speckle noise reduction (CST filter). *Canadian Journal of Remote Sensing*, **22**, 190–197.

Wald, L. (2002) *Data Fusion: Definitions and Architectures; Fusion of Images of Different Spatial Resolutions*, Les Presses de l'Ecole des Mines, Paris.

Walker, W.S., Kellndorfer, J.M. and Pierce, L.E. (2007) Quality assessment of SRTM C- and X-band interferometric data: Implications for the retrieval of vegetation canopy height. *Remote Sensing of Environment*, **106**, 428–448.

Wang, F. (1990) Fuzzy supervised classification of remote sensing images. *IEEE Transactions on Geoscience and Remote Sensing*, **28**, 194–201.

Wang, H. and Ellis, E.C. (2005) Spatial accuracy of orthorectified IKONOS imagery and historical aerial photographs across five sites in China. *International Journal of Remote Sensing*, **26**, 1893–1911.

Wang, J. (1993) LINDA – A system for automated linear feature detection. *Canadian Journal of Remote Sensing*, **19**, 9–21.

Wang, J. and Chang, C.-I. (2006) Independent Component Analysis-based dimensionality reduction with applications in hyperspectral image analysis. *IEEE Transactions on Geoscience and Remote Sensing*, **44**, 1586–1600.

Wang, L., Sousa, W.P., Gong, P. and Biging, G.S. (2004) Comparison of IKONOS and QuickBird images for mapping mangrove species on the Caribbean coast of Panama. *Remote Sensing of Environment*, **91**, 432–440.

Wardley, N.W. (1984) Vegetation index variability as a function of viewing geometry. *International Journal of Remote Sensing*, **5**, 861–870.

Warner, T.A., Nellis, M.D. and Foody, G.M. (eds) (2009) *The SAGE Handbook of Remote Sensing*, Sage Publications, Ltd, London.

Warrender, C.E. and Augusteijn, M.F. (1999) Fusion of image classifications using Bayesian techniques with Markov random fields. *International Journal of Remote Sensing*, **20**, 1987–2002.

Waske, B. and Braun, M. (2009) Classifier ensembles for land cover mapping using multitemporal SAR imagery. *ISPRS Journal of Photogrammetry and Remote Sensing*, **64**, 450–457.

Webster, R. and Oliver, M.A. (2007) *Geostatistics for Environmental Scientists*, John Wiley & Sons, Ltd, Chichester.

Wegener, M. (1990) Destriping multiple sensor imagery by improved histogram matching. *International Journal of Remote Sensing*, **11**, 859–975.

Wegman, E.J. (1990) Hyperdimensional data analysis using parallel coordinates. *Journal of the American Statistical Association*, **85**, 664–675.

Wegmüller U. (1997) Land applications of SAR interferometry, in Groll, H. and Nedkov, I. (eds.), *Microwave Physics and Techniques*. Kluwer Academic Publishers, Dordrecht, The Netherlands, pp. 235–250.

Wegmüller, U. and Werner, C.L. (1997) Retrieval of vegetation parameters with SAR interferometry. *IEEE Transactions on Geoscience and Remote Sensing*, **35**, 18–24.

Wegmüller, U., Werner, C.L., Nüesch, D. and Borgeaud, M. (1995) Forest mapping using ERS repeat-pass SAR interferometry. *ESA Earth Observation Quarterly*, **49**, 4–7.

Wegmüller, U., Werner, C.L., Wiesmann, A. and Strozzi, T. (2003) Radargrammetry and space triangulation for DEM generation and image ortho-rectification, *Proceedings of the IEEE International Geoscience and Remote Sensing Symposium, 2003 (IGARSS'03), Toulouse, 21–25 July 2003*, vol. 1, IEEE Press, Piscataway, pp. 179–181.

Wehr, A. and Lohr, U. (1999) Airborne laser scanning – an introduction and overview. *ISPRS Journal of Photogrammetry and Remote Sensing*, **54**, 68–82.

Weinand, H.C. (1974) Cosine theta in components analysis. *Annals of the Association of American Geographers*, **64**, 353.

Welch, R., Jordan, T.R., Lang, H. and Murakami, H. (1998) ASTER as a source of topographic data for the late 1990's. *IEEE Transactions on Geoscience and Remote Sensing*, **36**, 1282–1289.

Welford, B.P. (1962) Note on a method for calculating corrected sums of squares and products. *Technometrics*, **4**, 419–420.

Wemmert, C., Puissant, A., Forestier, G. and Gançarski, P. (2009) Multiresolution remote sensing image clustering. *IEEE Geoscience and Remote Sensing Letters*, **6**, 533–537.

Weng, Q. and Quattrochi, D.A. (eds) (2007) *Urban Remote Sensing*, CRC Press, Boca Raton.

Westin, T. (1990) Precision rectification of SPOT imagery. *Photogrammetric Engineering and Remote Sensing*, **56**, 247–253.

White, K. (1993) Image processing of Thematic Mapper data for discriminating piedmont surficial materials in the Tunisian Southern Atlas. *International Journal of Remote Sensing*, **14**, 961–977.

Wickham, J.D. and Riitters, K.H. (1995) Sensitivity of landscape metrics to pixel size. *International Journal of Remote Sensing*, **16**, 3585–3594.

Wilkinson, G. (2005) Results and implications of a study of fifteen years of satellite image classificatiom experiments. *IEEE Transactions on Geoscience and Remote Sensing*, **43**, 433–440.

Williams, J. (1995) *Geographic Information from Space*, John Wiley & Sons, Ltd and Praxis, Chichester.

Wilson, J.D. (1992) A comparison of procedures for classifying remotely sensed data using simulated data sets incorporating autocorrelation between spectral responses. *International Journal of Remote Sensing*, **13**, 2701–2725.

Wilson, J.P. and Fotheringham, A.S. (eds) (2008) *The Handbook of Geographic Information Science*, Blackwell Publishing, Ltd, Oxford.

Wise, S. (2002) *GIS Basics*, CRC Press, Boca Raton.

Wise, S.M. (2007) Effect of differing DEM creation methods on the results from a hydrological model. *Computers and Geosciences*, **33**, 1351–1365.

Witten, I.E. and Frank, E. (2005) *Data Mining: Practical Machine Learning Tools and Techniques*, 2nd edn. Amsterdam: Elsevier. The WEKA software is available from http://www.cs.waikato.ac.nz/ml/weka (accessed 9 June 2010).

Wolberg, G. (1990) *Digital Image Warping*.IEEE Computer Society Press, Los Alamitos, CA.

Wolf, P.R. and DeWitt, B.A. (2000) *Elements of Photogrammetry with Applications to GIS*, 3rd edn, McGraw-Hill, New York.

Wong, F., Orth, R. and Friedmann, D. (1981) The use of digital terrain models in the rectification of satellite-borne imagery. Proceedings of the 15th International Symposium on Remote Sensing of Environment, Environmental Research Institute of Michigan (ERIM), Ann Arbor, pp. 653–662.

Woodcock, C. and Harward, V.J. (1992) Nested-hierarchical scene models and image segmentation. *International Journal of Remote Sensing*, **16**, 3167–3187.

Woodcock, C.E. and Strahler, A.H. (1987) The factor of scale in remote sensing. *Remote Sensing of Environment*, **21**, 311–322.

Woodcock, C.E., Strahler, A.H. and Jupp, D.L.B. (1988a) The use of variograms in remote sensing. I: scene models and simulated images. *Remote Sensing of Environment*, **25**, 323–348.

Woodcock, C.E., Strahler, A.H. and Jupp, D.L.B. (1988b) The use of variograms in remote sensing. II: real digital images. *Remote Sensing of Environment*, **25**, 349–379.

Woodham, R.J. (1989) Determining intrinsic surface reflectance in rugged terrain and changing illumination, *Proceedings of the International Geoscience and Remote Sensing Symposium, IGARSS'89, 10–14 July 1989, Vancouver, British Columbia*, vol. 1, IEEE Press, Piscataway, pp. 1–5.

Woodhouse, I.H. (2006) *Introduction to Microwave Remote Sensing*, CRC Press, Boca Raton.

Wooding, M.G., Zmuda, A.D. and Griffiths, G.H. (1993) Crop discrimination using multi-temporal ERS-1 SAR data. Proceedings of the Second ERS-1 Symposium on Space at the Service of our Environment, Hamburg, European Space Agency, Paris, ESA SP-361, pp. 51–56.

Wooster, M. (2007) Progress report – remote sensing: sensors and systems. *Progress in Physical Geography*, **31**, 95–100.

Worboys, M. and Duckham, M. (2004) *GIS: A Computing Perspective*, 2nd edn, CRC Press, Boca Raton.

Wu, H.-H.P. and Schowengerdt, R.A. (1993) Improved fraction image estimation using image restoration. *IEEE Transactions on Geoscience and Remote Sensing*, **31**, 771–778.

Xiao, J., Li, J. and Moody, A. (2003) A detail-preserving and flexible adaptive filter for speckle suppression in SAR imagery. *International Journal of Remote Sensing*, **24**, 2451–2465.

Xie, H., Pierce, L.E. and Ulaby, F.T. (2002) SAR speckle reduction using wavelet denoising and Markov Random Fields. *IEEE Transactions on Geoscience and Remote Sensing*, **40**, 2196–2212.

Xu, H., Dvorkin, J. and Nur, A. (2001) Linking oil subsidence to surface subsidence from satellite SAR interferometry. *Geophysical Research Letters*, **28**, 1307–1310.

Xue, Y., Li, Y., Guang, J. *et al.* (2008) Small satellite remote sensing and applications – history, current and future. *International Journal of Remote Sensing*, **29**, 4339–4372.

Yamaguchi, Y., Kahle, A., Tsu, H. *et al.* (1998) An overview of advanced spaceborne thermal emission and reflection radiometer (ASTER). *IEEE Transactions on Geoscience and Remote Sensing*, **36**, 1062–1071.

Yocky, D.A. (1996) Multiresolution wavelet decomposition image merger of Landsat Thematic Mapper and SPOT panchromatic data. *Photogrammetric Engineering and Remote Sensing*, **62**, 1067–1084.

Yonezawa, C. (2007) Maximum likelihood classification combined with spectral angle mapper algorithm for high resolution satellite imagery. *International Journal of Remote Sensing*, **28**, 3729–3737.

Yool, S.R., Star, J.L., Estes, J.E. *et al.* (1986) Performance analysis of image processing algorithms for classification of natural vegetation in the mountains of Southern California. *International Journal of Remote Sensing*, **7**, 683–702.

Young, T.L. and Kaufman, Y.J. (1986) Non-Lambertian effects on remote sensing of surface reflectance and vegetation index. *IEEE Transactions on Geoscience and Remote Sensing*, **GE-24**, 699–707.

Yu, Q., Gong, P., Clinton, N. *et al.* (2006) Object-based detailed vegetation classification with airborne high spatial resolution remote sensing imagery. *Photogrammetric Engineering and Remote Sensing*, **72**, 799–811.

Yu, Q., Gong, P., Tian, Y.Q. *et al.* (2008) Factors affecting spatial variation of classification uncertainty in an image object-based vegetation mapping. *Photogrammetric Engineering and Remote Sensing*, **74**, 1007–1018.

Yu, S., De Backer, S. and Scheunders, P. (2002) Genetic feature selection combined with composite fuzzy nearest neighbour classifiers for hyperspectral satellite imagery. *Pattern Recognition Letters*, **23**, 183–190.

Yuan, D. and Elvidge, C.D. (1996) Comparison of radiometric normalization techniques. *ISPRS Journal of Photogrammetry and Remote Sensing*, **51**, 117–126.

Zebker, H.A. and Chen, K. (2005) Accurate estimation of correlation in INSAR observations. *IEEE Geoscience and Remote Sensing Letters*, **2**, 124–127.

Zebker, H.A. and Goldstein, R.M. (1986) Topographic mapping from interferometric synthetic aperture radar observations. *Journal of Geophysical Research*, **91**, 4993–4999.

Zervakis, M.E., Sundararajan, V. and Parhi, K.K. (2001) Vector processing of wavelet coefficients for robust image denoising. *Image and Vision Computing*, **19**, 435–450.

Zevenbergen, L.W. and Thorne, C.R. (1987) Quantitative analysis of land surface topography. *Earth Surface Processes and Landforms*, **12**, 47–56.

Zhan, X., Sohlberg, R.A., Townshend, J.R.G. *et al.* (2002) Detection of land cover changes using MODIS 250 m data. *Remote Sensing of Environment*, **83**, 336–350.

Zhang, K. and Whitman, D. (2005) Comparison of three algorithms for filtering airborne lidar data. *Photogrammetric Engineering and Remote Sensing*, **71**, 313–324.

Zhang, K., Chen, S., Whitman, D. *et al.* (2003) A progressive morphological filter for removing non-ground measurements from airborne LIDAR data. *IEEE Transactions on Geoscience and Remote Sensing*, **41**, 872–882.

Zhang, L., Liao, M., Yang, L. and Lin, H. (2007) Remote sensing change detection based on canonical correlation analysis and contextual bayes decision. *Photogrammetric Engineering and Remote Sensing*, **73**, 311–318.

Zhang, Q., Pavlic, G., Chen, W. *et al.* (2005) A semi-automatic segmentation procedure for feature extraction in remotely sensed imagery. *Computers and Geosciences*, **31**, 289–296.

Zhang, R. and Ma, J. (2008) An improved SVM method P-SVM for classification of remotely sensed data. *International Journal of Remote Sensing*, **29**, 6029–6036.

Zhang, X., Schaaf, C.B., Friedl, M.A. *et al.* (2002) MODIS tasselled cap transformation and its utility, *Proceedings of the International Geoscience and Remote Sensing Symposium, (IGARSS 2002), Toronto, Canada, 2428 June 2002*, Vol. 2, IEEE International, Piscataway, pp. 1063–1065.

Zhou, G., Ambrosia, V., Gasiewski, A.J. and Bland, G. (2009) Foreword to the special issue on Unmanned Airborne Vehicle (UAV) sensing systems for earth observations. *IEEE Transactions on Geoscience and Remote Sensing*, **47**, 687–689.

Zhou, J. and Civco, D.L. (1996) Using genetic learning neural networks for spatial decision making in GIS. *Photogrammetric Engineering and Remote Sensing*, **62**, 1287–1295.

Zhou, J., Civco, D.L. and Silander, J.A. (1998) A wavelet transform method to merge Landsat TM and SPOT panchromatic data. *International Journal of Remote Sensing*, **19**, 743–757.

Zhou, Q. and Liu, X. (2004) Analysis of errors of derived slope and aspect related to DEM data properties. *Computers and Geosciences*, **30**, 369–378.

Zhu, G. and Blumberg, D.G. (2002) Classification using ASTER data and SVM algorithms: the case study of Beer Sheva, Israel. *Remote Sensing of Environment*, **80**, 233–224.

Zortea, M. and Plaza, A. (2009) Spatial preprocessing for endmember extraction. *IEEE Transactions on Geoscience and Remote Sensing*, **47**, 2679–2693.

Zwally, H.J., Schutz, B., Abdalati, W. *et al.* (2002) ICESat's laser measurements of polar ice, atmosphere, ocean, and land. *Journal of Geodynamics*, **34**, 405–445.

Index

Computer Processing of Remotely-Sensed Images: An Introduction, Fourth Edition Paul M. Mather and Magaly Koch